THEORY AND EXAMPLES OF ORDINARY DIFFERENTIAL EQUATIONS

SERIES ON CONCRETE AND APPLICABLE MATHEMATICS

ISSN: 1793-1142

Series Editor: Professor George A. Anastassiou
Department of Mathematical Sciences
The University of Memphis
Memphis, TN 38152, USA

Series on Concrete and Applicable Mathematics – Vol.10

THEORY AND EXAMPLES OF ORDINARY DIFFERENTIAL EQUATIONS

Chin-Yuan Lin

National Central University, Taiwan

World Scientific

NEW JERSEY · LONDON · SINGAPORE · BEIJING · SHANGHAI · HONG KONG · TAIPEI · CHENNAI

Published by

World Scientific Publishing Co. Pte. Ltd.

5 Toh Tuck Link, Singapore 596224

USA office: 27 Warren Street, Suite 401-402, Hackensack, NJ 07601

UK office: 57 Shelton Street, Covent Garden, London WC2H 9HE

British Library Cataloguing-in-Publication Data
A catalogue record for this book is available from the British Library.

THEORY AND EXAMPLES OF ORDINARY DIFFERENTIAL EQUATIONS
Series on Concrete and Applicable Mathematics — Vol. 10

ISBN-13 978-981-4307-12-3
ISBN-10 981-4307-12-2

Printed in Singapore.

To my wife and brothers,

and

in memory of my mother, Liu Gim

Preface

This book is intended for undergraduate students majoring in mathematics who have studied calculus and are now reading advanced calculus and linear algebra. It would also be of interest to mathematics researchers, since it covers the theory as completely as possible. Because of its many examples and detailed solutions, this book might be found useful as well by students with their majors other than mathematics.

As is revealed by its contents, this book consists of eight chapters. Chapter 1 is on linear equations and Chapter 2 on systems of linear first order equations. Chapter 3 is on power series solutions and Chapter 4 on adjoint operators and nonhomogeneous boundary value problems. Chapter 5 is on Green functions and Chapter 6 on eigenfunction expansions. Finally, Chapter 7 is on long time behavior of systems of differential equations and Chapter 8 on existence and uniqueness theorems.

With the exception of Chapter 8, each chapter is basically divided into seven sections. The first section begins with the Introduction which describes the background and the goals. The second section states the main results and the third illustrates them by many examples. The fourth section proves the main results and the fifth extends them. The final two sections contain a problems set and its solutions set, respectively.

The proof of the main results requires mostly the knowledge of advanced calculus and linear algebra, and partly the knowledge of a complex variable. As a handy reference, such a knowledge is always provided where it is needed. This applies to the section about the extension of the main results where sometimes requires more advanced materials. Such a section might be skipped in the first reading of this book.

The author is indebted to Professors Jerome A. Goldstein and Gisele R. Goldstein at University of Memphis, and Professor Julio G. Dix at Texas State University, for their teaching and friendship.

Finally, the author wishes to thank his wife and son for their support and .encouragement.

Contents

List of Figures

CHAPTER 1

Linear Equations

1. Introduction

Consider the linear ordinary differential equations of nth order

$$Ly = f(x), \quad a < x < b, \tag{1.1}$$

where Ly is defined by

$$Ly \equiv p_0(x)y^{(n)} + p_1(x)y^{(n-1)} + \cdots + p_{n-1}(x)y' + p_n(x)y,$$

$n \in \mathbb{N}$. Here the complex-valued functions $f(x), p_i(x), i = 0, 1, \ldots, n$, are assumed to be continuous, and $p_0(x)$ is also restricted by the condition that there is a $\delta_0 > 0$, such that

$$|p_0(x)| \geq \delta_0 > 0$$

for all $x \in [a, b]$.

Thus, when $n = 2$, we have the linear second order ordinary differential equations

$$Ly = p_0(x)y'' + p_1(x)y' + p_2(x)y = 0, \quad a < x < b.$$

DEFINITION 1.1. *The equation (1.1) is called a nonhomogeneous equation if the function $f(x)$ is not identical to zero. Otherwise, it reduces to*

$$Ly = 0, \quad a < x < b \tag{1.2}$$

and is called a homogeneous equation.

DEFINITION 1.2. *A complex-valued function $z(x)$ defined on (a, b), is called a solution to the nonhomogeneous equation (1.1) if*

$$Lz = f.$$

DEFINITION 1.3. *A set $\{z_1, z_2, \ldots, z_k\}$ of k solutions $z_i(x), i = 1, 2, \ldots, k, k \in \mathbb{N}$, of the homogeneous equation (1.2), is called linearly dependent, if there exist k complex numbers $c_1, c_2, \ldots, c_k$, not all zero, such that*

$$\sum_{i=1}^{k} c_i z_i(x) = 0$$

for all $x \in (a, b)$.

Conversely, if the equation

$$\sum_{i=1}^{k} c_i z_i(x) = 0$$

for all $x \in (a, b)$, has only the zero solutions $0 = c_1 = c_2 = \cdots = c_k$, then the set

$$\{z_1(x), z_2(x), \ldots, z_k(x)\}$$

is called linearly independent.

DEFINITION 1.4. *The Wronskian* $W = W(z_1, z_2, \ldots, z_k)$ *of the k solutions*

$$z_i, \quad i = 1, \ldots, k$$

of (1.2), *is defined by the determinant*

$$W(x) = \begin{vmatrix} z_1(x) & z_2(x) & \cdots & z_k(x) \\ z_1'(x) & z_2'(x) & \cdots & z_k'(x) \\ \vdots & \vdots & \vdots & \vdots \\ z_1^{(k-1)}(x) & z_2^{(k-1)}(x) & \cdots & z_k^{(k-1)}(x) \end{vmatrix}.$$

The concept of linear independence is closely related to the Wronskian $W(x)$ (see Section 2).

DEFINITION 1.5. *A linearly independent set* $\{z_1, z_2, \ldots, z_n\}$ *of n solutions*

$$z_i, \quad i = 1, \ldots, n$$

of (1.2), *is called a basis or a fundamental set of solutions for* (1.2). *Furthermore, the matrix*

$$\Phi(x) \equiv \begin{pmatrix} z_1 & z_2 & \cdots & z_n \\ z_1' & z_2' & \cdots & z_n' \\ \vdots & \vdots & \vdots & \vdots \\ z_1^{(n-1)} & z_2^{(n-1)} & \cdots & z_n^{(n-1)} \end{pmatrix},$$

is called a fundamental matrix for the homogeneous equation (1.2).

Thus the Wronskian $W(x)$ is the determinant $|\Phi(x)|$ of the fundamental matrix $\Phi(x)$.

Theoretically speaking, the homogeneous equation (1.2) has a nontrivial solution by the theorem of existence and uniqueness in Chapter 8. To see this, consider the initial value problem for (1.2):

$$Ly = 0, \quad x \in (a, b)$$

$$x_0 \in (a, b), \tag{1.3}$$

$$y(x_0) = y_0, y'(x_0) = y_1, \ldots, y^{(n-1)}(x_0) = y_{n-1},$$

where $y_0, \ldots, y_{n-1}$ are n given nonzero complex numbers. The theorem of existence and uniqueness in Chapter 8 assures a unique nontrivial solution $z(x)$ for this problem, and thus $z(x)$ is a nontrivial solution to (1.2). This solution $z(x)$ is a zero solution if the initial values $y_i, i = 0, \ldots, n-1$, are zeros.

Likewise, the nonhomogeneous equation (1.1) also has a nontrivial solution theoretically, by the theorem of existence and uniqueness in Chapter 8. To see this, consider the initial value problem for (1.1):

$$Ly = f(x), \quad x \in (a, b)$$

$$y(x_0) = y_0, y'(x_0) = y_1, \ldots, y^{(n-1)}(x_0) = y_{n-1}, \tag{1.4}$$

$$x_0 \in (a, b),$$

where $y_1, y_2, \ldots, y_{n-1}$, are n given nonzero complex numbers. It follows from the existence and uniqueness theorem in Chapter 8 that this problem has a unique nontrivial solution $z(x)$, and thus $z(x)$ is a nontrivial solution to (1.1).

It is the purpose of this chapter to show further that (1.2) has a fundamental set $\{z_1, z_2, \ldots, z_n\}$ of solutions, and that every solution $z(x)$ to (1.2) is of the form

$$z = \sum_{i=1}^{n} c_i z_i,$$

a linear combination of the elements in the fundamental set $\{z_1, z_2, \ldots, z_n\}$, where $c_i, i = 1, \ldots, n$, are constant complex numbers. Furthermore, the initial value problem (1.3) has a unique solution

$$z(x) = \sum_{i=1}^{n} c_i z_i(x),$$

where the constants $c_i, i = 1, \ldots, n$, are uniquely determined by the initial values $y_i, i = 0, \ldots, n-1$. When the coefficient functions $p_i(x)$ are all constants, a fundamental set of solutions will be demonstrated explicitly.

It is also the purpose of this chapter to show further that every solution z to (1.1) takes the form

$$z = \sum_{i=1}^{n} c_i z_i + g(x),$$

where $g(x)$ is an arbitrary solution of (1.1), and that the initial value problem (1.4) has a unique solution

$$z(x) = \sum_{i=1}^{n} c_i z_i(x) + g(x),$$

where the constants c_i are uniquely determined by the initial values $y_i, i = 0, \ldots, n-1$. Methods for finding such a particular solution $g(x)$ are given.

The third purpose of this chapter is to study a few equations of special type, including nonlinear first order ones. This needs a separate treatment.

The final purpose of this chapter is to demonstrate the technique of the Laplace transform for computing explicitly the solution of the initial value problem (1.4) with constant coefficients. This technique is particularly useful when the nonhomogeneous term $f(x)$ is a piecewise continuous function or an impulsive forcing function.

The rest of this chapter is organized as follows. Section 2 states the main results, and Section 3 studies examples for illustration. Section 4 deals with the proof of the main results, and Section 5 examines a few equations of special type, including nonlinear first order ones. Section 6 demonstrates the technique of the Laplace transform for computing the solution of the initial value problem (1.4). Finally, Section 7 contains a problems set, and their solutions are presented in Section 8.

2. Main Results

Firstly, we consider the case where the coefficient functions $p_i(x), i = 0, 1, \ldots, n$ in (1.1), are constants a_i's with $a_0 \neq 0$.

THEOREM 2.1. *Suppose that the algebraic equation*

$$p(\lambda) \equiv a_0 \lambda^n + a_1 \lambda^{n-1} + \cdots + a_{n-1}\lambda + a_n = 0$$

has s distinct solutions (or called roots) $\lambda_1, \lambda_2, \ldots, \lambda_s$ *where* $1 \leq s \leq n$, *and that the solution* λ_i *is repeated* $m_i (\in \mathbb{N})$ *times. In other words, we assume that*

$$p(\lambda) = a_0(\lambda - \lambda_1)^{m_1}(\lambda - \lambda_2)^{m_2} \cdots (\lambda - \lambda_s)^{m_s} = 0,$$

that $\lambda_i \neq \lambda_j$ *for* $i \neq j$, *and that*

$$m_1 + m_2 + \cdots + m_s = n.$$

Then the set of functions

$$\{x^k e^{x\lambda_i}, i = 1, 2, \ldots, s; k = 0, 1, \ldots, (m_i - 1)\}$$

is a fundamental set of solutions for the homogeneous equation (1.2).

Thus if $p(\lambda) = (\lambda - 1)^2(\lambda - 2) = 0$, then

$$\{e^x, xe^x, e^{2x}\}$$

is a fundamental set of solutions.

REMARK 2.2. *In fact, the expression* $p(\lambda) = 0$ *in Theorem 2.1 is always true. This follows from the fact:*

FACT 2.3 (Fundamental Theorem of Algebra). *Every polynomial equation* $p(\lambda) = 0$ *has a complex root.*

Using this fact, we have that

$$0 = p(\lambda) = a_0(\lambda - \mu_1)q_1(\lambda),$$

where μ_1 *is a complex root of* $p(\lambda) = 0$ *and* $q_1(\lambda)$ *is a polynomial of degree* $(n-1)$. *Using the above fact again, we have the polynomial equation*

$$0 = q_1(\lambda) = (\lambda - \mu_2)q_2(\lambda),$$

where μ_1 *is a complex root of* $q_1(\lambda) = 0$ *and* $q_2(\lambda)$ *is polynomial of degree* $(n-2)$. *Repeating this process, we obtain that* $p(\lambda) = 0$ *takes the form above* [**6**, *pages 17–18*].

REMARK 2.4. *An intuitive idea for obtaining the equation* $p(\lambda) = 0$ *in Theorem 2.1 is to seek a solution* z *to the homogeneous equation* (1.2), *of the form*

$$z = e^{x\lambda}.$$

Calculations shows that

$$Lz = p(\lambda)e^{x\lambda},$$

and so $z = e^{x\lambda}$ *is a solution to the equation* (1.2) *if* $p(\lambda) = 0$. *Here note that* $e^{x\lambda}$ *is never zero.*

The fundamental set of solutions in Theorem 2.1 is, in general, a set of complex-valued functions. Under a certain circumstance, it can be a set of real-valued functions as the following shows:

COROLLARY 2.5. *Suppose that the* $p(\lambda)$ *in Theorem 2.1 is either of the form*

$$p(\lambda) = a_0[(\lambda - \lambda_1)^{m_1}(\lambda - \overline{\lambda_1})^{m_1}] \cdots [(\lambda - \lambda_s)^{m_s}(\lambda - \overline{\lambda_s})^{m_s}](\lambda - \lambda_{s+1})^{m_{s+1}} \quad (2.1)$$

or of the form

$$p(\lambda) = a_0[(\lambda - \lambda_1)^{m_1}(\lambda - \overline{\lambda_1})^{m_1}] \cdots [(\lambda - \lambda_s)^{m_s}(\lambda - \overline{\lambda_s})^{m_s}], \quad (2.2)$$

to which corresponds either

$$2(m_1 + m_2 + \cdots + m_s) + m_{s+1} = n$$

or

$$2(m_1 + m_2 + \cdots + m_s) = n.$$

Here

$$\lambda_j = \alpha_j + i\beta_j, j = 1, 2, \ldots, s,$$

$$\alpha_j, \beta_j \in \mathbb{R},$$

$i \equiv \sqrt{-1},$ *the complex unit, with* $i^2 = -1,$

$\overline{\lambda_j} = \alpha_j + i\beta_j,$ *the complex conjugate of* $\lambda_j,$

$\lambda_{s+1} \in \mathbb{R},$ *and*

$\lambda_{j_1} \neq \lambda_{j_2}$ *for* $j_1 \neq j_2,$ *where* $j_i, j_2 = 1, 2, \ldots, (s+1).$

Then according as (2.1) or (2.2), the set

$$\{x^l e^{x\lambda_{s+1}}, x^k e^{x\alpha_j} \cos(x\beta_j), x^k e^{x\alpha_j} \sin(x\beta_j), l = 0, 1, \ldots, (m_{s+1} - 1);$$
$$j = 1, 2, \ldots, s; k = 0, 1, \ldots, (m_s - 1)\}$$

or the set

$$\{x^k e^{x\alpha_j} \cos(x\beta_j), x^k e^{x\alpha_j} \sin(x\beta_j), j = 1, 2, \ldots, s; k = 0, 1, \ldots, (m_s - 1)\}$$

is a fundamental set of real-valued solutions for the homogeneous equation (1.2).

Thus if $p(\lambda) = [\lambda - (1 + 2i)]^2 [\lambda - (1 - 2i)]^2 (\lambda - 3) = 0$, then

$$\{e^x \cos(2x), xe^x \cos(2x), e^x \sin(2x), xe^x \sin(2x), e^{3x}\}$$

is a fundamental set of real-valued solutions.

Corollary 2.5 assumes that the roots of the algebraic equation $p(\lambda) = 0$ appear in complex conjugate pairs, and it is this case when the coefficients $a_0, a_1, \ldots, a_n$ are real constants:

LEMMA 2.6. *Suppose that the coefficient constants in the polynomial* $p(\lambda)$ *are real numbers. If* λ *is a root of the algebraic equation* $p(\lambda) = 0$, *then the complex conjugate* $\overline{\lambda}$ *of* λ *is also a root.*

PROOF. See the Problems set in Section 8.   □

DEFINITION 2.7. *The following are defined:*
- $p(\lambda)$ *is called the characteristic polynomial,*
- $p(\lambda) = 0$ *is called the characteristic equation, and*
- λ_i *is called a characteristic value of multiplicity* m_i.

REMARK 2.8. $p(\lambda)$ *can be obtained symbolically by setting the ith term* $p_i(x)y^{(n-i)}$ *in* Ly *equal to* $a_i \lambda^{n-i}$. *In this correspondence, we have*

$$Ly = p(D)y, \tag{2.3}$$

where $p(D)$ *is obtained symbolically from* $p(\lambda)$, *by replacing* λ *by* D *and* D *is defined as the first order differential operator* $\frac{d}{dx}$ *having the properties:*

$$D^2 = \frac{d^2}{dx^2}, D^3 = \frac{d^3}{dx^3}, \ldots,$$

and so on. Thus, in the case of Theorem 2.1, we have

$$Ly = p(D)y = a_0(D - \lambda_1)^{m_1}(D - \lambda_2)^{m_2} \cdots (D - \lambda_s)^{m_s} y, \qquad (2.4)$$

and in the case of Corollary 2.5, we have

$$Ly = a_0[(D - \lambda_1)^{m_1}(D - \overline{\lambda_1})^{m_1}] \cdots [(D - \lambda_s)^{m_s}(D - \overline{\lambda_s})^{m_s}](D - \lambda_{s+1})^{m_{s+1}}$$
$$(2.5)$$

or

$$Ly = a_0[(D - \lambda_1)^{m_1}(D - \overline{\lambda_1})^{m_1}] \cdots [(D - \lambda_s)^{m_s}(D - \overline{\lambda_s})^{m_s}]. \qquad (2.6)$$

DEFINITION 2.9. *Combined with Theorem 2.1, (2.4) shows, for example, that*

$$[(D - \lambda_1)^{m_1}(D - \lambda_2)^{m_2}][q_1(x)e^{x\lambda_1} + q_2(x)e^{x\lambda_2}] = 0.$$

In that case, we call

$$[(D - \lambda_1)^{m_1}(D - \lambda_2)^{m_2}]$$

an annihilator of the function

$$q_1(x)e^{x\lambda_1} + q_2(x)e^{x\lambda_2}.$$

Here $q_1(x)$ is a polynomial in x, of degree $k, k = 0, 1, \ldots, (m_1 - 1)$, and $q_2(x)$ is a polynomial in x, of degree $k, k = 0, 1, \ldots, (m_2 - 1)$.

Similarly, Corollary 2.5 and (2.5) show, for example, that

$$(D^2 + \beta_1^2)^{m_1}[q_1(x)\cos(x\beta_1) + q_2(x)\sin(x\beta_1)] = 0,$$

and so $(D^2 + \beta_1^2)^{m_1}$ is an annihilator of the function

$$[Q_1(x)\cos(x\beta_1) + Q_2(x)\sin(x\beta_1)].$$

Here $Q_1(x)$ and $Q_2(x)$ are two polynomials in x, of degree $k, k = 0, 1, \ldots, (m_1 - 1)$.

Secondly, we consider the case where the coefficient functions $p_i(x)$'s in (1.2) are not constants. In this case, there still exists theoretically a fundamental set of solutions for (1.2) but such a set is not as explicit as that in Theorem 2.1:

THEOREM 2.10. *There exists a fundamental set of solutions for the homogeneous equation (1.2).*

A fundamental set of solutions for (1.2) is very useful, in that every solution to (1.2) can be expressed as a linear combination of the elements in the fundamental set:

THEOREM 2.11. *If z is a solution to the homogeneous equation (1.2), then z is of the form*

$$z = \sum_{i=1}^{n} c_i z_i,$$

where the set $\{z_1, z_2, \ldots, z_n\}$ is a fundamental set of solutions for the equation (1.2) and $c_i, i = 1, 2, \ldots, c_n$, are constant complex numbers.

Furthermore, the unique solution for the initial value problem (1.3) is given by

$$z(x) = \sum_{i=1}^{n} c_i z_i(x),$$

where the constants c_i's are uniquely determined by the initial values y_j, $j = 0, \ldots, n - 1$.

DEFINITION 2.12. *The function*

$$z(x) = \sum_{i=1}^{n} c_i z_i$$

in Theorem 2.1 is called the general solution to the homogeneous equation (1.2).

The existence of a fundamental set of solutions for the homogeneous equation (1.2) is also very useful, in that every solution to the nonhomogeneous equation (1.1) can be expressed as a linear combination of the elements in the fundamental set, plus a particular solution to (1.1):

THEOREM 2.13. *If z is a solution to the nonhomogeneous equation* (1.1), *then z is of the form*

$$z = \sum_{i=1}^{n} c_i z_i + g(x),$$

where $g(x)$ is a particular solution to the equation (1.1), *the set $\{z_1, z_2, \ldots, z_n\}$, is a fundamental set of solutions for the equation* (1.2), *and c_i's are constant complex numbers.*

Furthermore, the unique solution for the initial value problem (1.4) *is given by*

$$z(x) = \sum_{i=1}^{n} c_i z_i(x) + g(x),$$

where the constants c_i's are uniquely determined by the initial values y_j, $j = 0, \ldots, n-1$.

REMARK 2.14. *A nontrivial solution $z(x)$ and a particular solution $g(x)$ to the nonhomogeneous equation* (1.1) *always exist by the theorem of existence and uniqueness from Chapter 8.*

DEFINITION 2.15. *The function*

$$z = \sum_{i=1}^{n} c_i z_i + g(x)$$

in Theorem 2.13 is called the general solution to the nonhomogeneous equation (1.1).

Another use of a fundamental set of solutions is that it can yield a particular solution explicitly for the nonhomogeneous equation (1.1) through a formula of constants variation:

THEOREM 2.16 (Variation of Constants). *A particular solution $g(x)$ to the nonhomogeneous equation* (1.1) *exists and is of the form*

$$g(x) = \sum_{i=1}^{n} \varphi_i(x) z_i(x),$$

where $\{z_i, z_2, \ldots, z_n\}$, is a fundamental set of solutions for (1.2) *and $\varphi_i(x)$ equals some indefinite integral*

$$\varphi_i(x) = \int \frac{W_i(x)}{W(x)} [f(x)/p_0(x)] \, dx,$$

for which the integration constant can be taken to be zero. Here $W(x)$ is the Wronskian of the n solutions $z_1, z_2, \ldots, z_n$ to the equation (1.2), and W_i is obtained from $W(x)$ by replacing the ith column by the column

$$\begin{pmatrix} 0 \\ \vdots \\ 0 \\ 1 \end{pmatrix}.$$

REMARK 2.17. *A heuristic argument for obtaining the particular solution $g(x)$ of the form in Theorem 2.16 is made in subsection 4.5.*

A different method for obtaining a particular solution explicitly to the nonhomogeneous equation (1.1) exists and is called the method of undetermined coefficients, if the coefficient functions $p_i(x)$ are constants and $f(x)$ is either a polynomial function, an exponential function, a sine function, a cosine function, or combinations of them.

THEOREM 2.18 (Method of Undetermined Coefficients). *Suppose that the coefficient functions $p_i(x)$'s in (1.1) are constants, that the corresponding characteristic equation $p(\lambda) = 0$ is of the form*

$$p(\lambda) = a_0 (\lambda - \lambda_1)^{m_1} \cdots (\lambda - \lambda_s)^{m_s}$$

as in Theorem 2.1, and finally that the nonhomogeneous term $f(x)$ in (1.1) is of the form

$$f(x) = q_1(x) e^{\mu_1 x},$$

where $q_1(x) = b_0 x^{n_1} + b_1 x^{n_1 - 1} + \cdots + b_{n_1 - 1} x + b_{n_1}$, a polynomial of degree $n_1 (\in \{0\} \cup \mathbb{N})$ with $b_0 \neq 0$.
    If μ_1 is not in the set

$$\{\lambda_1, \lambda_2, \ldots, \lambda_s\},$$

then a particular solution $g(x)$ to the nonhomogeneous equation (1.1) is of the form

$$g(x) = Q_1(x) e^{x \mu_1}.$$

Conversely, if $\mu_1 = \lambda_1$, for example, then a particular solution $g(x)$ to the nonhomogeneous equation (1.1) is of the form

$$g(x) = x^{m_1} Q_1(x) e^{x \lambda_1}.$$

Here $Q_1(x)$ is a polynomial of degree $n_1 (\in \{0\} \cup \mathbb{N})$, the same degree as that of $q_1(x)$.
    Thus if $p(\lambda) = (\lambda - 1)(\lambda - 2) = 0$ and $f(x) = (x^2 + 1)e^{2x}$, then a particular solution $g(x)$ is of the form:

$$g(x) = x(c_0 + c_1 x + c_2 x^2) e^{2x},$$

where c_0, c_1, and c_2 are three constants to be determined.
    It is a useful property that linear operations can be performed on the nonhomogeneous equation (1.1):

LEMMA 2.19 (Superposition Principle). *If $y_i, i = 1, 2$, is a solution to the equation*

$$Ly = f_i(x), \quad i = 1, 2,$$

then $(\alpha_1 y_1 + \alpha_2 y_2)$ is a solution to the equation

$$Ly = \alpha_1 f_1(x) + \alpha_2 f_2(x).$$

PROOF. See the Problems set in Section 8. □

REMARK 2.20. *With the aid of Lemma 2.19, Theorem 2.18 applies to nonhomogeneous equations of the form*

$$Ly = q_1(x)e^{\mu_1 x} + q_2(x)e^{\mu_2 x} + \cdots + q_r(x)e^{\mu_r x}. \tag{2.7}$$

This form has $r(\in \mathbb{N})$ similar nonhomogeneous terms, as opposed to one single nonhomogeneous term $q_1(x)e^{\mu_1 x}$ that Theorem 2.18 has. In applying, we solve, separately, the nonhomogeneous equation

$$Ly = q_i(x)e^{\mu_i x}$$

for a particular solution $g_i(x), i = 1, 2, \ldots, r$, and then sum up those solutions $g_i(x)$. The sum $\sum_{i=1}^{r} g_i(x)$ will be a particular solution to the nonhomogeneous equation (2.7) by the superposition principle in Lemma 2.19.

COROLLARY 2.21. *Theorem 2.18 also applies to the case when the nonhomogeneous term $f(x)$ is of the form*

$$f(x) = q_1(x)e^{x\alpha_1}\sin(\beta_1 x) \quad or$$
$$= q_1(x)e^{x\alpha_1}\cos(\beta_1 x) \quad or$$
$$= q_1(x)e^{x\alpha_1}\cos(x\beta_1) + q_2(x)e^{x\alpha_1}\sin(x\beta_1).$$

Here $\alpha_1, \beta_1 \in \mathbb{R}$, $q_1(x)$ is a polynomial of degree n_1, and $q_2(x)$ is a polynomial of degree n_2.

If the complex conjugate pair $\{\alpha_1 + i\beta_1, \alpha_1 - i\beta_1\}$ is not in the set

$$\{\lambda_1, \lambda_2, \ldots, \lambda_s\},$$

then the corresponding particular solutions $g(x)$'s are of the form

$$g(x) = Q_1(x)e^{x\alpha_1}\cos(x\beta_1) + Q_2(x)e^{x\alpha_1}\sin(x\beta_1),$$
$$= the\ same\ as\ above, \quad and$$
$$= R_1(x)e^{x\alpha_1}\cos(x\beta_1) + R_2(x)e^{x\alpha_1}\sin(x\beta_1), \quad respectively.$$

Here $Q_1(x)$ and $Q_2(x)$ are polynomials of degree n_1, and $R_1(x)$ and $R_2(x)$ are polynomials of degree $n_3 \equiv \max\{n_1, n_2\}$.

Conversely, if

$$\alpha_1 + i\beta_1 = \lambda_1$$
$$\alpha_1 - i\beta_1 = \lambda_2,$$

for example, for which there must

$$m_1 = m_2,$$

then the particular solutions $g(x)$'s to the nonhomogeneous equation (1.1) take the form

$$g(x) = x^{m_1} S_1(x)e^{x\alpha_1}\cos(x\beta_1) + x^{m_1} S_2(x)e^{x\alpha_1}\sin(x\beta_1),$$
$$= the\ same\ as\ obove, \quad and$$
$$= x^{m_1} T_1(x)e^{x\alpha_1}\cos(x\beta_1) + x^{m_1} T_2 e^{x\alpha_1}\sin(x\beta_1), \quad respectively.$$

Here $S_1(x)$ and S_2 are polynomials of degree n_1, and $T_1(x)$ and $T_2(x)$ are polynomials of degree

$$n_3 \equiv \max\{n_1, n_2\}.$$

PROOF. Observe that

$$e^{x\alpha_1} \sin(\beta_1 x) = e^{\alpha_1} \frac{e^{i\beta_1 x} - e^{-i\beta_1 x}}{2i} = \frac{1}{2i}[e^{x(\alpha_1+i\beta_1)} - e^{x(\alpha_1-i\beta_1)}] \quad \text{and}$$

$$e^{x\alpha_1} \cos(\beta_1 x) = e^{x\alpha_1} \frac{e^{i\beta_1 x} + e^{-i\beta_1 x}}{2} = \frac{1}{2}[e^{x(\alpha_1+i\beta_1)} + e^{x(\alpha_1-i\beta_1)}],$$

and that the above Remark 2.20 can be used. Here $i \equiv \sqrt{-1}$, the complex unit, and $i^2 = -1$. For the remaining details, see the Problems set in Section 8. □

Thus if $p(\lambda) = [\lambda - (1 + 2i)]^2[\lambda - (1 - 2i)]^2 = 0$ and $f(x) = xe^x \cos(2x) + e^x \sin(2x)$, then a particular solution $g(x)$ takes the form

$$g(x) = x^2(c_0 + c_1 x)e^x \cos(2x) + x^2(d_0 + d_1 x)e^x \sin(2x),$$

where c_0, c_1, d_0, and d_1 are four constants to be determined.

As is stated in Section 1, the concept of linear independence of n solutions to the homogeneous equation (1.2) is closely related to its associated quantity, the Wronskian $W(x)$:

PROPOSITION 2.22. *Suppose that $z_i, i = 1, 2, \ldots, k, 1 \leq k \leq n$, are k solutions to the homogeneous equation (1.2). Then the following statements (i) and (ii) are equivalent, and they are equivalent to the following statement (iii) if $k = n$:*

(i) *The set $\{z_i, z_2, \ldots, z_k\}$ is linearly independent.*
(ii) *The associated Wronskian $W(x)$ with the k solutions $z_1, z_2, \ldots, z_n$ is nonzero somewhere.*
(iii) *The associated Wronskian $W(x)$ with $k = n$ is nowhere zero.*

3. Examples

Before we prove the main results in Section 2, we look at some examples.

EXAMPLE 3.1. Solve the homogeneous equation

$$y'' - 3y' + 2y = 0, \quad 0 < x < 1$$

and the initial value problem

$$y'' - 3y' + 2y = 0, \quad 0 < x < 1$$
$$y(0) = 3, y'(0) = 4.$$

Solution. The characteristic equation is

$$p(\lambda) = \lambda^2 - 3\lambda + 2$$
$$= (\lambda - 1)(\lambda - 2) = 0,$$

and so it follows from Theorem 2.1 that a fundamental set of solutions is

$$\{e^x, e^{2x}\}$$

and that the general solution is

$$z(x) = c_1 e^x + c_2 e^{2x}.$$

Here c_1 and c_2 are two constants.

Using the initial values: $y(0) = 3, \quad y'(0) = 4$, we have

$$3 = z(0) = c_1 + c_2,$$
$$4 = y'(0) = c_1 + 2c_2,$$

and so

$$c_1 = 2,$$
$$c_2 = 1.$$

Thus the unique solution for the initial value problem, by Theorem 2.11, is:

$$z(x) = 2e^x + e^{2x}.$$

EXAMPLE 3.2. Solve the homogeneous equation

$$y'' - 6y' + 25y = 0, \quad 0 < x < 1.$$

Solution. The characteristic equation is

$$p(\lambda) = \lambda^2 - 6\lambda + 25$$
$$= [\lambda - (3 + 4i)][\lambda - (3 - 4i)]$$
$$= 0,$$

and so it follows from Corollary 2.5 that a fundamental set of solutions is

$$\{e^{3x}\cos(4x), e^{3x}\sin(4x)\}$$

and that the general solution is

$$z(x) = c_1 e^{3x}\cos(4x) + c_2 e^{3x}\sin(4x).$$

Here c_1 and c_2 are two constants.

EXAMPLE 3.3. Solve the homogeneous equation

$$y'' - 12y' + 36y = 0, \quad 0 < x < 1.$$

Solution. The characteristic equation is

$$p(\lambda) = \lambda^2 - 12\lambda + 36$$
$$= (\lambda - 6)^2 = 0,$$

and so it follows from Theorem 2.1 that a fundamental set of solutions is

$$\{e^{6x}, xe^{6x}\}$$

and that the general solution is

$$z(x) = c_1 e^{6x} + c_2 x e^{6x}.$$

Here c_1 and c_2 are two constants.

EXAMPLE 3.4. Solve the homogeneous equation

$$y''' - 11y'' + 55y' - 125y = 0, \quad 0 < x < 1.$$

Solution. The characteristic equation is

$$p(\lambda) = \lambda^3 - 11\lambda^2 + 55\lambda - 125$$
$$= (\lambda^2 - 6\lambda + 25)(\lambda - 5)$$
$$= [\lambda - (3 + 4i)][\lambda - (3 - 4i)](\lambda - 5)$$
$$= 0,$$

and so it follows from Corollary 2.5 that a fundamental set of solutions is
$$\{e^{3x}\cos(4x), e^{3x}\sin(4x), e^{5x}\}$$
and that the general solution is
$$z(x) = c_1 e^{3x}\cos(4x) + c_2 e^{3x}\sin(4x) + c_3 e^{5x}.$$
Here $c_i, i = 1, 2, 3$, are three constants.

EXAMPLE 3.5. Solve the nonhomogeneous equation
$$y'' + y = \sec(x), \quad 0 < x < 1.$$

Solution. The characteristic equation is
$$\begin{aligned}
p(\lambda) &= \lambda^2 + \lambda \\
&= (\lambda - i)[\lambda - (-i)] \\
&= 0,
\end{aligned}$$
and so it follows from Corollary 2.5 that a fundamental set of solutions is
$$\{\cos(x), \sin(x)\}.$$

The Wronskian $W(x)$ is:
$$\begin{aligned}
W(x) &= \begin{vmatrix} \cos(x) & \sin(x) \\ -\sin(x) & \cos(x) \end{vmatrix} \\
&= 1.
\end{aligned}$$

Other calculations show that
$$\begin{aligned}
W_1(x) &= \begin{vmatrix} 0 & \sin(x) \\ 1 & \cos(x) \end{vmatrix} \\
&= -\sin(x), \\
W_2(x) &= \begin{vmatrix} \cos(x) & 0 \\ -\sin(x) & 1 \end{vmatrix} \\
&= 1, \\
\varphi_1(x) &= \int \frac{W_1(x)}{W(x)} \sec(x)\, dx \\
&= -\int \tan(x)\, dx \\
&= \ln|\cos(x)|, \\
\varphi_2(x) &= \int \frac{W_2(x)}{W(x)} \sec(x)\, dx \\
&= \int 1\, dx \\
&= x, \quad \text{and a particular solution is} \\
g(x) &= \varphi_1(x)\cos(x) + \varphi_2(x)\sin(x) \\
&= \cos(x)\ln|\cos(x)| + x\sin(x).
\end{aligned}$$

Thus it follows from Theorem 2.16 (Variation of Constants) that the general solution is
$$z(x) = c_1\cos(x) + c_2\sin(x) + g(x)$$

$$= c_1 \cos(x) + c_2 \sin(x) + \cos(x) \ln |\cos(x)| + x \sin(x),$$

where c_1 and c_2 are two constants.

EXAMPLE 3.6. Solve the nonhomogeneous equation

$$y'' - 3y' + 2y = xe^{2x} + e^{3x}, \quad 0 < x < 1.$$

Solution. It follows from Example 3.1 that a fundamental set of solutions is

$$\{e^x, e^{2x}\}.$$

Next, a particular solution $g_1(x)$ to the nonhomogeneous equation

$$y'' - 3y' + 2y = e^{3x}, \tag{3.1}$$

by Theorem 2.18 (Method of Undetermined Coefficients), takes the form

$$g_1(x) = ce^{3x},$$

where the constant c is to be determined by the condition that $g_1(x)$ satisfy the equation (3.1). Calculations show that

$$e^{3x} = g_1'' - 3g_1' + 2g_1$$
$$= 2ce^{3x},$$

and so $c = \frac{1}{2}$ and

$$g_1(x) = \frac{1}{2}e^{3x}.$$

Similarly, a particular solution $g_2(x)$ to the nonhomogeneous equation

$$y'' - 3y' + 2y = xe^{2x}, \tag{3.2}$$

by Theorem 2.18, takes the form

$$g_2(x) = x(c_1 x + c_0)e^{2x},$$

where constants c_1 and c_0 are to be determined by the condition that $g_2(x)$ satisfy the equation (3.2). Calculations show that

$$xe^{2x} = g_2'' - 3g_2' + 2g_2$$
$$= [2c_1 x + (2c_1 + c_0)]e^{2x},$$

and so $c_1 = \frac{1}{2}, c_0 = -1$, and

$$g_2(x) = x(\frac{1}{2}x - 1)e^{2x}.$$

Applying Remark 2.20, we see that

$$g(x) = g_1(x) + g_2(x)$$

is a particular solution to the nonhomogeneous equation

$$y'' - 3y' + 2y = x^{2x} + e^{3x}.$$

It follows from Theorem 2.13 that the general solution is

$$z(x) = d_1 e^x + d_2 e^{2x} + g(x),$$

where d_1 and d_2 are two constants.

4. Proof of the Main Results

In this section, we shall prove the main results in Section 2. The reader is referred to Sections 1 and 2 for related definitions and assumptions.

4.1. The Wronskian $W(x)$. The quantity, the Wronskian $W(x)$, is closely related to the concept of linear independence:

Proposition 2.22. *Suppose that $z_i, i = 1, 2, \ldots, k, 1 \le k \le n$, are k solutions to the homogeneous equation (1.2). Then the following statements (i) and (ii) are equivalent, and they are equivalent to the following statement (iii) if $k = n$:*

  (i) *The set $\{z_i, z_2, \ldots, z_k\}$ is linearly independent.*
  (ii) *The associated Wronskian $W(x)$ with the k solutions $z_1, z_2, \ldots, z_n$ is nonzero somewhere.*
  (iii) *The associated Wronskian $W(x)$ with $k = n$ is nowhere zero.*

PROOF. We divide the proof into three steps.

Step 1. Assume that (ii) holds, and we shall show that (i) is true. Let complex constants $c_i, i = 1, 2, \ldots, k$, be such that

$$c_1 z_1(x) + c_2 z_2(x) + \cdots + c_k z_k(x) = 0 \tag{4.1}$$

for all $t \in [a, b]$. Differentiating the expression (4.1) $(k-1)$ times, we obtain

$$c_1 z_1(x) + c_2 z_2(x) + \cdots + c_k z_k(x) = 0,$$
$$c_1 z_1'(x) + c_2 z_2'(x) + \cdots + c_k z_k'(x) = 0,$$

$$\vdots$$

$$c_1 z_1^{(k-1)}(x) + c_2 z_2^{(k-1)}(x) + \cdots + c_k z_k^{(k-1)}(x) = 0$$

which is the same as

$$\begin{pmatrix} z_1(x) & \cdots & z_k(x) \\ z_1'(x) & \cdots & z_k'(x) \\ \vdots & \vdots & \vdots \\ z_1^{(k-1)} & \cdots & z_k^{(k-1)} \end{pmatrix} \begin{pmatrix} c_1 \\ c_2 \\ \vdots \\ c_k \end{pmatrix} = \begin{pmatrix} 0 \\ 0 \\ \vdots \\ 0 \end{pmatrix}. \tag{4.2}$$

Since $W(x_0) \ne 0$ for some $x_0 \in [a, b]$, we have

$$c_1 = c_2 = \cdots = c_k = 0,$$

and so the set $\{z_1, \ldots, z_k\}$ is linearly independent. This is (i). Here we used the fact from linear algebra:

FACT 4.1. *Suppose that A is a matrix of $m \times m, m \in \mathbb{N}$, that the determinant $|A|$ of A is not zero, and that $\vec{b}$ is an m-dimensional column vector. Then the equation $A\vec{b} = \vec{0}$ has only the zero solution $\vec{b} = \vec{0}$.*

Step 2. Assume that (i) holds, and we shall show that (ii) is true. Claim that if (ii) is not true, then a contradiction results. To this end, assume that the Wronskian $W(x) \equiv 0$ for all $x \in [a, b]$. Then the expression (4.2) implies that $c_i, i = 1, 2, \ldots, k$, are not all zero, a contradiction to the fact that the set $\{z_1(x), \ldots, z_k(x)\}$ is linearly independent by (i). Thus (ii) is true. Here we used the fact from linear algebra:

FACT 4.2. *Suppose that A is a matrix of $m \times m$, $m \in \mathbb{N}$, that the determinant $|A|$ of A is zero, and that $\vec{b}$ is an m-dimensional column vector. Then the equation $A\vec{b} = \vec{0}$ has a nonzero solution $\vec{b}$.*

Step 3. Suppose that $k = n$, and we shall show that (ii) is equivalent to (iii). Differentiating the Wronskian $W(x)$, we have

$$
\frac{d}{dx}W(x) = \begin{vmatrix} z_1'(x) & \cdots & z_k'(x) \\ z_1'(x) & \cdots & z_n'(x) \\ z_1''(x) & \cdots & z_n''(x) \\ \vdots & \vdots & \vdots \\ z_1^{(n-1)}(x) & \cdots & z_n^{(n-1)}(x) \end{vmatrix} + \begin{vmatrix} z_1(x) & \cdots & z_n(x) \\ z_1''(x) & \cdots & z_n''(x) \\ z_1''(x) & \cdots & z_n''(x) \\ z_1'''(x) & \cdots & z_n'''(x) \\ \vdots & \vdots & \vdots \\ z_1^{(n-1)}(x) & \cdots & z_n^{(n-1)}(x) \end{vmatrix}
$$
$$
+ \cdots
$$
$$
+ \begin{vmatrix} z_1(x) & \cdots & z_n(x) \\ \vdots & \vdots & \vdots \\ z_1^{(n-1)}(x) & \cdots & z_n^{(n-1)}(x) \\ z_1^{(n-1)}(x) & \cdots & z_n^{(n-1)}(x) \end{vmatrix} + \begin{vmatrix} z_1(x) & \cdots & z_n(x) \\ \vdots & \vdots & \vdots \\ z_1^{(n-2)}(x) & \cdots & z_n^{(n-2)}(x) \\ z_1^{(n)}(x) & \cdots & z_n^{(n)}(x) \end{vmatrix}, \tag{4.3}
$$

which equals

$$
0 + \begin{vmatrix} z_1(x) & \cdots & z_n(x) \\ \vdots & \vdots & \vdots \\ z_1^{(n-2)}(x) & \cdots & z_n^{(n-2)}(x) \\ \sum_{i=1}^n \frac{p_i(x)}{p_0(x)} z_1^{(n-i)}(x) & \cdots & \sum_{i=1}^n \frac{p_i(x)}{p_0(x)} z_n^{(n-i)}(x) \end{vmatrix},
$$

$$
= \begin{vmatrix} z_1(x) & \cdots & z_n(x) \\ \vdots & \vdots & \vdots \\ z_1^{(n-2)}(x) & \cdots & z_n^{(n-2)}(x) \\ \frac{p_1(x)}{p_0(x)} z_1^{(n-1)}(x) & \cdots & \frac{p_1(x)}{p_0(x)} z_n^{(n-1)}(x) \end{vmatrix}
$$

$$
= \frac{p_1(x)}{p_0(x)} W(x).
$$

Thus, solving the first order differential equation

$$
\frac{d}{dx}W(x) = \frac{p_1(x)}{p_0(x)} W(x)
$$

for $W(x)$, we obtain, for $x_0 \in [a, b]$,

$$
W(x) = W(x_0) e^{\int_{t=x_0}^{x} \frac{p_1(t)}{p_0(t)} dt},
$$

and this shows that the statement (ii) is the same as the statement (i) since the exponential function e^t, $t \in [a, b]$, is never zero. Here for a proof of (4.3), see the Problems set in Section 7; in addition, we used the fact from linear algebra:

FACT 4.3. *Suppose that*

$$
A(x) = \begin{pmatrix} \vec{a_1}(x) \\ \vdots \\ \vec{a_m}(x) \end{pmatrix}
$$

is a matrix of $m \times m, m \in \mathbb{N}$, where $a_i(\vec{x}), i = 1, 2, \ldots, m$, are m-dimensional row vectors that are differentiable functions of the variable $x \in [a, b]$. Then

- *the deterivative $\frac{d}{dx}|A(x)|$ of the determinant $|A(x)|$ equals*

$$|\frac{d}{dx}A(x)| = \sum_{i=1}^{m} \begin{vmatrix} a_1(\vec{x}) \\ \vdots \\ \frac{d}{dt}a_i(\vec{x}) \\ \vdots \\ a_m(\vec{x}) \end{vmatrix},$$

- *the determinant $|A(x)|$ of $A(x)$ equals zero when two rows in $A(x)$ are the same.*
- *the determinant $|A(x)|$ of $A(x)$ is unchanged when the rows in $A(x)$ multiplied by constants are added to other rows in $A(x)$, and*
- *if $a_{i_0}(\vec{x}) = c_{i_0}(x)b_{i_0}(\vec{x})$ for some i_0, where $c_{i_0}(x)$ is a scalar function of t, then the determinant $|A(x)|$ of $A(x)$ satisfies*

$$|A(x)| = c_{i_0}(x) \begin{vmatrix} a_1(\vec{x}) \\ \vdots \\ b_{i_0}(\vec{x}) \\ \vdots \\ a_m(\vec{x}) \end{vmatrix}.$$

$\square$

4.2. Fundamental Sets of Solutions. When the coefficient functions $p_i(x)$ are constants $a_i, i = 0, \ldots, n$, with $a_0 \neq 0$, the homogeneous equation (1.2) has a fundamental set of solutions of the form

$$\{x^k e^{x\lambda_i}, i = 1, 2, \ldots, s; \quad k = 0, 1, \ldots, (m_i - 1)\}.$$

Theorem 2.1. *Suppose that the algebraic equation*

$$p(\lambda) \equiv a_0 \lambda^n + a_1 \lambda^{n-1} + \cdots + a_{n-1}\lambda + a_n = 0$$

has s distinct solutions (or called roots) $\lambda_1, \lambda_2, \ldots, \lambda_s$, where $1 \leq s \leq n$, and that the solution λ_i is repeated $m_i(\in \mathbb{N})$ times. In other words, we assume that

$$p(\lambda) = a_0(\lambda - \lambda_1)^{m_1}(\lambda - \lambda_2)^{m_2} \cdots (\lambda - \lambda_s)^{m_s} = 0,$$

that $\lambda_i \neq \lambda_j$ for $i \neq j$, and that

$$m_1 + m_2 + \cdots + m_s = n.$$

Then the set of functions

$$\{x^k e^{x\lambda_i}, i = 1, 2, \ldots, s; \quad k = 0, 1, \ldots, (m_i - 1)\}$$

is a fundamental set of solutions for the homogeneous equation (1.2).

PROOF. We divide the proof into three steps.
Step 1. Claim that the functions

$$x^k e^{\lambda_i x}, i = 1, 2, \ldots, s; \quad k = 0, 1, \ldots, (m_i - 1),$$

are solutions to the homogeneous equations (1.2). To this end, we note that

$$De^{\lambda x} = \frac{d}{dx}e^{\lambda x} = \lambda e^{\lambda x},$$

$$D^2 e^{\lambda x} = \frac{d}{dx}(\lambda e^{\lambda x}) = \lambda^2 e^{\lambda x},$$

$$\vdots$$

$$D^n e^{\lambda x} = \lambda^n e^{\lambda x},$$

and so

$$Le^{\lambda x} = P(D)e^{\lambda x} \quad \text{by the equation (2.3)}$$
$$= [a_0\lambda^n + a_1\lambda^{n-1} + \cdots + a_{n-1}\lambda + a_n]e^{\lambda x}$$
$$= p(\lambda)e^{\lambda x}.$$

It follows that if λ_1 is a root to the algebraic equation $p(\lambda) = 0$ with multiciplicity m_1, then

$$x^k e^{\lambda_1 x}, \quad k = 0, 1, \ldots, (m_1 - 1)$$

are solutions to the equation (1.2). This is because

$$Le^{\lambda_1 x} = p(\lambda_1)e^{\lambda_1 x} = 0 \quad \text{obviously,}$$

$$p(\lambda) = a_0(\lambda - \lambda_1)^{m_1}q(\lambda) \quad \text{for some polynomial} \quad q(\lambda) \quad \text{of degree} \quad (n - m_1),$$

$$p^{(k)}(\lambda_1) = \frac{d^k}{d\lambda^k}p(\lambda)|_{\lambda=\lambda_1} = 0, k = 1, 2, \ldots, (m_1 - 1),$$

and

$$L(x^k e^{x\lambda_1}) = L(\frac{\partial^k}{\partial\lambda^k}e^{x\lambda}|_{\lambda=\lambda_1}) = \frac{\partial^k}{\partial\lambda^k}Le^{x\lambda}|_{\lambda=\lambda_1} = \frac{\partial^k}{\partial\lambda^k}(p(\lambda)e^{x\lambda})|_{\lambda=\lambda_1} \qquad (4.4)$$

$$= [p^{(k)}(\lambda_1) + kp^{(k-1)}(\lambda_1)x + \frac{k(k-1)}{2!}p^{(k-2)}(\lambda_1)x^2 + \cdots$$

$$+ kp'(\lambda_1)x^{k-1} + p(\lambda_1)x^k]e^{x\lambda_1} \qquad (4.5)$$

$$= 0, \quad k = 1, 2, \ldots, (m_1 - 1).$$

Here (4.5) can be proved by the mathematical induction, for which see the Problems set in Section 7; however, (4.4) results from the fact in advanced calculus [2, page 360]:

FACT 4.4. *If a real-valued function $h(x, \lambda)$ of x, λ, is such that $Lh(x, \lambda) = P(D)h(x, \lambda)$ and $\frac{\partial}{\partial\lambda}h(x, \lambda)$ exist at (x, λ_1), and $L\frac{\partial}{\partial\lambda}h(x, \lambda)$ and $\frac{\partial}{\partial\lambda}Lh(x, \lambda)$ are continuous at (x, λ_1), then*

$$L\frac{\partial}{\partial\lambda}h(x, \lambda_1) = \frac{\partial}{\partial\lambda}Lh(x, \lambda_1).$$

Apply this fact to the real part $e^{x\alpha}\cos(x\beta)$ and the imaginary part $e^{x\alpha}\sin(x\beta)$, separately, of the complex-valued function

$$e^{x\lambda} = e^{x\alpha}\cos(\beta x) + ie^{x\alpha}\sin(x\beta),$$

where $\lambda = \alpha + i\beta$, $\alpha, \beta \in \mathbb{R}$, and $i \equiv \sqrt{-1}$, the complex unit. Then (4.4) follows.

REMARK 4.5. *A heuristic way to see* (4.5) *is comparing them to the binomial theorem:*

$$(\alpha + \beta)^k = \alpha^k + k\alpha^{k-1}\beta + \frac{k(k-1)}{2!}\alpha^{k-2}\beta^2 + \cdots + k\alpha\beta^{k-1} + \beta^k$$

for $\alpha, \beta \in \mathbb{R}$. *It follows that*

$$\frac{\partial^k}{\partial\lambda^k}(p(\lambda)e^{x\lambda}) = (p(\lambda)e^{x\lambda})^{(k)}$$

$$= p^{(k)}(\lambda) + kp^{(k-1)}(\lambda)\frac{\partial}{\partial\lambda}e^{x\lambda} + \frac{k(k-1)}{2!}p^{(k-2)}\frac{\partial^2}{\partial\lambda^2}e^{x\lambda} + \cdots$$

$$+ kp'(\lambda)\frac{\partial^{k-1}}{\partial\lambda^{k-1}}e^{x\lambda} + p(\lambda)\frac{\partial^k}{\partial\lambda^k}e^{x\lambda} \qquad (4.6)$$

$$= [p^{(k)}(\lambda) + kp^{(k-1)}(\lambda)x + \frac{k(k-1)}{2!}p^{(k-2)}(\lambda)x^2 + \cdots$$

$$+ kp'(\lambda)x^{k-1} + p(\lambda)x^k]e^{x\lambda},$$

which, when $\lambda = \lambda_1$, *are* (4.4) *and* (4.5).

The above arguments can be repeated to the case of other roots $\lambda_i, i = 2, 3, \ldots, s$, and this completes the proof of the claim in this Step 1.

Step 2. It follows from (4.6) that

$$\frac{\partial^k}{\partial\lambda^k}(q(\lambda)e^{x\lambda}) = [q^{(k)}(\lambda) + kq^{(k-1)}(\lambda)x + \frac{k(k-1)}{2!}q^{(k-2)}(\lambda)x^2 + \cdots$$

$$+ kq'(\lambda)x^{k-1} + q(\lambda)x^k]e^{x\lambda} \qquad (4.7)$$

is not identical to zero if $q(\lambda)$ is not a zero polynomial of λ. To see this, let

$$q(\lambda) = b_0\lambda^m + b_1\lambda^{m-1} + \cdots + b_m$$

be a nonzero polynomial of λ with degree $m, m \in \{0\} \cup \mathbb{N}$, for which $b_0 \neq 0$. Then the term of highest degree in the polynomial of λ on the right side of (4.7) is the term $q(\lambda)x^k$, which has the degree m. Thus the polynomial on the right side of (4.6) is a nonzero polynomial of degree m.

Step 3. Claim that the set

$$\{x^k e^{x\lambda_i}, i = 1, 2, \ldots, s; \quad k = 0, 1, \ldots, (m_i - 1)\}$$

is a fundamental set for the homogeneous equation (1.2).

To prove the claim, we suppose that there exist constants $c_{i,k}, i = 1, 2, \ldots, s$, and $k = 0, 1, \ldots, (m_i - 1)$, not all zero, such that

$$\sum_{i=1}^{s}\sum_{k=0}^{m_i-1} c_{i,k}x^k e^{x\lambda_i} = \sum_{i=1}^{s} q_i(x)e^{x\lambda_i}$$

$$= q_1(x)e^{x\lambda_1} + q_2(x)e^{x\lambda_2} + \cdots + q_s(x)e^{x\lambda_s} = 0, \qquad (4.8)$$

and we seek a contradiction. Here

$$q_i(x) = \sum_{k=0}^{m_i-1} c_{i,k}x^k$$

$$= c_{i,0} + c_{i,1}x + \cdots + c_{i,m_i-1}x^{m_i-1}$$

is a polynomial in x of degree at most $(m_i - 1)$, and it might be a zero polynomial since $c_{i,k}$ might be zero for this i and for all $k = 0, 1, \ldots, (m_i - 1)$.

Suppose that we delete the zero polynomials, if any, in the set of polynomials $\{q_i(x), i = 1, 2, \ldots, s\}$ and rearrange the order of $q_i(x)$ so that (4.8) becomes

$$\sum_{i=1}^{s_0} q_i(x)e^{x\lambda_i} = q_1(x)e^{x\lambda_1} + q_2(x)e^{x\lambda_2} + \cdots + q_{s_0}(x)e^{x\lambda_{s_0}}$$

$$= 0, \tag{4.9}$$

where $1 \leq s_0 \leq s$ and $q_i(x)$ is not identical to zero for all $1 \leq i \leq s_0 \leq s$. Here note that if there exist exactly one zero polynomial, then $s_0 = n - 1$, and that if exactly two zero polynomials, then $s_0 = n - 2$, and that, $\ldots$, and so on.

Now, to eliminate $q_1(x)$ from (4.9), we divide (4.9) by $e^{x\lambda_1} \neq 0$ and differentiate the result with respect to the variable t, m_1 times, if necessary. It follows from Step 2 that

$$\sum_{i=2}^{s_0} Q_i(x)e^{x(\lambda_i-\lambda_1)} = 0 = \sum_{i=2}^{s_0} Q_i(x)e^{x\lambda_i}$$

$$= Q_2(x)e^{x\lambda_2} + Q_3(x)e^{x\lambda_3} + \cdots + Q_{s_0}(x)e^{x\lambda_{s_0}} \tag{4.10}$$

for some polynomials $Q_i(x), i = 2, 3, \ldots, s_0$, in x with degree at most $(m_i - 1)$, where these polynomials $Q_i(x)$'s are all not zero polynomials since the polynomials $q_i(x)$'s are not zero polynomials by Step 2.

Repeating this elimination, we end up with

$$R_{s_0}(x)e^{x\lambda_{s_0}} = 0 \tag{4.11}$$

for some nonzero polynomial $R_{s_0}(x)$. But this is a contradiction since (4.11) implies

$$R_{s_0}(x) \equiv 0,$$

a zero polynomial. Thus the claim in this Step 3 is proved.   □

The fundamental set of solutions in Theorem 2.1 is, in general, a set of complex-valued functions. Under a certain circumstance, it can be a set of real-valued functions as the following shows:

Corollary 2.5. *Suppose that the $p(\lambda)$ in Theorem 2.1 is either of the form*

$$p(\lambda) = a_0[(\lambda - \lambda_1)^{m_1}(\lambda - \overline{\lambda_1})^{m_1}] \cdots [(\lambda - \lambda_s)^{m_s}(\lambda - \overline{\lambda_s})^{m_s}](\lambda - \lambda_{s+1})^{m_{s+1}} \tag{2.1}$$

or of the form

$$p(\lambda) = a_0[(\lambda - \lambda_1)^{m_1}(\lambda - \overline{\lambda_1})^{m_1}] \cdots [(\lambda - \lambda_s)^{m_s}(\lambda - \overline{\lambda_s})^{m_s}], \tag{2.2}$$

to which corresponds either

$$2(m_1 + m_2 + \cdots + m_s) + m_{s+1} = n$$

or

$$2(m_1 + m_2 + \cdots + m_s) = n.$$

Here

$\lambda_j = \alpha_j + i\beta_j, j = 1, 2, \ldots, s,$

$\alpha_j, \beta_j \in \mathbb{R},$

$i \equiv \sqrt{-1}, \quad$ *the complex unit with* $\quad i^2 = -1,$

$\overline{\lambda_j} = \alpha_j + i\beta_j, \quad$ *the complex conjugate of* $\lambda_j,$

$\lambda_{s+1} \in \mathbb{R}, \quad$ *and*

$$\lambda_{j_1} \neq \lambda_{j_2} \quad for \quad j_1 \neq j_2, \quad where \quad j_i, j_2 = 1, 2, \dots, (s+1).$$

Then according as (2.1) or (2.2), the set

$$\{x^l e^{x\lambda_{s+1}}, x^k e^{x\alpha_j} \cos(x\beta_j), x^k e^{x\alpha_j} \sin(x\beta_j), \quad l = 0, 1, \dots, (m_{s+1} - 1);$$
$$j = 1, 2, \dots, s; \quad k = 0, 1, \dots, (m_s - 1)\}$$

or the set

$$\{x^k e^{x\alpha_j} \cos(x\beta_j), x^k e^{x\alpha_j} \sin(x\beta_j), \quad j = 1, 2, \dots, s; \quad k = 0, 1, \dots, (m_s - 1)\}$$

is a fundamental set of real-valued solutions for the homogeneous equation (1.2).

PROOF. It suffices to assume the case where (2.1) holds, since the other case where (2.2) holds is similar.

We divide the proof into two steps.

Step 1. Claim that the functions in the set

$$S \equiv \{x^l e^{x\lambda_{s+1}}, \quad l = 0, 1, \dots, (m_{s+1} - 1);$$
$$x^k e^{x\alpha_j} \cos(x\beta_j), \quad j = 1, 2, \dots, s, \quad k = 0, 1, \dots, (m_s - 1);$$
$$x^k e^{x\alpha_j} \sin(x\beta_j), \quad j = 1, 2, \dots, s, \quad k = 0, 1, \dots, (m_s - 1)\},$$

are solutions to (1.2). According to Theorem 2.1, we have that

$$L(x^l e^{x\lambda_{s+1}}) = 0, \quad l = 0, 1, \dots, (m_{s+1} - 1),$$
$$L(x^k e^{x\lambda_j}) = 0, \quad j = 1, 2, \dots, s; \quad k = 0, 1, \dots, (m_s - 1), \quad and$$
$$L(x^k e^{\overline{\lambda_s}}) = 0, \quad j = 1, 2, \dots, s; \quad k = 0, 1, \dots, (m_s - 1).$$

Since

$$e^{x\lambda_j} = e^{x\alpha_j + ix\beta_j} = e^{x\alpha_j} e^{ix\beta_j}$$
$$= e^{x\alpha_j} [\cos(x\beta_j) + i\sin(x\beta_j)],$$
$$e^{x\overline{\lambda_j}} = e^{x\alpha_j - ix\beta_j}$$
$$= e^{x\alpha_j} [\cos(x\beta_j) - i\sin(x\beta_j)],$$

$$x^k e^{x\alpha_j} \cos(x\beta_j) = x^k \frac{e^{x\lambda_j} + e^{x\overline{\lambda_j}}}{2}, \quad and \qquad (4.12)$$

$$x^k e^{x\alpha_j} \sin(x\beta_j) = x^k \frac{e^{x\lambda_j} - e^{x\overline{\lambda_j}}}{2i}, \qquad (4.13)$$

it follows that

$$L(x^k e^{x\alpha_j} \cos(x\beta_j)) = \frac{1}{2}[L(x^k e^{x\lambda_j}) + L(x^k e^{x\overline{\lambda_j}})] = 0 \quad and$$

$$L(x^k e^{x\alpha_j} \sin(x\beta_j)) = \frac{1}{2i}[L(x^k e^{x\lambda_j}) - L(x^k e^{x\overline{\lambda_j}})] = 0.$$

Thus the claim in this Step 1 is proved.

Step 2. Claim that the set S in Step 1 is linearly independent. To see this, suppose that there are constants

$$c_l, \quad l = 0, 1, \dots, (m_{s+1} - 1),$$
$$d_{k,j}, \quad j = 1, 2, \dots, s; \quad k = 0, 1, \dots, (m_s - 1),$$
$$b_{k,j}, \quad j = 1, 2, \dots, s; \quad k = 0, 1, \dots, (m_s - 1),$$

such that

$$\sum_{l=0}^{m_{s+1}-1} c_l x^l e^{x\lambda_{s+1}} + \sum_{k=0}^{m_s-1}\sum_{j=1}^{s} d_{k,j} x^k e^{x\alpha_j}\cos(x\beta_j)$$

$$+ \sum_{k=0}^{m_s-1}\sum_{j=1}^{s} b_{k,j} x^k e^{x\alpha_j}\sin(x\beta_j) = 0 \tag{4.14}$$

for all $x \in [a, b]$. We need to show that all those constants $c_l, d_{k,j}$, and $b_{k,j}$ are zero. Using (4.12) and (4.13), we have that (4.14) reduces to

$$\sum_{l=0}^{m_{s+1}-1} x^l e^{x\lambda_{s+1}} + \sum_{k=0}^{m_s-1}\sum_{j=1}^{s} [\frac{1}{2}d_{k,j} + \frac{1}{2i}b_{k,j}]x^k e^{x\lambda_j}$$

$$+ \sum_{k=0}^{m_s-1}\sum_{j=1}^{s} [\frac{1}{2}d_{k,j} - \frac{1}{2i}b_{k,j}]x^k e^{x\overline{\lambda_j}} = 0,$$

and so

$$c_l = 0,$$

$$\frac{1}{2}d_{k,j} + \frac{1}{2i}b_{k,j} = 0, \quad \text{and}$$

$$\frac{1}{2}d_{k,j} - \frac{1}{2i}b_{k,j} = 0,$$

for all l, k, and j since the set

$$\{x^l e^{x\lambda_{s+1}}, \quad l = 0, \dots, (m_{s+1} - 1),$$

$$x^k e^{x\alpha_j}\cos(x\beta_j), \quad j = 1, \dots, s; \quad k = 0, \dots, (m_s - 1),$$

$$x^k e^{x\alpha_j}\sin(x\alpha_j), \quad j = 1, \dots, s; \quad k = 0, \dots, (m_s - 1)\}$$

is linearly independent by Theorem 2.1. Thus all the constants $c_l, d_{k,j}$, and $b_{k,j}$ are zero, and the claim in Step 2 is proved. $\square$

When the coefficient functions $p_i(x)$'s in (1.2) are not constants, there still exists theoretically a fundamental set of solutions. But such a set is not as explicit as that in Theorem 2.1.

Theorem 2.10. *There exists a fundamental set of solutions for the homogeneous equation* (1.2).

PROOF. The theorem of existence and uniqueness in Chapter 8 implies that the homogeneous equation (1.2) with each prescribed initial data $\vec{\xi_i}$ has a unique solution z_i. Here $\tau_0 \in (a, b)$, and

$$\vec{\xi_i} = (y(\tau_0), y'(\tau_0), \dots, y^{(n-2)}(\tau_0), y^{(n-1)}(\tau_0))$$

$$= (0, 0, \dots, 0, 1, 0, \dots, 0), \quad i = 1, 2, \dots, n,$$

a row vector that has every component being 0 except for the ith component, which is 1. It follows from Proposition 2.22 that the set

$$\{z_1, z_2, \dots, z_n\}$$

of n solutions $z_1, z_2, \ldots, z_n$ to the homogeneous equation (1.2) is a fundamental set of solutions for the equation (1.2), since the Wronskian $W(x)$ at $x = \tau_0$ is the determinant of the identity matrix, which is not zero:

$$W(\tau_0) = \begin{vmatrix} 1 & 0 & \cdots & 0 \\ 0 & 1 & \cdots & 0 \\ \vdots & \vdots & \vdots & \vdots \\ 0 & 0 & \cdots & 1 \end{vmatrix} = 1 \neq 0.$$

$\square$

4.3. Homogeneous Equations. One use of a fundamental set of solutions for the homogeneous equation (1.2) is that every solution to the equation (1.2) is a linear combination of the solutions in the fundamental set:

Theorem 2.11. *If z is a solution to the homogeneous equation (1.2), then z is of the form*

$$z = \sum_{i=1}^{n} c_i z_i,$$

where the set $\{z_1, z_2, \ldots, z_n\}$, is a fundamental set of solutions for the equation (1.2) and c_i's are constant complex numbers.

Furthermore, the unique solution for the initial value problem (1.3) is given by

$$z(x) = \sum_{i=1}^{n} c_i z_i(x),$$

where the constants c_i's are uniquely determined by the initial values y_j, $j = 0, \ldots, n - 1$.

PROOF. Let z be a solution to the equation (1.2), and let $\{z_1, z_2, \ldots, z_n\}$ be a fundamental set of solutions for the equation (1.2).

We divide the proof into four steps.

Step 1. It is easy to see that $z(x)$ is a solution to the initial value problem

$$Ly(x) = 0, \quad x \in (a, b)$$
$$y(x_0) = z(x_0), \quad y'(x_0) = z'(x_0), \ldots, \quad y^{(n-1)}(x_0) = z^{(n-1)}(x_0),$$

where x_0 is a given point in (a, b).

Step 2. Claim that

$$u \equiv \sum_{i=1}^{n} c_i z_i(x)$$

for suitably chosen constants c_i's is also a solution to the initial value problem in Step 1. To this end, choose c_i so that

$$u(x_0) = c_1 z_1(x_0) + \cdots + c_n z_n(x_0) = z(x_0),$$

$$\vdots$$

$$u^{(n-1)}(x_0) = c_1 z^{(n-1)}(x_0) + \cdots + c_n z_n^{(n-1)}(x_0) = z^{(n-1)}(x_0),$$

or equivalently,

$$
\begin{pmatrix} z_1(x_0) & \cdots & z_n(x_0) \\ \vdots & \vdots & \vdots \\ z_1^{(n-1)}(x_0) & \cdots & z_n^{(n-1)}(x_0) \end{pmatrix} \begin{pmatrix} c_1 \\ \vdots \\ c_n \end{pmatrix} = \begin{pmatrix} z(x_0) \\ \vdots \\ z^{(n-1)}(x_0) \end{pmatrix}.
$$

Such a choice of c_i's is allowed, since the Wronskian $W(x_0)$ at x_0 is not zero by Proposition 2.22. It is then readily seen that the $u(x)$ with such a choice of c_i's is a solution to the initial value problem in Step 1, and that the claim in this Step 2 is proved.

Step 3. It follows from the theorem of uniqueness and existence in Chapter 8 that the two solutions z and u to the initial value problem in Step 1 are, in fact, the same, and so

$$
z = u = c_1 z_1 + \cdots + c_n z_n,
$$

a linear combination of z_i's.

Step 4. As in Step 2, the constants c_i's are unique, and so

$$
z(x) = \sum_{i=1}^{n} c_i z_i(x)
$$

is the unique solution for the initial value problem (1.3). $\quad\square$

4.4. Nonhomogeneous Equations. The existence of a fundamental set of solutions for the equation (1.2) is also useful, in that every solution to the nonhomogeneous equation (1.1) is the sum of a particular solution to the equation (1.1) and a linear combination of the elements in the fundamental set:

Theorem 2.13. *If z is a solution to the nonhomogeneous equation (1.1), then z is of the form*

$$
z = \sum_{i=1}^{n} c_i z_i + g(x),
$$

where $g(x)$ is a particular solution to the equation (1.1), the set $\{z_1, z_2, \ldots, z_n\}$, is a fundamental set of solutions for the equation (1.2), and c_i's are constant complex numbers.

Furthermore, the unique solution for the initial value problem (1.4) is given by

$$
z(x) = \sum_{i=1}^{n} c_i z_i(x) + f(x),
$$

where the constants c_i's are uniquely determined by the initial values y_j, $\quad j = 0, \ldots, n-1$.

PROOF. Let z be a solution to the nonhomogeneous equation (1.1). Since $g(x)$ is also a solution to the equation (1.1) by assumption, it follows from the superpostion principle in Lemma 2.19 that

$$
(z - g)
$$

is a solution to the equation $Ly = f(x) - f(x) = 0$. Using Theorem 2.10, we thus have that $(z - g)$ takes the form

$$
(z - g) = \sum_{i=1}^{n} c_i z_i,
$$

where $\{z_1, z_2, \ldots, z_n\}$ is a fundamental set of solutions for the equation (1.2) and c_i's are constant complex numbers. Hence

$$z = \sum_{i=1}^{n} c_i z_i + g.$$

Furthermore, as in Step 2 in the proof of Theorem 2.11, we readily see that the constants c_i's are unique, and so

$$z(x) = \sum_{i=1}^{n} c_i z_i(x) + g(x)$$

is the uniquely solution for the initial value problem (1.4). $\qquad\square$

4.5. A Particular Solution. In this subsection, we show that a particular solution to the nonhomogeneous equation (1.1) can be found by using the method of undetermined coefficients and the method of constants variation. We treat the latter method, first.

Theorem 2.16 (Variation of Constants). *A particular solution $g(x)$ to the non-homogeneous equation (1.1) exists and is of the form*

$$g(x) = \sum_{i=1}^{n} \varphi_i(x) z_i(x),$$

where $\{z_i, z_2, \ldots, z_n\}$ is a fundamental set of solutions for the homogeneous equation (1.2) and $\varphi_i(x)$ equals some indefinite integral

$$\varphi_i(x) = \int \frac{W_i(x)}{W(x)} [f(x)/p_0(x)]\, dx,$$

for which the integration constant can be taken to be zero. Here $W(x)$ is the Wronskian of the n solutions $z_1, z_2, \ldots, z_n$ to the equation (1.2), and W_i is obtained from $W(x)$ by replacing the ith column by the column

$$\begin{pmatrix} 0 \\ \vdots \\ 0 \\ 1 \end{pmatrix}.$$

PROOF. We need to show that the function

$$u(x) \equiv \sum_{i=1}^{n} z_i(x) c_i(x)$$

is a solution to the nonhomogeneous equation (1.1), where

$$c_i(x) = \int \frac{W_i(x)}{W(x)} [f(x)/p_0(x)]\, dx,$$

$W(x)$ is the Wronskian of the fundamental set

$$\{z_1, \ldots, z_n\},$$

and $W_i(x)$ is obtained from $W(x)$ by replacing the ith column by the column

$$\begin{pmatrix} 0 \\ \vdots \\ 0 \\ 1 \end{pmatrix}.$$

We divide the proof into three steps.

Step 1. Claim that

$$\sum_{i=1}^{n} z_i^{(k)}(x) W_i(x) = z_1^{(k)} W_1 + z_2^{(k)} W_2 + \cdots + z_n^{(k)} W_n$$

$$= 0, \quad k = 0, 1, \ldots, (n-2), \quad \text{and}$$

$$\sum_{i=1}^{n} z_i^{(n-1)}(x) W_i(x) = z_1^{(n-1)} W_1 + z_2^{(n-1)} W_2 + \cdots + Z_n^{(n-1)} W_n$$

$$= W(x),$$

or equivalently,

$$z_1 W_1 + z_2 W_2 + \cdots + z_n W_n = 0,$$
$$z_1' W_1 + z_2' W_2 + \cdots + z_n' W_n = 0,$$

$$\vdots$$

$$z_1^{(n-2)} W_1 + z_2^{(n-2)} W_2 + \cdots + z_n^{(n-2)} W_2 = 0,$$
$$z_1^{(n-1)} W_1 + z_2^{(n-1)} W_2 + \cdots + z_n^{(n-1)} W_n = W(x).$$

To show the claim, let

$$\Omega(x, k), \quad k = 0, 1, \ldots, (n-1),$$

be obtained from $W(x)$ by replacing the nth row vector by the row vector

$$\begin{pmatrix} z_1^{(k)} & z_2^{(k)} & \cdots & z_n^{(k)} \end{pmatrix}.$$

Also, let

$$\omega(x, i), \quad i = 1, \ldots, n$$

be the determinant, obtained from the determinant $W_i(x)$ by deleting both the ith column and the n row. It follows that

$$\Omega(x, k) = \begin{vmatrix} z_1 & z_2 & \cdots & z_n \\ z_1' & z_2' & \cdots & z_n' \\ \vdots & \vdots & \vdots & \vdots \\ z_1^{(n-2)} & z_2^{(n-2)} & \cdots & z_n^{(n-2)} \\ z_1^{(k)} & z_2^{(k)} & \cdots & z_n^{(k)} \end{vmatrix}$$

$$= \begin{cases} 0, & \text{if } k = 0, 1, \ldots, (n-2); \\ W(x), & \text{if } k = (n-1) \end{cases}$$

by the fact from linear algebra:

FACT 4.6. *The determinant of a square matrix is equal to zero if two rows in that matrix are identical.*

It also follows that

$$\Omega(x,k) = z_1^{(k)}\omega(x,1) + z_2^{(k)}\omega(x,2) + \cdots + z_n^{(k)}\omega(x,n), \quad k = 0,1,\ldots,(n-1),$$

and that

$$\sum_{i=1}^{n} z_i^{(k)} W_i = z_1^{(k)} W_1 + z_2^{(k)} W_2 + \cdots + z_n^{(k)} W_n$$

$$= z_1^{(k)}\omega(x,1) + z_2^{(k)}\omega(x,2) + \cdots + z_n^{(k)}\omega(x,n)$$

$$= \Omega(x,k), \quad k = 0,1,\ldots,(n-1).$$

This is the result of expanding the determinant $\Omega(x,k)$ along its nth row and expanding the determinant W_i along its ith column. Here we used the fact from linear algebra about expanding the determinant of a square matrix:

FACT 4.7. *Let $|A|$ be the determinant*

$$|A| = \begin{vmatrix} a_{11} & \cdots & a_{1n} \\ \vdots & \vdots & \vdots \\ a_{n1} & \cdots & a_{nn} \end{vmatrix},$$

where $m = 2,3,\ldots$. Let $|A_{ij}|$ be the determinant obtained from $|A|$ by deleting both the ith row and the jth column. Then

$$|A| = \sum_{i=1}^{m} a_{ij}|A_{ij}|$$

$$= a_{1j}|A_{ij}| + \cdots + a_{mj}|A_{m,j}|, \quad \text{the expansion of } |A|$$

$$\text{along the } j\text{th column}, \quad j = 1,\ldots,m$$

$$= \sum_{j=1}^{m} a_{ij}|A_{ij}|$$

$$= a_{i1}|A_{i1}| + \cdots + a_{im}|A_{im}|, \quad \text{the expansion of } |A|$$

$$\text{along the } i\text{th row}, \quad i = 1,\ldots,m.$$

Thus the claim in Step 1 is proved.

Step 2. Step 1 implies that

$$\sum_{i=1}^{n} z_i^{(k)}(x)c_i'(x) = \sum_{i=1}^{n} z_i^{(k)}(x)\frac{W_i(x)}{W(x)}[f(x)/p_0(x)]$$

$$= \begin{cases} 0, & \text{if } k = 0,1,\ldots,(n-2); \\ \frac{f(x)}{p_0(x)}, & \text{if } k = (n-1), \end{cases}$$

and so we have

$$u(x) = \sum_{i=1}^{n} z_i(x)c_i(x),$$

$$u'(x) = \sum_{i=1}^{n} z_i'(x)c_i(x) + \sum_{i=1}^{n} z_i(x)c_i'(x)$$

$$= \sum_{i=1}^{n} z_i'(x)c_i(x) + 0,$$

$$u''(x) = \sum_{i=1}^{n} z_i''(x)c_i(x) + \sum_{i=1}^{n} z_i'(x)c_i(x)$$

$$= \sum_{i=1}^{n} z_i''(x)c_i(x) + 0,$$

$$\vdots,$$

$$u^{(n-1)}(x) = \sum_{i=1}^{n} z_i^{(n-1)}(x)c_i(x) + \sum_{i=1}^{n} z_i^{(n-2)}(x)c_i'(x)$$

$$= \sum_{i=1}^{n} z_i^{(n-1)}(x)c_i(x) + 0, \quad \text{and}$$

$$u^{(n)}(x) = \sum_{i=1}^{n} z_i^{(n)}(x)c_i(x) + \sum_{i=1}^{n} z_i^{(n-1)}(x)c_i'(x)$$

$$= \sum_{i=1}^{n} z_i^{(n)}(x)c_i(x) + \frac{f(x)}{p_0(x)}.$$

It follows that

$$Lu = p_0 u^{(n)} + p_1(x)u^{(n-1)} + \cdots + p_{n-1}(x)u' + p_n(x)u$$

$$= \sum_{i=1}^{n} [p_0(x)z_i^{(n)} + p_1(x)z_i^{(n-1)} + \cdots + p_{n-1}(x)z_i' + p_n(x)z_i]c_i(x)$$

$$\quad + p_0(x)\frac{f(x)}{p_0(x)}$$

$$= \sum_{i=1}^{n} c_i(x)Lz_i + f(x)$$

$$= 0 + f(x),$$

which proves that $u(x)$ is a particular solution to the nonhomogeneous equation (1.1). $\qquad\square$

REMARK 4.8. *A heuristic argument for obtaining the particular solution*

$$u(x) = \sum_{i=1}^{n} z_i(x) \int \frac{W_i(x)}{W(x)} [f(x)/p_0(x)] \, dx$$

can be made as follows.

We seek a particular solution $g(x)$ *of the form*

$$g(x) = \sum_{i=1}^{n} c_i(x)z_i(x),$$

where the functions z_i's *are the elements in the fundamental set*

$$\{z_i, \ldots, z_n\}$$

of solutions to the homogeneous equation (1.2) and $c_i(x)$'s are n unknown functions to be determined by the condition that $g(x)$ satisfies the nonhomogeneous equation (1.1)

$$Ly = f(x).$$

The form that $g(x)$ takes is the form that the general solution in Theorem 2.11, to the homogeneous equation (1.2) takes, except that here, $c_i(x)$'s are functions instead of constants. This delivers the name, Variation of Constants, of this method.

Calculations show that

$$g = \sum_{i=1}^{n} z_i c_i,$$

$$g' = \sum_{i=1}^{n} z_i' c_i + \sum_{i=1}^{n} z_i c_i'$$

$$= \sum_{i=1}^{n} z_i' c_i + 0 \quad if \ \sum_{i=1}^{n} z_i c_i' = 0,$$

$$g'' = \sum_{i=1}^{n} z_i'' c_i + \sum_{i=1}^{n} z_i' c_i' \quad if \ \sum_{i=1}^{n} z_i c_i' = 0$$

$$= \sum_{i=1}^{n} z_i'' c_i + 0 \quad if \ \sum_{i=1}^{n} z_i' c_i' = 0 = \sum_{i=1}^{n} z_i c_i',$$

$$\vdots,$$

$$g^{(n-1)} = \sum_{i=1}^{n} z_i^{(n-1)} c_i + \sum_{i=1}^{n} z_i^{(n-2)} c_i' \quad if \ \sum_{i=1}^{n} z_i^{(k)} c_i' = 0, k = 0, \ldots, (n-3),$$

$$= \sum_{i=1}^{n} z_i^{(n-1)} c_i + 0 \quad if \ \sum_{i=1}^{n} z_i^{(k)} c_i' = 0, \quad k = 0, 1, \ldots, (n-2), \ and$$

$$g^{(n)} = \sum_{i=1}^{n} z_i^{(n)} c_i + \sum_{i=1}^{n} z_i^{(n-1)} c_i' \quad if \ \sum_{i=1}^{n} z_i^{(k)} c_i' = 0, \quad k = 0, 1, \ldots, (n-2).$$

Thus, if

$$\sum_{i=1}^{n} z_i^{(k)} c_i' = 0, k = 0, 1, \ldots, (n-2),$$

then

$$Lg = p_0(x) g^{(n)} + \cdots + p_{n-1}(x) g' + p_n g$$

$$= \sum_{i=1}^{n} [p_0(x) z_i + \cdots + p_{n-1} z_i' + p_n z] c_i + p_0(x) \sum_{i=1}^{n} z_i^{(n-1)} c_i'$$

$$= \sum_{i=1}^{n} c_i L z_i + p_0(x) \sum_{i=1}^{n} z_i^{(n-1)} c_i'$$

$$= 0 + p_0(x) \sum_{i=1}^{n} z_i^{(n-1)} c_i',$$

which equals $f(x)$ if additionally,

$$p_0(x) \sum_{i=1}^{n} z_i^{(n-1)} c_i' = f(x).$$

It follows that $g(x)$ is a particular solution to the nonhomogeneous equation (1.1)

$$Ly = f(x),$$

if

$$\begin{pmatrix} z_1 & \cdots & z_n \\ z_1' & \cdots & z_n' \\ \vdots & \vdots & \vdots \\ z_1^{(n-1)} & \cdots & z_n^{(n-1)} \end{pmatrix} \begin{pmatrix} c_1' \\ c_2' \\ \vdots \\ c_n' \end{pmatrix} = \begin{pmatrix} 0 \\ \vdots \\ 0 \\ \frac{f(x)}{p_0(x)} \end{pmatrix},$$

that is, if

$$c_i'(x) = \frac{W_i(x)}{W(x)}[f(x)/p_0(x)], \quad i = 1, \ldots, n.$$

We now treat the method of undetermined coefficients:

Theorem 2.18 (Method of Undetermined Coefficients). *Suppose that the coefficient functions $p_i(x)$'s in (1.1) are constants, that the corresponding characteristic equation $p(\lambda) = 0$ is of the form*

$$p(\lambda) = a_0(\lambda - \lambda_1)^{m_1} \cdots (\lambda - \lambda_s)^{m_s}$$

as in Theorem 2.1, and finally that the nonhomogeneous term $f(x)$ in (1.1) is of the form

$$f(x) = q_1(x)e^{\mu_1 x},$$

where $q_1(x) = b_0 x^{n_1} + b_1 x^{n_1-1} + \cdots + b_{n_1-1}x + b_{n_1}$ is a polynomial of degree $n_1(\in \{0\} \cup \mathbb{N})$, for which $b_0 \neq 0$.
   If μ_1 is not in the set

$$\{\lambda_1, \lambda_2, \ldots, \lambda_s\},$$

then a particular solution $g(x)$ to the nonhomogeneous equation (1.1) is of the form

$$g(x) = Q_1(x)e^{x\mu_1}.$$

Conversely, if $\mu_1 = \lambda_1$, for example, then a particular solution $g(x)$ to the nonhomogeneous equation (1.1) is of the form

$$g(x) = x^{m_1}Q_1(x)e^{x\lambda_1}.$$

Here $Q_1(x)$ is a polynomial of degree $n_1 \in \{0\} \cup \mathbb{N}$, the same degree as that of $q_1(x)$.

PROOF. Note that from Theorem 2.1, we have that

$$(D - \mu_1)^{n_1+1}[q_1(x)e^{x\mu_1}] = 0,$$

that is, the function $q_1(x)e^{x\mu_1}$ is annihilated by the differential operator

$$(D - \mu_1)^{n_1+1}.$$

Here $D = \frac{d}{dx}$ is defined in the Remark 2.8.
   Let $g_p(x)$ be a particular solution to the nonhomogeneous equation (1.1)

$$Ly = f(x) = q_1(x)e^{x\mu_1},$$

which exists by Remark 2.14. Since

$$Lg_p(x) = p(D)g_p(x) = a_0(D - \lambda_1)^{m_1} \cdots (D - \lambda_s)^{m_s} g_p(x)$$

by Remark 2.8, we have

$$(D - \mu_1)^{n_1+1} Lg_p(x) = a_0(D - \mu_1)^{n_1+1}(D - \lambda_1)^{m_1} \cdots (D - \lambda_s)^{m_s} g_p(x)$$
$$= (D - \mu_1)^{n_1+1} q_1(x)e^{x\mu_1}$$
$$= 0,$$

and the corresponding characteristic equation is

$$P(\lambda) \equiv a_0(\lambda - \mu_1)^{n_1+1}(\lambda - \lambda_1)^{m_1} \cdots (\lambda - \lambda_s)^{m_s} = 0.$$

Consider two cases: Case 1 and Case 2.

Case 1. Suppose that λ_1 is not in the set

$$\{\lambda_1, \lambda_2, \ldots, \lambda_s\}.$$

Then Theorems 2.1 and 2.11 show that $g_p(x)$ is of the form

$$g_p(x) = \sum_{l=0}^{n_1} d_l x^l e^{x\mu_1} + \sum_{k=0}^{m_i-1} \sum_{i=1}^{m_s} c_{k,i} x^k e^{x\lambda_i},$$

where $d_l, c_{k,i}$ are complex constants.

On the other hand, the function

$$z(x) \equiv \sum_{k=0}^{m_i-1} \sum_{i=1}^{m_s} c_{k,i} x^k e^{x\lambda_i}$$

is a solution to the homogeneous equation (1.2)

$$Ly = 0,$$

and so the superposition principle in Lemma 2.19 implies that the function

$$g(x) \equiv (g_p(x) - z(x))$$

$$= [\sum_{l=0}^{n_1} d_l x^l] e^{x\mu_1}$$

is a solution to the nonhomogeneous equation

$$Ly = f(x) - 0 = q_1(x)e^{x\mu_1},$$

as desired.

Case 2. Suppose that $\mu_1 = \lambda_1$. Then the corresponding characteristic equation is

$$P(\lambda) \equiv a_0(\lambda - \lambda_1)^{m_1+n_1+1}(\lambda - \lambda_2)^{m_2} \cdots (\lambda - \lambda_s)^{m_s} = 0,$$

and so the particular solution $g_p(x)$ is of the form

$$g_p(x) = \sum_{k=0}^{m_1+n_1} c_{k,1} x^k e^{x\lambda_1} + \sum_{k=0}^{m_i-1} \sum_{i=2}^{s} c_{k,i} x^k e^{x\lambda_i}$$

$$= \sum_{k=0}^{m_1-1} c_{k,1} x^k e^{x\lambda_1} + \sum_{k=m_1}^{m_1+n_1} c_{k,1} x^k e^{x\lambda_1}$$

$$+ \sum_{k=0}^{m_i-1} \sum_{i=2}^{s} c_{k,i} x^k e^{x\lambda_i}$$

$$= \sum_{k=0}^{m_i-1} \sum_{i=1}^{s} c_{k,i} x^k e^{x\lambda_i} + x^{m_1} \sum_{k=0}^{n_1} c_{k,1} x^k e^{x\lambda_1},$$

by Theorems 2.1 and 2.11. Since the function $z(x)$ in Case 1 above is a solution to the homogeneous equation

$$Ly = 0,$$

and so the superposition principle in Lemma 2.19 implies that the function

$$g(x) \equiv (g_p(x) - z(x))$$

$$= x^{m_1} [\sum_{k=0}^{n_1} c_{k,1} x^k] e^{x\lambda_1}$$

$$= x^{m_1} [\sum_{k=0}^{n_1} c_{k.1} x^k] e^{x\mu_1}$$

is a solution to the nonhomogeneous equation

$$Ly = f(x) - 0 = q_1(x) e^{x\mu_1},$$

as desired. □

The material of this Section 4 is based on Coddington and Levinson [5] and Coddington [6, page 84–86, 94–97].

5. Equations of Special Type

In this section, we study some special types of nonlinear first order equations, together with Cauchy equations of second order. Nonlinear equations are, in general, difficult to solve, and their solutions for initial value problems, when existing, might not be unique.

5.1. First Order Equations. To begin with, we consider the first order linear equation

$$\frac{d}{dx} y(x) + p(x) y(x) = q(x), \quad a < x < b, \tag{5.1}$$

in which a and b are two constants, and $p(x)$ and $q(x)$ are two continuous functions on $[a, b]$.

THEOREM 5.1. *The equation (5.1) has the general solution*

$$y(x) = e^{-\int p(x)\,dx} \int e^{\int p(x)\,dx} q(x)\,dx + c e^{-\int p(x)\,dx},$$

where c is a constant and the integration constants in the above indefinite integrals can be taken to be zeros.

PROOF. Multiplying the equation (5.1) by the function $e^{\int p(x)\,dx}$, we have

$$\frac{d}{dx} [e^{\int p(x)\,dx} y(x)] = e^{\int p(x)\,dx} q(x),$$

and so integrating both sides gives

$$y(x) = e^{-\int p(x)\,dx} \int e^{\int p(x)\,dx} q(x)\,dx + ce^{-\int p(x)\,dx}.$$

Here c is a constant, and integration constants are taken to be zero. $\square$

EXAMPLE 5.2. Solve the equation

$$\frac{d}{dx}y(x) + 2xy(x) = x, \quad 3 < x < 8$$

for $y = y(x)$.

Solution. Compute

$$\int 2x\,dx = x^2,$$

$$\int e^{x^2} x\,dx = \frac{1}{2}e^{x^2}.$$

It follows from Theorem 5.1 that the general solution is

$$y(x) = e^{-x^2}[\frac{1}{2}e^{x^2}] + ce^{-x^2}$$

$$= \frac{1}{2} + ce^{-x^2},$$

where c is a constant.

Next we consider the separable equations:

DEFINITION 5.3. *A separable equation is an equation of the form*

$$g(y)\frac{d}{dx}y(x) = f(x), \tag{5.2}$$

$$a < x < b, \quad c < y < d,$$

where $f(x)$ is a continuous function on $[a,b]$ and $g(y)$ is a continuous function on $[c,d]$.

THEOREM 5.4. *The separable equation has the general solution $y = y(x)$ defined by the equation*

$$\int g(y)\,dy = \int f(x)\,dx + c,$$

where c is a constant and the integration constants in the above indefinite integrals can be taken to be zero.

PROOF. Integrating both sides of (5.2), we have

$$\int_{t=x_0}^{x} g(y(t))\frac{d}{dt}y(t)\,dt = \int_{t=x_0}^{x} f(t)\,dt,$$

which is the same as

$$\int_{y(t)=y(x_0)}^{y(x)} g(y(t))\,dy(t) = \int_{t=x_0}^{x} f(t)\,dt \tag{5.3}$$

by the formula of variable change for definite integrals, Apostol [**2**, page 164]:

FACT 5.5.

$$\int_{t=x_0}^{x} g(y(t)) \frac{d}{dt} y(t)\, dt = \int_{y(t)=y(x_0)}^{y(x)} g(y(t))\, dy(t)$$

for the above $g(y)$, where $a \le x_0 \le b$.

Since, in (5.3), the contributions from the substitution $t = x_0$ are just constants, we for convenience symbolically write the result in (5.3) in terms of indefinite indefinite integrals as

$$\int g(y)\, dy = \int f(x)\, dx + c.$$

Here c is a constant, and the integration constants in the above indefinite integrals can be taken to be zero.   □

EXAMPLE 5.6. Solve the equation

$$\frac{d}{dx} y(x) = (1 + y^2)[\cos(x) + \tan(x)], \quad 2 < x < 4$$

for $y = y(x)$.

Solution. Using Theorem 5.4, we have, by separating the variables x and y and integrating both sides, that

$$\int \frac{1}{1 + y^2}\, dy = \int [\cos(x) + \tan(x)]\, dx + c,$$

and so

$$\tan^{-1}(y) = \sin(x) + \ln|\sec(x)| + c$$

or

$$y(x) = \tan[\sin(x) + \ln|\sec(x)| + c].$$

EXAMPLE 5.7. Solve the initial value problem, Coddington and Levinson [5, page 7]:

$$\frac{d}{dx} y(x) = y(x)^{\frac{1}{3}}, \quad 0 < x < 1$$
$$y(0) = 0.$$

Solution. There are an infinite number of solutions for this initial value problem, and so solutions are not unique. Those solutions are, for any $0 \le x_0 \le 1$, given by

$$y(x) = [\frac{2(x - x_0)}{3}]^{\frac{3}{2}}, \quad 0 < x_0 < x \le 1,$$
$$y(x) = 0, \quad 0 \le x \le x_0.$$

To see that they are solutions, we can check that

$$y(0) = 0,$$
$$\frac{d}{dx} y = 0 = y^{\frac{1}{3}} \quad \text{for } 0 \le x < x_0,$$
$$\frac{d}{dx} y = (x - x_0)^{\frac{1}{2}} = y^{\frac{1}{3}} \quad \text{for } 0 < x_0 < x \le 1,$$
$$\frac{d}{dx} y(x)|_{x=x_0} = 0 = y^{\frac{1}{3}}(x_0).$$

To partially derive the solutions, separate the variables x and y and integrate both sides. It follows from Theorem 5.4 that

$$\int y^{-\frac{1}{3}}\, dy = \int 1\, dx + c,$$

and so

$$\frac{3}{2}y^{\frac{2}{3}} = x + c.$$

Using the initial value $y(0) = 0$, we obtain

$$0 = c,$$

and so

$$y(x) = [\frac{2}{3}x]^{\frac{3}{2}}.$$

This is only one of the solutions.

We now consider homogeneous equations:

DEFINITION 5.8. *A homogeneous equation is an equation of the form*

$$\frac{d}{dx}y(x) = f(\frac{y}{x}),$$
$$a < x < b, \tag{5.4}$$
$$c < y < d,$$

where $f(z)$ is a continuous function of z when $z = \frac{y}{x}$ is defined.

THEOREM 5.9. *The homogeneous equation (5.4) can be, by the change $v = \frac{y}{x}$ of variable, transformed into the solvable separable equation*

$$\frac{1}{f(v) - v}\frac{d}{dx}v(x) = \frac{1}{x},$$

if $\frac{1}{f(v)-v}$ is a continuous function of v when $v = \frac{y}{x}$ is defined.

PROOF. Using the change of variable $v = \frac{y}{x}$, we have

$$y = vx,$$
$$\frac{d}{dx}y = \frac{d}{dx}v(x)x + v(x),$$

which, when substituted into the equation (5.4), yields

$$x\frac{d}{dx}v(x) + v = f(v).$$

Thus

$$\frac{1}{f(v) - v}\frac{d}{dx}v(x) = \frac{1}{x},$$

a separable equation that can be solved by Theorem 5.4 if $\frac{1}{f(v)-v}$ is a continuous function of v when $v = \frac{y}{x}$ is defined.   □

EXAMPLE 5.10. Solve the equation

$$\frac{d}{dx}y(x) = \frac{6xy + 2y^2}{y^2 - 3xy}, \quad 3 < x < 8$$

for $y = y(x)$.

Solution. The equation can be rewritten as

$$\frac{d}{dx}y(x) = \frac{6 + 2\frac{y}{x}}{\frac{y}{x} - 3},$$

a homogeneous equation. Thus, using the change $v = \frac{y}{x}$ of variable, it follows from Theorem 5.9 that

$$\frac{1}{x} = \frac{1}{f(v) - v}\frac{d}{dx}v$$

$$= \frac{3 - v}{v^2 - 5v - 6}\frac{d}{dx}v$$

$$= [\frac{\frac{-3}{7}}{v - 6} + \frac{\frac{-4}{7}}{v + 1}]\frac{d}{dx}v.$$

Integrating both sides, it follows from Theorem 5.4 that

$$\frac{-3}{7}\ln|v - 6| - \frac{4}{7}\ln|v + 1| = \ln|x| + c_1,$$

and so

$$(v - 6)^3(v + 1)^4 x^7 = c_2.$$

Changing $v = \frac{y}{x}$ back to y yields

$$(y - 6x)^3(y + x)^4 = c_2.$$

Here c_1 and c_2 are constants.

EXAMPLE 5.11. Solve the equation

$$\frac{d}{dx}y(x) = \frac{2x + 3y}{x - y}, \quad 3 < x < 8$$

for $y = y(x)$.

Solution. The equation in the example can be rewritten as the homogeneous equation

$$\frac{d}{dx}y = \frac{2 + \frac{y}{x}}{1 - \frac{y}{x}} = f(\frac{y}{x}),$$

and so, using the change $v = \frac{y}{x}$ of variable, it follows from Theorem 5.9 that

$$f(v) - v = \frac{2 + 3v}{1 - v} - v$$

$$= \frac{v^2 + 2v + 2}{1 - v},$$

$$\frac{1}{x} = \frac{1}{f(v) - v}\frac{d}{dx}v$$

$$= \frac{1 - v}{v^2 + 2v + 2}\frac{d}{dx}v$$

$$= [\frac{2}{(v + 1)^2 + 1} - \frac{v + 1}{(v + 1)^2 + 1}]\frac{d}{dx}v.$$

Integrating both sides, it follows from Theorem 5.4 that

$$2\tan^{-1}(v + 1) - \frac{1}{2}\ln[(v + 1)^2 + 1] = \ln|x| + c,$$

where c is a constant. Changing $v = \frac{y}{x}$ back to y, we have

$$2\tan^{-1}(\frac{y}{x} + 1) - \frac{1}{2}\ln[(\frac{y}{x} + 1)^2 + 1] + \ln|x| + c,$$

which defines $y = y(x)$ as a function of x implicitly.

EXAMPLE 5.12. Solve the equation

$$\frac{d}{dx}y(x) = \frac{2x + 3y + 5}{x - y - 2}, \quad 3 < x < 8,$$

that is similar to the equation in the previous example, for $y = y(x)$.

Solution. We can transform our equation into the equation in the previous example, by using the linear transformation

$$x = u + \alpha,$$
$$y = v + \beta,$$

where the constants α and β are to be determined from the relations

$$2x + 3y + 5 = (2u + 3v) + (2\alpha + 3\beta + 5),$$
$$x - y - 2 = (u - v) + (\alpha - \beta - 2)$$

by requiring

$$2\alpha + 3\beta + 5 = 0,$$
$$\alpha - \beta - 2 = 0.$$

Solving for α and β, we obtain

$$\alpha = \frac{1}{5}, \quad \beta = -\frac{9}{5}.$$

Since

$$\frac{d}{dx}y = \frac{dy}{dv}\frac{dv}{du}\frac{du}{dx} = \frac{dv}{du}$$

by chain rule, it follows that our equation becomes

$$\frac{d}{du}v = \frac{2u + 3v}{u - v},$$

the solved separable equation in the previous example. Thus we finially obtain

$$2\tan^{-1}(\frac{y + \frac{9}{5}}{x - \frac{1}{5}} + 1) - \frac{1}{2}\ln[(\frac{y + \frac{9}{5}}{x - \frac{1}{5}} + 1)^2 + 1] = \ln|x - \frac{1}{5}| + c,$$

where c is a constant.

Next we have

THEOREM 5.13. *The equation*

$$\frac{d}{dx}y(x) = f(\alpha x + \beta y),$$
$$a < x < b, \quad c < y < d,$$

can be transformed into a solvable separable equation

$$\frac{1}{\beta f(u) + \alpha}\frac{d}{dx}u(x) = 1$$

by the change $u = \alpha x + \beta y$ of variable. Here $f(u)$ is a continuous function of $u = \alpha x + \beta y$, such that $\frac{1}{\beta f(u) + \alpha}$ is defined.

PROOF. The change $u = \alpha x + \beta y$ of variable yields

$$\frac{d}{dx}u = \alpha + \beta\frac{d}{dx}y,$$

and so

$$\frac{1}{\beta f(u) + \alpha}\frac{d}{dx}u(x) = 1,$$

a separable equation that can be solved by Theorem 5.4.   □

EXAMPLE 5.14. Solve the equation

$$\frac{d}{dx}y(x) = \sin(x+y), \quad 3 < x < 8$$

for $y = y(x)$.

Solution. Using the change $u = x + y$ of variable, it follows from Theorem 5.13 that

$$1 = \frac{1}{\sin(u) + 1}\frac{d}{dx}u(x)$$

$$= \frac{1 - \sin(u)}{(\cos(u))^2}\frac{d}{dx}u$$

$$= [(\sec(u))^2 - (\cos^{-2}(u))\sin(u)]\frac{d}{dx}u.$$

Integrating both sides, it follows from Theorem 5.4 that

$$x + c = \tan(u) - \frac{1}{\cos(u)}$$

$$= \tan(x+y) - \sec(x+y),$$

where c is a constant.

Next we consider Bernoulli equations:

DEFINITION 5.15. *A Bernoulli equation is an equation of the form*

$$\frac{d}{dx}y(x) + p(x)y = q(x)y^r, \quad a < x < b, \tag{5.5}$$

where $r \neq 0, \neq 1$, is a real number, and $p(x)$ and $q(x)$ are continuous functions on $[a, b]$. Here note that when $r = 0$ or 1, the equation is a solvable linear equation.

THEOREM 5.16. *The Bernoulli equation (5.5) can be, by the change $u = y^{1-r}$ of variable when $y^{1-r} = y^{1-r}(x)$ is defined, transformed into the solvable linear equation*

$$\frac{d}{dx}u(x) + (1 - r)p(x)u(x) = (1 - r)q(x).$$

PROOF. Using the change $u = y^{1-r}$ of variable when y^{1-r} is defined, we have

$$\frac{d}{dx}u(x) = (1 - r)y^{-r}\frac{d}{dx}y(x),$$

and so the Bernoulli equation becomes

$$\frac{d}{dx}v(x) + (1 - r)p(x)v(x) = (1 - r)q(x),$$

a solvable linear equation.   □

EXAMPLE 5.17. Solve the equation

$$\frac{d}{dx}y(x) + 2y = -3y^3, \quad 3 < x < 8$$

for $y = y(x)$.

Solution. Using the change $u = y^{1-3} = y^{-2}$ of variable, it follows from Theorem 5.16 that

$$\frac{d}{dx}u(x) - 4u(x) = 6,$$

a solvable linear equation that has the solution

$$u(x) = ce^{4x} - \frac{3}{2}$$

by Theorem 5.1. Here c is a constant. Changing $v = y^{-2}$ back to y, we have

$$y^2(x) = \frac{1}{ce^{4x} - \frac{3}{2}}.$$

Finally, we consider in this subsection, Riccati equations:

DEFINITION 5.18. *A Riccati equation is an equation of the form*

$$\frac{d}{dx}y(x) = r(x) + q(x)y + p(x)y^2, a < x < b, \tag{5.6}$$

where $p(x), q(x)$, and $r(x)$ are continuous functions on $[a, b]$.

THEOREM 5.19. *The Riccati equation (5.6) can be, by the change $y(x) = y_p(x) + \frac{1}{v(x)}$ of variable, transformed into the solvable linear equation*

$$\frac{d}{x}v(x) + [2p(x)y_p(x) + q(x)]v(x) = -p(x),$$

where $y_p(x)$ is a particular solution to the Riccati equation that is found in advance.

PROOF. By the change $y(x) = y_p(x) + \frac{1}{v(x)}$ of variable, we have

$$y'(x) = y_p'(x) - \frac{1}{v^2(x)}v'(x),$$

and so the Riccati equation (5.6) becomes

$$v'(x) + [2p(x)y_p(x) + q(x)]v(x) = -p(x).$$

Here note that $y_p(x)$ satisfies the Riccati equation (5.6) for which

$$\frac{d}{dx}y_p(x) = r(x) + q(x)y_p(x) + p(x)(y_p)^2.$$

□

EXAMPLE 5.20. Solve the differential equation

$$y'(x) = 2x + x^4 - 2x^2y + y^2, \quad 3 < x < 8$$

for $y = y(x)$.

Solution. We can find a particular solution $y_p = x^2$ by inspection. Using the change of variable

$$y = y_p + \frac{1}{v}$$
$$= x^2 + \frac{1}{v},$$

it follows from Theorem 5.19 that $v = v(x)$ satisfies the linear equation

$$-1 = v' + [2x^2 - 2x^2]v = -1$$
$$= v'.$$

Solving for v, we obtain $v = -x + c$, where c is a constant. Thus the general solution is

$$y = x^2 + \frac{1}{-x + c}.$$

5.2. Euler Equations. We consider the Euler equations or the equi-dimensional equations, Boyce and DiPrima [**3**, pages 239–248]:

$$x^2 y'' + a_1 x y' + a_0 y = 0, \quad a < x < b, x \neq 0, \tag{5.7}$$

where a_1 and a_0 are two complex constants.

For our use in this subsection, we recall two definitions from calculus:

DEFINITION 5.21. *For $x \in \mathbb{R}$, define*

$$e^x = \sum_{k=0}^{\infty} \frac{x^k}{k!}.$$

DEFINITION 5.22. *For $x > 0$ and $\lambda \in \mathbb{R}$, define*

$$x^\lambda = e^{\lambda \ln(x)}.$$

Following Definition 5.21, we define:

DEFINITION 5.23. *For $z \in \mathbb{C}$, define*

$$e^z = \sum_{k=0}^{\infty} \frac{z^k}{k!}.$$

In particular,

$$e^{\mu + i\theta} = e^\mu e^{i\theta}$$
$$= e^\mu [\cos(\theta) + i \sin(\theta)],$$

where $i \equiv \sqrt{-1}, i^2 = 1$, and $\mu, \theta \in \mathbb{R}$.

Similarly, we use Definition 5.22 to define:

DEFINITION 5.24. *For $x > 0$ and $\lambda = \alpha + i\beta \in \mathbb{C}$, where α and β are in $\mathbb{R}$, define*

$$x^\lambda = e^{\lambda \ln(x)}$$
$$= e^{(\alpha + i\beta) \ln(x)}$$
$$= e^{\alpha \ln(x)} e^{i\beta \ln(x)}$$
$$= x^\alpha [\cos(\beta \ln(x)) + i \sin(\beta \ln(x))].$$

DEFINITION 5.25. *The characteristic equation for the Euler equation (5.7) is*

$$p(\lambda) = \lambda(\lambda - 1) + a_1\lambda + a_0 = 0.$$

REMARK 5.26. From Section 2, we know that $p(\lambda) = 0$ takes the form

$$p(\lambda) = (\lambda - \lambda_1)(\lambda - \lambda_2) = 0$$

for two complex numbers λ_1 and λ_2.

THEOREM 5.27. *For $x > 0$, the Euler equation (5.7) has the general solution $y = y(x)$:*

Case 1. *If $\lambda_1 \neq \lambda_2$, then*

$$y(x) = c_1 x^{\lambda_1} + c_2 x^{\lambda_2}.$$

Case 2. *If $\lambda_1 = \lambda_2$, then*

$$y(x) = c_1 x^{\lambda_1} + c_2[\ln(x)]x^{\lambda_1}.$$

Here c_1 and c_2 are two constants.

Furthermore, under Case 1, if $\lambda_1 = \alpha + i\beta$ and $\lambda_2 = \alpha - i\beta$ in which $\alpha, \beta \in \mathbb{R}, i = \sqrt{-1}$, and λ_1 is the complex conjugate of λ_2, then the real-valued general solution with $x > 0$, is:

$$y(x) = c_1 x^\alpha \cos(\beta \ln(x)) + c_2 x^\alpha \sin(\beta \ln(x)),$$

where c_1 and c_2 are two real constants.

Finally, for $x < 0$, the general solution is obtained by replacing the x in the above $y(x)$ by $-x$.

We look at some examples first.

EXAMPLE 5.28. *Solve the equation*

$$x^2 y'' + 3xy' - 8y = 0, \quad 3 < x < 8$$

for $y = y(x)$.

Solution. The characteristic equation is

$$p(\lambda) = \lambda(\lambda - 1) + 3\lambda - 8$$
$$= \lambda^2 + 2\lambda - 8 = 0,$$

and so $\lambda = 2$ or -4. It follows from Theorem 5.27 that the general solution is

$$y(x) = c_1 x^2 + c_2 x^{-4},$$

where c_1 and c_2 are two constants.

EXAMPLE 5.29. *Solve the equation*

$$x^2 y'' - 2xy' + 3 = 0, \quad 3 < x < 8$$

for $y = y(x)$.

Solution. The characteristic equation is

$$p(\lambda) = \lambda(\lambda - 1) - 2\lambda + 3$$
$$= \lambda^2 - 3\lambda + 3 = 0,$$

and so $\lambda = \frac{3}{2} + i\frac{\sqrt{3}}{2}$ or $\frac{3}{2} - i\frac{\sqrt{3}}{2}$. It follows from Theorem 5.27 that the real-valued general solution is

$$y(x) = c_1 x^{\frac{3}{2}} \cos(\frac{\sqrt{3}}{2} \ln(x)) + c_2 x^{\frac{3}{2}} \sin(\frac{\sqrt{3}}{2} \ln(x)),$$

where c_1 and c_2 are two real constants.

EXAMPLE 5.30. *Solve the equation*

$$x^2 y'' - xy' + y = 0, \quad 3 < x < 8$$

for $y = y(x)$.

Solution. The characteristic equation is

$$\begin{aligned} p(\lambda) &= \lambda(\lambda - 1) - \lambda + 1 \\ &= \lambda^2 - 2\lambda + 1 = 0, \end{aligned}$$

and so $\lambda = 1$ or 1. It follows from Theorem 5.27 that the general solution is

$$y(x) = c_1 x + c_1 x \ln(x),$$

where c_1 and c_2 are two constants.

We now prove Theorem 5.27:

PROOF. By defining

$$Ly \equiv x^2 y'' + a_1 xy' + a_0 y,$$

and by noting that x^λ is well-defined for $x > 0$, it follows that

$$\begin{aligned} Lx^\lambda &= x^\lambda [\lambda(\lambda - 1) + a_1\lambda + a_0] \\ &= x^\lambda p(\lambda). \end{aligned}$$

Thus x^λ for $x > 0$ is a solution to the Euler equation (5.7) if and only if

$$p(\lambda) = (\lambda - \lambda_1)(\lambda - \lambda_2) = 0.$$

Case 1. If $\lambda_1 \neq \lambda_2$, then

$$\{x^{\lambda_1}, x^{\lambda_2}\}$$

is a fundamental set of solutions, and so the general solution is

$$y(x) = c_1 x^{\lambda_1} + c_2 x^{\lambda_2},$$

where c_1 and c_2 are two constants. This is because the Wronskian

$$\begin{aligned} W(x^{\lambda_1}, x^{\lambda_2}) &= \begin{vmatrix} x^{\lambda_1} & x^{\lambda_2} \\ \lambda_1 x^{\lambda_1 - 1} & \lambda_2 x^{\lambda_2 - 1} \end{vmatrix} \\ &= x^{\lambda_1 + \lambda_2 - 1}(\lambda_2 - \lambda_1) \\ &\neq 0. \end{aligned}$$

Case 2. If $\lambda_1 = \lambda_2$, then not only x^{λ_1} is a solution, but also $x^{\lambda_1} \ln(x)$ is a solution. This is because, on one hand,

$$\begin{aligned} \frac{\partial}{\partial\lambda}[Lx^\lambda] &= L\frac{\partial}{\partial\lambda}x^\lambda \\ &= L[x^\lambda \ln(x)], \end{aligned}$$

and on the other hand,

$$\frac{\partial}{\partial \lambda}[Lx^\lambda] = \frac{\partial}{\partial \lambda}[x^\lambda p(\lambda)]$$

$$= \frac{\partial}{\partial \lambda}[x^\lambda(\lambda - \lambda_1)^2]$$

$$= x^\lambda(\lambda - \lambda_1)^2 \ln(x) + x^\lambda 2(\lambda - \lambda_1),$$

which is zero when $\lambda = \lambda_1$. Thus

$$L[x^{\lambda_1} \ln(x)] = 0,$$

and so $x^{\lambda_1} \ln(x)$ is a second solution. We claim that the Wronskian $W(x^{\lambda_1}, x^{\lambda_1} \ln(x))$ is not zero, so that

$$\{x^{\lambda_1}, x^{\lambda_1} \ln(x)\}$$

is a fundamental set of solutions, and that the general solution is

$$y(x) = c_1 x^{\lambda_1} + c_2 x^{\lambda_1} \ln(x).$$

To see the claim, we compute

$$W(x^{\lambda_1}, x^{\lambda_1} \ln(x)) = \begin{vmatrix} x^{\lambda_1} & x^{\lambda_1} \ln(x) \\ \lambda_1 x^{\lambda_1-1} & (1 + \lambda_1 \ln(x))x^{\lambda_1-1} \end{vmatrix}$$

$$= x^{2\lambda_1-1}$$

$$\neq 0.$$

For $x < 0$, x^λ is not defined but this problem can be solved by using the change of variable : $\xi = -x > 0$. It follows that

$$\frac{d}{dx}y = \frac{dy}{d\xi}\frac{d\xi}{dx} = (-1)\frac{dy}{d\xi},$$

$$\frac{d^2y}{dx} = \frac{d}{dx}(\frac{dy}{dx}) = \frac{d}{d\xi}(\frac{dy}{dx})\frac{d\xi}{dx}$$

$$= \frac{d^2y}{d\xi^2},$$

and so the Euler equation (5.7) becomes

$$\xi^2 \frac{d^2y}{d\xi^2} + a_1 \xi \frac{dy}{d\xi} + a_0 y = 0. \tag{5.8}$$

Again, this is an Euler equation but now the independent variable ξ is positive, and so the results above for the Euler equation (5.7) with $x > 0$ can be applied. It results that for $x < 0$, the general solution is obtained by changing the x in the general solution $y(x)$ for $x > 0$ into $-x$.

Now suppose that $x > 0$ and that $\lambda_1 = \alpha + i\beta$ is the complex conjugate of $\lambda_2 = \alpha - i\beta$, where $i = \sqrt{-1}$, and α and $\beta \neq 0$ are real numbers. Since

$$x^{\lambda_1} = x^\alpha[\cos(\beta \ln(x)) + i \sin(\beta \ln(x))],$$

$$x^{\lambda_2} = x^\alpha[\cos(\beta \ln(x)) - i \sin(\beta \ln(x))],$$

it follows from superposition principle that

$$y_1 = x^\alpha \cos(\beta \ln(x)) = \frac{x^{\lambda_1} + x^{\lambda_2}}{2} \quad \text{and}$$

$$y_2 = x^\alpha \sin(\beta \ln(x)) = \frac{x^{\lambda_1} - x^{\lambda_2}}{2i}$$

are two real-valued solutions to the Euler equation (5.7) with $x > 0$. It follows that $\{y_1, y_2\}$ is a fundamental set and that the general real-valued solution to Euler equation (5.7) with $x > 0$ is

$$y(x) = c_1 y_1 + c_2 y_2,$$

where c_1 and c_2 are two real constants. To see this, we compute the Wronskian

$W(y_1, y_2)$

$$= \begin{vmatrix} x^\alpha \cos(\beta \ln(x)) & x^\alpha \sin(\beta \ln(x)) \\ \alpha x^{\alpha-1} \cos(\beta \ln(x)) - x^{\alpha-1}\beta \sin(\beta \ln(x)) & \alpha x^{\alpha-1} \sin(\beta \ln(x)) + x^{\alpha-1}\beta \cos(\beta \ln(x)) \end{vmatrix}$$

$$= x^{2\alpha-1}\beta,$$

which is not zero.   □

Another method for solving the Euler equation (5.7) is to use the change of variable, Boyce and DiPrima [**3**, pages 246–247]: $t = \ln(x)$.

PROPOSITION 5.31. *The change $t = \ln(x)$ of variable transforms the Euler equation (5.7) with $x > 0$ into the solvable linear second order equation with constant coefficients:*

$$\frac{d^2 y}{dt^2} + (a_1 - 1)\frac{dy}{dt} + a_0 y = 0.$$

PROOF. The change $t = \ln(x)$ of variable gives

$$\frac{dy}{dx} = \frac{dy}{dt}\frac{dt}{dx} = \frac{dy}{dt}\frac{1}{x},$$

$$\frac{d^2 y}{dx^2} = \frac{d}{dt}\left(\frac{dy}{dx}\right)\frac{dt}{dx}$$

$$= [\frac{d^2 y}{dt^2}\frac{1}{x} + \frac{dy}{dt}\frac{-1}{x^2}\frac{dx}{dt}]\frac{dt}{dx}$$

$$= \frac{1}{x^2}\frac{d^2 y}{dt^2}(-\frac{1}{x^2})\frac{dy}{dt},$$

and so the Euler equation (5.7) with $x > 0$, becomes

$$\frac{d^2 y}{dt^2} + (a_1 - 1)\frac{dy}{dt} + a_0 y = 0,$$

a solvable linear second order equation with constant coefficients.   □

EXAMPLE 5.32. Solve the equation

$$x^2 y'' + 3xy' - 8 = x + \ln(x)$$

for $y = y(x)$.

Solution. Using the change $t = \ln(x)$ of variable, we have from Proposition 5.31 that

$$\frac{d^2 y}{dt^2} + 2\frac{dy}{dt} - 8 = e^t + t.$$

The characteristic equation is

$$p(\lambda) = \lambda^2 + 2\lambda - 8 = 0,$$

which has the two roots: $2, -4$. Thus a fundamental set of solutions is

$$\{e^{2t}, e^{-4t}\}.$$

To obtain a particular solution, we consider the two equations separately:

$$\frac{d^2y}{dt^2} + 2\frac{dy}{dt} - 8y = e^t, \tag{5.9}$$

$$\frac{d^2y}{dt^2} + 2\frac{dy}{dt} - 8y = t. \tag{5.10}$$

For the equation (5.9), a particular solution takes the form

$$g_{p_1} = \alpha e^t,$$

where constant α is determined by the condition that g_{p_1} satisfies the equation (5.9). Calculations show that $\alpha = \frac{-1}{5}$.

For the equation (5.10), a particular solution takes the form

$$g_{p_2} = ct + d,$$

where constants c and d are determined by the condition that g_{p_1} satisfies the equation (5.10). Calculations show that $c = \frac{-1}{8}$ and $d = \frac{-1}{32}$.
Thus the general solution is

$$y(t) = c_1 e^{2t} + c_2 e^{-4t} + g_{p_1} + g_{p_2},$$

where c_1 and c_2 are two constants. Changing t back to x, we have

$$y(x) = c_1 x^2 + c_2 x^{-4} + \frac{-1}{5}x + \frac{-1}{8}\ln(x) + \frac{-1}{32}.$$

A change of variable that is more general than the one in Proposition 5.31 is given next, which transforms a linear second order equation with variable coefficients into a linear second order equation with constant coefficients, or into a linear second order equation without the term of first order derivative, Boyce and DiPrima [**3**, pages 142–143]:

PROPOSITION 5.33. *The change $t = f(x)$ variable transforms the equation*

$$y''(x) + p(x)y' + q(x)y = 0, \quad a < x < b \tag{5.11}$$

into the equation

$$\frac{d^2y}{dt^2} + \frac{f''(x) + p(x)f'(x)}{(f'(x))^2}\frac{dy}{dt} + \frac{q(x)}{(f'(x))^2}y = 0, \tag{5.12}$$

provided that $p(x)$ is a continuous function on $[a, b]$, $q(x)$ is a continuously differentiable function on $[a, b]$, and $f(x)$ is a twice continuously differentiable function on $[a, b]$ that has the inverse f^{-1}.

In particular, if $f(x)$ is such that

$$f'(x) = e^{-\int p(x)\,dx},$$

then the equation (5.12) becomes

$$\frac{d^2y(t)}{dt^2} + \frac{q(x)}{(f'(x))^2}y(t) = 0,$$

an equation without the term of first order derivative where x is evaluated at $x = F^{-1}(t)$.

Furthermore, suppose $f(x)$ is such that $\frac{q(x)}{(f'(x))^2} = c_1$ where c_1 is a constant making $\frac{q(x)}{c_1} > 0$, and such that

$$\frac{f''(x) + p(x)f'(x)}{(f'(x))^2} = c_2$$

is a constant for which

$$c_2 = \frac{f''(x) + 2p(x)f'(x)}{(f'(x))^2} \tag{5.13}$$

$$= \frac{q'(x) + 2p(x)q(x)}{2q(x)\sqrt{\frac{q(x)}{c_1}}}. \tag{5.14}$$

Then the equation (5.12) becomes

$$\frac{d^2y(t)}{dt^2} + c_2\frac{dy(t)}{dt} + c_1y(t) = 0,$$

an equation with constant coefficients.

PROOF. The change $t = f(x)$ of variable gives

$$\frac{dy}{dx} = \frac{dy}{dt}\frac{dt}{dx} = f'(x)\frac{dy}{dt},$$

$$\frac{d^2y}{dx^2} = \frac{d}{dt}(f'(x)\frac{dy}{dt})\frac{dt}{dx}$$

$$= [f'(x)\frac{d^2y}{dt^2} + \frac{dy}{dt}\frac{d}{dt}]\frac{dt}{dx}$$

$$= (f'(x))^2\frac{d^2y}{dt^2} + f''(x)\frac{dy}{dt},$$

and so (5.11) becomes the equation (5.12)

$$\frac{d^2y}{dt^2} + \frac{f''(x) + p(x)f'(x)}{(f'(x))^2}\frac{dy}{dt} + \frac{q(x)}{(f'(x))^2}y = 0.$$

If $f(x)$ is such that

$$f'(x) = e^{-\int p(x)\,dx},$$

then the equation (5.12) becomes

$$\frac{d^2y}{dt^2} + \frac{q(x)}{(f'(x))^2}y = 0,$$

an equation without the term of first order derivative where x is evaluated at $x = f^{-1}(t)$.

Suppose that there are two constants c_1 and c_2, such that

$$\frac{q(x)}{(f'(x))^2} = c_1,$$

$$\frac{f''(x) + p(x)f'(x)}{(f'(x))^2} = c_2,$$

where c_1 is such that

$$\frac{q(x)}{c_1} > 0.$$

Then

$$c_2 = \frac{f''(x) + p(x)f'(x)}{(f'(x))^2}$$
$$= \frac{q'(x) + 2p(x)q(x)}{2q(x)\sqrt{\frac{q(x)}{c_1}}},$$

and (5.12) becomes

$$\frac{d^2y}{dt^2} + c_2\frac{dy}{dt} + c_1 y = 0,$$

an equation with constant coefficients.   □

EXAMPLE 5.34. Solve the equation

$$y''(x) + 2(e^{2x} - 1)y' + e^{4x}y = 0, \quad 3 < x < 8$$

for $y = y(x)$.

Solution. We use Proposition 5.33. Choose $c_1 = 1$ and compute

$$f(x) = \int \sqrt{\frac{q(x)}{c_1}}\,dx = \int e^{2x}\,dx$$
$$= \frac{e^{2x}}{2}.$$

Now check if the condition (5.13) is satisfied:

$$\frac{q'(x) + 2p(x)q(x)}{q(x)\sqrt{\frac{q(x)}{c_1}}} = \frac{4e^{4x} + 4(e^{2x} - 1)e^{4x}}{e^{6x}}$$
$$= 4,$$

which is a constant, as desired. It follows from Proposition 5.33 that we have the transformed equation

$$y''(t) + 4y'(t) + y(t) = 0,$$

and the general solution is by Section 2

$$y(t) = c_1 e^{(-2+\sqrt{3})t} + c_2 e^{(-2-\sqrt{3})t},$$

where c_1 and c_2 are two constants. Changing t back to x through $t = f(x) = \frac{e^{2x}}{2}$, we have the general solution

$$y(x) = c_1 e^{(-2+\sqrt{3})\frac{e^{2x}}{2}} + c_2 e^{(-2-\sqrt{3})\frac{e^{2x}}{2}}.$$

EXAMPLE 5.35. Transform the equation

$$y'' - \frac{2x}{x^2 + 1}y' + \sin(x)y = 0, \quad 3 < x < 8$$

into an equation that has no term of the first order derivative.

Solution. We use Proposition 5.33. The condition $f''(x) + p(x)f'(x) = 0$ gives

$$f'(x) = e^{\int \frac{2x}{x^2+1}\,dx}$$
$$= e^{\ln(x^2+1)}$$
$$= x^2 + 1,$$

and so $f(x) = \frac{x^3}{3} + x$. It follows from Proposition 5.33 that, using the transform

$$t = f(x) = \frac{x^3}{3} + x,$$

we obtain the equation

$$y''(t) + \frac{\sin(f^{-1}(t))}{[(f^{-1}(t))^2 + 1]^2} y(t) = 0,$$

that has no first order derivative term.

5.3. Exact Equations. We consider a nonlinear first order equation of the form

$$M(x, y) + N(x, y) \frac{d}{dx} y = 0, \quad a < x < b. \tag{5.15}$$

where $M(x, y)$ and $N(x, y)$ are two continuous functions defined on the region

$$R \equiv \{(x, y) : a < x < b, c < y < d\}.$$

Here a, b, c, and d are four given real numbers.

THEOREM 5.36. *Suppose that functions $M(x, y)$ and $N(x, y)$, together with their partial derivatives $\frac{\partial}{\partial y} M(x, y)$ and $\frac{\partial}{\partial x} N(x, y)$, are continuous on the region R. Then the following two statements (i) and (ii) are equivalent:*

(i) *$\frac{\partial}{\partial y} M = \frac{\partial}{\partial x} N$ on R.*
(ii) *There is a continuous function $F(x, y)$ on R*

$$F(x, y) = \int_{t=x_0}^{x} M(t, y) \, dt + g(y), \tag{5.16}$$

such that

$$\frac{\partial}{\partial x} F = M,$$

$$\frac{\partial}{\partial y} F = N.$$

Here $x_0 \in (a, b)$, and the continuously differentiable function $g(y)$ on (c, d) is determined by the condition

$$\frac{\partial}{\partial y} F = N. \tag{5.17}$$

REMARK 5.37. In the calculation of $F(x, y)$, we can drop the contribution from the lower limit of the definite integral

$$\int_{t=x_0}^{x} M(t, y) \, dt$$

since it, after the substitution of $t = x_0$, becomes a function of y only but not x and can be incorporated into $g(y)$ to make a new $g(y)$ depending only on y. In other words, we can take the integration constant in the indefinite integral

$$G(x, y) \equiv \int M(x, y) \, dx$$

to be zero, where $G(x, y)$ is a function such that

$$\frac{\partial}{\partial x} G(x, y) = M(x, y).$$

This is always true since, for example, we can take

$$G(x, y) = \int_{t=x_0}^{x} M(t, y)\, dt.$$

DEFINITION 5.38. *The equation* (5.15)

$$M(x, y) + N(x, y)\frac{d}{dx}y = 0, \quad a < x < b,$$

where $M(x, y)$ and $N(x, y)$, together with their partial derivatives $\frac{\partial}{\partial y}M(x, y)$ and $\frac{\partial}{\partial x}N(x, y)$, are continuous on R, is exact if the condition

$$\frac{\partial}{\partial y}M = \frac{\partial}{\partial x}N.$$

Here is our main result in this subsection:

THEOREM 5.39. *If $y = y(x)$ is a solution to an exact equation* (5.15)*, then $y(x)$ is of the form*

$$F(x, y) = c$$

for some constant c where $F(x, y)$ is defined by equations (5.16) *and* (5.17)*.*
Conversely, the function $y = y(x)$ defined by

$$F(x, y) = c,$$

where $F(x, y)$ is conditioned by (5.16) *and* (5.17)*, is a solution to an exact equation* (5.15)*.*

DEFINITION 5.40. *The function $y = y(x)$ defined by*

$$F(x, y) = c,$$

is called the general solution to an exact equation (5.15)*.*

In general, the first order equation (5.15) is not exact but it might become exact when multiplied by some function:

DEFINITION 5.41. *A function $\mu = \mu(x, y)$ is called an integrating factor for the equation* (5.15) *if the equation* (5.15) *multiplied by μ becomes exact.*

LEMMA 5.42. *Suppose that $\mu = \mu(x, y)$ is an integrating factor for the equation* (5.15) *and that μ has partial derivatives $\frac{\partial}{\partial x}\mu(x, y)$ and $\frac{\partial}{\partial y}\mu(x, y)$. Then $\mu(x, y)$ satisfies the equation*

$$M(x, y)\frac{\partial}{\partial y}\mu(x, y) - N(x, y)\frac{\partial}{\partial x}\mu(x, y) = [\frac{\partial}{\partial x}N(x, y) - \frac{\partial}{\partial y}M(x, y)]\mu(x, y). \quad (5.18)$$

The equation (5.18) is generally difficult to solve for $\mu(x, y)$. However,

COROLLARY 5.43. *If the $\mu(x, y)$ in Lemma 5.42 depends only on x, then $\mu = \mu(x)$ is determined by the separable equation*

$$\frac{d}{dx}\mu(x) = \frac{\frac{\partial}{\partial y}M(x, y) - \frac{\partial}{\partial x}N(x, y)}{N(x, y)}\mu(x). \quad (5.19)$$

COROLLARY 5.44. *If the $\mu = \mu(y)$ in Lemma 5.42 depends only on y, then $\mu(y)$ is determined by the separable equation*

$$\frac{d}{dy}\mu(y) = \frac{\frac{\partial}{\partial x}N(x, y) - \frac{\partial}{\partial y}M(x, y)}{M(x, y)}\mu(y). \quad (5.20)$$

Before we prove the main results, we first look at some examples.

EXAMPLE 5.45. Solve the equation

$$(2xy + y^3) + (x^2 + 3xy^2)\frac{d}{dx}y(x) = 0$$

for $y = y(x)$.

Solution. Since

$$\frac{\partial}{\partial y}(2xy + y^3) = 2x + 3y^2$$

$$= \frac{\partial}{\partial x}(x^2 + 3xy^2),$$

the equation in the example is exact. It follows from equations (5.16) and (5.17) that

$$F(x, y) = \int_{t=x_0}^{x} (2ty + y^3)\, dt + g(y)$$

$$= [x^2 y + xy^3] - [x_0^2 y + x_0 y^3] + g(y),$$

$$x^2 + 3xy^2 = \frac{\partial}{\partial y}F(x, y)$$

$$= [x^2 + 3xy^2] - [x_0^2 + 3x_0 y^2] + g'(y),$$

and so

$$g'(y) = x_0^2 + 3x_0 y^2.$$

Integrating both sides of this separable equation, we obtain

$$g(y) = x_0^2 y + x_0 y^3,$$

in which we take the integration constant to be zero for simplicity. Substituting the obtained $g(y)$ into the above $F(x, y)$, we have the final

$$F(x, y) = [x^2 y + xy^3] - [x_0^2 y + x_0 y^3] + x_0^2 y + x_0 y^3$$
$$= x^2 y + xy^3.$$

It follows from Theorem 5.39 that the function $y = y(x)$ determined by the equation

$$F(x, y) = c,$$

is the general solution. Here c is a constant.

If combined with Remark 5.37, we can write

$$F(x, y) = x^2 y + xy^3 + g(y)$$
$$g'(y) = 0,$$

and so, taking $g(y) = 0$, we obtain the same $F(x, y)$ as above.

EXAMPLE 5.46. Solve the equation

$$(e^y + 2xy^2) + (xe^y + 2x^2 y)\frac{d}{dx}y = 0$$

for $y = y(x)$.

Solution. Since

$$\frac{\partial}{\partial y}(e^y + 2xy^2) = e^y + 4xy$$

$$= \frac{\partial}{\partial x}(xe^y + 2x^2y),$$

the equation in the example is exact. It follows from (5.16), (5.17), and Remark 5.37 that

$$F(x,y) = \int (e^y + 2xy^2)\, dy + g(y)$$

$$= xe^y + x^2y^2 + g(y),$$

$$xe^y + 2x^2y = \frac{\partial}{\partial y}F(x,y)$$

$$= xe^y + 2x^2y + g'(y),$$

and so

$$g'(y) = 0.$$

Integrating both sides of this separable equation, we obtain

$$g(y) = 0,$$

in which we take the integration constant to be zero for simplicity. Substituting the obtained $g(y)$ into the above $F(x,y)$, we have the final

$$F(x,y) = xe^y + x^2y^2 + 0.$$

It follows from Theorem 5.39 that the function $y = y(x)$ determined by the equation

$$F(x,y) = c,$$

is the general solution. Here c is a constant.

EXAMPLE 5.47. Solve the equation

$$(-y + 2x^3y^2) + (x + 2x^4y)\frac{d}{dx}y = 0$$

for $y = y(x)$.

Solution. It is readily checked that

$$\frac{\partial}{\partial y}(-y + 2x^3y^2) = (-1 + 4x^3y)$$

$$\neq \frac{\partial}{\partial x}(x + 2x^4y)(= 1 + 8x^3y),$$

and so the equation in the example is not exact. However, using the formula (5.19) in Corollary 5.43, we have

$$\frac{d}{dx}\mu(x) = \frac{(-1 + 4x^3y) - (1 + 8x^3y)}{x + 2x^4y}\mu(x)$$

$$= \frac{-2}{x}\mu(x).$$

Integrating both sides of this separable equation, we obtain the integrating factor

$$\mu(x) = x^{-2},$$

in which we take the integration constant to be zero for simplicity. Multiplying the equation in this example by $\mu(x)$, we have the exact equation

$$(\frac{-y}{x^2} + 2xy^2) + (\frac{1}{x} + 2x^2y)\frac{d}{dx}y = 0.$$

It follows from (5.16), (5.17), and Remark 5.37 that

$$F(x, y) = \int (\frac{-y}{x^2} + 2xy^2)\, dx + g(y)$$

$$= x^{-1}y + x^2y^2 + g(y),$$

$$(x^{-1} + 2x^2y) = \frac{\partial}{\partial y}F(x, y)$$

$$= x^{-1} + x^2y^2 + g'(y),$$

and so

$$g'(y) = 0.$$

Integrating both sides of this separable equation, we derive

$$g(y) = 0,$$

in which we take the integration constant to be zero for simplicity. Substituting the obtained $g(y)$ into the above $F(x, y)$, it follows that

$$F(x, y) = x^{-1}y + x^2y^2 + 0.$$

Thus Theorem 5.39 says that the function $y = y(x)$ determined by the equation

$$F(x, y) = c,$$

is the general solution. Here c is a constant.

Proof of Theorem 5.36.

PROOF. We divide the proof into two steps.

Step 1. Suppose that (i) holds, and we need to show that (ii) is true. Define the function $F(x, y)$ by

$$F(x, y) = \int_{t=x_0}^{x} M(t, y)\, dt + g(y),$$

where $x_0 \in (a, b)$ and the function $g(y)$ is to be determined by the condition

$$\frac{\partial}{\partial y}F = N.$$

Once g can be found, (ii) follows. This is because

$$\frac{\partial}{\partial x}F = \frac{\partial}{\partial x}[\int_{t=x_0}^{x} M(t, y)\, dt + g(y)]$$

$$= M(x, y) + 0$$

by the fundamental theorem of calculus, Apostol [**2**, pages 161–162]:

FACT 5.48. *Suppose that $f(x)$ is a continuous function defined on $[a, b]$ for which the Riemann integral*

$$G(x) \equiv \int_{t=a}^{x} f(t)\, dt$$

exists for each $x \in [a, b]$. Then for each $x \in [a, b]$, $G'(x)$ exists, and equals $f(x)$.

To see that $g(x)$ can be found, compute the condition

$$N(x, y) = \frac{\partial}{\partial y} F(x, y)$$

$$= \frac{\partial}{\partial y} \left[\int_{t=x_0}^{x} M(t, y) \, dt + g(y) \right]$$

$$= \int_{t=x_0}^{x} \frac{\partial}{\partial y} M(t, y) \, dt + g'(y)$$

$$= \int_{t=x_0}^{x} \frac{\partial}{\partial t} N(t, y) \, dt + g'(y) \quad \text{by the assumption (i),}$$

and so

$$g'(y) = N(x, y) - \int_{t=x_0}^{x} \frac{\partial}{\partial t} N(t, y) \, dt$$

$$\equiv G(x, y).$$

It follows that $g(y)$ is solvable if $G(x, y)$ does not depend on x, since then it is a separable equation. But this is true, since

$$\frac{\partial}{\partial x} G(x, y) = \frac{\partial}{\partial x} N(x, y) - \frac{\partial}{\partial x} \int_{t=x_0}^{x} \frac{\partial}{\partial t} N(t, y) \, dt$$

$$= \frac{\partial}{\partial x} N(x, y) - \frac{\partial}{\partial x} N(x, y) \quad \text{by the Fundamental Theorem of Calculus}$$

$$= 0,$$

which shows that $G(x, y)$ does not depend on x.

Step 2. Suppose that (ii) holds, and we need to show that (i) is true. Notice that

$$\frac{\partial^2}{\partial y \partial x} F(x, y) = \frac{\partial}{\partial y} \left[\frac{\partial}{\partial x} F(x, y) \right]$$

$$= \frac{\partial}{\partial y} M(x, y), \quad \text{and}$$

$$\frac{\partial^2}{\partial x \partial y} F(x, y) = \frac{\partial}{\partial x} \left[\frac{\partial}{\partial y} F(x, y) \right]$$

$$= \frac{\partial}{\partial x} N(x, y),$$

and that $\frac{\partial}{\partial y} M(x, y)$ and $\frac{\partial}{\partial x} N(x, y)$ are continuous by assumption. It follows that

$$\frac{\partial^2}{\partial x \partial y} F(x, y) = \frac{\partial^2}{\partial y \partial x} F(x, y),$$

or equivalently

$$\frac{\partial}{\partial y} M(x, y) = \frac{\partial}{\partial x} N(x, y),$$

and this is (i). Here we used the fact about equality of mixed derivatives from Advanced Calculus, Apostol [**2**, page 360]:

FACT 5.49. *Suppose that $f(x, y)$ is a function on the region*

$$R = \{(x, y) : a < x < b, c < y < d\},$$

such that both partial derivatives $\frac{\partial}{\partial x} f(x, y)$ and $\frac{\partial}{\partial y} f(x, y)$ exist on R and that both $\frac{\partial^2}{\partial x \partial y} f(x, y)$ and $\frac{\partial^2}{\partial y \partial x} f(x, y)$ are continuous on R. Then

$$\frac{\partial^2}{\partial x \partial y} f(x, y) = \frac{\partial^2}{\partial y \partial x} f(x, y).$$

$\square$

Proof of Theorem 5.39.

PROOF. We divide the proof into two steps.

Step 1. Suppose that $y = y(x)$ is a solution to an exact equation (5.15). We need to show that y is of the form

$$F(x, y) = c$$

for some constant c where $F(x, y)$ is determined by equations (5.16) and (5.17).

Since equation (5.15) is exact, it follows from Theorem 5.36 that there is a continuous function $F(x, y)$ on R determined by (5.16) and (5.17), such that

$$M(x, y) = \frac{\partial}{\partial x} F(x, y) \quad \text{and}$$

$$N(x, y) = \frac{\partial}{\partial y} F(x, y).$$

Substituting the obtained $M(x, y)$ and $N(x, y)$ into equation (5.15), we have

$$\begin{aligned}
0 &= M(x, y) + N(x, y) \frac{d}{dx} y \\
&= \frac{\partial}{\partial x} F(x, y) + [\frac{\partial}{\partial y} F(x, y)] \frac{d}{dx} y \\
&= \frac{d}{dx} F(x, y),
\end{aligned}$$

and so

$$F(x, y) = c$$

for some constant c. This completes Step 1. Here we used the fact from Advanced Calculus, Apostol [**2**, pages 353–354]:

FACT 5.50. *Suppose that $f(x, y)$ is a differentiable function on the region*

$$R = \{(x, y) : a < x < b, c < y < d\}$$

and that $y = y(x)$ is a differentiable function on

$$\{x : a < x < b\}.$$

Then

$$\frac{d}{dx} f(x, y) = \frac{\partial}{\partial x} f(x, y) + [\frac{\partial}{\partial y} f(x, y)] \frac{d}{dx} y.$$

Step 2. Conversely, suppose that $y = y(x)$ is a function defined by the equation

$$F(x, y) = c$$

where $F(x, y)$ is conditioned by (5.16) and (5.17). We need to show that y is a solution to the equation (5.15).

Notice that

$$M(x, y) = \frac{\partial}{\partial x} F(x, y) \quad \text{and}$$

$$N(x, y) = \frac{\partial}{\partial y} F(x, y)$$

under our assumption. It follows from differentiation and the Fact in Step 1 that

$$0 = \frac{d}{dx} F(x, y)$$

$$= \frac{\partial}{\partial x} F(x, y) + [\frac{\partial}{\partial y} F(x, y)] \frac{d}{dx} y$$

$$= M(x, y) + N(x, y) \frac{d}{dx} y,$$

and so $y = y(x)$ is a solution to the equation (5.15). $\qquad\qquad\square$

6. The Technique of the Laplace Transform

In this section, we shall use the technique of the Laplace transform to compute explicitly the unique continuous solution of the initial value problem with constant coefficients $a, b,$ and c:

$$ay''(t) + by'(t) + cy(t) = g(t), \quad 0 < t < \infty,$$
$$y(0) = y_0, \qquad\qquad\qquad\qquad\qquad\quad (6.1)$$
$$y'(0) = y_1.$$

This technique is particularly useful when $g(t)$ is a piecewise continuous function on $[0, \infty)$ or a so-called delta function $\delta(t - t_0)$ where $0 \le t_0 < \infty$. In the former case, the unique continuous solution of (6.1) is assured by the existence theorem in Chapter 8. However, in the latter case, the equation (6.1) should be interpreted appropriately, for which see Chapter 5 for the meaning of $\delta(t - t_0)$. Basically, the delta function $\delta(t - t_0)$ is symbolically characterized by the two properties:

- $\delta(t - t_0) = 0$ for $t \ne t_0$.
- For a piecewise continuous function $f(t)$ on (α, β) where (α, β) is any subset of $(-\infty, \infty)$, it is true that

$$\int_\alpha^\beta \delta(t - t_0) f(t) \, dt = \begin{cases} 0, & \text{if } t_0 \notin (\alpha, \beta); \\ f(t_0), & \text{if } t_0 \in (\alpha, \beta). \end{cases}$$

At $t = t_0$, $\delta(t - t_0)$ is thought of as ∞.

DEFINITION 6.1. *The Laplace transform of a piecewise continuous function $f(t)$ on $[0, \infty)$ is the function*

$$F(s) = \mathcal{L}\{f(t)\} \equiv \int_{t=0}^\infty e^{-st} f(t) \, dt$$

$$\equiv \lim_{A \to \infty} \int_{t=0}^A e^{-st} f(t) \, dt, \quad \text{an improper integral,}$$

where $s > 0$.

While $\mathcal{L}\{f(t)\}$ does not necessarily exist for an arbitrary f, a sufficient condition for it to exist is this:

THEOREM 6.2. *Suppose that* $f(t)$ *is a piecewise continuous function on* $[0, \infty)$ *and that there are constants* $M > 0, \alpha \in \mathbb{R}$, *and* $t_0 \in [0, \infty)$, *such that*

$$|f(t)| \le M e^{\alpha t} \quad \text{for all } t \ge t_0. \tag{6.2}$$

Then the Laplace transform $F(s) = \mathcal{L}\{f(t)\}$ *exists as a function of* $s \in [\alpha, \infty)$.

DEFINITION 6.3. *A piecewise continuous function that satisfies* (6.2) *is called of exponential growth with order* α.

The proof of this theorem will be given at the end of this section, and some examples will be examined first:

EXAMPLE 6.4.

$$\mathcal{L}\{t\} = \frac{1}{s^2} \quad \text{for } s > 0, \text{ and in general,}$$

$$\mathcal{L}\{t^n\} = \frac{n!}{s^{n+1}} \quad \text{for } s > 0, \text{ where } n = 0, 1, 2, \ldots.$$

Solution. By the formula of integration by parts, we have

$$\int_{t=0}^{A} e^{-st} t \, dt = -\frac{1}{s} [te^{-st}|_{t=0}^{A} - \int_{0}^{A} e^{-st} \, dt]$$

$$= -\frac{1}{s} [(Ae^{-sA} - 0) + \frac{1}{s}(e^{-sA} - 1)]$$

$$\longrightarrow \frac{1}{s^2}$$

as $A \longrightarrow +\infty$.

The other formula is proved by mathematical induction.

EXAMPLE 6.5.

$$\mathcal{L}\{\cos(at)\} = \frac{s}{s^2 + a^2} \quad \text{for } s > 0, \text{ and}$$

$$\mathcal{L}\{\sin(at)\} = \frac{a}{s^2 + a^2} \quad \text{for } s > 0.$$

Here a is a real constant.

Solution. By the formula of integration by parts, we have

$$I \equiv \int_{t=0}^{A} e^{-st} \cos(at) \, dt$$

$$= \frac{1}{a} [e^{-st} \sin(at)|_{t=0}^{A} + s \int_{0}^{A} e^{-st} \sin(at) \, dt]$$

$$= [\frac{1}{a} e^{-st} \sin(at) - \frac{s}{a^2} e^{-st} \cos(at)]|_{t=0}^{A}$$

$$- \frac{s^2}{a^2} \int_{t=0}^{A} e^{-st} \cos(at) \, dt,$$

and so

$$I = \frac{a^2}{s^2 + a^2}[\frac{1}{a}e^{-st}\sin(at) - \frac{s}{a^2}e^{-st}\cos(at)]|_{t=0}^{A}$$

$$= \frac{a^2}{s^2 + a^2}\{[\frac{1}{a}e^{-sA}\sin(aA) - \frac{s}{a^2}e^{-sA}\cos(aA)] + \frac{s}{a^2}\}$$

$$\longrightarrow \frac{a^2}{s^2 + a^2}(0 + \frac{s}{a^2}) = \frac{s}{s^2 + a^2} \quad \text{for } s > 0$$

as $A \longrightarrow +\infty$. Here we used

$$|e^{-sA}\cos(aA)| \le e^{-sA} \longrightarrow 0 \quad \text{and}$$

$$|e^{-sA}\sin(aA)| \le e^{-sA} \longrightarrow 0.$$

The other formula is similarly proved.

From Example 6.4, we see that the Laplace transform of the function t is $\frac{1}{s^2}$. Are there other functions that also have $\frac{1}{s^2}$ as their Laplace transform? This is answered by the theorem whose proof will be omitted, Widder [**22**]:

THEOREM 6.6 (Lerch). *If $f(t)$ and $g(t)$ are two continuous functions on $[0, \infty)$ that have the same Laplace transform, then*

$$f = g \quad identically.$$

Thus we can define

DEFINITION 6.7. *If $f(t)$ is continuous on $[0, \infty)$ and $\pounds\{f(t)\} = F(s)$, then the inverse $\pounds^{-1}$ of $\pounds$ is defined by*

$$\pounds^{-1}\{F(s)\} = f(t).$$

Since the function $f(t) = t$ is continuous and $\pounds\{f(t)\} = \frac{1}{s^2}$, we have by Theorem 6.6 that $f(t) = t$ is the only continuous function that has $\frac{1}{s^2}$ as its Laplace transform. Therefore, we have

$$\pounds^{-1}\{\frac{1}{s^2}\} = t.$$

The technique for solving the initial value problem in this section consists in taking the Laplace transform on both sides of (6.1), and therefore we need to know the computing of $\pounds\{y'\}$ and $\pounds\{y''\}$ of the derivatives y'' and y' of y, Boyce and DiPrima [**3**]:

THEOREM 6.8. *Assume that y is continuous and y' is piecewise continuous on $[0, \infty)$ (or more generally, y' is Riemann integrable on any $[0, A]$). Further assume that y is of exponential growth with order α. Then $\pounds\{y(t)\}$ and $\pounds\{y'(t)\}$ exist for $s > \alpha$, and*

$$\pounds\{y'(t)\} = s\pounds\{y(t)\} - y(0).$$

In general, $\pounds\{y^{(n)}(t)\}, \pounds\{y^{(n-1)}(t)\}, \dots, \pounds\{y'(t)\}$, and $\pounds\{y(t)\}$ exist for $s > \alpha$, and

$$\pounds\{y^{(n)}(t)\} = s^n \pounds\{y(t)\} - s^{n-1}y(0) - s^{n-2}y'(0) - \cdots - sy^{(n-2)}(0) - y^{(n-1)}(0),$$

provided that $y, y', \dots$, and $y^{(n-1)}$ are continuous and $y^{(n)}$ is piecewise continuous on $[0, \infty)$ (or more generally, $y^{(n)}$ is Riemann integrable on any $[0, A]$) and that

$y, y', \ldots,$ and $y^{(n-1)}$ are of exponential growth with the order α. Here $n \in \mathbb{N}$. In particular,

$$\mathcal{L}\{y''(t)\} = s^2 \mathcal{L}\{y(t)\} - sy(0) - y'(0).$$

The proof of this Theorem 6.8 will be given at the end of this section. We now illustrate the technique of the Laplace transform:

EXAMPLE 6.9. Solve for y:

$$y'' + 4y = t, \quad 0 < t < \infty,$$
$$y(0) = 0,$$
$$y'(0) = 0.$$

Solution. Assume that each Laplace transform of $y, y',$ and y'' exists and that y is continuous on $[0, \infty)$. Then by letting $\mathcal{L}\{y(t)\} = Y(s)$, we have

$$\mathcal{L}\{t\} = \frac{1}{s}$$
$$= \mathcal{L}\{y''\} + 4\mathcal{L}\{y\}$$
$$= [s^2 Y - sy(0) - y'(0)] + 4Y,$$

and so

$$Y = \frac{1}{s^2(s^2 + 4)} = \frac{1}{4}\left(\frac{1}{s^2} - \frac{1}{s^2 + 4}\right).$$

Thus

$$y(t) = \mathcal{L}^{-1}\{Y(s)\} = \frac{1}{4}[t - \frac{1}{2}\sin(2t)].$$

From this, we readily check that y is continuous, and that each Laplace transform of $y, y',$ and y'' exists.

To solve larger class of equations, we need more formulas for the Laplace transform, Churchill [4]:

THEOREM 6.10. *The following are true:*

$(i)\, \mathcal{L}\{e^{at}\} = \dfrac{1}{s-a} \quad$ *for $s > a$.*

$(ii)\, \mathcal{L}\{e^{at} f(t)\} = F(s-a) \quad$ *for $s > (a+b)$, if $\mathcal{L}\{f(t)\} = F(s)$ for $s > a$.*

$(iii)\, \mathcal{L}\{t^n f(t)\} = (-1)^n \dfrac{d^n}{ds^n} F(s), n \in \mathbb{N} \quad$ *for $s > a$, if $\mathcal{L}\{f(t)\} = F(s)$ for $s > a$.*

$(iv)\, \mathcal{L}\{u_c(t)\} = \dfrac{e^{-cs}}{s} \quad$ *for $s > 0$.*

$(v)\, \mathcal{L}\{u_c(t) f(t-c)\} = e^{-cs} F(s) \quad$ *for $s > a$, if $\mathcal{L}\{f(t)\} = F(s)$ for $s > a$.*

$(vi)\, \mathcal{L}\{\delta(t-c)\} = e^{-cs} \quad$ *for $s > 0$.*

$(vii)\, \mathcal{L}\{(f * g)(t)\} = \mathcal{L}\{(g * f)(t)\} = F(s)G(s) \quad$ *for $s > a$,*

if $\mathcal{L}\{f(t)\} = F(s)$ and $\mathcal{L}\{g(t)\} = G(s)$ for $s > a$.

Here

$$u_c(t) = \begin{cases} 0, & \text{if } 0 \le t < c; \\ 1, & \text{if } c \le t < \infty, \end{cases} \quad \text{the step function at } t = c,$$

$$(f * g)(t) = \int_{\tau=0}^{t} f(t-\tau)g(\tau)\, d\tau, \quad \text{the convolution of } f \text{ with } g,$$

and $\delta(t-c)$ is the delta function at $t = c$.

By direct computations, we can prove these formulas, except for formulas (*iii*) and (*vii*). A proof of these exceptions is a consequence of applying Advanced Calculus, Apostol [**2**], and is referred to the book, Churchill [**4**].

We next examine more equations by using these formulas:

EXAMPLE 6.11. Solve for y:

$$y'' + 2y' + 2y = e^{-t}\cos(t), \quad 0 < t < \infty,$$
$$y(0) = 1,$$
$$y'(0) = -1.$$

Solution. Assume that y is continuous on $[0, \infty)$ and that each Laplace transform of y, y' and y'' exists. Then by letting $Y(s) = \mathcal{L}\{y(t)\}$, we have

$$\mathcal{L}\{e^{-t}\cos(t)\} = \frac{(s+1)}{(s+1)^2 + 1}$$
$$= \mathcal{L}\{y'' + 2y' + 2y\}$$
$$= [s^2 Y - sy(0) - y'(0)] + 2[sY - y(0)] + 2Y,$$

and so

$$Y = \frac{s+1}{s^2 + 2s + 2} + \frac{s+1}{[(s+1)^2 + 1]^2}$$
$$= \frac{(s+1)}{(s+1)^2 + 1} + \frac{1}{2}(-1)\frac{d}{ds}[\frac{1}{(s+1)^2 + 1}].$$

Thus

$$y(t) = \mathcal{L}^{-1}\{Y(s)\} = e^{-t}\cos(t) + \frac{1}{2}te^{-t}\sin(t).$$

From this, we readily check that y is continuous on $[0, \infty)$ and that each Laplace transform of y, y', and y'' exist.

EXAMPLE 6.12. Solve for y:

$$y'' + 9y = g(t), \quad 0 < t < \infty,$$
$$y(0) = 2,$$
$$y'(0) = 3,$$

where

$$g(t) = \begin{cases} 0, & \text{if } 0 \le t < 1; \\ t - 2, & \text{if } 1 \le t < \pi; \\ \sin(t), & \text{if } \pi \le t < \infty. \end{cases}$$

Solution. Assume that y is continuous on $[0, \infty)$ and that each Laplace transform of y, y', and y'' exist. Rewriting $g(t)$ in terms of step functions, we have

$$g(t) = (t-2)u_1(t) + [\sin(t) - (t-2)]u_\pi(t)$$
$$= -u_1(t) + (t-1)u_1(t) - u_\pi(t)\sin(t-\pi) - (t-\pi)u_\pi(t) - (\pi - 2)u_\pi(t).$$

Then by letting $Y(s) = \mathcal{L}\{y(t)\}$, we have

$$\mathcal{L}\{g(t)\} = -\frac{e^{-s}}{s} + \frac{e^{-s}}{s^2} - \frac{e^{-\pi s}}{s^2 + 1} - \frac{e^{-\pi s}}{s^2} - (\pi - 2)\frac{e^{-\pi s}}{s}$$

$$= \mathcal{L}\{y'' + 9y\}$$
$$= [s^2 Y - sy(0) - y'(0)] + 9Y,$$

and so

$$Y = \frac{2s+3}{s^2+9} + e^{-s}[\frac{1}{s^2(s^2+9)} - \frac{1}{s(s^2+9)}] -$$

$$e^{-\pi s}[\frac{1}{(s^2+1)(s^2+9)} + \frac{1}{s^2(s^2+9)} + (\pi-2)\frac{1}{s(s^2+9)}].$$

Since

$$\frac{1}{s(s^2+9)} = \frac{\frac{1}{9}}{s} + \frac{-\frac{1}{9}s}{s^2+9},$$

$$\frac{1}{s^2(s^2+9)} = \frac{1}{9}(\frac{1}{s^2} - \frac{1}{s^2+9}),$$

$$\frac{1}{(s^2+1)(s^2+9)} = \frac{1}{8}(\frac{1}{s^2+1} - \frac{1}{s^2+9}),$$

we have

$$y(t) = 2\cos(3t) + \sin(3t) + u_1(t)[\frac{1}{9}(t-1) - \frac{1}{27}\sin 3(t-1)]$$

$$- u_1(t)[\frac{1}{9} - \frac{1}{9}\cos 3(t-1)] - u_\pi(t)[\frac{1}{8}\sin(t-\pi) - \frac{1}{24}\sin 3(t-\pi)]$$

$$- u_\pi(t)[\frac{1}{9}(t-\pi) - \frac{1}{27}\sin 3(t-\pi)] - (\pi-2)u_\pi(t)[\frac{1}{9} - \frac{1}{9}\cos 3(t-\pi)]$$

or

$$y(t) = 2\cos(3t) + \sin(3t) + \frac{1}{3}\int_{\tau=0}^{t} u_1(t-\tau)(t-1-\tau)\sin(3\tau)\,d\tau$$

$$- \frac{1}{3}\int_{\tau=0}^{t} u_1(t-\tau)\sin(3\tau)\,d\tau - \frac{1}{3}\int_{\tau=0}^{t} u_\pi(t-\tau)\sin(t-\pi-\tau)\sin(3\tau)\,d\tau$$

$$- \frac{1}{3}\int_{\tau=0}^{t} u_\pi(t-\tau)(t-\pi-\tau)\sin(3\tau)\,d\tau - \frac{\pi-2}{3}\int_{\tau=0}^{t} u_\pi(t-\tau)\sin(3\tau)\,d\tau$$

in terms of convolutions. In this example, the first expression of y is better than the second one, although situations might vary.

EXAMPLE 6.13. Solve for y:

$$y'' + 2y' + 2y = \delta(t-1), \quad 0 < t < \infty,$$
$$y(0) = 1,$$
$$y'(0) = -1.$$

Solution. Assume that y is continuous on $[0, \infty)$ and that each Laplace transform of y, y', and y'' exist. Then by letting $\mathcal{L}\{y(t)\} = Y(s)$, we have

$$\mathcal{L}\{\delta(t-1)\} = e^{-s}$$
$$= \mathcal{L}\{y'' + 2y' + 2y\}$$
$$= [s^2 Y - sy(0) - y'(0)] + 2[sY - y(0)] + 2y$$
$$= (s^2 + 2s + 2)Y - (s+1),$$

and so

$$Y(s) = \frac{(s+1)}{(s+1)^2 + 1} + e^{-s}\frac{1}{(s+1)^2 + 1}.$$

Thus

$$y(t) = \mathcal{L}^{-1}\{Y(s)\}$$
$$= e^{-t}\cos(t) + u_1(t)e^{(t-1)}\sin(t-1).$$

From this, we readily check that y is continuous on $[0, \infty)$ and that each Laplace transform of y, y', and y'' exist.

EXAMPLE 6.14. Solve for y:

$$y'' + 2y' + 2y = g(t), \quad 0 < t < \infty,$$
$$y(0) = 1,$$
$$y'(0) = -1,$$

where $g(t)$ is a piecewise continuous function on $[0, \infty)$ or an impulsive forcing function involving delta functions whose Laplace transform exist.

Solution. Assume that y is continuous on $[0, \infty)$ and that each Laplace transform of y, y', and y'' exist. Then by letting $\mathcal{L}\{y(t)\} = F(s)$ and $\mathcal{L}\{g(t)\} = G(s)$, we have

$$\mathcal{L}\{g(t)\} = G(s)$$
$$= \mathcal{L}\{y'' + 2y' + 2y\}$$
$$= (s^2 + 2s + 2)Y - (s + 1),$$

and so

$$Y(s) = \frac{(s+1)}{(s+1)^2 + 1} + \frac{1}{(s+1)^2 + 1}G(s).$$

Thus

$$y(t) = e^{-t}\cos(t) + \int_{\tau=0}^{t} e^{-(t-\tau)}\cos(t-\tau)g(\tau)\,d\tau.$$

From this, we readily check that y is continuous on $[0, \infty)$ and that each Laplace transform of y, y', and y'' exist.

Finally, we prove Theorems 6.2 and 6.8:

Proof of Theorem 6.2. Since $f(t)$ is of exponential growth with order α, there are constants $M > 0$ and $t_0 \in [0, \infty)$, such that

$$|e^{-st}f(t)| \le e^{-st}Me^{\alpha t} \quad \text{for all } t \ge t_0.$$

Since

$$\int_{t=0}^{\infty} e^{-(s-\alpha)}\,dt = \lim_{A \to \infty} \frac{1}{s-\alpha}e^{-(s-\alpha)t}\big|_{t=A}^{0}$$
$$= \frac{1}{s-\alpha}, \quad \text{convergent for } s > \alpha,$$

we have

$$\mathcal{L}\{f(t)\} = \int_{t=0}^{\infty} e^{-st}f(t)\,dt$$

converges and exists by the comparison test in Advanced Calculus, Apostol [2, Page 276]. □

Proof of Theorem 6.8. We begin with the proof of the first statement. Since y' is piecewise continuous on $[0, \infty)$ and so Riemann integrable on any $[0, A]$, we have

$$\int_{t=0}^{A} e^{-st} y'(t)\, dt = \int_{t=0}^{A} e^{-st}\, dy$$

by the result in Advanced Calculus, Apostol [**2**, Page 163]. The right side of the above, by the formula of integration by parts, Apostol [**2**, Page 144], equals

$$e^{-s} y(t)|_{t=0}^{A} + s \int_{t=0}^{A} e^{-st} y(t)\, dt,$$

and this converges to $-y(0) + s\mathcal{L}\{y(t)\}$. This is because $\mathcal{L}\{y(t)\}$ exists by Theorem 6.2 and because

$$|e^{-sA} y(A)| \le e^{-sA} M e^{\alpha A}$$

by the assumption of exponential growth on y, which coverges to zero for $s > \alpha$ as $A \longrightarrow +\infty$.

The other statement about the general case is proved by mathematical induction. $\square$

7. Problems

Find a fundamental set of solutions and then the general solution for the homogeneous equations, where $y = y(x)$, $\quad 3 < x < 8$:

1. $y'' - 4y' + 3y = 0$.

2. $y'' - 2y' + y = 0$.

3. $y'' - 4y' + 5y = 0$.

4. $y'' - (1 + 2i)y' + (3 - i) = 0$.

5. $y''' + 4y'' + y' - 6y = 0$.

6. $y''' + 2y'' - 2y' + 24y = 0$.

7. $y''' - 6y'' + 12y' - 8y = 0$.

8. $y''' - 9y'' + 19y' - 35y = 0$.

9. $y^{(4)} - 4y''' + 14y'' - 20y' + 25y = 0$.

Find a particular solution by using the method of undetermined coefficients and the method of constants variation and then find the general solution for the nonhomogeneous equations, where $y = y(x), x > 0$:

10. $y'' - 4y' + 3y = g(x)$, where

$$g(x) = (3x^2 + 5x + 2) \quad \text{or}$$
$$= e^{5x} x \quad \text{or}$$
$$= (3x^2 + 5x + 2) + e^{5x} x.$$

11. $y'' + 4y = \tan(2x)$.

12. $y'' - 2y' + y = g(x)$, where

$$g(x) = e^x (2x + 3) \quad \text{or}$$
$$= e^{3x} \sin(2x) \quad \text{or}$$
$$= e^x (2x + 3) + e^{3x} \sin(2x).$$

13. $y'' - 4y' + 5y = e^{2x} \cos(x)$.

Find an annihilator for the functions:

14.

$$(6x^3 + 2x^2), (x^5 + x)e^{3x}, (3x^3 + 1)e^{(2+3i)x},$$

$$(x^4 + 2)\cos(3x), (2x^3 + 3x + 5)e^{2x}\cos(3x),$$

$$(2x + 3)e^x \cos(2x), (x^2 + 1)e^x \sin(2x).$$

15. Prove Lemma 2.6.

16. Prove Lemma 2.19.

17. Prove Corollary 2.21.

18. Prove the expression (4.5) by mathematical induction.

19. Suppose that $z(x)$ is a nonzero solution to the homogeneous equation (1.2). Use Theorem 2.1 to show that $z(x)$ tends to zero as $x \longrightarrow +\infty$ if all the characteristic values for the characteristic equation $p(\lambda) = 0$ has negative real parts. Conversely, show that $z(x)$ tends to positive infinity as $x \longrightarrow +\infty$ if all the real parts are positive. Finally, apply them to Problems **1** to **9**.

20. Prove the expression (4.3) by mathematical induction.

21. (Reduction of Order) (Coddington [**6**, pages 118–122]) Consider the homogeneous equation (1.2) when $n = 2$. Suppose that ϕ_1 is a nontrivial solution to the equation (1.2), and that there is a $\delta_1 > 0$ such that

$$|\phi_1(x)| \geq \delta_1 > 0$$

for all $x \in [a, b]$, that is, $|\phi_1(x)|$ on $[a, b]$ is bounded away from zero. Follow the steps below to show that a second solution ϕ_2 to the equation (1.2), of the form

$$\phi_2 = u(x)\phi_1$$

for some continuously differentiable function $u(x)$ on $[a, b]$, exists, and that the set $\{\phi_1, \phi_2\}$ is a fundamental set of solutions for the equation (1.2):

(i) Seek ϕ_2 of the form $\phi_2 = u(x)\phi_1$, and substitute it into the equation (1.2)

$$Ly = p_0(x)y'' + p_1(x)y' + p_2(x)y = 0$$

to obtain the equation of first order

$$(p_0\phi_1)v' + (2p_0\phi_1' + p_1\phi_1)v = 0,$$

where $v(x) = u'(x)$.

(ii) Since

$$|p_0(x)\phi_1(x)| \geq \delta_0\delta_1,$$

bounded away from zero, it follows from Theorem 2.10 that this first order equation has a nontrivial solution $z(x)$. Thus

$$u = \int z(x)\, dx,$$

where the integration constant can be taken to zero. In this first order case, we have explicitly

$$z(x) = Ce^{\int \frac{2p_0(x)\phi_1'(x) + p_1(x)\phi_1(x)}{p_0(x)\phi_1(x)}\, dx},$$

and we can take the constant $C = 1$ and the integration constant in the exponent to be zero.

(iii) Use the technique of differentiation to show that the set $\{\phi_1, \phi_2\}$ is linearly independent.

22. Solve the first order differential equation for $y = y(x)$, where $3 < x < 8$:

(a) $(2x \sin(y) + e^x y^2 + y) + (x^2 \cos(y) + 2e^x y + x) \frac{d}{dx} y(x) = 0.$

(b) $(2x - y^2 \sin(x)) + (x^2 + y^2 \cos(x) + 2y \cos(x)) \frac{d}{dx} y(x) = 0.$

(c) $\frac{d}{dx} y(x) = \frac{y^2}{1+x^2}.$

(d) $\frac{d}{dx} y(x) = (\frac{y}{x})^2 + \frac{y}{x} + 1.$

(e) $\frac{d}{dx} y(x) = (2x + y - 3)^2 + (2x + y - 5).$

(f) $-(3x + y) + (x + 3y) \frac{d}{dx} y(x) = 0.$

(g) $-(3x + y + 7) + (x + 3y + 9) \frac{d}{dx} y(x) = 0.$

(h) $\frac{d}{dx} y(x) + \frac{1}{2} \cos(x) y = \frac{1}{2} \cos(x) y^{\frac{1}{2}}.$

(i) $y' = e^x + e^{2x} - 2e^x y + y^2.$

23. Solve the second order equations for $3 < x < 8$:

(a) $x^2 y'' + 5xy' - 24y = 0.$

(b) $x^2 y'' + 3xy' - 8y = \cos(\ln(x)).$

(c) $y'' - \cot(x)y' + [\sin(x)]^2 y = 0$ (Use Proposition 5.33).

24. Use Proposition 5.33 to transform the second order equation

$$y''(x) + \frac{2x}{x^2 + 1} y'(x) + \cos(x)y(x) = 0, \quad 3 < x < 8$$

into an equation that has no first order derivative term.

25. Using the technique of the Laplace transform to solve for y:

$$y'' + 9y = g(t), \quad 0 < t < \infty,$$
$$y(0) = 2,$$
$$y'(0) = 3,$$

where

$$g(t) = \begin{cases} 0, & \text{if } 0 \leq t < 2; \\ t^2, & \text{if } 2 \leq t < \infty, \end{cases}$$
$$= u_2(t)[(t - 2)^2 + 4(t - 2) + 4].$$

26. Use the technique of the Laplace transform to solve for y:

$$y'' + 9y = \delta(t - 2), \quad 0 < t < \infty,$$
$$y(0) = 2,$$
$$y'(0) = 3.$$

8. Solutions

1. $\{e^x, e^{3x}\}$; $c_1 e^x + c_2 e^{3x}$.

2. $\{e^x, xe^x\}$; $c_1 e^x + c_2 xe^x$.

3. $\{e^{2x} \cos(x), e^{2x} \sin(x)\}$; $c_1 e^{2x} \cos(x) + c_2 e^{2x} \sin(x)$.

4. $\{e^{-ix}, e^{(1+3i)x}\}$; $c_1 e^{-ix} + c_2 e^{(1+3i)x}$.

5. $\{e^x, e^{-2x}, e^{-3x}\}$; $c_1 e^x + c_2 e^{-2x} + c_3 e^{-3x}$.

6. $\{e^{2x}, xe^{2x}, e^{-6x}\}$; $c_1 e^{2x} + c_2 xe^{2x} + c_3 e^{-6x}$.

7. $\{e^{2x}, xe^{2x}, x^2 e^{2x}\}$; $c_1 e^{2x} + c_2 xe^{2x} + c_3 x^2 e^{2x}$.

8. $\{e^x \cos(2x), e^x \sin(2x), e^{7x}\}$; $c_1 e^x \cos(2x) + c_2 e^x \sin(2x) + c_3 e^{7x}$.

9.

$$\{e^x \cos(2x), e^x \sin(2x), xe^x \cos(2x), xe^x \sin(x)\};$$
$$c_1 e^x \cos(2x) + c_2 e^x \sin(2x) + c_3 xe^x \cos(2x) + c_4 xe^x \sin(2x).$$

10.

$$y_p = x^2 + \frac{13}{3}x + \frac{52}{9}, y = c_1 e^x + c_2 e^{3x} + y_p;$$

$$y_p = e^{5x}\left(\frac{1}{8}x - \frac{3}{32}\right), y = c_1 e^x + c_2 e^{3x} + y_p;$$

$$y_p = [x^2 + \frac{13}{3}x + \frac{52}{9}] + [e^{5x}\left(\frac{1}{8}x + \frac{3}{32}\right)], y = c_1 e^x + c_2 e^{3x} + y_p.$$

11.

$$y_p = -\frac{1}{4}[\cos(2x)] \ln|\sec(2x) + \tan(2x)|,$$
$$y = c_1 \cos(2x) + c_2 \sin(2x) + y_p.$$

12.

$$y_{p_1} = x^2 e^x \left(\frac{1}{3}x + \frac{3}{2}\right), y = c_1 e^x + c_2 xe^x + y_{p_1};$$

$$y_{p_2} = e^{3x}\left[-\frac{1}{8}\cos(2x)\right], y = c_1 e^x + c_2 xe^x + y_{p_2};$$

$$y_p = y_{p_1} + y_{p_2}, y = c_1 e^x + c_2 xe^x + y_p.$$

13.

$$y_p = xe^{2x}\left[\frac{1}{2}\sin(2x)\right],$$
$$y = c_1 e^{2x} \cos(2x) + c_2 e^{2x} \sin(2x) + y_p.$$

14.

$$D^4; (D-3)^6; [D-(2+3i)]^4; (D^2+9)^5;$$
$$[(D-2)^2 + 9]^4; [(D-1)^2 + 4]^2; [(D-1)^2 + 4]^3.$$

15. Suppose that λ is a root of the algebraic equation

$$p(\lambda) = a_0\lambda^n + a_1\lambda^{n-1} + \cdots + a_{n-1}\lambda + a_n = 0.$$

Since $a_k, k = 0, \ldots, n$, are real, we have

$$0 = \overline{p(\lambda)}$$
$$= \overline{a_0\lambda^n + \cdots + a_{n-1}\lambda + a_n}$$
$$= a_0\overline{\lambda}^n + \cdots + a_{n-1}\overline{\lambda} + a_n,$$

and so $\overline{\lambda}$ is also a root.

16. Since

$$Ly_1 = f_1(x) \quad \text{and}$$
$$Ly_2 = f_2(x),$$

we have

$$L(\alpha_1 y_1 + \alpha_2 y_2) = L(\alpha_1 y_1) + L(\alpha_2 y_2)$$
$$= \alpha_1 L y_1 + \alpha_2 L y_2$$
$$= \alpha_1 f_1(x) + \alpha_2 f_2(x).$$

17.
Since

$$e^{x\alpha_1} \sin(\beta_1 x) = e^{\alpha_1} \frac{e^{i\beta_1 x} - e^{-i\beta_1 x}}{2i} = \frac{1}{2i}[e^{x(\alpha_1 + i\beta_1)} - e^{x(\alpha_1 - i\beta_1)}] \quad \text{and}$$
$$e^{x\alpha_1} \cos(\beta_1 x) = e^{x\alpha_1} \frac{e^{i\beta_1 x} + e^{-i\beta_1 x}}{2} = \frac{1}{2}[e^{x(\alpha_1 + i\beta_1)} + e^{x(\alpha_1 - i\beta_1)}],$$

the proof is complete by applying Remark 2.20. Here $i \equiv \sqrt{-1}$, the complex unit, and $i^2 = -1$.

18. For $k = 1$, we have

$$\frac{\partial}{\partial \lambda}[p(\lambda)e^{x\lambda}] = p'(\lambda)e^{x\lambda} + p(\lambda)xe^{x\lambda}$$
$$= [p'(\lambda) + p(\lambda)x]e^{x\lambda},$$

and so (4.5) is true.

Next assume (4.5) is true for $k = j \in \mathbb{N}$, and show it is also true for $k = j + 1$. This will complete the induction proof.

Now for $k = j + 1$, we have

$$\frac{\partial}{\partial \lambda}\{\frac{\partial^j}{\partial \lambda^j}[p(\lambda)e^{x\lambda}]\} = \frac{\partial}{\partial \lambda}\{[p^{(j)}(\lambda) + jp^{(j-1)}(\lambda)x + \frac{j(j-1)}{2!}p^{(j-2)}(\lambda)x^2 + \cdots$$
$$+ jp'(\lambda)x^{j-1} + p(\lambda)x^j]e^{x\lambda}\}$$
$$= [p^{(j+1)}(\lambda) + jp^{(j)}(\lambda)x + \frac{j(j-1)}{2!}p^{(j-1)}(\lambda)x^2 + \cdots$$
$$+ p'(\lambda)x^j]e^{x\lambda} +$$
$$[p^{(j)}(\lambda) + jp^{(j-1)}(\lambda)x + \frac{j(j-1)}{2!}p^{(j-2)}(\lambda)x^2 + \cdots$$
$$+ jp'(\lambda)x^{j-1} + p(\lambda)x^j]xe^{x\lambda}$$
$$= [p^{j+1}(\lambda) + (j+1)p^{(j)}(\lambda)x + \frac{j(j+1)}{2!}p^{(j-1)}(\lambda)x^2 + \cdots$$
$$+ (j+1)p'(\lambda)x^j + p(\lambda)x^{j+1}]e^{x\lambda}.$$

This is (4.5).

19. Since

$$e^{a+ib} = e^a e^{ib} = e^a[\cos(b) + i\sin(b)]$$

for $a, b \in \mathbb{R}$ and $|e^{ib}| = 1$, the results follow. The solutions in Problems **1** to **3** and in Problems **7** to **9** all go to $+\infty$ as $x \longrightarrow +\infty$. The solution $c_1 e^{-ix} + c_2 e^{(1+3i)x}$ in Problem **4** goes to a constant or $+\infty$ according as $c_2 = 0$ or $c_1 = 0$. The solution $c_1 e^x + c_2 e^{-2x} + c_3 e^{-3x}$ in Problem **5** goes to $+\infty$ or 0 according as $c_1 \neq 0$ or $c_1 = 0$. The solution $c_1 e^{2x} + c_2 x e^{2x} + c_3 e^{-6x}$ in Problem **6** goes to 0 or $+\infty$ according as $c_1 = 0 = c_2$ or not.

20. Let

$$A(t) = \begin{vmatrix} a_{11}(t) & a_{12}(t) & \cdots & a_{1n}(t) \\ a_{21}(t) & a_{22}(t) & \vdots & a_{2n}(t) \\ & \vdots & & \\ a_{n1}(t) & a_{n2}(t) & \cdots & a_{nn}(t) \end{vmatrix}$$

$$\equiv \begin{vmatrix} a_1(t) \\ a_2(t) \\ \vdots \\ a_n(t) \end{vmatrix}.$$

It suffices to show that

$$\frac{d}{dt}A(t) = \begin{vmatrix} a_1'(t) \\ a_2(t) \\ \vdots \\ a_{n-1}(t) \\ a_n(t) \end{vmatrix} + \begin{vmatrix} a_1(t) \\ a_2'(t) \\ a_3(t) \\ \vdots \\ a_n(t) \end{vmatrix} + \cdots + \begin{vmatrix} a_1(t) \\ a_2(t) \\ \vdots \\ a_{n-1}(t) \\ a_n'(t) \end{vmatrix}. \qquad (8.1)$$

For $n = 2$, we have

$$A(t) = \begin{vmatrix} a_{11}(t) & a_{12}(t) \\ a_{21}(t) & a_{22}(t) \end{vmatrix}$$

$$= a_{11}(t)a_{22}(t) - a_{12}(t)a_{21}(t),$$

and so

$$\frac{d}{dt}A(t) = a_{11}'(t)a_{22}(t) + a_{11}(t)a_{22}'(t) - [a_{12}'(t)a_{21}(t) + a_{12}(t)a_{21}'(t)]$$

$$= [a_{11}'(t)a_{22}(t) - a_{12}'(t)a_{21}(t)] + [a_{11}(t)a_{22}'(t) - a_{12}(t)a_{21}'(t)]$$

$$= \begin{vmatrix} a_1'(t) \\ a_2(t) \end{vmatrix} + \begin{vmatrix} a_1(t) \\ a_2'(t) \end{vmatrix}.$$

This is (8.1).

Next assume (8.1) is true for $n = k \in \mathbb{N}$, and show that it is also true for $n = k + 1$. This will complete the induction proof.

Now for $n = k + 1$, we have

$$A(t) = \begin{vmatrix} a_1(t) \\ \vdots \\ a_{k+1}(t) \end{vmatrix},$$

where $a_j(t) = \begin{pmatrix} a_{j1}(t) & \cdots & a_{j(k+1)}(t) \end{pmatrix}$ for $j = 1, \ldots, (k + 1)$. To reduce the order of $A(t)$ from $(k + 1)$ to k, we expand the determinant $A(t)$ along the first row, and we have

$$A(t) = a_{11}(t)A_{11}(t) - a_{12}(t)A_{12}(t) + a_{13}(t)A_{13}(t) - \cdots$$

$$+ (-1)^{1+(k+1)}a_{1(k+1)}(t)A_{1(k+1)}(t).$$

Here the $A_{1j}(t)$ is the determinant of the $k \times k$ matrix, obtained from the $(k + 1) \times (k + 1)$ matrix, corresponding to $A(t)$, by deleting row 1 and column j for $j = 1, \ldots, (k + 1)$.

Hence

$$\frac{d}{dt}A(t) = [a'_{11}(t)A_{11}(t) - a'_{12}(t)A_{12}(t) + a'_{13}(t)A_{13}(t) - \cdots$$
$$+ (-1)^{1+(k+1)}a'_{1(k+1)}(t)A_{1(k+1)}(t)]+$$
$$[a_{11}(t)A'_{11}(t) - a_{12}(t)A'_{12}(t) + a_{13}(t)A'_{13}(t) - \cdots$$
$$+ (-1)^{1+(k+1)}a_{1(k+1)}(t)A'_{1(k+1)}(t)]$$

$$= \begin{vmatrix} a'_1(t) \\ a_2(t) \\ \vdots \\ a_k(t) \\ a_{k+1}(t) \end{vmatrix} + [a_{11}(t)A'_{11}(t) - a_{12}(t)A'_{12}(t) + a_{13}(t)A'_{13}(t) - \cdots$$
$$+ (-1)^{1+(k+1)}a_{1(k+1)}(t)A'_{1(k+1)}(t)]$$

$$= \begin{vmatrix} a'_1(t) \\ a_2 \\ \vdots \\ a_k(t) \\ a_{k+1}(t) \end{vmatrix} + \begin{vmatrix} a_1 \\ a'_2(t) \\ a_3(t) \\ \vdots \\ a_{k+1}(t) \end{vmatrix} + \cdots + \begin{vmatrix} a_1(t) \\ a_2(t) \\ \vdots \\ a_k(t) \\ a'_{k+1}(t) \end{vmatrix}.$$

This is (8.1). Here we reversed the process of expanding a determinant and used the induction assumption that (8.1) is true for $n = k$.

22.

(a) $F(x, y) = x^2 \sin(y) + e^x y^2 + xy = c.$

(b) An integrating factor $\mu(y) = e^y$; $F(x, y) = e^y x^2 + e^y y^2 \cos(x) = c.$

(c) $y = \frac{-1}{\tan^{-1}(x)+c}.$

(d) $\tan^{-1}(\frac{y}{x}) = \ln|x| + c.$

(e) $\frac{2x+y-6}{2x+y-5} = ce^x.$

(f) $c(y+x)(y-x)^2 = 1.$

(g) $(\alpha, \beta) = (\frac{-3}{2}, \frac{-5}{2}); c(y+x+4)(y-x+1)^2 = 1.$

(h) $y = (1 + ce^{\frac{-1}{2}\sin(x)})^2.$

(i) $y_p = e^x; y = e^x + \frac{1}{c-(e^x+\frac{e^{2x}}{2})}.$

23.

(a) $y = c_1 x^3 + c_2 x^{-8}.$

(b) $y = c_1 x + c_2 x \ln(x) + \frac{-1}{2}\sin(\ln(x)).$

(c) $y = c_1 \cos(-\cos(x)) + c_2 \sin(-\cos(x)).$

24. $t = f(x) = \tan^{-1}(x), y''(t) + \frac{\cos(\tan(t))}{[(\tan(t))^2+1]^2}y(t) = 0.$

25.

$$y(t) = [2\cos(t) + \sin(t)] + u_2(t)[\frac{(t-2)^2}{9} + \frac{(t-2)}{2} - \frac{34}{81}$$
$$- \frac{1}{6}\sin(3(t-2)) - \frac{38}{81}\cos 3(t-2)].$$

26. $y(t) = [2\cos(t) + \sin(t)] + u_2(t)[\frac{1}{3}\sin 3(t-2)].$

CHAPTER 2

Systems of Linear First Order Equations

1. Introduction

Consider a system of $n(\in \mathbb{N})$ linear first order differential equations

$$\frac{d}{dt}u(t) = A(t)u(t) + f(t), \quad a < t < b, \tag{1.1}$$

where a and b are real numbers and

$$u(t) = \begin{pmatrix} u_1(t) \\ u_2(t) \\ \vdots \\ u_n \end{pmatrix}, \quad \frac{d}{dt}u(t) = \begin{pmatrix} \frac{d}{dt}u_1(t) \\ \frac{d}{dt}u_2(t) \\ \vdots \\ \frac{d}{dt}u_n(t) \end{pmatrix},$$

$$f(t) = \begin{pmatrix} f_1(t) \\ f_2(t) \\ \vdots \\ f_n(t) \end{pmatrix}, \quad \text{and} \quad A(t) = \begin{pmatrix} p_{11}(t) & p_{12}(t) & \cdots & p_{1n}(t) \\ p_{21}(t) & p_{22}(t) & \cdots & p_{2n}(t) \\ \vdots & \vdots & \vdots & \vdots \\ p_{n1}(t) & p_{n2}(t) & \cdots & p_{nn}(t) \end{pmatrix}.$$

Here $u(t)$ is the column vector of n continuous complex-valued functions

$$u_i(t), i = 1, 2, \ldots, n,$$

on $[a, b]$, and $\frac{d}{dt}u$ is the derivative of $u(t)$; $f(t)$ is the column vector of n continuous complex-valued functions

$$f_i(t), i = 1, 2, \ldots, n,$$

on $[a, b]$, and $A(t)$ is the square matrix of $n \times n$ continuous complex-valued functions

$$p_{ij}(t), i, j = 1, 2, \ldots, n,$$

on $[a, b]$.

A special case of the system (1.1) is the linear ordinary differential equations of nth order

$$Ly = f(t), \quad a < t < b, \tag{1.2}$$

where Ly is defined by

$$Ly \equiv p_0(t)y^{(n)} + p_1(t)y^{(n-1)} + \cdots + p_{n-1}(t)y' + p_n(t)y$$

for $n \in \mathbb{N}$ and the complex-valued functions $f(t), p_i(t), i = 0, 1, \ldots, n$, are assumed to be continuous. Here $p_0(t)$ is also restricted by the condition that there is a $\delta_0 > 0$ such that

$$|p_0(t)| \geq \delta_0 > 0$$

for all $t \in [a, b]$. To see that (1.2) is a special case of (1.1), use the substitutions

$$u_1 = y,$$

$$u_2 = y',$$

$$\vdots$$

$$u_n = y^{(n-1)}.$$

It follows from differentiation with respect to t that

$$\frac{d}{dt}\begin{pmatrix} u_1 \\ u_2 \\ \vdots \\ u_{n-1} \\ u_n \end{pmatrix} = \begin{pmatrix} y' \\ y'' \\ \vdots \\ y^{(n-1)} \\ y^{(n)} \end{pmatrix} = \begin{pmatrix} u_2 \\ u_3 \\ \vdots \\ u_n \\ -\frac{p_1}{p_0}y^{(n-1)} - \frac{p_1}{p_0}y^{(n-2)} - \cdots - \frac{p_n}{p_0}y + \frac{h(t)}{p_0} \end{pmatrix}$$

$$= \begin{pmatrix} u_2 \\ u_3 \\ \vdots \\ u_n \\ -\frac{p_1}{p_0}u_n - \frac{p_2}{p_0}u_{n-1} - \cdots - \frac{p_n}{p_0}u_1 \end{pmatrix} + \begin{pmatrix} 0 \\ 0 \\ \vdots \\ 0 \\ h(t) \end{pmatrix}$$

$$= \begin{pmatrix} 0 & 1 & 0 & 0 & \cdots & 0 \\ 0 & 0 & 1 & 0 & \cdots & 0 \\ \vdots & \vdots & \vdots & \vdots & \vdots & \vdots \\ 0 & 0 & 0 & \cdots & 0 & 1 \\ -\frac{p_n}{p_0} & -\frac{p_{n-1}}{p_0} & \cdots & \cdots & \cdots & -\frac{p_1}{p_0} \end{pmatrix}\begin{pmatrix} u_1 \\ u_2 \\ \vdots \\ u_{n-1} \\ u_n \end{pmatrix} + \begin{pmatrix} 0 \\ 0 \\ \vdots \\ 0 \\ h(t) \end{pmatrix},$$

and this is the desired system.

Thus if $n = 2$, we have the equation

$$Ly(t) = p_0(t)y'' + p_1(t)y' + p_0(t)y = h(t), \quad a < t < b,$$

and the substitutions

$$\begin{cases} u_1 = y \\ u_2 = y' \end{cases}$$

reduces this equation to the system

$$\frac{d}{dt}\begin{pmatrix} u_1 \\ u_2 \end{pmatrix} = \begin{pmatrix} 0 & 1 \\ -\frac{p_1}{p_0} & -\frac{p_2}{p_0} \end{pmatrix}\begin{pmatrix} u_1 \\ u_2 \end{pmatrix} + \begin{pmatrix} 0 \\ h(t) \end{pmatrix}.$$

The above technique of substitutions also applies to systems of linear higher order equations (see Section 3 about examples).

DEFINITION 1.1. *The system* (1.1) *is called a nonhomogeneous system if* $f(t)$ *is not identical to zero. Otherwise, it reduces to*

$$\frac{d}{dt}u(t) = A(t)u, \tag{1.3}$$

and is called a homogeneous system.

A special system that is closely related to (1.3) is, Boyce and DiPrima [**3**, page 357]:

DEFINITION 1.2. *The system*

$$t\frac{d}{dt}u(t) = Au(t), \quad a < t < b, t \neq 0 \tag{1.4}$$

is called the Euler system.

DEFINITION 1.3. *A complex-valued column vector function $z(t)$ defined on (a,b) is called a solution to the nonhomogeneous system (1.1) if $z(t)$ satisfies*

$$\frac{d}{dt}u = A(t)u + f.$$

It is easy to see that if $z = \begin{pmatrix} z_1 \\ \vdots \\ z_n \end{pmatrix}$ is a solution to the nonhomogeneous system (1.1), then z_1 is a solution to the nonhomogeneous equation (1.2), and that if y is a solution to the nonhomogeneous equation (1.2), then $z \equiv \begin{pmatrix} y \\ y' \\ \vdots \\ y^{(n-1)} \end{pmatrix}$ is a solution to the nonhomogeneous system (1.1).

DEFINITION 1.4. *A set $\{z_1, z_2, \ldots, z_k\}$ of k solutions $z_i(t), i = 1, 2, \ldots, k, k \in \mathbb{N}$, of (1.3), is called linearly dependent, if there exist k complex numbers $c_1, c_2, \ldots, c_k$, not all zero, such that*

$$\sum_{i=1}^{k} c_i z_i(t) = 0$$

for all $t \in [a, b]$.

Conversely, if the equation

$$\sum_{i=1}^{k} c_i z_i(t) = 0$$

for all $t \in [a, b]$, has only the zero solutions $0 = c_1 = c_2 = \cdots = c_k$, then the set

$$\{z_1(t), z_2(t), \ldots, z_k(t)\}$$

is called linearly independent.

DEFINITION 1.5. *The Wronskian $W = W(z_1, z_2, \ldots, z_n)$ of n solutions*

$$z_i(t), \quad i = 1, \ldots, n,$$

of (1.3), is the determinant

$$W(t) = \begin{vmatrix} z_1(t) & z_2(t) & \cdots & z_n(t) \end{vmatrix}.$$

The concept of linear independence is closely related to the Wronskian $W(t)$ (see Section 2).

DEFINITION 1.6. *A linearly independent set $\{z_1, z_2, \ldots, z_n\}$ of n solutions*

$$z_1(t), \quad i = 1, \ldots, n,$$

of (1.3), is called a basis or a fundamental set of solutions for (1.3).
Furthermore, the matrix

$$\Phi \equiv \begin{pmatrix} z_1 & z_2 & \cdots & z_n \end{pmatrix},$$

is called a fundamental matrix for (1.3).
Thus, the Wronskian $W(t)$ is the determinant of the fundamental matrix $\Phi(t)$.

Theoretically speaking, the homogeneous system (1.3) has a nontrivial solution by the theorem of existence and uniqueness. To see this, consider the initial value problem for (1.3):

$$\frac{d}{dt}u(t) = A(t)u, \quad a < t < b$$
$$t_0 \in (a, b) \tag{1.5}$$
$$u(t_0) = u_0,$$

where u_0 is a given nonzero constant complex column vector. The theorem of existence and uniqueness in Chapter 8 assures the existence of a unique nontrivial solution $z(t)$ to this problem, and thus $z(t)$ is a nontrivial solution to (1.3).

Likewise, the nonhomogeneous system (1.1) also has a nontrivial solution theoretically, by the theorem of existence and uniqueness in Chapter 8. To see this, consider the initial value problem for (1.1):

$$\frac{d}{dt}u(t) = A(t)u(t) + f(t), \quad t < a < b$$
$$t_0 \in (a, b) \tag{1.6}$$
$$u(t_0) = u_0,$$

where u_0 is a given nonzero constant complex column vector. It follows from the theorem of existence and uniqueness in Chapter 8 that this problem has a unique nontrivial solution $z(t)$, and thus $z(t)$ is a nontrivial solution to (1.1).

It is the purpose of this chapter to show further that (1.3) has a fundamental set $\{z_1, z_2, \ldots, z_n\}$ of solutions, and that every solution $z(t)$ to (1.3) is of the form

$$z = \sum_{i=1}^{n} c_i z_i,$$

a linear combination of the elements in the fundamental set $\{z_1, z_2, \ldots, z_n\}$, where $c_i, i = 1, 2, \ldots, n$, are constant complex numbers. Or, in terms of the fundamental matrix $\Phi(t)$ associated with the fundamental set $\{z_1, \ldots, z_n\}$, $z(t)$ takes the form

$$z(t) = \Phi(t)c,$$

where $c = \begin{pmatrix} c_1 \\ \vdots \\ c_n \end{pmatrix}$, a constant column vector. Those results will be applied to the Euler system (1.4).

It is also the purpose of this chapter to show further that every solution z to (1.1) takes the form

$$z = \sum_{i=1}^{n} c_i z_i + g(t)$$
$$= \Phi c + g(t),$$

where $g(t)$ is an arbitrary solution to the system (1.1). Methods for finding such a solution $g(t)$ explicitly are given.

When $A(t) = A$ is independent of t, the exponential function of a matrix, denoted by e^{tA}, will be defined and shown to be a fundamental matrix for (1.3). Methods for computing e^{tA} and other fundamental matrices explicitly, will be demonstrated. A method of elimination, Nagel and Saff [19, pages 211–212], for computing the solution u directly, will also be illustrated (see Section 3). Compared to other methods of computing fundamental matrices, this method is more efficient when the order $n \times n$ of the matrix A is small, e.g. 2×2 or 3×3.

The rest of this chapter is organized as follows. Section 2 states the main results, and Section 3 studies examples for illustration. Section 4 deals with the proof of the main results, and Section 5 provides an additional method for computing e^{tA}. This additional method involves more advanced linear algebra. Finally, Section 6 contains a problems set, and their solutions are presented in Section 7.

2. Main Results

Firstly, we consider the case where $A(t) = A$ is independent of t.

The following theorem defines the exponential function e^{tA} of tA, and derives its useful properties:

THEOREM 2.1. *For each $m \in \mathbb{N}$, let S_m be the square matrix defined by*

$$S_m \equiv \sum_{k=0}^{m} \frac{(tA)^k}{k!},$$

the partial sum of the symbolic infinite series of matrices

$$\sum_{k=0}^{\infty} \frac{(tA)^k}{k!}.$$

Then the following are true:

(i) *The limit $\lim_{m \to \infty} (S_m)_{\alpha\beta}$ of the element $(S_m)_{\alpha\beta}$ of the matrix S_m at the position $(\alpha, \beta), \alpha, \beta = 1, \ldots, n$, as $m \longrightarrow \infty$, exists absolutely and uniformly for bounded t.*

(ii) *The square matrix e^{tA}, whose element $(e^{tA})_{\alpha\beta}$ at the position (α, β) is defined by the complex number $\lim_{m \to \infty} (S_m)_{\alpha\beta}$ from (i), has the properties: for $t, s \in \mathbb{R}$,*

(a) $e^{tA}|_{t=0} = I$, *the identity matrix.*

(b) $e^{(t+s)A} = e^{tA}e^{sA} = e^{sA}e^{tA}$.

(c) $e^{-tA} = (e^{tA})^{-1}$, *the inverse of the matrix e^{tA}.*

(d) $\frac{d}{dt}e^{tA} = Ae^{tA} = e^{tA}A$, *and $e^{(t-t_0)A}u_0$ is the unique solution to the initial value problem (1.5).*

(e) e^{tA} *is a fundamental matrix for the homogeneous system (1.3).*

Theorem 2.1 can be applied to (1.4):

COROLLARY 2.2. *The Euler system (1.4) has a fundamental matrix e^{xA}, where x is evaluated at*

$$x = \ln(t).$$

PROOF. By the change of variable $x = \ln(t)$, we have

$$\frac{d}{dt}u = \frac{du}{dx}\frac{dx}{dt}$$
$$= \frac{1}{t}\frac{du}{dx},$$

and so the Euler system (1.4) becomes the type of (1.3)

$$\frac{du}{dx} = Au.$$

By Theorem 2.1, this system has a fundamental matrix e^{xA}. Thus changing x back to t, the Euler system has a fundamental matrix e^{xA}, where x is evaluated at

$$x = \ln(t).$$

$\square$

For $\lambda \in \mathbb{C}$, let $p(\lambda)$ denote the determinant $|A - \lambda I|$ of the square matrix $(A - \lambda I)$. $p(\lambda)$ is a polynomial in λ, of degree n.

Here is a method of computing e^{tA}:

THEOREM 2.3. *Suppose that the algebraic equation*

$$p(\lambda) \equiv a_0 \lambda^n + a_1 \lambda^{n-1} + \cdots + a_{n-1}\lambda + a_n = 0$$

has s distinct solutions (or called roots) $\lambda_1, \lambda_2, \ldots, \lambda_s$ where $1 \le s \le n$, and that the solution λ_i is repeated $m_i (\in \mathbb{N})$ times. In other words, we assume that

$$p(\lambda) = a_0 (\lambda - \lambda_1)^{m_1} (\lambda - \lambda_2)^{m_2} \cdots (\lambda - \lambda_s)^{m_s} = 0,$$

$\lambda_i \ne \lambda_j$ *for $i \ne j$, and that*

$$m_1 + m_2 + \cdots + m_s = n.$$

Then there exist n unique square matrices

$$M_{i,k}, \quad i = 1, 2, \ldots, s; \quad k = 0, 1, \ldots, (m_i - 1),$$

such that

$$e^{tA} = \sum_{k=0}^{m_i - 1} \sum_{i=1}^{s} t^k e^{t\lambda_i} M_{i,k},$$

or that

$$e^{tA} = \sum_{j=1}^{n} z_j(t) N_j,$$

if we relabel the n functions

$$t^k e^{t\lambda_i}, \quad i = 1, 2, \ldots, s; \quad k = 0, 1, \ldots, (m_i - 1),$$

as the n functions

$$z_i(t), \quad i = 1, 2, \ldots, n,$$

and relabel the n square matrices

$$M_{i,k}, \quad i = 1, 2, \ldots, s; \quad k = 0, 1, \ldots, (m_i - 1),$$

as the n square matrices

$$N_j, \quad j = 1, 2, \ldots, n.$$

Here the n matrices $N_j, j = 1, \ldots, n$, are the unique n solutions of the system of algebraic equations:

$$\begin{pmatrix} z_1(0) & z_2(0) & \cdots & z_n(0) \\ z_1'(0) & z_2'(0) & \cdots & z_n'(0) \\ \vdots & & & \\ z_1^{(n-1)}(0) & z_2^{(n-1)}(0) & \cdots & z_n^{(n-1)}(0) \end{pmatrix} \begin{pmatrix} N_1 \\ N_2 \\ \vdots \\ N_n \end{pmatrix} = \begin{pmatrix} I \\ A \\ \vdots \\ A^{n-1} \end{pmatrix}.$$

Thus if $p(\lambda) = (\lambda - 1)^2(\lambda - 2) = 0$, then

$$z_1 = e^t, z_2 = te^t, z_3 = e^{2t},$$

$$\begin{pmatrix} 1 & 0 & 1 \\ 1 & 1 & 2 \\ 1 & 2 & 4 \end{pmatrix} \begin{pmatrix} N_1 \\ N_2 \\ N_3 \end{pmatrix} = \begin{pmatrix} I \\ A \\ A^2 \end{pmatrix},$$

and $N_i, i = 1, 2, 3$, are uniquely solvable.

REMARK 2.4. In fact, the form

$$p(\lambda) = a_0(\lambda - \lambda_1)^{m_1} \cdots (\lambda - \lambda_s)^{m_s} = 0$$

in Theorem 2.3 is always true. A proof of this was given in Chapter 1.

DEFINITION 2.5. *The following are defined:*

- $p(\lambda) \equiv |A - \lambda I|$ *is called the characteristic polynomial,*
- $p(\lambda) = 0$ *is called the characteristic equation,*
- λ_i *is called a characteristic value or an eigenvalue of multiplicity m_i, and*
- *a nonzero constant vector u is called an eigenvector corresponding to λ_i, if $(A - \lambda_i)u = 0$.*

The problem of finding a fundamental set can also be studied from the view point of eigenvalues and eigenvectors. It will follow that a fundamental set of solutions (or a fundamental matrix) for the system (1.3) can be computed easily when the eigenvalue λ_i has m_i linearly independent eigenvectors

$$u_1^{(i)}, u_2^{(i)}, \ldots, u_{m_i}^{(i)} :$$

THEOREM 2.6. *Assume that to each eigenvalue λ_i of multiplicity m_i, there correspond m_i linearly independent eigenvectors*

$$u_1^{(i)}, u_2^{(i)}, \ldots, u_{m_i}^{(i)}.$$

Then the set

$$\{e^{t\lambda_i} u_k^{(i)} : i = 1, 2, \ldots, s; k = 1, 2, \ldots, m_i\}$$

is a fundamental set of solutions for the homogeneous system (1.3). Here $u_k^{(i)}, k = 1, \ldots, m_i$, are to be solved from the system of algebraic equations

$$(A - \lambda_i I)u = 0.$$

PROOF. Since $u_k^{(i)}$ satisfies

$$(A - \lambda_i)u = 0,$$

we have

$$\frac{d}{dt} e^{t\lambda_i} u_k^{(i)} = \lambda_i e^{t\lambda_i} u_k^{(i)}$$
$$= A(e^{t\lambda_i} u_k^{(i)}),$$

and so $e^{t\lambda_i} u_k^{(i)}$ is a solution of (1.3).

To see that

$$\{e^{t\lambda_i} u_k^{(i)} : i = 1, \ldots, s; k = 1, \ldots, m_i\}$$

is linearly independent, we suppose

$$\sum_{k=1}^{m_i} \sum_{i=1}^{s} c_{ik} e^{t\lambda_i} u_k^{(i)} = 0 \qquad (2.1)$$

for all t, and show that $c_{ik} = 0$ for all i, k.

Since (2.1) is the same as the matrix equation

$$\Phi(t)c \equiv \left(e^{t\lambda_1} u_1^{(1)} \quad \cdots \quad e^{t\lambda_2} u_1^{(2)} \quad \cdots \quad e^{t\lambda_s} u_1^{(s)} \quad \cdots \right) \begin{pmatrix} c_{11} \\ \vdots \\ c_{21} \\ \vdots \\ c_{s1} \\ \vdots \end{pmatrix}$$

$$= 0$$

and the determinant $|\Phi(t)|$ of $\Phi(t)$ is not zero for all t, we have $c = 0$. This completes the proof. Here, to see $|\Phi(t)| \neq 0$, we calculate

$$|\Phi(t)| = e^{m_1 t\lambda_1} e^{m_2 t\lambda_2} \cdots e^{m_s t\lambda_s} \left| u_1^{(1)} \quad \cdots \quad u_1^{(2)} \quad \cdots \quad u_1^{(s)} \quad \cdots \right|$$

$$\equiv e^{\sum_{i=1}^{s} t\lambda_i} |u|.$$

Now the result follows, since the determinant $|u|$ is not zero by the linear independence of $\{u_k^{(i)}\}$. Here we used the fact from linear algebra that multiplying a column of a square matrix B by a constant α gives that the determinant of the resultant matrix equals $\alpha |B|$. □

A general form of Theorem 2.6 is presented in Section 5.

REMARK 2.7. *If A is a real matrix, where, for example,*

$$s = 2,$$
$$\lambda_1 = a + ib, \quad a, b \in \mathbb{R},$$
$$\lambda_2 = \overline{\lambda_1} = a - ib,$$
$$u_1^{(1)} = c + id, \quad c, d \in \mathbb{R}^2,$$
$$u_1^{(2)} = \overline{u_1^{(1)}} = c - id,$$

then the set of the real part and the imaginary part

$$\{e^{ta}[c\cos(tb) - d\sin(tb)], e^{ta}[d\cos(tb) + c\sin(tb)]\}$$

of $e^{t\lambda_1} u_1^{(1)} = e^{t(a+ib)}(c + id)$, is a fundamental set of real-valued solutions to the homogeneous system (1.3). This should be compared to the fundamental set

$$\{e^{t(a+ib)}(c + id), e^{t(a-ib)}(c - id)\}$$

of complex-valued solutions from Theorem 2.6.

Thus if $p(\lambda) = (\lambda - \sqrt{2})^2 (\lambda - \sqrt{3})^3 = 0$, and if $u_1^{(1)}, u_2^{(1)}$ are two linearly independent eigenvectors corresponding to the eigenvalue $\sqrt{2}$, and if $u_1^{(2)}, u_2^{(2)}, u_3^{(2)}$ are three linearly independent eigenvectors corresponding to the eigenvalue $\sqrt{3}$, then the set

$$\{e^{\sqrt{2}t} u_1^{(1)}, e^{\sqrt{2}t} u_2^{(1)}, e^{\sqrt{3}t} u_1^{(2)}, e^{\sqrt{3}t} u_2^{(2)}, e^{\sqrt{3}t} u_3^{(2)}\}$$

is a fundamental set.

Not every matrix A satisfies the assumption in Theorem 2.6, but it is this case if the multiplicity of each eigenvalue of A equals one, or if A is a real symmetric matrix:

DEFINITION 2.8. *Recall that the matrix A is real symmetric if each entry in A is real and A equals its tranpose A^t. Here A^t is obtained from A by interchanging the rows (in A) with columns (in A).*

COROLLARY 2.9. *If A has n distinct eigenvalues, or if A is real symmetric, then the assumption in Theorem 2.6 is satisfied and the results in Theorem 2.6 hold.*

Thus if $p(\lambda) = (\lambda - 1)(\lambda - 2)(\lambda - 3) = 0$, for which the numbers $1, 2$, and 3 are eigenvalues, each of them being of multiplicity one, and if the vectors

$$u_1 = \begin{pmatrix} 1 \\ 0 \\ 1 \end{pmatrix}, u_2 = \begin{pmatrix} -2 \\ 1 \\ 0 \end{pmatrix}, u_3 = \begin{pmatrix} 0 \\ -3 \\ 2 \end{pmatrix},$$

are found eigenvectors, respectively, then

$$\{e^t u_1, e^{2t} u_2, e^{3t} u_3\}$$

is a fundamental set of solutions.

REMARK 2.10. *An intuitive idea for obtaining the fundamental set of solutions in Theorem 2.6, is to seek a solution z to the homogeneous system (1.3), of the form*

$$z = e^{t\lambda} u,$$

where the complex number λ and the complex vector u are to be determined. Calculations show that, in order for z to be a solution to (1.3), we must have

$$\frac{d}{dt} z = \lambda e^{t\lambda} u$$

$$= e^{t\lambda} Au,$$

and so

$$(A - \lambda I)u = 0.$$

Thus u is nonzero and $z = e^{t\lambda} u$ is a nontrivial solution to (1.3), if $p(\lambda) = |A - \lambda I| = 0$. Here note that $e^{t\lambda}$ is never zero.

The computed e^{tA} in Theorem 2.3 is, in general, complex-valued. However, under a certain circumstance, it can be real-valued as the following shows:

COROLLARY 2.11. *Suppose that the $p(\lambda)$ in Theorem 2.3 is either of the form*

$$p(\lambda) = a_0[(\lambda - \lambda_1)^{m_1}(\lambda - \overline{\lambda_1})^{m_1}] \cdots [(\lambda - \lambda_r)^{m_r}(\lambda - \overline{\lambda_r})^{m_r}](\lambda - \lambda_{r+1})^{m_{r+1}} \quad (2.2)$$

or of the form

$$p(\lambda) = a_0[(\lambda - \lambda_1)^{m_1}(\lambda - \overline{\lambda_1})^{m_1}] \cdots [(\lambda - \lambda_r)^{m_r}(\lambda - \overline{\lambda_r})^{m_r}], \quad (2.3)$$

to which corresponds either

$$2(m_1 + m_2 + \cdots + m_r) + m_{r+1} = n$$

or

$$2(m_1 + m_2 + \cdots + m_r) = n.$$

Here

$$\lambda_j = \alpha_j + i\beta_j, j = 1, 2, \ldots, r, \quad \text{some } r \in \mathbb{N}$$

$$\alpha_j, \beta_j \in \mathbb{R}, i \equiv \sqrt{-1}, \quad \text{the complex unit, with} \quad i^2 = -1,$$

$$\overline{\lambda_j} = \alpha_j + i\beta_j, \quad \text{the complex conjugate of } \lambda_j,$$

$$\lambda_{r+1} \in \mathbb{R}, \quad \text{and}$$

$$\lambda_i \neq \lambda_j \quad \text{for} \quad i \neq j, \quad \text{where} \quad i, j = 1, 2, \ldots, (r+1).$$

Then according as (2.2) or (2.3), there exist n real square matrices

$$L_{r+1,l}, l = 0, 1, \ldots, (m_{r+1} - 1);$$
$$M_{i,k}, N_{i,k}, i = 1, 2, \ldots, r; k = 0, 1, \ldots, (m_i - 1) \tag{2.4}$$

or n real square matrices

$$M_{i,k}, N_{i,k}, i = 1, \ldots, s; k = 0, 1, \ldots, (m_i - 1), \tag{2.5}$$

such that

$$e^{tA} = \sum_{l=0}^{m_{r+1}-1} t^l e^{t\lambda_{r+1}} L_{(r+1),l} \ +$$

$$\sum_{i=1}^{r} \sum_{k=0}^{m_i-1} x^k e^{t\alpha_i} \cos(t\beta_i) M_{i,k} + \sum_{i=1}^{r} \sum_{k=0}^{m_i-1} x^k e^{t\alpha_i} \sin(t\beta_i) N_{i,k} \tag{2.6}$$

or

$$e^{tA} = \sum_{i=1}^{r} \sum_{k=0}^{m_i-1} t^k e^{t\alpha_i} \cos(t\beta_i) M_{i,k} + \sum_{i=1}^{r} \sum_{k=0}^{m_i-1} x^k e^{t\alpha_i} \sin(t\beta_i) N_{i,k}. \tag{2.7}$$

Equivalently,

$$e^{tA} = \sum_{j=1}^{n} z_j(t) P_j,$$

if we relabel the n real-valued functions

$$t^l e^{t\lambda_{r+1}}, l = 0, \ldots, (m_{r+1} - 1);$$

$$t^k e^{t\alpha_i} \cos(t\beta_i), t^k e^{t\alpha_i} \sin(t\beta_i), i = 1, \ldots, r; k = 0, \ldots, (m_i - 1),$$

in (2.6) or the n real-valued functions

$$t^k e^{t\alpha_i} \cos(t\beta_i), t^k e^{t\alpha_i} \sin(t\beta_i), i = 1, \ldots, s; k = 0, \ldots, (m_i - 1),$$

in (2.7), as the n functions

$$z_1(t), \ldots, z_n(t),$$

and relabel the n real square matrices in (2.4) or the n real square matrices in (2.5), as the n real square matrices

$$P_1, \ldots, P_n.$$

Here the n real square matrices $P_j, j = 1, \ldots, n$, are the unique n solutions of the system of algebraic equations:

$$\begin{pmatrix} z_1(0) & z_2(0) & \cdots & z_n(0) \\ z_1'(0) & z_2'(0) & \cdots & z_n'(0) \\ \vdots & & & \\ z_1^{(n-1)}(0) & z_2^{(n-1)}(0) & \cdots & z_n^{(n-1)}(0) \end{pmatrix} \begin{pmatrix} P_1 \\ P_2 \\ \vdots \\ P_n \end{pmatrix} = \begin{pmatrix} I \\ A \\ \vdots \\ A^{n-1} \end{pmatrix}.$$

Thus if $p(\lambda) = (\lambda - (1 + 2i))(\lambda - (1 - 2i)) = 0$, then

$$z_1 = e^t \cos(2t), z_2 = e^t \sin(2t),$$

$$\begin{pmatrix} 1 & 0 \\ 1 & 2 \end{pmatrix} \begin{pmatrix} P_1 \\ P_2 \end{pmatrix} = \begin{pmatrix} I \\ A \end{pmatrix},$$

and $P_i, i = 1, 2$, are uniquely solvable.

Corollary 2.11 assumes that the roots of $p(\lambda) = |A - \lambda I| = 0$ appear in complex conjugate pairs. This assumption is satisfied when A is real:

LEMMA 2.12. *Suppose that A is a real matrix. If λ is a root of the algebraic equation*

$$p(\mu) = |A - \mu I| = 0,$$

then the complex conjugate $\overline{\lambda}$ of λ is also a root.

PROOF. See the Problems set in Section 6. □

Secondly, we consider the case where $A(t)$ is not a constant matrix but depends on t. In this case, there still exists theoretically a fundamental set of solutions for the homogeneous system (1.3). But such a fundamental set is not as explicit as that, e^{tA}, in Theorem 2.3:

THEOREM 2.13. *There exists a fundamental set of solutions for the homogeneous system (1.3).*

A fundamental set of solutions for (1.3) is very useful, in that every solution to (1.3) can be expressed as a linear combination of the elements in the fundamental set:

THEOREM 2.14. *If z is a solution to (1.3), then z is of the form*

$$z = \sum_{i=1}^{n} c_i z_i, \quad \textit{a linear combination of } z_i$$
$$= \Phi(t)c,$$

where the set $\{z_1, z_2, \ldots, z_n\}$ is a fundamental set of solutions for (1.3), and $c_i, i = 1, 2, \ldots, c_n$, are constant complex numbers; $\Phi(t)$ is the fundamental matrix associated with the fundamental set $\{z_1, \ldots, z_n\}$, and $c = \begin{pmatrix} c_1 \\ \vdots \\ c_n \end{pmatrix}$ is a constant vector.

DEFINITION 2.15. *The function*

$$z(t) = \sum_{i=1}^{n} c_i z_i$$
$$= \Phi(t)c$$

in Theorem 2.3 is called the general solution of (1.3).

A fundamental set of solutions for the homogeneous system (1.3) is also very useful, in that every solution to the nonhomogeneous system (1.1) is a linear combination of the elements in the fundamental set, plus a particular solution to (1.1):

THEOREM 2.16. *If z is a solution to the nonhomogeneous system* (1.1), *then z is of the form*

$$z = \sum_{i=1}^{n} c_i z_i + g(t)$$

$$= \Phi(t)c + g(t),$$

where $g(t)$ is a particular solution to (1.1), *and the set $\{z_1, z_2, \ldots, z_n\}$ is a fundamental set of solutions for* (1.3); $c_i, i = 1, 2, \ldots, n$, *are constant complex numbers, $\Phi(t)$ is the fundamental matrix associated with the fundamental set $\{z_1, \ldots, z_n\}$, and $c = \begin{pmatrix} c_1 \\ \vdots \\ c_n \end{pmatrix}$ is a constant vector.*

DEFINITION 2.17. *The function*

$$z = \sum_{i=1}^{n} c_i z_i + g(t)$$

$$= \Phi(t)c + g(t)$$

in Theorem 2.16 is called the general solution to the nonhomogeneous equation (1.1).

Another use of a fundamental set of solutions is that it can yield a particular solution explicitly for the nonhomogeneous system (1.1), through a formula of parameters variation:

THEOREM 2.18 (Variation of Constants). *A particular solution $g(t)$ to the nonhomogeneous system* (1.1) *exists and is of the form*

$$g(t) = \Phi(t)\varphi(t),$$

where $\Phi(t)$ is the fundamental matrix associated with the fundamental set $\{z_i, z_2, \ldots, z_n\}$ of solutions for (1.3) *and $\varphi(t)$ equals some indefinite integral*

$$\varphi(t) = \int \Phi^{-1}(t) f(t) \, dt + c \quad \text{or}$$

$$= \int_{\tau=t_1}^{t} \Phi^{-1}(\tau) f(\tau) \, d\tau$$

for any constant c or for any constant $a < t_1 < b$. Here $\Phi^{-1}(t)$ is the inverse of the fundamental matrix $\Phi(t)$.

Thus the particular solution $g(t)$ is of the form

$$g(t) = \Phi(t) [\int \Phi^{-1}(t) f(t) \, dt + c] \quad \text{or}$$

$$= \int_{\tau=t_1}^{t} \Phi(t)\Phi^{-1}(\tau) f(\tau) \, d\tau.$$

COROLLARY 2.19. *The unique solution to the initial value problem* (1.6) *is*

$$\Phi(t)\Phi^{-1}(t_0) u_0 + \int_{\tau=t_0}^{t} \Phi(t)\Phi^{-1}(\tau) f(\tau) \, d\tau,$$

where $\Phi(t)$ is a fundamental matrix for the homogeneous system (1.3) and $\Phi^{-1}(t)$ is the inverse of $\Phi(t)$.

PROOF. See the Problems set in Section 6. □

COROLLARY 2.20. *When $A(t) = A$ is a constant matrix, it follows that*

$$e^{(t-t_0)A}u_0 + \int_{\tau=t_0}^{t} e^{(t-\tau)A}f(\tau)\,d\tau$$

is the unique solution to the initial value problem (1.6).

PROOF. See the Problems set in Section 6. □

A heuristic argument for obtaining the particular solution $g(t)$ of the form in Theorem 2.18 is to seek $g(t)$ of the form

$$g(t) = \Phi(t)c(t),$$

where $\Phi(t)$ is the fundamental matrix associated with the fundamental set

$$\{z_1(t), z_2(t), \ldots, z_n(t)\}$$

and $c(t)$ is an unknown vector function to be determined.

The form that $g(t)$ takes is the form that the general solution in Theorem 2.14 to the homogeneous equation (1.3) takes, except that here, the $c(t)$ is a vector function, instead of a constant vector. This delivers the name, Variation of Constants, of this method.

Calculations show that in order for $g(t)$ to be a solution to the nonhomogeneous system (1.1), we must have

$$\begin{aligned}
\frac{d}{dt}g(t) &= \Phi'(t)c(t) + \Phi(t)c'(t) \\
&= \begin{pmatrix} z_1'(t) & \cdots & z_n'(t) \end{pmatrix} c(t) + \Phi(t)c'(t) \\
&= \begin{pmatrix} A(t)z_1(t) & \cdots & A(t)z_n(t) \end{pmatrix} c(t) + \Phi(t)c'(t) \\
&= A(t)\Phi(t)c(t) + \Phi(t)c'(t) \\
&= A(t)[\Phi(t)c(t)] + f(t).
\end{aligned}$$

Thus $\Phi(t)c'(t) = f(t)$, and so

$$c(t) = \int \Phi^{-1}(t)f(t)\,dt.$$

Here $\Phi^{-1}(t)$ is the inverse matrix of $\Phi(t)$.

A different method for obtaining a particular solution explicitly to the nonhomogeneous system (1.1) exists and is called the method of undetermined coefficients, if $A(t) = A$ is a constant matrix and $f(t)$ is a vector function with components being either polynomial functions, exponential functions, sine, cosine, or combinations of them.

DEFINITION 2.21. *A vector polynomial $q(t)$ in the variable t, of degree $m \in \mathbb{N} \cup \{0\}$, is an expression of the form*

$$q(t) = c_m t^m + c_{m-1}t^{m-1} + \cdots + c_1 t + c_0,$$

where $c_i, i = 0, \ldots, m$, are constant vectors.

THEOREM 2.22 (Method of Undetermined Coefficients). *Suppose that $A(t) = A$ is a constant matrix, that its corresponding characteristic equation $p(\lambda) = 0$ is of the form*

$$p(\lambda) = a_0(\lambda - \lambda_1)^{m_1} \cdots (\lambda - \lambda_s)^{m_s}$$

as in Theorem 2.3, and finally that the nonhomogeneous term $f(t)$ in the system (1.1) is of the form

$$f(t) = q_1(t)e^{\mu_1 t},$$

where $q_1(t) = b_0 t^{n_1} + b_1 t^{n_1-1} + \cdots + b_{n_1-1}t + b_{n_1}$, a vector polynomial of degree $n_1 \in \{0\} \cup \mathbb{N}$, with $b_0 \neq 0$.

If μ_1 is not in the set

$$\{\lambda_1, \lambda_2, \ldots, \lambda_s\},$$

then a particular solution $g(t)$ to the nonhomogeneous system (1.1) is of the form

$$g(t) = Q_1(t)e^{t\mu_1},$$

where $Q_1(t)$ is a vector polynomial of degree n_1.

Conversely, if $\mu_1 = \lambda_1$, for example, then a particular solution $g(x)$ to the nonhomogeneous system (1.1) is of the form

$$g(t) = R_1(t)e^{t\lambda_1},$$

where $R_1(t)$ is a vector polynomial of degree $n_1 + m_1$.

REMARK 2.23. The proof of Theorem 2.22 given in Subsection 4.7 seems new.

Thus if $p(\lambda) = (\lambda - 2)(\lambda - 3) = 0$ and $f(t) = e^{2t}\begin{pmatrix} t^2 \\ 5 \end{pmatrix}$, then a particular solution $g(t)$ is of the form

$$g(t) = [c_3 t^3 + c_2 t^2 + c_1 t + c_0]e^{2t},$$

where $c_i, i = 0, \ldots, 3$, are four constant vectors to be determined.

It is a useful property that linear operations can be performed on the nonhomogeneous system (1.1):

LEMMA 2.24 (Superposition Principle). *If $z_i, i = 1, 2$, is a solution to the system*

$$\frac{d}{dt}u = A(t)u + f_i(t), \quad i = 1, 2,$$

then $(\alpha_1 z_1 + \alpha_2 z_2)$ is a solution to the system

$$\frac{d}{dt}u = A(t)u + \alpha_1 f_1(t) + \alpha_2 f_2(t).$$

PROOF. See the Problems set in Section 6. □

REMARK 2.25. *With the aid of Lemma 2.24, Theorem 2.22 applies to nonhomogeneous systems of the form*

$$\frac{d}{dt}u = A(t)u + q_1(t)e^{\mu_1 t} + q_2(t)e^{\mu_2 t} + \cdots + q_r(t)e^{\mu_r t}. \tag{2.8}$$

This form has $r(\in \mathbb{N})$ similar nonhomogeneous terms, as opposed to one single nonhomogeneous term $q_1(t)e^{\mu_1 t}$ that Theorem 2.22 has. In applying, we solve, separately, the nonhomogeneous system

$$\frac{d}{dt}u = A(t) + q_i(t)e^{\mu_i t}$$

for a particular solution $g_i(t), i = 1, 2, \ldots, r$, and then sum up those solutions $g_i(t)$'s. The sum $\sum_{i=1}^{r} g_i(t)$ will be a particular solution to the nonhomogeneous system (2.8), by the superposition principle in Lemma 2.24.

COROLLARY 2.26. *Theorem 2.22 also applies to the case where the nonhomogeneous term $f(t)$ is of the form*

$$f(t) = q_1(t)e^{t\alpha_1}\sin(\beta_1 t) \quad or$$
$$= q_1(t)e^{t\alpha_1}\cos(\beta_1 t) \quad or$$
$$= q_1(t)e^{t\alpha_1}\cos(t\beta_1) + q_2(t)e^{t\alpha_1}\sin(t\beta_1),$$

where $\alpha_1, \beta_1 \in \mathbb{R}$, $q_1(t)$ is a vector polynomial of degree n_1, and $q_2(x)$ is a vector polynomial of degree n_2. The results are:
If the complex conjugate pair

$$\{\alpha_1 + i\beta_1, \alpha_1 - i\beta_1\}$$

is not in the set

$$\{\lambda_1, \lambda_2, \ldots, \lambda_s\},$$

then the corresponding particular solutions $g(t)$'s are, repectively, of the form

$$g(t) = Q_1(t)e^{t\alpha_1}\cos(t\beta_1) + Q_2(t)e^{t\alpha_1}\sin(t\beta_1),$$
$$= the\ same\ as\ above, \quad and$$
$$= R_1(t)e^{t\alpha_1}\cos(t\beta_1) + R_2(t)e^{t\alpha_1}\sin(t\beta_1),$$

where $Q_1(t)$ and $Q_2(t)$ are vector polynomials of degree n_1, and $R_1(t)$ and $R_2(t)$ are polynomials of degree

$$n_3 \equiv \max\{n_1, n_2\}.$$

Conversely, if

$$\alpha_1 = i\beta_1 = \lambda_1$$
$$\alpha_1 - i\beta_1 = \lambda_2,$$

for example, for which there must

$$\lambda_1 = \lambda_2,$$

then the particular solutions $g(t)$'s to the nonhomogeneous system (1.1) take, respectively, the form

$$g(t) = S_1(t)e^{t\alpha_1}\cos(t\beta_1) + S_2(t)e^{t\alpha_1}\sin(t\beta_1),$$
$$= the\ same\ as\ above, \quad and$$
$$= T_1(t)e^{t\alpha_1}\cos(t\beta_1) + T_2(t)e^{t\alpha_1}\sin(t\beta_1),$$

where $S_1(t)$ and $S_2(t)$ are vector polynomials of degree $(n_1 + m_1)$, and $T_1(t)$ and $T_2(t)$ are vector polynomials of degree

$$n_3 \equiv \max\{n_1 + m_1, n_2 + m_1\}.$$

PROOF. See the Problems set in Section 6.   □

Thus if $p(\lambda) = (\lambda - (1 + 2i))(\lambda - (1 - 2i)) = 0$ and $f(t) = e^t \cos(2t)\begin{pmatrix} t^2 \\ 1 \end{pmatrix}$, then a particular solution $g(t)$ is of the form

$$g(t) = [c_3 t^3 + c_2 t^2 + c_1 t + c_0]e^t \cos(2t)$$

$$+ [d_3 t^3 + d_2 t^2 + d_1 t + d_0] e^t \sin(2t)$$

where $c_i, d_i, i = 0, \ldots, 3$, are eight constant vectors to be determined.

Although they are not unique, fundamental matrices for the homogeneous system (1.3) are related:

LEMMA 2.27. *Suppose that $\Phi(t)$ and $\Psi(t)$ are two fundamental matrices for the homogeneous system (1.3) with associated fundamental solution sets, $\{\phi_1, \ldots, \phi_n\}$ and $\{\psi_1, \ldots, \psi_n\}$, respectively. Then*

$$\Phi = \Psi C$$

for some nonsingular matrix C.

COROLLARY 2.28. *If $\Phi(t)$ is a fundamental matrix for the homogeneous system (1.3), then*

$$e^{tA} = \Phi(t)\Phi(0)^{-1}.$$

PROOF. See the Problems set in Section 6.   □

As is stated in Section 1, the concept of linear independence of n solutions to the homogeneous system (1.3) is closely related to its associated quantity, the Wronskian $W(t)$:

PROPOSITION 2.29. *Suppose that $z_i, i = 1, 2, \ldots, n$, are n solutions to the homogeneous system (1.3). Then the following statements (i), (ii), and (iii) are equivalent:*

(i) *The set $\{z_i, z_2, \ldots, z_n\}$ is linearly independent.*
(ii) *The associated Wronskian $W(t)$ with the n solutions $z_1, z_2, \ldots, z_n$ is non-zero somewhere.*
(iii) *The associated Wronskian $W(t)$ is nowhere zero.*

3. Examples

Before we prove the main results in Section 2, we look at some examples.

EXAMPLE 3.1. Find the general solution to the homogeneous system, Nagel and Saff, [**19**, page 154]:

$$\frac{d}{dt} u(t) = Au(t),$$

and find the unique solution to the initial value problem:

$$\frac{d}{dt} u(t) = Au(t)$$
$$u(0) = u_0,$$

where

$$A = \begin{pmatrix} 1 & -2 & 2 \\ -2 & 1 & 2 \\ 2 & 2 & 1 \end{pmatrix} \quad \text{and} \quad u_0 = \begin{pmatrix} 1 \\ 2 \\ 3 \end{pmatrix}.$$

Solution. We use three methods to find the solution.

Method 1. Compute the characteristic equation $p(\lambda) = 0$ for A:

$$0 = p(\lambda)$$
$$= |A - \lambda I|$$

$$= \begin{vmatrix} (1-\lambda) & -2 & 2 \\ -2 & (1-\lambda) & 2 \\ 2 & 2 & (1-\lambda) \end{vmatrix}$$

$$= -(\lambda - 3)^2(\lambda + 3).$$

Thus the eigenvalues are 3 of multiplicity two, and -3 of multiplicity one, respectively. The exponential function e^{tA} by Theorem 2.3 is:

$$e^{tA} = z_1(t)(t)N_1 + z_2(t)N_2 + z_3(t)N_3,$$

where $z_1(t) = e^{3t}$, $z_2(t) = te^{3t}$, and $z_3(t) = e^{-3t}$, and the 3×3 square matrices N_i, $i = 1, 2$, and 3, are unique solutions to the algebraic system of equations:

$$\begin{pmatrix} 1 & 0 & 1 \\ 3 & 1 & -3 \\ 9 & 6 & 9 \end{pmatrix} \begin{pmatrix} N_1 \\ N_2 \\ N_3 \end{pmatrix} = \begin{pmatrix} I \\ A \\ A^2 \end{pmatrix}.$$

Calculations show that

$$N_1 = \frac{1}{6} \begin{pmatrix} 4 & -2 & 2 \\ -2 & 4 & 2 \\ 2 & 2 & 4 \end{pmatrix},$$

$$N_2 = 0, \quad \text{and}$$

$$N_3 = \frac{1}{6} \begin{pmatrix} 2 & 2 & -2 \\ 2 & 2 & -2 \\ -2 & -2 & 2 \end{pmatrix}.$$

It follows by Theorem 2.3 that

$$e^{tA} = \frac{1}{6} \begin{pmatrix} (4e^{3t} + 2e^{-3t}) & (-2e^{3t} + 2e^{-3t}) & (2e^{3t} - 2e^{-3t}) \\ (-2e^{3t} + 2e^{-3t}) & (4e^{3t} + 2e^{-3t}) & (2e^{3t} - 2e^{-3t}) \\ (2e^{3t} - 2e^{-3t}) & (2e^{3t} - 2e^{-3t}) & (4e^{3t} + 2e^{-3t}) \end{pmatrix},$$

and so the general solution is

$$e^{tA}c$$

by Theorem 2.14, where $c = \begin{pmatrix} c_1 \\ c_2 \\ c_3 \end{pmatrix}$ is a constant column vector. Thus the unique solution to the initial value problem is

$$e^{tA}u_0$$

by Theorem 2.1.

Method 2. We use the results in Method 1 that

$$p(\lambda) = |A - \lambda I| = -(\lambda - 3)^2(\lambda + 3) = 0$$

and that the eigenvalues are 3 of multiplicity two and -3 of multiplicity one, respectively. Since A is real symmetric, it follows from Corollary 2.9 that there are three linearly independent eigenvectors, in which the eigenvalue 3 has two linearly independent eigenvectors, and in which the eigenvalue -3 has one linearly independent eigenvector. These eigenvectors can be calculated as follows.

For the eigenvalue $\lambda_1 = 3$, the equation $(A - 3I)u = 0$ where $u = \begin{pmatrix} x_1 \\ x_2 \\ x_3 \end{pmatrix}$, gives

$$\begin{pmatrix} -2 & -2 & 2 \\ -2 & -2 & 2 \\ 2 & 2 & -2 \end{pmatrix} \begin{pmatrix} x_1 \\ x_2 \\ x_3 \end{pmatrix} = \begin{pmatrix} 0 \\ 0 \\ 0 \end{pmatrix}.$$

Solving for x_1, x_2, and x_3, we have

$$-x_1 - x_2 + x_3 = 0,$$

and so $x_3 = x_1 + x_3$ and

$$u = \begin{pmatrix} x_1 \\ x_2 \\ x_1 + x_2 \end{pmatrix}$$

$$= \begin{pmatrix} x_1 \\ 0 \\ x_1 \end{pmatrix} + \begin{pmatrix} 0 \\ x_2 \\ x_2 \end{pmatrix}$$

$$= x_1 \begin{pmatrix} 1 \\ 0 \\ 1 \end{pmatrix} + x_2 \begin{pmatrix} 0 \\ 1 \\ 1 \end{pmatrix}.$$

Thus the corresponding two linearly independent eigenvectors can be chosen as

$$u_1^{(1)} = \begin{pmatrix} 1 \\ 0 \\ 1 \end{pmatrix}, u_2^{(1)} = \begin{pmatrix} 0 \\ 1 \\ 1 \end{pmatrix}.$$

For the eigenvalue $\lambda_2 = -3$, the equation $(A - (-3)I)u = 0$ where $u = \begin{pmatrix} x_1 \\ x_2 \\ x_3 \end{pmatrix}$, gives

$$\begin{pmatrix} 4 & -2 & 2 \\ -2 & 4 & 2 \\ 2 & 2 & 4 \end{pmatrix} \begin{pmatrix} x_1 \\ x_2 \\ x_3 \end{pmatrix} = \begin{pmatrix} 0 \\ 0 \\ 0 \end{pmatrix}.$$

Solving for x_1, x_2, and x_3, we have

$$x_2 = x_1, x_3 = -x_1,$$

and so

$$u = \begin{pmatrix} x_1 \\ x_1 \\ -x_1 \end{pmatrix}$$

$$= x_1 \begin{pmatrix} 1 \\ 1 \\ -1 \end{pmatrix}.$$

Thus the corresponding eigenvector can be chosen as

$$u_1^{(2)} = \begin{pmatrix} 1 \\ 1 \\ -1 \end{pmatrix}.$$

It follows from Corollary 2.9 that the set

$$\{e^{3t}u_1^{(1)}, e^{3t}u_2^{(1)}, e^{-3t}u_1^{(2)}\}$$

is a fundamental set of solutions, and then it follows from Theorem 2.14 that the general solution is

$$z(t) = c_1 e^{3t} u_1^{(1)} + c_2 e^{3t} u_2^{(1)} + c_3 e^{-3t} u_1^{(2)},$$

where c_i, $i = 1, 2$, and 3, are complex constants.

The unique solution to the initial value problem can be found by solving the algebraic system of equations for c_1, c_2, and c_3:

$$u_0 = \begin{pmatrix} 1 \\ 2 \\ 3 \end{pmatrix}$$

$$= z(0) = c_1 u_1^{(1)} + c_2 u_2^{(1)} + c_3 u_1^{(2)}$$

$$= \begin{pmatrix} 1 & 0 & 1 \\ 0 & 1 & 1 \\ 1 & 1 & -1 \end{pmatrix} \begin{pmatrix} c_1 \\ c_2 \\ c_3 \end{pmatrix}.$$

The result that is found is:

$$c_1 = 1, c_2 = 2, c_3 = 0,$$

and so the unique solution is:

$$z(t) = e^{3t} u_1^{(1)} + 2e^{3t} u_2^{(1)},$$

which is consistent with the unique solution $e^{tA} u_0$ that was found in Method 1.

Method 3. Rewrite the equations in the example as

$$(D - 1)u_1 + 2u_2 - 2u_3 = 0 \qquad (3.1)$$

$$2u_1 + (D - 1)u_2 - 2u_3 = 0 \qquad (3.2)$$

$$-2u_1 - 2u_2 + (D - 1)u_3 = 0, \qquad (3.3)$$

where $D = \frac{d}{dt}$ is a first order differential operator.

Apply the differential operator $(D - 1)$ to the equation (3.1), and multiply the equation (3.2) by 2. Adding the two results together, we obtain

$$(D = 1)(D - 3)u_1 - 2(D - 3)u_2 = 0. \qquad (3.4)$$

Adding equation (3.1) to equation (3.3), we have

$$(D - 3)u_1 + (D - 3)u_3 = 0$$

$$(D - 3)(u_1 + u_2) = 0. \qquad (3.5)$$

Multiplying equation (3.5) by 2 and adding the result to equation (3.4), we derive

$$(D - 3)(D + 3)u_1 = 0.$$

This equation has the solution

$$u_1 = c_1 e^{3t} + c_1 e^{-3t}$$

by the theory of Chapter 1, where c_1 and c_2 are two constants.

The equation (3.5) has the solution

$$u_1 + u_3 = c_3 e^{3t}$$

by the theory of Chapter 1, which, together with the solved u_1 above, gives

$$u_3 = (c_3 - c_1)e^{3t} - c_2 e^{-3t}.$$

The solved u_3 and u_1, when substituted into equation (3.1), gives

$$u_2 = (c_3 - 2c_1)e^{3t} + c_2 e^{-3t}.$$

Thus the general solution is

$$u = \begin{pmatrix} u_1 \\ u_2 \\ u_3 \end{pmatrix}$$

$$= e^{3t} \begin{pmatrix} c_1 \\ c_3 - 2c_1 \\ c_3 - c_1 \end{pmatrix} + e^{-3t} \begin{pmatrix} c_2 \\ c_2 \\ -c_2 \end{pmatrix}$$

$$= e^{3t} \left[\begin{pmatrix} 0 \\ c_3 \\ c_3 \end{pmatrix} + \begin{pmatrix} c_1 \\ -2c_1 \\ -c_1 \end{pmatrix} \right] + c_2 e^{-3t} \begin{pmatrix} 1 \\ 1 \\ -1 \end{pmatrix}$$

$$= c_3 e^{3t} \begin{pmatrix} 0 \\ 1 \\ 1 \end{pmatrix} + c_1 e^{3t} \begin{pmatrix} 1 \\ -2 \\ -1 \end{pmatrix} + c_2 e^{-3t} \begin{pmatrix} 1 \\ 1 \\ -1 \end{pmatrix}.$$

Here note that the set

$$\{ v_1 = \begin{pmatrix} 1 \\ -2 \\ -1 \end{pmatrix}, v_2 = \begin{pmatrix} 0 \\ 1 \\ 1 \end{pmatrix}, v_3 = \begin{pmatrix} 1 \\ 1 \\ -1 \end{pmatrix} \}$$

of three vectors is a fundamental set, since

$$u_2^{(1)} = v_2, u_1^{(2)} = v_3, \quad \text{and} \quad 2u_2^{(1)} + u_1^{(1)} = v_1,$$

where the fundamental set $\{u_1^{(1)}, u_2^{(1)}, u_1^{(2)}\}$ comes from Method 2. Here we used the two facts:

FACT 3.2. *If column operations are applied to a matrix B of $n \times n$, then B becomes a matrix of the form BC, where C is an invertible matrix. (Friedberg, Insel, and Spence* [**9**, *page* 130])

FACT 3.3. *If $\Phi(t)$ is a fundamental matrix for the homogeneous system* (1.3) *and C is an invertible matrix. then $\Phi(t)C$ is also a fundamental matrix,* [**5**, *page* 70]. *See the Problems set in Section 6.*

As in Method 2, the initial condition $u(0) = u_0 = \begin{pmatrix} 1 \\ 2 \\ 3 \end{pmatrix}$ determines the constants

c_1, c_2, c_3 and then the corresponding unique solution $u = \begin{pmatrix} u_1 \\ u_2 \\ u_3 \end{pmatrix}$.

EXAMPLE 3.4. Find the general solution to the Euler system

$$t\frac{du}{dt} = \begin{pmatrix} 1 & -2 & 2 \\ -2 & 1 & 2 \\ 2 & 2 & 1 \end{pmatrix} u.$$

Solution. It follows from Corollary 2.2 and Theorem 2.14 that the general solution is $e^{xA}c$, where e^{xA} is calculated in Example 3.1 and x is evaluated at $x = \ln(t)$.

EXAMPLE 3.5. Find the general solution to the nonhomogeneous system

$$\frac{d}{dt}u(t) = Au(t) + f(t),$$

and find the unique solution to the initial value problem for the nonhomogeneous system:

$$\frac{d}{dt}u(t) = Au(t) + f(t)$$
$$u(0) = u_0,$$

where

$$A = \begin{pmatrix} 1 & 3 \\ -1 & 5 \end{pmatrix},$$

$$f(t) = \begin{pmatrix} \cos(t) \\ t \end{pmatrix}, \quad \text{and}$$

$$u_0 = \begin{pmatrix} 1 \\ 2 \end{pmatrix}.$$

Solution. We use three methods to solve the problem.
Method 1. Computing the characteristic equation

$$p(\lambda) = |A - \lambda I|$$
$$= \begin{vmatrix} (1 - \lambda) & 3 \\ -1 & (5 - \lambda) \end{vmatrix}$$
$$= (\lambda - 2)(\lambda - 4) = 0,$$

we obtain that the eigenvalues are 2 of multiplicity one and 4 of multiplicity one, respectively. It follows from Corollary 2.9 that there exist two linearly independent eigenvectors, in which the eigenvalue 2 has one linearly independent eigenvector and so does the eigenvalue 4. These eigenvectors can be calculated as follows.

For the eigenvalue $\lambda_1 = 2$, the equation $(A - 2I)u = 0$ where $u = \begin{pmatrix} x_1 \\ x_2 \end{pmatrix}$, gives

$$\begin{pmatrix} -1 & 3 \\ -1 & 3 \end{pmatrix} \begin{pmatrix} x_1 \\ x_2 \end{pmatrix} = \begin{pmatrix} 0 \\ 0 \end{pmatrix}.$$

Solving for x_1 and x_2, we have

$$x_1 = 3x_2,$$

and so

$$u = \begin{pmatrix} 3x_2 \\ x_2 \end{pmatrix}$$
$$= x_2 \begin{pmatrix} 3 \\ 1 \end{pmatrix}.$$

Thus the corresponding eigenvector can be chosen as

$$u_1^{(1)} = \begin{pmatrix} 3 \\ 1 \end{pmatrix}.$$

For the eigenvalue $\lambda_2 = 4$, the equation $(A - 4I)u = 0$ where $u = \begin{pmatrix} x_1 \\ x_2 \end{pmatrix}$, gives

$$\begin{pmatrix} -3 & 3 \\ -1 & 1 \end{pmatrix} \begin{pmatrix} x_1 \\ x_2 \end{pmatrix} = \begin{pmatrix} 0 \\ 0 \end{pmatrix}.$$

Solving for x_1 and x_2, we have

$$x_1 = x_2,$$

and so

$$u = \begin{pmatrix} x_2 \\ x_2 \end{pmatrix}$$

$$= x_2 \begin{pmatrix} 1 \\ 1 \end{pmatrix}.$$

Thus the corresponding eigenvector can be chosen as

$$u_1^{(2)} = \begin{pmatrix} 1 \\ 1 \end{pmatrix}.$$

It follows from Corollary 2.9 that the set

$$\{e^{2t}u_1^{(1)}, e^{4t}u_1^{(2)}\}$$

is a fundamental set and then the associated matrix

$$\Phi(t) = \begin{pmatrix} e^{2t}u_1^{(1)} & e^{4t}u_1^{(2)} \end{pmatrix}$$

$$= \begin{pmatrix} 3e^{2t} & e^{4t} \\ e^{2t} & e^{4t} \end{pmatrix},$$

is a fundamental matrix.

The inverse of $\Phi(t)$ is

$$\Phi^{-1}(t) = \frac{1}{2} \begin{pmatrix} e^{-2t} & -e^{-2t} \\ -e^{-4t} & 3e^{-4t} \end{pmatrix},$$

which is obtained by augmenting $\Phi(t)$ with the identity $I = \begin{pmatrix} 1 & 0 \\ 0 & 1 \end{pmatrix}$:

$$\begin{pmatrix} \Phi(t) & | & I \end{pmatrix}$$

and then applying row operations to the augmented matrix to reach a matrix of the form

$$\begin{pmatrix} I & | & \Psi(t) \end{pmatrix}.$$

The resulting matrix $\Psi(t)$ is the inverse $\Phi^{-1}(t)$ of $\Phi(t)$.

A particular solution is

$$g(t) = \Phi(t)\varphi(t)$$

$$= \begin{pmatrix} \frac{3}{8}t + \frac{9}{32} - \frac{41}{85}\cos(t) + \frac{23}{85}\sin(t) \\ -\frac{1}{8}t + \frac{1}{32} - \frac{7}{85}\cos(t) + \frac{6}{85}\sin(t) \end{pmatrix} \quad \text{where}$$

$$\varphi(t) = \int \Phi^{-1}(t)f(t)\, dt$$

$$= \frac{1}{2} \begin{pmatrix} \frac{1}{5}(e^{-2t}\sin(t) - 2e^{-2t}\cos(t)) - (\frac{-1}{2})(te^{-2t} + \frac{1}{2}e^{-2t}) \\ -\frac{1}{17}(e^{-4t}\sin(t) - 4e^{-4t}\cos(t)) + 3(\frac{-1}{4})(te^{-4t} + \frac{1}{4}e^{-4t}) \end{pmatrix},$$

which is obtained by applying Theorem 2.18.

It follows from Theorem 2.16 that the general solution to the nonhomogeneous system is:

$$z(t) = \Phi(t)c + g(t),$$

where $c = \begin{pmatrix} c_1 \\ c_2 \end{pmatrix}$ is a constant vector. The unique solution for the initial value problem by Corollary 2.19 is

$$z(t) = \Phi(t)\Phi(0)^{-1}u_0 + \Phi(t)\varphi(\tau)|_{\tau=0}^{t}.$$

Method 2. This is a method of elimination. Rewrite the equations in the example as

$$(D-1)u_1 - 3u_2 = \cos(t) \tag{3.6}$$
$$u_1 + (D-5)u_2 = t, \tag{3.7}$$

where $D = \frac{d}{dt}$ is a first order differential operator and $D^2 = DD = \frac{d^2}{dt^2}$ is the second order differential operator.

Applying the differential operator $(D-1)$ to equation (3.7), from which subtract equation (3.6), we obtain the equation

$$(D^2 - 6D + 8)u_2 = 1 - t - \cos(t). \tag{3.8}$$

This equation has, by the theory in Chapter 1, the general solution

$$u_2 = c_1 e^{2t} + c_2 e^{4t} + [-\frac{1}{8}t + \frac{1}{32} - \frac{7}{85}\cos(t) + \frac{6}{85}\sin(t)].$$

Substituting the solved u_2 into equation (3.7), we obtain

$$u_1 = 3c_1 e^{2t} + c_2 e^{4t} + [\frac{3}{8}t + \frac{9}{32} - \frac{41}{85}\cos(t) + \frac{23}{85}\sin(t)].$$

Thus the solved $u(t) = \begin{pmatrix} u_1 \\ u_2 \end{pmatrix}$ gives the general solution.

As in Method 1, the initial condition $u(0) = u_0 = \begin{pmatrix} 1 \\ 2 \end{pmatrix}$ determines the constants c_1 and c_2 and then the unique solution for the initial value problem.

Method 3. This is the method of undetermined coefficients for finding a particular solution. Here note that the nonhomogeneous term $f(t)$ can be rewrittens as the sum of two terms:

$$f(t) = \begin{pmatrix} \cos(t) \\ t \end{pmatrix} = \begin{pmatrix} \cos(t) \\ 0 \end{pmatrix} + \begin{pmatrix} 0 \\ t \end{pmatrix}$$
$$= \cos(t)\begin{pmatrix} 1 \\ 0 \end{pmatrix} + t\begin{pmatrix} 0 \\ 1 \end{pmatrix}.$$

We will apply the method of undetermined coefficients in Theorem 2.22, Remark 2.25, and Corollary 2.26, to the above two terms, separately.

For the equation

$$\frac{d}{dt}u = Au + \cos(t)\begin{pmatrix} 1 \\ 0 \end{pmatrix},$$

we assume that a particular solution is of the form

$$g_1(t) = a\cos(t) + b\sin(t),$$

where a and b are two constant vectors to be determined. Since $g_1(t)$ satisfies the equation by assumption, we have

$$-a\sin(t) + b\cos(t) = A(a\cos(t) + b\sin(t)) + \cos(t)\begin{pmatrix} 1 \\ 0 \end{pmatrix},$$

and so

$$Ab = -a$$

$$Aa + \begin{pmatrix} 1 \\ 0 \end{pmatrix} = b.$$

It follows that

$$A^2 a + A\begin{pmatrix} 1 \\ 0 \end{pmatrix} = Ab = -a$$

$$(A^2 + I)a = -A\begin{pmatrix} 1 \\ 0 \end{pmatrix} = \begin{pmatrix} -1 \\ 1 \end{pmatrix}$$

$$a = \begin{pmatrix} -\frac{41}{85} \\ -\frac{7}{85} \end{pmatrix}$$

$$b = \begin{pmatrix} \frac{23}{85} \\ \frac{6}{85} \end{pmatrix}.$$

For the equation

$$\frac{d}{dt}u = Au + t\begin{pmatrix} 0 \\ 1 \end{pmatrix},$$

we assume that a particular solution is of the form

$$g_2(t) = ct + d,$$

where c and d are two constant vectors to be determined. Since $g_2(t)$ satisfies the equation by assumption, we have

$$c = A(ct + d) + t\begin{pmatrix} 0 \\ 1 \end{pmatrix},$$

and so

$$c = Ad$$

$$Ac = \begin{pmatrix} 0 \\ -1 \end{pmatrix}.$$

Solving for c and d, we have

$$c = \begin{pmatrix} \frac{3}{8} \\ -\frac{1}{8} \end{pmatrix}$$

$$d = \begin{pmatrix} \frac{9}{32} \\ \frac{1}{32} \end{pmatrix}.$$

Thus a particular solution for the equation

$$\frac{d}{dt}u = Au + f(t)$$

is

$$g(t) = g_1(t) + g_2(t),$$

and the general solution is

$$u(t) = \Phi(t)c + g(t)$$

where the fundamental matrix $\Phi(t)$ was obtained in Method 1.

EXAMPLE 3.6. Find a particular solution to the nonhomogeneous system

$$\frac{d}{dt}u = Au + f(t),$$

where

$$A = \begin{pmatrix} 1 & 3 \\ -1 & 5 \end{pmatrix} \quad \text{and}$$

$$f(t) = \begin{pmatrix} e^{2t} \\ e^{4t} + e^t \end{pmatrix}.$$

Solution. We will use the method of undetermined coefficients in Theorem 2.22, Remark 2.25, and Corollary 2.26, to find a particular solution.

Note that the matrix A here is the same as that in the previous Example 3.5, and so we have from there that 2 and 4 are the two eigenvalues and that $x_2 \begin{pmatrix} 3 \\ 1 \end{pmatrix}$ and $x_1 \begin{pmatrix} 1 \\ 1 \end{pmatrix}$ are the two corresponding eigenvectors.

Note also that the nonhomogeneous term $f(t)$ can be rewritten as the sum of two terms:

$$f(t) = e^{2t} \begin{pmatrix} 1 \\ 0 \end{pmatrix} + e^{4t} \begin{pmatrix} 0 \\ 1 \end{pmatrix},$$

and we will apply the method of undetermined coefficients to these two terms, separately.

For the equation

$$\frac{d}{dt}u = Au + e^{2t} \begin{pmatrix} 1 \\ 0 \end{pmatrix},$$

in which the number 2 in the exponent of the function e^{2t} also appears to be an eigenvalue of A, we assume that a particular solution is of the form

$$g_1(t) = ae^{2t} + bte^{2t},$$

where $a = \begin{pmatrix} a_1 \\ a_2 \end{pmatrix}$ and b are two constant vectors to be determined. Since $g_1(t)$ satisfies the equation by assumption, we have

$$2ae^{2t} + be^{2t} + 2bte^{2t} = A(ae^{2t} + bte^{2t}) + e^{2t} \begin{pmatrix} 1 \\ 0 \end{pmatrix},$$

and so

$$2a + b = Aa + \begin{pmatrix} 1 \\ 0 \end{pmatrix}$$

$$2b = Ab.$$

b is the eigenvector of A corresponding to the eigenvalue 2, and was already obtained above as

$$x_2 \begin{pmatrix} 3 \\ 1 \end{pmatrix}.$$

Using this obtained b, we derive

$$(A-2)a = \begin{pmatrix} (-a_1 + 3a_2) \\ (-a_1 + 3a_2) \end{pmatrix}$$

$$= b - \begin{pmatrix} 1 \\ 0 \end{pmatrix}$$

$$= \begin{pmatrix} 3x_2 - 1 \\ x_2 \end{pmatrix},$$

and so

$$3x_2 - 1 = x_2 \quad \text{or} \quad x_2 = \frac{1}{2},$$

and then

$$a = \begin{pmatrix} a_1 \\ a_2 \end{pmatrix} = \begin{pmatrix} 3a_2 - \frac{1}{2} \\ a_2 \end{pmatrix} = a_2 \begin{pmatrix} 3 \\ 1 \end{pmatrix} + \begin{pmatrix} -\frac{1}{2} \\ 0 \end{pmatrix}.$$

We can choose any a_2 or the simplest case where $a_2 = 0$, for which

$$a = \begin{pmatrix} -\frac{1}{2} \\ 0 \end{pmatrix}.$$

For the equation

$$\frac{d}{dt}u = Au + e^{4t}\begin{pmatrix} 0 \\ 1 \end{pmatrix},$$

in which the number 4 in the exponent of the exponential function e^{4t} appears to be an eigenvalue of A, we assume that a particular solution is of the form

$$g_2(t) = ce^{4t} + dte^{4t},$$

where $c = \begin{pmatrix} c_1 \\ c_2 \end{pmatrix}$ and d are two constant vectors to be determined. Since $g_2(t)$ satisfies the equation by assumption, we have

$$4ce^{4t} + de^{4t} + 4dte^{4t} = A(ce^{4t} + dte^{4t}) + e^{4t}\begin{pmatrix} 0 \\ 1 \end{pmatrix},$$

and so

$$4c + d = Ac + \begin{pmatrix} 0 \\ 1 \end{pmatrix}$$

$$4d = Ad,$$

d is the eigenvector of A corresponding to the eigenvalue 4, and was already obtained above as

$$x_1 \begin{pmatrix} 1 \\ 1 \end{pmatrix}.$$

Using the obtained d, we derive

$$\begin{pmatrix} -3 & 3 \\ -1 & 1 \end{pmatrix} c = (A - 4)c$$

$$= d - \begin{pmatrix} 0 \\ 1 \end{pmatrix}$$

$$= \begin{pmatrix} x_1 \\ x_1 - 1 \end{pmatrix}.$$

It follows that

$$3(x_1 - 1) = x_1 \quad \text{or} \quad x_1 = \frac{3}{2},$$

$$c = \begin{pmatrix} c_1 \\ c_2 \end{pmatrix} = \begin{pmatrix} c_2 - \frac{1}{2} \\ c_2 \end{pmatrix} = c_2 \begin{pmatrix} 1 \\ 1 \end{pmatrix} + \begin{pmatrix} -\frac{1}{2} \\ 0 \end{pmatrix}.$$

We can choose any c_2 or the simplest case where $c_2 = 0$, for which

$$c = \begin{pmatrix} -\frac{1}{2} \\ 0 \end{pmatrix}.$$

For the equation

$$\frac{d}{dt} u = Au + e^t \begin{pmatrix} 0 \\ 1 \end{pmatrix},$$

we assume that a particular solution is of the form

$$g_3(t) = \alpha e^t,$$

where α is a constant vector to be determined. Since $g_3(t)$ satisfies the equation by assumption, we have

$$\alpha e^t = A\alpha e^t + e^t \begin{pmatrix} 0 \\ 1 \end{pmatrix},$$

and so

$$(A - I)\alpha = \begin{pmatrix} 0 \\ -1 \end{pmatrix}.$$

It is easy to see that

$$\alpha = \begin{pmatrix} 1 \\ 0 \end{pmatrix}.$$

Thus a particular solution for the equation

$$\frac{d}{dt} u = Au + f(t)$$

is

$$g(t) = g_1(t) + g_2(t) + g_3(t).$$

EXAMPLE 3.7. Find the general solution to the homogeneous system

$$\frac{d}{dt} u(t) = Au(t), \quad \text{where}$$

$$A = \begin{pmatrix} -3 & -2 \\ 9 & 5 \end{pmatrix}.$$

Solution.
Method 1. Computing the characteristic equation

$$p(\lambda) = |A - \lambda I|$$
$$= \begin{vmatrix} -3 - \lambda & -2 \\ 9 & 5 - \lambda \end{vmatrix}$$
$$= \lambda^2 - 2\lambda + 3 = 0,$$

we obtain that the eigenvalues are $(1 + i\sqrt{2})$ of multiplicity one and $(1 - i\sqrt{2})$ of multiplicity one, respectively. The exponential function e^{tA} by Theorem 2.3 is:

$$e^{tA} = z_1(t)N_1 + z_2(t)N_2,$$

where $z_1(t) = e^t \cos(\sqrt{2}t)$, $z_2(t) = e^t \sin(\sqrt{2}t)$, and the 2×2 square matrices $N_i, i = 1, 2$, are unique solutions to the algebraic system of equations:

$$\begin{pmatrix} 1 & 0 \\ 1 & \sqrt{2} \end{pmatrix} \begin{pmatrix} N_1 \\ N_2 \end{pmatrix} = \begin{pmatrix} I \\ A \end{pmatrix}.$$

Calculations show that

$$N_1 = \begin{pmatrix} 1 & 0 \\ 0 & 1 \end{pmatrix} \quad \text{and} \quad N_2 = \frac{1}{\sqrt{2}} \begin{pmatrix} -4 & -2 \\ 9 & 4 \end{pmatrix}.$$

It follows by Theorem 2.3 that

$$e^{tA} = \begin{pmatrix} e^t \cos(\sqrt{2}t) - \frac{4}{\sqrt{2}} e^t \sin(\sqrt{2}t) & \frac{-2}{\sqrt{2}} e^t \sin(\sqrt{2}t) \\ \frac{9}{\sqrt{2}} e^t \sin(\sqrt{2}t) & e^t \cos(\sqrt{2}t) + \frac{4}{\sqrt{2}} e^t \sin(\sqrt{2}t) \end{pmatrix},$$

and so the general solution is

$$e^{tA} c$$

by Theorem 2.14, where $c = \begin{pmatrix} c_1 \\ c_2 \end{pmatrix}$ is a constant column vector.

Method 2. From Method 1 above, A has the two eigenvalues $\{1 + \sqrt{2}i, 1 - \sqrt{2}i\}$, a conjugate pair. For the eigenvalue $\lambda_1 = 1 + \sqrt{2}i$, a corresponding eigenvector $u = \begin{pmatrix} x_1 \\ x_2 \end{pmatrix}$ satisfies

$$\begin{pmatrix} 0 \\ 0 \end{pmatrix} = (A - \lambda_1 I)u$$

$$= \begin{pmatrix} -4 - \sqrt{2}i & -2 \\ 9 & (4 - \sqrt{2}i) \end{pmatrix} \begin{pmatrix} x_1 \\ x_2 \end{pmatrix}.$$

This gives

$$-(4 + \sqrt{2}i)x_1 - 2x_2 = 0,$$

and so we can take $u = \begin{pmatrix} 2 \\ -4 - \sqrt{2}i \end{pmatrix}$. The calculation

$$e^{t\lambda_1} u = e^t [\cos(\sqrt{2}t) + i \sin(\sqrt{2}t)] \begin{pmatrix} 2 \\ -4 - \sqrt{2}i \end{pmatrix}$$

$$= e^t \begin{pmatrix} 2\cos(\sqrt{2}t) \\ -4\cos(\sqrt{2}t) + \sqrt{2}\sin(\sqrt{2}t) \end{pmatrix} + i e^t \begin{pmatrix} 2\sin(\sqrt{2}t) \\ -4\sin(\sqrt{2}t) - \sqrt{2}\cos(\sqrt{2}t) \end{pmatrix}$$

$$\equiv z_1(t) + i z_2(t)$$

shows by using Remark 2.7 that $\{z_1, z_2\}$ is a fundamental set of solutions and that $c_1 z_1 + c_2 z_2$ is the general solution. Here c_1 and c_2 are two scalar constants.

EXAMPLE 3.8. Find the general solution to the homogeneous system

$$\frac{d}{dt} u(t) = Au(t),$$

and the general solution to the Euler system

$$t \frac{d}{dt} u(t) = Au(t),$$

where

$$A = \begin{pmatrix} 1 & -1 \\ 1 & 3 \end{pmatrix}.$$

Solution. Compute the characteristic equation

$$p(\lambda) = |A - \lambda I|$$

$$= \begin{vmatrix} 1 - \lambda & -1 \\ 1 & 3 - \lambda \end{vmatrix}$$

$$= (\lambda - 2)^2 = 0,$$

we obtain that the only eigenvalue is 2 of multiplicity two. It follows from Theorem 2.3 that the exponential function e^{tA} is:

$$e^{tA} = e^{2t} N_1 + te^{2t} N_2,$$

where the 2×2 matrices N_1, N_2 are unique solutions to the system of algebraic equations:

$$\begin{pmatrix} 1 & 0 \\ 2 & 1 \end{pmatrix} \begin{pmatrix} N_1 \\ N_2 \end{pmatrix} = \begin{pmatrix} I \\ A \end{pmatrix}.$$

Calculations show that

$$N_1 = \begin{pmatrix} 1 & 0 \\ 0 & 1 \end{pmatrix}, N_2 = \begin{pmatrix} -1 & -1 \\ 1 & 1 \end{pmatrix},$$

and so

$$e^{tA} = \begin{pmatrix} e^{2t}(1 - t) & -te^{2t} \\ t^{2t} & e^{2t}(1 + t) \end{pmatrix}.$$

It results that the general solution to the homogeneous system is

$$e^{tA} c,$$

and that the general solution to the Euler system is, by Corollary 2.2,

$$e^{xA},$$

where $c = \begin{pmatrix} c_1 \\ c_2 \end{pmatrix}$ is a constant column vector and x is evaluated at $x = \ln(t)$.

REMARK 3.9. *We cannot solve the above Example by using the Method 2 in the Solution to Example 3.1, since we only have one eigenvector for the eigenvalue 2 of multiplicity two. To overcome this difficulty, we can use the method in Section 5.*

EXAMPLE 3.10. Find the general solution to the system of two linear second order equations, Nagel and Saff [**19**, pages 211–212], Cullen [**7**, pages 323–328]:

$$x''(t) + y'(t) - x(t) + y(t) = t$$
$$x'(t) + y'(t) - x(t) = t^2.$$

Solution. We use two methods to solve the problem, in which method 2, a method of elimination, is more efficient in the current case where only two equations are involved.

Method 1. Using the substitutions

$$u_1 = x$$
$$u_2 = x'$$
$$u_3 = y,$$

we have

$$u_1' = x' = u_2$$

$$u_2' = x'' = -y' + x - y + t = -u_3' + u_1 - u_3 + t$$
$$u_3' = -x' + x + t^2 = -u_2 + u_1 + t^2,$$

which is the same as

$$u_1' = u_2$$
$$u_2' + u_3' = u_1 - u_3 + t$$
$$u_3' = u_1 - u_2 + t^2.$$

Written in terms of matrices, the above is equal to

$$F \begin{pmatrix} u_1' \\ u_2' \\ u_3' \end{pmatrix} = G \begin{pmatrix} u_1 \\ u_2 \\ u_3 \end{pmatrix} + h(t),$$

where the matrices

$$F = \begin{pmatrix} 1 & 0 & 0 \\ 0 & 1 & 1 \\ 0 & 0 & 1 \end{pmatrix}, G = \begin{pmatrix} 0 & 1 & 0 \\ 1 & 0 & -1 \\ 1 & -1 & 0 \end{pmatrix}, \quad \text{and} \quad h(t) = \begin{pmatrix} 0 \\ t \\ t^2 \end{pmatrix}.$$

Using row operations, the above system is equivalent to the system of linear first order equations

$$\begin{pmatrix} u_1' \\ u_2' \\ u_3' \end{pmatrix} = \frac{d}{dt} \begin{pmatrix} u_1 \\ u_2 \\ u_3 \end{pmatrix} = A \begin{pmatrix} u_1 \\ u_2 \\ u_3 \end{pmatrix} + f(t),$$

and can be solved by the theory of Section 2. Here the matricies

$$A = \begin{pmatrix} 0 & 1 & 0 \\ 0 & 1 & -1 \\ 1 & -1 & 0 \end{pmatrix} \quad \text{and} \quad f(t) = \begin{pmatrix} 0 \\ t - t^2 \\ t^2 \end{pmatrix}$$

are obtained by augmenting the matrix F with the matrix G, and $h(t)$:

$$\begin{pmatrix} F & | & G & | & h(t) \end{pmatrix}$$

and then by applying the row operations to the augmented matrix

$$\begin{pmatrix} F & | & G & | & h(t) \end{pmatrix}$$

to reach a matrix of the form

$$\begin{pmatrix} I & | & B & | & c(t) \end{pmatrix},$$

where I is the identity matrix. The resulting matrix B is the matrix A, and the resulting $c(t)$ is the $f(t)$. Here we used the fact from linear algebra, Friedberg, Insel, and Spence [**9**, pages 141–142]:

FACT 3.11. *Let $U, V,$ and W be three square matrices of order $m \times m, m \in \mathbb{N}$, with U being invertible. If row operations are performed on the augmented matrix*

$$\begin{pmatrix} U & | & V & | & W \end{pmatrix}$$

and this matrix becomes a matrix of the form

$$\begin{pmatrix} I & | & X & | & Y \end{pmatrix},$$

then

$$X = U^{-1}V$$
$$Y = U^{-1}W.$$

Here I is the identity matrix.

As in solving Example 3.1, we have

$$e^{tA} = \frac{1}{2} \begin{pmatrix} \frac{3}{2}e^t - te^t + \frac{1}{2}e^{-t} & te^t + \frac{1}{2}e^t - \frac{1}{2}e^{-t} & \frac{1}{2}e^t - te^t - \frac{1}{2}e^{-t} \\ \frac{1}{2}e^t - te^t - \frac{1}{2}e^{-t} & \frac{3}{2}e^t + te^t + \frac{1}{2}e^{-t} & -\frac{1}{2}e^t - te^t + \frac{1}{2}e^{-t} \\ e^t - e^{-t} & -e^t + e^{-t} & e^t + e^{-t} \end{pmatrix},$$

which results from the calculations:

$1, 1,$ and -1 are eigenvalues of A;

$e^{tA} = e^t N_1 + te^t N_2 + e^{-t} N_3$,

$N_1, N_2,$ and N_3 are unique solution to the algebraic system :

$I = N_1 + N_3$

$A = N_1 + N_2 - N_3$

$A^2 = N_1 + 2N_2 + N_3,$ and are given by

$$N_1 = \frac{1}{2} \begin{pmatrix} \frac{3}{2} & \frac{1}{2} & \frac{1}{2} \\ \frac{1}{2} & \frac{3}{2} & -\frac{1}{2} \\ 1 & -1 & 1 \end{pmatrix},$$

$$N_2 = \frac{1}{2} \begin{pmatrix} -1 & 1 & -1 \\ -1 & 1 & -1 \\ 0 & 0 & 0 \end{pmatrix}, \quad \text{and}$$

$$N_3 = \frac{1}{2} \begin{pmatrix} \frac{1}{2} & -\frac{1}{2} & -\frac{1}{2} \\ -\frac{1}{2} & \frac{1}{2} & \frac{1}{2} \\ -1 & 1 & 1 \end{pmatrix}, \quad \text{respectively.}$$

It follows from the remark after Theorem 2.18 that a particular solution $g(t)$ is given by

$$g(t) = \int_{\tau=0}^t e^{tA} e^{-\tau A} f(\tau)\, d\tau$$

$$= \int_{\tau=0}^t e^{(t-\tau)A} f(\tau)\, d\tau$$

$$= \frac{1}{2} \int_{\tau=0}^t \begin{pmatrix} [-(2t+1)\tau^2 + 2\tau^3 + (t+\frac{1}{2})\tau]e^{(t-\tau)} - \frac{1}{2}\tau e^{-(t-\tau)} \\ [2\tau^3 - (3+2t)\tau^2 + (\frac{3}{2}+t)\tau]e^{(t-\tau)} \\ [2\tau^2 - \tau]e^{(t-\tau)} + \tau e^{-(t-\tau)} \end{pmatrix} d\tau$$

$$= \frac{1}{2} \begin{pmatrix} (-2t^2 - 8t - 10) + (-3te^t + \frac{21}{2}e^t - \frac{1}{2}e^{-t}) \\ (-4t - 8) + (\frac{15}{2} - 3t)e^t + \frac{1}{2}e^{-t} \\ (-2t^2 - 2t - 4) + (3e^t + e^{-t}) \end{pmatrix}.$$

Here we used the calculations:

$$(C_1 \quad C_2 \quad C_3) \begin{pmatrix} 0 \\ \tau - \tau^2 \\ \tau^2 \end{pmatrix} = C_1 \times 0 + C_2 \times (\tau - \tau^2) + C_3 \times \tau^2,$$

where $C_1, C_2,$ and C_3 are the three columns of the matrix $e^{(t-\tau)A}$;

$$\int_{\tau=0}^t \tau e^{(t-\tau)}\, d\tau = -t - 1 + e^t;$$

$$\int_{\tau=0}^{t} \tau^2 e^{(t-\tau)} \, d\tau = -(t^2 + 2t + 2 - 2e^t);$$

$$\int_{\tau=0}^{t} \tau^3 e^{(t-\tau)} \, d\tau = -(t^3 + 3t^2 + 6t + 6 - 6e^t); \quad \text{and}$$

$$\int_{\tau=0}^{t} \tau e^{-(t-\tau)} \, d\tau = t - 1 + e^{-t}.$$

As a consequence, the general solution is

$$\begin{pmatrix} x(t) \\ x'(t) \\ y(t) \end{pmatrix} = \begin{pmatrix} u_1 \\ u_2 \\ u_3 \end{pmatrix}$$

$$= e^{tA} \begin{pmatrix} c_1 \\ c_2 \\ c_3 \end{pmatrix} + g(t)$$

$$= \frac{1}{2} \begin{pmatrix} e^t(\frac{3}{2}c_1 + \frac{1}{2}c_2 + \frac{1}{2}c_3 + \frac{21}{2}) + te^t(-c_1 + c_2 - c_3 - 3) + e^{-t}(\frac{1}{2}c_1 - \frac{1}{2}c_2 - \frac{1}{2}c_3 - \frac{1}{2}) + (-2t^2 - 8t - 10) \\ e^t(\frac{1}{2}c_1 + \frac{3}{2}c_2 - \frac{1}{2}c_3 + \frac{15}{2}) + te^t(-c_1 + c_2 - c_3 - 3) + e^{-t}(-\frac{1}{2}c_1 + \frac{1}{2}c_2 + \frac{1}{2}c_3 + \frac{1}{2}) + (-4t - 8) \\ e^t(c_1 - c_2 + c_3 + 3) + e^{-t}(-c_1 + c_2 + c_3 + 1) + (-2t^2 - 2t - 4) \end{pmatrix}.$$

Thus we have

$$x(t) = d_3 e^t - d_1 t e^t - \frac{1}{2} d_2 e^{-t} - t^2 - 4t - 5$$

$$y(t) = d_1 e^t + d_2 e^{-t} - t^2 - t - 2,$$

where

$$d_1 = \frac{1}{2}(c_1 - c_2 + c_3 + 3),$$

$$d_2 = \frac{1}{2}(-c_1 + c_2 + c_3 + 1), \quad \text{and}$$

$$d_3 = \frac{1}{2}(\frac{3}{2}c_1 + \frac{1}{2}c_2 + \frac{1}{2}c_3 + \frac{21}{2}),$$

three constants.

Remark. Using the method in Section 5, we will obtain a fundamental matrix $\Phi(t)$ given by

$$\Phi(t) = \begin{pmatrix} z_1(t) & z_2(t) & z_3(t) \end{pmatrix},$$

where the three linearly independent solutions z_1, z_2, and z_3 are given by

$$z_1(t) = e^t \begin{pmatrix} 1 \\ 1 \\ 0 \end{pmatrix}, z_2(t) = e^t [\begin{pmatrix} 1 \\ 0 \\ 1 \end{pmatrix} + t \begin{pmatrix} -1 \\ -1 \\ 0 \end{pmatrix}], \quad \text{and}$$

$$z_3(t) = e^{-t} \begin{pmatrix} -1 \\ 1 \\ 2 \end{pmatrix}, \quad \text{respectively.}$$

Here we observe that

$$z_1(t) = C_1 + C_2$$

$$z_2(t) = C_1 + C_3$$

$$z_3(t) = 2(C_2 + C_3),$$

where C_1, C_2, and C_3 are column one, two, and three of the matrix e^{tA}, respectively.

Method 2. This is a method of elimination. Rewrite the two equations in the example as

$$(D^2 - 1)x + (D + 1)y = t \tag{3.9}$$
$$(D - 1)x + Dy = t^2, \tag{3.10}$$

where $D = \frac{d}{dt}$, is the first order differential operator and $D^2 = DD = \frac{d^2}{dt^2}$, is the second order differential operator. Applying the differential operator

$$(D + 1)$$

to the equation (3.10), we obtain the equation

$$(D^2 - 1)x + (D + 1)Dy = 2t + t^2. \tag{3.11}$$

To eliminate the function x, we subtract the equation (3.9) from the equation (3.11). It follows that

$$(D + 1)(D - 1)y = (D^2 - 1)y = y'' - y = t^2 + t, \tag{3.12}$$

a second order nonhomogeneous equation with the nonhomogeneous term of a polynomial function of degree two.

Using the theory of Undetermined Coefficients Method in Chapter 1, (3.12) has the solution

$$y = c_1 e^t + c_2 e^{-t} - t^2 - t - 2,$$

in which $\{e^t, e^{-t}\}$ is a fundamental set of solutions, resulting from the homogeneous equation $y'' - y = 0$; c_1 and c_2 are two constants; $g(t) \equiv -t^2 - t - 2$ is a particular solution, resulting from the conditions that $g(t)$ is of the form

$$g(t) = at^2 + bt + c$$

and that $g(t)$ satisfies the equation (3.11), for which the constants a, b, and c are then determined.

Substituting the y into the equation (3.10), we obtain

$$(D - 1)x = x' - x$$
$$= t^2 - y' = 2t^2 + t + 2 - c_1 e^t + c_2 e^{-t}, \tag{3.13}$$

a first order nonhomogeneous equation with the nonhomogeneous term of a polynomial function of degree two and two exponential functions, one of which satisfies the homogeneous equation

$$x' - x = 0.$$

Using the theory of Undetermined Coefficients Method in Chapter 1, the equation (3.13) has the solution

$$x = c_3 e^t - c_1 t e^t - \frac{1}{2} c_2 e^{-t} - t^2 - 4t - 5,$$

in which $\{e^t\}$ is a fundamental set of solution, resulting from the homogeneous equation $x' - x = 0$; c_3 is a constant; $g(t) \equiv -c_1 t e^t - \frac{1}{2} c_2 e^{-t} - t^2 - 4t - 5$ is a particular solution, resulting from the conditions that $g(t)$ is of the form

$$g(t) = \alpha t^2 + \beta t + \gamma + A t e^t + B e^{-t}$$

and that $g(t)$ satisfies the equation (3.13), for which the constants α, β, γ, A, and B are then determined.

4. Proof of the Main Results

In this section, we will prove the main results in Section 2. The reader is referred to Sections 1 and 2 for related definitions and assumptions.

4.1. The Exponential Function exp(tA). In this subsection, we firstly assume that the matrix $A(t) = A$ does not depend on t, for simplicity. Under this assumption, the exponential function e^{tA} of (tA) can be defined as a square matrix. This function has many useful properties, and is a fundamental matrix for the homogeneous system (1.3). These results are stated in:

THEOREM 2.1. *For each $m \in \mathbb{N}$, let S_m be the square matrix defined by*

$$S_m \equiv \sum_{k=0}^{m} \frac{(tA)^k}{k!},$$

the partial sum of the symbolic infinite series of matrices

$$\sum_{k=0}^{\infty} \frac{(tA)^k}{k!}.$$

Then the following are true:

(i) *The limit $\lim_{m \to \infty}(S_m)_{\alpha\beta}$ of the element $(S_m)_{\alpha\beta}$ of the matrix S_m at the position $(\alpha, \beta), \alpha, \beta = 1, \ldots, n$, as $m \longrightarrow \infty$, exists absolutely and uniformly for bounded t.*

(ii) *The square matrix e^{tA}, whose element $(e^{tA})_{\alpha\beta}$ at the position (α, β) is defined by the complex number $\lim_{m \to \infty}(S_m)_{\alpha\beta}$ from (i), has the properties: for $t, s \in \mathbb{R}$,*

    (a) $e^{tA}|_{t=0} = I,$ *the identity matrix.*

    (b) $e^{(t+s)A} = e^{tA}e^{sA} = e^{sA}e^{tA}.$

    (c) $e^{-tA} = (e^{tA})^{-1},$ *the inverse of the matrix e^{tA}.*

    (d) $\frac{d}{dt}e^{tA} = Ae^{tA} = e^{tA}A$, *and $e^{(t-t_0)A}u_0$ is the unique solution to the initial value problem (1.5).*

    (e) e^{tA} *is a fundamental matrix for the homogeneous system (1.3).*

To prove Theorem 2.1, we need some preparations.

DEFINITION 4.1. *The norm $\|B\|$ of a square matrix B is defined by*

$$\|B\| \equiv \max_{1 \le \alpha, \beta \le n} |(B)_{\alpha\beta}|,$$

a nonnegative number.

LEMMA 4.2. *For two square matrices U and V, the inequality*

$$\|UV\| \le n\|U\|\|V\|$$

is true.

PROOF. Calculations show

$$(UV)_{\alpha\beta} = \sum_{l=1}^{n}(U)_{\alpha l}(V)_{l\beta}$$

$$\|UV\| = \max_{1 \le \alpha\beta \le n} |\sum_{l=1}^{n}(U)_{\alpha l}(V)_{l\beta}|$$

$$\leq \sum_{l=1}^{n} \|U\|\|V\|$$
$$= n\|U\|\|V\|,$$

and so the Lemma is proved. □

Proof of Theorem 2.1.

PROOF. We divide the proof into six steps.

Step 1. To obtain (i), we note that $A^0 = I$, the identity matrix, and that

$$\left(\sum_{k=0}^{m} \frac{(tA)^k}{k!}\right)_{\alpha\beta} = \sum_{k=0}^{m} \frac{t^k(A^k)_{\alpha\beta}}{k!}.$$

We now estimate the following:

$$\|A^2\| = \|AA\| \leq n\|A\|^2,$$
$$\|A^3\| = \|A^2 A\| \leq n\|A^2\|\|A\| \leq n^2\|A\|^3,$$

$$\vdots$$

$$\|A^k\| \leq n^{k-1}\|A\|^k,$$

$$0 \leq \left|\frac{t^k(A^k)_{\alpha\beta}}{k!}\right| \leq \frac{|t^k|}{k!}\|A^k\|,$$

$$\leq \frac{|t|^k}{k!} n^{k-1}\|A\|^k, \quad k \geq 2$$

$$= \left(\frac{|t|\|A\|^k}{k!}\right)\frac{1}{n}, \quad k \geq 2$$

$$\sum_{k=2}^{\infty} \frac{(|t|\|A\|)^k}{k!}\frac{1}{n} = \frac{1}{n}(e^{|t|\|A\|} - 1 - |t|\|A\|),$$

which converges uniformly and absolutely for bounded $|t|$. It follows from comparison test that

$$\sum_{k=0}^{\infty} \frac{t(A^k)_{\alpha\beta}}{k!} = 1 + t(A)_{\alpha\beta} + \sum_{k=2}^{\infty} \frac{t^k(A^k)_{\alpha\beta}}{k!}$$

converges uniformly and absolutely for bounded $|t|$. This is (i). Here we used Lemma 4.2 and the fact of comparison test from advanced calculus, Apostol [2]:

FACT 4.3. *Suppose that a_m and b_m are two sequences of complex numbers. If*

$$0 \leq |a_m| \leq |b_m|$$

and

$$\sum_{m=2}^{\infty} |b_m|$$

converges, then the infinite series

$$\sum_{m=2}^{\infty} |a_m|$$

converges also.

Step 2. To obtain (a), let $t = 0$, from which S_m becomes I, the identity matrix. (a) results as $m \longrightarrow \infty$.

Step 3. To obtain (b), we calculate

$$
(e^{tA}e^{sA})_{\alpha\beta} = \sum_{l=1}^{n}(e^{tA})_{\alpha l}(e^{sA})_{l\beta}
$$

$$
= \sum_{l=1}^{n}[\sum_{k=0}^{\infty}\frac{t^k(A^k)_{\alpha l}}{k!}][\sum_{k=0}^{\infty}\frac{s^k(A^k)_{l\beta}}{k!}]
$$

$$
= \sum_{l=1}^{n}\sum_{m=0}^{\infty}\sum_{k=0}^{m}[\frac{t^{m-k}(A^{m-k})_{\alpha l}}{(m-k)!}\frac{s^k(A^k)_{l\beta}}{k!}]
$$

$$
= \sum_{m=0}^{\infty}\sum_{k=0}^{m}[\sum_{l=1}^{n}(A^{m-k})_{\alpha l}(A^k)_{l\beta}]\frac{t^{m-k}s^k}{(m-k)!k!}\frac{m!}{m!}
$$

$$
= \sum_{m=0}^{\infty}\sum_{k=0}^{m}(A^{(m-k)+k})_{\alpha\beta}\binom{m}{k}t^{m-k}s^k\frac{1}{m!}
$$

$$
= \sum_{m=0}^{\infty}\frac{(A^m)_{\alpha\beta}}{m!}(t+s)^m
$$

$$
= (e^{(t+s)A})_{\alpha\beta}
$$

$$
= (e^{(s+t)A})_{\alpha\beta}
$$

$$
= (e^{sA}e^{tA})_{\alpha\beta},
$$

and so, (b) follows. Here we used the facts from binomial theorem and advanced calculus, Apostol [2]:

FACT 4.4. *If a and b are two complex numbers, then*

$$
(a+b)^m = \sum_{k=0}^{m}\binom{m}{k}a^{m-k}b^k
$$

$$
= \sum_{k=0}^{m}\frac{m!}{(m-k)!k!}a^{m-k}b^k,
$$

where $m \in \mathbb{N}$.

FACT 4.5. *If $\sum_{m=0}^{\infty}a_m$ and $\sum_{m=0}^{\infty}b_m$ are two absolutely convergent series of complex numbers, then the Cauchy product*

$$
\sum_{m=0}^{\infty}c_m,
$$

where $c_m = \sum_{k=0}^{m}a_{m-k}b_k$, converges absolutely to

$$
\sum_{m=0}^{\infty}a_m \sum_{m=0}^{\infty}b_m.
$$

Step 4. To obtain (c), let $s = -t$ in (b), from which we have

$$
I = e^{0A} = e^{tA}e^{-tA} = e^{-tA}e^{tA}.
$$

It follows that

$$(e^{tA})^{-1} = e^{-tA},$$

which is (c).

Step 5. To obtain (d), note that

$$(e^{tA})_{\alpha\beta} = \sum_{k=0}^{\infty} \frac{t^k (A^k)_{\alpha\beta}}{k!}$$

converges for bounded $|t|$ by (i), and so can be term by term differentiated with respect to t. It follows that

$$\frac{d}{dt}(e^{tA})_{\alpha\beta} = \sum_{k=0}^{\infty} \frac{d}{dt} \frac{t^k (A^k)_{\alpha\beta}}{k!}$$

$$= \sum_{k=1}^{\infty} \frac{t^{k-1}(A^k)_{\alpha\beta}}{(k-1)!}$$

$$= \sum_{k=1}^{\infty} [\frac{t^{k-1}}{(k-1)!} \sum_{l=1}^{n} (A^{k-1})_{\alpha l}(A)_{l\beta}] (= \sum_{k=1}^{\infty} [\frac{t^{k-1}}{(k-1)!} \sum_{l=1}^{n} (A)_{\alpha l}(A^{k-1})_{l\beta}] \text{ also})$$

$$= \sum_{l=1}^{n} [\sum_{k=1}^{\infty} \frac{t^{k-1}(A^{k-1})_{\alpha l}}{(k-1)!} (A)_{l\beta}] (= \sum_{l=1}^{n} [(A)_{\alpha l} \sum_{k=1}^{\infty} \frac{t^{k-1}(A^{k-1})_{l\beta}}{(k-1)!}] \text{ also})$$

$$= \sum_{l=1}^{n} (e^{tA})_{\alpha l}(A)_{l\beta} (= \sum_{l=1}^{n} (A)_{\alpha l}(e^{tA})_{l\beta} \text{ also})$$

$$= (e^{tA}A)_{\alpha\beta} (= (Ae^{tA})_{\alpha\beta} \text{ also}),$$

which is (d). Here we used the fact from advanced calculus, Apostol [2, page 236]:

FACT 4.6. *If the power series $\sum_{k=0}^{\infty} a_k z^k$ converges to a complex-valued function $f(z)$ for z in an open disk of radius $r_0 > 0$:*

$$\{z \in \mathbb{C} : |z| < r_0\},$$

then the derivative $f'(z)$ exists and equals

$$\sum_{k=1}^{\infty} \frac{d}{dz}(a_k z^k).$$

That $e^{(t-t_0)A}u_0$ is the unique solution to the initial value problem (1.5) follows from the simple calculation:

$$e^{(t-t_0)A}u_0|_{t=t_0} = Iu_0$$

$$= u_0,$$

in which uniqueness follows from existence and uniqueness theorem in Chapter 8.

Step 6. To obtain (e), let

$$e^{tA} = \begin{pmatrix} z_1(t) & z_2(t) & \cdots & z_n(t) \end{pmatrix},$$

in which $z_i(t), i = 1, \ldots, n$, is the ith column of the matrix e^{tA}. It follows from (d) that

$$\begin{pmatrix} z_1'(t) & \cdots & z_n'(t) \end{pmatrix} = \frac{d}{dt}e^{tA}$$

$$= Ae^{tA} = A\begin{pmatrix} z_1(t) & \cdots & z_n(t) \end{pmatrix}$$

$$= \begin{pmatrix} Az_1(t) & \cdots & Az_n(t) \end{pmatrix}.$$

This shows that

$$z_i'(t) = Az_i(t), \quad i = 1, \ldots, n,$$

and so $z_i(t), i = 1, \ldots, n$, are solutions to the homogeneous system (1.3). The set

$$\{z_1, \ldots, z_n\}$$

is linearly independent because the matrix e^{tA} is nonsingular for all t, by (c). Here we used the fact from linear algebra:

FACT 4.7. *Suppose that* $B = \begin{pmatrix} b_1 & \cdots & c_n \end{pmatrix}$ *is a square matrix, where* $b_i, i = 1, \ldots, n$, *are* n *column vectors. If* B *is nonsingular, then the set*

$$\{b_1, \ldots, b_n\}$$

is a linearly independent set of n *vectors.*

□

4.2. The Wronskian $W(t)$. As is stated in Section 1, the concept of linear independence of n solutions to the homogeneous system (1.3) is closely related to its associated quantity, the Wronskian $W(t)$. Here is this relation:

Proposition 2.29. *Suppose that* $z_i, i = 1, 2, \ldots, n$, *are* n *solutions to the homogeneous system* (1.3). *Then the following statements* (i), (ii), *and* (iii) *are equivalent:*

   (i) *The set* $\{z_i, z_2, \ldots, z_n\}$ *is linearly independent.*
   (ii) *The associated Wronskian* $W(t)$ *with the* n *solutions* $z_1, z_2, \ldots, z_n$, *is nonzero somewhere.*
   (iii) *The associated Wronskian* $W(t)$ *is nowhere zero.*

PROOF. We divide the proof into three steps.

Step 1. Assume that (ii) holds, and we shall show that (i) is true. Let complex constants $c_i, i = 1, 2, \ldots, n$, be such that

$$c_1 z_1(t) + c_2 z_2(t) + \cdots + c_n z_n(t) = 0, \quad \text{the zero column vector} \qquad (4.1)$$

for all $t \in [a, b]$. This is the same as

$$\begin{pmatrix} z_1(t) & \cdots & z_n(t) \end{pmatrix} \begin{pmatrix} c_1 \\ c_2 \\ \vdots \\ c_n \end{pmatrix} = 0, \quad \text{the zero vector.} \qquad (4.2)$$

Since $W(t_0) \neq 0$ for some $t_0 \in [a, b]$, we have

$$c_1 = c_2 = \cdots = c_n = 0,$$

and so the set $\{z_1, \ldots, z_n\}$ is linearly independent. This is (i). Here we used the fact from linear algebra:

FACT 4.8. *Suppose that* A *is a matrix of* $m \times m, m \in \mathbb{N}$, *that the determinant* $|A|$ *of* A *is not zero, and that* $\vec{b}$ *is an* m-*dimensional column vector. Then the equation* $A\vec{b} = \vec{0}$ *has only the zero solution* $\vec{b} = \vec{0}$.

Step 2. Assume that (i) holds, and we shall show that (ii) is true. Claim that if (ii) is not true, then a contradiction results. To this end, assume that the Wronskian $W(t) \equiv 0$ for all $t \in [a, b]$. Then the expression (4.2) implies that $c_i, i = 1, 2, \ldots, n$, are not all zero, a contradiction to the fact that the set $\{z_1(t), \ldots, z_n(t)\}$ is linearly independent by (i). Thus (ii) is true. Here we used the fact from linear algebra:

FACT 4.9. *Suppose that A is a matrix of $m \times m, m \in \mathbb{N}$, that the determinant $|A|$ of A is zero, and that $\vec{B}$ is an m-dimensional column vector. Then the equation $A\vec{b} = \vec{0}$ has a nonzero solution $\vec{b}$.*

Step 3. We shall show that (ii) is equivalent to (iii), [5, pages 28-29]: Let $\Phi(t)$ be the matrix
$$\begin{pmatrix} z_1 & \cdots & z_n \end{pmatrix}.$$

(a) Claim that
$$\Phi'(t) = A(t)\Phi(t).$$

To see this, compute the derivative
$$\begin{aligned} \frac{d}{dt}\Phi(t) &= \begin{pmatrix} z_1' & \cdots & z_n' \end{pmatrix} \\ &= \begin{pmatrix} A(t)z_1 & \cdots & A(t)z_n \end{pmatrix} \\ &= A(t) \begin{pmatrix} z_1 & \cdots & z_n \end{pmatrix} \\ &= A(t)\Phi(t), \end{aligned}$$

and the claim is proved. Here note that
$$z_i = \begin{pmatrix} z_{1i} \\ \vdots \\ z_{ni} \end{pmatrix}, i = 1, \ldots, n,$$

are n solutions to the homogeneous system (1.3).

(b) The above claim (a) implies that
$$\frac{d}{dt}z_{ik} = z_{ik}' = \sum_{l=1}^{n} p_{il}(t)z_{lk}$$

for $i, k = 1, \ldots, n$.

(c) Next, differentiating the Wronskian $W(t)$ where
$$W(t) = \begin{vmatrix} z_1(t) & \cdots & z_n(t) \end{vmatrix}, \quad \text{and}$$
$$z_i(t) = \begin{pmatrix} z_{1i}(t) \\ \vdots \\ z_{ni}(t) \end{pmatrix}, i = 1, 2, \ldots, n, \quad n \text{ solutions,}$$

we have
$$\frac{d}{dx}W(t) = \begin{vmatrix} z_{11}'(t) & \cdots & z_{1n}'(t) \\ z_{21}(t) & \cdots & z_{2n}(t) \\ z_{31}(t) & \cdots & z_{3n}(t) \\ \vdots & \vdots & \vdots \\ z_{n1}(t) & \cdots & z_{nn}(t) \end{vmatrix} + \begin{vmatrix} z_{11}(t) & \cdots & z_{1n}(t) \\ z_{21}'(t) & \cdots & z_{2n}'(t) \\ z_{31}(t) & \cdots & z_{3n}(t) \\ \vdots & \vdots & \vdots \\ z_{n1}(t) & \cdots & z_{nn}(t) \end{vmatrix}$$
$$+ \cdots$$

$$+ \begin{vmatrix} z_{11}(t) & \cdots & z_{1n}(t) \\ \vdots & \vdots & \vdots \\ z'_{(n-1)1}(t) & \cdots & z'_{(n-1)n}(t) \\ z_{n1}(t) & \cdots & z_{nn}(t) \end{vmatrix} + \begin{vmatrix} z_{11}(t) & \cdots & z_{1n}(t) \\ \vdots & \vdots & \vdots \\ z_{(n-1)1}(t) & \cdots & z_{(n-1)n}(t) \\ z'_{n1}(t) & \cdots & z'_{nn}(t) \end{vmatrix}.$$

The right side of the above expression has n determinants, which we denote by $D_1, D_2, \ldots, D_n$, respectively. Applying (b) and row operations to each determinant D_i, we have, for example.

$$D_1 = \begin{vmatrix} \sum_{l=1}^{n} p_{1l} z_{l1} & \sum_{l=1}^{n} p_{1l} z_{l2} & \cdots & \sum_{l=1}^{n} p_{1l} z_{ln} \\ z_{21} & z_{22} & \cdots & z_{2n} \\ \vdots & & & \\ z_{n1} & z_{n2} & \cdots & z_{nn} \end{vmatrix}$$

$$= \begin{vmatrix} p_{11} z_{11} & p_{11} z_{12} & \cdots & p_{11} z_{1n} \\ z_{21} & z_{22} & \cdots & z_{2n} \\ \cdots & & & \\ z_{n1} & z_{n2} & \cdots & z_{nn} \end{vmatrix}$$

$$= p_{11} \begin{vmatrix} z_{11} & z_{12} & \cdots & z_{1n} \\ \vdots & & & \\ z_{n1} & z_{n2} & \cdots & z_{nn} \end{vmatrix}$$

$$= p_{11} W(t),$$

in which the sum of $-p_{12}$ times row 2, $-p_{13}$ times row 3, ..., and $-p_{1n}$ times row n, is added into row 1. Similarly, we have

$$D_2 = p_{22} W(t), \ldots, D_n = p_{nn} W(t),$$

and then it follows that

$$\frac{d}{dt} W(t) = [p_{11} + p_{22} + \cdots + p_{nn}] W(t).$$

Solving this first order differential equation for $W(t)$, we obtain, for $t_0 \in [a, b]$,

$$W(t) = W(t_0) e^{\int_{t=t_0}^{t} [p_{11} + \cdots + p_{nn}] \, dt},$$

and this shows that the statement (ii) is the same as the statement (i), since the exponential function $e^t, t \in [a, b]$, is never zero. Here we used the fact from linear algebra:

FACT 4.10. *Suppose that*

$$A(t) = \begin{pmatrix} \vec{a_1}(t) \\ \vdots \\ \vec{a_m}(t) \end{pmatrix}$$

is a matrix of $m \times m$, $m \in \mathbb{N}$, where $\vec{a_i}(t), i = 1, 2, \ldots, m$, are m-dimensional row vectors, which are differentiable functions of the variable $t \in [a, b]$. Then

- the derivative $\frac{d}{dt}|A(t)|$ of the determinant $|A(t)|$ equals

$$\frac{d}{dx}|A(t)| = \sum_{i=1}^{m} \begin{vmatrix} \vec{a_1}(t) \\ \vdots \\ \frac{d}{dt}\vec{a_i}(t) \\ \vdots \\ \vec{a_m}(t) \end{vmatrix},$$

- the determinant $|A(t)|$ of $A(t)$ is unchanged when the rows in $A(t)$ multiplied by constants are added to other rows in $A(t)$, and
- if $\vec{a_{i_0}}(t) = c_{i_0}(t)\vec{b_{i_0}}(t)$ for some i_0, where $c_{i_0}(t)$ is a scalar function of t, then the determinant $|A(t)|$ of $A(t)$ satisfies

$$|A(t)| = c_{i_0}(t) \begin{vmatrix} \vec{a_1}(t) \\ \vdots \\ \vec{b_{i_0}}(t) \\ \vdots \\ \vec{a_m}(t) \end{vmatrix}.$$

$\square$

4.3. Methods of Computing the Fundamental Matrix exp(tA). Here is a method of computing the fundamental matrix e^{tA}:

Theorem 2.3. *Suppose that the algebraic equation*

$$p(\lambda) \equiv a_0\lambda^n + a_1\lambda^{n-1} + \cdots + a_{n-1}\lambda + a_n = 0$$

has s distinct solutions (or called roots) $\lambda_1, \lambda_2, \ldots, \lambda_s$, where $1 \le s \le n$, and that the solution λ_i is repeated $m_i(\in \mathbb{N})$ times. In other words, we assume that

$$p(\lambda) = a_0(\lambda - \lambda_1)^{m_1}(\lambda - \lambda_2)^{m_2}\cdots(\lambda - \lambda_s)^{m_s} = 0,$$

$\lambda_i \ne \lambda_j$ *for $i \ne j$, and that*

$$m_1 + m_2 + \cdots + m_s = n.$$

Then there exist n unique square matrices

$$M_{i,k}, \quad i = 1, 2, \ldots, s; \quad k = 0, 1, \ldots, (m_i - 1),$$

such that

$$e^{tA} = \sum_{k=0}^{m_i-1}\sum_{i=1}^{s} t^k e^{t\lambda_i} M_{i,k},$$

or that

$$e^{tA} = \sum_{j=1}^{n} z_j(t)N_j,$$

if we relabel the n functions

$$t^k e^{t\lambda_i}, \quad i = 1, 2, \ldots, s; \quad k = 0, 1, \ldots, (m_i - 1),$$

as the n functions

$$z_i(t), \quad i = 1, 2, \ldots, n,$$

and relabel the n square matrices

$$M_{i,k}, \quad i = 1, 2, \ldots, s; \quad k = 0, 1, \ldots, (m_i - 1),$$

as the n square matrices

$$N_j, \quad j = 1, 2, \ldots, n.$$

Here the n matrices $N_j, j = 1, \ldots, n$, are the n unique solutions of the system of algebraic equations:

$$\begin{pmatrix} z_1(0) & z_2(0) & \cdots & z_n(0) \\ z_1'(0) & z_2'(0) & \cdots & z_n'(0) \\ \vdots & & & \\ z_1^{(n-1)}(0) & z_2^{(n-1)}(0) & \cdots & z_n^{(n-1)}(0) \end{pmatrix} \begin{pmatrix} N_1 \\ N_2 \\ \vdots \\ N_n \end{pmatrix} = \begin{pmatrix} I \\ A \\ \vdots \\ A^{n-1} \end{pmatrix}.$$

PROOF. We divide the proof into three steps.

Step 1. Claim that

$$p(D)e^{tA} = 0,$$

where D is the differential operator $\frac{d}{dt}$ and $p(D)$ is obtained from the characteristic polynomial $p(\lambda) \equiv |A - \lambda I|$ of A, by replacing the λ by D, for which

$$p(D) = a_0 D^n + a_1 D^{n-1} + \cdots + a_{n-1} D + a_0$$

$$= a_0 \frac{d^n}{dt^n} + a_1 \frac{d^{n-1}}{dt^{n-1}} + \cdots + a_{n-1} \frac{d}{dt} + a_n.$$

To see the claim, note by Theorem 2.1 that

$$De^{tA} = Ae^{tA},$$

$$D^2 e^{tA} = A^2 e^{tA},$$

$$\vdots$$

$$D^n e^{tA} = A^n e^{tA}.$$

It follows that

$$p(D)e^{tA} = [a_0 A^n + a_1 A^{n-1} + \cdots + a_{n-1} A + a_n]e^{tA}$$

$$= p(A)e^{tA}$$

$$= 0, \quad \text{the zero matrix.}$$

Here we used the fact from linear algebra [**9**, page 285]:

FACT 4.11 (Cayley-Hamilton Theorem). *If B is a square matrix and $q(\lambda)$ is the characteristic polynomial $|B - \lambda I|$ of B, then*

$$q(A) = 0.$$

Step 2. Claim that there are n square matrices

$$M_{i,k}, \quad i = 1, \ldots, s; \quad k = 0, 1, \ldots, (m_i-),$$

such that

$$e^{tA} = \sum_{i=1}^{s} \sum_{k=0}^{m_i - 1} t^k e^{\lambda_i t} M_{i,k}.$$

To see the claim, note that Step 1 shows that each entry $z(t)$ of a function of t, in the matrix A, is a solution to the nth order linear differential equation

$$p(D)y(t) = a_0 y^{(n)} + a_1 y^{(n-1)} + \cdots + a_{n-1} y' + a_n y = 0,$$

and that this equation has, by Theorem 2.1 in Chapter 1, the set

$$\{t^k e^{t\lambda_i}, i = 1, 2, \ldots, s; k = 0, 1, \ldots, (m_i - 1)\}$$

as a fundamental set. Thus each entry $z(t)$ in A is of the form

$$z(t) = \sum_{i=1}^{s} \sum_{k=0}^{m_i - 1} c_{i,k} t^k e^{t\lambda_i}$$

for some constants $c_{i,k}$, and then the claim follows.

Step 3. From Step 2, we have that

$$e^{tA} = \sum_{j=1}^{n} z_j(t) N_j, \tag{4.3}$$

if we relabel the n functions

$$t^k e^{t\lambda_i}, \quad i = 1, 2, \ldots, s; \quad k = 0, 1, \ldots, (m_i - 1),$$

as the n functions

$$z_i(t), \quad i = 1, 2, \ldots, n,$$

and relabel the n square matrices

$$M_{i,k}, \quad i = 1, 2, \ldots, s; \quad k = 0, 1, \ldots, (m_i - 1),$$

as the n square matrices

$$N_j, \quad j = 1, 2, \ldots, n.$$

Here the matrices $N_j, j = 1, \ldots, n$, are the n unique solutions of the system of algebraic equations:

$$\begin{pmatrix} z_1(0) & z_2(0) & \cdots & z_n(0) \\ z_1'(0) & z_2'(0) & \cdots & z_n'(0) \\ \vdots & & & \\ z_1^{(n-1)}(0) & z_2^{(n-1)}(0) & \cdots & z_n^{(n-1)}(0) \end{pmatrix} \begin{pmatrix} N_1 \\ N_2 \\ \vdots \\ N_n \end{pmatrix} = \begin{pmatrix} I \\ A \\ \vdots \\ A^{n-1} \end{pmatrix}. \tag{4.4}$$

This is because differentiating the expression (4.3) for $(n-1)$ times and evaluating the results at $t = 0$ gives

$$I = \sum_{j=1}^{n} z_j(0) N_j,$$

$$A = \sum_{j=1}^{n} z_j'(0) N_j,$$

$$\vdots$$

$$A^{n-1} = \sum_{j=1}^{n} z_j^{(n-1)} N_j,$$

which is the same as (4.4). $\square$

The proof of Theorem 2.3 is taken from Cullen [**7**, pages 307–313].

The computed e^{tA} in Theorem 2.3 is, in general, complex-valued, but under a certain circumstance, it can be real-valued as the following shows:

Corollary 2.11. *Suppose that the $p(\lambda)$ in Theorem 2.3 is either of the form*

$$p(\lambda) = a_0[(\lambda - \lambda_1)^{m_1}(\lambda - \overline{\lambda_1})^{m_1}]\cdots[(\lambda - \lambda_s)^{m_r}(\lambda - \overline{\lambda_r})^{m_r}](\lambda - \lambda_{r+1})^{m_{r+1}} \quad (2.2)$$

or of the form

$$p(\lambda) = a_0[(\lambda - \lambda_1)^{m_1}(\lambda - \overline{\lambda_1})^{m_1}]\cdots[(\lambda - \lambda_r)^{m_r}(\lambda - \overline{\lambda_r})^{m_r}], \quad (2.3)$$

to which correspondings either

$$2(m_1 + m_2 + \cdots + m_r) + m_{r+1} = n$$

or

$$2(m_1 + m_2 + \cdots + m_r) = n.$$

Here

$$\lambda_j = \alpha_j + i\beta_j, j = 1, 2, \ldots, r, \quad \text{some } r \in \mathbb{N}$$
$$\alpha_j, \beta_j \in \mathbb{R},$$
$$i \equiv \sqrt{-1}, \quad \text{the complex unit with } i^2 = -1,$$
$$\overline{\lambda_j} = \alpha_j + i\beta_j, \quad \text{the complex conjugate of } \lambda_j,$$
$$\lambda_{r+1} \in \mathbb{R}, \quad \text{and}$$
$$\lambda_i \neq \lambda_j \quad \text{for } i \neq j, \quad \text{where } i, j = 1, 2, \ldots, (r+1).$$

Then according as (2.2) or (2.3), there exist n real square matrices

$$L_{r+1,l}, \quad l = 0, 1, \ldots, (m_{r+1} - 1);$$
$$M_{i,k}, N_{i,k}, \quad i = 1, 2, \ldots, r; \quad k = 0, 1, \ldots, (m_i - 1) \quad (4.5)$$

or n real square matrices

$$M_{i,k}, N_{i,k}, \quad i = 1, \ldots, r; \quad k = 0, 1, \ldots, (m_i - 1) \quad (4.6)$$

such that

$$e^{tA} = \sum_{l=0}^{m_{r+1}-1} t^l e^{t\lambda_{r+1}} L_{(r+1),l} \; +$$

$$\sum_{i=1}^{r}\sum_{k=0}^{m_i-1} x^k e^{t\alpha_i}\cos(t\beta_i)M_{i,k} + \sum_{i=1}^{r}\sum_{k=0}^{m_i-1} x^k e^{t\alpha_i}\sin(t\beta_i)N_{i,k} \quad (4.7)$$

or that

$$e^{tA} = \sum_{i=1}^{r}\sum_{k=0}^{m_i-1} t^k e^{t\alpha_i}\cos(t\beta_i)M_{i,k} + \sum_{i=1}^{r}\sum_{k=0}^{m_i-1} x^k e^{t\alpha_i}\sin(t\beta_i)N_{i,k}. \quad (4.8)$$

Equivalently,

$$e^{tA} = \sum_{j=1}^{n} z_j(t)P_j,$$

if we relabel the n real-valued functions

$$t^l e^{t\lambda_{r+1}}, \quad l = 0, \ldots, (m_{r+1} - 1);$$
$$t^k e^{t\alpha_i}\cos(t\beta_i), t^k e^{t\alpha_i}\sin(t\beta_i), \quad i = 1, \ldots, r; \quad k = 0, \ldots, (m_i - 1)$$

in (4.7) or the n real-valued functions

$$t^k e^{t\alpha_i} \cos(t\beta_i), t^k e^{t\alpha_i} \sin(t\beta_i), \quad i = 1, \ldots, r; \quad k = 0, \ldots, (m_i - 1),$$

in (4.8), as the n functions

$$z_1(t), \ldots, z_n(t),$$

and relabel the n real square matrices in (4.5) or (4.6) as the n real square matrices

$$P_1, \ldots, P_n.$$

Here the n square matrices $P_j, j = 1, \ldots, n$, are the unique n solutions of the system of algebraic equations:

$$\begin{pmatrix} z_1(0) & z_2(0) & \cdots & z_n(0) \\ z_1'(0) & z_2'(0) & \cdots & z_n'(0) \\ \vdots & & & \\ z_1^{(n-1)}(0) & z_2^{(n-1)}(0) & \cdots & z_n^{(n-1)}(0) \end{pmatrix} \begin{pmatrix} P_1 \\ P_2 \\ \vdots \\ P_n \end{pmatrix} = \begin{pmatrix} I \\ A \\ \vdots \\ A^{n-1} \end{pmatrix}.$$

PROOF. It suffices to assume the case that

$$p(\lambda) = a_0[(\lambda - \lambda_1)^{m_1}(\lambda - \overline{\lambda_1})^{m_1}] \cdots [(\lambda - \lambda_r)^{m_r}(\lambda - \overline{\lambda_r})^{m_r}](\lambda - \lambda_{r+1})^{m_{r+1}},$$

since the other case is similar.

We divide the proof into two steps.

Step 1. Claim that there are n square matrices

$$L_{r+1,l}, \quad l = 0, 1, \ldots, (m_{r+1} - 1);$$

$$M_{i,k}, N_{i,k}, \quad i = 1, \ldots, r; \quad k = 0, 1, \ldots, (m_i - 1),$$

such that

$$e^{tA} = \sum_{l=0}^{m_{r+1}-1} t^l e^{t\lambda_{r+1}} +$$

$$\sum_{i=1}^{r} \sum_{k=0}^{m_i-1} t^k e^{t\alpha_i} \cos(t\beta_i) M_{i,k} +$$

$$\sum_{i=1}^{r} \sum_{k=0}^{m_i-1} t^k e^{t\alpha_i} \sin(t\beta_i) N_{i,k}.$$

To prove the claim, we use the Step 1 of the proof of Theorem 2.3, and it follows that

$$p(D)e^{tA} = p(A)e^{tA} = 0, \quad \text{the zero matrix.}$$

Thus each entry $z(t)$ of a function of t, in the matrix A, is a solution to the nth order linear differential equations

$$p(D)y = a_0 y^{(n)} + a_1 y^{(n-1)} + \cdots + a_{n-1} y' + a_n = 0,$$

and this equation has, by Corollary 2.5 in Chapter 1, the set

$$S \equiv \{x^l e^{t\lambda_{r+1}}, l = 0, 1, \ldots, (m_{r+1} - 1),$$

$$x^k e^{t\alpha_j} \cos(t\beta_j), j = 1, 2, \ldots, r; k = 0, 1, \ldots, (m_r - 1), \quad \text{and}$$

$$x^k e^{t\alpha_j} \sin(t\beta_j), j = 1, 2, \ldots, r; k = 0, 1, \ldots, (m_r - 1)\},$$

as a fundamental set. It results that each entry $z(t)$ in the matrix A is of the form

$$z(t) = \sum_{l=0}^{m_{r+1}-1} t^l e^{t\lambda_{r+1}} c_l +$$

$$\sum_{i=1}^{r} \sum_{k=0}^{m_i-1} t^k e^{t\alpha_i} \cos(t\beta_i) b_{i,k} +$$

$$\sum_{i=1}^{r} \sum_{k=0}^{m_i-1} t^k e^{t\alpha_i} \sin(t\beta_i) d_{i,k}$$

for some n constants $c_l, b_{i,k}$, and $d_{i,k}$, which is a linar combination of the elements in the fundamental set S. Thus the claim is proved.

Step 2. From Step 1, we have

$$e^{tA} = \sum_{j=1}^{n} z_j(t) P_j,$$

if we relabel the n real-valued functions

$$t^l e^{t\lambda_{r+1}}, l = 0, \ldots, (m_{r+1} - 1);$$

$$t^k e^{t\alpha_i} \cos(t\beta_i), t^k e^{t\alpha_i} \sin(t\beta_i), i = 1, \ldots, r; k = 0, \ldots, (m_i - 1),$$

as the n functions

$$z_1(t), \ldots, z_n(t),$$

and relabel the n real square matrices

$$L_{r+1,l}, l = 0, \ldots, (m_{r+1} - 1);$$

$$M_{i,k}, N_{i,k}, i = 1, \ldots, r; k = 0, \ldots, (m_i - 1),$$

as the n real square matrices

$$P_1, \ldots, P_n.$$

Here the matrices $P_j, j = 1, \ldots, n$, are the unique n solutions of the system of algebraic equations:

$$\begin{pmatrix} z_1(0) & z_2(0) & \cdots & z_n(0) \\ z_1'(0) & z_2'(0) & \cdots & z_n'(0) \\ \vdots & & & \\ z_1^{(n-1)}(0) & z_2^{(n-1)}(0) & \cdots & z_n^{(n-1)}(0) \end{pmatrix} \begin{pmatrix} P_1 \\ P_2 \\ \vdots \\ P_n \end{pmatrix} = \begin{pmatrix} I \\ A \\ \vdots \\ A^{n-1} \end{pmatrix}. \tag{4.9}$$

This is because differentiating the expression (4.9) for $(n-1)$ times and evaluating the results at $t = 0$ gives

$$I = \sum_{j=1}^{n} z_j(0) P_j,$$

$$A = \sum_{j=1}^{n} z_j'(0) P_j,$$

$$\vdots$$

$$A^{n-1} = \sum_{j=1}^{n} z_j^{(n-1)} P_j,$$

which is the same as (4.9).   □

4.4. Fundamental Sets of Solutions. When the matrix $A(t) = A$ does not depend on t, the fundamental matrix e^{tA} exists by Theorem 2.1, and can be computed explicitly by Theorem 2.3 and Corollary 2.11. Furthermore, even when the matrix $A(t)$ does depend on t, there still exists theoretically a fundamental set of solutions for (1.3) but such a set is not as explicit as e^{tA}.

THEOREM 2.13. *There exists a fundamental set of solutions for the homogeneous system* (1.3).

PROOF. The theorem of existence and uniqueness in Chapter 8 implies that the homogeneous system (1.3) has a unique vector-valued solution z_i, when equipped with each initial data $u_i(t_0), i = 1, \ldots, n$:

$$t_0 \in (a, b),$$

$$u_i(t_0) = \begin{pmatrix} 0 \\ 0 \\ \vdots \\ 0 \\ 1 \\ 0 \\ \vdots \\ 0 \end{pmatrix}$$

Here $u_i(t_0)$ is a column vector that has every component being 0, except for the ith component, which is 1. It follows that the set

$$\{z_1, z_2, \ldots, z_n\}$$

is a fundamental set of vector-valued solutions for the system (1.3), since the Wronskian $W(t)$ at $t = t_0$ is the determinant of the identity matrix, which is not zero:

$$W(t_0) = \begin{vmatrix} 1 & 0 & \cdots & 0 \\ 0 & 1 & \cdots & 0 \\ \vdots & \vdots & \vdots & \vdots \\ 0 & 0 & \cdots & 1 \end{vmatrix} = 1 \neq 0.$$

□

4.5. Homogeneous Systems. One use of a fundamental set of solutions for the homogeneous system (1.3) is that every solution to the system (1.3) is some linear combination of the solutions in the fundamental set, as the following shows:

THEOREM 2.13. *If z is a solution to* (1.3), *then z is of the form*

$$z = \sum_{i=1}^{n} c_i z_i, \quad \text{some linear combination of } z_i$$

$$= \Phi(t)c,$$

where the set $\{z_1, z_2, \ldots, z_n\}$ *is a fundamental set of solutions for* (1.3), *and* $c_i, i = 1, 2, \ldots, c_n$, *are constant complex numbers;* $\Phi(t)$ *is the fundamental matrix associated with the fundamental set* $\{z_1, \ldots, z_n\}$, *and* $c = \begin{pmatrix} c_1 \\ \vdots \\ c_n \end{pmatrix}$ *is a constant vector.*

PROOF. Let z be a solution to the system (1.3), and let $\{z_1, z_2, \ldots, z_n\}$ be a fundamental set of solutions for the system (1.3).

Now we divide the proof into three steps.

Step 1. It is easy to see that $z(t)$ is a solution to the initial value problem

$$\frac{d}{dt} u(t) = A(t)u(t), \quad t \in (a, b)$$
$$u(t_0) = z(t_0),$$

where t_0 is a given point in (a, b).

Step 2. Claim that $u \equiv \sum_{i=1}^n c_i z_i(t)$ for suitably chosen constants $c_i, i = 1, \ldots, n$, is also a solution to the initial value problem in Step 1. To this end, choose c_i so that

$$u(t_0) = c_1 z_1(t_0) + \cdots + c_n z_n(t_0) = z(t_0),$$

or equivalently, that

$$\begin{pmatrix} z_1(t_0) & \cdots & z_n(t_0) \end{pmatrix} \begin{pmatrix} c_1 \\ \vdots \\ c_n \end{pmatrix} = z(t_0).$$

Such a choice of $c_i, i = 1, \ldots, n$, is allowed, since the Wronskian $W(t_0)$ at t_0 is not zero by Proposition 2.29. It is then readily seen that the $u(t)$ with such a choice of $c_i, i = 1, \ldots, n$, is a solution to the initial value problem in Step 1, and that the claim in this Step 2 is proved.

Step 3. It follows from the theorem of uniqueness and existence in Chapter 8 that the two solutions z and u to the initial value problem in Step 1 are, in fact, the same, and so

$$z = u = c_1 z_1 + \cdots + c_n z_n,$$

some linear combination. □

Although they are not unique, fundamental matrices for the homogeneous system (1.3) are related, as the following shows :

Lemma 2.27. *Suppose that* $\Phi(t)$ *and* $\Psi(t)$ *are two fundamental matrices for the homogeneous system* (1.3), *with associated fundamental solution sets* $\{\phi_1, \ldots, \phi_n\}$ *and* $\{\psi_1, \ldots, \psi_n\}$, *respectively. Then*

$$\Phi = \Psi C$$

for some nonsingular matrix C.

PROOF. Since $\phi_i, i = 1, \ldots, n$, is a solution to the homogeneous system (1.3), and since Ψ is a fundamental matrix for the homogeneous system (1.3), it follows from Theorem 2.13 that for each i,

$$\phi_i = \Psi c_i$$

is true for some constant column vector c_i. From this, we have

$$\Phi = \begin{pmatrix} \phi_1 & \cdots & \phi_n \end{pmatrix}$$
$$= \begin{pmatrix} \Psi c_1 & \cdots & \Psi c_n \end{pmatrix}$$
$$= \Psi C,$$

where the square matrix

$$C = \begin{pmatrix} c_1 & \cdots & c_n \end{pmatrix}$$

is nonsingular since Φ and Ψ are nonsingular by Proposition 2.29. $\qquad\square$

REMARK 4.12. *Another proof of Lemma 2.27 can be proceeded as follows, in which Theorem 2.13 is not used, [5, page 70]: Let a square matrix $C(t)$ be defined by*

$$C(t) \equiv \Psi^{-1}(t)\Phi(t),$$

in which $\Psi(t)$ is nonsingular and $\Psi^{-1}(t)$ exists by Proposition 2.29, and we show that $C(t)$ is a constant matrix. To see this, note that

$$\Phi(t) = \Psi(t)C(t),$$

and so, by differentiating $\Phi(t)$ with respect to t, we have

$$\Phi'(t) = \Psi'(t)C(t) + \Psi(t)C'(t).$$

Since

$$\Phi'(t) = A(t)\Phi(t)$$
$$\Psi'(t) = A(t)\Psi(t),$$

as is proved in Step 3 in the proof of Proposition 2.29, it follows that

$$A(t)\Phi(t) = A(t)\Psi(t)C(t)$$
$$= A(t)\Psi(t)C(t) + \Psi(t)C'(t),$$

and so

$$\Psi(t)C'(t) = 0,$$

the zero square matrix. Since $\Psi(t)$ is invertible by Proposition 2.29, we have that $C'(t) = 0$, and so $C(t)$ is a constant matrix. That $C(t)$ is nonsingular follows from $C(t) = \Psi^{-1}(t)\Phi(t)$, where $\Psi^{-1}(t)$ and $\Phi(t)$ are nonsingular by Proposition 2.29.

4.6. Nonhomogeneous Systems. The existence of a fundamental set of solutions for the system (1.3) is also useful, in that every solution to the nonhomogeneous system (1.1) is the sum of a particular solution to the system (1.1) and some linear combination of the elements in the fundamental set. This fact is stated in:

Theorem 2.16. *If z is a vector-valued solution to the nonhomogeneous system (1.1), then z is of the form*

$$z = \sum_{i=1}^{n} c_i z_i + g(t)$$
$$= \Phi(t)c + g(t),$$

where $g(t)$ is a particular vector-valued solution to the system (1.1), the set $\{z_1, z_2, \ldots, z_n\}$, which exists by Theorem 2.13, is a fundamental set of vector-valued solutions for the system (1.3), and $c_i, i = 1, 2, \ldots, n$, are constant complex numbers;

$\Phi(t)$ is the fundamental matrix associated with the fundamental set $\{z_1, \ldots, z_n\}$, and c is the constant column vector $\begin{pmatrix} c_1 \\ \vdots \\ c_n \end{pmatrix}$.

PROOF. Let z be a vector solution to the nonhomogeneous system (1.1). Since $g(t)$ is, by assumption, also a solution to the system (1.1), it follows from the superposition principle in Lemma 2.24 that

$$(z - g)$$

is a solution to the system $Ly = f(t) - f(t) = 0$. Using Theorem 2.13, we thus have that $(z - g)$ takes the form

$$(z - g) = \sum_{i=1}^{n} c_i z_i,$$

where $\{z_1, z_2, \ldots, z_n\}$ is a fundamental set of solutions for the equation (1.3), and $c_i, i = 1, 2, \ldots, n$, are constant complex numbers. Thus

$$z = \sum_{i=1}^{n} c_i z_i + g,$$

which completes the proof.   □

4.7. A Particular Solution. Another use of a fundamental set of vector-valued solutions is that it can yield a particular solution explicitly for the nonhomogeneous system (1.1), through a formula of parameters variation. This fact is stated in:

Theorem 2.18. *A particular solution $g(t)$ to the nonhomogeneous system (1.1) exists, and is of the form*

$$g(t) = \Phi(t)\varphi(t),$$

where $\Phi(t)$ is the fundamental matrix associated with the fundamental set $\{z_i, z_2, \ldots, z_n\}$ of solutions for the homogeneous system (1.3), and $\varphi(t)$ equals some indefinite integral,

$$\varphi(t) = \int \Phi^{-1}(t) f(t) \, dt + c \quad or$$

$$= \int_{\tau = t_1}^{t} \Phi^{-1}(\tau) f(\tau) \, d\tau$$

for any constant c or for any constant $a < t_1 < b$. Here $\Phi^{-1}(t)$ is the inverse of the fundamental matrix $\Phi(t)$, which exists since the fundamental set $\{z_1, \ldots, z_n\}$ is linearly independent, for which the determinant $|\Phi(t)|$ is not zero.

PROOF. We need to show that the function

$$u(t) \equiv \Phi(t)\varphi(t)$$

is a solution to the nonhomogeneous system (1.1), where $\Phi(t)$ is the fundamental matrix associated with the fundamental set $\{z_1, z_2, \ldots, z_n\}$ of vector-valued solutions for the homogeneous system (1.3), and $\varphi(t)$ is defined by

$$\varphi(t)(t) = \int \Phi^{-1}(t) f(t) \, dt.$$

Compute the derivative of $u(t)$ gives

$$\frac{d}{dt}u(t) = \Phi'(t)\varphi(t) + \Phi(t)\Phi^{-1}(t)f(t)$$
$$= A(t)\Phi(t)\varphi(t) + f(t)$$
$$= A(t)u(t) + f(t),$$

and so $u(t)$ is a solution to the nonhomogeneous system (1.1). Here note

$$\Phi'(t) = \begin{pmatrix} z_1' & \cdots & z_n' \end{pmatrix}$$
$$= \begin{pmatrix} A(t)z_1 & \cdots & A(t)z_n \end{pmatrix}$$
$$= A(t)\begin{pmatrix} z_1 & \cdots & z_n \end{pmatrix}$$
$$= A(t)\Phi(t).$$

$\square$

A different method for obtaining a particular solution explicitly to the nonhomogeneous system (1.1) exists, and is called the method of undetermined coefficients, provided that $A(t) = A$ is a constant matrix and that $f(t)$ is a vector function with components being either polynomial functions, exponential functions, sine, cosine, or combinations of them. This is stated in:

Theorem 2.22. *Suppose that the function elements $a_{ij}(t), i, j = 1, \ldots, n$, of the matrix A in the nonhomogeneous system (1.1) are all identical to constants, that the corresponding characteristic system $p(\lambda) = 0$ is of the form*

$$p(\lambda) = a_0(\lambda - \lambda_1)^{m_1} \cdots (\lambda - \lambda_s)^{m_s}$$

as in Theorem 2.3, and finally that the nonhomogeneous term $f(t)$ in the system (1.1) is of the form

$$f(t) = q_1(t)e^{\mu_1 t},$$

where $q_1(t) = b_0 t^{n_1} + b_1 t^{n_1-1} + \cdots + b_{n_1-1}t + b_{n_1}$, a vector polynomial of degree $n_1 \in \{0\} \cup \mathbb{N}$, for which $b_0 \neq 0$.
  If μ_1 is not in the set

$$\{\lambda_1, \lambda_2, \ldots, \lambda_s\},$$

then a particular solution $g(t)$ to the nonhomogeneous system (1.1) is of the form

$$g(t) = Q_1(t)e^{t\mu_1},$$

where $Q_1(t)$ is a vector polynomial of degree n_1.
  Conversely, if $\mu_1 = \lambda_1$, for example, then a particular solution $g(x)$ to the nonhomogeneous system (1.1) is of the form

$$g(t) = R_1(t)e^{t\lambda_1},$$

where $R_1(t)$ is a vector polynomial of degree $(n_1 + m_1)$.

Here

DEFINITION 4.13. *A vector polynomial $q(t)$ in the variable t, of degree $m \in \mathbb{N} \cup \{0\}$, is an expression of the form*

$$q(t) = c_m t^m + c_{m-1}t^{m-1} + \cdots + c_1 t + c_0,$$

where $c_i, i = 0, \ldots, m$, are constant vectors.

To prove Theorem 2.22, we need some preparations.

LEMMA 4.14. *Suppose that the formula*

$$e^{tA} = \sum_{k=0}^{m_i-1} \sum_{i=1}^{s} t^k e^{t\lambda_i} M_{i,k}$$

in Theorem 2.3 is rewritten as

$$e^{tA} = \sum_{i=1}^{s} M^{(i)}(t),$$

where

$$M^{(1)}(t) = \sum_{k=0}^{m_1-1} t^k e^{t\lambda_1} M_{1,k}$$

$$= e^{t\lambda_1}[M_{1,0} + tM_{1,1} + \cdots + t^{m_1-1}M_{1,m_1-1}],$$

$$M^{(2)}(t) = \sum_{k=0}^{m_2-1} t^k e^{t\lambda_2} M_{2,k},$$

$$= e^{t\lambda_2}[M_{2,0} + tM_{2,1} + \cdots + t^{m_2-1}M_{2,m_2-1}]$$

$$\cdots$$

$$M^{(s)}(t) = \sum_{k=0}^{m_s-1} t^k e^{t\lambda_s} M_{s,k}$$

$$= e^{t\lambda_s}[M_{s,0} + tM_{s,1} + \cdots + t^{m_s-1}M_{s,m_s-1}].$$

Then each matrix $M^{(i)}(t), i = 1, \ldots, s$, is a matrix solution to the homogeneous system (1.3), in the sense that

$$\frac{d}{dt}M^{(i)}(t) = AM^{(i)}(t), \quad i = 1, \ldots, s.$$

PROOF. We divide the proof into two steps.

Step 1. Note that

$$Ae^{tA} = \sum_{i=1}^{s}[\sum_{k=0}^{m_i-1} t^k e^{t\lambda_i} AM_{i,k}] \quad \text{and}$$

$$\frac{d}{dt}e^{tA} = \sum_{i=1}^{s}[\sum_{k=0}^{m_i-1} \frac{d}{dt}(t^k e^{t\lambda_i}) M_{i,k}]$$

$$= \sum_{i=1}^{s}[\sum_{k=0}^{m_i-1} (kt^{k-1} + \lambda_i t^k)e^{t\lambda_i} M_{i,k}].$$

Using the property

$$\frac{d}{dt}e^{tA} = Ae^{tA}$$

in Theorem 2.1, we have

$$0 = \frac{d}{dt}e^{tA} - Ae^{tA}$$

$$= \sum_{i=1}^{s} \sum_{k=0}^{m_i-1} [t^k e^{t\lambda_i} AM_{i,k} - (kt^{k-1} + \lambda_i t^k)e^{t\lambda_i} M_{i,k}$$

$$= \sum_{i=1}^{s} [(AM_{i,0} - \lambda_i M_{i,0} - M_{i,1}) + (AM_{i,1} - \lambda_i M_{i,1} - 2M_{i,2}) + \cdots$$
$$+ (AM_{i,m_i-2} - \lambda_i M_{i,m_i-2} - (m_i - 1)M_{i,m_i-1}) +$$
$$(AM_{i,m_i-1} - \lambda_i M_{i,m_i-1}].$$

Since the set

$$\{t^k e^{t\lambda_i} : i = 1, \ldots, s; k = 1, \ldots, (m_i - 1)\}$$

is linearly independent from Chapter 1, it follows that

$$AM_{i,0} = \lambda_i M_{i,0} + M_{i,1},$$
$$AM_{i,1} = \lambda_i M_{i,1} + 2M_{i,2},$$
$$\vdots$$
$$AM_{i,m_i-2} = \lambda_i M_{i,m_i-2} + (m_i - 1)M_{i,m_i-1},$$
$$AM_{i,m_i-1} = \lambda_i M_{i,m_i-1}.$$

Step 2. Compute

$$\frac{d}{dt} M^{(i)}(t) = \sum_{k=0}^{m_i-1} \frac{d}{dt}(t^k e^{t\lambda_i}) M_{i,k}$$
$$= \sum_{k=0}^{m_i-1} (kt^{k-1} + \lambda_i t^k) e^{t\lambda_i} M_{i,k}$$
$$= [(\lambda_i M_{i,0} + M_{i,1}) + t(\lambda_i M_{i,1} + 2M_{i,2}) + \cdots$$
$$+ t^{m_i-2}(\lambda_i M_{i,m_i-1} + (m_i - 1)M_{i,m_i-2}) + t^{m_i-1}\lambda_i M_{i,m_i-1}]e^{t\lambda_i}$$

and

$$AM^{(i)}(t) = \sum_{k=0}^{m_i-1} t^k e^{t\lambda_i} AM_{i,k}$$
$$= [AM_{i,0} + tAM_{i,1} + \cdots + t^{m_i-1}AM_{i,m_i-1}]e^{t\lambda_i}.$$

It follows from Step 1 that

$$\frac{d}{dt} M^{(i)}(t) = AM^{(i)}(t).$$

$\square$

LEMMA 4.15. *Suppose that $y(t)$ is a continuously differentiable function on $[a, b]$, that $h(t)$ is a continuous function on $[a, b]$, and that α is a given complex constant. Then the differential equation*

$$y'(t) - \alpha y = h(t) \tag{4.10}$$

has the general solution

$$y(t) = e^{t\alpha} \int e^{-t\alpha} h(t) \, dt + ce^{t\alpha},$$

where c is a complex constant, and the integration constant for the indefinite integral $\int e^{-t\alpha} h(t) \, dt$ is taken to be zero.

PROOF. Multiplying the equation (4.10) by $e^{-t\alpha}$, we have

$$(e^{-t\alpha}y)' = e^{-t\alpha}(y' - \alpha y)$$
$$= e^{-t\alpha}h(t),$$

and so integrating both sides give

$$y(t) = e^{t\alpha} \int e^{-t\alpha}h(t)\,dt + ce^{t\alpha}.$$

Here c is a complex constant, and

$$\int e^{-t\alpha}h(t)\,dt$$

is the indefinite integral of the function

$$e^{-t\alpha}h(t),$$

in which the integration constant is taken to be zero.   □

LEMMA 4.16. *Suppose that y and h are two differentiable functions on (a,b), and that y and h are also continuous at end points a and b. If*

$$y'(t) = h'(t), \quad t \in (a,b),$$

then, by taking out, if any, the constants in y and h, and denoting the resultant y and h by $\tilde{y}$ and $\tilde{h}$, respectively, we have

$$\tilde{y} = \tilde{h}.$$

PROOF. It follows from the mean value theorem in calculus that

$$y = h + c,$$

where c is a constant. By taking out, if any, the constants in y and h, we have

$$\tilde{y} = \tilde{h}.$$

□

REMARK 4.17. *It is easy to see that Lemmas 4.15 and 4.16 also hold for vector-valued $y(t)$ and $h(t)$.*

LEMMA 4.18. *If $v(t)$ is a vector-valued solution to the nonhomogeneous system*

$$\frac{d}{dt}u(t) = Au(t) + f(t) \tag{4.11}$$
$$= Au(t) + q_1(t)e^{t\mu_1}$$

in Theorem 2.22, then $v(t)$ satisfies the homogeneous system

$$0 = (D - \mu_1)^{n_1+1}[(D - A)u] \tag{4.12}$$
$$= (D - A)[(D - \mu_1)^{n_1+1}u].$$

Here $D = \frac{d}{dt}$.

PROOF. From Chapter 1, we know that

$$(D - \mu_1)^{n_1+1}(t^{n_1} e^{t\mu_1}) = 0,$$

the differential operator $(D - \mu_1)^{n_1+1}$ annihilating the complex-valued function $t^{n_1} e^{t\mu_1}$. It results that

$$(D - \mu_1)^{n_1+1}(q_1(t) e^{t\mu_1}) = 0,$$

the differential operator $(D - \mu_1)^{n_1+1}$ annihilating the vector-valued function $q_1(t) e^{t\mu_1}$. Thus applying the differential operator $(D - \mu_1)^{n_1+1}$ to the nonhomogeneous system

$$(D - A)u = q_1(t) e^{t\mu_1},$$

we obtain the homogeneous system (4.12), which completes the proof. $\square$

LEMMA 4.19.

$$\int e^{t(A-\mu_1)} c \, dt = e^{t(A-\mu_1)} (A - \mu_1)^{-1} c$$

is true, where the complex number μ_1 is such that $(A - \mu_1)^{-1}$ exists, and the integration constant is taken to be zero.

PROOF. Since

$$\frac{d}{dt} e^{t(A-\mu_1)} = e^{t(A-\mu_1)} (A - \mu_1)$$

by Theorem 2.1, it results that

$$e^{t(A-\mu_1)} = \int e^{t(A-\mu_1)} (A - \mu_1),$$

and so

$$\int e^{t(A-\mu_1)} c \, dt = \int e^{t(A-\mu_1)} (A - \mu_1)(A - \mu_1)^{-1} c \, dt$$

$$= e^{t(A-\mu_1)} (A - \mu_1)^{-1} c.$$

This completes the proof. Here the integration constants are all taken to be zero. $\square$

LEMMA 4.20. If

$$Q(t) \equiv \int e^{-t\mu} M^{(i)}(t) \, dt,$$

then

$$Q(t) = -\frac{1}{\mu}[e^{-t\mu} M^{(i)}(t) - AQ(t)],$$

where the matrices $M^{(i)}(t)$ come from Lemma 4.14, and the complex number μ is such that $e^{-t\mu} M^{(i)}(t)$ has no constant term. Here the integration constant for $Q(t)$ is taken to be zero.

Furthermore, $e^{t\mu} Q(t)$ satisfies

$$\frac{d}{dt} e^{t\mu} Q(t) = A(e^{t\mu} Q(t)),$$

that is, $e^{t\mu} Q(t)$ is a matrix solution to the homogeneous system (1.3).

REMARK 4.21. Thus, the case where $\mu = \lambda_1$ and $M^{(i)}(t) = M^{(1)}(t)$ does not satisfy the assumptions in Lemma 4.20, because $e^{-t\lambda_1} M^{(1)}(t)$ has a constant term $M_{1,0}$. However, the case where $\mu = \lambda_1$ and $M^{(i)}(t), i = 2, 3, \ldots, s$, does.

PROOF. Using integration by parts, Lemma 4.14, and (the Lemma 4.16 in) Remark 4.17, we have

$$Q(t) \equiv \int e^{-t\mu} M^{(i)}(t)\, dt$$

$$= \frac{-1}{\mu} \int M^{(i)}(t)\, de^{-t\mu}$$

$$= \frac{-1}{\mu} [e^{-t\mu} M^{(i)}(t) - \int e^{-t\mu} \frac{d}{dt} M^{(i)}(t)\, dt]$$

$$= \frac{-1}{\mu} [e^{-t\mu} M^{(i)}(t) - \int e^{-t\mu} A M^{(i)}(t)\, dt]$$

$$= \frac{-1}{\mu} [e^{-t\mu} M^{(i)}(t) - A Q(t)],$$

where used was the assumption that $e^{-t\mu} M^{(i)}(t)$ has no constant term. Furthermore, since $\frac{d}{dt} Q(t) = e^{-t\mu} M^{(i)}(t)$, we have

$$A(e^{t\mu} Q(t)) = e^{t\mu} [e^{-t\mu} M^{(i)}(t)] + \mu e^{t\mu} Q(t)$$

$$= e^{t\mu} \frac{d}{dt} Q(t) + \mu e^{t\mu} Q(t)$$

$$= \frac{d}{dt} [e^{t\mu} Q(t)].$$

This completes the proof.   □

We are now ready to prove theorem 2.22:

PROOF. Suppose that $h(t)$ is a solution to the nonhomogeneous system (4.11). Then $h(t)$ satisfies the homogeneous system (4.12) by Lemma 4.18, which by Theorem 2.1, has the general solution $(D - \mu_1)^{n_1+1} h(t)$ given by

$$(D - \mu_1)^{n_1+1} h(t) = e^{tA} c$$

$$= \sum_{i=1}^{s} M^{(i)}(t) c \tag{4.13}$$

for some constant vector c, where the matrices $M^{(i)}(t)$ come from Lemma 4.14.

We consider two cases.

Case 1. Suppose that μ_1 is not in the set $\{\lambda_1, \lambda_2, \ldots, \lambda_s\}$, for which

$$(A - \mu_1)^{-1}$$

exists.

Applying (the Lemma 4.15 in) Remark 4.17 and Lemma 4.19 to equation (4.13), we have

$$(D - \mu_1)^{n_1} h(t) = e^{t\mu_1} \int e^{t(A - \mu_1)} c\, dt + e^{t\mu_1} d_0$$

$$= e^{t\mu_1} e^{t(A - \mu_1)} (A - \mu_1)^{-1} c + e^{t\mu_1} d_0 \tag{4.14}$$

$$= e^{tA} (A - \mu_1)^{-1} c + e^{t\mu_1} d_0.$$

Applying this argument again, to (4.14) instead of (4.13), we have

$$(D - \mu_1)^{n_1-1} h(t) = e^{t\mu_1} \int e^{-t\mu_1} [e^{tA} (A - \mu_1)^{-1} c + e^{t\mu_1} d_0]\, dt + e^{t\mu_1} d_1$$

$$= e^{t\mu_1}[e^{t(A-\mu_1)}(A-\mu_1)^{-1}(A-\mu_1)^{-1}c + td_0] + e^{t\mu_1}d_1$$
$$= e^{tA}(A-\mu_1)^{-2}c + te^{t\mu_1}d_0 + e^{t\mu_1}d_1.$$

By repeating this argument, we have eventually

$$h(t) = e^{tA}(A-\mu_1)^{-(n_1+1)}c + e^{t\mu_1}[t^{n_1}d_0 + t^{n_1-1}d_1 + \cdots + td_{n_1-1} + d_{n_1}]$$

for some constant vectors $d_0, \ldots, d_{n_1}$.

Since

$$e^{tA}(A-\mu_1)^{-(n_1+1)}c$$

is clearly a solution to the homogeneous system (1.3), it follows from superposition principle, Lemma 2.24, that

$$\begin{aligned} g(t) &\equiv h(t) - e^{tA}(A-\mu_1)^{-(n_1+1)}c \\ &= [t^{n_1}d_0 + \cdots + td_{n_1-1} + d_{n_1}]e^{t\mu_1} \\ &= Q_1(t)e^{t\mu_1} \end{aligned}$$

is a particular solution to the nonhomogeneous system (4.13). Here $Q_1(t)$ is a vector polynomial of degree n_1.

Case 2. Suppose that $\mu_1 = \lambda_1$, for which

$$(A-\mu_1)^{-1} = (A-\lambda_1)^{-1}$$

does not exist.

Applying Lemma 4.15 and Remark 4.17 to the equation (4.13), we have

$$\begin{aligned} &(D-\mu_1)^{n_1}h(t) \\ &= e^{t\mu_1}\int e^{-t\mu_1}\sum_{i=1}^{s}M^{(i)}(t)c\,dt + e^{t\mu_1}d_0 \\ &= e^{t\mu_1}\int \left[\sum_{k=0}^{m_1-1}t^k M_{1,k}c + \sum_{i=2}^{s}M^{(i)}(t)c\right]dt + e^{t\mu_1}d_0 \qquad (4.15) \\ &= Q(t)c + e^{t\mu_1}\sum_{k=0}^{m_1-1}\frac{1}{k+1}t^{k+1}M_{1,k}c + e^{t\mu_1}d_0, \end{aligned}$$

where

$$Q(t)c = \sum_{i=2}^{s}e^{t\mu_1}\int e^{-t\mu_1}M^{(i)}(t)c\,dt.$$

Since $Q(t)c$ is a solution to the homogeneous system (1.3) by Lemma 4.20, we have

$$\begin{aligned} Q(t)c &= e^{tA}c_1 \\ &= \sum_{i=1}^{s}M^{(i)}(t)c_1 \\ &= \sum_{k=0}^{m_1-1}t^k e^{t\lambda_1}M_{1,k}c_1 + \sum_{i=2}^{s}M^{(i)}(t)c_1 \end{aligned}$$

for some constant c_1.

It is easy to see that $Q(t)c$ has no terms involving $e^{t\mu_1} = e^{t\lambda_1}$, and so $M_{1,k}c_1 = 0$, and then

$$Q(t)c = \sum_{i=2}^{s} M^{(i)}(t)c_1.$$

Therefore, the above arguments can be applied again to (4.15) instead of (4.13), and we have

$(D - \mu_1)^{n_1-1} h(t)$

$$= e^{t\mu_1} \int e^{-t\mu_1} [\sum_{i=2}^{s} M^{(i)}(t)c_1 + \sum_{k=0}^{m_1-1} \frac{1}{k+1} t^{k+1} e^{t\mu_1} M_{1,k}c + e^{t\mu_1} d_0] \, dt + e^{t\mu_1} d_1$$

$$= Q(t)c_1 + [\sum_{k=0}^{m_1-1} \frac{1}{(k+2)(k+1)} t^{k+2} e^{t\mu_1} M_{1,k}c + t e^{t\mu_1} d_0 + e^{t\mu_1} d_1],$$

where

$$Q(t)c_1 = \sum_{i=2}^{s} e^{t\mu_1} \int e^{-t\mu_1} M^{(i)}(t)c_1 \, dt$$

$$= e^{tA} c_2$$

$$= \sum_{i=1}^{s} M^{(i)}(t)c_2$$

$$= \sum_{i=2}^{s} M^{(i)}(t)c_2 \quad \text{as before.}$$

Again, the above arguments can be repeated, and eventually, we have

$h(t)$

$$= Q(t)c_{n_1} + \sum_{k=0}^{m_1-1} \frac{1}{(k+n_1+1)(k+n_1)\cdots(k+2)(k+1)} t^{k+n_1+1} e^{t\mu_1} M_{1,k}c$$

$$+ (t^{n_1} d_0 + t^{n_1-1} d_1 + \cdots + t d_{n_1-1} + d_{n_1}) e^{t\mu_1}$$

for some constant vectors $c_1, \ldots, c_{n_1+1}$ and $d_0, \ldots, d_{n_1}$, where

$$Q(t)c_{n_1} = \sum_{i=2}^{s} M^{(i)} c_{n_1+1}$$

is a solution to the homogeneous system (1.3) by Lemma 4.14.

It follows from superposition principle, Lemma 2.24 that

$$g(t) \equiv h(t) - \sum_{i=2}^{s} M^{(i)}(t)c_{n_1+1}$$

$$= \sum_{k=0}^{m_1-1} \frac{1}{(k+n_1+1)(k+n_1)\cdots(k+2)(k+1)} t^{k+n_1+1} e^{t\mu_1} M_{1,k}c$$

$$+ (t^{n_1} d_0 + t^{n_1-1} d_1 + \cdots + t d_{n_1-1} + d_{n_1}) e^{t\mu_1}$$

$$\equiv R_1(t) e^{t\mu_1}$$

is a particular solution to the nonhomogeneous system (4.13). Here $R_1(t)$ is a vector polynomial of degree $(n_1 + m_1)$.   □

REMARK 4.22. The above proof seems new.

5. An Additional Method for Computing a Fundamental Matrix

The method in Theorem 2.3 is used to compute the special fundamental matrix e^{tA}. In addition to this method, there is another one [**19**, pages 560–563], which computes a fundamental set of solutions first. Once this is done, computing a fundamental matrix and then the e^{tA} follows easily.

For this purpose, we prepare the following definitions.

DEFINITION 5.1. *We recall from linear algebra* [**9**, *pages 418-419*] *that*

- *A nonzero constant vector u in $\mathbb{C}^n$ is called a generalized eigenvector of A, if there is a complex number λ such that*

$$(A - \lambda I)^p u = 0 \quad \text{the zero vector, but} \quad (A - \lambda I)^{p-1} u \neq 0$$

  for some positive integer $p \in \mathbb{N}$.
- *The λ is called a generalized eigenvalue.*
- *When $p = 1$, the λ is called an eigenvalue and u is called an eigenvector, corresponding to λ.*
- *The set*

$$S_u^{(p)} \equiv \{v_1, v_2, \ldots, v_p\}$$
$$\equiv \{(A - \lambda I)^{p-1} u, (A - \lambda I)^{p-2} u, \ldots, (A - \lambda I) u, u\}$$

  is called a cycle of generalized eigenvectors of A, corresponding to the generalized eigenvalue λ.
- *The p is called the length of the cycle $S_u^{(p)}$, and u is called the generator of the cycle $S_u^{(p)}$.*

It will follow from the above definition that

LEMMA 5.2. *The following*

$$(A - \lambda I) v_1^{(p)} = 0,$$
$$(A - \lambda I)^2 v_2^{(p)} = 0,$$
$$(A - \lambda I)^3 v_3^{(p)} = 0,$$
$$\vdots,$$
$$(A - \lambda I)^p v_p^{(p)} = 0$$

are true, where

$$v_k^{(p)} = (A - \lambda I)^{p-k} u, k = 1, 2, \ldots, p.$$

PROOF. See the Problems set in Section 6.   □

DEFINITION 5.3. *Define the p vector-valued functions*

$$z_k^{(p)}(t) \equiv e^{t\lambda} \sum_{l=0}^{k-1} \frac{t^l (A - \lambda I)^l}{l!} v_k^{(p)}$$

$$= e^{t\lambda}[v_k^{(p)} + \frac{t(A - \lambda I)}{1!}v_k^{(p)} + \cdots + \frac{t^{k-1}(A - \lambda I)^{k-1}}{(k-1)!}v_k^{(p)}],$$

$$= e^{t\lambda}[v_k^{(p)} + \frac{t}{1!}v_{k-1}^{(p)} + \cdots + \frac{t^{k-1}}{(k-1)!}v_1^{(p)}],$$

$$k = 1, 2, \ldots p.$$

Then the set

$$Z_u^{(p)}(t) \equiv \{z_1^{(p)}(t), z_2^{(p)}(t), \ldots, z_p^{(p)}(t)\}$$

of p vector-valued functions is called the cycle of vector-valued functions, associated with the cycle $S_u^{(p)}$.

The above p vector-valued functions $z_k^{(p)}(t), k = 1, \ldots, p$, are related to the exponential function e^{tA}:

LEMMA 5.4. *The following*

$$z_k^{(p)}(t) = e^{tA}v_k^{(p)}, k = 1, 2, \ldots, p,$$

is true.

PROOF. We divide the proof into two steps.
Step 1. Claim that

$$e^{tA} = e^{t(A - \mu I) + t\mu I}$$

$$= e^{t(A - \mu I)}e^{t\mu}$$

$$= e^{t\mu}e^{t(A - \mu I)},$$

where $\mu \in \mathbb{C}$ is a complex number. For a proof, see the Problems set in Section 6.

Step 2. It follows from Step 1, the definition of e^{tA}, and Lemma 5.2 that the αth component $(e^{tA}v_k^{(p)})_\alpha$ of the vector $e^{tA}v_k^{(p)}$, where $1 \le \alpha \le n$, satisfies:

$$(e^{tA}v_k^{(p)})_\alpha = (e^{t\lambda}e^{t(A - \lambda I)}v_k^{(p)})_\alpha$$

$$= e^{t\lambda}\sum_{\beta=1}^{n}(e^{t(A - \lambda I)})_{\alpha\beta}(v_k^{(p)})_\beta$$

$$= e^{t\lambda}\sum_{\beta=1}^{n}\lim_{m\to\infty}(\sum_{j=0}^{m}\frac{t^j(A - \lambda I)^j}{j!})_{\alpha\beta}(v_k^{(p)})_\beta$$

$$= e^{t\lambda}\lim_{m\to\infty}\sum_{\beta=1}^{n}(\sum_{j=0}^{m}\frac{t^j(A - \lambda I)^j}{j!})_{\alpha\beta}(v_k^{(p)})_\beta$$

$$= e^{t\lambda}\lim_{m\to\infty, m\ge k}(\sum_{j=0}^{m}\frac{t^j(A - \lambda I)^j}{j!}v_k^{(p)})_\alpha$$

$$= e^{t\lambda}\lim_{m\to\infty}(\sum_{j=0}^{k-1}\frac{t^j(A - \lambda I)^j}{j!}v_k^{(p)})_\alpha$$

$$= (e^{t\lambda}\sum_{j=0}^{k-1}\frac{t^j(A - \lambda I)^j}{j!}v_k^{(p)})_\alpha$$

$$= (z_k^{(p)}(t))_\alpha.$$

Thus

$$e^{tA}v_k^{(p)} = z_k^{(p)}(t), k = 1, \ldots, p.$$

$\square$

REMARK 5.5. *We observe that*

- *Each cycle $S_u^{(p)}$, corresponding to a generalized eigenvalue λ of the matrix A, has a unique element $(A - \lambda I)^{p-1}u$ that is an eigenvector of A. This is because*

$$(A - \lambda I)[(A - \lambda I)^{p-1}u] = (A - \lambda I)^p u = 0.$$

- *Since $v \equiv (A - \lambda I)^{p-1}u$ is an eigenvector, for which*

$$(A - \lambda I)v = 0,$$

λ is not only a generalized eigenvalue, but also an eigenvalue of A.

DEFINITION 5.6. *Recall that the set*

$$E_\lambda \equiv \{v \in \mathbb{C} : v \quad is \ an \ eigenvector \ of \ A, \ corresponding \ to \ the \ eigenvalue \ \lambda\}$$
$$= \{v \in \mathbb{C} : (\lambda I - A)v = 0\},$$

which is a vector space, is called the eigenspace, corresponding to the eigenvalue λ.

DEFINITION 5.7. *The dimension of E_λ is the maximal number of linearly independent eigenvectors in E_λ.*

We are now in a position to describe, what we call, another method of computing a fundamental set of solutions:

THEOREM 5.8. *As in Theorem 2.1, assume that*

$$p(\lambda) = a_0(\lambda - \lambda_1)^{m_1}(\lambda - \lambda_2)^{m_2} \cdots (\lambda - \lambda_s)^{m_s} = 0,$$

where $\lambda_i \neq \lambda_j$ for $1 \leq i \neq j \leq s$ and $1 \leq s \leq n$, and

$$m_1 + m_2 + \cdots + m_s = n.$$

Then for each $1 \leq i \leq s$, the complex number λ_i is a generalzed eigenvalue (also an eigenvalue) of A, and corresponding to this generalized eigenvalue λ_i, there are n_i cycles

$$S_{u_l^{(i)}}^{(p_l^{(i)})}, l = 1, 2, \ldots, n_i,$$

of generalized eigenvectors (see Definition 5.1) and there are n_i associated cycles

$$Z_{u_l^{(i)}}^{(p_l^{(i)})}(t), l = 1, 2, \ldots, n_i,$$

of vector-valued functions (see Definition 5.3), each cycle $Z_{u_l^{(i)}}^{(p_l^{(i)})}(t)$ having $p_l^{(i)}$ vector-valued functions, such that the union $Y_i(t)$ of those n_i cycles

$$Z_{u_l^{(i)}}^{(p_l^{(i)})}(t), l = 1, 2, \ldots, n_i,$$

is a set of

$$m_i = p_1^{(i)} + p_2^{(i)} + \cdots + p_{n_i}^{(i)}$$

linearly independent vector-valued solutions to the homogeneous system (1.3).

Here the positive integer n_i is the dimension of the eigenspace E_{λ_i} of the eigenvalue λ_i, and the generalized eigenvector $u_l^{(i)}$, which is the generator of the cycle $S_{u_l^{(i)}}^{(p_l^{(i)})}$, is a solution to the systems of algebraic equations:

$$(A - \lambda_i I)^{p_l^{(i)}} w = 0 \quad \text{and}$$
$$(A - \lambda_i I)^{p_l^{(i)} - 1} w \neq 0$$

for some $1 \leq p_l^{(i)} \leq m_i$.

Furthermore the union

$$Z(t) = Y_1(t) \cup Y_2(t) \cdots \cup Y_s(t)$$

of the s $Y_i(t)'s$ is a fundamental set of

$$n = m_1 + m_2 + \cdots + m_s$$

vector-valued solutions to the homogeneous system (1.3).

Thus if $m_i = 3$ and $n_i = 1$, then $l = 1$ only, and $p_l^{(i)} = p_1^{(i)} = 3$. Or if $m_i = 3$ and $n_i = 2$, then $l = 1$ or 2, and $p_1^{(i)} = 1$ and $p_2^{(i)} = 2$. Or if $m_i = 3$ and $n_i = 3$, then $l = 1, 2, 3$, and $p_1^{(i)} = p_2^{(i)} = p_3^{(i)} = 1$.

Before we prove this theorem, we look at some concrete examples first.

EXAMPLE 5.9. Find the general solution to the homogeneous system

$$\frac{d}{dt} u(t) = Au(t), \quad \text{where}$$

$$A = \begin{pmatrix} 1 & -1 \\ 1 & 3 \end{pmatrix}.$$

Solution. This example was already studied in Section 3 by using Theorem 2.3. We now use the above Theorem 5.8 although using Theorem 2.3 is easier.

Computing the characteristic equation

$$p(\lambda) = |A - \lambda I|$$
$$= \begin{vmatrix} 1 - \lambda & -1 \\ 1 & 3 - \lambda \end{vmatrix}$$
$$= (\lambda - 2)^2 = 0,$$

we obtain that the only eigenvalue is $\lambda_1 = 2$ of multiplicity $m_1 = 2$. The corresponding eigenvectors can be calculated as follows:

The equation $(A - 2)u = 0$ where $u = \begin{pmatrix} x_1 \\ x_2 \end{pmatrix}$, gives

$$\begin{pmatrix} -1 & -1 \\ 1 & 1 \end{pmatrix} \begin{pmatrix} x_1 \\ x_2 \end{pmatrix} = \begin{pmatrix} 0 \\ 0 \end{pmatrix}.$$

Solving for x_1 and x_2, we have

$$x_1 = -x_2,$$

and so

$$u = \begin{pmatrix} x_1 \\ -x_1 \end{pmatrix}$$

$$= x_1 \begin{pmatrix} 1 \\ -1 \end{pmatrix}.$$

Thus the eigenspace E_{λ_1}, corresponding to the eigenvalue λ_1, is of dimension one, that is, $n_1 = 1$. It follows from Theorem 5.8 that

$$l = 1 \quad \text{only,}$$
$$p_l^{(1)} = m_1 = 2,$$

and that

$$(A - 2I)^2 u_1^{(1)} = \begin{pmatrix} 0 & 0 \\ 0 & 0 \end{pmatrix} u_1^{(1)} = \begin{pmatrix} 0 \\ 0 \end{pmatrix} \quad \text{and}$$
$$(A - 2I) u_1^{(1)} \neq 0,$$

for which the generator $u_1^{(1)}$ can be any two dimensional column vector, except for the eigenvector $u = \begin{pmatrix} 1 \\ -1 \end{pmatrix}$. We can take

$$u_1^{(1)} = \begin{pmatrix} 1 \\ 0 \end{pmatrix},$$

for example. Thus by Theorem 5.8, we have the fundamental set $Z(t)$ of solutions:

$$Z(t) = Y_1(t) = S_{u_1^{(1)}}^{(p_1^{(1)})}$$
$$= \{ z_1^{(p_1^{(1)})}(t), z_2^{(p_1^{(1)})}(t) \},$$

where

$$z_1^{(p_1^{(1)})}(t) = e^{2t} v_1^{(p_1^{(1)})},$$
$$z_2^{(p_1^{(1)})}(t) = e^{2t} [v_2^{(p_1^{(1)})} + t v_1^{(p_1^{(1)})}],$$
$$v_1^{(p_1^{(1)})} = (A - 2I)^{2-1} u_1^{(1)} = \begin{pmatrix} -1 \\ 1 \end{pmatrix}, \quad \text{and}$$
$$v_2^{(p_1^{(1)})} = (A - 2I)^{2-2} u_1^{(1)} = u_1^{(1)} = \begin{pmatrix} 1 \\ 0 \end{pmatrix},$$

and then we have the general solution

$$z(t) = c_1 z_1^{(p_1^{(1)})}(t) + c_2 z_2^{(p_1^{(1)})}(t),$$

by Theorem 2.14, where c_1 and c_2 are two complex constants.

EXAMPLE 5.10. Find the general solution for the homogeneous system

$$\frac{d}{dt} u(t) = A u(t), \quad \text{where}$$

$$A = \begin{pmatrix} 1 & 0 & 0 \\ 1 & 3 & 0 \\ 1 & 1 & 1 \end{pmatrix}.$$

Solution. Computing the characteristic equation

$$p(\lambda) = |A - \lambda I|$$

$$= \begin{vmatrix} 1-\lambda & 0 & 0 \\ 1 & 3-\lambda & 0 \\ 1 & 1 & 1-\lambda \end{vmatrix}$$

$$= (1-\lambda)^2(3-\lambda) = 0,$$

we obtain that the eigenvalues are $\lambda_1 = 1$ of multiplicity $m_1 = 2$, and $\lambda_2 = 3$ of multiplicity $m_2 = 1$, respectively. The corresponding eigenvectors can be calculated as follows.

For $\lambda_1 = 1$, the equation $(A - 1I)u = 0$ where $u = \begin{pmatrix} x_1 \\ x_2 \\ x_3 \end{pmatrix}$ gives

$$\begin{pmatrix} 0 & 0 & 0 \\ 1 & 2 & 0 \\ 1 & 1 & 0 \end{pmatrix} \begin{pmatrix} x_1 \\ x_2 \\ x_3 \end{pmatrix} = \begin{pmatrix} 0 \\ 0 \\ 0 \end{pmatrix}.$$

Solving for x_1, x_2, and x_3, we have

$$x_1 = x_2 = 0,$$

and so

$$u = \begin{pmatrix} 0 \\ 0 \\ x_3 \end{pmatrix}$$

$$= x_3 \begin{pmatrix} 0 \\ 0 \\ 1 \end{pmatrix}.$$

Thus the eigenspace E_{λ_1}, corresponding to the eigenvalue $\lambda_1 = 1$, is of dimension one, that is, $n_1 = 1$. It follows from Theorem 5.8 that

$$l = 1 = n_1 \quad \text{only}$$

$$p_l^{(1)} = m_1 = 2,$$

and that

$$(A - 1I)^2 u_1^{(1)} = \begin{pmatrix} 0 & 0 & 0 \\ 2 & 4 & 0 \\ 1 & 2 & 0 \end{pmatrix} \begin{pmatrix} x_1 \\ x_2 \\ x_3 \end{pmatrix} = \begin{pmatrix} 0 \\ 0 \\ 0 \end{pmatrix} \quad \text{and}$$

$$(A - 1I)u_1^{(1)} \neq 0.$$

Solving for x_1, x_2, and x_3, we obtain

$$x_1 = -2x_2,$$

and so

$$u_1^{(1)} = \begin{pmatrix} -2x_2 \\ x_2 \\ x_3 \end{pmatrix}$$

$$= \begin{pmatrix} -2x_2 \\ x_2 \\ 0 \end{pmatrix} + \begin{pmatrix} 0 \\ 0 \\ x_3 \end{pmatrix}$$

$$= x_2 \begin{pmatrix} -2 \\ 1 \\ 0 \end{pmatrix} = x_3 \begin{pmatrix} 0 \\ 0 \\ 1 \end{pmatrix}.$$

We can only take $u_1^{(1)} = \begin{pmatrix} -2 \\ 1 \\ 0 \end{pmatrix}$, but we cannot take $u_1^{(1)} = \begin{pmatrix} 0 \\ 0 \\ 1 \end{pmatrix}$ since $\begin{pmatrix} 0 \\ 0 \\ 1 \end{pmatrix}$ is an

eigenvector corresponding to the eigenvalue $\lambda_1 = 1$ and satisfies

$$(A - 1I)u_1^{(1)} = 0.$$

Thus by Theorem 5.8, we have

$$Y_1(t) = S_{u_1^{(1)}}^{(p_1^{(1)})}$$

$$= \{z_1^{(p_1^{(1)})}(t), z_2^{(p_1^{(1)})}(t)\},$$

where

$$z_1^{(p_1^{(1)})}(t) = e^t v_1^{(p_1^{(1)})},$$

$$z_2^{(p_1^{(1)})}(t) = e^t [v_2^{(p_1^{(1)})} + t v_1^{(p_1^{(1)})}],$$

$$v_1^{(p_1^{(1)})} = (A - 1I)^{2-1} u_1^{(1)} = \begin{pmatrix} 0 \\ 0 \\ 1 \end{pmatrix}, \quad \text{and}$$

$$v_2^{(p_1^{(1)})} = (A - 2I)^{2-2} u_1^{(1)} = u_1^{(1)} = \begin{pmatrix} -2 \\ 1 \\ 0 \end{pmatrix}.$$

For $\lambda_2 = 3$, the equation $(A - 3I)u = 0$ where $u = \begin{pmatrix} x_1 \\ x_2 \\ x_3 \end{pmatrix}$ gives

$$\begin{pmatrix} -2 & 0 & 0 \\ 1 & 0 & 0 \\ 1 & 1 & -2 \end{pmatrix} \begin{pmatrix} x_1 \\ x_2 \\ x_3 \end{pmatrix} = \begin{pmatrix} 0 \\ 0 \\ 0 \end{pmatrix}.$$

Solving for x_1, x_2, and x_3, we obtain

$$x_1 = 0, x_2 = 2x_3,$$

and so

$$u = \begin{pmatrix} 0 \\ 2x_3 \\ x_3 \end{pmatrix}$$

$$= x_3 \begin{pmatrix} 0 \\ 2 \\ 1 \end{pmatrix}.$$

Thus the eigenspace E_{λ_2}, corresponding to the eigenvalue $\lambda_2 = 3$, is of dimension one, that is, $n_2 = 1$. It follows from Theorem 5.8 that

$$l = 1 \quad \text{only}$$

$$p_l^{(2)} = m_2 = 1,$$

and that

$$(A - 3I)u_1^{(2)} = 0 \quad \text{and}$$
$$(A - 3I)^{1-1}u_1^{(2)} = u_1^{(2)} \neq 0.$$

Thus $u_1^{(2)}$ is just the eigenvector $\begin{pmatrix} 0 \\ 2 \\ 1 \end{pmatrix}$, obtained above, corresponding to the eigen-

value $\lambda_2 = 3$.

It follows from Theorem 5.8 that

$$Y_2(t) = S_{u_1^{(2)}}^{(p_1^{(2)})}$$
$$= \{z_1^{(p_1^{(2)})}(t)\},$$

where

$$z_1^{(p_1^{(2)})}(t) = e^{3t}[v_1^{(p_1^{(2)})}], \quad \text{and}$$
$$v_1^{(p_1^{(2)})} = (A - 3I)^{1-1}u_1^{(2)} = u_1^{(2)}$$
$$= \begin{pmatrix} 0 \\ 2 \\ 1 \end{pmatrix},$$

and that the fundamental set $Z(t)$ is

$$Z(t) = Y_1(t) \cup Y_2(t).$$

By Theorem 2.14, the general solution is

$$z(t) = c_1 z_1^{(p_1^{(1)})}(t) + c_2 z_2^{(p_1^{(1)})}(t) + c_3 z_1^{(p_1^{(2)})},$$

where c_1, c_2, and c_3 are three complex constants.

Proof of Theorem 5.8.

PROOF. We divide the proof into four steps.
Step 1. We first recall the theorem from linear algebra [**9**, pages 420–425],
[**13**]:

THEOREM 5.11 (Primary Decomposition Theorem). *As in Theorem 2.1, as-sume that*

$$p(\lambda) = a_0(\lambda - \lambda_1)^{m_1}(\lambda - \lambda_2)^{m_2} \cdots (\lambda - \lambda_s)^{m_s} = 0,$$

where $\lambda_i \neq \lambda_j$ *for* $1 \leq i \neq j \leq s$ *and* $1 \leq s \leq n$, *and*

$$m_1 + m_2 + \cdots + m_s = n.$$

Then for each $1 \leq i \leq s$, *the complex number* λ_i *is a generalzed eigenvalue (also an eigenvalue) of* A, *and corresponding to this generalized eigenvalue* λ_i, *there are* n_i *cycles*

$$S_{u_l^{(i)}}^{(p_l^{(i)})}, l = 1, 2, \ldots, n_i$$

*of generalized eigenvectors, each cycle $S_{u_l^{(i)}}^{(p_l^{(i)})}$ having $p_l^{(i)}$ generalized eigenvectors
with $u_l^{(i)}$ as the generator, such that the union R_i of those n_i cycles*

$$S_{u_l^{(i)}}^{(p_l^{(i)})}, l = 1, 2, \ldots, n_i,$$

is a set of

$$m_i = p_1^{(i)} + p_2^{(i)} + \cdots + p_{n_i}^{(i)}$$

linearly independent generalized eigenvectors.

*Here the positive integer n_i is the dimension of the eigenspace E_{λ_i} of the eigen-
value λ_i, and the generalized eigenvector $u_l^{(i)}$ as a generator, is a solution to the
systems of algebraic equations:*

$$(A - \lambda_i I)^{p_l^{(i)}} w = 0 \quad and$$

$$(A - \lambda_i I)^{p_l^{(i)} - 1} w \neq 0$$

for some $1 \leq p_l^{(i)} \leq m_i$.
    Furthermore, the union

$$S = R_1 \cup R_2 \cdots \cup R_s$$

of the s $R_i's$ is a set of

$$n = m_1 + m_2 + \cdots + m_s$$

linearly independent generalized eigenvectors.

Step 2. It follows from the Primary Decomposition Theorem above in Step 1
that Theorem 5.8 is almost proved, except for the part that

$$Z(t) = Y_1(t) \cup Y_2(t) \cup \cdots \cup Y_s(t)$$

is a fundamental set of

$$n = m_1 + m_2 + \cdots + m_s$$

vector-valued solutions to the homogeneous system (1.3), which will be proved in
Steps 3 and 4 below. Here the set

$$Y_i(t) = \cup_{l=1}^{n_i} Z_{u_l^{(i)}}^{p_l^{(i)}}(t)$$

$$= Z_{u_1^{(i)}}^{p_1^{(i)}}(t) \cup \cdots \cup Z_{u_{n_i}^{(i)}}^{p_{n_i}^{(i)}}(t)$$

is the union of those n_i cycles

$$Z_{u_l^{(i)}}^{(p_l^{(i)})}(t), l = 1, 2, \ldots, n_i$$

of vector-valued functions, that are associated with the n_i cycles

$$S_{u_l^{(i)}}^{(p_l^{(i)})}, l = 1, 2, \ldots, n_i$$

from Step 1, of generalized eigenvectors of the matrix A; the union of the n_i cycles
$S_{u_l^{(i)}}^{(p_l^{(i)})}, l = 1, 2, \ldots, n_i$, by Step 1, is a set of

$$m_i = p_1^{(i)} + p_2^{(i)} + \cdots + p_{n_i}^{(i)}$$

generalized eigenvectors of the matrix A, and the set $Y_i(t)$ is a set of

$$m_i = p_1^{(i)} + p_2^{(i)} + \cdots + p_{n_i}^{(i)}$$

vector-valued functions; the association of $Z_{u_l^{(i)}}^{(p_l^{(i)})}(t)$ with $S_{u_l^{(i)}}^{(p_l^{(i)})}$, for each $1 \leq l \leq n_i$, is made by Definitions 5.1 and 5.3 as the following shows:

$$S_{u_l^{(i)}}^{(p_l^{(i)})} = \{(A - \lambda_i)^{p_l^{(i)}-1}u_l^{(i)}, (A - \lambda_i)^{p_l^{(i)}-2}u_l^{(i)}, \ldots, (A - \lambda_i)u_l^{(i)}, u_l^{(i)}\}$$

$$= \{v_1^{(p_l^{(i)})}, v_2^{(p_l^{(i)})}, \ldots, v_{p_l^{(i)}}^{(p_l^{(i)})}\}, \quad \text{where}$$

$$v_k^{(p_l^{(i)})} = (A - \lambda_i)^{p_l^{(i)}-k}u_l^{(i)}, k = 1, 2, \ldots, p_l^{(i)};$$

$$Z_{u_l^{(i)}}^{p_l^{(i)}} = \{z_1^{(p_l^{(i)})}(t), z_2^{(p_l^{(i)})}(t), \ldots, z_{p_l^{(i)}}^{(p_l^{(i)})}(t)\}, \quad \text{where}$$

$$z_k^{(p_l^{(i)})}(t) \equiv e^{t\lambda} \sum_{l=0}^{k-1} \frac{t^l (A - \lambda I)^l}{l!} v_k^{(p_l^{(i)})}$$

$$= e^{t\lambda}[v_k^{(p_l^{(i)})} + \frac{t(A - \lambda I)}{1!}v_k^{(p_l^{(i)})} + \cdots + \frac{t^{k-1}(A - \lambda I)^{k-1}}{(k-1)!}v_k^{(p_l^{(i)})}],$$

$$= e^{t\lambda}[v_k^{(p_l^{(i)})} + \frac{t}{1!}v_{k-1}^{(p_l^{(i)})} + \cdots + \frac{t^{k-1}}{(k-1)!}v_1^{(p_l^{(i)})}],$$

$$k = 1, 2, \ldots, p_l^{(i)}.$$

Step 3. Since

$$z_k^{(p_l^{(i)})}(t) = e^{tA}v_k^{(p_l^{(i)})}, k = 1, 2, \ldots, p_l^{(i)}; l = 1, \ldots, n_i$$

by Lemma 5.4, and since

$$e^{tA}v_k^{(p_l^{(i)})}, k = 1, \ldots, p_l^{(i)}; l = 1, \ldots, n_i,$$

are solutions to the homogeneous system (1.3), by Theorem 2.1, it follows that

$$Z_{u_l^{(i)}}^{(p_l^{(i)})}(t) = \{z_1^{(p_l^{(i)})}(t), z_2^{(p_l^{(i)})}(t), \ldots, z_{p_l^{(i)}}^{(p_l^{(i)})}(t)\}$$

is a set of $p_l^{(i)}$ solutions to the homogeneous system (1.3), that

$$Y_i(t) = \cup_{l=1}^{n_i} Z_{u_l^{(i)}}^{(p_l^{(i)})}(t)$$

is set of

$$m_i = p_1^{(i)} + p_2^{(i)} + \cdots + p_{n_i}^{(i)}$$

solutions to the homogeneous system (1.3), and finally that

$$Z(t) = Y_1(t) \cup Y_2(t) \cup \cdots \cup Y_s(t)$$

is a set of

$$n = m_1 + m_2 + \cdots + m_s$$

solutions to the homogeneous system (1.3).

Step 4. Since $Z(t)$ is the set of these n solutions

$$z_k^{(p_l^{(i)})}(t) = e^{tA}v_k^{(p_l^{(i)})}, k = 1, \ldots, p_l^{(i)}; l = 1, \ldots, n_i; i = 1, \ldots, s,$$

to the homogeneous system (1.3), and since

$$Z(0) = \{(e^{tA}|_{t=0})v_k^{(p_l^{(i)})}, k = 1, \ldots, p_l^{(i)}; l = 1, \ldots, n_i; i = 1, \ldots, s\}$$
$$= \{v_k^{(p_l^{(i)})}, k = 1, \ldots, p_l^{(i)}; l = 1, \ldots, n_i; i = 1, \ldots, s\}$$
$$= R_1 \cup \cdots \cup R_s$$
$$= S,$$

which is linearly independent by Step 1, it follows from Proposition 2.29 that $Z(t)$ is linearly independent, so that it is a fundamental set of solutions for the homogeneous system (1.3). $\qquad\square$

6. Problems

Use each of the methods in this chapter to find a fundamental set of solutions, and then the general solution, for the homogeneous system:

$$\frac{d}{dt}u(t) = Au(t), \quad 3 < t < 8,$$

where A is given in Problems **1** through **7**.

1.
$$\begin{pmatrix} 7 & 3 \\ 1 & 5 \end{pmatrix}.$$

2.
$$\begin{pmatrix} 5 & 1 \\ -1 & 3 \end{pmatrix}.$$

3.
$$\begin{pmatrix} -1 & 5 \\ -2 & 5 \end{pmatrix}.$$

4.
$$A = \begin{pmatrix} 1+i & -3 \\ 1 & (1+3i) \end{pmatrix}.$$

5.
$$A = \begin{pmatrix} 2 & 1 & 1 \\ 1 & 2 & 1 \\ -2 & -2 & -1 \end{pmatrix}.$$

6.
$$\begin{pmatrix} -3 & 9 & 0 \\ 0 & -3 & 4 \\ -1 & 0 & 4 \end{pmatrix}.$$

7.
$$A = \begin{pmatrix} 10 & -5 & 0 \\ 0 & 10 & 8 \\ 20 & 0 & 8 \end{pmatrix}.$$

8. Find a particular solution and then the general solution, by using the method of constants variation, for the nonhomogeneous system:

$$\frac{d}{dt}u(t) = Au(t) + f(t), \quad 3 < t < 8,$$

where $A = \begin{pmatrix} 7 & 3 \\ 1 & 5 \end{pmatrix}$, the A in Problem **1**. and $f(t) = \begin{pmatrix} e^t \\ e^{2t} \end{pmatrix}.$

9. Find a particular solution and then the general solution, by using the method of undetermined coefficients, for the nonhomogeneous system:

$$\frac{d}{dt}u(t) = Au(t) + f(t), \quad 3 < t < 8,$$

where $A = \begin{pmatrix} 7 & 3 \\ 1 & 5 \end{pmatrix}$, the A in Problem **1**, and $f(t) = e^{4t}\begin{pmatrix} 1 \\ 2 \end{pmatrix}$.

10. Find a particular solution and then the general solution, by using the method of constants variation, for the nonhomogeneous system:

$$\frac{d}{dt}u(t) = Au(t) + f(t), \quad 3 < t < 8,$$

where $A = \begin{pmatrix} -1 & 5 \\ -2 & 5 \end{pmatrix}$, the A in Problem **3**, and $f(t) = \begin{pmatrix} e^{2t}\tan(t) \\ e^{2t} \end{pmatrix}$.

11. Use the concept of an annihilator in Chapter 1 to find an annihilator for each of the vector-valued functions:

(a) $\begin{pmatrix} 6t^3 \\ 2t^2 \end{pmatrix}$.

(b) $\begin{pmatrix} t^5 e^{3t} \\ te^{3t} \end{pmatrix}$.

(c) $\begin{pmatrix} 3t^3 e^{(2+3i)t} \\ e^{(2+3i)t} \end{pmatrix}$.

(d) $\begin{pmatrix} 2\cos(3t) \\ t^4\sin(3t) \end{pmatrix}$.

(e) $\begin{pmatrix} 2t^3 e^{2t}\cos(3t) \\ (3t+5)e^{2t}\sin(3t) \end{pmatrix}$.

(f) $\begin{pmatrix} 2te^t\cos(2t) \\ e^t\sin(2t) \end{pmatrix}$.

(g) $\begin{pmatrix} (t^2+1)e^t\sin(2t) \\ te^t\cos(2t) \end{pmatrix}$.

12. Prove Fact 3.3.

13. Prove Lemma 2.12.

14. Prove Lemma 2.24.

15. Prove Corollary 2.26.

16. Prove Corollary 2.19.

17. Prove Corollary 2.20.

18. Prove Corollary 2.28.

19. Prove that if $\Phi(t)$ is a fundamental matrix for the homogeneous system (1.3), then

$$\Phi(t)\Phi^{-1}(t_1) = e^{(t-t_1)A}$$

for any $a < t_1 < b$.

20. Prove Lemma 5.2.

21. Prove the Step 1 in the proof of Lemma 5.4

22. Suppose that $z(t)$ is a nonzero solution to the homogeneous system (1.3). Use Theorem 2.3 or Theorem 5.8 to show that $z(t)$ tends to zero as $t \longrightarrow +\infty$, if all the characteristic values for the characteristic equation $p(\lambda) = 0$ have negative real parts. Conversely, show that $z(t)$ tends to positive infinity as $t \longrightarrow +\infty$, if all the real parts are positive. Finally, apply them to Problems **1** to **7**.

7. Solutions

1. Method 1. From

$$(D - 7)u_1 = 3u_2,$$
$$(D - 5)u_2 = u_1,$$

where $D = \frac{d}{dt}$ and $u = \begin{pmatrix} u_1 \\ u_2 \end{pmatrix}$, we have

$$3u_1 = (D - 5)(D - 7)u_1$$
$$= u_1'' - 12u_1' + 35u_1.$$

Hence

$$u_1 = c_1 e^{4t} + c_2 e^{8t},$$

and then

$$u_2 = \frac{1}{3}(u_1' - 7u_1)$$
$$= -c_1 e^{4t} + \frac{1}{3}c_2 e^{8t}.$$

Since

$$u = \begin{pmatrix} u_1 \\ u_2 \end{pmatrix} = c_1 e^{4t} \begin{pmatrix} 1 \\ -1 \end{pmatrix} + c_2 e^{8t} \begin{pmatrix} 1 \\ \frac{1}{3} \end{pmatrix},$$

a fundamental set of solutions is

$$\left\{ e^{4t} \begin{pmatrix} 1 \\ -1 \end{pmatrix}, e^{8t} \begin{pmatrix} 1 \\ \frac{1}{3} \end{pmatrix} \right\}.$$

Method 2. Since A has the eigenpairs

$$\{\lambda_1, v_1\} = \{4, \begin{pmatrix} 1 \\ -1 \end{pmatrix}\} \quad \text{and} \quad \{\lambda_2, v_2\} = \{8, \begin{pmatrix} 3 \\ 1 \end{pmatrix}\},$$

we have the two linearly independent solutions

$$z_1(t) = e^{tA}v_1 = e^{t\lambda_1} e^{t(A-\lambda_1)} v_1 = e^{4t} v_1,$$
$$z_2 = e^{t\lambda_2} e^{t(A-\lambda_2)} v_2 = e^{8t} v_2,$$

and then the general solution is

$$y = c_1 z_1 + c_2 z_2.$$

Method 3. Since

$$e^{tA} = e^{4t} M_1 + e^{8t} M_2,$$
$$Ae^{tA} = 4e^{4t} M_1 + 8e^{8t} M_2,$$

we have

$$I = M_1 + M_2,$$
$$A = 4M_1 + 8M_2,$$

and so

$$M_1 = \begin{pmatrix} \frac{3}{4} & \frac{3}{4} \\ \frac{1}{4} & \frac{1}{4} \end{pmatrix},$$

$$M_2 = \begin{pmatrix} \frac{1}{4} & -\frac{3}{4} \\ -\frac{1}{4} & \frac{3}{4} \end{pmatrix}.$$

Thus the fundamental matrix

$$
e^{tA} = \begin{pmatrix} \frac{1}{4}e^{4t} + \frac{3}{4}e^{8t} & -\frac{3}{4}e^{4t} + \frac{3}{4}e^{8t} \\ -\frac{1}{4}e^{4t} + \frac{1}{4}e^{8t} & \frac{3}{4}e^{4t} + \frac{1}{4}e^{8t} \end{pmatrix},
$$

and the general solution is $y = e^{tA}c$, where c is a constant vector.

2. Method 1. Since

$$
(D - 5)u_1 = u_2,
$$
$$
(D - 3)u_2 = -u_1,
$$

where $D = \frac{d}{dt}$ and $u = \begin{pmatrix} u_1 \\ u_2 \end{pmatrix}$, we have

$$
-u_1 = (D - 3)(D - 5)u_1
$$
$$
= u_1'' - 8u_1' + 15u_1.
$$

Hence

$$
u_1 = c_1 e^{4t} + c_2 t e^{4t},
$$

and then

$$
u_2 = u_1' - 5u_1
$$
$$
= -c_1 e^{4t} + c_2 e^{4t}(1 - t).
$$

Thus the general solution is

$$
u = \begin{pmatrix} u_1 \\ u_2 \end{pmatrix} = c_1 e^{4t} \begin{pmatrix} 1 \\ -1 \end{pmatrix} + c_2 e^{4t} \begin{pmatrix} t \\ 1 - t \end{pmatrix},
$$

and

$$
\{ e^{4t} \begin{pmatrix} 1 \\ -1 \end{pmatrix}, e^{4t} \begin{pmatrix} t \\ 1 - t \end{pmatrix} \}
$$

is a fundamental set of solution.

Method 2. A has the eigenpair

$$
\{\lambda_1, v_1\} = \{4, \begin{pmatrix} 1 \\ -1 \end{pmatrix}\}
$$

and the generalized eigenpair

$$
\{\lambda_1, v_2\} = \{4, \begin{pmatrix} 0 \\ 1 \end{pmatrix}\},
$$

where v_2 satisfies $(A - \lambda_1)^2 v_2 = 0$. Thus the two linearly independent solutions are

$$
z_1 = e^{tA}v_1 = e^{t\lambda_1}e^{t(A - \lambda_1)}v_1
$$
$$
= e^{4t} \begin{pmatrix} 1 \\ -1 \end{pmatrix},
$$
$$
z_2 = e^{t\lambda_1}e^{t(A - \lambda_1)}v_2
$$
$$
= e^{4t}[I + t(A - \lambda_1)]v_2
$$
$$
= e^{4t}[\begin{pmatrix} 0 \\ 1 \end{pmatrix} + t \begin{pmatrix} 1 \\ -1 \end{pmatrix}],
$$

and the general solution is

$$
y = c_1 z_1 + c_2 z_2.
$$

Method 3. Since

$$e^{tA} = e^{4t} M_1 + te^{4t} M_2,$$
$$Ae^{tA} = 4e^{4t} M_1 + e^{4t} M_2 + 4te^{4t} M_2,$$

we have

$$I = M_1 + 0,$$
$$A = 4M_1 + M_2,$$

and so

$$M_1 = I \quad \text{and} \quad M_2 = \begin{pmatrix} 1 & 1 \\ -1 & -1 \end{pmatrix}.$$

Thus

$$e^{tA} = \begin{pmatrix} e^{4t} + te^{4t} & te^{4t} \\ -te^{4t} & e^{4t} - te^{4t} \end{pmatrix},$$

and $e^{tA}c$ is the general solution, where c is a constant vector.

3. Method 1. From

$$(D+1)u_1 = 5u_2,$$
$$(D-5)u_2 = -2u_1,$$

where $D = \frac{d}{dt}$ and $u = \begin{pmatrix} u_1 \\ u_2 \end{pmatrix}$, we have

$$-10u_1 = (D+1)(D-5)u_1$$
$$= u_1'' - 4u_1' - 5u_1.$$

Hence

$$u_1 = c_1 e^{2t} \cos(t) + c_2 e^{2t} \sin(t),$$

and then

$$u_2 = \frac{1}{5}(u_1' + u_1)$$
$$= \frac{1}{5}c_1[3e^{2t}\cos(t) - e^{2t}\sin(t)] + \frac{1}{5}c_2[3e^{2t}\sin(t) + e^{2t}\cos(t)].$$

Thus

$$u = \begin{pmatrix} u_1 \\ u_2 \end{pmatrix}$$
$$= c_1 \begin{pmatrix} e^{2t}\cos(t) \\ \frac{3}{5}e^{2t}\cos(t) - \frac{1}{5}e^{2t}\sin(t) \end{pmatrix} + c_2 \begin{pmatrix} e^{2t}\sin(t) \\ \frac{3}{5}e^{2t}\sin(t) + e^{2t}\cos(t) \end{pmatrix}$$
$$\equiv c_1 z_1(t) + c_2 z_2(t),$$

and $\{z_1, z_2\}$ is a fundamental set of solutions.

Method 2. A has the conjugate eigenpairs

$$\{\lambda_1, v_1\} = \{2+i, \begin{pmatrix} 3-i \\ 2 \end{pmatrix}\},$$

$$\{\lambda_2, v_2\} = \{2-i, \begin{pmatrix} 3+i \\ 2 \end{pmatrix}\}.$$

The calculation

$$e^{tA}v_1 = e^{t\lambda_1}e^{t(A-\lambda_1)}v_1$$

$$= e^{(2+i)t}v_1 = e^{2t}[\cos(t) + i\sin(t)]\begin{pmatrix} 3-i \\ 2 \end{pmatrix}$$

$$= e^{2t}\begin{pmatrix} 3\cos(t) + \sin(t) \\ 2\cos(t) \end{pmatrix} + ie^{2t}\begin{pmatrix} 3\sin(t) - \cos(t) \\ 2\sin(t) \end{pmatrix}$$

$$\equiv z_1(t) + iz_2(t)$$

gives that z_1 and z_2 are two linearly independent real-valued solutions, and that $u = c_1z_1 + c_2z_2$ is the general solution.

Method 3. Since

$$e^{tA} = e^{2t}\cos(t)M_1 + e^{2t}\sin(t)M_2,$$

$$Ae^{tA} = [2e^{2t}\cos(t) - e^{2t}\sin(t)]M_1 + [2e^{2t}\sin(t) + e^{2t}\cos(t)]M_2,$$

we have

$$I = M_1 + 0,$$
$$A = 2M_1 + M_2,$$

and so

$$M_1 = I \quad \text{and} \quad M_2 = \begin{pmatrix} -3 & 5 \\ -2 & 3 \end{pmatrix}.$$

Thus

$$e^{tA} = \begin{pmatrix} e^{2t}\cos(t) - 3e^{2t}\sin(t) & 5e^2\sin(t) \\ -2e^{2t}\sin(t) & e^{2t}\cos(t) + 3e^{2t}\sin(t) \end{pmatrix},$$

and $e^{tA}c$ is the general solution, where c is a constant vector.

4. Method 1. Since

$$[(D - (1+i)]u_1 + 3u_2 = 0,$$
$$-u_1 + [D - (1+3i)]u_2 = 0,$$

where $D = \frac{d}{dt}$ and $u = \begin{pmatrix} u_1 \\ u_2 \end{pmatrix}$, we have

$$0 = [D - (1+i)][D - (1+3i)]u_2 + 3u_2$$
$$= u_2'' - (2+4i)u_2' + (1+4i)u_2.$$

Hence

$$u_2 = c_1e^t + c_2e^{(1+4i)t},$$
$$u_1 = u_2' - (1+3i)u_2$$
$$= c_1(-3i)e^t + c_2(i)e^{(1+4i)t}.$$

The general solution is

$$u = \begin{pmatrix} u_1 \\ u_2 \end{pmatrix} = c_1e^t\begin{pmatrix} -3i \\ 1 \end{pmatrix} + c_2e^{(1+4i)t}\begin{pmatrix} i \\ 1 \end{pmatrix}$$

$$\equiv c_1z_1(t) + c_2z_2(t),$$

and $\{z_1, z_2\}$ is a fundamental set of solutions.

Method 2. Since A has the eigenpairs

$$\{\lambda_1, v_1\} = \{1, \begin{pmatrix} -3i \\ 1 \end{pmatrix}\},$$

$$\{\lambda_2, v_2\} = \{1 + 4i, \begin{pmatrix} i \\ 1 \end{pmatrix}\},$$

we have the two linearly independent solutions

$$z_1(t) = e^{tA} v_1 = e^{t\lambda_1} e^{t(A-\lambda_1)} v_1$$

$$= e^t \begin{pmatrix} -3i \\ 1 \end{pmatrix} = e^t \begin{pmatrix} 0 \\ 1 \end{pmatrix} + i e^t \begin{pmatrix} -3 \\ 0 \end{pmatrix},$$

$$z_2(t) = e^{t\lambda_2} e^{t(A-\lambda_2)} v_2$$

$$= e^{(1+4i)t} \begin{pmatrix} i \\ 1 \end{pmatrix} = e^t \begin{pmatrix} -\sin(4t) \\ \cos(4t) \end{pmatrix} + i e^t \begin{pmatrix} \cos(4t) \\ \sin(4t) \end{pmatrix},$$

and $\{z_1, z_2\}$ is a fundamental set of solutions.

Method 3. Since

$$e^{tA} = e^t M_1 + e^{(1+4i)t} M_2,$$

$$A e^{tA} = e^t M_1 + (1 + 4i) e^{(1+4i)t} M_2,$$

we have

$$I = M_1 + M_2,$$

$$A = M_1 + (1 + 4i) M_2,$$

and so

$$M_1 = \begin{pmatrix} \frac{3}{4} & -\frac{3i}{4} \\ \frac{i}{4} & \frac{1}{4} \end{pmatrix} = \left(\tfrac{i}{4} v_1 \quad \tfrac{1}{4} v_1 \right),$$

$$M_2 = \begin{pmatrix} \frac{1}{4} & \frac{3i}{4} \\ \frac{-i}{4} & \frac{3}{4} \end{pmatrix} = \left(-\tfrac{i}{4} v_2 \quad \tfrac{3}{4} v_2 \right).$$

Here v_1 and v_2 are the two eigenvectors in Method 2 above. Thus the fundamental matrix

$$e^{tA} = \left(e^t (\tfrac{i}{4}) v_1 + e^{(1+4i)t} (\tfrac{-i}{4}) v_2 \quad e^t (\tfrac{1}{4}) v_1 + e^{(1+4i)t} (\tfrac{3}{4}) v_2 \right),$$

and $e^{tA} c$ is the general solution, where c is a constant vector.

5. Method 1. Since

$$(D - 2) u_1 - u_2 - u_3 = 0,$$

$$- u_1 + (D - 2) u_2 - u_3 = 0,$$

$$2 u_1 + 2 u_2 + (D + 1) u_3 = 0,$$

where $D = \frac{d}{dt}$ and $u = \begin{pmatrix} u_1 \\ u_2 \\ u_3 \end{pmatrix}$, we have

$$(D - 1) u_1 - (D - 1) u_2 = 0 = (D - 1)(u_1 - u_2),$$

$$- (D - 1) u_1 + [(D + 1)(D - 2) + 2] u_2 = 0,$$

and then

$$u_1 - u_2 = c_1 e^t,$$
$$0 = [(D+1)(D-2) + 2]u_2 - (D-1)u_2$$
$$= u_2'' - 2u_2' + u_2.$$

Hence

$$u_2 = c_2 e^t + c_3 t e^t,$$

and then

$$u_1 = u_2 + c_1 e^t$$
$$= c_1 e^t + c_2 e^t + c_3 t e^t,$$
$$u_3 = -u_1 + u_2' - 2u_2$$
$$= -c_1 e^t - 2c_2 e^t + c_3 e^t (1 - 2t).$$

Thus the general solution is

$$u = \begin{pmatrix} u_1 \\ u_2 \\ u_3 \end{pmatrix}$$

$$= c_1 e^t \begin{pmatrix} 1 \\ 0 \\ -1 \end{pmatrix} + c_2 e^t \begin{pmatrix} 1 \\ 1 \\ -2 \end{pmatrix} + c_3 e^t [\begin{pmatrix} 0 \\ 0 \\ 1 \end{pmatrix} + t \begin{pmatrix} 1 \\ 1 \\ -2 \end{pmatrix}]$$

$$\equiv c_1 z_1(t) + c_2 z_2(t) + c_3 z_3(t),$$

and $\{z_1, z_2, z_3\}$ is a fundamental set of solutions.

Method 2. A has the eigenpairs

$$\{\lambda_1, v_1\} = \{1, \begin{pmatrix} 1 \\ 0 \\ -1 \end{pmatrix}\},$$

$$\{\lambda_2, v_2\} = \{1, \begin{pmatrix} 0 \\ 1 \\ -1 \end{pmatrix}\},$$

and the generalized eigenpair

$$\{\lambda_3, v_3\} = \{1, \begin{pmatrix} 0 \\ 0 \\ 1 \end{pmatrix}\},$$

where v_3 satisfies $(A-1)^2 v_3 = 0$. Hence we have the three linearly independent solutions

$$z_1(t) = e^{tA} v_1 + e^{t\lambda_1} e^{t(A-\lambda_1)} v_1$$
$$= e^t v_1,$$
$$z_2(t) = e^{t\lambda_2} e^{t(A-\lambda_2)} v_2$$
$$= e^t v_2,$$
$$z_3(t) = e^{t\lambda_3} e^{t(A-\lambda_3)} v_3$$
$$= e^t [v_3 + t(A-1)v_3]$$

$$= e^t[v_3 + t \begin{pmatrix} 1 \\ 1 \\ -2 \end{pmatrix}],$$

and the general solution is

$$u = c_1 z_1 + c_2 z_2 + c_3 z_3.$$

Method 3. Since

$$e^{tA} = e^t M_1 + t e^t M_2 + t^2 e^t M_3,$$

we have

$$M_1 + t M_2 + t^2 M_3 = e^{t(A-1)}$$
$$= I + t(A - 1),$$

and so

$$M_1 = I,$$

$$M_2 = A - 1 = \begin{pmatrix} 1 & 1 & 1 \\ 1 & 1 & 1 \\ -2 & -2 & -2 \end{pmatrix},$$

$$M_3 = 0.$$

Thus the fundamental matrix

$$e^{tA} = \begin{pmatrix} e^t + t e^t & t e^t & t e^t \\ t e^t & e^t + t e^t & t e^t \\ -2t e^t & -2t e^t & e^t - 2t e^t \end{pmatrix},$$

and $e^{tA}c$ is the general solution, where c is a constant vector.

6. Method 1. From

$$(D + 3)u_1 - 9u_2 = 0,$$
$$(D + 3)u_2 - 4u_3 = 0,$$
$$(D - 4)u_3 + u_1 = 0,$$

where $D = \frac{d}{dt}$ and $u = \begin{pmatrix} u_1 \\ u_2 \\ u_3 \end{pmatrix}$, we have

$$0 = (D - 4)(D + 3)^2 u_1 + 36 u_1$$
$$= u_1''' + 2u_1'' - 15u_1'.$$

Hence

$$u_1 = c_1 + c_2 e^{3t} + c_3 e^{-5t},$$

and then

$$u_2 = \frac{1}{9}(u_1' + 3u_1)$$
$$= \frac{1}{3}c_1 + \frac{2}{3}c_2 e^{3t} - \frac{2}{9}c_3 e^{-5t},$$

$$u_3 = \frac{1}{4}(u_2' + 3u_2)$$

$$= \frac{1}{4}c_1 + c_2 e^{3t} + \frac{1}{9}c_3 e^{-5t}.$$

Thus the general solution

$$u = \begin{pmatrix} u_1 \\ u_2 \\ u_3 \end{pmatrix}$$

$$= c_1 \begin{pmatrix} 1 \\ \frac{1}{3} \\ \frac{1}{4} \end{pmatrix} + c_2 e^{3t} \begin{pmatrix} 1 \\ \frac{2}{3} \\ 1 \end{pmatrix} + c_3 e^{-5t} \begin{pmatrix} 1 \\ -\frac{2}{9} \\ \frac{1}{9} \end{pmatrix}$$

$$= c_1 z_1(t) + c_2 z_2(t) + c_3 z_3(t),$$

and $\{z_1, z_2, z_3\}$ is a fundamental set of solutions.

Method 2. Since A has the eigenpairs

$$\{\lambda_1, v_1\} = \{0, \begin{pmatrix} 1 \\ \frac{1}{3} \\ \frac{1}{4} \end{pmatrix}\},$$

$$\{\lambda_2, v_2\} = \{3, \begin{pmatrix} 1 \\ \frac{2}{3} \\ 1 \end{pmatrix}\},$$

$$\{\lambda_3, v_3\} = \{-5, \begin{pmatrix} 1 \\ -\frac{2}{9} \\ \frac{1}{9} \end{pmatrix}\},$$

we have the three linearly independent solutions

$$z_1(t) = e^{tA}v_1 = e^{t\lambda_1}e^{t(A-\lambda_1)}v_1 = v_1,$$
$$z_2(t) = e^{t\lambda_2}e^{t(A-\lambda_2)}v_2 = e^{3t}v_2,$$
$$z_3(t) = e^{t\lambda_3}e^{t(A-\lambda_3)}v_3 = e^{-5t}v_3,$$

and then the general solution is

$$u = c_1 z_1 + c_2 z_2 + c_3 z_3.$$

Method 3. Since

$$e^{tA} = M_1 + e^{3t}M_2 + e^{-5t}M_3,$$
$$Ae^{tA} = 3e^{3t}M_2 - 5e^{5t}M_3,$$
$$A^2 e^{tA} = 9e^{3t}M_2 + 25e^{-5t}M_3,$$

we have

$$M_1 = \frac{1}{15} \begin{pmatrix} 12 & 36 & -36 \\ 4 & 12 & -12 \\ 3 & 9 & -9 \end{pmatrix} = \left(\tfrac{4}{5}v_1 \quad \tfrac{12}{5}v_1 \quad -\tfrac{12}{5}v_1 \right),$$

$$M_2 = \frac{1}{24} \begin{pmatrix} -6 & -9 & 36 \\ -4 & -6 & 24 \\ -6 & -9 & 36 \end{pmatrix} = \left(-\tfrac{1}{4}v_2 \quad -\tfrac{3}{8}v_2 \quad \tfrac{3}{2}v_2 \right),$$

$$M_3 = \frac{1}{40}\begin{pmatrix} 18 & -81 & 36 \\ -4 & 18 & -8 \\ 2 & -9 & 4 \end{pmatrix} = \left(\frac{9}{20}v_3 \quad -\frac{81}{40}v_3 \quad \frac{9}{10}v_3 \right).$$

Here v_1, v_2, and v_3 are the three eigenvectors from Method 2 above. Thus the fundamental matrix

$$e^{tA} = \left(\frac{4}{5}v_1 - \frac{1}{4}e^{3t}v_2 + \frac{9}{20}e^{-5t}v_3 \quad \frac{12}{5}v_1 - \frac{3}{8}e^{3t}v_2 - \frac{81}{40}e^{-5t}v_3 \quad -\frac{12}{5}v_1 + \frac{3}{2}e^{3t}v_2 + \frac{9}{10}e^{-5t}v_3 \right),$$

and the general solutions is $e^{tA}c$ where c is a constant vector.

7. Method 1. Since

$$(D - 10)u_1 + 5u_2 = 0,$$
$$(D - 10)u_2 - 8u_3 = 0,$$
$$- 20u_1 + (D - 8)u_3 = 0,$$

where $D = \frac{d}{dt}$ and $u = \begin{pmatrix} u_1 \\ u_2 \\ u_3 \end{pmatrix}$, we have

$$(D - 10)^2 u_1 + 40u_3 = 0,$$
$$- 20u_1 + (D - 8)u_3 = 0,$$

and then

$$0 = (D - 8)(D - 10)^2 u_1 + 800u_1$$
$$= u_1''' - 28u_1'' + 260u_1',$$

Hence

$$u_1 = c_1 + c_2 e^{14y}\cos(8t) + c_3 e^{14t}\sin(8t),$$

and then

$$u_2 = -\frac{1}{5}(u_1' - 10u_1)$$
$$= 2c_1 + c_2[-\frac{4}{5}e^{14t}\cos(8t) + \frac{8}{5}e^{14t}\sin(8t)] + c_3[-\frac{4}{5}e^{14t}\sin(8t) - \frac{8}{5}e^{14t}\cos(8t)],$$

$$u_3 = \frac{1}{8}(u_2' - 10u_2)$$
$$= -\frac{5}{2}c_1 + c_2[\frac{6}{5}e^{14t}\cos(8t) + \frac{8}{5}e^{14t}\sin(8t)] + c_3[-\frac{8}{5}e^{14t}\cos(8t) + \frac{6}{5}e^{14t}\sin(8t)].$$

Thus the general solution is

$$u = \begin{pmatrix} u_1 \\ u_2 \\ u_3 \end{pmatrix} = c_1\begin{pmatrix} 1 \\ 2 \\ -\frac{5}{2} \end{pmatrix} + c_2\begin{pmatrix} e^{14t}\cos(8t) \\ -\frac{4}{5}e^{14t}\cos(8t) + \frac{8}{5}e^{14t}\sin(8t) \\ \frac{6}{5}e^{14t}\cos(8t) + \frac{8}{5}e^{14t}\sin(8t) \end{pmatrix}$$

$$+ c_3\begin{pmatrix} e^{14t}\sin(8t) \\ -\frac{8}{5}e^{14t}\cos(8t) - \frac{4}{5}e^{14t}\sin(8t) \\ -\frac{8}{5}e^{14t}\cos(8t) + \frac{6}{5}e^{14t}\sin(8t) \end{pmatrix}$$

$$= c_1\begin{pmatrix} 1 \\ 2 \\ -\frac{5}{2} \end{pmatrix} + c_2[e^{14t}\cos(8t)\begin{pmatrix} 1 \\ -\frac{4}{5} \\ \frac{6}{5} \end{pmatrix} - e^{14t}\sin(8t)\begin{pmatrix} 0 \\ -\frac{8}{5} \\ -\frac{8}{5} \end{pmatrix}]$$

$$+ c_3[e^{14t}\cos(8t)\begin{pmatrix} 0 \\ -\frac{8}{5} \\ -\frac{8}{5} \end{pmatrix} + e^{14t}\sin(8t)\begin{pmatrix} 1 \\ -\frac{4}{5} \\ \frac{6}{5} \end{pmatrix}]$$

$$\equiv c_1 z_1(t) + c_2 z_2(t) + c_3 z_3(t),$$

and $\{z_1, z_2, z_3\}$ is a fundamental set of solutions.

Method 2. A has the eigenpair

$$\{\lambda_1, v_1\} = \{0, \begin{pmatrix} 1 \\ 2 \\ -\frac{5}{2} \end{pmatrix}\},$$

and the conjugate eigenpairs

$$\{\lambda_2, v_2\} = \{14 + 8i, \begin{pmatrix} 1 \\ -\frac{4}{5} \\ \frac{6}{5} \end{pmatrix} + i\begin{pmatrix} 0 \\ -\frac{8}{5} \\ -\frac{8}{5} \end{pmatrix}\},$$

$$\{\lambda_3, v_3\} = \{\overline{\lambda_2}, \overline{v_2}\}.$$

The calculation

$$e^{tA}v_2 = e^{t\lambda_2}e^{t(A-\lambda_2)}v_2$$

$$= e^{14t+i8t}v_2 = e^{14t}[\cos(8t) + i\sin(8t)][\begin{pmatrix} 1 \\ -\frac{4}{5} \\ \frac{6}{5} \end{pmatrix} + i\begin{pmatrix} 0 \\ -\frac{8}{5} \\ -\frac{8}{5} \end{pmatrix}]$$

$$= e^{14t}[\begin{pmatrix} 1 \\ -\frac{4}{5} \\ \frac{6}{5} \end{pmatrix}\cos(8t) - \begin{pmatrix} 0 \\ -\frac{8}{5} \\ -\frac{8}{5} \end{pmatrix}\sin(8t)] + ie^{14t}[\begin{pmatrix} 0 \\ -\frac{8}{5} \\ -\frac{8}{5} \end{pmatrix}\cos(8t) + \begin{pmatrix} 1 \\ -\frac{4}{5} \\ \frac{6}{5} \end{pmatrix}\sin(8t)]$$

$$\equiv z_2(t) + iz_3(t)$$

gives that z_2, and z_3, plus $z_1 = e^{t\lambda_1}v_1$, are three linearly independent real-valued solutions, and that $c_1 z_1 + c_2 z_2 + c_3 z_3$ is the general solution.

Method 3. From

$$e^{tA} = M_1 + e^{14t}\cos(8t)M_2 + e^{14t}\sin(8t)M_3,$$

$$Ae^{tA} = [14e^{14t}\cos(8t) - 8e^{14t}\sin(8t)]M_2 + [14e^{14}\sin(8t) + 8e^{14t}\cos(8t)]M_3,$$

$$A^2 e^{tA} = [132e^{14t}\cos(8t) - 224e^{1t}\sin(8t)]M_2$$
$$+ [132e^{14t}\sin(8t) + 224e^{14t}\cos(8t)]M_3,$$

we have

$$I = M_1 + M_2,$$
$$A = 14M_2 + 8M_3,$$
$$A^2 = 132M_2 + 224M_3,$$

and so

$$M_1 = \frac{1}{260}\begin{pmatrix} 80 & 40 & -40 \\ 160 & 80 & -80 \\ -200 & -100 & 100 \end{pmatrix}$$

$$= \left(\frac{4}{13} \begin{pmatrix} 1 \\ 2 \\ -\frac{5}{2} \end{pmatrix} \quad \frac{2}{13} \begin{pmatrix} 1 \\ 2 \\ -\frac{5}{2} \end{pmatrix} \quad -\frac{2}{13} \begin{pmatrix} 1 \\ 2 \\ -\frac{5}{2} \end{pmatrix} \right)$$

$$\equiv \begin{pmatrix} p_1 & p_2 & p_3 \end{pmatrix},$$

$$M_2 = \frac{1}{260} \begin{pmatrix} 180 & -40 & 40 \\ -160 & 180 & 80 \\ 200 & 100 & 160 \end{pmatrix}$$

$$= \left(\frac{1}{13} \begin{pmatrix} 9 \\ -8 \\ 10 \end{pmatrix} \quad \frac{1}{13} \begin{pmatrix} -2 \\ 9 \\ 5 \end{pmatrix} \quad \frac{2}{13} \begin{pmatrix} 1 \\ 2 \\ 4 \end{pmatrix} \right)$$

$$\equiv \begin{pmatrix} q_1 & q_2 & q_3 \end{pmatrix},$$

$$M_3 = \frac{1}{1040} \begin{pmatrix} 40 & -370 & -280 \\ 1120 & 40 & 480 \\ 1200 & -700 & -80 \end{pmatrix}$$

$$= \left(\frac{1}{26} \begin{pmatrix} 1 \\ 28 \\ 30 \end{pmatrix} \quad \frac{1}{104} \begin{pmatrix} -37 \\ 4 \\ -70 \end{pmatrix} \quad \frac{1}{26} \begin{pmatrix} -7 \\ 12 \\ -2 \end{pmatrix} \right)$$

$$\equiv \begin{pmatrix} r_1 & r_2 & r_3 \end{pmatrix}.$$

Thus the fundamental matrix

$$e^{tA} = \begin{pmatrix} s_1(t) & s_2(t) & s_3(t) \end{pmatrix},$$

where

$$s_1(t) = p_1 + e^{14t} \cos(8t) q_1 + e^{14t} \sin(8t) r_1,$$
$$s_2(t) = p_2 + e^{14t} \cos(8t) q_2 + e^{14t} \sin(8t) r_2,$$
$$s_3(t) = p_3 + e^{14t} \cos(8t) q_3 + e^{14t} \sin(8t) r_3.$$

$e^{tA} c$ is the general solution, where c is a constant vector.

8. From the solution to Problem **1**, we have the fundamental matrix

$$\Phi(t) = \begin{pmatrix} e^{4t} & e^{8t} \\ -e^{4t} & \frac{1}{3} e^{8t} \end{pmatrix},$$

and so its inverse is

$$\Phi(t)^{-1} = \frac{3}{4} \begin{pmatrix} \frac{1}{3} e^{-4t} & -e^{-4t} \\ e^{-8t} & e^{-8t} \end{pmatrix}.$$

A particular solution by the constants variation formula is

$$u_p = \Phi(t) \int^t \Phi(\tau)^{-1} f(\tau) \, d\tau$$

$$= \Phi(t) \frac{3}{4} \begin{pmatrix} -\frac{1}{9} e^{-3t} + \frac{1}{2} e^{-2t} \\ -\frac{1}{7} e^{-7t} - \frac{1}{6} e^{-6t} \end{pmatrix}$$

$$= \begin{pmatrix} -\frac{4}{21} e^t + \frac{1}{4} e^{2t} \\ \frac{1}{21} e^t - \frac{5}{12} e^{2t} \end{pmatrix},$$

and then the general solution is

$$u = \Phi(t) c + u_p.$$

Here c is a constant vector.

9. From the solution to Problem **1**, we have the fundamental set of solutions:

$$\{z_1(t), z_2(t)\} = \{e^{4t}\begin{pmatrix} 1 \\ -1 \end{pmatrix}, e^{8t}\begin{pmatrix} 3 \\ 1 \end{pmatrix}\}.$$

Hence a particular solution u_p, by the method of undetermined coefficients, takes the form:

$$u_p = e^{4t}a + te^{4t}b,$$

where a and b are two constant vectors to be determined. Since u_p satisfies

$$\frac{d}{dt}u = Au + f,$$

we have

$$4a + b = Aa + \begin{pmatrix} 1 \\ 2 \end{pmatrix},$$

$$4b = Ab.$$

Being solved, $b = \alpha\begin{pmatrix} 1 \\ -1 \end{pmatrix}$, eigenvectors of A, corresponding to the eigenvalue 4.

Here α are constants. It follows that $a = \begin{pmatrix} a_1 \\ a_2 \end{pmatrix}$ satisfies

$$(A - 4)a = \begin{pmatrix} 3(a_1 + a_2) \\ (a_1 + a_2) \end{pmatrix}$$

$$= b - \begin{pmatrix} 1 \\ 2 \end{pmatrix} = \begin{pmatrix} \alpha - 1 \\ -\alpha - 2 \end{pmatrix}.$$

Hence

$$\alpha - 1 = 3(-\alpha - 2) \quad \text{or} \quad \alpha = -\frac{5}{4},$$

and then

$$b = \begin{pmatrix} -\frac{5}{4} \\ \frac{5}{4} \end{pmatrix},$$

$$a = \begin{pmatrix} a_1 \\ -\frac{3}{4} - a_1 \end{pmatrix}$$

$$= a_1\begin{pmatrix} 1 \\ -1 \end{pmatrix} + \begin{pmatrix} 0 \\ -\frac{3}{4} \end{pmatrix}.$$

We can choose any a_1 or choose $a_1 = 0$, the simplest case, for which

$$a = \begin{pmatrix} 0 \\ -\frac{3}{4} \end{pmatrix}.$$

Therefore, a particular solution is

$$u_p = e^{4t}\begin{pmatrix} 0 \\ -\frac{3}{4} \end{pmatrix} + te^{4t}\begin{pmatrix} -\frac{5}{4} \\ \frac{5}{4} \end{pmatrix},$$

and the general solution is

$$u = c_1 z_1(t) + c_2 z_2(t) + u_p.$$

Here c_1 and c_2 are two constants.

10. From the solution to Problem **3**, we have the fundamental matrix

$$\Phi(t) \equiv e^{tA} = e^{2t} \begin{pmatrix} \cos(t) - 3\sin(t) & 5\sin(t) \\ -2\sin(t) & \cos(t) + 3\sin(t) \end{pmatrix},$$

and so its inverse is

$$\Phi(t)^{-1} = e^{-2t} \begin{pmatrix} \cos(t) + 3\sin(t) & -5\sin(t) \\ 2\sin(t) & \cos(t) - 3\sin(t) \end{pmatrix}.$$

A particular solution, by the constants variation formula, is

$$u_p = \Phi(t) \int_t \Phi(\tau)^{-1} f(\tau)\, d\tau$$

$$= \Phi(t) \begin{pmatrix} (4\cos(t) - 3\sin(t)) + 3\ln|\sec(t) + \tan(t)| \\ (3\cos(t) - \sin(t)) + 2\ln|\sec(t) + \tan(t)| \end{pmatrix}$$

$$= e^{2t} \begin{pmatrix} 4 + (3\cos(t) + \sin(t))\ln|\sec(t) + \tan(t)| \\ 3 + (2\cos(t))\ln|\sec(t) + \tan(t)| \end{pmatrix},$$

and the general solution is

$$u = e^{tA} c + u_p.$$

Here c is a constant vector.

11.

(a) D^4.

(b) $(D-3)^6$.

(c) $[D - (2+3i)]^4$.

(d) $(D^2 + 9)^5$.

(e) $[(D-2)^2 + 9]^4$.

(f) $[(D-1)^2 + 4]^2$.

(g) $[(D-1)^2 + 4]^3$.

12. Since $\Phi(t)$ is a fundamental matrix, we have $|\Phi(t)| \neq 0$ and

$$\frac{d}{dt}\Phi(t) = A(t)\Phi(t).$$

Hence

$$\frac{d}{dt}(\Phi(t)C) = \frac{d}{dt}\Phi(t)C = A(t)\Phi(t)C$$

$$= A(t)(\Phi(t)C).$$

Since $|\Phi(t)C| = |\Phi(t)||C| \neq 0$, $\Phi(t)C$ is a fundamental matrix.

13. Since λ satisfies $Au = \lambda u$ for some nonzero vector u and A is real, we have, by taking complex conjugate on both sides,

$$\overline{\lambda}\overline{u} = A\overline{u}.$$

Thus $\overline{\lambda}$ is also a root of the equation

$$p(\mu) = |A - \mu|,$$

and the corresponding eigenpair is $\{\overline{\lambda}, \overline{u}\}$.

14. Since

$$\frac{d}{dt}z_1 = A(t)z_1 + f_1(t),$$

$$\frac{d}{dt}z_2 = A(t)z_2 + f_2(t),$$

we complete the proof by multiplying the top equation by α_1, multiplying the bottom equation by α_2, and finally, adding them together.

15. Since

$$e^{t\alpha_1}\sin(\beta_1 t) = e^{\alpha_1}\frac{e^{i\beta_1 t} - e^{-i\beta_1 t}}{2i} = \frac{1}{2i}[e^{t(\alpha_1+i\beta_1)} - e^{t(\alpha_1-i\beta_1)}], \quad \text{and}$$

$$e^{t\alpha_1}\cos(\beta_1 t) = e^{t\alpha_1}\frac{e^{i\beta_1 t} + e^{-i\beta_1 t}}{2} = \frac{1}{2}[e^{t(\alpha_1+i\beta_1)} + e^{t(\alpha_1-i\beta_1)}],$$

the proof is complete by applying Remark 2.25. Here $i \equiv \sqrt{-1}$, the complex unit, and $i^2 = -1$.

16. Since the solution u takes the form

$$u(t) = \Phi(t)c + \int_{\tau=t_0}^{t}\Phi(t)\Phi(\tau)^{-1}f(\tau)\,d\tau,$$

we have, by using the initial condition $u(t_0) = u_0$,

$$u_0 = \Phi(t_0)c + 0,$$

and so $c = \Phi(t_0)^{-1}u_0$. Thus the unique solution u is

$$u(t) = \Phi(t)\Phi(t_0)^{-1}u_0 + \int_{\tau=t_0}^{t}\Phi(t)\Phi(\tau)^{-1}f(\tau)\,d\tau.$$

17. Taking $\Phi(t) = e^{tA}$ in the above solution to Problem **16** immediately gives the result.

18. Since, by Lemma 2.27,

$$e^{tA} = \Phi(t)C$$

for some nonsingular constant matrix C, we have, by taking $t = 0$,

$$I = \Phi(0)C.$$

Hence $C = \Phi(0)^{-1}$ and $e^{tA} = \Phi(t)\Phi(0)^{-1}$.

19. Since, by Lemma 2.27,

$$e^{tA} = \Phi(t)C$$

for some nonsingular constant matrix C, we have, by taking $t = t_1$,

$$e^{t_1 A} = \Phi(t_1)C.$$

Hence $C = \Phi(t_1)^{-1}e^{t_1 A}$ and

$$e^{(t-t_1)A} = \Phi(t)\Phi(t_1)^{-1}.$$

20. Since $(A - \lambda I)^p u = 0$, we have

$$(A - \lambda I)^k v_k^{(p)} = (A - \lambda I)^p u = 0.$$

21. As in the proof of Theorem 2.1, Part (*b*), we have

$$e^{tA} = e^{t(A-\mu I)+t\mu I}$$

$$= e^{t(A-\mu I)}e^{t\mu I}$$

$$= e^{t(A-\mu I)}e^{t\mu}.$$

Similarly, we have

$$e^{tA} = e^{t\mu I + t(A-\mu I)}$$

$$= e^{t\mu I}e^{t(A-\mu I)}$$

$$= e^{t\mu}e^{t(A-\mu I)}.$$

22. By Theorem 2.3, each solution u is of the form

$$u = e^{tA}c$$

$$= \sum_{k=0}^{m_i-1}\sum_{i=1}^{s} t^k e^{t\lambda_i} M_{i,k}c.$$

Since

$$\lim_{t\to\infty} t^k e^{t\lambda_i} = 0 \quad \text{or} \quad +\infty$$

according as all the real parts of $\lambda_i, i = 1, \ldots, s$ are negative or positive, the results follows.

For Problems **1**, to **5**, all solutions tend to $+\infty$. For Problem **6**, all solutions tend to either 0 or $+\infty$ or a constant. For Problem **7**, all solutions tend to either $+\infty$ or a constant.

CHAPTER 3

Power Series Solutions

1. Introduction

Let $A, B_k, k = 0, 1, 2, \ldots$, be constant complex matrices of order 2×2. Let $r_0 > 0$ be a given positive constant. On the open interval of x with $x = 0$ deleted:

$$\{x \in \mathbb{R} : 0 < |x| < r_0\},$$

consider the linear system of 2 first order ordinary differential equations of the form

$$\frac{d}{dx}u = (\frac{1}{x}A + \sum_{k=0}^{\infty} x^k B_k)u, \quad -r_0 < x < r_0, \quad \text{but} \quad x \neq 0$$

$$= (\frac{1}{x}A + B_0 + xB_1 + x^2B_2 + \cdots)u. \tag{1.1}$$

Here $u = u(x) \in \mathbb{C}^2$ is a complex vector-valued function of x, and the power series $\sum_{k=0}^{\infty} x^k B_k$ is assumed convergent absolutely for $|x| < r_0$. Another case can be similarly treated, where x is a complex number in the open disk of the complex plane:

$$\{x = a + ib \in \mathbb{C}, a, b \in \mathbb{R} : 0 < |x| < r_0\}$$

with the origin deleted. See Section 5.

DEFINITION 1.1. *The power series $\sum_{k=0}^{\infty} x^k B_k$ converges absolutely for each real x satisfying $|x| < r_0$, if the nonnegative number $\sum_{k=0}^{\infty} |x|^k |B_k|$ is finite for each such x. Here the nonnegative number $|B_k|$ is called the norm of the matrix B_k, and is defined as the sum of the absolute values of the elements $(B_k)_{lm}$ of B_k:*

$$|B_k| = \sum_{l,m=1}^{2} |(B_k)_{lm}|.$$

For example, the norm $|B|$ of the matrix $B = \begin{pmatrix} (1+2i) & 2 \\ -3 & 4 \end{pmatrix}$ equals

$$|B| = |1 + 2i| + 2 + |-3| + 4$$

$$= \sqrt{1^2 + 2^2} + 2 + 3 + 4 = 9 + \sqrt{5}.$$

DEFINITION 1.2. *The linear system (1.1) is called regular if $A = 0$. In this case, $B_{k_0} \neq 0$ for some k_0 is assumed, and $x = 0$ is allowed. Otherwise, it is called regular singular or weakly singular [23] or of a singularity of the first kind [5]. In the regular case, (1.1) is reduced to the equation*

$$\frac{d}{dx}u = (\sum_{k=0}^{\infty} x^k B_k)u, \quad -r_0 < x < r_0$$

$$= (B_0 + xB_1 + x^2B_2 + \cdots)u. \tag{1.2}$$

By integrating (1.2), we see that u satisfies the integral eqation

$$u(x) = u_0 + \int (\sum_{k=0}^{\infty} x^k B_k) u \, dx,$$

where u_0 is a constant vector $\in \mathbb{C}^2$. This equation has the successive approximations by indefinite integrals:

$$u_m(x) = u_0 + \int (\sum_{k=0}^{\infty} x^k B_k) u_{m-1}(x) \, dx, \quad m = 1, 2, \ldots,$$

$$= u_0 + \sum_{k=0}^{\infty} (\int x^k B_k u_{m-1}(x) \, dx). \tag{1.3}$$

One of the two purposes in this chapter is to show that, for each constant vector $u_0 \in \mathbb{C}^2$, the successive approximations (1.3) converge as $m \longrightarrow \infty$, and the limit u is a solution to (1.2). Here the integration constants are taken to be zero.

DEFINITION 1.3. *Such a solution u is called a power series solution to the system* (1.2).

In particular, the two solutions, corresponding to $u_0 = \begin{pmatrix} 1 \\ 0 \end{pmatrix}$ and $u_0 = \begin{pmatrix} 0 \\ 1 \end{pmatrix}$, respectively, are linearly independent, and thus constitute a fundamental set of solutions for (1.2). This result shall apply to the linear second order equation of the form:

$$y'' + (\sum_{k=0}^{\infty} p_k x^k) y' + (\sum_{k=0}^{\infty} q_k x^k) y = 0, \quad -r_0 < x < r_0, \tag{1.4}$$

where $y' = \frac{d}{dx} y, y'' = \frac{d^2}{dx^2} y$, and $p_k, q_k, k = 0, 1, 2, \ldots$, are complex constants. Here the power series $\sum_{k=0}^{\infty} p_k x^k$ and $\sum_{k=0}^{\infty} q_k x^k$ converge absolutely for each $-r_0 < x < r_0$. In this case,

DEFINITION 1.4. (1.4) *is also called a regular equation.*

See the main results in Section 2.

The other purpose of this chapter is to study the regularly or weakly singular system (1.1), where $A \neq 0$ and $x \neq 0$. Using the change of variable $x = e^t$, (1.1) is transformed into the system

$$\frac{d}{dx} v(t) = (A + e^t \sum_{k=0}^{\infty} e^{kt} B_k) v, \tag{1.5}$$

where $v(t) = u(x)|_{x=e^t}$ or $u(x) = v(t)|_{t=\ln(x)}$.

Here

DEFINITION 1.5. *The logrithmic function $\ln(x)$ of x is defined by*

$$t = \ln(x) = \begin{cases} \ln|x|, & \text{if } x > 0; \\ \ln|x| + i\pi, & \text{if } x < 0, \end{cases}$$

and the exponential function x^λ of x, for $\lambda \in \mathbb{C}$, is defined by

$$x^\lambda = e^{\lambda \ln(x)},$$

in which, for $\alpha, \beta \in \mathbb{R}$,

$$e^{\alpha + i\beta} = e^\alpha e^{i\beta}$$

$$= e^{\alpha}[\cos(\beta) + i\sin(\beta)].$$

Here $i = \sqrt{-1}$ with $i^2 = -1$.

(1.5) is further reduced to the system:

$$\frac{d}{dt}w(t) = e^{-tA}(\sum_{k=0}^{\infty} e^{(k+1)t}B_k)e^{tA}w$$

$$= \sum_{k=0}^{\infty}(e^{-tA}e^{(k+1)t}B_k e^{tA}w), \qquad (1.6)$$

by the substitution $v(t) = e^{tA}w(t)$, where the exponential function e^{tA} of A is defined by the matrix

$$\lim_{m\to\infty} \sum_{k=0}^{m} \frac{(tA)^k}{k!}$$

in a previous chapter. Notice that e^{tA} has many nice properties. We shall show that, for each generalized eigenvector $w_0 \in \mathbb{R}^2$, of A , the successive approximations

$$w_m(t) = w_0 + \int (\sum_{k=0}^{\infty} e^{-tA}e^{(k+1)t}B_k e^{tA}w_{m-1}(t))\, dt, \quad m = 1, 2, \ldots$$

$$= w_0 + \sum_{k=0}^{\infty}(\int e^{-tA}e^{(k+1)t}B_k e^{tA}w_{m-1}(t)\, dt) \qquad (1.7)$$

of (1.6) by indefinite integrals, converge as $m \longrightarrow \infty$, and their limit $w(t)$ is a solution to the system (1.6). Equivalently, $v(t) = e^{tA}w(t)$ is a solution to the system (1.5) or $u(x) = v(t)|_{t=\ln(x)}$ is a solution to the system (1.1). Here the integration constants are all taken to be zero, as before.

DEFINITION 1.6. *Such a solution* $u(x) = v(t)_{t=\ln(x)} = e^{tA}w(t)_{t=\ln(x)}$ *is called a power series solution to the system* (1.1).

Furthermore, to two chosen linearly independent generalized eigenvectors w_0's of A, the corresponding two solutions u's will be linearly independent also, and thus constitute a fundamental set of power series solutions for (1.1). Here

DEFINITION 1.7. *A complex number* $\lambda \in \mathbb{C}$ *ia a generalized eigenvalue of A if there is a positive integer* $k_0 = 1$ *or 2 and a nonzero vector* $\beta \in \mathbb{C}^2$, *such that*

$$(A - \lambda I)^{k_0}\beta = 0 \quad (or \quad \det((A - \lambda I)^{k_0}) = 0) \quad but \quad (A - \lambda I)^{k_0-1}\beta \neq 0.$$

In this case, β is called a generalized eigenvector of A, corresponding to λ. Here $\det(B)$ is the determinant of a matrix B, and I is the identity matrix. If $k_0 = 1$, λ is also called an eigenvalue, and β an eigenvector.

The above result shall apply to the linear second order equation of the form:

$$x^2 y'' + x(\sum_{k=0}^{\infty} p_k x^k)y' + (\sum_{k=0}^{\infty} q_k x^k)y = 0, \quad -r_0 < x < r_0 \quad but \quad x \neq 0, \qquad (1.8)$$

where p_k and q_k are the same as those in (1.1). In this case,

DEFINITION 1.8. *(1.8) is also called a regular singular or weakly singular, equation.*

See the main results in Section 2.

The rest of this chapter is organized as follows. Section 2 states the main results, and Section 3 illustrates these results by examples. Section 4 proves the main results, and Section 5 extends them to the case where $x = z$ is a complex number in an open disk with the origin deleted, of the complex plane, and the matrices A and B_k have the order $n \times n, n \in \mathbb{N}$. Finally, Section 6 presents a set of problems, and solutions of them are placed in Section 7.

The material in this chapter is based on our paper [**14**].

2. Main Results

We treat the regular system (1.2) first.

THEOREM 2.1. *On assuming that $\sum_{k=0}^{\infty} x^k B_k$ converges absolutely for each real x satisfying $|x| < r_0$, the successive approximations u_m in (1.3), for each constant vector $u_0 \in \mathbb{C}^2$, converges, and its limit $u(x) \equiv T(x)u_0$ is a power series solution to (1.2). Furthermore, the two solutions u's, corresponding to $u_0 = \begin{pmatrix} 1 \\ 0 \end{pmatrix}$ and $u_0 = \begin{pmatrix} 0 \\ 1 \end{pmatrix}$, respectively, constitute a fundamental set of power series solutions to (1.2). Being written out, $u_m, m = 1, 2, \ldots$, takes the form:*

$$u_1 = u_0 + \left(\sum_{k_1=0}^{\infty} \frac{x^{k_1+1}}{k_1 + 1} B_{k_1} \right) u_0$$

$$u_2 = u_1 + \sum_{k_2=0}^{\infty} \frac{x^{k_2+1}}{(k_2 + 1) + (k_1 + 1)} B_{k_2} \sum_{k_1=0}^{\infty} \frac{x^{k_1+1}}{k_1 + 1} B_{k_1} u_0$$

$$\vdots$$

$$u_m = u_{m-1} + \sum_{k_m=0}^{\infty} \frac{x^{k_m+1}}{(k_m + 1) + (k_{m-1} + 1) + \cdots + (k_1 + 1)} B_{k_m}$$

$$\sum_{k_{m-1}=0}^{\infty} \frac{x^{k_{m-1}+1}}{(k_{m-1} + 1) + (k_{m-2} + 1) + \cdots + (k_1 + 1)} B_{k_{m-1}}$$

$$\cdots \sum_{k_2=0}^{\infty} \frac{x^{k_2+1}}{(k_2 + 1) + k_1 + 1} B_{k_2} \sum_{k_1=0}^{\infty} \frac{x^{k_1+1}}{k_1 + 1} B_{k_1} u_0.$$

It follows from Theorem 2.1 that

COROLLARY 2.2. *A power series solution u to the system (1.2) is of the form*

$$u = \sum_{k=0}^{\infty} x^k c_k,$$

where the constant vectors $c_k \in \mathbb{C}^2$ can be determined by substituting this u into (1.2) and equating the coefficient of x^k on both sides.

The regular equation (1.4) can be reduced to the regular system (1.2) by the familiar substitution: $u_1 = y, u_2 = y'$, and the result becomes

$$\begin{cases} u_1' = y' = u_2 \\ u_2' = y'' = -\sum_{k=0}^{\infty} p_k x^k - \sum_{k=0}^{\infty} q_k x^k u_1 \end{cases}$$

or

$$\frac{d}{dx} u = [\begin{pmatrix} 0 & 1 \\ -q_0 & -p_0 \end{pmatrix} + \sum_{k=1}^{\infty} x^k \begin{pmatrix} 0 & 0 \\ -q_k & -p_k \end{pmatrix}] u, \tag{2.1}$$

where $u = \begin{pmatrix} u_1 \\ u_2 \end{pmatrix}$.

THEOREM 2.3. *The system* (2.1) *has the fundamental set of solutions:*

$$\{T(x) \begin{pmatrix} 1 \\ 0 \end{pmatrix}, T(x) \begin{pmatrix} 0 \\ 1 \end{pmatrix}\},$$

where $T(x)$ is the $T(x)$ in Theorem 2.1, and

$$B_0 = \begin{pmatrix} 0 & 1 \\ -q_0 & -p_0 \end{pmatrix}, \quad \text{and} \quad B_k = \begin{pmatrix} 0 & 0 \\ -q_k & -p_k \end{pmatrix}, k = 1, 2, \ldots.$$

Furthermore, the first components of the two vectors $T(x) \begin{pmatrix} 1 \\ 0 \end{pmatrix}$ and $T(x) \begin{pmatrix} 0 \\ 1 \end{pmatrix}$, constitute a fundamental set of solutions for the regular equation (1.4).

It follows from Theorems 2.1 and 2.3 that

COROLLARY 2.4. *A power series solution to the regular equation* (1.4) *takes the form*

$$y = \sum_{k=0}^{\infty} d_k x^k,$$

where the scalar constants $d_k \in \mathbb{C}$ can be determined by substituting this y into (1.4) *and equating the coefficient of x on both sides.*

Next, we examine the regularly singular system (1.1), where $A \neq 0$ and $x \neq 0$. The following lemmas are prepared for this purpose.

LEMMA 2.5. *If β is an eigenvalue of A, satisfying $(A - \lambda I)\beta = 0$, then*

$$e^{tA}\beta = e^{\lambda t}\beta. \tag{2.2}$$

If β is a generalized eigenvector of A, satisfying

$$(A - \lambda I)^2\beta = 0 \quad \text{but} \quad (A - \lambda I)\beta \neq 0,$$

then

$$e^{tA}\beta = e^{t\lambda}\{\beta + \frac{t(A - \lambda)}{1!}\beta\}. \tag{2.3}$$

PROOF. See Problems set in Section 7.   □

LEMMA 2.6. *Suppose that λ_1 and λ_2 are the two eigenvalues of A.*
If $\lambda_1 \neq \lambda_2$, then there exist, uniquely, two matrices $M_{1,0}$ and $M_{2,0}$, such that

$$e^{tA} = e^{\lambda_1 t} M_{1,0} + e^{\lambda_2 t} M_{2,0},$$

$$M_{1,0}^2 = M_{1,0}, \quad M_{2,0}^2 = M_{2,0}, \quad and \qquad (2.4)$$

$$M_{1,0} M_{2,0} = 0 = M_{2,0} M_{1,0},$$

where

$$\left\{ \begin{array}{l} I = M_{1,0} + M_{2,0}, \\ A = \lambda_1 M_{1,0} + \lambda_2 M_{2,0} \end{array} \right. \quad or \quad \left\{ \begin{array}{l} M_{1,0} = (\lambda_1 - \lambda_2)^{-1}(A - \lambda_2 I), \\ M_{2,0} = (\lambda_2 - \lambda_1)^{-1}(A - \lambda_1 I). \end{array} \right.$$

If $\lambda_1 = \lambda_2$, then there exist, uniquely, two matrices $N_{1,0}$ and $N_{1,1}$, such that

$$e^{tA} = e^{\lambda_1 t} N_{1,0} + t e^{\lambda_1 t} N_{1,1},$$

$$N_{1,0}^2 = N_{1,0}, \quad N_{1,1}^2 = 0, \quad and \qquad (2.5)$$

$$N_{1,0} N_{1,1} = N_{1,1} N_{1,0} = N_{1,1},$$

where

$$\left\{ \begin{array}{l} I = N_{1,0}, \\ A = \lambda_1 N_{1,0} + N_{1,1} \end{array} \right. \quad or \quad \left\{ \begin{array}{l} N_{1,0} = I, \\ N_{1,1} = A - \lambda_1 I. \end{array} \right.$$

PROOF. [**7**, Pages 307–313] Denote by $f(\lambda)$ the characteristic polynomial

$$det(A - \lambda I) = a_0(\lambda - \lambda_1)(\lambda - \lambda_2)$$

of A, where a_0 is a constant. Using the formula in a previous chapter:

$$\frac{d}{dt} e^{tA} = A e^{tA},$$

it follows that

$$\begin{aligned} f(\frac{d}{dt}) e^{tA} &= a_0(\frac{d}{dt} - \lambda_1)(\frac{d}{dt} - \lambda_2) e^{tA} \\ &= a_0(A - \lambda_1)(A - \lambda_2) e^{tA} \\ &= f(A) e^{tA} = 0. \end{aligned}$$

Here used was the Cayley-Hamilton theorem from linear algebra [**9**]:

$$f(A) = a_0(A - \lambda_1)(A - \lambda_2) = 0.$$

Thus each element of a function of t, in the matrix e^{tA}, is a solution to the ordinary differential equation with the unknown function $y(t)$:

$$a_0(\frac{d}{dt} - \lambda_1)(\frac{d}{dt} - \lambda_2) y(t) = 0.$$

Since this differential equation has obviously the fundamental set $\{e^{t\lambda_1}, e^{t\lambda_2}\}$ or $\{e^{t\lambda_1}, t e^{t\lambda_1}\}$ of solutions, according as $\lambda_1 \neq \lambda_2$ or $\lambda_1 = \lambda_2$, it follows accordingly that

$$e^{tA} = e^{t\lambda_1} M_{1,0} + e^{t\lambda_2} M_{2,0}$$

or

$$e^{tA} = e^{t\lambda_1} N_{1,0} + t e^{t\lambda_1} N_{1,1}$$

for some matrices $M_{1,0}, M_{2,0}, N_{1,0}$, and $N_{1,1}$.

These matrices can be uniquely determined by the formulas:

$$e^{tA}|_{t=0} = I, \quad \frac{d}{dt}e^{tA}|_{t=0} = Ae^{tA}|_{t=0} = A,$$

and these matrices thus determined will satisfy the required properties, if we use the fact that $f(A) = 0$. □

LEMMA 2.7. *The formulas of indefinite integrals*

$$\int e^{-tA}e^{\lambda t}\, dt = (\lambda I - A)^{-1}e^{t(\lambda - A)} \tag{2.6}$$

and

$$\int e^{-tA}te^{\lambda t}\, dt = t(\lambda I - A)^{-1}e^{t(\lambda - A)} - (\lambda I - A)^{-2}e^{t(\lambda - A)} \tag{2.7}$$

are true, if λ is not an eigenvalue of A. Here the integration constants are taken to be zero.

PROOF. See Problems set in Section 7. □

DEFINITION 2.8. *A Jordan basis J for A is a set $\{\beta_1, \beta_2\}$ of two linearly independent generalized eigenvectors β_1 and β_2 of A.*

Related to this is the following result, a fact from linear algebra [9]:

FACT 2.9. *Assume that λ_1 and λ_2 are the two eigenvalues of A, and that the real part $Re\lambda_1$ of λ_1 is greater than or equal to $Re\lambda_2$, the real part of λ_2. Then A always has such a Jordan basis $J = \{\beta_1, \beta_2\}$, which satisfies:*
either

(C1) $(A - \lambda_1 I)\beta_1 = 0$ *and* $(A - \lambda_2 I)\beta_2 = 0$, *where* $\lambda_1 = \lambda_2$ *is possible,*

or

(C2) $\lambda_1 = \lambda_2$, $(A - \lambda_1 I)\beta_1 = 0$, $(A - \lambda_1 I)\beta_2 = \beta_1$, *and*

$$(A - \lambda_1 I)\beta_2 \neq 0 \quad \text{but} \quad (A - \lambda_1 I)^2\beta_2 = 0.$$

THEOREM 2.10. *For each element w_0 in the Jordan basis $J = \{\beta_1, \beta_2\}$ of A, where the two eigenvalues λ_1 and λ_2 of A satisfies $Re\lambda_1 \geq Re\lambda_2$, the successive approximations $w_m = w_m[w_0]$ in (1.7) converge as $m \longrightarrow \infty$, and their limit $w \equiv S(t)w_0$ is a power series solution to (1.6). The two power series solutions w's, corresponding to the two w_0's, constitute a fundamental set of solutions for (1.6). Equivalently, the set of the two power series solutions $v(t) = e^{tA}w(t)$'s is a fundamental set of solutions for (1.5), or the set of the two solutions $u(x) = v(t)|_{t=\ln(x)}$'s is a fundamental set of solutions for the weakly singular system (1.1). Here notice that the substitution $t = \ln(x)$ is equivalent to the substitution $t = \ln|x|$, regardless of the sign of x.*

The $w_m[\beta_1]$ can be calculated by using (2.2) and (2.6), and has no logarithmic term.

If the difference $l = \lambda_1 - \lambda_2$ of the two eigenvalues λ_1 and λ_2 is not an integer, then $w_m[\beta_2]$ can be calculated as $w_m[\beta_1]$ can be, and has no logarithmic term.

If the above l is a zero integer, that is, if $\lambda_1 = \lambda_2$, then $w_m[\beta_2]$ is calculated by using

$$\begin{cases} \text{(2.2) and (2.6), as } w_m[\beta_1] \text{ can be, and has no logarithmic term, if (C1) holds,} \\ \text{(2.3), (2.6), and (2.7), and has a logarithmic term, if (C2) holds.} \end{cases}$$

If the above l is a positive integer and further equals a sum

$$\sum = [\sum_{i=1}^{q_0}(k_0 + 1)] + [\sum_{i=1}^{q_1}(k_1 + 1)] + \cdots + [\sum_{i=1}^{q_{s_0}}(k_{s_0} + 1)]$$

of some integers $\sum_{i=1}^{q_j}(k_j + 1), j = 0, 1, \ldots, s_0$ for some s_0, where k_j is the index of the term $B_{k_j} \neq 0$ and the number $(k_j + 1)$ is added up repeatedly for some q_j times, $q_j \in \mathbb{N}$, then $w_m[\beta_2]$ can be calculated by using (2.2), (2.4), (2.6), and (2.7), and has a logarithmic term.

However, even if the above l is a positive integer, but no sum $\sum$ of the above type exists, then $w_m[\beta_2]$ is calculated as $w_m[\beta_1]$ is, and has no logarithmic term.

Being written out, $w_m = w_m[\beta_1]$ takes the form:

$$w_1 = \beta_1 + \sum_{k_1=0}^{\infty}[(k_1 + 1) + (\lambda_1 - A)]^{-1}e^{t[(k_1+1)+(\lambda_1-A)]}B_{k_1}\beta_1,$$

$$w_2 = w_1 + \sum_{k_2=0}^{\infty}[(k_2 + 1) + (k_1 + 1) + (\lambda_1 - A)]^{-1}e^{t[(k_2+1)+(\lambda_1-A)]}B_{k_2}$$

$$\sum_{k_1=0}^{\infty}[(k_1 + 1) + (\lambda_1 - A)]^{-1}e^{t(k_1+1)}B_{k_1}\beta_1,$$

$$\vdots,$$

$$w_m = w_{m-1} +$$

$$\sum_{k_m=0}^{\infty}[(k_m+1)+(k_{m-1}+1)+\cdots+(k_1+1)+(\lambda_1-A)]^{-1}e^{t[(k_m+1)+(\lambda_1-A)]}B_{k_m}$$

$$\sum_{k_{m-1}=0}^{\infty}[(k_{m-1}+1)+(k_{m-2}+1)+\cdots+(k_1+1)+(\lambda_1-A)]^{-1}e^{t(k_{m-1}+1)}B_{k_{m-1}}\cdots$$

$$\sum_{k_1=0}^{\infty}[(k_1+1)+(\lambda_1-A)]^{-1}e^{t(k_1+1)}B_{k_1}\beta_1, \quad m=1,2,\ldots.$$

It will follow from the proof of Theorem 2.10 that

COROLLARY 2.11. *The power series solution $y_1 = e^{tA}S(t)\beta_1|_{t=\ln(x)}$ to the weakly singular system (1.1) takes the form*

$$y_1 = x^{\lambda_1}\sum_{k=0}^{\infty}x^k c_k,$$

where the constant vectors c_k can be determined by substituting the y_1 into (1.1) and equating the coefficients of x^k on both sides.

If either $l = \lambda_1 - \lambda_2$ is not an integer, or both $l = 0$ and (C1) , then the power series solution $y_2 = e^{tA}S(t)\beta_2|_{t=\ln(x)}$ to (1.1) takes the same form as y_1 does, and is similarly calculated as y_1 is.

If both $l = 0$ and (C2) , then the power series solution $y_2 = e^{tA}S(t)\beta_2|_{t=\ln(x)}$ takes the form

$$y_2 = y_1 \ln(x) + x^{\lambda_2}\sum_{k=0}^{\infty}x^k c_k.$$

If the above l is a positive integer and further equals a sum

$$\sum = [\sum_{i=1}^{q_0}(k_0+1)] + [\sum_{i=1}^{q_1}(k_1+1)] + \cdots + [\sum_{i=1}^{q_{s_0}}(k_{s_0}+1)]$$

of some integers $\sum_{i=1}^{q_j}(k_j+1), j = 0, 1, \ldots, s_0$ for some s_0, where k_j is the index of the term $B_{k_j} \neq 0$ and the number (k_j+1) is added up repeatedly for some q_j times, $q_j \in \mathbb{N}$, then the power series solution $y_2 = e^{tA}S(t)\beta_2|_{t=\ln(x)}$ takes the form

$$y_2 = \alpha y_1 \ln(x) + x^{\lambda_2} \sum_{k=0}^{\infty} x^k d_k.$$

However, even if the above l is a positive integer, but no sum $\sum$ of the above type exists, then the power series solution $y_2 = e^{tA}S(t)\beta_2|t = \ln(x)$ to (1.1) takes the same form as y_1 does, and is similarly calculated as y_1 is.

Here the scalar constant α and the constant vectors c_k, d_k can be determined by substituting the y_2 into (1.1) and equating the coefficients of x^k on both sides. Here also notice that the substitution $t = \ln(x)$ is equivalent to the substitution $t = \ln|x|$, regardless of the sign of x.

The regular singular equation (1.8) can be reduced to the weakly singular system (1.1) by the substitution [5, Page 124]: $u_1 = y, u_2 = xy'$, and the result becomes

$$\begin{cases} u_1' = y' = \frac{1}{x}u_2 \\ u_2' = y' + xy'' = \frac{1}{x}u_2 + \frac{-1}{x}[(\sum_{k=0}^{\infty} p_k x^k)u_2 + (\sum_{k=0}^{\infty} q_k x^k)u_1] \end{cases}$$

or

$$\frac{d}{dx}u = [\frac{1}{x}\begin{pmatrix} 0 & 1 \\ -q_0 & 1-p_0 \end{pmatrix} + \sum_{k=0}^{\infty} x^k \begin{pmatrix} 0 & 0 \\ -q_{k+1} & -p_{k+1} \end{pmatrix}]u, \qquad (2.8)$$

a form of (1.1), in which

$$u = \begin{pmatrix} u_1 \\ u_2 \end{pmatrix}, A = \begin{pmatrix} 0 & 1 \\ -q_0 & 1-p_0 \end{pmatrix}, \quad \text{and} \quad B_k = \begin{pmatrix} 0 & 0 \\ -q_{k+1} & -p_{k+1} \end{pmatrix}.$$

Thus Theorem 2.10 entails:

THEOREM 2.12. *The A in the weakly singular system (2.8) has the corresponding Jordan basis $J = \{\beta_1, \beta_2\}$, which either satisfies (C1) but $\lambda_1 \neq \lambda_2$, or satisfies (C2).*

The weakly regular system (2.8) has the fundamental set of solutions:

$$Q(x) = \{y_1(x), y_2(x)\}$$
$$= \{e^{tA}S(t)\beta_1, e^{tA}S(t)\beta_2\}|_{t=\ln(x)},$$

where y_1 and y_2 are calculated, according to Corollary 2.11, or $S(t)\beta_1$ and $S(t)\beta_2$ are calculated, according to Theorem 2.10.

Here notice that the substitution $t = \ln(x)$ is equivalent to the substitution $t = \ln|x|$, regardless of the sign of x.

Furthermore, the first components of the two vectors in $Q(x)$ constitute a fundamental set of solutions for the weakly singular equation (1.8).

Utilizing Corollary 2.11 and Theorem 2.12, we have the following results, except for the part: $d_0 = 0$:

COROLLARY 2.13 (The extension of the method of Frobenius [6], Page 167).
The power series solution $y_1 = e^{tA}S(t)\beta_1|_{t=\ln(x)}$ *to the weakly singular equation*
(1.8) *takes the form*

$$y_1 = x^{\lambda_1} \sum_{k=0}^{\infty} c_k x^k.$$

If the difference $l = \lambda_1 - \lambda_2$ *of the two eigenvalues* λ_1 *and* λ_2 *is not an integer,*
then the power series solution $y_2 = e^{tA}S(t)\beta_2|_{t=\ln(x)}$ *to the weakly singular equation*
(1.8) *takes the same form as* y_1 *does, and is similarly calculated as* y_1 *is.*
    If the above l *is a zero integer, that is, if* $\lambda_1 = \lambda_2$, *then* y_2 *takes the form*

$$y_2 = y_1 \ln(x) + x^{\lambda_1} \sum_{k=0}^{\infty} d_k x^k$$

$$= y_1 \ln(x) + x^{\lambda_1} \sum_{k=1}^{\infty} d_k x^k,$$

where $d_0 = 0$.
    If the above l *is a positive integer and further equals a sum*

$$\sum = [\sum_{i=1}^{q_0}(k_0 + 1)] + [\sum_{i=1}^{q_1}(k_1 + 1)] + \cdots + [\sum_{i=1}^{q_{s_0}}(k_{s_0} + 1)]$$

of some integers $\sum_{i=1}^{q_j}(k_j + 1), j = 0, 1, \ldots, s_0$ *for some* s_0, *where* k_j *is the index*
of the term $B_{k_j} \neq 0$ *and the number* $(k_j + 1)$ *is added up repeatedly for some* q_j
times, $q_j \in \mathbb{N}$, *then* y_2 *takes the form*

$$y_2 = \alpha y_1 \ln(x) + x^{\lambda_2} \sum_{k=0}^{\infty} e_k x^k.$$

    However, even if the above l *is a positive integer, but no sum* $\sum$ *of the above*
type exists, then y_2 *takes the same form as* y_1 *does, and is similarly calculated as*
y_1 *is.*
    Here the constants c_k, d_k, *and* α, e_k *can be determined by substituting the* y_1
and y_2 *into the equation* (1.8) *and equating the coefficients of* x^k *on both sides.*
Here also notice that the substitution $t = \ln(x)$ *is equivalent to the substitution*
$t = \ln|x|$, *regardless of the sign of* x.

PROOF. It remains to prove that $d_0 = 0$.
Compute the roots of the characteristic equation:

$$0 = |A - \lambda I|$$

$$= \lambda^2 + (p_0 - 1)\lambda + q_0,$$

which shows that, in order to have two equal roots, that is, $\lambda_1 = \lambda_2$, there must

$$0 = (p_0 - 1)^2 - 4q_0$$

and

$$\lambda_1 = \lambda_2 = \frac{1 - p_0}{2}.$$

Corresponding to λ_1, we can take the eigenvector $\beta_1 = \begin{pmatrix} 1 \\ \lambda_1 \end{pmatrix} = \begin{pmatrix} 1 \\ \frac{1-p_0}{2} \end{pmatrix}$ and the generalized eigenvector $\beta_2 = \begin{pmatrix} 0 \\ 1 \end{pmatrix}$. Here β_2 satisfies $(A - \lambda_1 I)\beta_2 = \beta_1$. Since the first component of β_2 equals zero, the Step 2 in the proof of Theorem 2.10 yields $d_0 = 0$. $\qquad\qquad\qquad\qquad\qquad\qquad\qquad\qquad\qquad\qquad\qquad\qquad\qquad$ $\square$

3. Examples

Before we prove the main results in Section 2, we look at some examples.

EXAMPLE 3.1. Consider the regular linear system

$$\frac{d}{dx}u = (A_0 + xA_1)u, \quad -r_0 < x < r_0$$

$$= \begin{pmatrix} 0 & 1 \\ x & 0 \end{pmatrix} u,$$

where

$$A_0 = \begin{pmatrix} 0 & 1 \\ 0 & 0 \end{pmatrix} \quad \text{and} \quad A_1 = \begin{pmatrix} 0 & 0 \\ 1 & 0 \end{pmatrix}.$$

Solution.

Method 1. By Corollary 2.2, the trial function

$$u = \sum_{k=0}^{\infty} c_k x^k$$

of the solution, after being substituted into the system, leads to

$$\frac{d}{dx}u = \sum_{k=0}^{\infty} k c_k x^{k-1} = \sum_{k=0}^{\infty} (k+1)c_{k+1} x^k$$

$$= (A_0 + xA_1)\left(\sum_{k=0}^{\infty} c_k x^k\right)$$

$$= \sum_{k=0}^{\infty} A_0 c_k x^k + \sum_{k=0}^{\infty} A_1 c_k x^{k+1}$$

$$= \sum_{k=0}^{\infty} A_0 c_k x^k + \sum_{k=1}^{\infty} A_1 c_{k-1} x^k,$$

or

$$c_1 = A_0 c_0,$$

$$(k+1)c_{k+1} = A_0 d_k + A_1 d_{k-1}, \quad k \geq 1.$$

A few terms of c_k are:

$$c_1 = A_0 c_0,$$

$$c_2 = \frac{1}{2}(A_0 c_1 + A_1 c_0),$$

$$c_3 = \frac{1}{3}(A_0 c_2 + A_1 c_1),$$

$$c_4 = \frac{1}{4}(A_0 c_3 + A_1 c_2),$$

$$\vdots$$

Corresponding to $c_0 = \begin{pmatrix} 1 \\ 0 \end{pmatrix}$ and $c_0 = \begin{pmatrix} 0 \\ 1 \end{pmatrix}$, respectively, the two solutions u's,

where

$$u = \sum_{k=0}^{\infty} c_k x^k,$$

are linearly independent.

Method 2. Using Theorem 2.1, it follows that

$$u_1 = u_0 + \int (A_0 + x A_1) u_0 \, dx$$

$$= u_0 + (x A_0 + \frac{x^2}{2} A_1) u_0 \quad \text{or}$$

$$u_1 = u_0 + \int \begin{pmatrix} 0 & 1 \\ x & 0 \end{pmatrix} u_0 \, dx$$

$$= u_0 + \begin{pmatrix} 0 & x \\ \frac{x^2}{2} & 0 \end{pmatrix} u_0,$$

$$u_2 = u_0 + \int (A_0 + x A_1) u_1 \, dx$$

$$= u_1 + \int (A_0 + x A_1)(x A_0 + \frac{x^2}{2} A_1) u_0 \, dx$$

$$= u_1 + \frac{x^3}{3} A_1 A_0 + \frac{x^3}{3 \times 2} A_0 A_1) u_0 \quad \text{or}$$

$$u_2 = u_1 + \int \begin{pmatrix} 0 & 1 \\ x & 0 \end{pmatrix} \begin{pmatrix} 0 & x \\ \frac{x^2}{2} & 0 \end{pmatrix} u_0 \, dx$$

$$= u_1 + \int \begin{pmatrix} \frac{x^2}{2} & 0 \\ 0 & x^2 \end{pmatrix} u_0 \, dx$$

$$= u_1 + \begin{pmatrix} \frac{x^3}{3 \times 2} & 0 \\ 0 & \frac{x^3}{3} \end{pmatrix} u_0,$$

$$u_3 = u_0 + \int (A_0 + x A_1) u_2 \, dx$$

$$= u_2 + \int (A_0 + x A_1) x^3 (\frac{1}{3} A_1 A_0 + \frac{1}{3 \times 2} A_0 A_1) u_0 \, dx$$

$$= u_2 + (\frac{x^4}{4 \times 3} A_0 A_1 A_0 + \frac{1}{5 \times 3 \times 2} A_1 A_0 A_1) u_0 \quad \text{or}$$

$$u_3 = u_2 + \int \begin{pmatrix} 0 & 1 \\ x & 0 \end{pmatrix} \begin{pmatrix} \frac{x^3}{3 \times 2} & 0 \\ 0 & \frac{x^3}{3} \end{pmatrix} u_0 \, dx$$

$$= u_2 + \int \begin{pmatrix} 0 & \frac{x^3}{3} \\ \frac{x^4}{3 \times 2} & 0 \end{pmatrix} u_0 \, dx$$

$$= u_2 + \begin{pmatrix} 0 & \frac{x^4}{4 \times 3} \\ \frac{x^5}{5 \times 3 \times 2} & 0 \end{pmatrix} u_0.$$

The other $u_j, j = 4, 5, \ldots$, are similarly calculated.

Corresponding to $u_0 = \begin{pmatrix} 1 \\ 0 \end{pmatrix}$ and $u_0 = \begin{pmatrix} 0 \\ 1 \end{pmatrix}$, respectively, the two solutions u's,

where

$$u = \lim_{j \to \infty} u_j,$$

are linearly independent.

EXAMPLE 3.2. Consider the regular linear equation (Airy equation)

$$y'' = xy, \quad -r_0 < x < r_0. \tag{3.1}$$

Solution.

Method 1. As (2.1) states it, the substitutions: $u_1 = y, u_2 = y'$, lead the above regular linear equation (3.1) to the regular linear system

$$\frac{d}{dx} u = (A_0 + xA_1)u,$$

where

$$A_0 = \begin{pmatrix} 0 & 1 \\ 0 & 0 \end{pmatrix} \quad \text{and} \quad A_1 = \begin{pmatrix} 0 & 0 \\ 1 & 0 \end{pmatrix}.$$

Observe that this system is the system in Example 3.1. Thus, employing Theorem 2.3 and Eample 3.1, it follows that, corresponding to $c_0 = \begin{pmatrix} 1 \\ 0 \end{pmatrix}$, there are

$$d_0 = \begin{pmatrix} 1 \\ 0 \end{pmatrix}, \quad d_1 = \begin{pmatrix} 0 \\ \frac{1}{2} \end{pmatrix}, \quad d_2 = \begin{pmatrix} 0 \\ 0 \end{pmatrix}, \quad d_3 = \begin{pmatrix} \frac{1}{3 \times 2} \\ 0 \end{pmatrix}, \quad d_4 = \begin{pmatrix} 0 \\ 0 \end{pmatrix}, \ldots,$$

and so the resulting solution y_1, which is the first component of $u = \sum_{k=0}^{\infty} c_k x^k$, becomes:

$$y_1 = 1 + \frac{1}{3 \times 2} x^3 + \cdots.$$

Similarly, corresponding to $c_0 = \begin{pmatrix} 0 \\ 1 \end{pmatrix}$, there are

$$c_0 = \begin{pmatrix} 0 \\ 1 \end{pmatrix}, \quad c_1 = \begin{pmatrix} 1 \\ 0 \end{pmatrix}, \quad d_2 = \begin{pmatrix} 0 \\ 0 \end{pmatrix}, \quad d_3 = \begin{pmatrix} 0 \\ \frac{1}{3} \end{pmatrix}, \quad d_4 = \begin{pmatrix} \frac{1}{4 \times 3} \\ 0 \end{pmatrix}, \ldots,$$

and so the resulting second linearly independent solutions y_2, which is the first component of $u = \sum_{k=0}^{\infty} c_k x^k$, becomes:

$$y_2 = x + \frac{1}{4 \times 3} x^4 + \cdots.$$

Or, corresponding to $u_0 = \begin{pmatrix} 1 \\ 0 \end{pmatrix}$, there are

$$u_1 = u_0 + (xA_0 + \frac{x^2}{2} A_1)u_0$$

$$= \begin{pmatrix} 1 \\ 0 \end{pmatrix} + x \begin{pmatrix} 0 \\ 0 \end{pmatrix} + \frac{x^2}{2} \begin{pmatrix} 0 \\ 1 \end{pmatrix} \quad \text{or}$$

$$u_1 = u_0 + \begin{pmatrix} 0 & x \\ \frac{x^2}{2} & 0 \end{pmatrix} u_0,$$

$$= \begin{pmatrix} 1 \\ 0 \end{pmatrix} + \begin{pmatrix} 0 & x \\ \frac{x^2}{2} & 0 \end{pmatrix} \begin{pmatrix} 1 \\ 0 \end{pmatrix},$$

$$u_2 = u_1 + \frac{x^3}{3} A_1 A_0 u_0 + \frac{x^3}{3 \times 2} A_0 A_1 u_0$$

$$= u_1 + \frac{x^3}{3} \begin{pmatrix} 0 \\ 0 \end{pmatrix} + \frac{x^3}{3 \times 2} \begin{pmatrix} 1 \\ 0 \end{pmatrix} \quad \text{or}$$

$$u_2 = u_1 + \begin{pmatrix} \frac{x^3}{3 \times 2} & 0 \\ 0 & \frac{x^3}{3} \end{pmatrix} u_0,$$

$$= u_1 + \begin{pmatrix} \frac{x^3}{3 \times 2} & 0 \\ 0 & \frac{x^3}{3} \end{pmatrix} \begin{pmatrix} 1 \\ 0 \end{pmatrix},$$

$$u_3 = u_2 + \frac{x^4}{4 \times 3} A_0 A_1 A_0 u_0 + \frac{x^5}{5 \times 3 \times 2} A_1 A_0 A_1 u_0$$

$$= u_2 + \frac{x^4}{4 \times 3} \begin{pmatrix} 0 \\ 0 \end{pmatrix} + \frac{x^5}{5 \times 3 \times 2} \begin{pmatrix} 0 \\ 1 \end{pmatrix} \quad \text{or}$$

$$u_3 = u_2 + \begin{pmatrix} 0 & \frac{x^4}{4 \times 3} \\ \frac{x^5}{5 \times 3 \times 2} & 0 \end{pmatrix} u_0$$

$$= u_2 + \begin{pmatrix} 0 & \frac{x^4}{4 \times 3} \\ \frac{x^5}{5 \times 3 \times 2} & 0 \end{pmatrix} \begin{pmatrix} 1 \\ 0 \end{pmatrix}.$$

The other $u_j, j = 4, 5, \ldots$, are similarly calculated. The resulting solution y_1, which is the first component of $\lim_{j \to \infty} u_j$, becomes

$$y_1 = 1 + \frac{x^3}{3 \times 2} + \cdots.$$

In the same way as above, corresponding to $u_0 = \begin{pmatrix} 0 \\ 1 \end{pmatrix}$, there are

$$u_1 = u_0 + x A_0 u_0 + \frac{x^2}{2} A_1 u_0$$

$$= \begin{pmatrix} 0 \\ 1 \end{pmatrix} + x \begin{pmatrix} 1 \\ 0 \end{pmatrix} + \frac{x^2}{2} \begin{pmatrix} 0 \\ 0 \end{pmatrix} \quad \text{or}$$

$$u_1 = u_0 + \begin{pmatrix} 0 & x \\ \frac{x^2}{2} & 0 \end{pmatrix} u_0$$

$$= \begin{pmatrix} 0 \\ 1 \end{pmatrix} + \begin{pmatrix} 0 & x \\ \frac{x^2}{2} & 0 \end{pmatrix} \begin{pmatrix} 0 \\ 1 \end{pmatrix},$$

$$u_2 = u_1 + \frac{x^3}{3} A_1 A_0 u_0 + \frac{x^3}{3 \times 2} A_0 A_1 u_0$$

$$= u_1 + \frac{x^3}{3} \begin{pmatrix} 0 \\ 1 \end{pmatrix} + \frac{x^3}{3 \times 2} \begin{pmatrix} 0 \\ 0 \end{pmatrix} \quad \text{or}$$

$$u_2 = u_1 + \begin{pmatrix} \frac{x^3}{3 \times 2} & 0 \\ 0 & \frac{x^3}{3} \end{pmatrix} u_0$$

$$= u_1 + \begin{pmatrix} \frac{x^3}{3 \times 2} & 0 \\ 0 & \frac{x^3}{3} \end{pmatrix} \begin{pmatrix} 0 \\ 1 \end{pmatrix},$$

$$u_3 = u_2 + \frac{x^4}{4 \times 3} A_0 A_1 A_0 u_0 + \frac{x^5}{5 \times 3 \times 2} A_1 A_0 A_1 u_0$$

$$= u_2 + \frac{x^4}{4 \times 3} \begin{pmatrix} 1 \\ 0 \end{pmatrix} + \frac{x^5}{5 \times 3 \times 2} \begin{pmatrix} 0 \\ 0 \end{pmatrix} \quad \text{or}$$

$$u_3 = u_2 + \begin{pmatrix} 0 & \frac{x^4}{4 \times 3} \\ \frac{x^5}{5 \times 3 \times 2} & 0 \end{pmatrix} u_0$$

$$= u_2 + \begin{pmatrix} 0 & \frac{x^4}{4 \times 3} \\ \frac{x^5}{5 \times 3 \times 2} & 0 \end{pmatrix} \begin{pmatrix} 0 \\ 1 \end{pmatrix}.$$

The other $u_j, j = 4, 5$, are similarly calculated. The resulting second linearly independent solution y_2, which is the first component of $\lim_{j \to \infty} u_j$, becomes:

$$y_2 = x + \frac{x^4}{4 \times 3} + \cdots .$$

Method 2. By Corollary 2.4, the solution y takes the form

$$y = \sum_{k=0}^{\infty} c_k x^k,$$

where the constants c_k can be determined by substituting this y into (3.1) and equating the coefficients of x^k on both sides. The result becomes:

$$y'' = \sum_{k=2}^{\infty} k(k-1)x^{k-2}$$

$$= \sum_{k=0}^{\infty} (k+2)(k+1)c_{k+2}x^k$$

$$= xy = \sum_{k=0}^{\infty} c_k x^{k+1}$$

$$= \sum_{k=1}^{\infty} c_{k-1}x^k,$$

or

$$c_2 = 0, \quad \text{and} \quad c_{k+2} = \frac{c_{k-1}}{(k+2)(k+1)}, k = 1, 2, \ldots .$$

A few more c_k's are:

$$c_3 = \frac{c_0}{3 \times 2},$$

$$c_4 = \frac{c_1}{4 \times 3},$$

$$c_5 = \frac{c_2}{5 \times 4} = 0,$$

$$c_6 = \frac{c_3}{6 \times 5} = \frac{c_0}{6 \times 5 \times 3 \times 2},$$

$$c_7 = \frac{c_4}{7 \times 6} = \frac{c_1}{7 \times 6 \times 4 \times 3},$$

$$\vdots$$

By choosing $\begin{pmatrix} c_0 \\ c_1 \end{pmatrix} = \begin{pmatrix} 1 \\ 0 \end{pmatrix}$ and $\begin{pmatrix} c_0 \\ c_1 \end{pmatrix} = \begin{pmatrix} 0 \\ 1 \end{pmatrix}$, respectively, the resulting two linearly independent solutions are:

$$y_1 = 1 + \frac{x^3}{3 \times 2} + \frac{x^6}{6 \times 5 \times 3 \times 2} + \cdots,$$

$$y_2 = x + \frac{x^4}{4 \times 3} + \frac{x^7}{7 \times 6 \times 4 \times 3} + \cdots.$$

EXAMPLE 3.3. Consider the regular linear system:

$$\frac{d}{dx} u = \begin{pmatrix} 0 & \frac{1}{1-x^2} \\ \beta & 0 \end{pmatrix} u, \quad |x| < 1$$

$$= \begin{pmatrix} 0 & 1 + x^2 + x^4 + \cdots \\ \beta & 0 \end{pmatrix} u,$$

where $\beta = -\alpha(\alpha + 1)$ with $\alpha > -1$.

Solution. Using Theorem 2.1, it follows that

$$u_1 = u_0 + \int \begin{pmatrix} 0 & 1 + x^2 + x^4 + \cdots \\ \beta & 0 \end{pmatrix} u_0 \, dx$$

$$= u_0 + \begin{pmatrix} 0 & x + \frac{x^3}{3} + \frac{x^5}{5} + \cdots \\ \beta x & 0 \end{pmatrix} u_0,$$

$$u_2 = u_0 + \int \begin{pmatrix} 0 & 1 + x^2 + x^4 + \cdots \\ \beta & 0 \end{pmatrix} u_1 \, dx$$

$$= u_1 + \int \begin{pmatrix} 0 & 1 + x^2 + x^4 + \cdots \\ \beta & 0 \end{pmatrix} \begin{pmatrix} 0 & x + \frac{x^3}{3} + \frac{x^5}{5} + \cdots \\ \beta x & 0 \end{pmatrix} u_0 \, dx$$

$$= u_1 + \int \begin{pmatrix} \beta(x + x^3 + x^5 + \cdots) & 0 \\ 0 & \beta(x + \frac{x^3}{3} + \frac{x^5}{5} + \cdots) \end{pmatrix} u_0 \, dx$$

$$= u_1 + \begin{pmatrix} \beta(\frac{x^2}{2} + \frac{x^4}{4} + \frac{x^6}{6} + \cdots) & 0 \\ 0 & \beta(\frac{x^2}{2} + \frac{x^4}{4 \times} + \frac{x^6}{6 \times 5} + \cdots) \end{pmatrix} u_0,$$

$$u_3 = u_0 + \int \begin{pmatrix} 0 & 1 + x^2 + x^4 + \cdots \\ \beta & 0 \end{pmatrix} u_2 \, dx$$

$$= u_2 + \int \begin{pmatrix} 0 & 1 + x^2 + x^4 + \cdots \\ \beta & 0 \end{pmatrix} \begin{pmatrix} \beta(\frac{x^2}{2} + \frac{x^4}{4} + \frac{x^6}{6} + \cdots) & 0 \\ 0 & \beta(\frac{x^2}{2} + \frac{x^4}{4 \times} + \frac{x^6}{6 \times 5} + \cdots) \end{pmatrix} u_0 \, dx$$

$$= u_2 + \int \begin{pmatrix} 0 & \beta[\frac{x^2}{2} + (\frac{1}{2} + \frac{1}{4 \times 3})x^4 + (\frac{1}{2} + \frac{1}{4 \times 3} + \frac{1}{6 \times 5})x^6 + \cdots] \\ \beta^2(\frac{x^2}{2} + \frac{x^4}{4 \times 3} + \frac{x^6}{6 \times 5} + \cdots) & 0 \end{pmatrix} u_0 \, dx$$

$$= u_2 + \begin{pmatrix} 0 & \beta[\frac{x^3}{3 \times 2} + (\frac{1}{2} + \frac{1}{4 \times 3})\frac{x^5}{5} + (\frac{1}{2} + \frac{1}{4 \times 3} + \frac{1}{6 \times 5})\frac{x^7}{7} + \cdots] \\ \beta^2(\frac{x^3}{3 \times 2} + \frac{x^5}{5 \times 4 \times 3} + \frac{x^7}{7 \times 6 \times 5} + \cdots) & 0 \end{pmatrix} u_0,$$

$$u_4 = u_0 + \int \begin{pmatrix} 0 & 1 + x^2 + x^4 + \cdots \\ \beta & 0 \end{pmatrix} u_3 \, dx$$

$$= u_3 + \int \begin{pmatrix} 0 & 1 + x^2 + x^4 + \cdots \\ \beta & 0 \end{pmatrix}$$

$$\begin{pmatrix} 0 & \beta[\frac{x^3}{3 \times 2} + (\frac{1}{2} + \frac{1}{4 \times 3})\frac{x^5}{5} + (\frac{1}{2} + \frac{1}{4 \times 3} + \frac{1}{6 \times 5})\frac{x^7}{7} + \cdots] \\ \beta^2(\frac{x^3}{3 \times 2} + \frac{x^5}{5 \times 4 \times 3} + \frac{x^7}{7 \times 6 \times 5} + \cdots) & 0 \end{pmatrix} u_0 \, dx$$

$$= u_3 +$$

$$\int \begin{pmatrix} \beta^2[\frac{x^3}{3\times2}+(\frac{1}{3\times2}+\frac{1}{5\times4\times3})x^5+\cdots) & 0 \\ 0 & \beta^2[\frac{x^3}{3\times2}+(\frac{1}{2}+\frac{1}{4\times3})\frac{x^5}{5}+(\frac{1}{2}+\frac{1}{4\times3}+\frac{1}{6\times5})\frac{x^7}{7}+\cdots] \end{pmatrix} u_0\, dx$$

$$= u_3 +$$

$$\begin{pmatrix} \beta^2[\frac{x^4}{4\times3\times2}+(\frac{1}{3\times2}+\frac{1}{5\times4\times3})\frac{x^6}{6}+\cdots] & 0 \\ 0 & \beta^2[\frac{x^4}{4\times3\times2}+(\frac{1}{2}+\frac{1}{4\times3})\frac{x^6}{6\times5}+(\frac{1}{2}+\frac{1}{4\times3}+\frac{1}{6\times5})\frac{x^8}{8\times7}+\cdots] \end{pmatrix} u_0.$$

The other $u_j, j = 5, 6, \ldots$, are similarly calculated. The resulting two solutions u's, corresponding to $u_0 = \begin{pmatrix} 1 \\ 0 \end{pmatrix}$ and $u_0 = \begin{pmatrix} 0 \\ 1 \end{pmatrix}$, where

$$u = \lim_{j\to\infty} u_j,$$

are linearly independent.

EXAMPLE 3.4. Consider the regular linear equation (the Legendre equation)

$$(1 - x^2)y'' - 2xy' + \alpha(\alpha + 1)y = 0, \quad |x| < 1. \tag{3.2}$$

Solution.
Method 1. The substitutions: $u_1 = y, u_2 = (1 - x^2)y'$ (or the substitutions: $u_1 = y, u_2 = y'$, might as well be used) lead the above (3.2) to the regular linear system

$$\frac{d}{dx} u = \begin{pmatrix} 0 & \frac{1}{1-x^2} \\ \beta & 0 \end{pmatrix} u$$

$$= \begin{pmatrix} 0 & 1 + x^2 + x^4 + \cdots \\ \beta & 0 \end{pmatrix} u,$$

where $\beta = -\alpha(\alpha + 1)$.

Observe that this system is the system in Example 3.3. Thus, employing Theorem 2.3 and Example 3.3, it follows that

$$u_1 = u_0 + \begin{pmatrix} 0 & x + \frac{x^3}{3} + \frac{x^5}{5} + \cdots \\ \beta x & 0 \end{pmatrix} u_0,$$

$$u_2 = u_1 + \begin{pmatrix} \beta(\frac{x^2}{2} + \frac{x^4}{4} + \frac{x^6}{6} + \cdots) & 0 \\ 0 & \beta(\frac{x^2}{2} + \frac{x^4}{4\times} + \frac{x^6}{6\times5} + \cdots) \end{pmatrix} u_0,$$

$$u_3 = u_2 +$$

$$\begin{pmatrix} 0 & \beta[\frac{x^3}{3\times2}+(\frac{1}{2}+\frac{1}{4\times3})\frac{x^5}{5}+(\frac{1}{2}+\frac{1}{4\times3}+\frac{1}{6\times5})\frac{x^7}{7}+\cdots] \\ \beta^2(\frac{x^3}{3\times2}+\frac{x^5}{5\times4\times3}+\frac{x^7}{7\times6\times5}+\cdots) & 0 \end{pmatrix} u_0,$$

$$u_4 = u_3 +$$

$$\begin{pmatrix} \beta^2[\frac{x^4}{4\times3\times2}+(\frac{1}{3\times2}+\frac{1}{5\times4\times3})\frac{x^6}{6}+\cdots] & 0 \\ 0 & \beta^2[\frac{x^4}{4\times3\times2}+(\frac{1}{2}+\frac{1}{4\times3})\frac{x^6}{6\times5}+(\frac{1}{2}+\frac{1}{4\times3}+\frac{1}{6\times5})\frac{x^8}{8\times7}+\cdots] \end{pmatrix} u_0,$$

and that the other $u_j, j = 5, 6, \ldots$, are similarly calculated.

Thus, corresponding to $u_0 = \begin{pmatrix} 1 \\ 0 \end{pmatrix}$ and $u_0 = \begin{pmatrix} 0 \\ 1 \end{pmatrix}$, respectively, the resulting two linearly independent solutions y_1 and y_2, which are the first components of the two u's, respectively, where

$$u = \lim_{j\to\infty} u_j,$$

become:

$$y_1 = 1 - \frac{\alpha(\alpha + 1)}{2!}x^2 + \frac{\alpha(\alpha - 2)(\alpha + 1)(\alpha + 3)}{4!}x^4 + \cdots,$$

$$y_2 = x - \frac{(\alpha - 1)(\alpha + 2)}{3!} x^3 + \cdots .$$

Here used were the calculations:

$$\frac{\beta}{4} + \frac{\beta^2}{4 \times 3 \times 2} = \frac{\alpha(\alpha + 1)(\alpha + 3)(\alpha - 2)}{4!},$$

$$\frac{\beta}{3 \times 2} + \frac{1}{3} = -\frac{(\alpha - 1)(\alpha + 2)}{3!},$$

$$\frac{\beta}{2} = -\frac{\alpha(\alpha + 1)}{2!},$$

$$1,$$

$$1,$$

which are the coefficients of x^4, x^3, x^2, x, and x^0, respectively.

Method 2. By Corollary 2.4, the solution y takes the form

$$y = \sum_{k=0}^{\infty} c_k x^k,$$

where the constants c_k can be determined by substituting this y into (3.2) and equating the coefficients of x^k on both sides. The result becomes:

$$y' = \sum_{k=0}^{\infty} k c_k x^{k-1},$$

$$y'' = \sum_{k=0}^{\infty} k(k - 1) c_k x^{k-2},$$

$$0 = (1 - x^2)y'' - 2xy' + \alpha(\alpha + 1)y$$

$$= [\sum_{k=0}^{\infty} k(k - 1)c_k x^{k-2} - \sum_{k=0}^{\infty} k(k - 1)c_k x^k] - \sum_{k=0}^{\infty} 2k c_k x^k + \sum_{k=0}^{\infty} \alpha(\alpha + 1)c_k x^k.$$

Since

$$\sum_{k=0}^{\infty} k(k - 1)c_k x^{k-2} = \sum_{k=0}^{\infty} (k + 2)(k + 1)c_{k+2} x^k,$$

it follows that

$$(k + 2)(k + 1)c_{k+2} = [k^2 + k - \alpha(\alpha + 1)]c_k$$
$$= -(\alpha - k)(\alpha + k + 1)c_k,$$

or that

$$c_{k+2} = -\frac{(\alpha - k)(\alpha + k + 1)}{(k + 1)(k + 2)}.$$

In terms of c_0 and c_1, a few c_k are:

$$c_2 = -\frac{\alpha(\alpha + 1)}{2!} c_0,$$

$$c_3 = -\frac{(\alpha - 1)(\alpha + 2)}{3!} c_1,$$

$$c_4 = \frac{\alpha(\alpha - 2)(\alpha + 1)(\alpha + 3)}{4!} c_0,$$

$$c_5 = \frac{(\alpha - 1)(\alpha - 3)(\alpha + 2)(\alpha + 4)}{5!} c_1,$$

$$\vdots$$

By choosing $\begin{pmatrix} c_0 \\ c_1 \end{pmatrix} = \begin{pmatrix} 1 \\ 0 \end{pmatrix}$, and $\begin{pmatrix} c_0 \\ c_1 \end{pmatrix} = \begin{pmatrix} 0 \\ 1 \end{pmatrix}$, respectively, the corresponding two linearly independent solutions are:

$$y_1 = 1 - \frac{\alpha(\alpha + 1)}{2!} x^2 + \frac{\alpha(\alpha - 2)(\alpha + 1)(\alpha + 3)}{4!} x^4 + \cdots,$$

$$y_2 = x - \frac{(\alpha - 1)(\alpha + 2)}{3!} x^3 + \frac{(\alpha - 1)(\alpha - 3)(\alpha + 2)(\alpha + 4)}{5!} x^5 + \cdots.$$

EXAMPLE 3.5. Consider the weakly singular linear system

$$\frac{d}{dx} u = (\frac{1}{x} A + x B_1) u, \quad x > 0,$$

where

$$A = \begin{pmatrix} 0 & 1 \\ 0 & -3 \end{pmatrix}, \quad \text{and} \quad B_1 = \begin{pmatrix} 0 & 0 \\ 1 & 0 \end{pmatrix}.$$

Solution.

Method 1. By Theorem 2.10, we have that

$$w_1 = w_0 + \int e^{-tA} e^{2t} B_1 w_0 \, dt$$

$$= w_0 + (2 - A)^{-1} e^{(2-A)t} B_1 w_0,$$

$$v_1 = e^{tA} w_1$$

$$= w_0 + (2 - A)^{-1} e^{2t} B_1 w_0,$$

$$u_1 = v_1|_{t=\ln(x)}$$

$$= w_0 + (2 - A)^{-1} x^2 B_1 w_0,$$

and that

$$w_2 = w_0 + \int e^{-tA} e^{2t} B_1 v_1 \, dt$$

$$= w_1 + (4 - A)^{-1} e^{(4-A)t} B_1 (2 - A)^{-1} B_1 w_0,$$

$$v_2 = e^{tA} w_2$$

$$= v_1 + (4 - A)^{-1} e^{4t} B_1 (2 - A)^{-1} B_1 w_0,$$

$$u_2 = v_2|_{t=\ln(x)}$$

$$= u_1 + (4 - A)^{-1} x^4 B_1 (2 - A)^{-1} B_1 w_0,$$

where $w_0 = \beta_1 = \begin{pmatrix} 1 \\ 0 \end{pmatrix}$ is an eigenvector for the eigenvalue $\lambda_1 = 0$ of A. The other w_j, v_j and $u_j, j = 3, 4, \ldots$, are similarly computed.

However, corresponding to the eigenvector $w_0 = \beta_2 = \begin{pmatrix} 1 \\ -3 \end{pmatrix}$ for the eigenvalue $\lambda_2 = -3$ of A, we have that

$$w_1 = w_0 + \int e^{-tA} e^{2t} B_1 e^{-3t} w_0 \, dt$$

$$= w_0 + (-1 - A)^{-1} e^{(-1-A)t} B_1 w_0,$$

$$v_1 = e^{tA} w_1$$

$$= e^{-3t} w_0 + (-1 - A)^{-1} e^{-t} B_1 w_0,$$

$$u_1 = v_1|_{t=\ln(x)}$$

$$= x^{-3} w_0 + (-1 - A)^{-1} x^{-1} B_1 w_0,$$

and that

$$w_2 = w_0 + \int e^{-tA} e^{2t} B_1 v_1 \, dt$$

$$= w_1 + (1 - A)^{-1} e^{t(1-A)} B_1 (-1 - A)^{-1} B_1 w_0,$$

$$v_2 = e^{tA} w_2$$

$$= v_1 + (1 - A)^{-1} e^{t} B_1 (-1 - A)^{-1} B_1 w_0,$$

$$u_2 = v_2|_{t=\ln(x)}$$

$$= u_1 + (1 - A)^{-1} x B_1 (-1 - A)^{-1} B_1 w_0.$$

The other w_j, v_j, and $u_j, j = 3, 4, \ldots$, are similarly calculated.

Here note that although the two eigenvalues $\lambda_1 = 0$ and $\lambda_2 = -3$ differ by the integer 3, no logarithmic terms are involved in the power series solutions. This is predicted by Theorem 2.10, as $l = 0 - (-3) = 3$ has no sum $\sum$ of the type in Theorem 2.10.

Method 2. The matrix A has the eigenvalues $\{\lambda_1, \lambda_2\} = \{0, -3\}$ as in Method 1. By Corollary 2.11, the first solution y_1 takes the form

$$y_1 = x^{\lambda_1} \sum_{k=0}^{\infty} x^k c_k = \sum_{k=0}^{\infty} x^k c_k,$$

and, after being substituted into the singular system in this example, it satisfies:

$$y_1' = \sum_{k=0}^{\infty} k x^{k-1} c_k = \sum_{k=0}^{\infty} (k+1) x^k c_{k+1}$$

$$= (\frac{1}{x} A + x B_1) \sum_{k=0}^{\infty} x^k c_k$$

$$= \sum_{k=0}^{\infty} x^{k-1} A c_k + \sum_{k=0}^{\infty} x^{k+1} B_1 c_k$$

$$= \sum_{k=-1}^{\infty} x^k A c_{k+1} + \sum_{k=1}^{\infty} x^k B_1 c_{k-1}.$$

It follows by comparing the coefficients of x^k on both sides that

$$k = -1: \quad A c_0 = 0,$$

$$k = 0: \quad c_1 = 0,$$

$$k \geq 1: \quad [(k+1) - A] c_{k+1} = B_1 c_{k-1}.$$

A few terms of c_k's are:

$$c_0 = \beta_1 = \begin{pmatrix} 1 \\ 0 \end{pmatrix}, \quad \text{as in Method 1,}$$

$$c_1 = 0,$$
$$c_2 = (2 - A)^{-1} B_1 c_0,$$
$$c_3 = 0 = c_5 = c_7 = \cdots,$$
$$c_4 = (4 - A)^{-1} B_1 c_2 = (4 - A)^{-1} B_1 (2 - A)^{-1} B_1 c_0,$$
$$\vdots$$

Thus

$$y_1 = c_0 + x^2 (2 - A)^{-1} B_1 c_0 + x^4 (4 - A)^{-1} B_1 (2 - A)^{-1} B_1 c_0 + \cdots.$$

Similarly, the second solution y_2 takes the form

$$y_2 = x^{\lambda_2} \sum_{k=0}^{\infty} x^k c_k = x^{-3} \sum_{k=0}^{\infty} x^k c_k,$$

and satisfies:

$$y_2' = \sum_{k=0}^{\infty} (k - 3) x^{k-4} c_k = \sum_{k=-4}^{\infty} (k + 1) x^k c_{k+4}$$

$$= (\frac{1}{x} A + x B_1) \sum_{k=0}^{\infty} x^{k-3} c_k$$

$$= \sum_{k=0}^{\infty} x^{k-4} A c_k + \sum_{k=0}^{\infty} x^{k-2} B_1 c_k$$

$$= \sum_{k=-4}^{\infty} x^k A c_{k+4} + \sum_{k=-2}^{\infty} x^k B_1 c_{k+2}.$$

This is because no sum $\sum$ of the type in Corollary 2.11 exists.
After comparing coefficients, one derives:

$$k = -4: \quad (3 + A) c_0 = 0,$$
$$k = -3: \quad (2 + A) c_1 = 0,$$
$$k \geq -2: \quad [(k + 1) - A] c_{k+4} = B_1 c_{k+2}.$$

A few terms of c_k's are:

$$c_0 = \beta_2 = \begin{pmatrix} 1 \\ -3 \end{pmatrix}, \quad \text{as in Method 1,}$$
$$c_1 = 0,$$
$$c_2 = (-1 - A)^{-1} B_1 c_0,$$
$$c_3 = 0 = c_5 = c_7 = \cdots,$$
$$c_4 = (1 - A)^{-1} B_1 c_2 = (1 - A)^{-1} B_1 (-1 - A)^{-1} B_1 c_0,$$
$$\vdots$$

Thus

$$y_2 = x^{-3} [c_0 + x^2 (-1 - A)^{-1} B_1 c_0 + x^4 (1 - A)^{-1} B_1 (-1 - A)^{-1} B_1 c_0 + \cdots].$$

EXAMPLE 3.6. Consider the weakly singular system

$$\frac{d}{dx}u = [\frac{1}{x}A + xB_1]u, \quad x > 0,$$

where

$$A = \begin{pmatrix} -3 & 0 \\ 0 & -3 \end{pmatrix} \quad B_1 = \begin{pmatrix} 0 & 0 \\ 1 & 0 \end{pmatrix}.$$

Solution.

Method 1. By Theorem 2.10, we have that

$$w_1 = w_0 + \int e^{-tA}e^{2t}B_1e^{-3t}w_0\, dt$$

$$= w_0 + (-1 - A)^{-1}e^{(-1-A)t}B_1w_0,$$

$$v_1 = e^{tA}w_1$$

$$= e^{-3t}w_0 + (-1 - A)^{-1}e^{-t}B_1w_0,$$

$$u_1 = v_1|_{t=\ln(x)}$$

$$= x^{-3}w_0 + (-1 - A)^{-1}x^{-1}B_1w_0,$$

and that

$$w_2 = w_0 + \int e^{-tA}e^{2t}B_1v_1\, dt$$

$$= w_1 + (1 - A)^{-1}e^{t(1-A)}B_1(-1 - A)^{-1}B_1w_0,$$

$$v_2 = e^{tA}w_2$$

$$= v_1 + (1 - A)^{-1}e^{t}B_1(-1 - A)^{-1}B_1w_0,$$

$$u_2 = v_2|_{t=\ln(x)}$$

$$= u_1 + (1 - A)^{-1}xB_1(-1 - A)^{-1}B_1w_0,$$

where $w_0 = \begin{pmatrix} 1 \\ 0 \end{pmatrix}$ and $w_0 = \begin{pmatrix} 0 \\ 1 \end{pmatrix}$, respectively, are two linearly independent eigenvectors, corresponding to the eigenvalue $\lambda_1 = \lambda_2 = -3$.

The other $w_j, v_j,$ and $u_j, j = 3, 4, \ldots,$ are similarly calculated.

Here note that although the two eigenvalues are identical: $\lambda_1 = \lambda_2 = -3$, no logarithmic terms are involved in the power series solutions. This is predicted by Theorem 2.10, as $l = -3 - (-3) = 0$ has no sum $\sum$ of the type in Theorem 2.10.

Method 2. The matrix A has the eigenvalues $\{\lambda_1, \lambda_2\} = \{-3, -3\}$ and the corresponding eigenvectors $\{\beta_1, \beta_2\} = \{\begin{pmatrix} 1 \\ 0 \end{pmatrix}, \begin{pmatrix} 0 \\ 1 \end{pmatrix}\}$, as in Method 1. By Corollary 2.11, both the first solution y_1 and the second solution y_2 take the same form

$$y = x^{\lambda_1}\sum_{k=0}^{\infty}x^kc_k = x^{-3}\sum_{k=0}^{\infty}x^kc_k,$$

and, after being substituted into the singular system in this example, it satisfies:

$$y' = \sum_{k=0}^{\infty}(k - 3)x^{k-4}c_k = \sum_{k=-4}^{\infty}(k + 1)x^kc_{k+4}$$

$$= (\frac{1}{x}A + xB_1)\sum_{k=0}^{\infty}x^{k-3}c_k$$

$$= \sum_{k=0}^{\infty} x^{k-4} A c_k + \sum_{k=0}^{\infty} x^{k-2} B_1 c_k$$

$$= \sum_{k=-4}^{\infty} x^k A c_{k+4} + \sum_{k=-2}^{\infty} x^k B_1 c_{k+2}.$$

It follows by comparing the coefficients of x^k on both sides that

$$k = -4: \quad (3 + A)c_0 = 0,$$
$$k = -3: \quad (2 + A)c_1 = 0,$$
$$k \geq -2: \quad [(k + 1) - A]c_{k+4} = B_1 c_{k+2}.$$

A few terms of c_k's are:

$$c_0 = \beta_1 = \begin{pmatrix} 1 \\ 0 \end{pmatrix} \quad \text{or} \quad c_0 = \beta_2 = \begin{pmatrix} 0 \\ 1 \end{pmatrix}, \text{ as in Method 1,}$$

$$c_1 = 0,$$
$$c_2 = (-1 - A)^{-1} B_1 c_0,$$
$$c_3 = 0 = c_5 = c_7 = \cdots,$$
$$c_4 = (1 - A)^{-1} B_1 c_2 = (1 - A)^{-1} B_1 (-1 - A)^{-1} B_1 c_0,$$

$$\vdots$$

Thus

$$y = x^{-3}[c_0 + x^2(-1 - A)^{-1} B_1 c_0 + x^4(1 - A)^{-1} B_1(-1 - A)^{-1} B_1 c_0 + \cdots],$$

which is y_1 and y_2, respectively, according as $c_0 = \beta_1$ or $c_0 = \beta_2$.

EXAMPLE 3.7. Consider the weakly singular system

$$\frac{d}{dx} u = [\frac{1}{x} A + x B_1]u, \quad x > 0,$$

where

$$A = \begin{pmatrix} 0 & 1 \\ 0 & 0 \end{pmatrix}, \quad \text{and} \quad B_1 = \begin{pmatrix} 0 & 0 \\ -1 & 0 \end{pmatrix}.$$

This system is associated with the Bessel equation of order zero.

Solution.

Method 1. Using Theorem 2.10, we have that

$$w_1 = w_0 + \int e^{-tA} e^{2t} B_1 w_0 \, dt$$

$$= w_0 + (2 - A)^{-1} e^{(2-A)t} B_1 w_0,$$

$$v_1 = e^{tA} w_1$$

$$= w_0 + e^{2t}(2 - A)^{-1} B_1 w_0,$$

$$u_1 = v_1|_{t=\ln(x)}$$

$$= w_0 + x^2(2 - A)^{-1} B_1 w_0,$$

and that

$$w_2 = w_0 + \int e^{-tA} e^{2t} B_1 v_1 \, dt$$

$$= w_1 + (4 - A)^{-1} e^{t(4-A)} B_1 (2 - A)^{-1} B_1 w_0,$$

$$v_2 = e^{tA} w_2$$

$$= v_1 + (4 - A)^{-1} e^{4t} B_1 (2 - A)^{-1} B_1 w_0,$$

$$u_2 = v_2|_{t=\ln(x)}$$

$$= u_1 + (4 - A)^{-1} x^4 B_1 (2 - A)^{-1} B_1 w_0,$$

where $w_0 = \beta_1 = \begin{pmatrix} 1 \\ 0 \end{pmatrix}$ is an eigenvector for the eigenvalue $\lambda_1 = 0$. The other w_j, v_j, and $u_j, j = 3, 4, \ldots$, are similarly calculated.

However, corresponding to the generalized eigenvector $w_0 = \beta_2 = \begin{pmatrix} 0 \\ 1 \end{pmatrix}$ satisfying both $A^2 \beta_2 = 0$ and $A\beta_2 = \beta_1$, for the generalized eigenvalue $\lambda_1 = \lambda_2 = 0$, we have that

$$w_1 = w_0 + \int e^{-tA} e^{2t} B_1 (w_0 + tAw_0) \, dt$$

$$= w_0 + (2 - A)^{-1} e^{(2-A)t} B_1 w_0 + (2 - A)^{-1} t e^{(2-A)t} B_1 A w_0$$

$$- (2 - A)^{-2} e^{(2-A)t} B_1 A w_0,$$

$$v_1 = e^{tA} w_1$$

$$= w_0 + tAw_0 + (2 - A)^{-1} e^{2t} B_1 w_0 +$$

$$+ (2 - A)^{-1} t e^{2t} B_1 A w_0 - (2 - A)^{-2} e^{2t} B_1 A w_0,$$

$$u_1 = v_1|_{t=\ln(x)}$$

$$= w_0 + Aw_0 \ln(x) + (2 - A)^{-1} x^2 B_1 w_0$$

$$+ (2 - A)^{-1} [x^2 \ln(x)] B_1 A w_0 - (2 - A)^{-2} x^2 B_1 A w_0,$$

and that

$$w_2 = w_0 + \int e^{-tA} e^{2t} B_1 v_1 \, dt$$

$$= w_1 + \int e^{-tA} e^{2t} B_1 [(2-A)^{-1} e^{2t} B_1 w_0$$

$$+ (2-A)^{-1} t e^{2t} B_1 A w_0 - (2-A)^{-2} e^{2t} B_1 A w_0] \, dt$$

$$= w_1 + (4-A)^{-1} e^{t(4-A)} B_1 (2-A)^{-1} B_1 w_0 + (4-A)^{-1} t e^{t(4-A)} B_1 (2-A)^{-1} B_1 A w_0$$

$$- (4-A)^{-2} e^{t(4-A)} B_1 (2-A)^{-1} B_1 A w_0 - (4-A)^{-1} e^{t(4-A)} B_1 (2-A)^{-2} B_1 A w_0,$$

$$v_2 = e^{tA} w_2$$

$$= v_1 + (4 - A)^{-1} e^{4t} B_1 (2 - A)^{-1} B_1 w_0 + (4 - A)^{-1} t e^{4t} B_1 (2 - A)^{-1} B_1 A w_0$$

$$- (4 - A)^{-2} e^{4t} B_1 (2 - A)^{-1} B_1 A w_0 - (4 - A)^{-1} e^{4t} B_1 (2 - A)^{-2} B_1 A w_0,$$

$$u_2 = v_2|_{t=\ln(x)}$$

$$= u_1 + (4 - A)^{-1} x^4 B_1 (2 - A)^{-1} B_1 w_0 + (4 - A)^{-1} [x^4 \ln(x)] B_1 (2 - A)^{-1} B_1 A w_0$$

$$- (4 - A)^{-2} x^4 B_1 (2 - A)^{-1} B_1 A w_0 - (4 - A)^{-1} x^4 B_1 (2 - A)^{-2} B_1 A w_0.$$

The other w_j, v_j, and $u_j, j = 3, 4, \ldots$, are similarly calculated.

A logarithmic term $t = \ln(x)$ is involved, as predicted by Theorem 2.10 since (C2) is satisfied.

Method 2. The matrix A has the eigenvalues $\{\lambda_1, \lambda_2\} = \{0, 0\}$ and the corresponding generalized eigenvectors $\{\beta_1, \beta_2\} = \{\begin{pmatrix} 1 \\ 0 \end{pmatrix}, \begin{pmatrix} 0 \\ 1 \end{pmatrix}\}$, as in Method 1. Here $A\beta_2 = \beta_1$. By Corollary 2.11, the first solution y_1 takes the form

$$y_1 = x^{\lambda_1} \sum_{k=0}^{\infty} x^k c_k = \sum_{k=0}^{\infty} x^k c_k,$$

and, after being substituted into the singular system in this example, it satisfies:

$$y_1' = \sum_{k=0}^{\infty} k x^{k-1} c_k = \sum_{k=0}^{\infty} (k+1) x^k c_{k+1}$$

$$= \left(\frac{1}{x} A + x B_1\right) \sum_{k=0}^{\infty} x^k c_k$$

$$= \sum_{k=0}^{\infty} x^{k-1} A c_k + \sum_{k=0}^{\infty} x^{k+1} B_1 c_k$$

$$= \sum_{k=-1}^{\infty} x^k A c_{k+1} + \sum_{k=1}^{\infty} x^k B_1 c_{k-1}.$$

It follows by comparing the coefficients of x^k on both sides that

$$k = -1: \quad A c_0 = 0,$$
$$k = 0: \quad c_1 = 0,$$
$$k \geq 1: \quad [(k+1) - A] c_{k+1} = B_1 c_{k-1}.$$

A few terms of c_k's are:

$$c_0 = \beta_1 = \begin{pmatrix} 1 \\ 0 \end{pmatrix}, \quad \text{as in Method 1,}$$

$$c_1 = 0,$$

$$c_2 = (2 - A)^{-1} B_1 c_0,$$

$$c_3 = 0 = c_5 = c_7 = \cdots,$$

$$c_4 = (4 - A)^{-1} B_1 c_2 = (4 - A)^{-1} B_1 (2 - A)^{-1} B_1 c_0,$$

$$\vdots$$

Thus

$$y_1 = c_0 + x^2 (2 - A)^{-1} B_1 c_0 + x^4 (4 - A)^{-1} B_1 (2 - A)^{-1} B_1 c_0 + \cdots$$

$$= \sum_{k=1}^{\infty} x^{2k} \prod_{j=1}^{k} [(2j - A)^{-1} B_1 c_0].$$

Similarly, it follows from Corollary 2.11 that the second solution y_2 takes the form

$$y_2 = y_1 \ln(x) + \sum_{k=0}^{\infty} x^k c_k,$$

corresponding to the generalized eigenpair $\{0, \beta_2\} = \{0, \begin{pmatrix} 0 \\ 1 \end{pmatrix}\}$, for which $A\beta_2 = \beta_1$.
After substitution into the singular system in this example, y_2 satisfies:

$$y_2' = y_1' \ln(x) + \frac{y_1}{x} + \sum_{k=0}^{\infty} (k+1)x^k c_{k+1}$$

$$= (\frac{1}{x}A + xB_1)(y_1 \ln(x) + \sum_{k=0}^{\infty} x^k c_k)$$

$$= (\frac{1}{x}A + xB_1)y_1 \ln(x) + \sum_{k=0}^{\infty} x^{k-1} A c_k + \sum_{k=0}^{\infty} x^{k+1} B_1 c_k.$$

Since y_1 is also a solution of the singular system, it follows that

$$\frac{y_1}{x} + \sum_{k=0}^{\infty}(k+1)x^k c_{k+1} = x^{-1}\beta_1 + \sum_{k=1}^{\infty} x^{2k-1} \prod_{j=1}^{k}[(2j-A)^{-1}B_1]\beta_1] + \sum_{k=0}^{\infty}(k+1)x^k c_{k+1}$$

$$= \sum_{k=-1,1,3,5,\ldots} x^k \prod_{j=1}^{\frac{k+1}{2}}[(2j-A)^{-1}B_1]\beta_1 + \sum_{k=0}^{\infty}(k+1)x^k c_{k+1}$$

$$= \sum_{k=-1}^{\infty} x^k A c_{k+1} + \sum_{k=1}^{\infty} x^k B_1 c_{k-1}.$$

On comparing the coefficients of x^k on both sides, one derives:

$k = -1:$ $\beta_1 = A c_0,$

$k = 0:$ $c_1 = 0,$

$k = 1, 3, 5, \ldots:$ $\prod_{j=1}^{\frac{k+1}{2}}[(2j-A)^{-1}B_1]\beta_1 + (k+1)c_{k+1} = A c_{k+1} + B_1 c_{k-1},$

$k = 2, 4, 6, \ldots:$ $(k+1)c_{k+1} = A c_{k+1} + B_1 c_{k-1}.$

A few terms of c_k's are:

$c_0 = \beta_2,$

$c_1 = 0,$

$c_3 = 0 = c_5 = c_7 = \cdots,$

$c_2 = (2 - A)^{-1}B_1 c_0 - (2 - A)^{-2}B_1\beta_1,$

$c_4 = (4 - A)^{-1}B_1 c_2 - (4 - A)^{-1}(4 - A)^{-1}B_1(2 - A)^{-1}B_1\beta_1$

$\quad = (4 - A)^{-1}B_1(2 - A)^{-1}B_1 c_0 - (4 - A)^{-1}B_1(2 - A)^{-2}B_1\beta_1$

$\quad - (4 - A)^{-2}B_1(2 - A)^{-1}B_1\beta_1,$

$\quad \vdots$

Thus

$$y_2 = y_1 \ln(x) + \beta_2 + x^2[(2 - A)^{-1}B_1 c_0 - (2 - A)^{-2}B_1\beta_1]$$

$$+ x^4[(4 - A)^{-1}B_1(2 - A)^{-1}B_1 c_0 - (4 - A)^{-1}B_1(2 - A)^{-2}B_1\beta_1$$

$$- (4 - A)^{-2}B_1(2 - A)^{-1}B_1\beta_1].$$

EXAMPLE 3.8. Consider the weakly singular system

$$\frac{d}{dx}u = [\frac{1}{x}A + xB_1]u,$$

where

$$A = \begin{pmatrix} 0 & 1 \\ 0 & -2 \end{pmatrix}, \quad \text{and} \quad B_1 = \begin{pmatrix} 0 & 0 \\ 1 & 0 \end{pmatrix}.$$

Solution.
Method 1. Using Theorem 2.10, we have that

$$w_1 = w_0 + \int e^{-tA}e^{2t}B_1w_0\,dt$$

$$= w_0 + (2 - A)^{-1}e^{(2-A)t}B_1w_0,$$

$$v_1 = e^{tA}w_1$$

$$= w_0 + (2 - A)^{-1}e^{2t}B_1w_0,$$

$$u_1 = v_1|_{t=\ln(x)}$$

$$= w_0 + (2 - A)^{-1}x^2B_1w_0,$$

and that

$$w_2 = w_0 + \int e^{-tA}e^{2t}B_1v_1\,dt$$

$$= w_1 + (4 - A)^{-1}e^{t(4-A)}B_1(2 - A)^{-1}B_1w_0,$$

$$v_2 = e^{tA}w_2$$

$$= v_1 + (4 - A)^{-1}e^{4t}B_1(2 - A)^{-1}B_1w_0,$$

$$u_2 = v_2|_{t=\ln(x)}$$

$$= u_1 + (4 - A)^{-1}x^4B_1(2 - A)^{-1}B_1w_0,$$

where $w_0 = \beta_1 = \begin{pmatrix} 1 \\ 0 \end{pmatrix}$ is an eigenvector for the eigenvalue $\lambda_1 = 0$. The other w_j, v_j, and $u_j, j = 3, 4, \ldots$, are similarly calculated.

However, corresponding to the eigenvector $w_0 = \beta_2 = \begin{pmatrix} 1 \\ -2 \end{pmatrix}$ for the eigenvalue $\lambda_2 = -2$, we have that

$$e^{tA} = M_{1,0} + e^{-2t}M_{2,0},$$

$$M_{1,0} = \begin{pmatrix} 1 & \frac{1}{2} \\ 0 & 0 \end{pmatrix}, \quad M_{2,0} = \begin{pmatrix} 0 & -\frac{1}{2} \\ 0 & 1 \end{pmatrix},$$

$$M_{1,0}^2 = M_{1,0}, \quad M_{2,0}^2 = M_{2,0}, \quad M_{1,0}M_{2,0} = 0 = M_{2,0}M_{1,0},$$

$$w_1 = w_0 + \int e^{-tA}e^{2t}B_1e^{-2t}w_0\,dt$$

$$= w_0 + tM_{1,0}B_1w_0 + \frac{e^{2t}}{2}M_{2,0}B_1w_0,$$

$$v_1 = e^{tA}w_1$$

$$= e^{-2t}w_0 + tM_{1,0}B_1w_0 + \frac{1}{2}M_{2,0}B_1w_0,$$

$$u_1 = v_1|_{t=\ln(x)}$$

$$= x^{-2}w_0 + [\ln(x)]M_{1,0}B_1w_0 + \frac{1}{2}M_{2,0}B_1w_0,$$

and that

$$w_2 = w_1 + \int e^{-tA}e^{2t}B_1(tM_{1,0}B_1w_0 + \frac{1}{2}M_{2,0}B_1w_0)\,dt$$

$$= w_1 + \frac{1}{2}(2-A)^{-1}e^{(2-A)t}B_1M_{2,0}B_1w_0$$

$$+ t(2-A)^{-1}e^{(2-A)t}B_1M_{1,0}B_1w_0 - (2-A)^{-2}e^{(2-A)t}B_1M_{1,0}B_{1,0}w_0,$$

$$v_2 = e^{tA}w_2$$

$$= v_1 + \frac{1}{2}(2-A)^{-1}e^{2t}B_1M_{2,0}B_1w_0$$

$$+ t(2-A)^{-1}e^{2t}B_1M_{1,0}B_1w_0 - (2-A)^{-2}e^{2t}B_1M_{1,0}B_1w_0,$$

$$u_2 = v_2|_{t=\ln(x)}$$

$$= u_1 + \frac{1}{2}(2-A)^{-1}x^2B_1M_{2,0}B_1w_0$$

$$+ [\ln(x)](2-A)^{-1}x^2B_1M_{1,0}B_1w_0 - (2-A)^{-2}x^2B_1M_{1,0}B_1w_0.$$

Here

$$M_{1,0}B_1w_0 = M_{1,0}B_1\beta_2 = \frac{1}{2}\beta_1 \quad \text{and} \quad M_{2,0}B_1w_0 = M_{2,0}B_1\beta_2 = -\frac{1}{2}\beta_2.$$

The other w_j, v_j, and $u_j, j = 3, 4, \ldots$, are similarly calculated.

A logarithmic term $t = \ln(x)$ is involved, as predicted by Theorem 2.10 since $l = 0 - (-2) = 2$ has the sum $\sum$ of the type in Theorem 2.10.

Method 2. The matrix A has the eigenpairs:

$$\{\lambda_1, \beta_1\} = \{0, \begin{pmatrix} 1 \\ 0 \end{pmatrix}\} \quad \text{and} \quad \{\lambda_2, \beta_2\} = \{-2, \begin{pmatrix} 1 \\ -2 \end{pmatrix}\},$$

as in Method 1. By Corollary 2.11, the first solution y_1 takes the form

$$y_1 = x^{\lambda_1}\sum_{k=0}^{\infty}x^k c_k = \sum_{k=0}^{\infty}x^k c_k,$$

and, after being substituted into the singular system in this example, it satisfies:

$$y_1' = \sum_{k=0}^{\infty}kx^{k-1}c_k = \sum_{k=0}^{\infty}(k+1)x^k c_{k+1}$$

$$= (\frac{1}{x}A + xB_1)\sum_{k=0}^{\infty}x^k c_k$$

$$= \sum_{k=0}^{\infty}x^{k-1}Ac_k + \sum_{k=0}^{\infty}x^{k+1}B_1c_k$$

$$= \sum_{k=-1}^{\infty}x^k Ac_{k+1} + \sum_{k=1}^{\infty}x^k B_1c_{k-1}.$$

It follows by comparing the coefficients of x^k on both sides that

$$k = -1: \quad Ac_0 = 0,$$

$$k = 0: \quad c_1 = 0,$$

$$k \geq 1: \quad [(k+1) - A]c_{k+1} = B_1 c_{k-1}.$$

A few terms of c_k's are:

$$c_0 = \beta_1 = \begin{pmatrix} 1 \\ 0 \end{pmatrix}, \quad \text{as in Method 1,}$$

$$c_1 = 0,$$

$$c_2 = (2 - A)^{-1} B_1 c_0,$$

$$c_3 = 0 = c_5 = c_7 = \cdots,$$

$$c_4 = (4 - A)^{-1} B_1 c_2 = (4 - A)^{-1} B_1 (2 - A)^{-1} B_1 c_0,$$

$$\vdots$$

Thus

$$y_1 = c_0 + x^2 (2 - A)^{-1} B_1 c_0 + x^4 (4 - A)^{-1} B_1 (2 - A)^{-1} B_1 c_0 + \cdots$$

$$= \sum_{k=1}^{\infty} x^{2k} \prod_{j=1}^{k} [(2j - A)^{-1} B_1 c_0].$$

Similarly, it follows from Corollary 2.11 that the second solution y_2 takes the form

$$y_2 = \alpha y_1 \ln(x) + x^{\lambda_2} \sum_{k=0}^{\infty} x^k c_k$$

$$= \alpha y_1 \ln(x) + x^{-2} \sum_{k=0}^{\infty} x^k c_k,$$

corresponding to the eigenpair $\{\lambda_2, \beta_2\} = \{-2, \begin{pmatrix} 1 \\ -2 \end{pmatrix}\}$. After substitution into the singular system in this example, y_2 satisfies:

$$y_2' = \alpha y_1' \ln(x) + \alpha \frac{y_1}{x} + \sum_{k=0}^{\infty} (k-2) x^{k-3} c_k$$

$$= (\frac{1}{x} A + x B_1)(\alpha y_1 \ln(x) + \sum_{k=0}^{\infty} x^{k-2} c_k)$$

$$= (\frac{1}{x} A + x B_1)\alpha y_1 \ln(x) + \sum_{k=0}^{\infty} x^{k-3} A c_k + \sum_{k=0}^{\infty} x^{k-1} B_1 c_k.$$

Since y_1 is also a solution of the singular system, it follows that

$$\alpha \frac{y_1}{x} + \sum_{k=-3}^{\infty} (k+1) x^k c_{k+3}$$

$$= \alpha x^{-1} \beta_1 + \alpha \sum_{k=1}^{\infty} x^{2k-1} \prod_{j=1}^{k} [(2j - A)^{-1} B_1] \beta_1] + \sum_{k=-3}^{\infty} (k+1) x^k c_{k+3}$$

$$= \alpha \sum_{k=-1,1,3,5,\ldots} x^k \prod_{j=1}^{\frac{k+1}{2}} [(2j - A)^{-1} B_1] \beta_1 + \sum_{k=-3}^{\infty} (k+1) x^k c_{k+3}$$

$$= \sum_{k=-3}^{\infty} x^k A c_{k+3} + \sum_{k=-1}^{\infty} x^k B_1 c_{k+1}.$$

On comparing the coefficients of x^k on both sides, one derives:

$$k = -3: \quad -2c_0 = Ac_0,$$
$$k = -2: \quad -c_1 = Ac_1,$$
$$k = -1: \quad \alpha\beta_1 = Ac_2 + B_1 c_0,$$
$$k = 0: \quad c_3 = Ac_3 + B_1 c_1,$$
$$k = 1: \quad 2c_4 + \alpha(2 - A)^{-1} B_1\beta_1 = Ac_4 + B_1 c_2,$$
$$k = 3, 5, 7, \ldots: \quad \alpha \prod_{j=1}^{\frac{k+1}{2}} [(2j - A)^{-1} B_1]\beta_1 + (k + 1)c_{k+3} = Ac_{k+3} + B_1 c_{k+1},$$
$$k = 2, 4, 6, \ldots: \quad (k + 1)c_{k+3} = Ac_{k+3} + B_1 c_{k+1}.$$

The α, and a few terms of c_k's involving a constant c, are:

$$c_0 = \beta_2,$$
$$c_1 = 0,$$
$$c_3 = 0 = c_5 = c_7 = \cdots,$$
$$\alpha = \frac{1}{2},$$
$$c_2 = \begin{pmatrix} c \\ \frac{1}{2} \end{pmatrix}, \quad \text{since } Ac_2 = \alpha\beta_1 - B_1 c_0 = \begin{pmatrix} \alpha \\ -1 \end{pmatrix},$$
$$c_4 = \begin{pmatrix} \frac{1}{2}(\frac{c}{4} - \frac{3}{32}) \\ \frac{1}{4}(c - \frac{1}{8}) \end{pmatrix},$$
$$\vdots$$

Assigning any value for the constant c will do, but assigning $c = -\frac{1}{4}$ is consistent with the results in Method 1, and gives

$$c_2 = \begin{pmatrix} -\frac{1}{4} \\ \frac{1}{2} \end{pmatrix}, c_4 = \begin{pmatrix} -\frac{5}{64} \\ -\frac{3}{32} \end{pmatrix}, \ldots.$$

Thus

$$y_2 = \frac{1}{2} y_1 \ln(x) + x^{-2}\beta_2 + \begin{pmatrix} -\frac{1}{4} \\ \frac{1}{2} \end{pmatrix} + x^2 \begin{pmatrix} -\frac{5}{64} \\ -\frac{3}{32} \end{pmatrix} + \cdots.$$

EXAMPLE 3.9. Consider the weakly singular equation

$$x^2 y'' + 4xy' - x^2 y = 0, \quad x > 0.$$

Solution.
Method 1. As (2.8) states it, the substitution: $u_k = x^{k-1} y^{(k-1)}, k = 1, 2$, leads the above weakly singular equation to the associated weakly singular linear system:

$$\frac{d}{dx} u = (\frac{1}{x} A + x B_1)u, \quad x > 0,$$

where

$$A = \begin{pmatrix} 0 & 1 \\ 0 & -3 \end{pmatrix}, \quad \text{and} \quad B_1 = \begin{pmatrix} 0 & 0 \\ 1 & 0 \end{pmatrix}.$$

Observe that this system is the system in Example 3.5. Thus, employing Theorem 2.12 and Example 3.5, it follows that, corresponding to the eigenpair

$$(w_0, \lambda_1) = (\begin{pmatrix} 1 \\ 0 \end{pmatrix}, 0),$$

there are

$$u_1 = w_0 + (2 - A)^{-1} x^2 B_1 w_0$$

$$= \begin{pmatrix} 1 \\ 0 \end{pmatrix} + \frac{x^2}{10} \begin{pmatrix} 1 \\ 2 \end{pmatrix},$$

$$u_2 = u_1 + (4 - A)^{-1} x^4 B_1 (2 - A)^{-1} B_1 w_0$$

$$= u_1 + \frac{x^4}{280} \begin{pmatrix} 1 \\ 4 \end{pmatrix}.$$

The other $u_j, j = 3, 4, \ldots$, are similarly calculated. The resulting solution y_1, which is the first component of $\lim_{j \to \infty} u_j$, becomes:

$$y_1 = 1 + \frac{x^2}{10} + \frac{x^4}{280} + \cdots.$$

In the same way as above, corresponding to the eigenpair

$$(w_0, \lambda_2) = (\begin{pmatrix} 1 \\ 3 \end{pmatrix}, -3),$$

there are

$$u_1 = x^{-3} w_0 + (-1 - A)^{-1} x^{-1} B_1 w_0$$

$$= x^{-3} \begin{pmatrix} 1 \\ 3 \end{pmatrix} + \frac{x^{-1}}{-2} \begin{pmatrix} 1 \\ -1 \end{pmatrix},$$

$$u_2 = u_1 + (1 - A)^{-1} x B_1 (-1 - A)^{-1} B_1 w_0$$

$$= u_1 + \frac{x}{-8} \begin{pmatrix} 1 \\ 1 \end{pmatrix}.$$

The other $u_j, j = 3, 4, \ldots$, are similarly calculated. The resulting second solution y_2, which is the first component of $\lim_{j \to \infty} u_j$, becomes:

$$y_2 = x^{-3} + \frac{x^{-1}}{-2} + \frac{x}{-8} + \cdots.$$

Method 2. The corresponding matrix A has the eigenvalues

$$\{\lambda_1, \lambda_2\} = \{0, -3\},$$

as in Method 1. By Corollary 2.13, the first solution y_1 takes the form:

$$y_1 = x^{\lambda_1} \sum_{k=0}^{\infty} x^k c_k = \sum_{k=0}^{\infty} x^k c_k,$$

and, after being substituted into the singular equation in this example, it satisfies:

$$y_1' = \sum_{k=0}^{\infty} k x^{k-1} c_k,$$

$$y_1'' = \sum_{k=0}^{\infty} k(k-1)x^{k-1}c_k,$$

$$0 = x^2 y_1'' + 4xy_1' - x^2 y_1$$

$$= \sum_{k=2}^{\infty} k(k-1)x^k c_k + \sum_{k=1}^{\infty} 4kx^k c_k - \sum_{k=0}^{\infty} x^{k+2} c_k,$$

where

$$\sum_{k=0}^{\infty} x^{k+2} c_k = \sum_{k=2}^{\infty} x^k c_{k-2}.$$

It follows by comparing the coefficients of x^k on both sides that

$$k = 1: \quad c_1 = 0,$$

$$k \geq 2: \quad c_k = \frac{1}{k(k+3)} c_{k-2}.$$

A few terms of c_k's are:

$$c_1 = 0,$$

$$c_2 = \frac{1}{10} c_0,$$

$$c_3 = 0 = c_1 = c_5 = c_7 = \cdots,$$

$$c_4 = \frac{1}{28} c_2 = \frac{1}{280} c_0,$$

$$\vdots$$

Thus by setting $c_0 = 1$, it follows that

$$y_1 = 1 + \frac{1}{10} x^2 + \frac{1}{280} x^4 + \cdots.$$

Similarly, by Corollary 2.13, the second solution y_2 takes the form:

$$y_2 = x^{\lambda_2} \sum_{k=0}^{\infty} x^k c_k = \sum_{k=0}^{\infty} x^{k-3} c_k,$$

and, after substitution into the singular equation in this example, it satisfies:

$$y_2' = \sum_{k=0}^{\infty} (k-3)x^{k-4} c_k,$$

$$y_2'' = \sum_{k=0}^{\infty} (k-3)(k-4)x^{k-5} c_k,$$

$$0 = x^2 y_2'' + 4xy_2' - x^2 u_2$$

$$= \sum_{k=0}^{\infty} (k-3)(k-4)x^{k-3} c_k + \sum_{k=0}^{\infty} 4(k-3)x^{k-3} c_k - \sum_{k=0}^{\infty} x^{k-1} c_k$$

$$= \sum_{k=-3}^{\infty} k(k-1)x^k c_{k+3} + \sum_{k=-3}^{\infty} 4kx^k c_{k+3} - \sum_{k=-1}^{\infty} x^k c_{k+1}.$$

By comparing the coefficients of x^k, it follows that

$$k = -3: \quad 12c_0 - 12c_0 = 0,$$
$$k = -2: \quad 6c_1 - 8c_1 = 0,$$
$$k \geq -1 \quad k(k+3)c_{k+3} = c_{k+1}.$$

A few terms of c_k's are:

$$c_1 = 0,$$
$$c_2 = -\frac{1}{2}c_0,$$
$$c_3 = 0 = c_5 = c_7 = \cdots,$$
$$c_4 = \frac{1}{4}c_2 = -\frac{1}{8}c_0,$$
$$\vdots$$

Thus by setting $c_0 = 1$, it follows that

$$y_2 = x^{-3} - \frac{1}{2}x^{-1} - \frac{1}{8}x + \cdots.$$

EXAMPLE 3.10. Consider the weakly singular equation, the Bessel equation of order zero:

$$x^2 y'' + xy' + x^2 y = 0, \quad x > 0.$$

Solution.

Method 1. As (2.8) states it, the substitution: $u_1 = y, u_2 = xy'$, leads the above weakly singular equation to the weakly singular system:

$$\frac{d}{dx}u = [\frac{1}{x}A + xB_1]u, \quad x > 0,$$

where

$$A = \begin{pmatrix} 0 & 1 \\ 0 & 0 \end{pmatrix}, \quad \text{and} \quad B_1 = \begin{pmatrix} 0 & 0 \\ -1 & 0 \end{pmatrix}.$$

Observe that this system is the system in Example 3.7. Thus, employing Theorem 2.12 and Example 3.7, it follows that, corresponding to the eigenpair

$$(w_0, \lambda_1) = (\begin{pmatrix} 1 \\ 0 \end{pmatrix}, 0),$$

there are

$$u_1 = w_0 + x^2(2 - A)^{-1}B_1 w_0$$
$$= \begin{pmatrix} 1 \\ 0 \end{pmatrix} + \frac{x^2}{-4}\begin{pmatrix} 1 \\ 2 \end{pmatrix},$$
$$u_2 = u_1 + (4 - A)^{-1}x^4 B_1(2 - A)^{-1}B_1 w_0$$
$$= u_1 + \frac{x^4}{16}\frac{1}{4}\begin{pmatrix} 1 \\ 4 \end{pmatrix}.$$

The other $u_j, j = 3, 4, \ldots$, are similarly calculated. The resuting solution y_1, which is the first component of $\lim_{j \to \infty} u_j$, becomes:

$$y_1 = 1 + \frac{x^2}{-4} + \frac{x^4}{64} + \cdots.$$

In the same way as above, corresponding to the eigenpair

$$(w_0, \lambda_2) = (\begin{pmatrix} 0 \\ 1 \end{pmatrix}, 0)$$

there are

$$u_1 = w_0 + A w_0 \ln(x) + (2 - A)^{-1} x^2 B_1 w_0 + (2 - A)^{-1} [x^2 \ln(x)] B_1 A w_0$$
$$- (2 - A)^{-2} x^2 B_1 A w_0$$

$$= \begin{pmatrix} 0 \\ 1 \end{pmatrix} + \begin{pmatrix} 1 \\ 0 \end{pmatrix} \ln(x) + \begin{pmatrix} 0 \\ 0 \end{pmatrix} + [x^2 \ln(x)] \frac{1}{-4} \begin{pmatrix} 1 \\ 2 \end{pmatrix} - \frac{x^2}{16} \begin{pmatrix} -4 \\ -4 \end{pmatrix},$$

$$u_2 = v_2|_{t=\ln(x)}$$
$$= u_1 + (4 - A)^{-1} x^4 B_1 (2 - A)^{-1} B_1 w_0 + (4 - A)^{-1} [x^4 \ln(x)] B_1 (2 - A)^{-1} B_1 A w_0$$
$$- (4 - A)^{-2} x^4 B_1 (2 - A)^{-1} B_1 A w_0 - (4 - A)^{-1} x^4 B_1 (2 - A)^{-2} B_1 A w_0$$

$$= u_1 + \begin{pmatrix} 0 \\ 0 \end{pmatrix} + [x^4 \ln(x)] \frac{1}{64} \begin{pmatrix} 1 \\ 4 \end{pmatrix} - x^4 \frac{1}{64} \begin{pmatrix} \frac{1}{2} \\ 1 \end{pmatrix} - x^4 \frac{1}{64} \begin{pmatrix} 1 \\ 4 \end{pmatrix}$$

$$= u_1 + [x^4 \ln(x)] \frac{1}{64} \begin{pmatrix} 1 \\ 4 \end{pmatrix} - x^4 \frac{1}{64} \begin{pmatrix} 1 + \frac{1}{2} \\ 5 \end{pmatrix}.$$

The other $u_j, j = 3, 4, \ldots$, are similarly calculated. The resulting second solution y_2, which is the first component of $\lim_{j \to \infty} u_j$, becomes:

$$y_2 = \ln(x) + \frac{x^2}{-4} \ln(x) + \frac{x^4}{64} \ln(x)$$

$$+ \frac{x^2}{4} - x^4 \frac{1}{64} (1 + \frac{1}{2}) + \cdots.$$

Method 2. The corresponding matrix A has the eigenvalues $\{\lambda_1, \lambda_2\} = \{0, 0\}$, as in Method 1. By Corollary 2.13, the first solution y_1 takes the form:

$$y_1 = x^{\lambda_1} \sum_{k=0}^{\infty} x^k c_k$$

$$= \sum_{k=0}^{\infty} x^k c_k,$$

and, after substitution into the regular singular equation in the example, it satisfies:

$$y_1' = \sum_{k=1}^{\infty} k x^{k-1} c_k,$$

$$y_1'' = \sum_{k=2}^{\infty} k(k-1) x^{k-2} c_k,$$

$$0 = x^2 y_1'' + x y_1' + x^2 y$$

$$= \sum_{k=2}^{\infty} k(k-1) x^k c_k + \sum_{k=1}^{\infty} k x^k c_k + \sum_{k=2}^{\infty} x^k c_{k-2}.$$

It follows by comparing the coefficients of x^k on both sides that

$$k = 1: \quad c_1 = 0,$$

$$k = 2: \quad c_2 = -\frac{1}{4}c_0,$$

$$k \geq 2: \quad c_k = -\frac{1}{k^2}c_{k-2}.$$

A few terms of c_k's are:

$$c_1 = 0,$$

$$c_2 = -\frac{1}{4}c_0,$$

$$c_3 = 0 = c_1 = c_5 = c_7 = \cdots,$$

$$c_4 = -\frac{1}{16}c_2 = \frac{1}{64}c_0,$$

$$\vdots$$

Thus, by setting $c_0 = 1$, it follows that

$$y_1 = 1 - x^2\frac{1}{4} + x^4\frac{1}{64} + \cdots$$

$$= \sum_{j=0}^{\infty} x^{2j}\frac{(-1)^j}{2^{2j}(j!)^2}.$$

Similarly, it follows from Corollary 2.13 that the second solution y_2 takes the form:

$$y_2 = y_1 \ln(x) + \sum_{k=1}^{\infty} x^k c_k.$$

After substitution into the singular equation in this example, y_2 satisfies:

$$y_2' = y_1' \ln(x) + \frac{y_1}{x} + \sum_{k=1}^{\infty} kx^{k-1}c_k,$$

$$y_2'' = y_1'' \ln(x) + 2\frac{y_1'}{x} - \frac{y_1}{x^2} + \sum_{k=2}^{\infty} k(k-1)x^{k-2}c_k,$$

$$0 = x^2 y_2'' + xy_2' + x^2 y_2$$

$$= x^2 y_1'' \ln(x) + 2xy_1' - y_1 + \sum_{k=2}^{\infty} k(k-1)x^{k-2}c_k$$

$$+ xy_1' \ln(x) + y_1 + \sum_{k=1}^{\infty} kx^k c_k$$

$$+ x^2 y_1 \ln(x) + \sum_{k=1}^{\infty} x^{k+2}c_k.$$

Since y_1 is also a solution of the singular equation, it follows that

$$0 = 2xy_1' + \sum_{k=2}^{\infty} k(k-1)x^k c_k + \sum_{k=1}^{\infty} kx^k c_k + \sum_{k=3}^{\infty} x^k c_{k-2},$$

where

$$2xy_1' = 2\sum_{j=1}^{\infty}(2j)x^{2j}\frac{(-1)^j}{2^{2j}(j!)^2}$$

$$= 2 \sum_{k=2,4,6,\ldots}^{\infty} k x^k \frac{(-1)^{\frac{k}{2}}}{2^k [(\frac{k}{2})!]^2}.$$

By comparing the coefficients of x^k, it follows that

$$k = 1: \quad c_1 = 0,$$

$$k = 2: \quad 2c_2 - 1 + 2c_2 = 0,$$

$$k = 3, 5, 7, \ldots: \quad c_k = -\frac{1}{k^2} c_{k-2},$$

$$k = 4, 6, 8, \ldots: \quad c_k = -\frac{1}{k^2} (c_{k-2} + 2 \frac{k(-1)^{\frac{k}{2}}}{2^k [(\frac{k}{2})!]^2}).$$

A few terms of c_k's are:

$$c_1 = 0,$$

$$c_2 = \frac{1}{4},$$

$$c_3 = 0 = c_1 = c_5 = c_7 + \cdots,$$

$$c_4 = -\frac{1}{16} (\frac{1}{4} + \frac{1}{8}),$$

$$\vdots$$

Thus

$$y_2 = y_1 \ln(x) + x^2 \frac{1}{4} - x^4 \frac{1}{16} (\frac{1}{4} + \frac{1}{8}) + \cdots.$$

EXAMPLE 3.11. Consider the weakly singular equation

$$x^2 y'' + 3xy' - x^2 y = 0, \quad x > 0.$$

Solution.
Method 1. As (2.8) states it, the substitution: $u_1 = y, u_2 = xy'$, leads the avobe weakly singular equation to the weakly singular system:

$$\frac{d}{dx} u = [\frac{1}{x} A + x B_1] u,$$

where

$$A = \begin{pmatrix} 0 & 1 \\ 0 & -2 \end{pmatrix}, \quad \text{and} \quad B_1 = \begin{pmatrix} 0 & 0 \\ 1 & 0 \end{pmatrix}.$$

Observe that this system is the system in Example 3.8. Thus, employing Theorem 2.12 and Example 3.8, it follows that, corresponding to the eigenpair

$$(w_0, \lambda_1) = (\begin{pmatrix} 1 \\ 0 \end{pmatrix}, 0),$$

there are

$$u_1 = w_0 + (2 - A)^{-1} x^2 B_1 w_0$$

$$= \begin{pmatrix} 1 \\ 0 \end{pmatrix} + \frac{x^2}{8} \begin{pmatrix} 1 \\ 2 \end{pmatrix},$$

$$u_2 = u_1 + (4 - A)^{-1} x^4 B_1 (2 - A)^{-1} B_1 w_0$$

$$= u_1 + x^4 \frac{1}{192} \begin{pmatrix} 1 \\ 4 \end{pmatrix}.$$

The other $u_j, j = 3, 4, \ldots$, are similarly calculated. The resulting solution y_1, which is the first component of $\lim_{j \to \infty} u_j$, becomes:

$$y_1 = 1 + \frac{x^2}{8} + \frac{x^4}{192} + \cdots .$$

In the same way as above, corresponding to the eigenpair

$$(w_0, \lambda_2) = (\begin{pmatrix} 1 \\ -2 \end{pmatrix}, -2),$$

there are

$$u_1 = x^{-2} w_0 + [\ln(x)] M_{1,0} B_1 w_0 + \frac{1}{2} M_{2,0} B_1 w_0$$

$$= x^{-2} \begin{pmatrix} 1 \\ -2 \end{pmatrix} + [\ln(x)] \begin{pmatrix} \frac{1}{2} \\ 0 \end{pmatrix} + \begin{pmatrix} -\frac{1}{2} \\ 1 \end{pmatrix},$$

$$u_2 = u_1 + \frac{1}{2} (2 - A)^{-1} x^2 B_1 M_{2,0} B_1 w_0$$

$$+ [\ln(x)](2 - A)^{-1} x^2 B_1 M_{1,0} B_1 w_0 - (2 - A)^{-2} x^2 B_1 M_{1,0} B_1 w_0$$

$$= u_1 + [x^2 \ln(x)] \frac{1}{8} \begin{pmatrix} \frac{1}{2} \\ 1 \end{pmatrix} - x^2 \begin{pmatrix} \frac{5}{64} \\ \frac{3}{32} \end{pmatrix}.$$

The other $u_j, j = 3, 4, \ldots$, are similarly calculated. The resultant second solution y_2, which is the first component of $\lim_{j \to \infty} u_j$, becomes:

$$y_2 = \frac{1}{2} \ln(x) + \frac{x^2}{16} \ln(x) + x^{-2} + (-\frac{1}{4}) + \frac{-5}{64} x^2 + \cdots .$$

Method 2. The corresponding matrix A has the eigenvalues

$$\{\lambda_1, \lambda_2\} = \{0, -2\},$$

as in Method 1. By Corollary 2.13, the first solution y_1 takes the form:

$$y_1 = x^{\lambda_1} \sum_{k=0}^{\infty} x^k c_k = \sum_{k=0}^{\infty} x^k c_k,$$

and, after being substituted into the singular equation in this example, it satisfies:

$$y_1' = \sum_{k=0}^{\infty} k x^{k-1} c_k,$$

$$y_1'' = \sum_{k=0}^{\infty} k(k-1) x^{k-1} c_k,$$

$$0 = x^2 y_1'' + 3xy_1' - x^2 y_1$$

$$= \sum_{k=2}^{\infty} k(k-1) x^k c_k + \sum_{k=1}^{\infty} 3k x^k c_k - \sum_{k=0}^{\infty} x^{k+2} c_k,$$

where

$$\sum_{k=0}^{\infty} x^{k+2} c_k = \sum_{k=2}^{\infty} x^k c_{k-2}.$$

It follows by comparing the coefficients of x^k on both sides that

$$k = 1 : \quad c_1 = 0,$$

$$k \geq 2: \quad c_k = \frac{1}{k(k+2)} c_{k-2}.$$

A few terms of c_k's are:

$$c_1 = 0,$$

$$c_2 = \frac{1}{8} c_0,$$

$$c_3 = 0 = c_1 = c_5 = c_7 = \cdots ,$$

$$c_4 = \frac{1}{24} c_2 = \frac{1}{192} c_0,$$

$$\vdots$$

Thus by setting $c_0 = 1$, it follows that

$$y_1 = 1 + \frac{1}{10} x^2 + \frac{1}{280} x^4 + \cdots$$

$$= \sum_{j=1}^{\infty} x^{2j} \frac{1}{(2^2)^j (j+1)! j!}.$$

Similarly, it follows from Corollary 2.13 that the second solution y_2 takes the form:

$$y_2 = \alpha y_1 \ln(x) + x^{\lambda_2} \sum_{k=1}^{\infty} x^k c_k,$$

where $\lambda_2 = -2$. After substitution into the singular equation in this example, y_2 satisfies:

$$y_2' = \alpha y_1' \ln(x) + \alpha \frac{y_1}{x} + \sum_{k=0}^{\infty} (k-2) x^{k-3} c_k,$$

$$y_2'' = \alpha y_1'' \ln(x) + 2\alpha \frac{y_1'}{x} - \alpha \frac{y_1}{x^2} + \sum_{k=0}^{\infty} (k-2)(k-3) x^{k-4} c_k,$$

$$0 = x^2 y_2'' + 3xy_2' - x^2 y_2$$

$$= \alpha x^2 y_1'' \ln(x) + 2\alpha x y_1' - \alpha y_1 + \sum_{k=0}^{\infty} (k-2)(k-3) x^{k-2} c_k$$

$$+ \alpha 3 x y_1' \ln(x) + 3\alpha y_1 + \sum_{k=0}^{\infty} 3(k-2) x^{k-2} c_k$$

$$- \alpha x^2 y_1 \ln(x) - \sum_{k=0}^{\infty} x^k c_k.$$

Since y_1 is also a solution of the singular equation, it follows that

$$0 = 2\alpha x y_1' + 2\alpha y_1 + \sum_{k=-2}^{\infty} k(k-1) x^k c_{k+2}$$

$$+ \sum_{k=-2}^{\infty} 3k x^k c_{k+2} - \sum_{k=0}^{\infty} x^k c_k,$$

where

$$2\alpha y_1 = 2\alpha \sum_{j=0}^{\infty} x^{2j} \frac{1}{(2^2)^j (j+1)! j!}$$

$$= 2\alpha \sum_{k=0,2,4,\ldots}^{\infty} x^k \frac{1}{2^k (\frac{k}{2}+1)! (\frac{k}{2})!}$$

and

$$2\alpha x y_1' = 2\alpha \sum_{j=1}^{\infty} (2j) x^{2j} \frac{1}{(2^2)^j (j+1)! j!}$$

$$= 2\alpha \sum_{k=2,4,6,\ldots}^{\infty} k x^k \frac{1}{2^k \frac{k}{2}+1)! (\frac{k}{2})!}.$$

By comparing the coefficients of x^k, it follows that

$$k = -2: \quad 6c_0 - 6c_0 = 0,$$
$$k = -1: \quad 2c_1 - 3c_1 = 0,$$
$$k = 0: \quad c_0 = 2\alpha,$$
$$k = 1,3,5,7,\ldots: \quad k(k+2)c_{k+2} = c_k,$$
$$k = 2,4,6,8,\ldots: \quad k(k+2)c_{k+2} = c_k - 2\alpha(k+1)\frac{1}{2^k(\frac{k}{2}+1)!(\frac{k}{2})!}.$$

A few terms of c_k's are:

$$c_0 = 2\alpha,$$
$$c_1 = 0,$$
$$c_2: \quad \text{arbitrary},$$
$$c_3 = 0 = c_1 = c_5 = c_7 = \cdots,$$
$$c_4 = \frac{1}{8}(c_2 - \frac{3}{4}\alpha),$$
$$\vdots$$

Assigning any value for c_2 and c_0 will do, but assigning $c_2 = -\frac{1}{4}$ and assigning $c_0 = 1$, for which $\alpha = \frac{1}{2}$, is consistent with the results in Method 1. Thus

$$c_0 = 1,$$
$$\alpha = \frac{1}{2},$$
$$c_2 = -\frac{1}{4},$$
$$c_3 = 0 = c_1 = c_5 = c_7 + \cdots,$$
$$c_4 = -\frac{5}{64},$$
$$\vdots$$

and so

$$y_2 = \frac{1}{2} y_1 \ln(x) + x^{-2} - \frac{1}{4} - x^2 \frac{5}{64} + \cdots.$$

4. Proof of the Main Results

We first prove

Theorem 2.1. *On assuming that $\sum_{k=0}^{\infty} x^k B_k$ converges absolutely for each real x satisfying $|x| < r_0$, the successive approximations u_m in (1.3), for each constant vector $u_0 \in \mathbb{C}^2$, converges, and its limit $u(x) \equiv T(x)u_0$ is a power series solution to (1.2). Furthermore, the two solutions u's, corresponding to $u_0 = \begin{pmatrix} 1 \\ 0 \end{pmatrix}$ and $u_0 = \begin{pmatrix} 0 \\ 1 \end{pmatrix}$, respectively, constitute a fundamental set of power series solutions to (1.2). Being written out, $u_m, m = 1, 2, \ldots$, takes the form:*

$$u_1 = u_0 + \left(\sum_{k_1=0}^{\infty} \frac{x^{k_1+1}}{k_1 + 1} B_{k_1} \right) u_0$$

$$u_2 = u_1 + \sum_{k_2=0}^{\infty} \frac{x^{k_2+1}}{(k_2+1) + (k_1+1)} B_{k_2} \sum_{k_1=0}^{\infty} \frac{x^{k_1+1}}{k_1 + 1} B_{k_1} u_0$$

$$\vdots$$

$$u_m = u_{m-1} + \sum_{k_m=0}^{\infty} \frac{x^{k_m+1}}{(k_m+1) + (k_{m-1}+1) + \cdots + (k_1+1)} B_{k_m}$$

$$\sum_{k_{m-1}=0}^{\infty} \frac{x^{k_{m-1}+1}}{(k_{m-1}+1) + (k_{m-2}+1) + \cdots + (k_1+1)} B_{k_{m-1}}$$

$$\cdots \sum_{k_2=0}^{\infty} \frac{x^{k_2+1}}{(k_2+1) + k_1 + 1)} B_{k_2} \sum_{k_1=0}^{\infty} \frac{x^{k_1+1}}{k_1 + 1} B_{k_1} u_0.$$

PROOF. We divide the proof into five steps.

Step 1. For a sequence $f_m = \begin{pmatrix} (f_m)_1 \\ (f_m)_2 \end{pmatrix} \in \mathbb{C}^2$ of vectors that converges to $f = \begin{pmatrix} (f)_1 \\ (f)_2 \end{pmatrix} \in \mathbb{C}^2$, it is true that

$$\begin{cases} |f_1 + f_2| \leq |f_1| + |f_2|, \\ |f| = |\lim_{m \to \infty} f_m| = \lim_{m \to \infty} |f_m|, \end{cases}$$

where $|f| \equiv |(f)_1| + |(f)_2|$. This follows from:

$$|f_1 + f_2| = |(f_1)_1 + (f_2)_1| + |(f_1)_2 + (f_2)_2|$$
$$\leq |f_1| + |f_2|, \quad \text{and}$$
$$|f_m| = |(f_m)_1| + |(f_m)_2| \longrightarrow |(f)_1| + |(f)_2|,$$

as $m \longrightarrow \infty$.

Step 2. For a matrix $A = (a_{jk})_{2 \times 2} = \begin{pmatrix} a_{11} & a_{12} \\ a_{21} & a_{22} \end{pmatrix}$, define

$$|A| \equiv \sum_{j,k=1}^{2} |a_{jk}|.$$

Claim that
$$|Af| \le |A||f|$$
is true, where vector $f = \begin{pmatrix} (f)_1 \\ (f)_2 \end{pmatrix} \in \mathbb{C}^2$. This follows from:

$$|(f)_j| \le |f|, j = 1, 2, Af = \begin{pmatrix} \sum_{j=1}^{2} a_{1j}(f)_j \\ \sum_{j=1}^{2} a_{2j}(f)_j \end{pmatrix},$$

and

$$|Af| = \sum_{k=1}^{2} \left| \sum_{j=1}^{2} a_{kj}(f)_j \right| \le \sum_{k=1}^{2} \sum_{j=1}^{2} |a_{kj}||f| = |A||f|.$$

Step 3. The series form that the u_m assumes, as put in the Theorem 2.1, is formally derived from the successive approximations (1.3), in which integration is performed but integration constants are all taken to be zero.

Step 4. We now show that the series u_m converges for each x satisfying $|x| < r_0$. Applying Steps 1 and 2 to the expression:

$$u_m = u_{m-1} + \sum_{k_m=0}^{\infty} \frac{x^{k_m+1}}{(k_m+1) + (k_{m-1}+1) + \cdots + (k_1+1)} B_{k_m}$$

$$\sum_{k_{m-1}=0}^{\infty} \frac{x^{k_{m-1}+1}}{(k_{m-1}+1) + (k_{m-2}+1) + \cdots + (k_1+1)} B_{k_{m-1}}$$

$$\cdots \sum_{k_2=0}^{\infty} \frac{x^{k_2+1}}{(k_2+1) + k_1+1} B_{k_2} \sum_{k_1=0}^{\infty} \frac{x^{k_1+1}}{k_1+1} B_{k_1} u_0,$$

it follows that
$$|u_m - u_{m-1}| \le a_m,$$

where a_m is defined by

$$a_m = \sum_{k_m=0}^{\infty} \frac{|x^{k_m+1}|}{(k_m+1) + (k_{m-1}+1) + \cdots + (k_1+1)} |B_{k_m}|$$

$$\sum_{k_{m-1}=0}^{\infty} \frac{|x^{k_{m-1}+1}|}{(k_{m-1}+1) + (k_{m-2}+1) + \cdots + (k_1+1)} |B_{k_{m-1}}|$$

$$\cdots \sum_{k_2=0}^{\infty} \frac{|x^{k_2+1}|}{(k_2+1) + k_1+1} |B_{k_2}| \sum_{k_1=0}^{\infty} \frac{|x^{k_1+1}|}{k_1+1} |B_{k_1}||u_0|.$$

Here note that $\sum_{k=0}^{\infty} |x^k||B_k|$ converges for each real x satisfying $|x| \le r_0$, by assumption.

Calculations show that, for each real x satisfying $0 < |x| \le r_0$, we have

$$\frac{a_m}{a_{m-1}} \le \sum_{k_m=0}^{\infty} \frac{|x^{k_m+1}|}{(k_m+1) + (k_{m-1}+1) + \cdots + (k_1+1)} |B_{k_m}||u_0|$$

$$\le \frac{|x|}{m} \sum_{k_m=0}^{\infty} |x^{k_m}||B_{k_m}|$$

$$\longrightarrow 0 < 1,$$

as $m \longrightarrow \infty$. Thus $\sum_{m=1}^{\infty} a_m$ converges by the ratio test in calculus. Using

$$|u_k - u_{k-1}| \leq a_k, k = 1, 2, \ldots$$

from above, the comparison test in calculus, and the fact in calculus also that absolute convergence implies the usual convergence, it follows that

$$u_m = u_0 + \sum_{k=1}^{m} (u_k - u_{k-1})$$

converges for each x satisfying $0 < |x| \leq r_0$. Convergence holds also for $x = 0$ since, then, $u_m = u_0$ converges.

Step 5. Since term by term differentiation is allowed for a convergent power series, a result in calculus, and further since $u_m, m = 1, 2, \ldots$, are obtained as indefinite integrals, it follows that the limit $T(x)u_0$ of u_m is a power series solution to the regular system (1.2). The set $\{T(x) \begin{pmatrix} 1 \\ 0 \end{pmatrix}, T(x) \begin{pmatrix} 0 \\ 1 \end{pmatrix}$ is linearly independent, since

at $x = 0$, it is $\{\begin{pmatrix} 1 \\ 0 \end{pmatrix}, \begin{pmatrix} 0 \\ 1 \end{pmatrix}\}$, which is so.

The proof is complete. $\square$

Next we prove

Theorem 2.10. *For each element w_0 in the Jordan basis $J = \{\beta_1, \beta_2\}$ of A, where the two eigenvalues λ_1 and λ_2 of A satisfies $Re\lambda_1 \geq Re\lambda_2$, the successive approximations $w_m = w_m[w_0]$ in (1.7) converge as $m \longrightarrow \infty$, and their limit $w \equiv S(t)w_0$ is a power series solution to (1.6). The two power series solutions w's, corresponding to the two w_0's, constitute a fundamental set of solutions for (1.6). Equivalently, the set of the two power series solutions $v(t) = e^{tA}w(t)$'s is a fundamental set of solutions for (1.5), or the set of the two solutions $u(x) = v(t)_{t=\ln(x)}$'s is a fundamental set of solutions for the weakly singular system (1.1). Here notice that the substitution $t = \ln(x)$ is equivalent to the substitution $t = \ln|x|$, regardless of the sign of x.*

The $w_m[\beta_1]$ can be calculated by using (2.2) and (2.6), and has no logarithmic term.

If the difference $l = \lambda_1 - \lambda_2$ of the two eigenvalues λ_1 and λ_2 is not an integer, then $w_m[\beta_2]$ can be calculated as $w_m[\beta_1]$ can be, and has no logarithmic term.

If the above l is a zero integer, that is, if $\lambda_1 = \lambda_2$, then $w_m[\beta_2]$ is calculated by using

$$\left\{ \begin{array}{l} \text{(2.2) and (2.6), as } w_m[\beta_1] \text{ can be, and has no logarithmic term, if (C1) holds,} \\ \text{(2.3), (2.6), and (2.7), and has a logarithmic term, if (C2) holds.} \end{array} \right.$$

If the above l is a positive integer and further equals a sum

$$\sum = [\sum_{i=1}^{q_0} (k_0 + 1)] + [\sum_{i=1}^{q_1} (k_1 + 1)] + \cdots + [\sum_{i=1}^{q_{s_0}} (k_{s_0} + 1)]$$

of some integers $\sum_{i=1}^{q_j} (k_j + 1), j = 0, 1, \ldots, s_0$ for some s_0, where k_j is the index of the term $B_{k_j} \neq 0$ and the number $(k_j + 1)$ is added up repeatedly for some q_j times, $q_j \in \mathbb{N}$, then $w_m[\beta_2]$ can be calculated by using (2.2), (2.4), (2.6), and (2.7), and has a logarithmic term.

However, even if the above l is a positive integer, but no sum $\sum$ of the above type exists, then $w_m[\beta_2]$ is calculated as $w_m[\beta_1]$ is, and has no logarithmic term. Being written out, $w_m = w_m[\beta_1]$ takes the form:

$$w_1 = \beta_1 + \sum_{k_1=0}^{\infty} [(k_1 + 1) + (\lambda_1 - A)]^{-1} e^{t[(k_1+1)+(\lambda_1-A)]} B_{k_1}\beta_1,$$

$$w_2 = w_1 + \sum_{k_2=0}^{\infty} [(k_2 + 1) + (k_1 + 1) + (\lambda_1 - A)]^{-1} e^{t[(k_2+1)+(\lambda_1-A)]} B_{k_2}$$

$$\sum_{k_1=0}^{\infty} [(k_1 + 1) + (\lambda_1 - A)]^{-1} e^{t(k_1+1)} B_{k_1}\beta_1,$$

$$\vdots\,,$$

$$w_m = w_{m-1} +$$

$$\sum_{k_m=0}^{\infty} [(k_m + 1) + (k_{m-1} + 1) + \cdots + (k_1 + 1) + (\lambda_1 - A)]^{-1} e^{t[(k_m+1)+(\lambda_1-A)]} B_{k_m}$$

$$\sum_{k_{m-1}=0}^{\infty} [(k_{m-1} + 1) + (k_{m-2} + 1) + \cdots + (k_1 + 1) + (\lambda_1 - A)]^{-1} e^{t(k_{m-1}+1)} B_{k_{m-1}}$$

$$\cdots \sum_{k_1=0}^{\infty} [(k_1 + 1) + (\lambda_1 - A)]^{-1} e^{t(k_1+1)} B_{k_1}\beta_1, \quad m = 1, 2, \ldots.$$

PROOF. We divide the proof into seventeen steps.

Step 1. By the (2.2) in Lemma 2.5, we have $e^{tA}\beta_1 = e^{\lambda_1 t}\beta_1$. It follows from the equation (1.7) that

$$w_1 = \beta_1 + \sum_{k_1=0}^{\infty} [(k_1 + 1) + (\lambda_1 - A)]^{-1} e^{t[(k_1+1)+(\lambda_1-A)]} B_{k_1}\beta_1,$$

$$w_2 = w_1 + \sum_{k_2=0}^{\infty} [(k_2 + 1) + (k_1 + 1) + (\lambda_1 - A)]^{-1} e^{t[(k_2+1)+(\lambda_1-A)]} B_{k_2}$$

$$\sum_{k_1=0}^{\infty} [(k_1 + 1) + (\lambda_1 - A)]^{-1} e^{t(k_1+1)} B_{k_1}\beta_1,$$

and, in general, that

$$w_j = w_{j-1} + \sum_{k_j=0}^{\infty} [(k_j + 1) + (k_{j-1} + 1) + \cdots + (k_1 + 1)$$

$$+ (\lambda_1 - A)]^{-1} e^{t[(k_j+1)+(\lambda_1-A)]} B_{k_j}$$

$$\sum_{k_{j-1}=0}^{\infty} [(k_{j-1} + 1) + (k_{j-2} + 1) + \cdots + (k_1 + 1) + (\lambda_1 - A)]^{-1} e^{t(k_{j-1}+1)} B_{k_{j-1}} \cdots$$

$$\sum_{k_1=0}^{\infty} [(k_1 + 1) + (\lambda_1 - A)]^{-1} e^{t(k_1+1)} B_{k_1}\beta_1, \quad j = 1, 2, \ldots.$$

Here used were that $e^{at}e^{-tA} = e^{(a-A)t}$ for $a \in \mathbb{C}$, that $(\lambda_1 + b)$ is not an eigenvalue of A for which $((\lambda_1 + b) - A)^{-1}$ exists, where $b \geq 1$, and that the formula (2.6) in Lemma 2.7 is applicable.

Thus the series takes the form, as desired. We next show convergence of $w_j = w_j[\beta_1]$.

Step 2. Let

$$\sigma_0 = \sup_{\lambda \in \sigma(A)} |-\lambda_1 + \lambda|,$$

where $\sigma(A) = \{\lambda_1, \lambda_2\}$ is the spectrum set of A. Then $|-\lambda_1 + A| \leq \sigma_0$ [**21**, Page 277].

Step 3. Choose a $j_0 \in \mathbb{N}$, such that $j_0 > \sigma_0$. It follows [**21**, Page 277] that, for $k = 1, 2, \ldots$, we have $[k + j_0 - (-\lambda_1 + A)]^{-1}$ exists and equals its Neumann series:

$$(k + j_0)^{-1}[1 + (-\lambda_1 + A) + (-\lambda_1 + A)^2 + \cdots].$$

Furthermore, it is true [**21**, Page 193] that, for $k = 1, 2, \ldots$, we have

$$|[k + j_0 - (\lambda_1 + A)]^{-1}| \leq (k + j_0)^{-1}[1 + \frac{\sigma_0}{k + j_0} + (\frac{\sigma_0}{k + j_0})^2 + \cdots]$$
$$= (k + j_0 - \sigma_0)^{-1}.$$

Here used was the fact that a geometric series converges if the absolute value of its common ratio is less than one.

Step 4. For $j = 1, 2, \ldots$, let

$$a_j \equiv \sum_{k_j=0}^{\infty} |[(k_j + 1) + (k_{j-1} + 1) + \cdots + (k_1 + 1) + (\lambda_1 - A)]^{-1}||e^{t[(k_j+1)+(\lambda_1-A)]}|||B_{k_j}|$$

$$\sum_{k_{j-1}=0}^{\infty} (|[(k_{j-1} + 1) + (k_{j-2} + 1) + \cdots + (k_1 + 1) + (\lambda_1 - A)]^{-1}|)e^{t(k_{j-1}+1)}|B_{k_{j-1}}| \cdots$$

$$\sum_{k_1=0}^{\infty} (|[(k_1 + 1) + (\lambda_1 - A)]^{-1}|)e^{t(k_1+1)}|B_{k_1}||\beta|.$$

Here note that, by assumption, $\sum_{k=0}^{\infty} e^{t(k+1)}|B_k|$ converges for each complex $t = \ln|x|$ or $\ln|x| + i\pi$, satisfying $0 < |x| = |e^t| < r_0$. It follows that we have, for $j = 2, 3, \ldots$,

$$|w_{j_0+j} - w_{j_0+j-1}| \leq a_{j_0+j},$$

as the Step 4 in the proof of Theorem 2.1.

Step 5. Calculations show that

$$a_{j_0+1} \leq (1 + j_0 - \sigma_0)^{-1} \sum_{k_{j_0+1}=0}^{\infty} (|e^{t(\lambda_1-A)}|)e^{t(k_{j_0+1}+1)}|B_{k_{j_0+1}}|a_{j_0},$$

$$a_{j_0+2} \leq (2 + j_0 - \sigma_0)^{-1} \sum_{k_{j_0+2}=0}^{\infty} (|e^{t(\lambda_1-A)}|)e^{t(k_{j_0+2}+1)}|B_{k_{j_0+2}}|a_{j_0+1},$$

and, in general, that

$$a_{j_0+j} \leq (j + j_0 - \sigma_0)^{-1} \sum_{k_{j_0+j}=0}^{\infty} (|e^{t(\lambda_1-A)}|)e^{t(k_{j_0+j}+1)}|B_{k_{j_0+j}}|a_{j_0+j-1}, \quad j = 1, 2, \ldots.$$

Since, for each complex $t = \ln|x|$ or $\ln|x| + i\pi$, satisfying $0 < |x| = |e^t| < r_0$,

$$\frac{a_{j_0+j}}{a_{j_0+j-1}} \leq (j + j_0 - \sigma_0)^{-1} \sum_{k=0}^{\infty} (|e^{t(\lambda_1-A)}|)e^{t(k+1)}\|B_k\| \longrightarrow 0 < 1$$

as j goes to infinity, the number series $\sum_{j=1}^{\infty} a_{j_0+j}$ converges by the ratio test. Thus, from

$$|w_{j_0+j} - w_{j_0+j-1}| \leq a_{j_0+j}$$

above, from the comparison test in calculus, and from the fact in calculus also that absolute convergence implies the usual convergence, it follows that the vector series

$$\sum_{j=1}^{\infty} (w_{j_0+j} - w_{j_0+j-1}) = \lim_{j\to\infty} (w_{j_0+j} - w_{j_0})$$

converges, and then so does the series $\lim_{j\to\infty} w_j$.

Step 6. The convergence of $\lim_{j\to\infty} w_j$ in Step 5, combined with Step 1, shows the estimates:

$$v(t) = e^{tA} \lim_{j\to\infty} w_j$$

$$= e^{t\lambda_1}[\beta_1 + \quad \text{terms of the order and higher orders of} \quad e^t],$$

$$u_1(x) = v(t)|_{t=\ln(x)}$$

$$= x^{\lambda_1}[\beta_1 + \quad \text{terms of the order and higher orders of} \quad x].$$

We next turn to the calculation of $w_m[\beta_2]$.

Step 7. If $\lambda_1 = \lambda_2$, and $(C1)$ holds, then $w_m[\beta_2]$ is calculated as $w_m[\beta_1]$ is, in which β_1 is replaced by β_2. Furthermore, we have the estimate:

$$u_2 = \{e^{tA} \lim_{m\to\infty} w_m[\beta_2]\}|_{t=\ln(x)}$$

$$= x^{\lambda_2}[\beta_2 + \quad \text{terms of the order and higher orders of} \quad x].$$

Step 8. If $\lambda_1 = \lambda_2$, and $(C2)$ holds, then

$$e^{tA}\beta_2 = e^{\lambda_1 t}[\beta_2 + t(A - \lambda_1)\beta_2]$$

$$= e^{\lambda_1 t}(\beta_2 + t\beta_1)$$

by (2.3) in Lemma 2.5. Using Lemma 2.7, it follows from (1.7) that

$$w_1 = \phi_1 + \psi_1 + \varphi_1,$$

where

$$\phi_1 = \beta_2 + \sum_{k_1=0}^{\infty} [(k_1 + 1) + (\lambda_1 - A)]^{-1} e^{t[(k_1+1)+(\lambda_1-A)]} B_{k_1}\beta_2$$

$$\psi_1 = t \sum_{k_1=0}^{\infty} [(k_1 + 1) + (\lambda_1 - A)]^{-1} e^{t[(k_1+1)+(\lambda_1-A)]} B_{k_1}\beta_1$$

$$\varphi_1 = - \sum_{k_1=0}^{\infty} [(k_1 + 1) + (\lambda_1 - A)]^{-2} e^{t[(k_1+1)+(\lambda_1-A)]} B_{k_1} \beta_1,$$

and that $w_2 = \phi_2 + \psi_2 + \varphi_2$, where

$$\phi_2 = \phi_1 + \sum_{k_2=0}^{\infty} [(k_2 + 1) + (k_1 + 1) + (\lambda_1 - A)]^{-1} e^{t[(k_2+1)+(\lambda_1-A)]} B_{k_2}$$

$$\sum_{k_1=0}^{\infty} [(k_1 + 1) + (\lambda_1 - A)]^{-1} e^{t(k_1+1)} B_{k_1} \beta_2,$$

$$\psi_2 = \psi_1 + t \sum_{k_2=0}^{\infty} [(k_2 + 1) + (k_1 + 1) + (\lambda_1 - A)]^{-1} e^{t[(k_2+1)+(\lambda_1-A)]} B_{k_2}$$

$$\sum_{k_1=0}^{\infty} [(k_1 + 1) + (\lambda_1 - A)]^{-1} e^{t(k_1+1)} B_{k_1} \beta_1,$$

$$\varphi_2 = \varphi_1 - \sum_{k_2=0}^{\infty} [(k_2 + 1) + (k_1 + 1) + (\lambda_1 - A)]^{-2} e^{t[(k_2+1)+(k_1+1)+(\lambda_1-A)]} B_{k_2}$$

$$\sum_{k_1=0}^{\infty} [(k_1 + 1) + (\lambda_1 - A)]^{-1} B_{k_1} \beta_1$$

$$- \sum_{k_2=0}^{\infty} [(k_2 + 1) + (k_1 + 1) + (\lambda_1 - A)]^{-1} e^{t[(k_2+1)+(k_1+1)+(\lambda_1-A)]} B_{k_2}$$

$$\sum_{k_1=0}^{\infty} [(k_1 + 1) + (\lambda_1 - A)]^{-2} B_{k_1} \beta_1.$$

Here notice that, in calculating w_2, the term involving $t\beta_1$ in w_1 generates two terms, one term involving $t\beta_1$ also and the other term involving β_1 without t.

Continued in this way, $w_j = \phi_j + \psi_j + \varphi_j, j = 1, 2, \ldots$, are similarly calculated, and satisfy the estimates, as in Steps 2 to 5:

$$\left\{ \begin{array}{l} |\phi_{j_0+j} - \phi_{j_0+j-1}| \leq b_{j_0+j}; \\ \text{For each complex } t = \ln|x| \text{ or } \ln|x| + i\pi, \text{ satisfying } 0 < |x| = |e^t| < r_0, \text{ we have} \\ \frac{b_{j_0+j}}{b_{j_0+j-1}} \leq (j + j_0 - \sigma_0)^{-1} \sum_{k=0}^{\infty} (|e^{t(\lambda_1-A)}|) e^{t(k+1)} |B_k| \\ \longrightarrow 0 < 1 \quad \text{as } j \text{ goes to infinity;} \end{array} \right.$$

$$\left\{ \begin{array}{l} |\psi_{j_0+j} - \psi_{j_0+j-1}| \leq |t| c_{j_0+j}; \\ \text{For each complex } t \text{ satisfying } 0 < |x| = |e^t| < r_0, \text{ we have} \\ \frac{c_{j_0+j}}{c_{j_0+j-1}} \leq (j + j_0 - \sigma_0)^{-1} \sum_{k=0}^{\infty} (|e^{t(\lambda_1-A)}|) e^{t(k+1)} |B_k| \\ \longrightarrow 0 < 1 \quad \text{as } j \text{ goes to infinity;} \end{array} \right.$$

$$\left\{ \begin{array}{l} |\varphi_{j_0+j} - \varphi_{j_0+j-1}| \leq (j_0 + j) d_{j_0+j}; \\ \text{For each complex } t \text{ satisfying } 0 < |x| = |e^t| < r_0, \text{ we have} \\ \frac{(j_0+j)\varphi_{j_0+j}}{(j_0+j-1)\varphi_{j_0+j-1}} \leq \frac{j_0+j}{j_0+j-1} (j + j_0 - \sigma_0)^{-1} \sum_{k=0}^{\infty} (|e^{t(\lambda_1-A)}|) e^{t(k+1)} |B_k| \\ \longrightarrow 0 < 1 \quad \text{as } j \text{ goes to infinity.} \end{array} \right.$$

Here are that $b_j, j = 1, 2, \ldots$, equals the a_j in Step 4, that $c_j, j = 2, 3, \ldots$, equals the a_j in Step 4, that σ_0 is as in Step 2, that j_0 is as in Step 3, and that

$$d_j \equiv \sum_{k_j=0}^{\infty} |[(k_j+1)+(k_{j-1}+1)+\cdots+(k_1+1)+(\lambda_1-A)]^{-1}||e^{t[(k_j+1)+(\lambda_1-A)]}|||B_{k_j}|$$

$$\sum_{k_{j-1}=0}^{\infty} (|[(k_{j-1}+1)+(k_{j-2}+1)+\cdots+(k_1+1)+(\lambda_1-A)]^{-1}|)e^{t(k_{j-1}+1)}|B_{k_{j-1}}|$$

$$\cdots \sum_{k_1=0}^{\infty} |[(k_1+1)+(\lambda_1-A)]^{-2}|e^{t(k_1+1)}|B_{k_1}|||\beta_1|$$

is true, and finally that

$$|[(l_2+j_0)+(\lambda_1-A)]^{-1}| \le [(l_1+j_0)-\sigma_0]^{-1}$$

is true for $0 \le l_1 < l_2$.

Thus, as in Step 4, using the ratio test, the comparison test in calculus, and the fact in calculus also that absolute convergence implies the usual convergence, it follows that the vector series $\phi_j, \psi_j, \varphi_j$, and then w_j all converges for each complex t satisfying $0 < |x| = |e^t| < r_0$.

Step 9. By Step 8, we have the estimate:

$$u_2 = \{e^{tA} \lim_{m\to\infty} w_m[\beta_2]\}|_{t=\ln(x)}$$

$$= x^{\lambda_1}[\beta_2 + \ln(x)\beta_1 + \quad \text{terms of the order and higher order of} \quad x].$$

Step 10. Denoting $\lim_{j\to\infty} \psi_j$ by ψ, it is easy to see that

$$e^{tA}(w_0 + \psi) = e^{tA}(\beta_2 + \psi)$$

$$= e^{\lambda_1 t}\beta_2 + e^{\lambda_1 t}t\beta_1 + e^{tA}\psi$$

$$= e^{\lambda_1 t}\beta_2 + tS(t)\beta_1,$$

where $S(t)\beta_1 = \lim_{m\to\infty} w_m[\beta_1]$ in Step 1.

Step 11. Suppose that $l = \lambda_1 - \lambda_2$ equals a sum

$$\sum = [\sum_{i=1}^{q_0}(k_0+1)] + [\sum_{i=1}^{q_1}(k_1+1)] + \cdots + [\sum_{i=1}^{q_{s_0}}(k_{s_0}+1)]$$

of some integers $\sum_{i=1}^{q_j}(k_j+1), j = 0, 1, \ldots, s_0$ for some s_0, where k_j is the index of the term $B_{k_j} \ne 0$ and the number (k_j+1) is added up repeatedly for some q_j times, $q_j \in \mathbb{N}$.

We further assume that $\sum = (k_0+1)$; the geneal case is similar and will be considered in Step 14 after this case.

Applying (2.2), (2.4), and Lemma 2.7 to (1.7) for computing $w_m = w_m[\beta_2]$, it follows that

$$w_1 = \beta_2 + \int e^{-tA} \sum_{k_1 \ne k_0} e^{(k_1+1)t} B_{k_1} e^{t\lambda_2}\beta_2\, dt$$

$$+ \int [e^{-t\lambda_1}M_{1,0} + e^{-t\lambda_2}M_{2,0}]e^{(k_0+1)t} B_{k_0} e^{\lambda_2 t}\beta_2\, dt$$

$$= \beta_2 + \sum_{k_1 \neq k_0}^{\infty} [(k_1 + 1) + (\lambda_2 - A)]^{-1} e^{t[(k_1+1)+(\lambda_2-A)]} B_{k_1} \beta_2$$

$$+ t M_{1,0} B_{k_0} \beta_2 + e^t M_{2,0} B_{k_0} \beta_2,$$

that

$$v_1 = e^{tA} w_1$$

$$= e^{t\lambda_2} \beta_2 + \sum_{k_1 \neq k_0}^{\infty} [(k_1 + 1) + (\lambda_2 - A)]^{-1} e^{t[(k_1+1)+\lambda_2]} B_{k_1} \beta_2$$

$$+ [e^{t\lambda_1} M_{1,0} + e^{t\lambda_2} M_{2,0}][t M_{1,0} B_{k_0} \beta_2 + e^t M_{2,0} B_{k_0} \beta_2],$$

where the last term equals

$$t e^{t\lambda_1} M_{1,0} B_{k_0} \beta_2 + e^{t(\lambda_2+1)} M_{2,0} B_{k_0} \beta_2$$

with $\lambda_1 = \lambda_2 + 1 + k_0$, and finally that

$$w_2 = w_1 +$$

$$\int [e^{-tA} \sum_{k_2=0}^{\infty} e^{t(k_2+1)} B_{k_2} e^{tA}] \sum_{k_1 \neq k_2} [(k_1+1)+(\lambda_2 - A)]^{-1} e^{t[(k_1+1)+(\lambda_2-A)]} B_{k_1} \beta_2 \, dt$$

$$+ \int e^{-tA} \sum_{k_2=0}^{\infty} e^{t(k_2+1)} B_{k_2} [e^{t\lambda_1} M_{1,0} + e^{t\lambda_2} M_{2,0}][t M_{1,0} B_{k_0} \beta_2 + e^t M_{2,0} B_{k_0} \beta_2] \, dt$$

$$= w_1 + \sum_{k_2=0}^{\infty} [(k_2 + 1) + (k_1 + 1) + (\lambda_2 - A)]^{-1} e^{t[(k_2+1)+(\lambda_2-A)]} B_{k_2}$$

$$\sum_{k_1 \neq k_0}^{\infty} [(k_1 + 1) + (\lambda_2 - A)]^{-1} e^{t(k_1+1)} B_{k_1} \beta_2$$

$$+ \sum_{k_2=0}^{\infty} [(k_2 + 2) + (\lambda_2 - A)]^{-1} e^{t[(k_2+2)+(\lambda_2-A)]} M_{2,0} B_{k_0} \beta_2$$

$$+ t \sum_{k_2=0}^{\infty} [(k_2 + 1) + (\lambda_1 - A)]^{-1} e^{t[(k_2+1)+(\lambda_1-A)]} B_{k_2} M_{1,0} B_{k_0} \beta_2$$

$$- \sum_{k_2=0}^{\infty} [(k_2 + 1) + (\lambda_1 - A)]^{-2} e^{t[(k_2+1)+(\lambda_1-A)]} B_{k_2} M_{1,0} B_{k_0} \beta_2.$$

Here notice that, in calculating w_2, the term involving $t M_{1,0} B_{k_0} \beta_2$ generates two terms, one term involving $t M_{1,0} B_{k_0} \beta_2$ also but the other term involving $M_{1,0} B_{k_0} \beta_2$ without t.

Continued in this way, other $w_m, m \geq 3$, are similarly calculated, in which (2.4) is no longer used. Thus the proof of the convergence of the w_m resembles Step 8.

Step 12. By Step 11, we have the estimate:

$$u_2 = \{e^{tA} \lim_{m \to \infty} w_m[\beta_2]\}|_{t=\ln(x)}$$

$$= x^{\lambda_2}[\beta_2 + \quad \text{terms of the order and higher orders of} \quad x$$

$$+ \quad (\text{terms of the order and higher orders of} \quad x^{1+k_0}) \ln(x)].$$

Step 13. Denoting by ψ the series involving the above $tM_{1,0}B_{K_0}\beta_2$, we have

$$e^{tA}\psi = \alpha t S(t)\beta_1$$

for some constant α, where $S(t)\beta_1 = \lim_{m\to\infty} w_m[\beta_1]$ in Step 1. This follows, because, by (2.4),

$$(A - \lambda_1)M_{1,0}B_{k_0}\beta_2 = (\lambda_2 - \lambda_1)M_{2,0}M_{1,0}B_{k_0}\beta_2 = 0,$$

which implies $M_{1,0}B_{k_0}\beta_2 = \alpha\beta_1$ for some constant α.

Step 14. Under the general case where

$$l = \sum = [\sum_{i=1}^{q_0}(k_0 + 1)] + [\sum_{i=1}^{q_1}(k_1 + 1)] + \cdots + [\sum_{i=1}^{q_{s_0}}(k_{s_0} + 1)],$$

$w_m[\beta_2]$ is calculated as in Step 11, in which (2.4) is applied repeatedly. Furthermore, we have the estimate:

$$u_2 = \{e^{tA}\lim_{m\to\infty} w_m[\beta_2]\}|_{t=\ln(x)}$$

$$= x^{\lambda_2}[\beta_2 + \quad \text{terms of the order and higher orders of} \quad x$$

$$+ \quad (\text{terms of the order and higher orders of} \quad x^l)\ln(x)].$$

Step 15. Since term by term differentiation is allowed for a convergent power series, a fact in calculus, and since $w_m[w_0], m = 1, 2, \ldots$, are obtained as indefinite integrals, it follows that $w^{(1)} = \lim_{m\to\infty} w_m[\beta_1]$ and $w^{(2)} = \lim_{m\to\infty} w_m[\beta_2]$ are two power series solutions to (1.6).

Step 16. We show that the calculated

$$y_1(x) = e^{tA}w^{(1)}|_{t=\ln(x)} \quad \text{and} \quad y_2(x) = e^{tA}w^{(2)}|_{t=\ln(x)}$$

are linearly independent, by considering

$$c_1 y_1 + c_2 y_2 = 0 \tag{4.1}$$

for all $0 < |x| < r_0$ and demonstrating that the constants c_1 and c_2 are zero.

In the case where either $l = \lambda_1 - \lambda_2$ is not an integer, or else l is a positive integer but no such a sum of the type $\sum$ (as stated in the Theorem) exists, or else both $l = 0$ and (C1) hold, substitute both the $y_1 = u_1$ in Step 6 and the $y_2 = u_2$ in Step 7 into (4.1), and divide the resultant (4.1) by x^{λ_2}. Letting $x \longrightarrow 0$, it follows that $c_1\beta_1 + c_2\beta_2 = 0$ or $c_2\beta_2 = 0$ according as $\lambda_1 = \lambda_2$ or not. In either case, we have $c_1 = 0 = c_2$, which proves linear independence. Here note that $\lim_{x\to 0} x^{\lambda_1-\lambda_2} = 0$ if $Re(\lambda_1) > Re(\lambda_2)$, and that β_1 and β_2 are linearly independent.

In the case where both $l = 0$ and (C2) hold, substitute both the $y_1 = u_1$ in Step 6 and the $y_2 = u_2$ in Step 9 into (4.1), and divide the resultant (4.1) by $x^{\lambda_2}\ln(x)$. Letting $x \longrightarrow 0$, it follows that $c_2\beta_1 = 0$. Hence $c_2 = 0$, and then $c_1 = 0$.

In the case where l is a positive integer and equals a sum $\sum$ of some integers, as stated in the Theorem, substitute the $y_1 = u_1$ in Step 6 and the $y_2 = u_2$ in Step 12 or more generally, the $y_2 = u_2$ in Step 4, into (4.1), and then divide the resultant (4.1) by x^{λ_2}. Letting $x \longrightarrow 0$, it follows that $c_2\beta_2 = 0$. Hence $c_2 = 0$, and then $c_1 = 0$.

Step 17. Finally, we claim that the substitution $t = \ln(x)$ is equivalent to the substitution $t = \ln|x|$, irrespective of the sign of x.

Using Steps 10 and 13, it follows that, if $x < 0$, then the constant $i\pi$ in $\ln(x) = \ln|x| + i\pi$ for $y_2 = u_2$, contributes to a term of a multiple of y_1. This multiple of $y_1 = u_1$ can be incorporated into $y_1 = u_1$, and the remaining $y_2 = u_2$ is still linearly independent of y_1. This is because the orginal y_2 and y_1 are linearly independent by Step 16. Thus the claim is proved.

The proof is now complete. $\square$

5. The Extension of the Main Results

In this section, we extend the main results in Section 2 to the case where the weakly singular system consists of n equations.

Let $A \neq 0$ and $B_k, k = 0, 1, 2, \ldots$, be $n \times n$ matrices with their entries being complex numbers, where $n \in \mathbb{N}$. Consider the singular linear system of n first order differential equations

$$\frac{d}{dz}u(z) = (\frac{1}{z}A + \sum_{k=0}^{\infty} z^k B_k)u(z) \tag{5.1}$$

in a punctured disk $\{z \in \mathbb{C} : 0 < |z| < r_0\}$, where r_0 is a given positive number. This system is called weakly singular [**23**], or a system with a singularity of the first kind at 0 [**5**]. Here the series $\sum_{k=0}^{\infty} z^k B_k$ is assumed convergent absolutely for z satisfying $0 < |z| < r_0$, and one example of such a series is an analytic matrix-valued function of z.

A special case of (5.1) is the nth order regular singular equation:

$$z^n y^{(n)} + z^{n-1}b_1(z)y^{(n-1)} + \cdots + zb_{n-1}(z)y' + b_n(z)y = 0, \tag{5.2}$$

where

$$b_k(z) = \sum_{j=0}^{\infty} b_k^{(j)} z^j$$

for each $k = 1, \ldots, n$, is an analytic function in the disk $\{z \in \mathbb{C} : 0 \leq |z| < r_0\}$. To see this, using the substitution [**5**, page 124]:

$$u_k = z^{k-1}y^{(k-1)}, \quad k = 1, \ldots, n, \tag{5.3}$$

it follows that

$$u_1 = y,$$
$$u_2 = zy',$$
$$u_3 = z^2 y'',$$
$$\vdots,$$
$$u_n = z^{n-1}y^{(n-1)},$$

and that

$$u_1' = y' = \frac{1}{z}u_2,$$
$$u_2' = zy'' + y' = \frac{1}{z}u_3 + \frac{1}{z}u_2,$$
$$u_3 = z^2 y''' + 2zy'' = \frac{1}{z}u_4 + \frac{1}{z}2u_3,$$

$$u_4 = z^3 y^{(4)} + 3z^2 y''' = \frac{1}{z} u_5 + \frac{1}{z} 3 u_4,$$

$$\vdots ,$$

$$u'_{n-1} = z^{n-2} y^{(n-1)} + (n-2) z^{n-3} y^{(n-2)} = \frac{1}{z} u_n + \frac{1}{z}(n-2) u_{n-1},$$

$$u'_n = z^{n-1} y^{(n)} + (n-1) z^{n-2} y^{(n-1)}$$

$$= -\frac{1}{z}(b_1 u_n + b_2 u_{n-1} + \cdots + b_{n-1} u_2 + b_n u_1) + \frac{1}{z}(n-1) u_n.$$

Or equivalently,

$$\frac{d}{dz} u = \frac{1}{z}
\begin{pmatrix}
0 & 1 & 0 & 0 & 0 & 0 & \cdots & & 0 \\
0 & 1 & 1 & 0 & 0 & 0 & \cdots & & 0 \\
0 & 0 & 2 & 1 & 0 & 0 & \cdots & & 0 \\
0 & 0 & 0 & 3 & 1 & 0 & \cdots & & 0 \\
& & & \vdots & & & & & \\
0 & 0 & 0 & 0 & \cdots & 0 & (n-2) & & 1 \\
-b_n & -b_{n-1} & \cdots & \cdots & \cdots & \cdots & & -b_2 & (n-1) - b_1
\end{pmatrix} u \qquad (5.4)$$

$$= (\frac{1}{z} A + \sum_{k=0}^{\infty} z^k B_k) u,$$

where

$$A =
\begin{pmatrix}
0 & 1 & 0 & 0 & 0 & 0 & \cdots & & 0 \\
0 & 1 & 1 & 0 & 0 & 0 & \cdots & & 0 \\
0 & 0 & 2 & 1 & 0 & 0 & \cdots & & 0 \\
0 & 0 & 0 & 3 & 1 & 0 & \cdots & & 0 \\
& & & \vdots & & & & & \\
0 & 0 & 0 & 0 & \cdots & 0 & (n-2) & & 1 \\
-b_n^{(0)} & -b_{n-1}^{(0)} & \cdots & \cdots & \cdots & \cdots & & -b_2^{(0)} & (n-1) - b_1^{(0)}
\end{pmatrix}$$

and

$$B_k =
\begin{pmatrix}
0 & 0 & \cdots & 0 \\
0 & 0 & \cdots & 0 \\
& \vdots & & \\
0 & 0 & \cdots & 0 \\
-b_n^{(k+1)} & -b_{n-1}^{(k+1)} & \cdots & -b_1^{(k+1)}
\end{pmatrix} .$$

In this case, the Frobenius's method is usually used to obtain power series solutions [5, pages 132–135].

We shall solve (5.1) by using the exponential function

$$e^{tA} \equiv \sum_{k=0}^{\infty} \frac{(tA)^k}{k!},$$

defined in a previous chapter and the results of a Jordan canonical form of a matrix [7, 9, 18]. We can compute explicitly the power series solutions and prove their convergence, both simultaneously. The proof is patterned after that for Section 2, and the reader is referred to [13] for the details. Compare this method with the existing methods of obtaining such power series solutions [5, 23].

To this end, we reduce (5.1), by the substitution $z = e^t$, to the system

$$\frac{d}{dt}v(t) = (A + e^t \sum_{k=0}^{\infty} e^{kt} B_k)v(t),$$ (5.5)

where $v(t) = u(z)|_{z=e^t}$, and $u(z) = v(t)|_{t=\ln(z)}$.

DEFINITION 5.1. *For a complex number* $z = a + ib = |z|e^{i\theta}$, *where* $a, b \in \mathbb{R}$, $|z| = \sqrt{a^2 + b^2}$ *is the magnitude or the length of* z, $i = \sqrt{-1}, i^2 = -1, -\pi < \theta \leq \pi$, *and* $e^{i\theta} = \sum_{k=0}^{\infty} \frac{(i\theta)^k}{k!} = \cos(\theta) + i\sin(\theta)$, *define*

$$\ln(z) = \ln|z| + i\theta.$$

We further reduce (5.5) to the system

$$\frac{d}{dt}w(t) = e^{-tA}(e^t \sum_{k=0}^{\infty} e^{kt} B_k)e^{tA}w(t),$$ (5.6)

by the substitution $v(t) = e^{tA}w(t)$. Here is the main result of this section:

THEOREM 5.2. *For each element (or generalized eigenvector)* w_0 *in the Jordan canonical basis* J, *the limit* $w(t)$, *as* $j \to \infty$, *of the successive approximations*

$$w_j(t) = w_0 + \int e^{-tA}(e^t \sum_{k=0}^{\infty} e^{kt} B_k)e^{tA}w_{j-1}\, dt, j = 1, 2, \ldots,$$ (5.7)

of (5.6) *by indefinite integrals, is a power series solution to* (5.6), *and the* n *power series solutions* $w(t)$, *corresponding to the* n *elements* w_0 *in the Jordan canonical basis* J, *is a fundamental set of solutions for* (5.6). *Thus, the set of the* n

$$u(z) \equiv v(t)|_{t=\ln(z)} = e^{tA}w(t)|_{t=\ln(z)}$$

is a fundamental set of solutions for (5.1). *Here the integration constants in the indefinite integrals are all taken to be zero.*

The calculations of the indefinite integral in (5.7) *are performed by using the known formula*

$$e^{tA}w_0 = e^{t\lambda_0}\{w_0 + \frac{t(A - \lambda_0)}{1!}w_0 + \cdots + \frac{[t(A - \lambda_0)]^{m_0-1}}{(m_0 - 1)!}w_0\},$$

if w_0 *is a generalized eigenvector satisfying* $(A-\lambda_0)^{m_0}w_0 = 0$ *but* $(A-\lambda_0)^{m_0-1}w_0 \neq 0$ *for some positive integer* m_0. *Furthermore, if the difference* $l = \lambda - \mu$ *of two eigenvalues* λ *and* μ *of* A *is a positive integer and equals a sum*

$$\sum = [\sum_{i=1}^{q_0}(k_0 + 1)] + [\sum_{i=1}^{q_1}(k_1 + 1)] + \cdots + [\sum_{i=1}^{q_{s_0}}(k_{s_0} + 1)]$$

of some integers $\sum_{i=1}^{q_j}(k_j + 1), j = 0, 1, \ldots, s_0$ *for some* s_0, *where* k_j *is the index of the term* $B_{k_j} \neq 0$ *and the number* $(k_j + 1)$ *is added up repeatedly for some* q_j *times,* $q_j \in \mathbb{N}$, *then also used is the additional known formula* [7, *Pages 307–313*] *(See Proposition 5.5 below):*

$$e^{tA} = \sum_{k=0}^{m_i-1} \sum_{i=1}^{s} t^k e^{t\lambda_i} M_{i,k},$$ (5.8)

and the additional properties [**14**] *that* $M_{i,k}$ *satisfies:*

$$M_{i,l}M_{j,m} = 0 \quad \text{if } i \neq j;$$

$$M_{i,m}M_{i,l} = M_{i,l}M_{i,m} = \frac{(A - \lambda_i)^l}{l!}M_{i,m} = \frac{(A - \lambda_i)^{m+l}}{l!m!}M_{i,0} \tag{5.9}$$

$$\text{if} \quad 0 \leq l, m \leq m_i - 1 \quad \text{and} \quad l + m \leq m_i - 1;$$

$$M_{i,m}M_{i,l} = 0 \quad \text{if} \quad l + m > m_i - 1.$$

Here $\lambda_i, i = 1, \ldots, s$, *are the all eigenvalues of* A *with respective multiplicity* m_i. *However, even if the above* l *is a positive integer but no sum* $\sum$ *of the above type exists, then there is no need to use the additional known formula* (5.8) *and the additional property* (5.9).

In particular,

COROLLARY 5.3. *Under counting multiplicity, if no eigenvalues of* A *differ by nonnegative integers, then no logarithmic terms are involved in the power series solutions. In this case, all solutions are of the form*

$$z^{\lambda_0}\sum_{k=0}^{\infty} c_k z^k,$$

where λ_0 *is an eigenvalue of* A, *and* c_k *are constant vectors.*

In fact,

COROLLARY 5.4. *The power series solutions have logarithmic terms only if the* m_0 *in Theorem 5.2, is greater than or equal to 2, or if the difference* $l = \lambda - \mu$ *of two eigenvalues* λ *and* μ *of* A *is a positive integer and further equals a sum*

$$\sum = [\sum_{i=1}^{q_0}(k_0 + 1)] + [\sum_{i=1}^{q_1}(k_1 + 1)] + \cdots + [\sum_{i=1}^{q_{s_0}}(k_{s_0} + 1)]$$

of some integers $\sum_{i=1}^{q_j}(k_j + 1), j = 0, 1, \ldots, s_0$ *for some* s_0, *where* k_j *is the index of the term* $B_{k_j} \neq 0$ *and the number* $(k_j + 1)$ *is added up repeatedly for some* q_j *times,* $q_j \in \mathbb{N}$.

Here notice that even if the above l *is a positive integer, but no sum* $\sum$ *of the above type exists, then no logarithmic terms are involved.*

Before illustrating Theorem 5.2 by examples, we need to show how a Jordan canonical basis can be found.

PROPOSITION 5.5. *Assume that the characteristic equation* $|A - \lambda I| = 0$ *of* A *takes the form*

$$a_0(\lambda - \lambda_1)^{m_1}(\lambda - \lambda_2)^{m_2} \cdots (\lambda - \lambda_s)^{m_s} = 0, \tag{5.10}$$

where $s \in \mathbb{N}$, $\lambda_i \neq \lambda_j$ *for* $i \neq j$, *and* $m_1 + m_2 + \cdots + m_s = n$. *This is always true* [**6**, *pages 17–18*].

Then there exist n *unique square matrices*

$$M_{i,k}, \quad i = 1, 2, \ldots, s; \quad k = 0, 1, \ldots, (m_i - 1),$$

such that

$$e^{tA} = \sum_{k=0}^{m_i-1}\sum_{i=1}^{s} t^k e^{t\lambda_i} M_{i,k}.$$

Furthermore, after relabelling the n functions

$$t^k e^{t\lambda_i}, i = 1, 2, \ldots, s; k = 0, 1, \ldots, (m_i - 1),$$

as the n functions

$$z_i(t), i = 1, 2, \ldots, n,$$

and relabelling the n matrices

$$M_{i,k}, i = 1, 2, \ldots, s; k = 0, 1, \ldots, (m_i - 1),$$

as the n matrices

$$N_j, j = 1, 2, \ldots, n,$$

it follows that the n matrices $N_j, j = 1, \ldots, n$, can be computed as the n unique solutions of the algebraic equations:

$$\begin{pmatrix} z_1(0) & z_2(0) & \cdots & z_n(0) \\ z_1'(0) & z_2'(0) & \cdots & z_n'(0) \\ \vdots & & & \\ z_1^{(n-1)}(0) & z_2^{(n-1)}(0) & \cdots & z_n^{(n-1)}(0) \end{pmatrix} \begin{pmatrix} N_1 \\ N_2 \\ \vdots \\ N_n \end{pmatrix} = \begin{pmatrix} I \\ A \\ \vdots \\ A^{n-1} \end{pmatrix}.$$

PROOF. See Chapter 2, [**7**, pages 307–313], where the Cayley-Hamilton Theorem [**9**] is used. $\square$

PROPOSITION 5.6. *The $M_{i,k}$ satisfies:*

$$M_{i,l}M_{j,m} = 0 \quad \text{if } i \neq j;$$

$$M_{i,m}M_{i,l} = M_{i,l}M_{i,m} = \frac{(A - \lambda_i)^l}{l!}M_{i,m} = \frac{(A - \lambda_i)^{m+l}}{l!m!}M_{i,0}$$

if $0 \leq l, m \leq m_i - 1$ and $l + m \leq m_i - 1$;

$$M_{i,m}M_{i,l} = 0 \quad \text{if } l + m > m_i - 1; \quad \text{and}$$

$$M_{i,k}M_{i,0} = M_{i,k}, \quad i = 1, \ldots, s.$$

PROOF. See our article [**14**] or Chapter 7. $\square$

DEFINITION 5.7. *The eigenvalue λ_i of A has the multiplicity m_i, if the characteristic equation*

$$|A - \lambda I| = 0$$

has the root $\lambda = \lambda_i$, repeated exactly for m_i times. Here $|A - \lambda I|$ is the determinant of the matrix $(A - \lambda I)$.

A Jordan canonical basis J for A will be defined after the theorem from linear algebra [**9**, pages 420–425]:

THEOREM 5.8 (Primary Decomposition Theorem). *As in Proposition 5.5, assume*

$$p(\lambda) = a_0(\lambda - \lambda_1)^{m_1}(\lambda - \lambda_2)^{m_2} \cdots (\lambda - \lambda_s)^{m_s} = 0,$$

where $\lambda_i \neq \lambda_j$ for $i \neq j$, $1 \leq s \leq n$, and $m_1 + m_2 + \cdots + m_s = n$. Then for each $1 \leq i \leq s$, the complex number λ_i is a generalzed eigenvalue (also an eigenvalue) of A, and corresponding to this generalized eigenvalue λ_i, there are n_i cycles

$$S_{u_l^{(i)}}^{(p_l^{(i)})}, l = 1, 2, \ldots, n_i$$

of generalized eigenvectors, each cycle $S_{u_l^{(i)}}^{(p_l^{(i)})}$ *having* $p_l^{(i)}$ *generalized eigenvectors with* $u_l^{(i)}$ *as the generator, such that the union* R_i *of those* n_i *cycles*

$$S_{u_l^{(i)}}^{(p_l^{(i)})}, l = 1, 2, \ldots, n_i,$$

is a set of

$$m_i = p_1^{(i)} + p_2^{(i)} + \cdots + p_{n_i}^{(i)}$$

linearly independent generalized eigenvectors. Here the positive integer n_i *is the dimension of the eigenspace* E_{λ_i} *of the eigenvalue* λ_i, *and the generalized eigenvector* $u_l^{(i)}$ *as a generator, is a solution to the systems of algebraic equations:*

$$(A - \lambda_i I)^{p_l^{(i)}} w = 0 \quad and$$

$$(A - \lambda_i I)^{p_l^{(i)} - 1} w \neq 0$$

for some $1 \leq p_l^{(i)} \leq m_i$.
   Furthermore, the union

$$S = R_1 \cup R_2 \cdots \cup R_s$$

of the s $R_i's$ *is a set of*

$$n = m_1 + m_2 + \cdots + m_s$$

linearly independent generalized eigenvectors.

   Here

   DEFINITION 5.9. *The eigenspace* E_{λ_i} *of the eigenvalue* λ_i *is the number of the linearly independent eigenvectors, corresponding to* λ_i.

   Thus, If $n = 3, p(\lambda) = 3(\lambda - 1)(\lambda - 2)(\lambda - 3) = 0$, then $n_1 = n_2 = n_3 = 1, \lambda_1 = 1, \lambda_2 = 2, \lambda_3 = 3$, and

$$p_1^{(1)} = 1, p_1^{(2)} = 1, p_1^{(3)} = 1.$$

If $n = 4, p(\lambda) = 2(\lambda - 1)^4 = 0$ and $n_1 = 2$, then $\lambda_1 = 1, m_1 = 4$, and either

$$p_1^{(1)} = 1 \quad and \quad p_2^{(1)} = 3$$

or

$$p_1^{(1)} = 2 \quad and \quad p_2^{(1)} = 2.$$

   DEFINITION 5.10. *The set* S *in the above Theorem 5.8 is called a Jordan canonical basis* J *for the matrix* A.

   While Theorem 5.8 is the usual method for finding a Jordan canonical basis, here we will also present a recent method from our article [13]. For brevity, we write the cycle $S_{u_l^{(i)}}^{p_l^{(i)}}$ in the above Theorem 5.8, as $S_p^{(l)}$, and thus

$$S_p^{(l)} = \{p, (A - \lambda_i)p, \ldots, (A - \lambda_i)^{l-2}p, (A - \lambda_i)^{l-1}p\},$$

which has l generalized eigenvectors, and has the generator p.

   DEFINITION 5.11. l *is called the length of the cycle* $S_p^{(l)}$, *and* $(A - \lambda_i)^{l-1}p$ *is called the initial vector for* $S_p^{(l)}$.

DEFINITION 5.12. $S_p^{(l)}$ is said to have the subcycles:

$$\{(A - \lambda_i)^{l-1}p, 0, \ldots\ldots\ldots\ldots\ldots\ldots\ldots\ldots\ldots\ldots\ldots, 0\}$$
$$\{(A - \lambda_i)^{l-2}p, (A - \lambda_i)^{l-1}p, 0, \ldots\ldots\ldots\ldots\ldots\ldots, 0\}$$
$$\{(A - \lambda_i)^{l-3}p, (A - \lambda_i)^{l-2}p, (A - \lambda_i)^{l-1}p, 0, \ldots\ldots\ldots, 0\}$$
$$\vdots$$
$$\{(A - \lambda_i)p, (A - \lambda_i)^2 p, (A - \lambda_i)^3 p, \ldots, (A - \lambda_i)^{l-1}p, 0\}.$$

DEFINITION 5.13. For each $1 \le i \le s$, the matrices $M_{i,k}, k = 0, 1, \ldots, (m_i - 1)$, in Proposition 5.5 is said to have the augmented matrix

$$F_i = \begin{pmatrix} M_{i,0} \\ 1!M_{i,1} \\ 2!M_{i,2} \\ \vdots \\ (m_i - 1)!M_{i,m_i-1} \end{pmatrix}.$$

By [13], each nonzero column in F_i is a cycle of generalized eigenvectors for A, corresponding to the generalized eigenvalue λ_i, that has its corresponding column in the component matrix $M_{i,0}$ of F_i, as the generator, and has the number $1 \le l \le m_i$, as the length. The subcycles for each cycle column in F_i, can be regarded as additional columns in F_i, relative to the regular columns in F_i.

Here is our recent method for computing a Jordan basis J, [13]:

THEOREM 5.14. For each $1 \le i \le s$, perform column operations on the augmented matrix F_i, in such a way that any one of the initial vectors for the regular column cyles in F_i that is linearly dependent with the others, shall be reduced to zero. Continue column operations until all the initial vectors for the regular column cycles in F_i are linearly independent. Here multiplying the additional columns in F_i by constants, and adding them to regular columns in F_i are allowed; however, multiplying the regular columns by constants, and adding them to additional columns are not allowed. Then, by discarding the additional columns (that come from subcyles) in F_i, we have that the remaining column vectors in the component matrices $k! \times M_{i,k}, k = 0, 1, \ldots, (m_i - 1), i = 1, \ldots, s$, of the augmented matrices $F_i, i = 1, \ldots, s$, is a Jordan canonical basis for A.

Remark. At each stage of column operations, if we are able to pick regular columns in $F_i, i = 1, \ldots, s$, that constitute exactly n linearly independent vectors in $\mathbb{C}^n$, then we have obtained a Jordan canonical basis for A.

As illustration of Theorem 5.14, we consider two examples. We first take the example from [10, page 172]:

EXAMPLE 5.15.

$$A = \begin{pmatrix} 1 & 0 & 0 & 1 & -1 \\ 0 & 1 & -2 & 3 & -3 \\ 0 & 0 & -1 & 2 & -2 \\ 1 & -1 & 1 & 0 & 1 \\ 1 & -1 & 1 & -1 & 2 \end{pmatrix}.$$

Calculations show that the characteristic equation is

$$|A - \lambda I| = (\lambda - 1)^4(\lambda + 1) = 0,$$

and that the exponential function is, by Proposition 5.5,

$$e^{tA} = e^t M_{1,0} + te^t M_{1,1} + t^2 e^t M_{1,2} + t^3 e^t M_{1,3} + e^{-t} M_{2,0},$$

where

$$M_{1,0} = \begin{pmatrix} 1 & 0 & 0 & 0 & 0 \\ 0 & 1 & -1 & 1 & -1 \\ 0 & 0 & 0 & 1 & -1 \\ 0 & 0 & 0 & 1 & 0 \\ 0 & 0 & 0 & 0 & 1 \end{pmatrix}, \quad M_{1,1} = \begin{pmatrix} 0 & 0 & 0 & 1 & -1 \\ 0 & 0 & 0 & 1 & -1 \\ 0 & 0 & 0 & 0 & 0 \\ 1 & -1 & 1 & -1 & 1 \\ 1 & -1 & 1 & -1 & 1 \end{pmatrix}$$

$$M_{1,2} = 0, \quad M_{1,3} = 0$$

$$M_{2,0} = \begin{pmatrix} 0 & 0 & 0 & 0 & 0 \\ 0 & 0 & 1 & -1 & 1 \\ 0 & 0 & 1 & -1 & 1 \\ 0 & 0 & 0 & 0 & 0 \\ 0 & 0 & 0 & 0 & 0 \end{pmatrix}.$$

Forming the augmented matrix

$$F = \begin{pmatrix} M_{1,0} \\ M_{1,1} \end{pmatrix},$$

and performing column operations on F, we have

$$G = \begin{pmatrix} 1 & 0 & 0 & 0 & 0 \\ 1 & 1 & 0 & 1 & 0 \\ 0 & 0 & 0 & 1 & 0 \\ 0 & 0 & 0 & 1 & 1 \\ 0 & 0 & 0 & 0 & 1 \\ * & * & * & * & * \\ 0 & 0 & 0 & 1 & 0 \\ 0 & 0 & 0 & 1 & 0 \\ 0 & 0 & 0 & 0 & 0 \\ 0 & -1 & 0 & -1 & 0 \\ 0 & -1 & 0 & -1 & 0 \end{pmatrix}.$$

Here we added column 2 to columns 1 and 3, and added column 4 to column 5. Now a Jordan canonical basis $J = \{C_i\}_{i=1}^5$ for A is the set of columns 2 and 4 of the augmented matrix G, and column 3 of $M_{2,0}$, that is,

$$J = \{C_1, C_2, C_3, C_4, C_5\} = \left\{ \begin{pmatrix} 0 \\ 0 \\ 0 \\ -1 \\ -1 \end{pmatrix}, \begin{pmatrix} 0 \\ 1 \\ 0 \\ 0 \\ 0 \end{pmatrix}, \begin{pmatrix} 1 \\ 1 \\ 0 \\ -1 \\ -1 \end{pmatrix}, \begin{pmatrix} 0 \\ 1 \\ 1 \\ 1 \\ 0 \end{pmatrix}, \begin{pmatrix} 0 \\ 1 \\ 0 \\ 0 \\ 0 \end{pmatrix} \right\},$$

and $AC = CD$, where

$$C = \begin{pmatrix} C_1 & C_2 & C_3 & C_4 & C_5 \end{pmatrix}, \quad D = \begin{pmatrix} 1 & 1 & 0 & 0 & 0 \\ 0 & 1 & 0 & 0 & 0 \\ 0 & 0 & 1 & 1 & 0 \\ 0 & 0 & 0 & 1 & 0 \\ 0 & 0 & 0 & 0 & -1 \end{pmatrix}.$$

EXAMPLE 5.16. Consider the matrix in [**9**, page 438]:

$$A = \begin{pmatrix} 2 & -4 & 2 & 2 \\ -2 & 0 & 1 & 3 \\ -2 & -2 & 3 & 3 \\ -2 & -6 & 3 & 7 \end{pmatrix}.$$

Calculations show that the characteristic equation is

$$|A - \lambda I| = (\lambda - 2)^2(\lambda - 4)^2 = 0,$$

and that the exponential function is, by Proposition 5.5,

$$e^{tA} = e^{2t}M_{1,0} + te^{2t}M_{1,1} + e^{4t}M_{2,0} + te^{4t}M_{2,1},$$

where

$$M_{1,0} = \begin{pmatrix} 1 & 2 & -1 & -1 \\ 1 & 1 & 0 & -1 \\ 1 & 0 & 1 & -1 \\ 1 & 2 & -1 & -1 \end{pmatrix}, \quad M_{1,1} = 0,$$

$$M_{2,0} = \begin{pmatrix} 0 & -2 & 1 & 1 \\ -1 & 0 & 0 & 1 \\ -1 & 0 & 0 & 1 \\ -1 & -2 & 1 & 2 \end{pmatrix}, \quad M_{2,1} = \begin{pmatrix} 0 & 0 & 0 & 0 \\ 0 & -2 & 1 & 1 \\ 0 & -2 & 1 & 1 \\ 0 & -2 & 1 & 1 \end{pmatrix}.$$

Form the augmented matrix

$$F = \begin{pmatrix} M_{2,0} \\ M_{2,1} \end{pmatrix} = \begin{pmatrix} 0 & -2 & 1 & 1 \\ -1 & 0 & 0 & 1 \\ -1 & 0 & 0 & 1 \\ -1 & -2 & 1 & 2 \\ * & * & * & * \\ 0 & 0 & 0 & 0 \\ 0 & -2 & 1 & 1 \\ 0 & -2 & 1 & 1 \\ 0 & -2 & 1 & 1 \end{pmatrix}.$$

Now a Jordan canonical basis $J = \{C_i\}_{i=1}^4$ for A is the set of columns 1 and 2 of $M_{1,0}$ and column 3 of F, that is,

$$\{C_1, C_2, C_3, C_4\} = \{\begin{pmatrix} 1 \\ 1 \\ 1 \\ 1 \end{pmatrix}, \begin{pmatrix} 2 \\ 1 \\ 0 \\ 2 \end{pmatrix}, \begin{pmatrix} 0 \\ 1 \\ 1 \\ 1 \end{pmatrix}, \begin{pmatrix} 1 \\ 0 \\ 0 \\ 1 \end{pmatrix}\},$$

and $AC = CD$, where

$$C = (C_1 \quad C_2 \quad C_3 \quad C_4), \quad D = \begin{pmatrix} 2 & 0 & 0 & 0 \\ 0 & 2 & 0 & 0 \\ 0 & 0 & 4 & 1 \\ 0 & 0 & 0 & 4 \end{pmatrix}.$$

We now illustrate Theorem 5.2 by the following examples:

EXAMPLE 5.17. Consider the weakly singular system

$$\frac{d}{dz}u = [\frac{1}{x}A + z^4 B_4]u, \quad 0 < |z| < r_0,$$

where

$$A = \begin{pmatrix} 0 & 1 & 0 \\ 0 & 1 & 1 \\ 0 & 15 & 3 \end{pmatrix}, \quad \text{and} \quad B_4 = \begin{pmatrix} 0 & 0 & 0 \\ 0 & 0 & 0 \\ -4 & -5 & -6 \end{pmatrix}.$$

Solution. The usual calculations by Theorem 5.8 show that the characteristic equation

$$|A - \lambda I| = 0$$

has the eigenvalues λ's: $6, 0, -2$, and that the corresponding eigenpairs $\{\lambda, w_0\}$'s are

$$\{6, \begin{pmatrix} 1 \\ 6 \\ 30 \end{pmatrix}\}, \{0, \begin{pmatrix} 1 \\ 0 \\ 0 \end{pmatrix}\}, \quad \text{and} \quad \{-2, \begin{pmatrix} 1 \\ -2 \\ 6 \end{pmatrix}\}.$$

This can also be obtained by Theorem 5.14, as follows. By Proposition 5.5, the exponential function is

$$e^{tA} = e^{6t}M_{1,0} + M_{2,0} + e^{-2t}M_{3,0},$$

where

$$M_{1,0} = \frac{1}{48}(2A + A^2) = \frac{1}{48}\begin{pmatrix} 0 & 3 & 1 \\ 0 & 18 & 6 \\ 0 & 90 & 30 \end{pmatrix},$$

$$M_{2,0} = I - \frac{1}{12}(A^2 - 4A) = \begin{pmatrix} 1 & \frac{1}{4} & -\frac{1}{12} \\ 0 & 0 & 0 \\ 0 & 0 & 0 \end{pmatrix},$$

$$M_{3,0} = \frac{1}{16}(A^2 - 6A) = \frac{1}{16}\begin{pmatrix} 0 & -5 & 1 \\ 0 & 10 & -2 \\ 0 & -30 & 6 \end{pmatrix}.$$

Thus a Jordan canonical basis is, by Theorem 5.14,

$$J = \{\frac{1}{48}\begin{pmatrix} 1 \\ 6 \\ 30 \end{pmatrix}, \begin{pmatrix} 1 \\ 0 \\ 0 \end{pmatrix}, \frac{1}{16}\begin{pmatrix} 1 \\ -2 \\ 6 \end{pmatrix}\},$$

that is, the third column of $M_{1,0}$, the first column of $M_{2,0}$, and the third column of $M_{3,0}$. Although this method is more tedious for this example, compared to the usual method, it is more efficient for large matrix.

Using Theorem 5.2, it follows that, for each eigenpair $\{\lambda, w_0\}$, there are

$$w_1 = w_0 + \int e^{-tA}e^{5t}B_4 e^{tA}w_0 \, dt$$

$$= w_0 + (\lambda + 5 - A)^{-1} e^{t(\lambda+5-A)} B_4 w_0,$$

$$v_1 = e^{tA} w_1$$

$$= e^{t\lambda} w_0 + (\lambda + 5 - A)^{-1} e^{t(\lambda+5)} B_4 w_0,$$

$$u_1 = v_1|_{t=\ln(z)}$$

$$= z^{\lambda} w_0 + (\lambda + 5 - A)^{-1} z^{\lambda+5} B_4 w_0,$$

and

$$w_2 = w_0 + \int e^{-tA} e^{5t} B_4 v_1 \, dt$$

$$= w_1 + (\lambda + 10 - A)^{-1} e^{t(\lambda+10-A)} B_4 (\lambda + 5 - A)^{-1} B_4 w_0,$$

$$v_2 = e^{tA} w_2$$

$$= v_1 + (\lambda + 10 - A)^{-1} e^{t(\lambda+10)} B_4 (\lambda + 5 - A)^{-1} B_4 w_0,$$

$$u_2 = v_2|_{t=\ln(z)}$$

$$= u_1 + (\lambda + 10 - A)^{-1} z^{\lambda+10} B_4 (\lambda + 5 - A)^{-1} B_4 w_0.$$

The other $u_j, j = 3, 4, \ldots$, are similarly calculated.

EXAMPLE 5.18. Consider the weakly singular system

$$\frac{d}{dz} u = [\frac{1}{z} A + z^3 B_3] u, \quad 0 < |z| < r_0,$$

where

$$A = \begin{pmatrix} 0 & 1 & 0 \\ 0 & 1 & 1 \\ 0 & -1 & 3 \end{pmatrix}, \quad \text{and} \quad B_3 = \begin{pmatrix} 0 & 0 & 0 \\ 0 & 0 & 0 \\ -3 & -4 & -5 \end{pmatrix}.$$

Solution. The usual calculations by Theorem 5.8 show that the characteristic equation

$$|A - \lambda I| = 0$$

has the eigenvalues λ's: $2, 0, 2$, that the eigenpairs $\{\lambda, w_0\}$'s are

$$\{2, \beta_1\} = \{2, \begin{pmatrix} 1 \\ 2 \\ 2 \end{pmatrix}\},$$

$$\{0, \beta_3\} = \{0, \begin{pmatrix} 1 \\ 0 \\ 0 \end{pmatrix}\},$$

and that the generalized eigenpair is

$$\{2, \beta_2\} = \{2, \begin{pmatrix} -\frac{1}{2} \\ 0 \\ 2 \end{pmatrix}\}.$$

Here $\beta_2 = \begin{pmatrix} a \\ b \\ c \end{pmatrix}$ satisfies

$$0 = (A - 2)^2 \beta_2$$

$$= \begin{pmatrix} 4 & -3 & 1 \\ 0 & 0 & 0 \\ 0 & 0 & 0 \end{pmatrix} \beta_2 = \begin{pmatrix} 4a - 3b + c \\ 0 \\ 0 \end{pmatrix},$$

$$(A - 2)\beta_2 = \beta_1,$$

and, being solved, is a vector of the form

$$\beta_2 = \begin{pmatrix} a \\ b \\ 3b - 4a \end{pmatrix} = a \begin{pmatrix} 1 \\ 0 \\ -4 \end{pmatrix} + b \begin{pmatrix} 0 \\ 1 \\ 3 \end{pmatrix}$$

satisfying $\beta_1 = (A - 2)\beta_2 = (b - 2a) \begin{pmatrix} 1 \\ 2 \\ 2 \end{pmatrix}$, and is linearly independent of β_1. We

can arbitrarily choose the values of a and b that satisfy $b - 2a = 1$. Here we chose
$a = -\frac{1}{2}$ and $b = 0$.

This above result can also be obtained by Theorem 5.14 as follows. By Proposition 5.5, the exponential function is

$$e^{tA} = e^{2t} M_{1,0} + t e^{2t} M_{1,1} + M_{2,0},$$

where

$$M_{1,0} = A - \frac{A^2}{4} = \begin{pmatrix} 0 & \frac{3}{4} & -\frac{1}{4} \\ 0 & 1 & 0 \\ 0 & 0 & 1 \end{pmatrix},$$

$$M_{1,1} = -A + \frac{A^2}{2} = \begin{pmatrix} 0 & -\frac{1}{2} & \frac{1}{2} \\ 0 & -1 & 1 \\ 0 & -1 & 1 \end{pmatrix},$$

$$M_{2,0} = I - \frac{1}{4}(4A - A^2) = \begin{pmatrix} 1 & -\frac{3}{4} & \frac{1}{4} \\ 0 & 0 & 0 \\ 0 & 0 & 0 \end{pmatrix}.$$

Thus a Jordan canonical basis is

$$J = \{ \frac{1}{2} \begin{pmatrix} 1 \\ 2 \\ 2 \end{pmatrix}, \frac{1}{2} \begin{pmatrix} -\frac{1}{2} \\ 0 \\ 2 \end{pmatrix}, \begin{pmatrix} 1 \\ 0 \\ 0 \end{pmatrix} \},$$

that is, the third column of $M_{1,1}$, the third column of $M_{1,0}$, and the first column
of $M_{2,0}$.

Using Theorem 5.2, it follows that, for each eigenpair $\{\lambda, w_0\} \neq \{2, \beta_2\}$, there
are

$$w_1 = w_0 + \int e^{-tA} e^{4t} B_3 e^{t\lambda} w_0 \, dt$$

$$= w_0 + (\lambda + 4 - A)^{-1} e^{t(\lambda + 4 - A)} B_3 w_0,$$

$$v_1 = e^{tA} w_1$$

$$= e^{t\lambda} w_0 + (\lambda + 4 - A)^{-1} e^{t(\lambda + 4)} B_3 w_0,$$

$$u_1 = v_1 |_{t = \ln(z)}$$

$$= z^\lambda w_0 + (\lambda + 4 - A)^{-1} z^{\lambda + 4} B_3 w_0,$$

and

$$w_2 = w_0 + \int e^{-tA} e^{4t} B_3 v_1 \, dt$$

$$= w_1 + (\lambda + 8 - A)^{-1} e^{t(\lambda+8-A)} B_3 (\lambda + 4 - A)^{-1} B_3 w_0,$$

$$v_2 = e^{tA} w_2$$

$$= v_1 + (\lambda + 8 - A)^{-1} e^{t(\lambda+8)} B_3 (\lambda + 4 - A)^{-1} B_3 w_0,$$

$$u_2 = v_2|_{t=\ln(z)}$$

$$= u_1 + (\lambda + 8 - A)^{-1} z^{\lambda+8} B_3 (\lambda + 4 - A)^{-1} B_3 w_0.$$

The other $u_j, j = 3, 4, \ldots$, are similarly calculated.

Using Theorem 5.2, it also follows that, for the generalized eigenpair $\{2, \beta_2\}$, there are

$$w_1 = \beta_2 + \int e^{-tA} e^{4t} B_3 e^{2t} (\beta_2 + t(A-2)\beta_2) \, dt$$

$$= \beta_2 + (2 + 4 - A)^{-1} e^{(2+4-A)t} B_3 \beta_2 + (2 + 4 - A)^{-1} t e^{(2+4-A)t} B_3 \beta_1$$

$$- (2 + 4 - A)^{-2} e^{(2+4-A)t} B_3 \beta_1,$$

$$v_1 = e^{tA} w_1$$

$$= e^{2t} (\beta_2 + t\beta_1) + (2 + 4 - A)^{-1} e^{(2+4)t} B_3 \beta_2 +$$

$$+ (2 + 4 - A)^{-1} t e^{(2+4)t} B_3 \beta_1 - (2 + 4 - A)^{-2} e^{(2+4)t} B_3 \beta_1,$$

$$u_1 = v_1|_{t=\ln(z)}$$

$$= z^2 (\beta_2 + \beta_1 \ln(z)) + (2 + 4 - A)^{-1} z^{2+4} B_3 \beta_2$$

$$+ (2 + 4 - A)^{-1} [z^{2+4} \ln(z)] B_3 \beta_1 - (2 + 4 - A)^{-2} z^{2+4} B_3 \beta_1,$$

and that

$$w_2 = w_0 + \int e^{-tA} e^{4t} B_3 v_1 \, dt$$

$$= w_1 + (2 + 8 - A)^{-1} e^{t(2+8-A)} B_3 (2 + 4 - A)^{-1} B_3 \beta_2$$

$$+ (2 + 8 - A)^{-1} t e^{t(2+8-A)} B_3 (2 + 4 - A)^{-1} B_3 \beta_1$$

$$- (2 + 8 - A)^{-2} e^{t(2+8-A)} B_3 (2 + 4 - A)^{-1} B_3 \beta_1$$

$$- (2 + 8 - A)^{-1} e^{t(2+8-A)} B_3 (2 + 4 - A)^{-2} B_3 \beta_1,$$

$$v_2 = e^{tA} w_2$$

$$= v_1 + (2 + 8 - A)^{-1} e^{(2+8)t} B_3 (2 + 4 - A)^{-1} B_3 \beta_2$$

$$+ (2 + 8 - A)^{-1} t e^{(2+8)t} B_3 (2 + 4 - A)^{-1} B_3 \beta_1$$

$$- (2 + 8 - A)^{-2} e^{(2+8)t} B_3 (2 + 4 - A)^{-1} B_3 \beta_1$$

$$- (2 + 8 - A)^{-1} e^{(2+8)t} B_3 (2 + 4 - A)^{-2} B_3 \beta_1,$$

$$u_2 = v_2|_{t=\ln(x)}$$

$$= u_1 + (2 + 8 - A)^{-1} z^{(2+8)} B_3 (2 + 4 - A)^{-1} B_3 \beta_2$$

$$+ (2 + 8 - A)^{-1} [z^{(2+8)} \ln(z)] B_3 (2 + 4 - A)^{-1} B_3 \beta_1$$

$$- (2 + 8 - A)^{-2} z^{(2+8)} B_3 (2 + 4 - A)^{-1} B_3 \beta_1$$
$$- (2 + 8 - A)^{-1} z^{(2+8)} B_3 (2 + 4 - A)^{-2} B_3 \beta_1.$$

The other w_j, v_j, and $u_j, j = 3, 4, \ldots$, are similarly calculated.

EXAMPLE 5.19. Consider the weakly singular system

$$\frac{d}{dz} u = [\frac{1}{z} A + z^5 B_5] u, \quad 0 < |z| < r_0,$$

where A is the A in Example 5.17, and

$$B_5 = \begin{pmatrix} 0 & 0 & 0 \\ 0 & 0 & 0 \\ -5 & -6 & -7 \end{pmatrix}.$$

Solution. As in Example 5.17, the eigenpairs $\{\lambda_j, \beta_j\}, j = 1, 2, 3$, are:

$$\{6, \begin{pmatrix} 1 \\ 6 \\ 30 \end{pmatrix}\}, \{0, \begin{pmatrix} 1 \\ 0 \\ 0 \end{pmatrix}\}, \quad \text{and} \quad \{-2, \begin{pmatrix} 1 \\ -2 \\ 6 \end{pmatrix}\},$$

respectively. Although the eigenvalues are mutually different, the solutions have logarithmic terms. This will be seen soon.

For the eigenpair $\{\lambda, w_0\} = \{\lambda_1, \beta_1\}$ or $\{\lambda_3, \beta_3\}$, there are

$$w_1 = w_0 + \int e^{-tA} e^{6t} B_5 e^{t\lambda} w_0 \, dt$$
$$= w_0 + (\lambda + 6 - A)^{-1} e^{t(\lambda + 6 - A)} B_5 w_0,$$
$$v_1 = e^{tA} w_1$$
$$= e^{t\lambda} w_0 + (\lambda + 6 - A)^{-1} e^{t(\lambda + 6)} B_5 w_0,$$
$$u_1 = v_1|_{t = \ln(z)}$$
$$= z^\lambda w_0 + (\lambda + 6 - A)^{-1} z^{\lambda + 6} B_5 w_0,$$

and

$$w_2 = w_1 + (\lambda + 12 - A)^{-1} e^{t(\lambda + 12 - A)} B_5 (\lambda + 6 - A)^{-1} B_5 w_0,$$
$$v_2 = e^{tA} w_2$$
$$= v_1 + (\lambda + 12 - A)^{-1} e^{t(\lambda + 12)} B_5 (\lambda + 6 - A)^{-1} B_5 w_0,$$
$$u_2 = v_2|_{t = \ln(z)}$$
$$= u_1 + (\lambda + 12 - A)^{-1} z^{\lambda + 12} B_5 (\lambda + 6 - A)^{-1} B_5 w_0.$$

The other $u_j, j = 3, 4, \ldots$, are similarly calculated.

Applying Proposition 5.5, it follows that

$$e^{tA} = e^{6t} M_{1,0} + M_{2,0} + e^{-2t} M_{3,0},$$

where $M_{j,0}, j = 1, 2, 3$, satisfy

$$M_{1,0}^2 = M_{1,0}, M_{1,0} M_{2,0} = 0 = M_{1,0} M_{3,0},$$
$$M_{2,0}^2 = M_{2,0}, M_{2,0} M_{1,0} = 0 = M_{2,0} M_{3,0},$$
$$M_{3,0}^2 = M_{3,0}, M_{3,0} M_{1,0} = 0 = M_{3,0} M_{2,0}$$

by Proposition 5.6, and can be calculated by

$$I = M_{1,0} + M_{2,0} + M_{3,0},$$
$$A = 6M_{1,0} - 2M_{3,0},$$
$$A^2 = 36M_{1,0} + 4M_{3,0}.$$

Calculations show that

$$M_{1,0} = \frac{1}{48}(2A + A^2) = \frac{1}{48}\begin{pmatrix} 0 & 3 & 1 \\ 0 & 18 & 6 \\ 0 & 90 & 30 \end{pmatrix},$$

$$M_{2,0} = I - \frac{1}{12}(A^2 - 4A) = \begin{pmatrix} 1 & \frac{1}{4} & -\frac{1}{12} \\ 0 & 0 & 0 \\ 0 & 0 & 0 \end{pmatrix},$$

$$M_{3,0} = \frac{1}{16}(A^2 - 6A) = \frac{1}{16}\begin{pmatrix} 0 & -5 & 1 \\ 0 & 10 & -2 \\ 0 & -30 & 6 \end{pmatrix}.$$

For the eigenpair $\{\lambda_2, \beta_2\}$, it follows from Theorem 5.2 that

$$w_1 = \beta_2 + \int e^{-tA} e^{6t} B_5 e^{tA} \beta_2 \, dt$$

$$= \beta_2 + \int (e^{-6t} M_{1,0} + M_{2,0} + e^{2t} M_{3,0}) e^{6t} B_5 \beta_2 \, dt$$

$$= \beta_2 + t M_{1,0} B_5 \beta_2 + \frac{1}{6} e^{6t} M_{2,0} B_5 \beta_2 + \frac{1}{8} e^{8t} M_{3,0} B_5 \beta_2,$$

$$v_1 = e^{tA} w_1$$

$$= \beta_2 + t e^{6t} M_{1,0} B_5 \beta_2 + \frac{1}{6} e^{6t} M_{2,0} B_5 \beta_2 + \frac{1}{8} e^{6t} M_{3,0} B_5 \beta_2.$$

$$u_1 = v_1|_{t=\ln(z)}$$

$$= \beta_2 + [z^6 \ln(z)] M_{1,0} B_5 \beta_2 + \frac{1}{6} z^6 M_{2,0} B_5 \beta_2 + \frac{1}{8} z^6 M_{3,0} B_5 \beta_2,$$

$$w_2 = w_0 + \int e^{-tA} e^{6t} B_5 v_1 \, dt$$

$$= w_1 + t(12 - A)^{-1} e^{t(12-A)} B_5 (M_{1,0} B_5 \beta_2)$$

$$- (12 - A)^{-2} e^{t(12-A)} B_5 (M_{1,0} B_5 \beta_2) + \frac{1}{6}(12 - A)^{-1} e^{t(12-A)} B_5 (M_{2,0} B_5 \beta_2)$$

$$+ \frac{1}{8}(12 - A)^{-1} e^{t(12-A)} B_5 M_{3,0} B_5 \beta_2,$$

$$v_2 = e^{tA} w_2$$

$$= v_1 + t(12 - A)^{-1} e^{12t} B_5 (M_{1,0} B_5 \beta_2)$$

$$- (12 - A)^{-2} e^{12t} B_5 (M_{1,0} B_5 \beta_2) + \frac{1}{6}(12 - A)^{-1} e^{12t} B_5 (M_{2,0} B_5 \beta_2)$$

$$+ \frac{1}{8}(12 - A)^{-1} e^{12t} B_5 M_{3,0} B_5 \beta_2,$$

$$u_2 = v_2|_{t=\ln(z)}$$

$$= u_1 + [\ln(z)](12 - A)^{-1} z^{12} B_5 (M_{1,0} B_5 \beta_2)$$

$$-(12-A)^{-2}z^{12}B_5(M_{1,0}B_5\beta_2)+\frac{1}{6}(12-A)^{-1}z^{12}B_5(M_{2,0}B_5\beta_2)$$

$$+\frac{1}{8}(12-A)^{-1}z^{12}B_5M_{3,0}B_5\beta_2.$$

The other $u_j, j = 3, 4, \ldots$, are similarly calculated. Those $u_j, j = 1, \ldots$, have logarithmic terms.

EXAMPLE 5.20. Consider the regular singular equation of order 3, where no logarithmic terms appear in the solutions:

$$z^3y''' + z^2(-1+6z^5)y'' + z(-15+5z^5)y' + 4z^5y = 0, \quad 0 < |z| < r_0.$$

Solution. Using the substitution (5.3), in conjunction with (5.4), it follows that this example is the Example 5.17, and so the solutions are the first components of the solutions in Example 5.17.

EXAMPLE 5.21. Consider the regular singular equation of order 3, where a logarithmic term appears in the solutions, caused by a repeated eigenvalue:

$$z^3y''' + z^2(-1+5z^4)y'' + z(1+4z^4)y' + 3z^4y = 0, \quad 0 < |z| < r_0.$$

Solution. Using the substitution (5.5), in conjunction with (5.6), it follows that this example is the Example 5.18, and the solutions are the first components of the solutions in Example 5.18.

EXAMPLE 5.22. Consider the regular singular equation of order 3, where a logarithmic term appears in the solutions, caused by the difference of two eigenvalues being a positive integer:

$$z^3y''' + z^2(-1+7z^6)y'' + z(-15+6z^6)y' + 5z^6y = 0, \quad 0 < |z| < r_0.$$

Solution. Using the substitution (5.3), in conjunction with (5.4), it follows that this example is the Example 5.19, and so the solutions are the first components of the solutions in Example 5.19.

EXAMPLE 5.23. Consider the weakly singular system

$$\frac{d}{dz}u = [\frac{1}{z}A + z^2B_2]u, \quad 0 < |z| < r_0,$$

where

$$A = \begin{pmatrix} 3 & -1 & -5 & 1 \\ 1 & 1 & -1 & 0 \\ 0 & 0 & -2 & -1 \\ 0 & 0 & 1 & 0 \end{pmatrix}, \quad \text{and} \quad B_2 = \begin{pmatrix} 2 & -4 & 2 & 2 \\ -2 & 0 & 1 & 3 \\ -2 & -2 & 3 & 3 \\ -2 & -6 & 3 & 7 \end{pmatrix}.$$

Here A comes from [18, page 270], and B_2 is from [9, page 438].

Solution. Using Propositions 5.5 and 5.6, we have the characteristic equation

$$0 = |A - \lambda I| = (\lambda - 2)^2(\lambda + 1)^2,$$

and

$$e^{tA} = e^{2t}M_{1,0} + te^{2t}M_{1,1} + e^{-t}M_{2,0} + te^{-t}M_{2,1},$$

where

$$M_{1,0} = \frac{1}{27}\begin{pmatrix} 27 & 0 & -25 & 14 \\ 0 & 27 & -4 & -10 \\ 0 & 0 & 0 & 0 \\ 0 & 0 & 0 & 0 \end{pmatrix}, \quad M_{1,1} = \frac{1}{9}\begin{pmatrix} 9 & -9 & -7 & 8 \\ 9 & -9 & -7 & 8 \\ 0 & 0 & 0 & 0 \\ 0 & 0 & 0 & 0 \end{pmatrix},$$

$$M_{2,0} = \frac{1}{27} \begin{pmatrix} 0 & 0 & 25 & -14 \\ 0 & 0 & 4 & 10 \\ 0 & 0 & 27 & 0 \\ 0 & 0 & 0 & 27 \end{pmatrix}, \quad M_{2,1} = \frac{1}{9} \begin{pmatrix} 0 & 0 & -13 & -13 \\ 0 & 0 & 2 & 2 \\ 0 & 0 & -9 & -9 \\ 0 & 0 & 9 & 9 \end{pmatrix},$$

$$M_{1,0}^2 = M_{1,0}, \quad M_{2,0}^2 = M_{2,0}, \quad M_{1,1}^2 = 0, \quad M_{2,1}^2 = 0,$$
$$M_{1,0}M_{1,1} = M_{1,1}M_{1,0} = M_{1,1}, \quad M_{2,0}M_{2,1} = M_{2,1}M_{2,0} = M_{2,1},$$
$$M_{1,k}M_{2,i} = M_{2,i}M_{1,k} = 0 \quad \text{for} \quad i, k = 0, 1.$$

It follows from Theorem 5.14 that a Jordan basis

$$J = \{\beta_{10}, \beta_{11}, \beta_{20}, \beta_{21}\}$$

is given by

$$\{\frac{1}{27}\begin{pmatrix} 14 \\ -10 \\ 0 \\ 0 \end{pmatrix}, \frac{1}{9}\begin{pmatrix} 8 \\ 8 \\ 0 \\ 0 \end{pmatrix}, \frac{1}{27}\begin{pmatrix} -14 \\ 10 \\ 0 \\ 27 \end{pmatrix}, \frac{1}{9}\begin{pmatrix} -13 \\ 2 \\ -9 \\ 9 \end{pmatrix}\},$$

that is, the fourth column of the augmented matrix $\begin{pmatrix} M_{1,0} \\ M_{1,1} \\ M_{2,0} \\ M_{2,1} \end{pmatrix}$.

We shall compute series solution only for $w_0 = \beta_{20}$, for which

$$e^{tA}w_0 = e^{-t}[w_0 + t(A + I)w_0] = e^{-t}(w_0 + t\beta_{21}).$$

The cases for w_0 equal to other elements in J will be similar.
Using Theorem 5.2, we have that

$$w_1 = w_0 + \int^{t} [(M_{1,0} - tM_{1,1})(B_2 w_0 + tB_2(A + I)w_0)$$

$$+ e^{3t}(M_{2,0} - tM_{2,1})(B_2 w_0 + tB_2(A + I)w_0)]\, dt$$

$$= tM_{1,0}B_2 w_0 + \frac{t^2}{2}[M_{1,0}B_2(A + I)w_0 - M_{1,1}B_2 w_0] - \frac{t^3}{3}M_{1,1}B_2(A + I)w_0$$

$$+ \frac{1}{3}e^{3t}M_{2,0}B_2 w_0 + (\frac{1}{3}te^{3t} - \frac{1}{9}e^{3t}[M_{2,0}B_2(A + I)w_0 - M_{2,1}B_2 w_0]$$

$$- (\frac{1}{3}t^2 e^{3t} - \frac{2}{9}te^{3t} + \frac{2}{27}e^{3t})M_{2,1}B_2(A + I)w_0,$$

$$v_1 = e^{tA}w_1$$

$$= e^{-t}[w_0 + t(A + I)w_0] + (-\frac{1}{3} + \frac{1}{2})t^3 e^{2t}M_{1,1}B_2(A + I)w_0$$

$$+ \quad \text{terms of orders less than or equal to} \quad t^2,$$

$$u_1 = v_1|_{t=\ln(z)}$$

$$= z^{-1}[w_0 + \ln(z)(A + I)w_0] + (-\frac{1}{3} + \frac{1}{2})z^2 \ln(z)]^3 M_{1,1}B_2(A + I)w_0$$

$$+ \quad \text{terms of orders less than or equal to} \quad [\ln(z)]^2,$$

and that

$$w_2 = w_0 + \int e^{-tA}e^{3t}B_2 v_1\, dt$$

$$= e^{t(5-A)}(5-A)^{-1}t^3(-\frac{1}{3}+\frac{1}{2})B_2M_{1,1}B_2(A+I)w_0$$

$$+ e^{-tA} \times \quad \text{terms of orders less than or equal to} \quad t^2,$$

$$v_2 = e^{tA}w_2$$

$$= e^{-t}[w_0 + t(A+I)w_0] + t^3e^{5t}(-\frac{1}{3}+\frac{1}{2})(5-A)^{-1}B_2M_{1,1}B_2(A+I)w_0$$

$$+ \quad \text{terms of lower orders} \quad t^2,$$

$$u_2 = v_2|_{t=\ln(z)}$$

$$= z^{-1}[w_0 + \ln(z)(A+I)w_0] + z^5[\ln(z)]^3(-\frac{1}{3}+\frac{1}{2})(5-A)^{-1}B_2M_{1,1}B_2(A+I)w_0$$

$$+ \quad \text{terms of orders less than or equal to} \quad [\ln(z)]^2,$$

where notice that

$$(A+I)w_0 = (A+I)\beta_{20} = \beta_{21}.$$

Thus, a logarithmic term of highest order $t^3 = [\ln(z)]^3$ is involved, as predicted.

6. Problems

1. Solve the regular linear system:

$$\frac{d}{dx}u = A(x)u, \quad -r_0 < x < r_0$$

$$= (A_0 + xA_1)u,$$

where

$$A(x) = \begin{pmatrix} 0 & 1 \\ -2\mu & 2x \end{pmatrix}, \quad A_0 = \begin{pmatrix} 0 & 1 \\ -2\mu & 0 \end{pmatrix}, \quad A_1 = \begin{pmatrix} 0 & 0 \\ 0 & 2 \end{pmatrix}, \quad \text{and } \mu \in \mathbb{R}.$$

2. Solve the regular linear equation (the Hermite equation of order $\mu \in \mathbb{R}$):

$$y'' - 2xy' + 2\mu y = 0.$$

Solve the weakly singular systems, where no logarithmic terms appear in the solutions:

3. $\frac{d}{dx}u = \frac{1}{x}A + B_0$, where

$$A = \begin{pmatrix} 0 & 1 \\ \frac{1}{2} & -\frac{1}{2} \end{pmatrix}, \quad \text{and} \quad B_0 = \begin{pmatrix} 0 & 0 \\ -1 & 0 \end{pmatrix}.$$

4. $\frac{d}{dx}u = \frac{1}{x}A + xB_1$, where A is the A in Problem **3**, and

$$B_1 = \begin{pmatrix} 0 & 0 \\ -2 & 0 \end{pmatrix}.$$

5. $\frac{d}{dx}u = \frac{1}{x}A + x^2B_2$, where A is the A in Problem **3**, and

$$B_2 = \begin{pmatrix} 0 & 0 \\ -3 & 0 \end{pmatrix}.$$

Solve the weakly singular systems, where a logarithmic term appears in the solutions, caused by a repeated eigenvalue:

6. $\frac{d}{dx}u = \frac{1}{x}A + B_0$, where

$$A = \begin{pmatrix} 0 & 1 \\ -1 & 2 \end{pmatrix}, \quad B_0 = \begin{pmatrix} 0 & 0 \\ -1 & 0 \end{pmatrix}.$$

7. $\frac{d}{dx}u = \frac{1}{x}A + xB_1$, where

$$A = \begin{pmatrix} 0 & 1 \\ -2 & 2\sqrt{2} \end{pmatrix}, \quad B_1 = \begin{pmatrix} 0 & 0 \\ -1 & 0 \end{pmatrix}.$$

8. $\frac{d}{dx}u = \frac{1}{x}A + x^3B_3$, where

$$A = \begin{pmatrix} 0 & 1 \\ -4 & 4 \end{pmatrix}, \quad B_3 = \begin{pmatrix} 0 & 0 \\ -1 & 0 \end{pmatrix}.$$

Solve the weakly singular systems, where a logarithmic term appears in the solutions, caused by the difference of two eigenvalues being a positive integer:

9. $\frac{d}{dx}u = \frac{1}{x}A + B_0$, where

$$A = \begin{pmatrix} 0 & 1 \\ -\frac{3}{4} & -2 \end{pmatrix}, \quad B_0 = \begin{pmatrix} 0 & 0 \\ -1 & 0 \end{pmatrix}.$$

10. $\frac{d}{dx}u = \frac{1}{x}A + B_0$, where

$$A = \begin{pmatrix} 0 & 1 \\ -2 & -3 \end{pmatrix}, \quad B_0 = \begin{pmatrix} 0 & 0 \\ -1 & 0 \end{pmatrix}.$$

11. $\frac{d}{dx}u = \frac{1}{x}A + xB_1$, where

$$A = \begin{pmatrix} 0 & 1 \\ -3 & -4 \end{pmatrix}, \quad B_1 = \begin{pmatrix} 0 & 0 \\ -1 & 0 \end{pmatrix}.$$

Solve the weakly singular equations, where no logarithmic terms appear in the solutions:

12. $x^2y'' + x(\frac{3}{2})y' + (-\frac{1}{2} + x)y = 0.$

13. $x^2y'' + x(\frac{3}{2})y' + (-\frac{1}{2} + 2x^2)y = 0.$

14. $x^2y'' + x(\frac{3}{2}y' + (-\frac{1}{2} + 3x^3)y = 0.$

Solve the weakly singular equations, where a logarithmic term appears in the solutions, caused by a repeated eigenvalue:

15. $x^2y'' + x(-1)y' + (x + 1)y = 0.$

16. $x^2y'' + x(1 - 2\sqrt{2})y' + (x^2 + 2)y = 0.$

17. $x^2y'' + x(-3)y' + (x^4 + 4)y = 0.$

Solve the weakly singular equation, where a logarithmic term appears in the solutions, caused by the difference of two eigenvalues being a positive integer:

18. $x^2y'' + x(3)y' + (x + \frac{3}{4})y = 0.$

19. $x^2y'' + x(4)y' + (x + 2)y = 0.$

20. $x^2y'' + x(5)y' + (x^2 + 3)y = 0.$

21. Prove Lemma 2.5.

22. Prove Lemma 2.7.

23. Solve the weakly singular system of order 3×3, where no logarithmic terms appear in the solutions :

$$\frac{d}{dz}u = (\frac{1}{z}A + z^3B_3)u, \quad 0 < |z| < r_0,$$

where

$$A = \begin{pmatrix} 0 & 1 & 0 \\ 0 & 1 & 1 \\ 0 & 8 & 3 \end{pmatrix}, \quad \text{and} \quad B_3 = \begin{pmatrix} 0 & 0 & 0 \\ 0 & 0 & 0 \\ -3 & -4 & -5 \end{pmatrix}.$$

24. Solve the weakly singular system of order 3×3, where a logarithmic term appears in the solutions, caused by a repeated eigenvalue:

$$\frac{d}{dz}u = (\frac{1}{z}A + zB_1)u, \quad 0 < |z| < r_0,$$

where

$$A = \begin{pmatrix} 0 & 1 & 0 \\ 0 & 1 & 1 \\ 0 & -\frac{1}{4} & 2 \end{pmatrix} \quad \text{and} \quad B_1 = \begin{pmatrix} 0 & 0 & 0 \\ 0 & 0 & 0 \\ -1 & -2 & -3 \end{pmatrix}.$$

25. Consider the weakly singular system of order 3×3, where a logarithmic term appears in the solutions, caused by the difference of two eigenvalues being a positive integer:

$$\frac{d}{dz}u = (\frac{1}{z}A + z^4 B_4)u, \quad 0 < |z| < r_0.$$

Here A is the A in Problem **23**, and

$$B_4 = \begin{pmatrix} 0 & 0 & 0 \\ 0 & 0 & 0 \\ -4 & -5 & -6 \end{pmatrix}.$$

26. Consider the regular singular equation of order 3, where no logarithmic terms appear in the solutions:

$$z^3 y''' + z^2(-1 + 5z^4)y'' + z(-8 + 4z^4)y' + 3z^4 y = 0, \quad 0 < |z| < r_0.$$

27. Consider the regular singular equation of order 3, where a logarithmic term appears in the solutions:

$$z^3 y''' + z^2(3z^2)y'' + z(\frac{1}{4} + 2z^2)y' + z^2 y = 0, \quad 0 < |z| < r_0.$$

28. Consider the regular singular equation of order 3, where a logarithmic term appears in the solutions:

$$z^3 y''' + z^2(-1 + 6z^5)y'' + z(-8 + 5z^5)y' + 4z^5 y = 0, 0 < |z| < r_0.$$

29. Tell, in advance, whether the power series solutions to the examples in Section 3, have logarithmic terms, or not.

30. Tell, in advance, whether the power series solutions to the examples in Section 5, have logarithmic terms.

31. Consider the Bessel equation of order ν [6, Page 172]:

$$z^2 y'' + zy' + (z^2 - \nu^2)y = 0, \quad \nu \in \mathbb{C} \quad \text{and } 0 < |z| < r_0,$$

where the complex number ν satisfies $\operatorname{Re}\nu \geq 0$. Tell, in advance, whether its power series solutions have logarithmic terms, or not.

7. Solutions

1. Method 1. Try the solution of the form

$$u(x) = \sum_{k=0}^{\infty} c_k x^k,$$

where the constant vectors c_k are to be determined. After substitution and calculations, $d_k, k = 0, \ldots,$ satisfy

$$d_1 = A_0 d_0,$$

$$(k+1)d_{k+1} = A_0 d_k + A_1 d_{k-1}, \quad k \geq 1.$$

A few terms of c_k are:

$$d_0 = \begin{pmatrix} 1 \\ 0 \end{pmatrix}, \quad d_1 = \begin{pmatrix} 0 \\ -2\mu \end{pmatrix}, \quad d_2 = \begin{pmatrix} -\mu \\ 0 \end{pmatrix}, \ldots,$$

or

$$d_0 = \begin{pmatrix} 0 \\ 1 \end{pmatrix}, \quad d_1 = \begin{pmatrix} 1 \\ 0 \end{pmatrix}, \quad d_2 = \begin{pmatrix} 0 \\ -\mu + 1 \end{pmatrix}, \quad d_3 = \begin{pmatrix} \frac{-\mu+1}{3} \\ 0 \end{pmatrix}, \ldots.$$

Notice that $\begin{pmatrix} 1 \\ 0 \end{pmatrix}$ and $\begin{pmatrix} 0 \\ 1 \end{pmatrix}$ are linearly independent vectors.

Method 2. By Theorem 2.1,

$$u_j = u_0 + \int A(x)u_{j-1}\, dx, \quad j = 1, 2, \ldots.$$

A few terms of u_j are:

$$u_1 = u_0 + \begin{pmatrix} 0 & x \\ -2\mu x & x^2 \end{pmatrix} u_0,$$

$$u_2 = u_1 + \begin{pmatrix} -\mu x^2 & \frac{x^3}{3} \\ \frac{-4\mu x^3}{3} & (\frac{x^4}{2} - \mu x^2) \end{pmatrix} u_0,$$

$$u_3 = u_2 + \begin{pmatrix} \frac{-\mu x^4}{3} & \frac{x^5}{10} - \frac{\mu x^3}{3} \\ \frac{2\mu^2 x^3}{3} - \frac{8\mu x^5}{15} & -\frac{\mu x^4}{6} + \frac{x^6}{6} - \frac{\mu x^4}{2} \end{pmatrix} u_0,$$

$$\vdots$$

The two vectors $u_0 = \begin{pmatrix} 1 \\ 0 \end{pmatrix}$ and $u_0 = \begin{pmatrix} 0 \\ 1 \end{pmatrix}$, respectively, result in two linearly independent power series solutions.

2. Method 1. This equation corresponds to the system in Problem **1**, by the substitution: $u_1 = y, u_2 = y'$, and so its two linearly independent solutions are the first components of the two solutions in Problem **1**.

Method 2. Try the solution of the form

$$y = \sum_{k=0}^{\infty} c_k x^k,$$

where the constants c_k are to be determined. After substitution and calculations, $c_k, k = 0, 1, \ldots,$ satisfy

$$c_{k+2} = \frac{2k - 2\mu}{(k+2)(k+1)} c_k.$$

A few terms of c_k are:

$$c_2 = \frac{-2\mu}{2 \times 1} c_0,$$

$$c_4 = \frac{4 - 2\mu}{4 \times 3} \times \frac{-2\mu}{2 \times 1} c_0,$$

$$\vdots,$$

or

$$c_3 = \frac{2 - 2\mu}{3 \times 2} c_1,$$

$$c_5 = \frac{6 - 2\mu}{5 \times 4} \times \frac{2 - 2\mu}{3 \times 2} c_1,$$

$$\vdots$$

3. The eigenpairs $\{\lambda, w_0\}$'s are:

$$\{\frac{1}{2}, \begin{pmatrix} 2 \\ 1 \end{pmatrix}\}, \quad \{-1, \begin{pmatrix} 1 \\ -1 \end{pmatrix}\}.$$

For the eigenpair $\{\lambda, w_0\}$, there are

$$w_1 = w_0 + (\lambda + 1 - A)^{-1} e^{t(\lambda + 1 - A)} B_0 w_0,$$

$$v_1 = e^{tA} w_1$$

$$= e^{t\lambda} w_0 + (\lambda + 1 - A)^{-1} e^{t(\lambda + 1)} B_0 w_0,$$

$$u_1 = v_1|_{t=\ln(x)}$$

$$= x^\lambda w_0 + (\lambda + 1 - A)^{-1} x^{\lambda + 1} B_0 w_0,$$

and

$$w_2 = w_1 + (\lambda + 2 - A)^{-1} e^{t(\lambda + 2 - A)} B_0 (\lambda + 1 - A)^{-1} B_0 w_0,$$

$$v_2 = e^{tA} w_2$$

$$= v_1 + (\lambda + 2 - A)^{-1} e^{t(\lambda + 2)} B_0 (\lambda + 1 - A)^{-1} B_0 w_0,$$

$$u_2 = v_2|_{t=\ln(x)}$$

$$= u_1 + (\lambda + 2 - A)^{-1} x^{\lambda + 2} B_0 (\lambda + 1 - A)^{-1} B_0 w_0.$$

The other $u_j, j = 3, 4, \ldots$, are similarly calculated.

4. The eigenpairs $\{\lambda, w_0\}$'s are:

$$\{\frac{1}{2}, \begin{pmatrix} 2 \\ 1 \end{pmatrix}\}, \quad \{-1, \begin{pmatrix} 1 \\ -1 \end{pmatrix}\}.$$

For the eigenpair $\{\lambda, w_0\}$, there are

$$w_1 = w_0 + (\lambda + 1 - A)^{-1} e^{t(\lambda + 2 - A)} B_1 w_0,$$

$$v_1 = e^{tA} w_1$$

$$= e^{t\lambda} w_0 + (\lambda + 2 - A)^{-1} e^{t(\lambda + 2)} B_1 w_0,$$

$$u_1 = v_1|_{t=\ln(x)}$$

$$= x^\lambda w_0 + (\lambda + 2 - A)^{-1} x^{\lambda + 2} B_1 w_0,$$

and

$$w_2 = w_1 + (\lambda + 4 - A)^{-1} e^{t(\lambda + 4 - A)} B_1 (\lambda + 2 - A)^{-1} B_1 w_0,$$

$$v_2 = e^{tA} w_2$$

$$= v_1 + (\lambda + 4 - A)^{-1} e^{t(\lambda + 4)} B_1 (\lambda + 2 - A)^{-1} B_1 w_0,$$

$$u_2 = v_2|_{t=\ln(x)}$$

$$= u_1 + (\lambda + 4 - A)^{-1} x^{\lambda+2} B_1 (\lambda + 2 - A)^{-1} B_1 w_0.$$

The other $u_j, j = 3, 4, \ldots$, are similarly calculated.

5. The eigenpairs $\{\lambda, w_0\}$'s are:

$$\{\frac{1}{2}, \begin{pmatrix} 2 \\ 1 \end{pmatrix}\}, \quad \{-1, \begin{pmatrix} 1 \\ -1 \end{pmatrix}\}.$$

For the eigenpair $\{\lambda, w_0\}$, there are

$$w_1 = w_0 + (\lambda + 3 - A)^{-1} e^{t(\lambda+3-A)} B_2 w_0,$$

$$v_1 = e^{tA} w_1$$

$$= e^{t\lambda} w_0 + (\lambda + 3 - A)^{-1} e^{t(\lambda+3)} B_2 w_0,$$

$$u_1 = v_1|_{t=\ln(x)}$$

$$= x^\lambda w_0 + (\lambda + 3 - A)^{-1} x^{\lambda \mid 3} B_2 w_0,$$

and

$$w_2 = w_1 + (\lambda + 6 - A)^{-1} e^{t(\lambda+6-A)} B_2 (\lambda + 3 - A)^{-1} B_2 w_0,$$

$$v_2 = e^{tA} w_2$$

$$= v_1 + (\lambda + 6 - A)^{-1} e^{t(\lambda+6)} B_2 (\lambda + 3 - A)^{-1} B_2 w_0,$$

$$u_2 = v_2|_{t=\ln(x)}$$

$$= u_1 + (\lambda + 6 - A)^{-1} x^{\lambda+6} B_2 (\lambda + 3 - A)^{-1} B_2 w_0.$$

The other $u_j, j = 3, 4, \ldots$, are similarly calculated.

6. For the eigenpair

$$\{\lambda, w_0\} = \{1, \beta_1\} = \{1, \begin{pmatrix} 1 \\ 1 \end{pmatrix}\},$$

there are

$$w_1 = w_0 + (\lambda + 1 - A)^{-1} e^{t(\lambda+1-A)} B_0 w_0,$$

$$v_1 = e^{tA} w_1$$

$$= e^{t\lambda} w_0 + (\lambda + 1 - A)^{-1} e^{t(\lambda+1)} B_0 w_0,$$

$$u_1 = v_1|_{t=\ln(x)}$$

$$= x^\lambda w_0 + (\lambda + 1 - A)^{-1} x^{\lambda+1} B_0 w_0,$$

and

$$w_2 = w_1 + (\lambda + 2 - A)^{-1} e^{t(\lambda+2-A)} B_0 (\lambda + 1 - A)^{-1} B_0 w_0,$$

$$v_2 = e^{tA} w_2$$

$$= v_1 + (\lambda + 2 - A)^{-1} e^{t(\lambda+2)} B_0 (\lambda + 1 - A)^{-1} B_0 w_0,$$

$$u_2 = v_2|_{t=\ln(x)}$$

$$= u_1 + (\lambda + 2 - A)^{-1} x^{\lambda+2} B_0 (\lambda + 1 - A)^{-1} B_0 w_0.$$

The other $u_j, j = 3, 4, \ldots$, are similarly calculated.

For the generalized eigenpair

$$\{\lambda, w_0\} = \{1, \beta_2\} = \{1, \begin{pmatrix} 0 \\ 1 \end{pmatrix}\},$$

in which $A^2\beta_2 = 0$ and $A\beta_2 = \beta_1$, there are

$$w_1 = \beta_2 + (2-A)^{-1}e^{(2-A)t}B_0\beta_2 + (2-A)^{-1}te^{(2-A)t}B_0\beta_1$$
$$- (2-A)^{-2}e^{(2-A)t}B_0\beta_1,$$

$$v_1 = e^{tA}w_1$$
$$= e^t(\beta_2 + t\beta_1) + (2-A)^{-1}e^{2t}B_0\beta_2 +$$
$$+ (2-A)^{-1}te^{2t}B_0\beta_1 - (2-A)^{-2}e^{2t}B_0\beta_1,$$

$$u_1 = v_1|_{t=\ln(x)}$$
$$= x(\beta_2 + \beta_1\ln(x)) + (2-A)^{-1}x^2B_0\beta_2$$
$$+ (2-A)^{-1}[x^2\ln(x)]B_0\beta_1 - (2-A)^{-2}x^2B_0\beta_1,$$

and

$$w_2 = w_1 + (4-A)^{-1}e^{t(4-A)}B_0(2-A)^{-1}B_0\beta_2 + (4-A)^{-1}te^{t(4-A)}B_0(2-A)^{-1}B_0\beta_1$$
$$- (4-A)^{-2}e^{t(4-A)}B_0(2-A)^{-1}B_0\beta_1 - (4-A)^{-1}e^{t(4-A)}B_0(2-A)^{-2}B_0\beta_1,$$

$$v_2 = e^{tA}w_2$$
$$= v_1 + (4-A)^{-1}e^{4t}B_0(2-A)^{-1}B_0\beta_2 + (4-A)^{-1}te^{4t}B_0(2-A)^{-1}B_0\beta_1$$
$$- (4-A)^{-2}e^{4t}B_0(2-A)^{-1}B_0\beta_1 - (4-A)^{-1}e^{4t}B_0(2-A)^{-2}B_0\beta_1,$$

$$u_2 = v_2|_{t=\ln(x)}$$
$$= u_1 + (4-A)^{-1}x^4B_0(2-A)^{-1}B_0\beta_2 + (4-A)^{-1}[x^4\ln(x)]B_0(2-A)^{-1}B_0\beta_1$$
$$- (4-A)^{-2}x^4B_0(2-A)^{-1}B_0\beta_1 - (4-A)^{-1}x^4B_0(2-A)^{-2}B_0\beta_1.$$

The other w_j, v_j, and $u_j, j = 3, 4, \ldots$, are similarly calculated.

7. For the eigenpair

$$\{\lambda, w_0\} = \{\sqrt{2}, \beta_1\} = \{\sqrt{2}, \begin{pmatrix} 1 \\ \sqrt{2} \end{pmatrix}\},$$

there are

$$w_1 = w_0 + (\lambda + 2 - A)^{-1}e^{t(\lambda+1-A)}B_1w_0,$$
$$v_1 = e^{tA}w_1$$
$$= e^{t\lambda}w_0 + (\lambda + 2 - A)^{-1}e^{t(\lambda+2)}B_1w_0,$$
$$u_1 = v_1|_{t=\ln(x)}$$
$$= x^\lambda w_0 + (\lambda + 2 - A)^{-1}x^{\lambda+2}B_1w_0,$$

and

$$w_2 = w_1 + (\lambda + 4 - A)^{-1}e^{t(\lambda+4-A)}B_1(\lambda + 2 - A)^{-1}B_1w_0,$$
$$v_2 = e^{tA}w_2$$
$$= v_1 + (\lambda + 4 - A)^{-1}e^{t(\lambda+4)}B_1(\lambda + 2 - A)^{-1}B_1w_0,$$

$$u_2 = v_2|_{t=\ln(x)}$$
$$= u_1 + (\lambda + 4 - A)^{-1}x^{\lambda+4}B_0(\lambda + 2 - A)^{-1}B_1w_0.$$

The other $u_j, j = 3, 4, \ldots$, are similarly calculated.

For the generalized eigenpair

$$\{\lambda, w_0\} = \{\sqrt{2}, \beta_2\} = \left\{\sqrt{2}, \begin{pmatrix} 0 \\ 1 \end{pmatrix}\right\},$$

in which $A^2\beta_2 = 0$ and $A\beta_2 = \beta_1$, there are

$$w_1 = \beta_2 + (\sqrt{2}+2-A)^{-1}e^{(\sqrt{2}+2-A)t}B_1\beta_2 + (\sqrt{2}+2-A)^{-1}te^{(\sqrt{2}+2-A)t}B_1\beta_1$$
$$- (\sqrt{2}+2-A)^{-2}e^{(\sqrt{2}+2-A)t}B_1\beta_1,$$

$$v_1 = e^{tA}w_1$$
$$= e^{\sqrt{2}t}(\beta_2 + t\beta_1) + (\sqrt{2}+2-A)^{-1}e^{t(\sqrt{2}+2)}B_1\beta_2 +$$
$$+ (\sqrt{2}+2-A)^{-1}te^{t(\sqrt{2}+2)}B_1\beta_1 - (\sqrt{2}+2-A)^{-2}e^{t(\sqrt{2}+2)}B_1\beta_1,$$

$$u_1 = v_1|_{t=\ln(x)}$$
$$= x^{\sqrt{2}}(\beta_2 + \beta_1\ln(x)) + (\sqrt{2}+2-A)^{-1}x^{\sqrt{2}+2}B_1\beta_2$$
$$+ (\sqrt{2}+2-A)^{-1}[x^{\sqrt{2}+2}\ln(x)]B_1\beta_1 - (\sqrt{2}+2-A)^{-2}x^{\sqrt{2}+2}B_1\beta_1,$$

and

$$w_2 = w_1 + (\sqrt{2}+4-A)^{-1}e^{t(\sqrt{2}+4-A)}B_1(\sqrt{2}+2-A)^{-1}B_1\beta_2$$
$$+ (\sqrt{2}+4-A)^{-1}te^{t(\sqrt{2}+4-A)}B_1(\sqrt{2}+2-A)^{-1}B_1\beta_1$$
$$- (\sqrt{2}+4-A)^{-2}e^{t(\sqrt{2}+4-A)}B_1(\sqrt{2}+2-A)^{-1}B_1\beta_1$$
$$- (\sqrt{2}+4-A)^{-1}e^{t(\sqrt{2}+4-A)}B_1(\sqrt{2}+2-A)^{-2}B_1\beta_1,$$

$$v_2 = e^{tA}w_2$$
$$= v_1 + (\sqrt{2}+4-A)^{-1}e^{(\sqrt{2}+4)t}B_1(\sqrt{2}+2-A)^{-1}B_1\beta_2$$
$$+ (\sqrt{2}+4-A)^{-1}te^{(\sqrt{2}+4)t}B_1(\sqrt{2}+2-A)^{-1}B_1\beta_1$$
$$- (\sqrt{2}+4-A)^{-2}e^{(\sqrt{2}+4)t}B_1(\sqrt{2}+2-A)^{-1}B_1\beta_1$$
$$- (\sqrt{2}+4-A)^{-1}e^{(\sqrt{2}+4)t}B_1(\sqrt{2}+2-A)^{-2}B_1\beta_1,$$

$$u_2 = v_2|_{t=\ln(x)}$$
$$= u_1 + (\sqrt{2}+4-A)^{-1}x^{\sqrt{2}+4}B_1(\sqrt{2}+2-A)^{-1}B_1\beta_2$$
$$+ (\sqrt{2}+4-A)^{-1}[x^{\sqrt{2}+4}\ln(x)]B_1(\sqrt{2}+2-A)^{-1}B_1\beta_1$$
$$- (\sqrt{2}+4-A)^{-2}x^{\sqrt{2}+4}B_1(\sqrt{2}+2-A)^{-1}B_1\beta_1$$
$$- (\sqrt{2}+4-A)^{-1}x^{\sqrt{2}+4}B_1(\sqrt{2}+2-A)^{-2}B_1\beta_1.$$

The other w_j, v_j, and $u_j, j = 3, 4, \ldots$, are similarly calculated.

8. For the eigenpair

$$\{\lambda, w_0\} = \{2, \beta_1\} = \left\{2, \begin{pmatrix} 1 \\ 2 \end{pmatrix}\right\},$$

there are

$$w_1 = w_0 + (\lambda + 4 - A)^{-1} e^{t(\lambda+4-A)} B_3 w_0,$$

$$v_1 = e^{tA} w_1$$

$$= e^{t\lambda} w_0 + (\lambda + 4 - A)^{-1} e^{t(\lambda+4)} B_3 w_0,$$

$$u_1 = v_1|_{t=\ln(x)}$$

$$= x^\lambda w_0 + (\lambda + 4 - A)^{-1} x^{\lambda+4} B_3 w_0,$$

and

$$w_2 = w_1 + (\lambda + 8 - A)^{-1} e^{t(\lambda+8-A)} B_3 (\lambda + 4 - A)^{-1} B_3 w_0,$$

$$v_2 = e^{tA} w_2$$

$$= v_1 + (\lambda + 8 - A)^{-1} e^{t(\lambda+8)} B_3 (\lambda + 4 - A)^{-1} B_3 w_0,$$

$$u_2 = v_2|_{t=\ln(x)}$$

$$= u_1 + (\lambda + 8 - A)^{-1} x^{\lambda+8} B_0 (\lambda + 4 - A)^{-1} B_3 w_0.$$

The other $u_j, j = 3, 4, \ldots$, are similarly calculated.

For the generalized eigenpair

$$\{\lambda, w_0\} = \{2, \beta_2\} = \left\{2, \begin{pmatrix} 0 \\ 1 \end{pmatrix}\right\},$$

in which $A^2 \beta_2 = 0$ and $A\beta_2 = \beta_1$, there are

$$w_1 = \beta_2 + (2 + 4 - A)^{-1} e^{(2+4-A)t} B_3 \beta_2 + (2 + 4 - A)^{-1} t e^{(2+4-A)t} B_3 \beta_1$$
$$- (2 + 4 - A)^{-2} e^{(2+4-A)t} B_3 \beta_1,$$

$$v_1 = e^{tA} w_1$$
$$= e^{2t}(\beta_2 + t\beta_1) + (2 + 4 - A)^{-1} e^{(2+4)t} B_3 \beta_2 +$$
$$+ (2 + 4 - A)^{-1} t e^{(2+4)t} B_1 \beta_1 - (2 + 4 - A)^{-2} e^{(2+4)t} B_0 \beta_1,$$

$$u_1 = v_1|_{t=\ln(x)}$$
$$= x^2(\beta_2 + \beta_1 \ln(x)) + (2 + 4 - A)^{-1} x^{2+4} B_3 \beta_2$$
$$+ (2 + 4 - A)^{-1}[x^{2+4} \ln(x)] B_3 \beta_1 - (2 + 4 - A)^{-2} x^{2+4} B_3 \beta_1,$$

and

$$w_2 = w_1 + (2 + 8 - A)^{-1} e^{t(2+8-A)} B_3 (2 + 4 - A)^{-1} B_3 \beta_2 +$$
$$(2 + 8 - A)^{-1} t e^{t(2+8-A)} B_3 (2 + 4 - A)^{-1} B_3 \beta_1$$
$$- (2 + 8 - A)^{-2} e^{t(2+8-A)} B_3 (2 + 4 - A)^{-1} B_3 \beta_1 -$$
$$(2 + 8 - A)^{-1} e^{t(2+8-A)} B_3 (2 + 4 - A)^{-2} B_3 \beta_1,$$

$$v_2 = e^{tA} w_2$$
$$= v_1 + (2 + 8 - A)^{-1} e^{(2+8)t} B_3 (2 + 4 - A)^{-1} B_3 \beta_2 +$$

$$(2 + 8 - A)^{-1} t e^{(2+8)t} B_3 (2 + 4 - A)^{-1} B_3 \beta_1$$
$$- (2 + 8 - A)^{-2} e^{(2+8)t} B_3 (2 + 4 - A)^{-1} B_3 \beta_1 -$$
$$(2 + 8 - A)^{-1} e^{(2+8)t} B_3 (2 + 4 - A)^{-2} B_3 \beta_1,$$

$$u_2 = v_2|_{t=\ln(x)}$$
$$= u_1 + (2 + 8 - A)^{-1} x^{(2+8)} B_3 (2 + 4 - A)^{-1} B_3 \beta_2 +$$
$$(2 + 8 - A)^{-1} [x^{(2+8)} \ln(x)] B_3 (2 + 4 - A)^{-1} B_3 \beta_1$$
$$- (2 + 8 - A)^{-2} x^{(2+8)} B_3 (2 + 4 - A)^{-1} B_3 \beta_1 -$$
$$(2 + 8 - A)^{-1} x^{(2+8)} B_3 (2 + 4 - A)^{-2} B_3 \beta_1.$$

The other w_j, v_j, and $u_j, j = 3, 4, \ldots$, are similarly calculated.

9. For the eigenpair

$$\{\lambda, w_0\} = \{\lambda_1, \beta_1\} = \{-\frac{1}{2}, \begin{pmatrix} 2 \\ -1 \end{pmatrix}\},$$

there are

$$w_1 = w_0 + (\lambda + 1 - A)^{-1} e^{t(\lambda + 1 - A)} B_0 w_0,$$
$$v_1 = e^{tA} w_1$$
$$= e^{t\lambda} w_0 + (\lambda + 1 - A)^{-1} e^{t(\lambda+1)} B_0 w_0,$$
$$u_1 = v_1|_{t=\ln(x)}$$
$$= x^{\lambda} w_0 + (\lambda + 1 - A)^{-1} x^{\lambda+1} B_0 w_0,$$

and

$$w_2 = w_1 + (\lambda + 2 - A)^{-1} e^{t(\lambda + 2 - A)} B_0 (\lambda + 1 - A)^{-1} B_0 w_0,$$
$$v_2 = e^{tA} w_2$$
$$= v_1 + (\lambda + 2 - A)^{-1} e^{t(\lambda+2)} B_0 (\lambda + 1 - A)^{-1} B_0 w_0,$$
$$u_2 = v_2|_{t=\ln(x)}$$
$$= u_1 + (\lambda + 2 - A)^{-1} x^{\lambda+2} B_0 (\lambda + 1 - A)^{-1} B_0 w_0.$$

The other $u_j, j = 3, 4, \ldots$, are similarly calculated.

For the eigenpair

$$\{\lambda_2, \beta_2\} = \{-\frac{3}{2}, \begin{pmatrix} 2 \\ -3 \end{pmatrix}\},$$

there are

$$e^{tA} = e^{\lambda_1 t} M_{1,0} + e^{\lambda_2 t} M_{2,0},$$
$$M_{1,0} = \begin{pmatrix} \frac{3}{2} & 1 \\ -\frac{3}{4} & -\frac{1}{2} \end{pmatrix}, \quad M_{2,0} = \begin{pmatrix} -\frac{1}{2} & -1 \\ \frac{3}{4} & \frac{3}{2} \end{pmatrix},$$
$$w_1 = \beta_2 + t M_{1,0} B_0 \beta_2 + e^t M_{2,0} B_0 \beta_2,$$
$$v_1 = e^{tA} w_1$$
$$= e^{\lambda_2 t} \beta_2 + t e^{\lambda_1 t} M_{1,0} B_0 \beta_2 + e^{(\lambda_2+1)t} M_{2,0} B_0 \beta_2.$$
$$u_1 = v_1|_{t=\ln(x)}$$

$$= x^{\lambda_2}\beta_2 + [x^{\lambda_1}\ln(x)]M_{1,0}B_0\beta_2 + x^{\lambda_2+1}M_{2,0}B_0\beta_2,$$

$$w_2 = w_1 + t(\lambda_1 + 1 - A)^{-1}e^{t(\lambda_1+1-A)}B_0(M_{1,0}B_0\beta_2)$$
$$- (\lambda_1 + 1 - A)^{-2}e^{t(\lambda_1+1-A)}B_0(M_{1,0}B_0\beta_2)+$$
$$(\lambda_2 + 2 - A)^{-1}e^{t(\lambda_2+2-A)}B_0(M_{2,0}B_0\beta_2),$$

$$v_2 = v_1 + t(\lambda_1 + 1 - A)^{-1}e^{t(\lambda_1+1)}B_0(M_{1,0}B_0\beta_2)$$
$$- (\lambda_1 + 1 - A)^{-2}e^{t(\lambda_1+1)}B_0(M_{1,0}B_0\beta_2)+$$
$$(\lambda_2 + 2 - A)^{-1}e^{t(\lambda_2+2)}B_0(M_{2,0}B_0\beta_2),$$

$$u_2 = u_1 + [\ln(x)](\lambda_1 + 1 - A)^{-1}x^{(\lambda_1+1)}B_0(M_{1,0}B_0\beta_2)$$
$$- (\lambda_1 + 1 - A)^{-2}x^{(\lambda_1+1)}B_0(M_{1,0}B_0\beta_2)+$$
$$(\lambda_2 + 2 - A)^{-1}x^{(\lambda_2+2)}B_0(M_{2,0}B_0\beta_2).$$

The other $u_j, j = 3, 4, \ldots$, are similarly calculated.

10. For the eigenpair

$$\{\lambda, w_0\} = \{\lambda_1, \beta_1\} = \left\{-1, \begin{pmatrix} 1 \\ -1 \end{pmatrix}\right\},$$

there are

$$w_1 = w_0 + (\lambda + 1 - A)^{-1}e^{t(\lambda+1-A)}B_0 w_0,$$
$$v_1 = e^{tA}w_1$$
$$= e^{t\lambda}w_0 + (\lambda + 1 - A)^{-1}e^{t(\lambda+1)}B_0 w_0,$$
$$u_1 = v_1|_{t=\ln(x)}$$
$$= x^{\lambda}w_0 + (\lambda + 1 - A)^{-1}x^{\lambda+1}B_0 w_0,$$

and

$$w_2 = w_1 + (\lambda + 2 - A)^{-1}e^{t(\lambda+2-A)}B_0(\lambda + 1 - A)^{-1}B_0 w_0,$$
$$v_2 = e^{tA}w_2$$
$$= v_1 + (\lambda + 2 - A)^{-1}e^{t(\lambda+2)}B_0(\lambda + 1 - A)^{-1}B_0 w_0,$$
$$u_2 = v_2|_{t=\ln(x)}$$
$$= u_1 + (\lambda + 2 - A)^{-1}x^{\lambda+2}B_0(\lambda + 1 - A)^{-1}B_0 w_0.$$

The other $u_j, j = 3, 4, \ldots$, are similarly calculated.

For the eigenpair

$$\{\lambda_2, \beta_2\} = \left\{-2, \begin{pmatrix} 1 \\ -2 \end{pmatrix}\right\},$$

there are

$$e^{tA} = e^{\lambda_1 t}M_{1,0} + e^{\lambda_2 t}M_{2,0},$$

$$M_{1,0} = \begin{pmatrix} 2 & 1 \\ -2 & -1 \end{pmatrix}, \quad M_{2,0} = \begin{pmatrix} -1 & -1 \\ 2 & 2 \end{pmatrix},$$

$$w_1 = \beta_2 + tM_{1,0}B_0\beta_2 + e^t M_{2,0}B_0\beta_2,$$
$$v_1 = e^{tA}w_1$$

$$= e^{\lambda_2 t}\beta_2 + te^{\lambda_1 t}M_{1,0}B_0\beta_2 + e^{(\lambda_2+1)t}M_{2,0}B_0\beta_2.$$

$$u_1 = v_1|_{t=\ln(x)}$$
$$= x^{\lambda_2}\beta_2 + [x^{\lambda_1}\ln(x)]M_{1,0}B_0\beta_2 + x^{\lambda_2+1}M_{2,0}B_0\beta_2,$$

$$w_2 = w_1 + t(\lambda_1 + 1 - A)^{-1}e^{t(\lambda_1+1-A)}B_0(M_{1,0}B_0\beta_2)$$
$$- (\lambda_1 + 1 - A)^{-2}e^{t(\lambda_1+1-A)}B_0(M_{1,0}B_0\beta_2)+$$
$$(\lambda_2 + 2 - A)^{-1}e^{t(\lambda_2+2-A)}B_0(M_{2,0}B_0\beta_2),$$

$$v_2 = v_1 + t(\lambda_1 + 1 - A)^{-1}e^{t(\lambda_1+1)}B_0(M_{1,0}B_0\beta_2)$$
$$- (\lambda_1 + 1 - A)^{-2}e^{t(\lambda_1+1)}B_0(M_{1,0}B_0\beta_2)+$$
$$(\lambda_2 + 2 - A)^{-1}e^{t(\lambda_2+2)}B_0(M_{2,0}B_0\beta_2),$$

$$u_2 = u_1 + [\ln(x)](\lambda_1 + 1 - A)^{-1}x^{(\lambda_1+1)}B_0(M_{1,0}B_0\beta_2)$$
$$- (\lambda_1 + 1 - A)^{-2}x^{(\lambda_1+1)}B_0(M_{1,0}B_0\beta_2)+$$
$$(\lambda_2 + 2 - A)^{-1}x^{(\lambda_2+2)}B_0(M_{2,0}B_0\beta_2).$$

The other $u_j, j = 3, 4, \ldots$, are similarly calculated.

11. For the eigenpair

$$\{\lambda, w_0\} = \{\lambda_1, \beta_1\} = \left\{-1, \begin{pmatrix} 1 \\ -1 \end{pmatrix}\right\},$$

there are

$$w_1 = w_0 + (\lambda + 1 - A)^{-1}e^{t(\lambda+1-A)}B_0w_0,$$
$$v_1 = e^{tA}w_1$$
$$= e^{t\lambda}w_0 + (\lambda + 1 - A)^{-1}e^{t(\lambda+1)}B_0w_0,$$
$$u_1 = v_1|_{t=\ln(x)}$$
$$= x^\lambda w_0 + (\lambda + 1 - A)^{-1}x^{\lambda+1}B_0w_0,$$

and

$$w_2 = w_1 + (\lambda + 2 - A)^{-1}e^{t(\lambda+2-A)}B_0(\lambda + 1 - A)^{-1}B_0w_0,$$
$$v_2 = e^{tA}w_2$$
$$= v_1 + (\lambda + 2 - A)^{-1}e^{t(\lambda+2)}B_0(\lambda + 1 - A)^{-1}B_0w_0,$$
$$u_2 = v_2|_{t=\ln(x)}$$
$$= u_1 + (\lambda + 2 - A)^{-1}x^{\lambda+2}B_0(\lambda + 1 - A)^{-1}B_0w_0.$$

The other $u_j, j = 3, 4, \ldots$, are similarly calculated.

For the eigenpair

$$\{\lambda_2, \beta_2\} = \left\{-3, \begin{pmatrix} 1 \\ -3 \end{pmatrix}\right\},$$

there are

$$e^{tA} = e^{\lambda_1 t}M_{1,0} + e^{\lambda_2 t}M_{2,0},$$

$$M_{1,0} = \begin{pmatrix} \frac{3}{2} & 1 \\ -\frac{3}{4} & -\frac{1}{2} \end{pmatrix}, \quad M_{2,0} = \begin{pmatrix} -\frac{1}{2} & -1 \\ \frac{3}{4} & \frac{3}{2} \end{pmatrix},$$

$$w_1 = \beta_2 + tM_{1,0}B_1\beta_2 + e^{2t}M_{2,0}B_0\beta_2,$$

$$v_1 = e^{tA}w_1$$

$$= e^{\lambda_2 t}\beta_2 + te^{\lambda_1 t}M_{1,0}B_1\beta_2 + e^{(\lambda_2+2)t}M_{2,0}B_1\beta_2.$$

$$u_1 = v_1|_{t=\ln(x)}$$

$$= x^{\lambda_2}\beta_2 + [x^{\lambda_1}\ln(x)]M_{1,0}B_1\beta_2 + x^{\lambda_2+2}M_{2,0}B_1\beta_2,$$

$$w_2 = w_1 + t(\lambda_1 + 2 - A)^{-1}e^{t(\lambda_1+2-A)}B_1(M_{1,0}B_1\beta_2)$$

$$- (\lambda_1 + 2 - A)^{-2}e^{t(\lambda_1+2-A)}B_1(M_{1,0}B_1\beta_2)+$$

$$(\lambda_2 + 4 - A)^{-1}e^{t(\lambda_2+4-A)}B_1(M_{2,0}B_1\beta_2),$$

$$v_2 = v_1 + t(\lambda_1 + 2 - A)^{-1}e^{t(\lambda_1+2)}B_1(M_{1,0}B_1\beta_2)$$

$$- (\lambda_1 + 2 - A)^{-2}e^{t(\lambda_1+2)}B_1(M_{1,0}B_1\beta_2)+$$

$$(\lambda_2 + 4 - A)^{-1}e^{t(\lambda_2+4)}B_1(M_{2,0}B_1\beta_2),$$

$$u_2 = u_1 + [\ln(x)](\lambda_1 + 2 - A)^{-1}x^{(\lambda_1+2)}B_1(M_{1,0}B_1\beta_2)$$

$$- (\lambda_1 + 2 - A)^{-2}x^{(\lambda_1+2)}B_1(M_{1,0}B_1\beta_2)+$$

$$(\lambda_2 + 4 - A)^{-1}x^{(\lambda_2+4)}B_1(M_{2,0}B_1\beta_2).$$

The other $u_j, j = 3, 4, \ldots$, are similarly calculated.

Using the substitution: $u_1 = y, u_2 = xy'$, in conjunction with (2.1), it follows that Problems **12** to **20** correspond to the weakly singular systems in Problems **3** to **11**, respectively, and so their solutions are the first components of the respective solutions in Problems **3** to **11**.

21. Using the formula in a previous chapter:

$$e^{tA} = e^{t\lambda}e^{t(A-\lambda I)}$$

$$= e^{t\lambda}\sum_{k=0}^{\infty}\frac{[t(A-\lambda I)]^k}{k!}$$

$$= e^{t\lambda}[I + t(A - \lambda I) + \frac{[t(A-\lambda I)]^2}{2!} + \cdots],$$

the results follow.

22. We complete the proof by differentiating the right side of (2.6) and (2.7), respectively, and using the property in a previous chapter, that

$$\frac{d}{dt}e^{t(\lambda-A)} = (\lambda - A)e^{t(\lambda-A)}.$$

23. The eigenpairs $\{\lambda, w_0\}$'s are

$$\{5, \begin{pmatrix} 1 \\ 5 \\ 20 \end{pmatrix}\}, \{0, \begin{pmatrix} 1 \\ 0 \\ 0 \end{pmatrix}\}, \quad \text{and} \quad \{-1, \begin{pmatrix} 1 \\ -1 \\ 2 \end{pmatrix}\}.$$

These can also be obtained by Theorem 5.14 as follows. By Proposition 5.5, the exponential function is

$$e^{tA} = e^{5t}M_{1,0} + M_{2,0} + e^{-t}M_{3,0},$$

where

$$M_{1,0} = \frac{1}{30}(A + A^2) = \frac{1}{30}\begin{pmatrix} 0 & 2 & 1 \\ 0 & 10 & 5 \\ 0 & 40 & 20 \end{pmatrix},$$

$$M_{2,0} = I - \frac{1}{5}(A^2 - 4A) = \begin{pmatrix} 1 & 0.6 & -0.2 \\ 0 & 0 & 0 \\ 0 & 0 & 0 \end{pmatrix},$$

$$M_{3,0} = \frac{1}{6}(A^2 - 5A) = \frac{1}{6}\begin{pmatrix} 0 & -4 & 1 \\ 0 & 4 & -1 \\ 0 & -8 & 2 \end{pmatrix}.$$

Thus by Theorem 5.14, a Jordan canonical basis is

$$J = \{\frac{1}{30}\begin{pmatrix} 1 \\ 5 \\ 20 \end{pmatrix}, \begin{pmatrix} 1 \\ 0 \\ 0 \end{pmatrix}, \frac{1}{6}\begin{pmatrix} 1 \\ -1 \\ 2 \end{pmatrix}\},$$

that is, the third column of $M_{1,0}$, the first column of $M_{2,0}$, and the third column of $M_{3,0}$.

For each eigenpair $\{\lambda, w_0\}$, there are

$$w_1 = w_0 + (\lambda + 4 - A)^{-1}e^{t(\lambda+4-A)}B_3w_0,$$
$$v_1 = e^{tA}w_1$$
$$= e^{t\lambda}w_0 + (\lambda + 4 - A)^{-1}e^{t(\lambda+4)}B_3w_0,$$
$$u_1 = v_1|_{t=\ln(z)}$$
$$= z^\lambda w_0 + (\lambda + 4 - A)^{-1}z^{\lambda+4}B_3w_0,$$

and

$$w_2 = w_1 + (\lambda + 8 - A)^{-1}e^{t(\lambda+8-A)}B_3(\lambda + 4 - A)^{-1}B_3w_0,$$
$$v_2 = e^{tA}w_2$$
$$= v_1 + (\lambda + 8 - A)^{-1}e^{t(\lambda+8)}B_3(\lambda + 4 - A)^{-1}B_3w_0,$$
$$u_2 = v_2|_{t=\ln(z)}$$
$$= u_1 + (\lambda + 8 - A)^{-1}z^{\lambda+8}B_3(\lambda + 4 - A)^{-1}B_3w_0.$$

The other $u_j, j = 3, 4, \ldots,$ are similarly calculated.

24. The eigenpairs $\{\lambda, w_0\}$'s are

$$\{\frac{3}{2}, \beta_1\} = \{\frac{3}{2}, \frac{1}{6}\begin{pmatrix} 4 \\ 6 \\ 3 \end{pmatrix}\},$$

$$\{0, \beta_3\} = \{0, \begin{pmatrix} 1 \\ 0 \\ 0 \end{pmatrix}\},$$

and the generalized eigenpair is

$$\{\frac{3}{2}, \beta_2\} = \{\frac{3}{2}, \begin{pmatrix} -\frac{4}{9} \\ 0 \\ 1 \end{pmatrix}\}.$$

Here $\beta_2 = \begin{pmatrix} a \\ b \\ c \end{pmatrix}$ satisfies

$$0 = (A - \frac{3}{2})^2 \beta_2$$

$$= \begin{pmatrix} 2.25 & -2 & 1 \\ 0 & 0 & 0 \\ 0 & 0 & 0 \end{pmatrix} \beta_2 = \begin{pmatrix} 2.25a - 2b + c \\ 0 \\ 0 \end{pmatrix},$$

$$(A - \frac{3}{2})\beta_2 = \beta_1,$$

and, being solved, is a vector of the form

$$\beta_2 = \begin{pmatrix} a \\ b \\ 2b - 2.25a \end{pmatrix} = a \begin{pmatrix} 1 \\ 0 \\ -2.25 \end{pmatrix} + b \begin{pmatrix} 0 \\ 1 \\ 2 \end{pmatrix}$$

satisfying $\beta_1 = (A - \frac{3}{2})\beta_2 = (\frac{b}{4} - \frac{3}{8}a) \begin{pmatrix} 4 \\ 6 \\ 3 \end{pmatrix}$, and is linearly independent of β_1. We

can arbitrarily choose the values of a and b that satisfy $\frac{1}{4}b - \frac{3}{8}a = \frac{1}{6}$. Here we chose
$a = -\frac{4}{9}$ and $b = 0$.

This above result can also be obtained by Theorem 5.14 as follows. By Proposition 5.5, the exponential function is

$$e^{tA} = e^{\frac{3}{2}t} M_{1,0} + t e^{\frac{3}{2}t} M_{1,1} + M_{2,0},$$

where

$$M_{1,0} = \frac{4}{9}(3A - A^2) = \begin{pmatrix} 0 & \frac{8}{9} & -\frac{4}{9} \\ 0 & 1 & 0 \\ 0 & 0 & 1 \end{pmatrix},$$

$$M_{1,1} = \frac{1}{3}(2A^2 - 3A) = \frac{1}{6} \begin{pmatrix} 0 & -2 & 4 \\ 0 & -3 & 6 \\ 0 & -1.5 & 3 \end{pmatrix},$$

$$M_{2,0} = I - \frac{4}{9}(3A - A^2) = \begin{pmatrix} 1 & -\frac{8}{9} & \frac{4}{9} \\ 0 & 0 & 0 \\ 0 & 0 & 0 \end{pmatrix}.$$

Thus a Jordan canonical basis is

$$J = \{ \frac{1}{6} \begin{pmatrix} 4 \\ 6 \\ 3 \end{pmatrix}, \begin{pmatrix} -\frac{4}{9} \\ 0 \\ 1 \end{pmatrix}, \begin{pmatrix} 1 \\ 0 \\ 0 \end{pmatrix} \},$$

that is, the third column of $M_{1,1}$, the third column of $M_{1,0}$, and the first column
of $M_{2,0}$.

For each eigenpair $\{\lambda, w_0\} \neq \{\frac{3}{2}, \beta_2\}$, there are

$$w_1 = w_0 + (\lambda + 2 - A)^{-1} e^{t(\lambda + 2 - A)} B_1 w_0,$$

$$v_1 = e^{tA} w_1$$

$$= e^{t\lambda} w_0 + (\lambda + 2 - A)^{-1} e^{t(\lambda + 2)} B_1 w_0,$$

$$u_1 = v_1 |_{t = \ln(z)}$$

$$= z^\lambda w_0 + (\lambda + 2 - A)^{-1} z^{\lambda+2} B_1 w_0,$$

and

$$w_2 = w_1 + (\lambda + 4 - A)^{-1} e^{t(\lambda+4-A)} B_1 (\lambda + 2 - A)^{-1} B_1 w_0,$$

$$v_2 = e^{tA} w_2$$

$$= v_1 + (\lambda + 4 - A)^{-1} e^{t(\lambda+4)} B_1 (\lambda + 2 - A)^{-1} B_1 w_0,$$

$$u_2 = v_2|_{t=\ln(z)}$$

$$= u_1 + (\lambda + 4 - A)^{-1} z^{\lambda+4} B_1 (\lambda + 2 - A)^{-1} B_1 w_0.$$

The other $u_j, j = 3, 4, \ldots$, are similarly calculated.

For the generalized eigenpair $\{\frac{3}{2}, \beta_2\}$, there are

$$w_1 = \beta_2 + (\frac{3}{2} + 2 - A)^{-1} e^{(\frac{3}{2}+2-A)t} B_1 \beta_2 + (\frac{3}{2} + 2 - A)^{-1} t e^{(\frac{3}{2}+2-A)t} B_1 \beta_1$$

$$- (\frac{3}{2} + 2 - A)^{-2} e^{(\frac{3}{2}+2-A)t} B_1 \beta_1,$$

$$v_1 = e^{tA} w_1$$

$$= e^{\frac{3}{2}t}(\beta_2 + t\beta_1) + (\frac{3}{2} + 2 - A)^{-1} e^{(\frac{3}{2}+2)t} B_1 \beta_2 +$$

$$+ (\frac{3}{2} + 2 - A)^{-1} t e^{(\frac{3}{2}+2)t} B_1 \beta_1 - (\frac{3}{2} + 2 - A)^{-2} e^{(\frac{3}{2}+2)t} B_1 \beta_1,$$

$$u_1 = v_1|_{t=\ln(z)}$$

$$= z^{\frac{3}{2}}(\beta_2 + \beta_1 \ln(z)) + (\frac{3}{2} + 2 - A)^{-1} z^{\frac{3}{2}+2} B_1 \beta_2$$

$$+ (\frac{3}{2} + 2 - A)^{-1} [z^{\frac{3}{2}+2} \ln(z)] B_1 \beta_1 - (\frac{3}{2} + 2 - A)^{-2} z^{\frac{3}{2}+2} B_1 \beta_1,$$

and

$$w_2 = w_1 + (\frac{3}{2} + 4 - A)^{-1} e^{t(\frac{3}{2}+4-A)} B_1 (\frac{3}{2} + 2 - A)^{-1} B_1 \beta_2$$

$$+ (\frac{3}{2} + 4 - A)^{-1} t e^{t(\frac{3}{2}+4-A)} B_1 (\frac{3}{2} + 2 - A)^{-1} B_1 \beta_1$$

$$- (\frac{3}{2} + 4 - A)^{-2} e^{t(\frac{3}{2}+4-A)} B_1 (\frac{3}{2} + 2 - A)^{-1} B_1 \beta_1$$

$$- (\frac{3}{2} + 4 - A)^{-1} e^{t(\frac{3}{2}+4-A)} B_1 (\frac{3}{2} + 2 - A)^{-2} B_1 \beta_1,$$

$$v_2 = e^{tA} w_2$$

$$= v_1 + (\frac{3}{2} + 4 - A)^{-1} e^{(\frac{3}{2}+4)t} B_1 (\frac{3}{2} + 2 - A)^{-1} B_1 \beta_2$$

$$+ (\frac{3}{2} + 4 - A)^{-1} t e^{(\frac{3}{2}+4)t} B_1 (\frac{3}{2} + 2 - A)^{-1} B_1 \beta_1$$

$$- (\frac{3}{2} + 4 - A)^{-2} e^{(\frac{3}{2}+4)t} B_1 (\frac{3}{2} + 2 - A)^{-1} B_1 \beta_1$$

$$- (\frac{3}{2} + 4 - A)^{-1} e^{(\frac{3}{2}+4)t} B_1 (\frac{3}{2} + 2 - A)^{-2} B_1 \beta_1,$$

$$u_2 = v_2|_{t=\ln(x)}$$

$$= u_1 + (\frac{3}{2} + 4 - A)^{-1} z^{(\frac{3}{2}+4)} B_1 (\frac{3}{2} + 2 - A)^{-1} B_1 \beta_2$$

$$+ (\frac{3}{2} + 4 - A)^{-1} [z^{(\frac{3}{2}+4)} \ln(z)] B_1 (\frac{3}{2} + 2 - A)^{-1} B_1 \beta_1$$

$$- (\frac{3}{2} + 4 - A)^{-2} z^{(\frac{3}{2}+4)} B_1 (\frac{3}{2} + 2 - A)^{-1} B_1 \beta_1$$

$$- (\frac{3}{2} + 4 - A)^{-1} z^{(\frac{3}{2}+4)} B_1 (\frac{3}{2} + 2 - A)^{-2} B_1 \beta_1.$$

The other w_j, v_j, and $u_j, j = 3, 4, \ldots$, are similarly calculated.

25. The eigenpairs $\{\lambda_j, \beta_j\}, j = 1, 2, 3$, are:

$$\{5, \begin{pmatrix} 1 \\ 5 \\ 20 \end{pmatrix}\}, \{0, \begin{pmatrix} 1 \\ 0 \\ 0 \end{pmatrix}\}, \quad \text{and} \quad \{-1, \begin{pmatrix} 1 \\ -1 \\ 2 \end{pmatrix}\},$$

respectively.

For the eigenpair $\{\lambda, w_0\} = \{\lambda_1, \beta_1\}$ or $\{\lambda_3, \beta_3\}$, there are

$$w_1 = w_0 + (\lambda + 5 - A)^{-1} e^{t(\lambda+5-A)} B_4 w_0,$$

$$v_1 = e^{tA} w_1$$

$$= e^{t\lambda} w_0 + (\lambda + 5 - A)^{-1} e^{t(\lambda+5)} B_4 w_0,$$

$$u_1 = v_1|_{t=\ln(z)}$$

$$= z^\lambda w_0 + (\lambda + 5 - A)^{-1} z^{\lambda+5} B_4 w_0,$$

and

$$w_2 = w_1 + (\lambda + 10 - A)^{-1} e^{t(\lambda+10-A)} B_4 (\lambda + 5 - A)^{-1} B_4 w_0,$$

$$v_2 = e^{tA} w_2$$

$$= v_1 + (\lambda + 10 - A)^{-1} e^{t(\lambda+10)} B_4 (\lambda + 5 - A)^{-1} B_4 w_0,$$

$$u_2 = v_2|_{t=\ln(z)}$$

$$= u_1 + (\lambda + 10 - A)^{-1} z^{\lambda+10} B_4 (\lambda + 5 - A)^{-1} B_4 w_0.$$

The other $u_j, j = 3, 4, \ldots$, are similarly calculated.

Applying Proposition 5.5, it follows that

$$e^{tA} = e^{5t} M_{1,0} + M_{2,0} + e^{-t} M_{3,0},$$

where

$$M_{1,0} = \frac{1}{30}(A + A^2) = \frac{1}{30} \begin{pmatrix} 0 & 2 & 1 \\ 0 & 10 & 5 \\ 0 & 40 & 20 \end{pmatrix},$$

$$M_{2,0} = I - \frac{1}{5}(-4A + A^2) = \begin{pmatrix} 1 & 0.6 & -0.2 \\ 0 & 0 & 0 \\ 0 & 0 & 0 \end{pmatrix},$$

$$M_{3,0} = \frac{1}{6}(-5A + A^2) = \frac{1}{6} \begin{pmatrix} 0 & -4 & 1 \\ 0 & 4 & -1 \\ 0 & -8 & 2 \end{pmatrix}.$$

Thus by Theorem 5.14, a Jordan canonical basis is

$$J = \{\frac{1}{30}\begin{pmatrix}1\\5\\20\end{pmatrix}, \begin{pmatrix}1\\0\\0\end{pmatrix}, \frac{1}{6}\begin{pmatrix}1\\-1\\2\end{pmatrix}\},$$

that is, the third column of $M_{1,0}$, the first column of $M_{2,0}$, and the third column of $M_{3,0}$.

For the eigenpair $\{\lambda_2, \beta_2\}$, there are

$$w_1 = \beta_2 + tM_{1,0}B_4\beta_2 + \frac{1}{5}e^{5t}M_{2,0}B_4\beta_2 + \frac{1}{6}e^{6t}M_{3,0}B_4\beta_2,$$

$$v_1 = e^{tA}w_1$$

$$= \beta_2 + te^{5t}M_{1,0}B_4\beta_2 + \frac{1}{5}e^{5t}M_{2,0}B_4\beta_2 + \frac{1}{6}e^{5t}M_{3,0}B_4\beta_2.$$

$$u_1 = v_1|_{t=\ln(z)}$$

$$= \beta_2 + [z^5\ln(z)]M_{1,0}B_4\beta_2 + \frac{1}{5}z^5M_{2,0}B_4\beta_2 + \frac{1}{6}z^5M_{3,0}B_4\beta_2,$$

$$w_2 = w_1 + t(10-A)^{-1}e^{t(10-A)}B_4(M_{1,0}B_4\beta_2)$$

$$- (10-A)^{-2}e^{t(10-A)}B_4(M_{1,0}B_4\beta_2) + \frac{1}{5}(10-A)^{-1}e^{t(10-A)}B_4(M_{2,0}B_4\beta_2)$$

$$+ \frac{1}{6}(10-A)^{-1}e^{t(10-A)}B_4M_{3,0}B_4\beta_2,$$

$$v_2 = e^{tA}w_2$$

$$= v_1 + t(10-A)^{-1}e^{10t}B_4(M_{1,0}B_4\beta_2)$$

$$- (10-A)^{-2}e^{10t}B_4(M_{1,0}B_4\beta_2) + \frac{1}{5}(10-A)^{-1}e^{10t}B_4(M_{2,0}B_4\beta_2)$$

$$+ \frac{1}{6}(10-A)^{-1}e^{10t}B_4M_{3,0}B_4\beta_2,$$

$$u_2 = v_2|_{t=\ln(z)}$$

$$= u_1 + [\ln(z)](10-A)^{-1}z^{10}B_4(M_{1,0}B_4\beta_2)$$

$$- (10-A)^{-2}z^{10}B_4(M_{1,0}B_4\beta_2) + \frac{1}{5}(10-A)^{-1}z^{10}B_4(M_{2,0}B_4\beta_2)$$

$$+ \frac{1}{6}(10-A)^{-1}z^{10}B_4M_{3,0}B_4\beta_2.$$

The other $u_j, j = 3, 4, \ldots$, are similarly calculated.

Using the substitution (5.3), in conjunction with (5.4), it follows that Problems **26** to **28** become Problems **23** to **25**, respectively, and so their solutions are the first components of the respective solutions in Problems **23** to **25**.

29. Apply Theorem 2.10, Corollary 2.11, and Corollary 2.13.

30. Apply Corollary 5.4.

31. The second power series solution has a logarithmic term only when $\nu = 0, 1, 2, 3, \ldots$. This follows from Corollary 2.13 because the term B_1 exists, because the eigenvalues are $\{\lambda_1, \lambda_2\} = \{\nu, -\nu\}$, which are identical when $\nu = 0$, and because $\lambda_1 - \lambda_2 = 2\nu = (1+1)^l = 2, 4, 6, \ldots$, for a finite positive integral l when $\nu = 1, 2, \ldots$.

CHAPTER 4

Adjoint Operators and Nonhomogeneous Boundary Value Problems

1. Introduction

Firstly, we consider the homogeneous boundary value problems with simple zero Dirichlet boundary condition:

$$p(x)y'' + q(x)y' + r(x)y = 0, \quad a < x < b$$
$$y(a) = 0 = y(b), \tag{1.1}$$

where $p(x), p'(x), q(x)$, and $r(x)$ are continuous complex-valued functions on $[a, b]$, $p(x) \geq \delta_0 > 0$ for some $\delta_0 > 0$, and $a < b$.

To rewrite (1.1) in a compact form, let $L : D(L) \subset L^2(a, b) \longrightarrow L^2(a, b)$ be a linear differential operator defined by

$$Lu \equiv p(x)u'' + q(x)u' + r(x)u$$

for

$$u \in D(L) \equiv \{v \in C^2[a, b] : v(a) = 0, v(b) = 0\}.$$

Here $D(L)$ is the domain of L, $C^2[a, b]$ is the vector space of twice continuously differentiable complex-valued functions on $[a, b]$, and $L^2(a, b)$ is the vector space of squarely Lebesgue integrable (for a moment, think of them as Riemann integrable) complex-valued functions on (a, b)(see Chapter 5, Subsection 5.2).

Thus the above boundary value problem can be written as the compact equation

$$Ly = 0, \quad y \in D(L). \tag{1.2}$$

We next derive the adjoint problem of (1.2).

Using the integration by parts formula, we have that, for y and $z \in C^2[a, b]$ and for

$$My \equiv py'' + qy' + ry,$$

the inner product $(My, z) \equiv \int_a^b (My) \bar{z} \, dx$ of My with z (see Chapter 6, Section 1) satisfies

$$
\begin{aligned}
(My, z) &\equiv \int_a^b (py'' + qy' + ry)\bar{z} \, dx \\
&= \int_a^b \bar{z} p \, dy' + \int_a^b \bar{z} q \, dy + \int_a^b \bar{z} ry \, dx \\
&= [\bar{z} p y'|_a^b - \int_a^b y'(\bar{z} p)' \, dx] + [\bar{z} q y|_a^b - \int_a^b y(\bar{z} q)' \, dx] + \int_a^b \bar{z} ry \, dx \\
&= (y, (\bar{p} z)'' - (\bar{q} z)' + \bar{r} z) + [p\bar{z} y' + q\bar{z} y - p\bar{z}' y - p'\bar{z} y]|_a^b,
\end{aligned}
$$

or equivalently,

$$(My, z) = (y, Nz) + [p\bar{z}y' + q\bar{z}y - p\bar{z}'y - p'\bar{z}y]|_a^b, \tag{1.3}$$

where $Nz = (\bar{p}z)'' - (\bar{q}z)' + \bar{r}z$. Here notice that $z(x)$ is of the form:

$$z(x) = z_1(x) + iz_2(x)$$

with $i \equiv \sqrt{-1}$, $i^2 = -1$, that z_1 and z_2 real-valued functions, and that

$$\bar{z} = z_1(x) - iz_2(x)$$

is the complex conjugate of z. Also notice that

$$\int_a^b y'(\bar{z}p)' \, dx = \int_a^b (\bar{z}p)' \, dy$$

$$= (\bar{z}p)'y|_a^b - \int_a^b y(\bar{z}p)'' \, dx$$

$$= (\bar{z}'p + \bar{z}p')y|_a^b - (y, (\bar{p}z)''),$$

and that

$$-\int_a^b y(\bar{z}q)' \, dx + \int_a^b y\bar{z}r \, dx = (y, -(\bar{q}z)' + \bar{r}z).$$

If the above y even lies in $D(L)$, then $y(a) = 0 = y(b)$ and (1.3) becomes

$$(Ly, z) = (y, Nz) + [\bar{z}(b)p(b)y'(b) - \bar{z}(a)p(a)y'(a)]. \tag{1.4}$$

This (1.4) can be simplified further if the z in (1.4) satisfies a certain condition. This leads to the concept of the adjoint of L, which we define now:

DEFINITION 1.1. *The adjoint operator*

$$L^\dagger : D(L^\dagger) \subset L^2(a, b) \longrightarrow L^2(a, b)$$

of L is defined by

$$L^\dagger z = Nz = (\bar{p}z)'' - (\bar{q}z)' + \bar{r}z$$

for $z \in D(L^\dagger)$, the domain of $L^\dagger$. Here $D(L^\dagger)$ (Nagel and Saff [19, page 684]) is the set of functions $z \in C^2[a, b]$ that satisfy

$$\bar{z}(b)p(b)y'(b) - \bar{z}(a)p(a)y'(a) = 0 \tag{1.5}$$

for all $y \in D(L)$; that is, the second term of the right side of (1.4) is zero.

Using $L^\dagger$, (1.4) reduces to

$$(Ly, z) = (y, L^\dagger z)$$

for all $y \in D(L)$.

To find $D(L^\dagger)$ more explicitly, we choose the $y \in D(L)$ that satisfies

$$\begin{pmatrix} y'(a) \\ y'(b) \end{pmatrix} = \begin{pmatrix} 1 \\ 0 \end{pmatrix} \quad \text{or} \quad \begin{pmatrix} 0 \\ 1 \end{pmatrix},$$

and substitute them into (1.5). It follows that

$$\bar{z}(a) = 0 = \bar{z}(b), \quad \text{or equivalently,} \quad z(a) = 0 = z(b).$$

Thus

$$D(L^\dagger) \subset \{v \in C^2[a, b] : v(a) = 0 = v(b)\}.$$

Conversely, if $z \in C^2[a, b]$ satisfies $z(a) = 0 = z(b)$, then clearly, for $y \in D(L)$, we have

$$\overline{z}(b)p(b)y'(b) - \overline{z}(a)p(a)y'(a) = 0,$$

(1.5) being satisfied, and so

$$D(L^\dagger) \supset \{v \in C^2[a, b] : v(a) = 0 = v(b)\}.$$

Therefore

$$D(L^\dagger) = \{z \in C^2[a, b] : z(a) = 0 = z(b)\},$$

and the adjoint problem of (1.2) is

$$L^\dagger z = 0, \quad z \in D(L^\dagger). \tag{1.6}$$

We now extend the definition of the adjoint operator $L^\dagger$ and the L to the case where L is equipped with more general boundary condition of the form:

$$B \begin{pmatrix} y(a) \\ y'(a) \\ y(b) \\ y'(b) \end{pmatrix} = \vec{0}_m,$$

where $\vec{0}_m, 1 \leq m \leq 4$, is the m-dimensional zero vector, and

$$B = (b_{ij})_{m \times 4} \equiv \begin{pmatrix} b_{11} & b_{12} & b_{13} & b_{14} \\ & \vdots & & \\ b_{m1} & b_{m2} & b_{m3} & b_{m4} \end{pmatrix}$$

for $1 \leq m \leq 4$, is the corresponding boundary operator, and is a matrix of $Rank(B)$ m. That is, the domain of the new L is:

$$D(L) \equiv \{y \in C^2[a, b] : B \begin{pmatrix} y(a) \\ y'(a) \\ y(b) \\ y'(b) \end{pmatrix} = \vec{0}_m\}.$$

Here $Rank(B) = m$ means that the m row vectors

$$B_i \equiv \begin{pmatrix} b_{i1} & b_{i2} & b_{i3} & b_{i4} \end{pmatrix}, 1 \leq i \leq m,$$

are linearly independent, that is, if $\sum_{i=1}^m c_i B_i = \vec{0}_4$ for $c_i \in \mathbb{C}$, then $c_i = 0$ for all i.

However, when $m = 0$, we define $B = \begin{pmatrix} 0 & 0 & 0 & 0 \end{pmatrix}$. In this case, the boundary condition

$$B \begin{pmatrix} y(a) \\ y'(a) \\ y(b) \\ y'(b) \end{pmatrix} = \vec{0}_m$$

is true for all $\tilde{y} \equiv \begin{pmatrix} y(a) \\ y'(a) \\ y(b) \\ y'(b) \end{pmatrix}$, and so there is no restriction on $\tilde{y}$ at all.

DEFINITION 1.2. *The general operator*

$$L : D(L) \subset L^2(a, b) \longrightarrow L^2(a, b),$$

is defined by

$$Ly = p(x)y'' + q(x)y' + r(x)y$$

for

$$y \in D(L) \equiv \{u \in C^2[a,b] : B \begin{pmatrix} u(a) \\ u'(a) \\ u(b) \\ u'(b) \end{pmatrix} = \vec{0}_m\},$$

and its adjoint operator

$$L^\dagger : D(L^\dagger) \subset L^2(a,b) \longrightarrow L^2(a,b),$$

is defined by

$$L^\dagger z = Nz = (\bar{p}z)'' - (\bar{q}z)' + \bar{r}z$$

for $z \in D(L^\dagger)$, the domain of $L^\dagger$. Here $D(L^\dagger)$ is the set of functions $z \in C^2[a,b]$ that satisfy

$$[p\bar{z}y' + q\bar{z}y - p\bar{z}'y - p'\bar{z}y|_a^b = 0$$

for all $y \in D(L)$, that is, z is such that the second term of the right side of (1.3) is zero. In this case, (1.3) reduces to

$$(Ly, z) = (y, L^\dagger z)$$

for all $y \in D(L)$.

DEFINITION 1.3. *The operator L is symmetric if*

$$(Ly, z) = (y, Lz)$$

for all $y, z \in D(L)$, or equivalently, if

$$[p\bar{z}y' + q\bar{z}y - p\bar{z}'y - p'\bar{z}y|_a^b = 0$$

for all $y, z \in D(L)$.

It follows readily from Definition 1.2 that

COROLLARY 1.4. *L is symmetric if and only if*

$$L^\dagger u = Lu$$

for $u \in D(L)$. In that case, $D(L^\dagger) \supset D(L)$, for which we say

$$L^\dagger \supset L.$$

It will be seen later in Section 4 that

PROPOSITION 1.5. *Suppose that $q = p', m = 2$, and*

$$B = (\alpha \ \ \beta), \quad where \quad \alpha = \begin{pmatrix} \alpha_{11} & \alpha_{12} \\ \alpha_{21} & \alpha_{22} \end{pmatrix}, \quad and \quad \beta = \begin{pmatrix} \beta_{11} & \beta_{12} \\ \beta_{21} & \beta_{22} \end{pmatrix},$$

and that the determinant $|\alpha|$ of α or the determinant $|\beta|$ of β is not zero. Then

$$My = (py')' + ry,$$
$$Nz = (\bar{p}z')' + \bar{r}z,$$

and (1.3) reduces to

$$(My, z) = (y, Nz) + p(y'\bar{z} - y\bar{z'})|_a^b.$$

In that case, L is symmetric if and only if

$$p(a) \begin{vmatrix} \beta_{11} & \beta_{12} \\ \beta_{21} & \beta_{22} \end{vmatrix} = p(b) \begin{vmatrix} \alpha_{11} & \alpha_{12} \\ \alpha_{21} & \alpha_{22} \end{vmatrix}.$$

However, if, instead,

$$\alpha = \begin{pmatrix} c_1 & c_2 \\ 0 & 0 \end{pmatrix} \quad and \quad \beta = \begin{pmatrix} 0 & 0 \\ d_1 & d_2 \end{pmatrix},$$

where c_1, c_2, d_1, and d_2 are complex numbers, such that

$$c_1^2 + c_2^2 \neq 0 \quad and \quad d_1^2 + d_2^2 \neq 0,$$

then L is still symmetric.

PROOF. See Section 4.   □

It will follow from the proof of Proposition 1.5 that Proposition 1.5 continues to hold for a relaxing L with absolutely continuous coefficients (see Section 4).

It is the purpose of this chapter to show that the nonhomogeneous boundary value problem

$$p(x)y'' + q(x)y' + r(x)y = f(x), a < x < b$$

$$B \begin{pmatrix} y(a) \\ y'(a) \\ y(b) \\ y'(b) \end{pmatrix} = \vec{\gamma} \tag{1.7}$$

where $f(x)$ is a given continuous complex-valued function on $[a,b]$, and $\vec{\gamma} \in \mathbb{C}^m, 1 \leq m \leq 4$, is a given constant $m-$dimensional complex vector, can be solved by studying the adjoint problem

$$\pi_m^\dagger : \qquad L^\dagger z = 0, z \in D(L^\dagger).$$

The problem $\pi_m^\dagger$ is the adjoint problem of the original problem

$$\pi_m : \qquad Ly = 0, y \in D(L).$$

Here observe that the case where $m = 0$, is trivial, since in this case, no restriction is put on $\tilde{y} = \begin{pmatrix} y(a) \\ y'(a) \\ y(b) \\ y'(b) \end{pmatrix}$, and so the problem π_m always has two linearly independent solutions. Thus there is no need to study the adjoint problem $\pi_m^\dagger$.

The rest of this chapter is organized as follows. Section 2 states the main results, and Section 3 studies examples for illustration. Section 4 deals with the proof of the main results, and Section 5 presents an abstract result that is close to one of the main results in Section 2. Finally, Section 6 contains a problems set, and their solutions are placed in Section 7.

The material in Sections 2 and 4 is based on Coddington and Levinson [5].

2. A Necessary and Sufficient Condition for Solvability

It is a fact from linear algebra that if $\{\alpha_i\}_{i=1}^m$ is a set of $0 \leq m \leq 4$ linearly independent four-dimensional vectors in $\mathbb{C}^4$, then there are $(4-m)$ four-dimensional vectors

$$\alpha_j, \quad j = (m+1), \dots, 4,$$

in $\mathbb{C}^4$, such that $\{\alpha_i\}_{i=1}^4$ are linearly independent. It follows from this fact that, for the boundary operator (see Section 1)

$$B = \begin{pmatrix} b_{11} & b_{12} & b_{13} & b_{14} \\ \vdots & & & \\ b_{m1} & b_{m2} & b_{m3} & b_{m4} \end{pmatrix},$$

a matrix of rank $1 \le m \le 3$, we can choose a matrix

$$B_c \equiv (b_{ij})_{(4-m)\times 4} = \begin{pmatrix} b_{(m+1)1} & b_{(m+1)2} & b_{(m+1)3} & b_{(m+1)4} \\ \vdots & & & \\ b_{41} & b_{42} & b_{43} & b_{44} \end{pmatrix}$$

so that the augmented matrix

$$(b_{ij})_{4\times 4} = \begin{pmatrix} B \\ B_c \end{pmatrix}$$

has the rank equal to 4. B_c is called a complementary matrix to B. For example, if

$$B = \begin{pmatrix} 1 & 0 & 0 & 0 \\ 0 & 0 & 1 & 0 \end{pmatrix},$$

we can choose

$$B_c = \begin{pmatrix} 0 & 1 & 0 & 0 \\ 0 & 0 & 0 & 1 \end{pmatrix} \quad \text{or} \quad \begin{pmatrix} 0 & 1 & 1 & 1 \\ 0 & 0 & 1 & 1 \end{pmatrix}.$$

If

$$B = \begin{pmatrix} 1 & 2 & 3 & 4 \end{pmatrix},$$

we can choose

$$B_c = \begin{pmatrix} 0 & 1 & 0 & 0 \\ 0 & 0 & 1 & 0 \\ 0 & 0 & 0 & 1 \end{pmatrix} \quad \text{or} \quad \begin{pmatrix} 0 & 1 & 2 & 3 \\ 0 & 0 & 1 & 2 \\ 0 & 0 & 0 & 1 \end{pmatrix}.$$

When $m = 0$, for which B is defined by $\begin{pmatrix} 0 & 0 & 0 & 0 \end{pmatrix}$ (see Section 1), there are many $B_c's$, such that B_c is a 4×4 matrix of rank 4. For instance, we can take

$$B_c = \begin{pmatrix} 1 & 0 & 0 & 0 \\ 0 & 1 & 0 & 0 \\ 0 & 0 & 1 & 0 \\ 0 & 0 & 0 & 1 \end{pmatrix} \quad \text{or} \quad \begin{pmatrix} 1 & 2 & 3 & 4 \\ 0 & 1 & 2 & 3 \\ 0 & 0 & 1 & 2 \\ 0 & 0 & 0 & 1 \end{pmatrix}.$$

In this case, we still call B_c a complementary matrix to B.

In general, B_c is not unique. However, when $m = 4$, B is already a 4×4 matrix of rank 4; in this case, we define $B_c = \begin{pmatrix} 0 & 0 & 0 & 0 \end{pmatrix}$, and call it the only complementary matrix to B.

For convenience, we write

$$\tilde{y} = \begin{pmatrix} y(a) \\ y'(a) \\ y(b) \\ y'(b) \end{pmatrix}$$

for $y \in C^2[a, b]$, and write

$$V\{\tilde{y}, \tilde{z}\} = [p\bar{z}y' + q\bar{z}y - p\bar{z}'y - p'\bar{z}y]\big|_a^b$$

for $y, z \in C^2[a, b]$, the second term of the right side of (1.3).

It follows that (1.3) can be rewritten as

LEMMA 2.1 (Green's formula).

$$(My, z) = (y, Nz) + V\{\tilde{y}, \tilde{z}\} \tag{2.1}$$

for all $y, z \in C^2[a, b]$, where

$$My = py'' + qy' + ry,$$
$$Nz = (\bar{p}z)'' - (\bar{q}z)' + \bar{r}z.$$

The quantity $V\{\tilde{y}, \tilde{z}\}$ can be written as a semibilinear form through a nonsingular 4×4 matrix S:

LEMMA 2.2. For $y, z \in C^2[a, b]$,

$$V\{\tilde{y}, \tilde{z}\} = (S\tilde{y}, \tilde{z}),$$

where

$$S = \begin{pmatrix} -(q(a) - p'(a)) & -p(a) & 0 & 0 \\ p(a) & 0 & 0 & 0 \\ 0 & 0 & (q(b) - p'(b)) & p(b) \\ 0 & 0 & -p(b) & 0 \end{pmatrix}.$$

PROOF. See Section 4.   □

The semibilinear form in Lemma 2.2 that represents the quantity $V\{\tilde{y}, \tilde{z}\}$, can be further decomposed into the sum of two semilinear forms:

LEMMA 2.3 (Boundary formula). Suppose that B is the boundary operator (see Sections 1 and 2), an $m \times 4$ matrix of rank $0 \le m \le 4$, and that B_c is a complementary $(4 - m) \times 4$ matrix to B, of rank $0 \le (4 - m) \le 4$. Then by letting

$$F = \begin{cases} B_c, & \text{if } m = 0; \\ \begin{pmatrix} B \\ B_c \end{pmatrix}, & \text{if } 1 \le m \le 3; \\ B, & \text{if } m = 4, \end{cases}$$

there is a unique matrix $G = (SF^{-1})^*$, the transposed complex conjugate of SF^{-1}, such that

$$V\{\tilde{y}, \tilde{z}\} = (F\tilde{y}, G\tilde{z})$$

for $y, z \in C^2[a, b]$. Here S is the S from Lemma 2.2, and the usual inner product

$$\left(\begin{pmatrix} \eta_1 \\ \eta_2 \\ \eta_3 \\ \eta_4 \end{pmatrix}, \begin{pmatrix} \zeta_1 \\ \zeta_2 \\ \zeta_3 \\ \zeta_4 \end{pmatrix} \right) \equiv \sum_{i=1}^{4} \eta_i \bar{\zeta}_i,$$

$\eta_i, \zeta_i \in \mathbb{C}$, in $\mathbb{C}^4$, was used.

Furthermore, if we write out G in the form

$$G = (SF^{-1})^* = \begin{cases} B^{\dagger}, & \text{for which } B_c^{\dagger} \equiv \begin{pmatrix} 0 & 0 & 0 & 0 \end{pmatrix}, & \text{if } m = 0; \\ \begin{pmatrix} B_c^{\dagger} \\ B^{\dagger} \end{pmatrix}, & & \text{if } 1 \le m \le 3; \\ B_c^{\dagger}, & \text{for which } B^{\dagger} \equiv \begin{pmatrix} 0 & 0 & 0 & 0 \end{pmatrix}, & \text{if } m = 4, \end{cases}$$

then

$$V\{\tilde{y}, \tilde{z}\} = (B\tilde{y}, B_c^\dagger \tilde{z}) + (B_c\tilde{y}, B^\dagger \tilde{z}) \tag{2.2}$$

for $y, z \in C^2[a, b]$. *Here for* $1 \le m \le 3$, $B^\dagger$ *is a* $(4-m) \times 4$ *matrix of rank* $(4-m)$, *and* $B_c^\dagger$ *is an* $m \times 4$ *matrix of rank* m.

PROOF. See Section 4. □

REMARK 2.4. *Here notice that* $B^\dagger$ *is not the adjoint of* B, *and that* $B_c^\dagger$ *is not the adjoint of* B_c, *either.*

The decomposition in Lemma 2.3 is unique in some sense:

LEMMA 2.5 (Boundary formula). *By following Lemma 2.3, consider the case where* $0 \le m \le 3$, *for which there are multiple complementary matrices* B_c's *to* B.

Suppose that $\hat{B}_c$ *is another matrix, complementary to* B, *and that* $\hat{B}_c^\dagger$ *and* $\hat{B}^\dagger$ *are the corresponding unique matrices, such that*

$$V\{\tilde{y}, \tilde{z}\} = (B\tilde{y}, \hat{B}_c^\dagger \tilde{z}) + (\hat{B}_c\tilde{y}, \hat{B}^\dagger \tilde{z}).$$

Then

$$\hat{B}^\dagger \tilde{z} = A^* B^\dagger \tilde{z}$$

or equivalently,

$$\hat{B}^\dagger \begin{pmatrix} z(a) \\ z'(a) \\ z(b) \\ z'(b) \end{pmatrix} = A^* B^\dagger \begin{pmatrix} z(a) \\ z'(a) \\ z(b) \\ z'(b) \end{pmatrix}$$

for some nonsingular square matrix $A = (a_{ij})_{(4-m) \times (4-m)}$.

PROOF. See Section 4. □

It will follow from Lemma 2.5 that an equivalent definition of the adjoint operator is:

DEFINITION 2.6. *The adjoint operator*

$$L^\dagger : D(L^\dagger) \subset L^2(a, b) \longrightarrow L^2(a, b)$$

is defined by

$$L^\dagger z = Nz = (\bar{p}z)'' - (\bar{q}z)' + \bar{r}z$$

for

$$z \in D(L^\dagger) \equiv \{u \in C^2[a, b] : B^\dagger \tilde{u} = \vec{0}_{4-m}\}$$
$$= \{u \in C^2[a, b] : V\{\tilde{y}, \tilde{z}\} = 0 \quad \text{for all } y \in D(L)\},$$

for which (2.1), *together with* (2.2), *reduces to*

$$(Ly, z) = (y, L^\dagger z)$$

for all $y \in D(L)$. *Here* $B^\dagger$ *is from Lemma 2.5.*

REMARK 2.7. *Since we could choose* $B^\dagger$ *or* $\hat{B}^\dagger$ *for the use in* $D(L^\dagger)$, *the definition of* $L^\dagger$ *in Definition 2.6 may not be well-defined. But since the* A *in Lemma 2.5 is invertible, we have*

$$\{u \in C^2[a, b] : B^\dagger \tilde{u} = \vec{0}\} = \{u \in C^2[a, b] : A^* B^\dagger \tilde{u} = \hat{B}^\dagger \tilde{u} = \vec{0}\},$$

and so the $L^\dagger$ *in Definition 2.6 is really well-defined.*

LEMMA 2.8. *Definitions 1.2 and 2.6 are equivalent.*

PROOF. See Section 4. □

Here is our main theorem:

THEOREM 2.9. *The nonhomogeneous boundary value problem*

$$p(x)y'' + q(x)y' + r(x)y = f(x), \quad a < x < b$$
$$B\tilde{y} = \vec{\gamma} \in \mathbb{C}^4 \tag{2.3}$$

has a solution if and only if

$$(f, z) = (\vec{\gamma}, B_c^\dagger \tilde{z}) \tag{2.4}$$

for every solution z of the adjoint problem:

$$\pi_m^\dagger : \quad L^\dagger z = 0, z \in D(L^\dagger).$$

Here $p(x), p'(x), q(x), r(x)$, and $f(x)$ are continuous complex-valued functions on $[a, b]$; $p(x) \geq \delta_0 > 0$ for some $\delta_0 > 0$, $a < b$, and $1 \leq m \leq 4$; $\vec{\gamma} \in \mathbb{C}^m$ is a constant m-dimensional complex vector, and B is a $m \times 4$ matrix of rank 4; finally B_c is a $(4 - m) \times 4$ matrix, complementary to B, of rank $(4 - m)$, and $B_c^\dagger$ is the unique matrix in Lemma 2.5.

PROOF. See Section 4. □

REMARK 2.10. *The left side (f, z) of (2.4) does not depend on the complementary matrix B_c but the right side $(\vec{\gamma}, B_c^\dagger \tilde{z})$ does, and so both sides of (2.4) seems inconsistent. However, they are, in fact, consistent, as the following Lemma shows:*

LEMMA 2.11. *For every solution z to the adjoint problem*

$$\pi_m^\dagger : \quad L^\dagger z = 0, z \in D(L^\dagger),$$

the quantity $B_c^\dagger \tilde{z}$ is independent of the complementary matrix B_c.

PROOF. By (2.1) and Lemma 2.5, we have

$$(My, z) - (y, Nz) = (B\tilde{y}, B_c^\dagger \tilde{z}) + (B_c \tilde{y}, B^\dagger \tilde{z})$$

for $y \in C^2[a, b]$ and z being a solution of the adjoint problem. This reduces to

$$(My, z) - 0 = (B\tilde{y}, B_c^\dagger \tilde{z}) + 0,$$

since $Nz = L^\dagger z = 0$ and $B^\dagger \tilde{z} = 0$. Thus $(B\tilde{y}, B_c^\dagger \tilde{z})$ is independent of the complementary matrix B_c as (My, z) is so.

We further show that $(\vec{a}, B_c^\dagger \tilde{z})$ is independent of B_c, not only for $\vec{a} = B\tilde{y}$ but even for each vector $\vec{a} \in \mathbb{C}^m$. In that case, by choosing $\vec{a} = B_c^\dagger \tilde{z}$, we have

$$(\vec{a}, B_c^\dagger \tilde{z}) = \|B_c^\dagger \tilde{z}\|^2,$$

and then $B_c^\dagger \tilde{z}$ is independent of B_c.

Since

$$\{B\vec{a} : \vec{a} \in \mathbb{C}^4\} = \{B\tilde{y} : y \in C^2[a, b]\},$$

the dimension of the vector space, the range of B, equals the rank of B by linear algebra, and since the rank of B is m by our assumption, it follows that the proof is complete. Here the range of B is the set of images of B. □

COROLLARY 2.12. *When $\vec{\gamma} = \vec{0}$, the zero vector, (2.3) has a solution if and only if*

$$(f, z) = 0,$$

that is, f is orthogonal to z for every solution z of the adjoint problem

$$\pi_m^{\dagger} : \quad L^{\dagger} z = 0, z \in D(L^{\dagger}).$$

PROOF. Letting $\vec{\gamma} = \vec{0}$ in (2.4), proves the corollary.   □

COROLLARY 2.13. *The equation (2.3) has a unique solution if $m = 2$ and if the boundary value problem*

$$\pi_m : \quad Ly = 0, y \in D(L)$$

only has the trivial zero solution.

PROOF. See Section 4   □

REMARK 2.14. *From the proof of Theorem 2.9 in Section 4, we will see that Theorem 2.9 can be extended to higher order equations:*

$$My = p_0(x)y^{(n)} + p_1(x)y^{(n-1)} + \cdots + p_{n-1}(x)y' + p_n(x)y.$$

For the details, we refer to the book by Coddington and Levinson [5].

3. Examples

Before we prove Theorem 2.9, we look at some examples.

EXAMPLE 3.1. Consider the equation with the Dirichlet boundary conditions:

$$y'' + 2y' + 2y = h(x), 0 < x < \pi$$
$$y(0) = \frac{1}{2} = y(\pi), \tag{3.1}$$

where $h(x)$ is a continuous function on $[0, \pi]$.

The corresponding boundary operator is the matrix

$$B = \begin{pmatrix} 1 & 0 & 0 & 0 \\ 0 & 0 & 1 & 0 \end{pmatrix}.$$

Here correspondingly, $m = 2$ and $\vec{\gamma} = \begin{pmatrix} \frac{1}{2} \\ \frac{1}{2} \end{pmatrix}$.

Denote $n = 2$, the order of the equation $y'' + 2y' + 2y = 0$. Find two complementary matrices $B_c, \hat{B}_c$ of B, and compute the corresponding matrices $B_c^{\dagger}, B^{\dagger}, \hat{B}_c^{\dagger}, \hat{B}^{\dagger}$.

Write out the boundary value problem $\pi_2 : Ly = 0, y \in D(L)$ and the adjoint problem $\pi_2^{\dagger} : L^{\dagger} z = 0, z \in D(L^{\dagger})$, and solve these two problems. Is L symmetric? Show that if the number of linearly independent solutions for the problem π_2 in our findings is exactly equal to k, then the adjoint problem $\pi_2^{\dagger}$ has exactly $l = (k+m-n)$ linearly independent solutions, as Proposition 4.11 in Section 4 predicts.

Show that $B_c^{\dagger}\tilde{z} = \hat{B}_c^{\dagger}\tilde{z}$ for z being a solution to the adjoint problem $\pi_2^{\dagger}$, as Lemma 2.11 predicts.

Use Theorem 2.9 to find the necessary and sufficient condition on h, so that (3.1) has a solution.

Solution. Let

$$B_c = \begin{pmatrix} 0 & 1 & 0 & 0 \\ 0 & 0 & 0 & 1 \end{pmatrix}, \hat{B}_c = \begin{pmatrix} 0 & 1 & 0 & 1 \\ 0 & 0 & 2 & 1 \end{pmatrix}.$$

Let $F = \begin{pmatrix} B \\ B_c \end{pmatrix}$. Augment F with the identity matrix I_4:

$$(F \mid I_4) = \begin{pmatrix} 1 & 0 & 0 & 0 & | & 1 & 0 & 0 & 0 \\ 0 & 0 & 1 & 0 & | & 0 & 1 & 0 & 0 \\ 0 & 1 & 0 & 0 & | & 0 & 0 & 1 & 0 \\ 0 & 0 & 0 & 1 & | & 0 & 0 & 0 & 1 \end{pmatrix}.$$

Applying row operations, in which we add row 3 to row 2, we obtain

$$(F \mid I_4) \longrightarrow \begin{pmatrix} 1 & 0 & 0 & 0 & | & 1 & 0 & 0 & 0 \\ 0 & 1 & 1 & 0 & | & 0 & 1 & 1 & 0 \\ 0 & 1 & 0 & 0 & | & 0 & 0 & 1 & 0 \\ 0 & 0 & 0 & 1 & | & 0 & 0 & 0 & 1 \end{pmatrix}.$$

Adding $(-1) \times$ row 2 to row 3, we have

$$\longrightarrow \begin{pmatrix} 1 & 0 & 0 & 0 & | & 1 & 0 & 0 & 0 \\ 0 & 1 & 1 & 0 & | & 0 & 1 & 1 & 0 \\ 0 & 0 & -1 & 0 & | & 0 & -1 & 0 & 0 \\ 0 & 0 & 0 & 1 & | & 0 & 0 & 0 & 1 \end{pmatrix},$$

which becomes, after adding row 3 to row 2 and multiplying row 3 by -1,

$$\longrightarrow \begin{pmatrix} 1 & 0 & 0 & 0 & | & 1 & 0 & 0 & 0 \\ 0 & 1 & 0 & 0 & | & 0 & 0 & 1 & 0 \\ 0 & 0 & 1 & 0 & | & 0 & 1 & 0 & 0 \\ 0 & 0 & 0 & 1 & | & 0 & 0 & 0 & 1 \end{pmatrix} = \left(I_4 \mid F^{-1} \right).$$

Applying similar row operations to $(\hat{F} \mid I_4)$, where $\hat{F} = \begin{pmatrix} B \\ \hat{B}_c \end{pmatrix}$, we have

$$(\hat{F} \mid I_4) \longrightarrow \begin{pmatrix} 1 & 0 & 0 & 0 & | & 1 & 0 & 0 & 0 \\ 0 & 1 & 0 & 0 & | & 0 & 2 & 1 & -1 \\ 0 & 0 & 1 & 0 & | & 0 & 1 & 0 & 0 \\ 0 & 0 & 0 & 1 & | & 0 & 0 & 0 & 1 \end{pmatrix} = \left(I_4 \mid \hat{F}^{-1} \right).$$

Here we used the fact from linear algebra that when row operations finally deliver

$$(\hat{F} \mid I_4) \longrightarrow (I_4 \mid W),$$

the matrix W is the inverse $\hat{F}^{-1}$ of $\hat{F}$.

Using Lemma 2.3, we have

$$S = \begin{pmatrix} -2 & -1 & 0 & 0 \\ 1 & 0 & 0 & 0 \\ 0 & 0 & 2 & 1 \\ 0 & 0 & -1 & 0 \end{pmatrix};$$

$$SF^{-1} = \begin{pmatrix} -2 & 0 & -1 & 0 \\ 1 & 0 & 0 & 0 \\ 0 & 2 & 0 & 1 \\ 0 & -1 & 0 & 0 \end{pmatrix};$$

$$\begin{pmatrix} B_c^\dagger \\ B^\dagger \end{pmatrix} = G = (SF^{-1})^* = \begin{pmatrix} -2 & 1 & 0 & 0 \\ 0 & 0 & 2 & -1 \\ -1 & 0 & 0 & 0 \\ 0 & 0 & 1 & 0 \end{pmatrix}, \quad \text{for which}$$

$$B_c^\dagger = \begin{pmatrix} -2 & 1 & 0 & 0 \\ 0 & 0 & 2 & -1 \end{pmatrix} \quad \text{and}$$

$$B^\dagger = \begin{pmatrix} -1 & 0 & 0 & 0 \\ 0 & 0 & 1 & 0 \end{pmatrix};$$

$$S\hat{F}^{-1} = \begin{pmatrix} -2 & -2 & -1 & 1 \\ 1 & 0 & 0 & 0 \\ 0 & 2 & 0 & 1 \\ 0 & -1 & 0 & 0 \end{pmatrix};$$

$$\begin{pmatrix} \hat{B}_c^\dagger \\ \hat{B}^\dagger \end{pmatrix} = \hat{G} = (S\hat{F}^{-1})^* = \begin{pmatrix} -2 & 1 & 0 & 0 \\ -2 & 0 & 2 & -1 \\ -1 & 0 & 0 & 0 \\ 1 & 0 & 1 & 0 \end{pmatrix}, \quad \text{for which}$$

$$\hat{B}_c^\dagger = \begin{pmatrix} -2 & 1 & 0 & 0 \\ -2 & 0 & 2 & -1 \end{pmatrix} \quad \text{and}$$

$$\hat{B}^\dagger = \begin{pmatrix} -1 & 0 & 0 & 0 \\ 1 & 0 & 1 & 0 \end{pmatrix}.$$

Using the Definition 2.6 of L and $L^\dagger$, we have

$$Ly = y'' + 2y' + 2y,$$
$$D(L) = \{y \in C^2[0, \pi] : y(0) = 0 = y(\pi)\};$$
$$L^\dagger z = z'' - 2z' + 2z,$$
$$D(L^\dagger) = \{z \in C^2[0, \pi] : B^\dagger \vec{z} = \vec{0}_2, \quad \text{or equivalently,}$$
$$\hat{B}^\dagger \vec{z} = \vec{0}_2, \quad \text{or equivalently,}$$
$$z(0) = 0 = z(\pi)\}.$$

Thus the problem π_2 becomes

$$\pi_2 : \quad y'' + 2y' + 2y = 0,$$
$$y(0) = 0 = y(\pi),$$

and the adjoint problem $\pi_2^\dagger$ becomes

$$\pi_2^\dagger : \quad z'' - 2z' + 2z = 0,$$
$$z(0) = 0 = z(\pi).$$

It follows that L is not symmetric, since

$$L^\dagger u \neq Lu$$

for $u \in D(L)$.

Here note that we can also obtain the $L^\dagger$, by using Definition 1.2. The expression

$$[p\bar{z}y' + q\bar{z}y - p\bar{z}'y - p'\bar{z}y]|_a^b = 0$$

for all $y \in D(L)$, in Definition 1.2, gives

$$\overline{z}(\pi)y'(\pi) - \overline{z}(0)y'(0) = 0.$$

Here $y \in C^2[0, \pi]$ and $y(0) = 0 = y(\pi)$. Thus choosing $y'(\pi) = 1$ and $y'(0) = 0$ leads to $\overline{z}(\pi) = 0 = z(\pi)$. But choosing $y'(\pi) = 0$ and $y'(0) = 1$ leads to $\overline{z}(0) = 0 = z(0)$. It follows that $D(L^\dagger) = \{z \in C^2[0, \pi] : z(0) = 0 = z(\pi)\}$.

Try $y = e^{rx}$ in the problem π_2, and we get the auxiliary equation $r^2 + 2r + 2 = 0$. The two roots for this equation are: $r = -1 \pm i$, and so the general solution for y is:

$$y = c_1 e^{-x}\cos(x) + c_2 e^{-x}\sin(x).$$

The boundary conditions $y(0) = 0 = y(\pi)$ give:

$$0 = y(0) = c_1, \quad 0 = -c_1 e^{-\pi}.$$

and so $y = c_2 e^{-x}\sin(x)$, where c_2 is an arbitrary constant. Thus the problem π_2 has exactly 1 linearly independent solution. Similar calculations show that $z = d_2 e^x \sin(x)$, where d_2 is an arbitrary constant, and that the adjoint problem $\pi_2^\dagger$ has also exactly 1 linearly independent solution. These two numbers 1 and 1 satisfy the relation:

$$1 = 1 + m - n = 1 + 2 - 2,$$

as desired.

Compute

$$B_c^\dagger \tilde{z} = \begin{pmatrix} -2 & 1 & 0 & 0 \\ 0 & 0 & 2 & -1 \end{pmatrix} d_2 \begin{pmatrix} 0 \\ 1 \\ 0 \\ -e^\pi \end{pmatrix}$$

$$= d_2 \begin{pmatrix} 1 \\ e^\pi \end{pmatrix},$$

$$\hat{B}_c^\dagger \tilde{z} = \begin{pmatrix} -2 & 1 & 0 & 0 \\ -2 & 0 & 2 & -1 \end{pmatrix} d_2 \begin{pmatrix} 0 \\ 1 \\ 0 \\ -e^\pi \end{pmatrix}$$

$$= d_2 \begin{pmatrix} 1 \\ e^\pi \end{pmatrix},$$

which are desired results.

By Theorem 2.9, (3.1) has a solution if and only if h satisfies

$$(h, z) = \int_0^\pi h(x)\overline{d_2}e^x \sin(x)\, dx$$

$$= (\vec{\gamma}, B_c^\dagger \tilde{z})$$

$$= \left(\begin{pmatrix} \frac{1}{2} \\ \frac{1}{2} \end{pmatrix}, d_2 \begin{pmatrix} 1 \\ e^\pi \end{pmatrix} \right)$$

$$= \overline{d_2}\frac{1}{2}(1 + e^\pi),$$

that is, h satisfies

$$\int_0^\pi h(x)e^x \sin(x)\, dx = \frac{1}{2}(1 + e^\pi).$$

An admissible choice for h is: $h(x) \equiv 1$. Choices: $h(x) = 2$ or $sin(x)$ or x, are wrong. In the case that $h(x) \equiv 1$, a family of solutions for (3.1) are given by

$$y = c_2 e^{-x} \sin(x) + \frac{1}{2},$$

using the method of undetermined coefficients or the method of variation of parameters in Chapter 1, where c_2 is an arbitrary constant. In the case that $h(x) = 2$ or $sin(x)$ or x, either method used, shows that no solution exists for (3.1). One advantage of using the method in this chapter, is that we can tell, in advance, whether the solutions for (3.1) exist or not, after we have solved the adjoint problem $\pi_2^\dagger$.

EXAMPLE 3.2. Consider the equation with the Dirichlet boundary condition of one side:

$$y'' + 2y' + 2y = h(x), \quad 0 < x < \pi,$$

$$y(0) = \frac{1}{2}, \tag{3.2}$$

where $h(x)$ is a continuous function on $[0, \pi]$.

The corresponding boundary operator is the matrix

$$B = \begin{pmatrix} 1 & 0 & 0 & 0 \end{pmatrix}.$$

Here, correspondingly, $m = 1, n = 2$, and $\vec{\gamma} = \frac{1}{2}$.

Do as in Example 3.1.

Solution. Let

$$B_c = \begin{pmatrix} 0 & 0 & 1 & 0 \\ 0 & 1 & 0 & 0 \\ 0 & 0 & 0 & 1 \end{pmatrix}, \hat{B}_c = \begin{pmatrix} 0 & 0 & 1 & 0 \\ 0 & 1 & 0 & 1 \\ 0 & 0 & 2 & 1 \end{pmatrix}.$$

Let $F = \begin{pmatrix} B \\ B_c \end{pmatrix}$.

The solution of Example 3.1 shows that

$$(F \mid I_4) \longrightarrow \begin{pmatrix} 1 & 0 & 0 & 0 & | & 1 & 0 & 0 & 0 \\ 0 & 1 & 0 & 0 & | & 0 & 0 & 1 & 0 \\ 0 & 0 & 1 & 0 & | & 0 & 1 & 0 & 0 \\ 0 & 0 & 0 & 1 & | & 0 & 0 & 0 & 1 \end{pmatrix} = (I_4 \mid F^{-1});$$

$$\hat{F} = \begin{pmatrix} B \\ \hat{B}_c \end{pmatrix};$$

$$(\hat{F} \mid I_4) \longrightarrow \begin{pmatrix} 1 & 0 & 0 & 0 & | & 1 & 0 & 0 & 0 \\ 0 & 1 & 0 & 0 & | & 0 & 2 & 1 & -1 \\ 0 & 0 & 1 & 0 & | & 0 & 1 & 0 & 0 \\ 0 & 0 & 0 & 1 & | & 0 & 0 & 0 & 1 \end{pmatrix} = (I_4 \mid \hat{F}^{-1});$$

$$S = \begin{pmatrix} -2 & -1 & 0 & 0 \\ 1 & 0 & 0 & 0 \\ 0 & 0 & 2 & 1 \\ 0 & 0 & -1 & 0 \end{pmatrix};$$

$$SF^{-1} = \begin{pmatrix} -2 & 0 & -1 & 0 \\ 1 & 0 & 0 & 0 \\ 0 & 2 & 0 & 1 \\ 0 & -1 & 0 & 0 \end{pmatrix};$$

$$\begin{pmatrix} B_c^\dagger \\ B^\dagger \end{pmatrix} = G = (SF^{-1})^* = \begin{pmatrix} -2 & 1 & 0 & 0 \\ 0 & 0 & 2 & -1 \\ -1 & 0 & 0 & 0 \\ 0 & 0 & 1 & 0 \end{pmatrix}, \quad \text{for which}$$

$$B_c^\dagger = \begin{pmatrix} -2 & 1 & 0 & 0 \end{pmatrix} \quad \text{and}$$

$$B^\dagger = \begin{pmatrix} 0 & 0 & 2 & -1 \\ -1 & 0 & 0 & 0 \\ 0 & 0 & 1 & 0 \end{pmatrix};$$

$$S\hat{F}^{-1} = \begin{pmatrix} -2 & -2 & -1 & 1 \\ 1 & 0 & 0 & 0 \\ 0 & 2 & 0 & 1 \\ 0 & -1 & 0 & 0 \end{pmatrix};$$

$$\begin{pmatrix} \hat{B}_c^\dagger \\ \hat{B}^\dagger \end{pmatrix} = \hat{G} = (S\hat{F}^{-1})^* = \begin{pmatrix} -2 & 1 & 0 & 0 \\ -2 & 0 & 2 & -1 \\ -1 & 0 & 0 & 0 \\ 1 & 0 & 1 & 0 \end{pmatrix}, \quad \text{for which}$$

$$\hat{B}_c^\dagger = \begin{pmatrix} -2 & 1 & 0 & 0 \end{pmatrix} \quad \text{and}$$

$$\hat{B}^\dagger = \begin{pmatrix} -2 & 0 & 2 & -1 \\ -1 & 0 & 0 & 0 \\ 1 & 0 & 1 & 0 \end{pmatrix}.$$

Using the Definition 2.6 of L and $L^\dagger$, we have

$$Ly = y'' + 2y' + 2y,$$
$$D(L) = \{y \in C^2[0, \pi] : y(0) = 0\};$$
$$L^\dagger z = z'' - 2z' + 2z,$$
$$D(L^\dagger) = \{z \in C^2[0, \pi] : B^\dagger \vec{z} = \vec{0}_3, \quad \text{or equivalently,}$$
$$\hat{B}^\dagger \vec{z} = \vec{0}_3, \quad \text{or equivalently,}$$
$$z(0) = 0 = z(\pi) = 2z(\pi) - z'(\pi)\}.$$

Thus the problem π_1 becomes

$$\pi_1 : \quad y'' + 2y' + 2y = 0,$$
$$y(0) = 0,$$

and the adjoint problem $\pi_1^\dagger$ becomes

$$\pi_1^\dagger : \quad z'' - 2z' + 2z = 0,$$
$$z(0) = 0 = z(\pi) = 2z(\pi) - z'(\pi).$$

It follows that L is not symmetric, since

$$L^\dagger u \neq Lu$$

for $u \in D(L)$.

Here note that we can also obtain the $L^\dagger$ by using Definition 1.2. The expression

$$[p\bar{z}y' + q\bar{z}y - p\bar{z}'y - p'\bar{z}y]\big|_a^b = 0$$

for all $y \in D(L)$, in Definition 1.2, gives

$$\overline{z}(\pi)y'(\pi) + [2\overline{z}(\pi) - \overline{z}'(\pi)]y(\pi) - \overline{z}(0)y'(0) = 0,$$

where $y \in C^2[0, \pi]$ and $y(0) = 0$. Thus choosing

$$\begin{pmatrix} y'(\pi) \\ y(\pi) \\ y'(0) \end{pmatrix} = \begin{pmatrix} 1 \\ 0 \\ 0 \end{pmatrix}, \begin{pmatrix} 0 \\ 1 \\ 0 \end{pmatrix}, \quad \text{and} \quad \begin{pmatrix} 0 \\ 0 \\ 1 \end{pmatrix}, \quad \text{respectively,}$$

leads to

$$\overline{z}(\pi) = 0 = z(\pi),$$
$$2\overline{z}(\pi) - \overline{z}'(\pi) = 0 = 2z(\pi) - z'(\pi), \quad \text{and}$$
$$\overline{z}(0) = 0 = z(0), \quad \text{respectively.}$$

It follows that

$$D(L^\dagger) = \{z \in C^2[0, \pi] : z(\pi) = 0 = z(0) = 2z(\pi) - z'(\pi)\}.$$

Try $y = e^{rx}$ in the problem π_2, and we get the auxiliary equation $r^2 + 2r + 2 = 0$. The two roots for this equation are: $r = -1 \pm i$, and so the general solution for y is:

$$y = c_1 e^{-x} \cos(x) + c_2 e^{-x} \sin(x).$$

The boundary conditions $y(0) = 0$ give:

$$0 = y(0) = c_1,$$

and so $y = c_2 e^{-x} \sin(x)$, where c_2 is an arbitrary constant. Thus the problem π_2 has exactly 1 linearly independent solution. Similar calculations show that $d_1 = 0 = d_2$, and $z \equiv 0$, and that the adjoint problem $\pi_2^\dagger$ has exactly 0 linearly independent solution. These two numbers 1 and 0 satisfy the relation:

$$0 = 1 + m - n = 1 + 1 - 2,$$

as desired.

Compute

$$B_c^\dagger \overline{z} = \begin{pmatrix} -2 & 1 & 0 & 0 \end{pmatrix} \begin{pmatrix} 0 \\ 0 \\ 0 \\ 0 \end{pmatrix}$$

$$= 0,$$

$$\hat{B}_c^\dagger \overline{z} = \begin{pmatrix} -2 & 1 & 0 & 0 \end{pmatrix} \begin{pmatrix} 0 \\ 0 \\ 0 \\ 0 \end{pmatrix}$$

$$= 0,$$

which are desired results.

By Theorem 2.9, (3.2) has a solution if and only if h satisfies

$$(h, z) = \int_0^\pi h(x) 0 \, dx = 0$$
$$= (\overline{\gamma}, B_c^\dagger \overline{z})$$

$$= (\frac{1}{2}, 0)$$
$$= 0.$$

But this is true for all h, and so (3.2) has a solution for each $h \in C[0, \pi]$.

EXAMPLE 3.3. Consider the equation with the Dirichlet boundary conditions, coupled with one sided Neumann boundary condition:

$$y'' + 2y' + 2y = h(x), \quad 0 < x < \pi,$$
$$y(0) = \frac{1}{2} = y(\pi), \tag{3.3}$$
$$y'(0) = \lambda_0 \in \mathbb{C},$$

where $h(x)$ is a continuous function on $[0, \pi]$.

The corresponding boundary operator is the matrix

$$B = \begin{pmatrix} 1 & 0 & 0 & 0 \\ 0 & 0 & 1 & 0 \\ 0 & 1 & 0 & 0 \end{pmatrix}.$$

Here, correspondingly, $m = 3, n = 2$ and $\vec{\gamma} = \begin{pmatrix} \frac{1}{2} \\ \frac{1}{2} \\ \lambda_0 \end{pmatrix}.$

Do as in Example 3.1.

Solution. Let

$$B_c = \begin{pmatrix} 0 & 0 & 0 & 1 \end{pmatrix}, \hat{B}_c = \begin{pmatrix} 0 & 0 & 2 & 1 \end{pmatrix}.$$

Let $F = \begin{pmatrix} B \\ B_c \end{pmatrix}.$

The solution of Example 3.1 shows that

$$(F \mid I_4) \longrightarrow \begin{pmatrix} 1 & 0 & 0 & 0 & | & 1 & 0 & 0 & 0 \\ 0 & 1 & 0 & 0 & | & 0 & 0 & 1 & 0 \\ 0 & 0 & 1 & 0 & | & 0 & 1 & 0 & 0 \\ 0 & 0 & 0 & 1 & | & 0 & 0 & 0 & 1 \end{pmatrix} = (I_4 \mid F^{-1}).$$

Letting $\hat{F} = \begin{pmatrix} B \\ \hat{B}_c \end{pmatrix}$, the familar row operations deliver

$$(\hat{F} \mid I_4) \longrightarrow \begin{pmatrix} 1 & 0 & 0 & 0 & | & 1 & 0 & 0 & 0 \\ 0 & 1 & 0 & 0 & | & 0 & 0 & 1 & 0 \\ 0 & 0 & 1 & 0 & | & 0 & 1 & 0 & 0 \\ 0 & 0 & 0 & 1 & | & 0 & -2 & 0 & 1 \end{pmatrix} = (I_4 \mid \hat{F}^{-1}).$$

As in the solution of Example 3.1, we have

$$S = \begin{pmatrix} -2 & -1 & 0 & 0 \\ 1 & 0 & 0 & 0 \\ 0 & 0 & 2 & 1 \\ 0 & 0 & -1 & 0 \end{pmatrix}.$$

It follows that

$$SF^{-1} = \begin{pmatrix} -2 & 0 & -1 & 0 \\ 1 & 0 & 0 & 0 \\ 0 & 2 & 0 & 1 \\ 0 & -1 & 0 & 0 \end{pmatrix};$$

$$\begin{pmatrix} B_c^\dagger \\ B^\dagger \end{pmatrix} = G = (SF^{-1})^* = \begin{pmatrix} -2 & 1 & 0 & 0 \\ 0 & 0 & 2 & -1 \\ -1 & 0 & 0 & 0 \\ 0 & 0 & 1 & 0 \end{pmatrix}, \quad \text{for which}$$

$$B_c^\dagger = \begin{pmatrix} -2 & 1 & 0 & 0 \\ 0 & 0 & 2 & -1 \\ -1 & 0 & 0 & 0 \end{pmatrix} \quad \text{and}$$

$$B^\dagger = \begin{pmatrix} 0 & 0 & 1 & 0 \end{pmatrix};$$

$$S\hat{F}^{-1} = \begin{pmatrix} -2 & 0 & -1 & 0 \\ 1 & 0 & 0 & 0 \\ 0 & 0 & 0 & 1 \\ 0 & -1 & 0 & 0 \end{pmatrix},$$

$$\begin{pmatrix} \hat{B}_c^\dagger \\ \hat{B}^\dagger \end{pmatrix} = \hat{G} = (S\hat{F}^{-1})^* = \begin{pmatrix} -2 & 1 & 0 & 0 \\ 0 & 0 & 0 & -1 \\ -1 & 0 & 0 & 0 \\ 0 & 0 & 1 & 0 \end{pmatrix}, \quad \text{for which}$$

$$\hat{B}_c^\dagger = \begin{pmatrix} -2 & 1 & 0 & 0 \\ 0 & 0 & 0 & -1 \\ -1 & 0 & 0 & 0 \end{pmatrix} \quad \text{and}$$

$$\hat{B}^\dagger = \begin{pmatrix} 0 & 0 & 1 & 0 \end{pmatrix}.$$

Using the Definition 2.6 of L and $L^\dagger$, we have

$$Ly = y'' + 2y' + 2y,$$
$$D(L) = \{y \in C^2[0, \pi] : y(0) = 0 = y(\pi) = y'(0)\};$$
$$L^\dagger z = z'' - 2z' + 2z,$$
$$D(L^\dagger) = \{z \in C^2[0, \pi] : B^\dagger \hat{z} = 0, \quad \text{or equivalently,}$$
$$\hat{B}^\dagger \hat{z} = 0, \quad \text{or equivalently,}$$
$$z(\pi) = 0\}.$$

Thus the problem π_3 becomes

$$\pi_3 : \quad y'' + 2y' + 2y = 0,$$
$$y(0) = 0 = y(\pi) = y'(0),$$

and the adjoint problem $\pi_3^\dagger$ becomes

$$\pi_3^\dagger : \quad z'' - 2z' + 2z = 0,$$
$$z(\pi) = 0.$$

It follows that L is not symmetric, since

$$L^\dagger u \neq Lu$$

for $u \in D(L)$.

Here note that we can also obtain the $L^\dagger$ by using the Definition 1.2. The expression

$$[p\bar{z}y' + q\bar{z}y - p\bar{z}'y - p'\bar{z}y]|_a^b = 0$$

for all $y \in D(L)$, in Definition 1.2, gives

$$\bar{z}(\pi)y'(\pi) = 0,$$

where $y \in C^l[0, \pi]$, and $y(0) = 0 = y(\pi) = y'(0)$. Thus choosing $y'(\pi) = 1$ leads to

$$\bar{z}(\pi) = 0 = z(\pi).$$

It follows that

$$D(L^\dagger) = \{z \in C^2[0, \pi] : z(\pi) = 0\}.$$

Try $y = e^{rx}$ in the problem π_2, and we get the auxiliary equation $r^2 + 2r + 2 = 0$. The two roots for this equation are: $r = -1 \pm i$, and so the general solution for y is:

$$y = c_1 e^{-x}\cos(x) + c_2 e^{-x}\sin(x).$$

The boundary conditions $y(0) = 0 = y(\pi) = y'(0)$ give:

$$0 = y(0) = c_1, \quad 0 = -c_1 e^{-\pi}, \quad 0 = c_2,$$

and so $y \equiv 0$. Thus the problem π_2 has exactly 0 linearly independent solution. Similar calculations show that $z = d_2 e^x \sin(x)$, where d_2 is an arbitrary constant, and that the adjoint problem $\pi_2^\dagger$ has exactly 1 linearly independent solution. These two numbers 0 and 1 satisfy the relation:

$$1 = 0 + m - n = 0 + 3 - 2,$$

as desired.

Compute

$$B_c^\dagger \bar{z} = \begin{pmatrix} -2 & 1 & 0 & 0 \\ 0 & 0 & 2 & -1 \\ -1 & 0 & 0 & 0 \end{pmatrix} d_2 \begin{pmatrix} 0 \\ 1 \\ 0 \\ -e^\pi \end{pmatrix}$$

$$= d_2 \begin{pmatrix} 1 \\ e^\pi \\ 0 \end{pmatrix},$$

$$\hat{B}_c^\dagger \bar{z} = \begin{pmatrix} -2 & 1 & 0 & 0 \\ 0 & 0 & 0 & -1 \\ -1 & 0 & 0 & 0 \end{pmatrix} d_2 \begin{pmatrix} 0 \\ 1 \\ 0 \\ -e^\pi \end{pmatrix}$$

$$= d_2 \begin{pmatrix} 1 \\ e^\pi \\ 0 \end{pmatrix},$$

which are desired results.

By Theorem 2.9, (3.3) has a solution if and only if h satisfies

$$(h, z) = \int_0^\pi h(x)\overline{d_2}e^x \sin(x)\,dx$$

$$= (\bar{\gamma}, B_c^\dagger \bar{z})$$

$$= (\begin{pmatrix} \frac{1}{2} \\ \frac{1}{2} \\ \lambda_0 \end{pmatrix}, d_2 \begin{pmatrix} 1 \\ e^\pi \\ 0 \end{pmatrix})$$

$$= \overline{d_2} \frac{1}{2}(1 + e^\pi),$$

that is, h satisfies

$$\int_0^\pi h(x) e^x \sin(x)\, dx = \frac{1}{2}(1 + e^\pi).$$

An admissible choice for h is: $h(x) \equiv 1$. Choices: $h(x) = 2$ or $sin(x)$ or x, are wrong. In the case that $h(x) \equiv 1$, the unique solution for (3.1) is given by

$$y = \lambda_0 e^{-x} \sin(x) + \frac{1}{2},$$

using the method of undetermined coefficients or the method of variation of parameters in Chapter 1. In the case that $h(x) = 2$ or $sin(x)$ or x, either method used, shows that no solution exists for (3.3). One advantage of using the method in this chapter is that we can tell, in advance, whether the solutions for (3.3) exist or not, when we have solved the adjoint problem $\pi_2^\dagger$.

EXAMPLE 3.4. Consider the equation with the Dirichlet boundary conditions, coupled with the two-sided Neumann boundary conditions:

$$y'' + 2y' + 2y = h(x), \quad 0 < x < \pi,$$
$$y(0) = a, y(\pi) = b, \qquad\qquad (3.4)$$
$$y'(0) = c, y'(\pi) = d,$$

where a, b, c and d are constants, and h is a continuous function on $[0, \pi]$.
Do as in Example 3.1.

Solution. Here the corresponding boundary operator is the matrix

$$B = \begin{pmatrix} 1 & 0 & 0 & 0 \\ 0 & 0 & 1 & 0 \\ 0 & 1 & 0 & 0 \\ 0 & 0 & 0 & 1 \end{pmatrix},$$

and $m = 4, n = 2$, and $\vec{\gamma} = \begin{pmatrix} a \\ b \\ c \\ d \end{pmatrix}$.

Let $B_c = 0_{0 \times 4}$, a zero matrix of order 0×4, or equivalently, a zero vector $\vec{0}_4$ of four dimensions. This is the only complementary matrix to B.

Let $F = \begin{pmatrix} B \\ B_c \end{pmatrix} = B$. Applying row operations to compute the inverse F^{-1} of F, we have

$$F^{-1} = \begin{pmatrix} 1 & 0 & 0 & 0 \\ 0 & 0 & 1 & 0 \\ 0 & 1 & 0 & 0 \\ 0 & 0 & 0 & 1 \end{pmatrix}.$$

Using Lemma 2.3, we have

$$S = \begin{pmatrix} -2 & -1 & 0 & 0 \\ 1 & 0 & 0 & 0 \\ 0 & 0 & 2 & 1 \\ 0 & 0 & -1 & 0 \end{pmatrix};$$

$$SF^{-1} = \begin{pmatrix} -2 & 0 & -1 & 0 \\ 1 & 0 & 0 & 0 \\ 0 & 2 & 0 & 1 \\ 0 & -1 & 0 & 0 \end{pmatrix};$$

$$\begin{pmatrix} B_c^\dagger \\ B^\dagger \end{pmatrix} = G = (SF^{-1})^* = \begin{pmatrix} -2 & 1 & 0 & 0 \\ 0 & 0 & 2 & -1 \\ -1 & 0 & 0 & 0 \\ 0 & 0 & 1 & 0 \end{pmatrix}, \quad \text{for which}$$

$$B_c^\dagger = \begin{pmatrix} -2 & 0 & 0 & 0 \\ 0 & 0 & 2 & -1 \\ -1 & 0 & 0 & 0 \\ 0 & 0 & 1 & 0 \end{pmatrix} \quad \text{and}$$

$B^\dagger = 0_{0 \times 4}$, a zero matrix of order 0×4, or equivalently,

a zero vector $\vec{0}_4$ of four dimensions.

Using the Definition 2.6 of L and $L^\dagger$, we have

$$Ly = y'' + 2y' + 2y,$$
$$D(L) = \{y \in C^2[0,1] : y(0) = 0, y(\pi) = 0, y'(0) = 0, y'(\pi) = 0\};$$
$$L^\dagger z = z'' - 2z' + 2z,$$
$$D(L^\dagger) = \{z \in C^2[0,1] : B^\dagger z = 0, \quad \text{or equivalently},$$
$$\text{no restriction on } \tilde{z}\}.$$

Thus the problem π_4 becomes

$$\pi_2 : \quad 2y'' + 5y' - 3y = 0,$$
$$y(0) = 0, y(\pi) = 0,$$
$$y'(0) = 0, y'(\pi) = 0,$$

and the adjoint problem $\pi_4^\dagger$ becomes

$$\pi_4^\dagger : \quad z'' - 2z' + 2z = 0.$$

It follows that L is not symmetric, since

$$L^\dagger u \neq Lu$$

for $u \in D(L)$.

Here note that we can also obtain the $L^\dagger$, by using Definition 1.2. The expression

$$[p\bar{z}y' + q\bar{z}y - p\bar{z}'y - p'\bar{z}y]|_a^b = 0$$

in Definition 1.2, is true for all $y \in D(L)$ and $z \in C^2[0,\pi]$, since

$$y(0) = 0 = y(\pi) = y'(0) = y'(\pi)$$

for $y \in D(L)$. It follows that

$$D(L^\dagger) = C^2[0, \pi].$$

Try $y = e^{rx}$ in the problem π_4, and we get the auxiliary equation $r^2 + 2r + 2 = 0$. The two roots for this equation are: $r = -1 + i, -1 - i$, and so the general solution for y is:

$$y = c_1 e^{-x} \cos(x) + c_2 e^{-x} \sin(x).$$

The boundary conditions $y(0) = 0 = y(\pi) = y'(0) = y'(\pi)$ give:

$$0 = c_1 = c_2,$$

and so $y \equiv 0$. Thus the problem π_4 has exactly 0 linearly independent solution. Similar calculations show that

$$z = c_1 e^x \cos(x) + c_2 e^x \sin(x),$$

and so the adjoint problem $\pi_4^\dagger$ has also exactly 2 linearly independent solution. These two numbers 0 and 2 satisfy the relation:

$$2 = 0 + m - n = 0 + 4 - 2,$$

as desired.

Compute

$$B_c^\dagger z = \begin{pmatrix} -2 & 1 & 0 & 0 \\ 0 & 0 & 2 & -1 \\ -1 & 0 & 0 & 0 \\ 0 & 0 & 1 & 0 \end{pmatrix} \begin{pmatrix} c_1 \\ c_1 + c_2 \\ -c_1 e^\pi \\ -(c_1 + c_2)e^\pi \end{pmatrix}$$

$$= \begin{pmatrix} c_2 - c_1 \\ (c_2 - c_1)e^\pi \\ -c_1 \\ -c_1 e^\pi \end{pmatrix}.$$

By Theorem 2.9, (3.4) has a solution if and only if h satisfies

$$(h, z) = \int_0^\pi h(x)(\overline{c_1} e^x \cos(x) + \overline{c_2} e^x \sin(x)) \, dx$$

$$= (\vec{\gamma}, B_c^\dagger z)$$

$$= (\begin{pmatrix} a \\ b \\ c \\ d \end{pmatrix}, \begin{pmatrix} c_2 - c_1 \\ (c_2 - c_1)e^\pi \\ -c_1 \\ -c_1 e^\pi \end{pmatrix})$$

$$= a\overline{(c_2 - c_1)} + b\overline{(c_2 - c_1)}e^\pi + c\overline{(-c_1)} + d\overline{(-c_1)}e^\pi.$$

When $h(x) \equiv 1$, many choices can be made for a, b, c, and d. For instance, we can choose

$$(a, b, c, d) = (\frac{1}{2}, \frac{1}{2}, 0, 0) \quad \text{or}$$

$$= (\frac{1}{2}(1 + \pi), 0, 0, 0).$$

Under the first choice, the unique solution of (3.4) is:

$$y = \frac{1}{2},$$

by using the method of undetermined coefficients or the method of parameters variation.

When $h(x) = \sin(2x)$, there is no solution of (3.4), no matter what $\vec{\gamma} = (a, b, c, d)$ is chosen. This can be checked by the method of undetermined coefficients or the method of parameters variation.

One advantage of using the method in this chapter is that we can tell, in advance, whether the solutions for (3.4) exist or not, when we have solved the adjoint problem $\pi_4^\dagger$.

EXAMPLE 3.5. Suppose that L is defined by

$$Ly = [(\cos(2x))y']' + \sin(x)y$$
$$D(L) = \{y \in C^2[0, \pi] : y \quad \text{satisfies} \quad B.C.\},$$

where the $B.C.$ is either the

- Zero Dirichlet Boundary Condition: $y(0) = 0 = y(\pi)$, or the
- Zero Neumann Boundary Condition: $y'(0) = 0 = y'(\pi)$, or the
- Zero Robin Boundary Condition:

$$y(0) = c_1 y'(\pi) \quad \text{and} \quad y(\pi) = -c_2 y'(\pi),$$

where c_1, c_2 are two positive constants, or the
- Periodic Boundary Condition:

$$y(0) = y(\pi) \quad \text{and} \quad y'(0) = y'(\pi).$$

Show that L is symmetric.

Solution. For the Dirichlet or Neumann or Robin Boundary conditions, we have $p(x) = \cos(2x), p(0) = 1, p(\pi) = 1$;

$$\alpha = \begin{pmatrix} 1 & 0 \\ 0 & 0 \end{pmatrix} \quad \text{or} \quad \begin{pmatrix} 0 & 1 \\ 0 & 0 \end{pmatrix} \quad \text{or} \quad \begin{pmatrix} 1 & -c_1 \\ 0 & 0 \end{pmatrix}, \quad \text{and}$$

$$\beta = \begin{pmatrix} 0 & 0 \\ 1 & 0 \end{pmatrix} \quad \text{or} \quad \begin{pmatrix} 0 & 0 \\ 0 & 1 \end{pmatrix} \quad \text{or} \quad \begin{pmatrix} 0 & 0 \\ 1 & c_2 \end{pmatrix}, \quad \text{correspondingly.}$$

It follows that L is symmetric by Proposition 1.5. Here $|\alpha|$ is the determinant of the matrix α.

However, for the Periodic Boundary Condition, we have

$$\alpha = \begin{pmatrix} 1 & 0 \\ 0 & 1 \end{pmatrix} \quad \text{and} \quad \beta = \begin{pmatrix} 1 & 0 \\ 0 & 1 \end{pmatrix}.$$

Thus L is still symmetric since $p(0)|\alpha| = 1 = p(\pi)|\beta|$.

4. Complementary Properties between Adjoint Problems and Original Problems

In order to prove Theorem 2.9, we need to prepare a few Lemmas and Propositions. Those Lemmas and propositions reflect the complementary properties between the adjoint problem and the original problem.

We first prove the equivalence of Definition 1.2 and Definition 2.6 in Lemma 2.8.

Proof of Lemma 2.8.

PROOF. Let

$$U = \{u \in C^2[a,b] : V\{\tilde{y}, \tilde{u}\} = 0 \quad \text{for all} \quad y \in D(L)\}$$

and

$$T = \{u \in C^2[a,b] : B^\dagger \tilde{u} = \vec{0}_{4-m}\}.$$

We need to show that $U = T$.

Step 1. Let $z \in T$, for which $B^\dagger \tilde{z} = \vec{0}_{4-m}$. We will show $z \in U$, so that $T \subset U$. By (2.2), we have

$$V\{\tilde{y}, \tilde{z}\} = (B\tilde{y}, B_c^\dagger \tilde{z}) + (B_c \tilde{y}, B^\dagger \tilde{z}),$$

which equals zero since the first term of the right side is zero by $y \in D(L)$ and since the second term of the right side is zero also by $z \in T$. Thus $z \in U$.

Step 2. Let $z \in U$, for which $V\{\tilde{y}, \tilde{z}\} = 0$. We will show $z \in T$, so that $U \subset T$. It follows from (2.2) in Lemma 2.5 again that

$$0 = V\{\tilde{y}, \tilde{z}\} = 0 + (B_c \tilde{y}, B^\dagger \tilde{z}).$$

Here used was $B\tilde{y} = \vec{0}_m$ by $y \in D(L)$. We claim that we can choose the $y \in D(L)$, for which $B\tilde{y} = \vec{0}_m$, such that

$$B_c \tilde{y} = \begin{pmatrix} 1 \\ 0 \\ \cdot \\ 0 \end{pmatrix},$$

a unit vector of $(4 - m)$ dimensions, with the first component equal to 1 and the remaining components equal to 0, or such that

$$B_c \tilde{y} = \begin{pmatrix} 0 \\ 1 \\ \cdot \\ 0 \end{pmatrix},$$

a unique vector of $(4 - m)$ dimensions, with the second component equal to 1 and the remaining components equal to 0, or such that ..., and so on. Here there are totally $(4 - m)$ such unit vectors. With these y, it follows that

$$B^\dagger \tilde{z} = 0,$$

for which $z \in T$.

To prove the claim, let $\hat{i}^{(k)}, 1 \leq k \leq (4 - m)$, be the $(4 - m)$- dimensional vector in $\mathbb{C}^{(4-m)}$, with the k-th component equal to 1 and the remaining components equal to 0. Since B_c has the rank of $(4 - m)$, it follows from linear algebra that vectors $\vec{\alpha}^{(k)}, 1 \leq k \leq (4 - m)$, of 4 dimensions in $\mathbb{C}^4$, exist to satisfy

$$B_c \vec{\alpha}^{(k)} = \hat{i}^{(k)}.$$

The claim follows, since we can choose the $y \in D(L)$, for which $y \in C^2[a,b]$ and $B\tilde{y} = \vec{0}_m$, such that

$$B_c \tilde{y} = \vec{\alpha}^{(k)}, k = 1, \cdots, (4 - m).$$

Here note that

$$A \equiv \begin{pmatrix} B \\ B_c \end{pmatrix}$$

is a matrix of 4×4 and has the rank equal to 4, so that A^{-1}, the inverse of A, exists, and so that

$$\begin{pmatrix} y(a) \\ y'(a) \\ y(b) \\ y'(b) \end{pmatrix} = \tilde{y} = A^{-1} \begin{pmatrix} \vec{0}_m \\ \vec{\alpha}^{(k)} \end{pmatrix}.$$

$\square$

Next, we prove Lemma 2.2 that the quantity $V\{\tilde{y}, \tilde{z}\}$ can be written as a semilinear form through a nonsingular matrix:

Proof of Lemma 2.2.

PROOF. For $y, z \in C^2[a, b]$, we have

$$V\{\tilde{y}, \tilde{z}\} = (p(x)\bar{z}y' + q(x)\bar{z}y - p\bar{z}'y - p'\bar{z}y)|_{x=a}^{b}$$
$$= (p\bar{z}y' + q\bar{z}y - p\bar{z}'y - p'\bar{z}y)(b) - (p\bar{z}y' + q\bar{z}y - p\bar{z}'y - p'\bar{z}y)(a)$$
$$= [-(q(a) - p'(a))y(a) - p(a)y'(a)]\bar{z}(a) + [p(a)y(a)]\bar{z}'(a)$$
$$+ [(q(b) - p'(b))y(b) + p(b)y'(b)]\bar{z}(b) + [-p(b)y(b)]\bar{z}'(b)$$
$$= \left(\begin{pmatrix} -(q(a) - p'(a))y(a) - p(a)y'(a) \\ p(a)y(a) \\ (q(b) - p'(b))y(b) + p(b)y'(b) \\ -p(b)y(b) \end{pmatrix}, \begin{pmatrix} z(a) \\ z'(a) \\ z(b) \\ z'(b) \end{pmatrix} \right)$$
$$= \left(\begin{pmatrix} -(q(a) - p'(a)) & -p(a) & 0 & 0 \\ p(a) & 0 & 0 & 0 \\ 0 & 0 & (q(b) - p'(b)) & p(b) \\ 0 & 0 & -p(b) & 0 \end{pmatrix} \begin{pmatrix} y(a) \\ y'(a) \\ y(b) \\ y'(b) \end{pmatrix}, \begin{pmatrix} z(a) \\ z'(a) \\ z(b) \\ z'(b) \end{pmatrix} \right).$$

$\square$

Recall the following definition:

DEFINITION 4.1. For a matrix $S = (s_{ij})_{4 \times 4}$, its transpose is $S^T \equiv (s_{ji})_{4 \times 4}$, its complex conjugate is $\bar{S} \equiv (\bar{s}_{ij})_{4 \times 4}$, and its transposed complex conjugate is $S^* \equiv \bar{S}^T$.

The following Lemmas 4.2 and 4.3 are two facts in linear algebra:

LEMMA 4.2. For $\vec{\alpha}, \vec{\beta} \in \mathbb{C}^4$, and S, a matrix of 4×4, we have

$$(S\alpha, \beta) = (\alpha, S^* \beta).$$

PROOF. Let

$$\vec{\alpha} = \begin{pmatrix} \alpha_1 \\ \alpha_2 \\ \alpha_3 \\ \alpha_4 \end{pmatrix}, \vec{\beta} = \begin{pmatrix} \beta_1 \\ \beta_2 \\ \beta_3 \\ \beta_4 \end{pmatrix},$$

and $S = (s_{ij})_{4 \times 4}$. By the definition of matrix multiplication, we have

$$S\vec{\alpha} = \sum_{j=1}^{4} s_{ij}\alpha_j,$$

and so

$$(S\vec{\alpha}, \vec{\beta}) = \sum_{i=1}^{4} (\sum_{j=1}^{4} s_{ij}\alpha_j)\overline{\beta_i}$$

$$= \sum_{i,j=1}^{4} \overline{s_{ij}\alpha_j\beta_i}$$

$$= \sum_{j=1}^{4} \alpha_j \overline{\sum_{i=1}^{4} \overline{s_{ij}}\beta_i}$$

$$= (\vec{\alpha}, S^*\vec{\beta}).$$

$\square$

LEMMA 4.3. *If S and F are two nonsingular matrices of 4×4, and if $\vec{\alpha}$ and $\vec{\beta}$ are two vectors of 4 dimensions in $\mathbb{C}^4$, then there is a unique nonsingular matrix $G = (SF^{-1})^*$ of 4×4, such that*

$$(S\vec{\alpha}, \vec{\beta}) = (F\vec{\alpha}, G\vec{\beta}).$$

PROOF.

$$(S\vec{\alpha}, \vec{\beta}) = (SF^{-1}F\vec{\alpha}, \vec{\beta})$$

$$= (F\vec{\alpha}, (SF^{-1})^*\vec{\beta}) \quad \text{by Lemma 4.2,}$$

and so $G = (SF^{-1})^*$ is unique and nonsingular. Here F^{-1} is the inverse of F, and that G is nonsingular means that G^{-1} exists or the determinant $det(G)$ of G is not zero. $\square$

Next, we prove Lemma 2.3 that the semilinear form in Lemma 2.2 can be further decomposed into the sum of two semilinear forms:

Proof of Lemma 2.3.

PROOF. Use the S in Lemma 2.2, and let

$$F = \begin{cases} B_c, & \text{if } m = 0; \\ \begin{pmatrix} B \\ B_c \end{pmatrix}, & \text{if } 1 \le m \le 3; \\ B, & \text{if } m = 4. \end{cases}$$

By Lemma 4.3, we have the unique nonsingular $G = (SF^{-1})^*$, such that

$$V\{\tilde{y}, \tilde{z}\} = (S\tilde{y}, \tilde{z})$$

$$= (F\tilde{y}, G\tilde{z}),$$

which equals

$$(B\tilde{y}, B_c^\dagger\tilde{z}) + (B_c\tilde{y}, B^\dagger\tilde{y}),$$

if we write out G in the form

$$G = (SF^{-1})^* = \begin{cases} B^\dagger, & \text{for which } B_c^\dagger \equiv \begin{pmatrix} 0 & 0 & 0 & 0 \end{pmatrix}, & \text{if } m = 0; \\ \begin{pmatrix} B_c^\dagger \\ B^\dagger \end{pmatrix}, & & \text{if } 1 \le m \le 3; \\ B_c^\dagger, & \text{for which } B^\dagger \equiv \begin{pmatrix} 0 & 0 & 0 & 0 \end{pmatrix}, & \text{if } m = 4. \end{cases}$$

This proves the lemma. Here note that $B_c^\dagger$ and $B^\dagger$ are unique. $\qquad\square$

Now we can prove the boundary formula in Lemma 2.5.

Proof of Lemma 2.5

PROOF. Note that $0 \le m \le 3$.
Let $\hat{B}_c$ be another matrix, complementary to B, and let

$$F = \begin{cases} B_c, & \text{if } m = 0; \\ \begin{pmatrix} B \\ B_c \end{pmatrix}, & \text{if } 1 \le m \le 3. \end{cases}$$

As in Lemma 2.3, the unique $\hat{G} = (S\hat{F}^{-1})^*$ exists to satisfy

$$(S\tilde{y}, \tilde{z}) = (\hat{F}\tilde{y}, \hat{G}\tilde{z}),$$

which equals

$$(B\tilde{y}, \hat{B}_c^\dagger \tilde{z}) + (\hat{B}_c\tilde{y}, \hat{B}^\dagger \tilde{z})$$

if we write out $\hat{G}$ in the form

$$\hat{G} = (S\hat{F}^{-1})^* = \begin{cases} \hat{B}^\dagger, & \text{for which } \hat{B}_c^\dagger \equiv \begin{pmatrix} 0 & 0 & 0 & 0 \end{pmatrix}, & \text{if } m = 0; \\ \begin{pmatrix} \hat{B}_c^\dagger \\ \hat{B}^\dagger \end{pmatrix}, & & \text{if } 1 \le m \le 3. \end{cases}$$

Here note that $\hat{B}_c^\dagger$ and $\hat{B}^\dagger$ are unique.
Rewrite $\hat{G}$ as

$$\begin{aligned} \hat{G} = (S\hat{F}^{-1})^* &= ((SF^{-1})(F\hat{F}^{-1}))^* \\ &= (F\hat{F}^{-1})^*(SF^{-1})^* = (F\hat{F}^{-1})^* G \\ &\equiv H^* G, \end{aligned}$$

which immediately proves the lemma for $m = 0$.
For $1 \le m \le 3$, observe that $H \equiv (F\hat{F}^{-1})$ satisfies

$$\begin{pmatrix} B \\ B_c \end{pmatrix} = F = H\hat{F} = H \begin{pmatrix} B \\ \hat{B}_c \end{pmatrix},$$

and so H must be of the form

$$\begin{pmatrix} I_{m \times m} & 0_{m \times (4-m)} \\ H_{(4-m) \times m} & H_{(4-m) \times (4-m)} \end{pmatrix}.$$

Here $I_{m \times m}$ is the identity matrix of $m \times m$, $0_{m \times (4-m)}$ is the zero matrix of $m \times (4 - m)$, $H_{(4-m) \times m}$ is a matrix of $(4 - m) \times m$, and $H_{(4-m) \times (4-m)}$ is a matrix of $(4 - m) \times (4 - m)$. Thus

$$\begin{pmatrix} I_{m \times m} & 0_{m \times (4-m)} \end{pmatrix} = \begin{pmatrix} 1 & 0 & 0 & 0 \end{pmatrix}$$

$$\text{or} \quad \begin{pmatrix} 1 & 0 & 0 & 0 \\ 0 & 1 & 0 & 0 \end{pmatrix}$$

$$\text{or} \quad \begin{pmatrix} 1 & 0 & 0 & 0 \\ 0 & 1 & 0 & 0 \\ 0 & 0 & 1 & 0 \end{pmatrix},$$

according as $m = 1$ or 2 or 3.

It follows that

$$\begin{pmatrix} \hat{B}_c^\dagger \\ \hat{B}^\dagger \end{pmatrix} = \hat{G} = H^*G$$

$$= \begin{pmatrix} I_{m \times m}^* & H_{(4-m) \times m}^* \\ 0_{m \times (4-m)}^* & H_{(4-m) \times (4-m)}^* \end{pmatrix} \begin{pmatrix} B_c^\dagger \\ B^\dagger \end{pmatrix},$$

from which

$$\hat{B}^\dagger = H_{(4-m) \times (4-m)}^* B^\dagger.$$

This proves the lemma. Here note that $H_{(4-m) \times (4-m)}$ is nonsingular, since the determinant $det(H_{(4-m) \times (4-m)})$ of $H_{(4-m) \times (4-m)}$ equals that of $H = (F\hat{F}^{-1})$, which is non-zero. $\qquad\square$

Recall from Chapter 1 the following three definitions and one lemma:

DEFINITION 4.4. *Two functions φ_1 and φ_2 on $[a, b]$ are linearly independent if*

$$c_1 \varphi_1(x) + c_2 \varphi_2(x) = 0$$

for all $x \in [a, b]$ implies the two constants c_1 and c_2 are zero.

DEFINITION 4.5. *$\{\varphi_1, \varphi_2\}$ is a fundamental set of solutions for the equation*

$$My = py'' + qy' + ry = 0$$

if φ_1 and φ_2 are two solutions to $My = 0$ and are linearly independent.

DEFINITION 4.6. *If $\{\varphi_1, \varphi_2\}$ is a fundamental set of solutions to the equation $My = 0$, then*

$$\Phi \equiv \begin{pmatrix} \varphi_1 & \varphi_2 \end{pmatrix}$$

is the associated matrix of 1×2, in connection with the equation $My = 0$.

For convenience, we write

$$\tilde{\Phi} = \begin{pmatrix} \tilde{\varphi}_1 & \tilde{\varphi}_2 \end{pmatrix} = \begin{pmatrix} \varphi_1(a) & \varphi_2(a) \\ \varphi_1'(a) & \varphi_2'(a) \\ \varphi_1(b) & \varphi_2(b) \\ \varphi_1'(a) & \varphi_2'(b) \end{pmatrix}.$$

LEMMA 4.7. *The equation $My = 0$ has at least one fundamental set of solutions.*

Next is the relation between two associated matrices:

LEMMA 4.8. *If Φ and Ψ are two associated matrices, in connection with equation $My = 0$, then*

$$\Psi = \Phi A$$

for some nonsingular matrix A of 2×2.

PROOF. Let $\{\phi_1, \phi_2\}$ and $\{\psi_1, \psi_2\}$ be two fundamental sets of solutions, with which Φ and Ψ are associated matrices, respectively, in connection with the equation $My = 0$. By Lemma 4.7, we have

$$\psi_1 = a_1\phi_1 + a_2\phi_2 = (\phi_1 \quad \phi_2)\begin{pmatrix} a_1 \\ a_2 \end{pmatrix}$$

$$\psi_2 = a_3\phi_1 + a_4\phi_2 = (\phi_1 \quad \phi_2)\begin{pmatrix} a_3 \\ a_4 \end{pmatrix}$$

(4.1)

for some constants a_1, a_2, a_3, and a_4, where

$$a_1^2 + a_2^2 \neq 0 \quad \text{and} \quad a_3^2 + a_4^2 \neq 0. \tag{4.2}$$

This is the same as

$$\Psi = (\psi_1 \quad \psi_2) = (\phi_1 \quad \phi_2) A,$$

where $A \equiv \begin{pmatrix} a_1 & a_3 \\ a_2 & a_4 \end{pmatrix}$. Claim that A is nonsingular, so that the proof is complete.

Suppose otherwise that A is singular, for which

$$0 = det(A) = a_1 a_4 - a_2 a_3, \tag{4.3}$$

and we seek a contradiction. This implies

$$a_2 \neq 0 \quad \text{or} \quad a_4 \neq 0, \tag{4.4}$$

since otherwise (4.1) gives

$$\psi_1 = a_1\phi_1 \quad \text{and} \quad \psi_2 = a_3\phi_1,$$

a contradiction to the linearly independence of $\{\psi_1, \psi_2\}$. Here (4.2) was used, so that $a_1 \neq 0$ and $a_3 \neq 0$. Therefore (4.4) is true.

But by (4.3), (4.1), and (4.4), we have

$$a_4\psi_1 = a_2\psi_2,$$

and again this is a contradiction since ψ_1 and ψ_2 are linearly independent. □

Now we are in a position to prove the following Lemma 4.9.

LEMMA 4.9. *The boundary value problem*

$$\pi_m : \quad Ly = 0, y \in D(L)$$

has exactly $0 \leq k \leq 2$ linearly independent solutions if and only if $B\tilde{\Phi}$ has the rank $(2 - k)$ for any associated matrix Φ, in connection with the equation $My = py'' + qy' + ry = 0$. Here the number 2 is the order n of the equation $Ly = 0$; since the equation $Ly = 0$ is a second order equation, the order n equals 2.

PROOF. We divide the proof into two steps.

Step 1. Let

$$\Phi = (\varphi_1 \quad \varphi_2)$$

be an associated matrix, in connection with $My = py'' + qy' + ry = 0$, where $\{\varphi_1, \varphi_2\}$ exists by Lemma 4.7, as a fundamental set of solutions for $My = py'' + qy' + ry = 0$. Then for c_1 and c_2 being constants, $\varphi \equiv c_1\varphi_1 + c_2\varphi_2$ is a solution to the problem

$$\pi_m : Ly = 0, y \in D(L)$$

if and only if

$$\vec{0}_m = B\tilde{\varphi} = B(c_1\tilde{\varphi}_1 + c_2\tilde{\varphi}_2)$$

$$= B\tilde{\Phi}\begin{pmatrix} c_1 \\ c_2 \end{pmatrix}.$$

has a solution

$$\begin{pmatrix} c_1 \\ c_2 \end{pmatrix},$$

that is, $\begin{pmatrix} c_1 \\ c_2 \end{pmatrix}$ is in the kernel $ker(B\widetilde{\Phi})$ of $B\Phi$. Thus the problem π_m has exactly k linearly independent solutions if and only if

$$rank(B\widetilde{\Phi}) = 2 - k.$$

Here used was the fact in linear algebra that the sum of $rank(B\widetilde{\Phi})$ and the dimension of the vector space of the kernel $ker(B\widetilde{\Phi})$ equals 2.

Step 2. Let Ψ be another associated matrix, in connection with $My = 0$. Then by Lemma 4.8, $\Psi = \Phi A$ for some nonsingular matrix

$$A = \begin{pmatrix} a_1 & b_1 \\ a_2 & b_2 \end{pmatrix}.$$

Since

$$\Psi = \Phi A = \begin{pmatrix} (a_1\varphi_1 + a_2\varphi_2) & (b_1\varphi_1 + b_2\varphi_2) \\ (a_1\varphi_1' + a_2\varphi_2') & (b_1\varphi_1' + b_2\varphi_2') \end{pmatrix},$$

we have

$$\tilde{\Psi} = \begin{pmatrix} (a_1\varphi_1(a) + a_2\varphi_2(a)) & (b_1\varphi_1(a) + b_2\varphi_2(a)) \\ (a_1\varphi_1'(a) + a_2\varphi_2'(a)) & (b_1\varphi_1'(a) + b_2\varphi_2'(a)) \\ (a_1\varphi_1(b) + a_2\varphi_2(b)) & (b_1\varphi_1(b) + b_2\varphi_2(b)) \\ (a_1\varphi_1'(b) + a_2\varphi_2'(b)) & (b_1\varphi_1'(b) + b_2\varphi_2'(b)) \end{pmatrix}$$

$$= \begin{pmatrix} \tilde{\Phi}\begin{pmatrix} a_1 \\ a_2 \end{pmatrix} & \tilde{\Phi}\begin{pmatrix} b_1 \\ b_2 \end{pmatrix} \end{pmatrix}$$

$$= \tilde{\Phi}\begin{pmatrix} a_1 & b_1 \\ a_2 & b_2 \end{pmatrix}$$

$$= \tilde{\Phi}A.$$

It follows that

$$Rank(B\tilde{\Psi}) = Rank(B(\tilde{\Phi}A)) = Rank((B\tilde{\Phi})A)$$

$$= Rank(B\tilde{\Phi}).$$

Here used was the fact in linear algebra that two matrices have the same rank if one of them equals the other multiplied by a nonsingular matrix.   □

If we start with the adjoint operator $L^\dagger$ of L and try to find its adjoint $(L^\dagger)^\dagger$, the result is

$$(L^\dagger)^\dagger = L,$$

as the following Lemma 4.10 shows.

LEMMA 4.10. *Use the two matrices $B^\dagger$ and $B_c^\dagger$ from Lemma 2.5, for which the adjoint $L^\dagger$ has its domain $D(L^\dagger)$ defined by*

$$D(L^\dagger) = \{u \in C^2[a,b] : B^\dagger\tilde{z} = \vec{0}_{4-m}\}.$$

It follows that the adjoint $(L^\dagger)^\dagger$ equals L.

PROOF. By taking the complex conjugate of (2.1) and (2.2), we have

$$(Nz, y) - (z, My) = \overline{(y, Nz) - (My, z)} = \overline{-V\{\tilde{y}, \tilde{z}\}}$$
$$= \overline{-(B\tilde{y}, B_c^\dagger \tilde{z}) - (B_c \tilde{y}, B^\dagger \tilde{z})}$$
$$= (B^\dagger \tilde{z}, -B_c \tilde{y}) + (B_c^\dagger \tilde{z}, -B \tilde{y}).$$

By the Definition 2.6 of adjoint, the adjoint

$$(L^\dagger)^\dagger : D((L^\dagger)^\dagger) \subset L^2(a, b) \longrightarrow L^2(a, b)$$

of $L^\dagger$ is defined by

$$(L^\dagger)^\dagger y = My = py'' + qy' + ry$$

for

$$y \in D((L^\dagger)^\dagger) \equiv \{u \in C^2[a, b] : B\tilde{y} = \vec{0}_m\}$$
$$= D(L).$$

Thus $(L^\dagger)^\dagger = L$. $\square$

There exists some duality between the number of linearly independent solutions of the boundary value problem

$$\pi_m : \quad Ly = 0, y \in D(L)$$

and that of linearly independent solutions to the adjoint problem

$$\pi_m^\dagger : \quad L^\dagger z = 0, z \in D(L^\dagger),$$

as the following Proposition 4.11 shows.

PROPOSITION 4.11. *Use the $B, B_c, B^\dagger$, and $B_c^\dagger$ in Lemma 2.5. Suppose the problem*

$$\pi_m : \quad Ly = 0, y \in D(L)$$

has exactly $0 \leq k \leq 2$ linearly independent solutions $\varphi_1, \ldots, \varphi_k$, where the number 2 is the order n of the equation $Ly = 0$. Since the equation $Ly = 0$ is a second order equation, the order n equals 2.
Then

- *$B_c \widetilde{\varphi_1}, \cdots, B_c \widetilde{\varphi_k}$ are linearly independent.*
- *$(B^\dagger \tilde{\Psi})^* \vec{b} = 0$ has exactly the k linearly independent solutions $B_c \widetilde{\varphi_1}, \ldots, B_c \widetilde{\varphi_k}$, where Ψ is any associated matrix, in connection with the equation $Nz = 0$.*
- *the adjoint problem*

$$\pi_m^\dagger : \quad L^\dagger z = 0, z \in D(L^\dagger)$$

  has exactly $l = (k + m - 2)$ linearly independent solutions, which are, for convenience, say $z_1, \ldots, z_{(k+m-2)}$.
- *$(B\tilde{\Phi})^* \vec{a} = 0$ has exactly the $(k + m - 2)$ linearly independent solutions $B_c^\dagger \widetilde{z_1}, \ldots, B_c^\dagger \widetilde{z_{(k+m-2)}}$, where Φ is any associated matrix, in connection with the equation $My = 0$.*

PROOF. We consider, separately, the cases: $m = 0, m = 4$, and $1 \leq m \leq 3$. When $m = 0$, clearly the problem

$$\pi_m : \quad Ly = My = py'' + qy' + ry = 0,$$
$$B\tilde{y} = \vec{0}_m,$$

has exactly 2 linearly independent solutions, and so $k = 2$. This is because $B = \begin{pmatrix} 0 & 0 & 0 & 0 \end{pmatrix}$, and so there is no restriction on $\tilde{y}$. From this, it follows that

$$y = c_1 y_1 + c_2 y_2,$$

where $\{y_1, y_2\}$ is a fundamental set of solutions for the equation

$$My = 0.$$

On the other hand, the adjoint problem

$$\pi_m^\dagger : \quad L^\dagger z = Nz = (\bar{p}z)'' - (\bar{q}z)' + \bar{r}z = 0,$$
$$B^\dagger \tilde{z} = \vec{0}_4,$$

only has the trivial solution, and so $l = 0$. To see this, let

$$z = d_1 z_1 + d_2 z_2,$$

where $\{z_1, z_2\}$ be a fundamental set of solutions for the equation

$$Nz = 0.$$

For z to satisfy $B^\dagger \tilde{z} = \vec{0}_4$, we must have

$$\vec{0}_4 = B^\dagger \tilde{z} = B^\dagger [(\tilde{z}_1 \quad \tilde{z}_2) \begin{pmatrix} d_1 \\ d_2 \end{pmatrix}].$$

Since the rank of $B^\dagger$ equals 4 by Lemma 2.3, there should

$$(\tilde{z}_1 \quad \tilde{z}_2) \begin{pmatrix} d_1 \\ d_2 \end{pmatrix} = \vec{0}_4.$$

This implies $d_1 = 0 = d_2$, and so $z = 0$, the trivial solution. This is because $\{z_1, z_2\}$ is a fundamental set of solution for $Nz = 0$, and then the determinant of $(\tilde{z}_1 \quad \tilde{z}_2)$ is nonzero.

Thus

$$0 = l = k + m - n = 2 + 0 - 2,$$

as desired.

Next, when $m = 4$, the problem π_4 only has trivial solution, and so $k = 0$. This is because the rank of B is 4 by assumption, and then the result follows from, as above, the case for $m = 0$. On the other hand, the adjoint problem $\pi_m^\dagger$ has exactly 2 linearly independent solutions, and so $l = 2$. This is because $B^\dagger = \begin{pmatrix} 0 & 0 & 0 & 0 \end{pmatrix}$ by Lemma 2.3, and then the result follows from the case for $m = 0$ again. Thus

$$2 = l = k + m - n = 0 + 4 - 2,$$

as desired.

Finally, consider the case where $1 \leq m \leq 3$. Let $\varphi_1, \ldots, \varphi_k$ be k linearly independent solutions of the boundary value problem π_m. Use the matrices $B, B_c, B^\dagger$, and $B_c^\dagger$ from Lemma 2.5, where note that B_c is complementary to B, and that $B_c^\dagger$ is complementary to $B^\dagger$. We divide the proof into four steps.

Step 1. We claim that $B_c \tilde{\varphi}_1, \ldots, B_c \tilde{\varphi}_k$ are linearly independent vectors. Suppose that they are not, and we seek a contradiction. Then there are constants $c_1, \ldots, c_k$, not all zero, such that

$$\sum_{i=1}^{k} c_i B_c \tilde{\varphi}_i = \vec{0}_{4-m}.$$

This implies

$$B_c \tilde{\varphi} = \vec{0}_{4-m},$$

where $\varphi \equiv \sum_{i=1}^{k} c_i \varphi_i$. But

$$B\tilde{\varphi} = \vec{0}_m$$

also, since $\varphi = \sum_{i=1}^{k} c_i \varphi_i$ is a solution of the problem π_m also and so lies in $D(L)$. Hence

$$\begin{pmatrix} B \\ B_c \end{pmatrix} \tilde{\varphi} = \begin{pmatrix} 0 \\ 0 \\ 0 \\ 0 \end{pmatrix},$$

and this implies

$$\tilde{\varphi} = \begin{pmatrix} \varphi(a) \\ \varphi'(a) \\ \varphi(b) \\ \varphi'(b) \end{pmatrix} = \begin{pmatrix} 0 \\ 0 \\ 0 \\ 0 \end{pmatrix}$$

since $\begin{pmatrix} B \\ B_c \end{pmatrix}$ has rank 4 and so is invertible. Thus

$$\varphi(a) = 0 = \varphi'(a) \quad \text{and} \quad \varphi(b) = 0 = \varphi'(b),$$

and then φ is a solution to the initial value problem

$$My = 0, y(a) = 0 = y'(a).$$

Since 0 is, clearly, also a solution, we have

$$\sum_{i=1}^{k} c_i \varphi_i = \varphi \equiv 0$$

by the theorem of existence and uniqueness in Chapter 8. Therefore $c_1 = \cdots = c_k = 0$, and this is a contradiction.

Step 2. Let ψ_1, ψ_2 be two linearly independent solutions of $Nz = 0$. This is possible by Chapter 1. We claim that the equation

$$(B^{\dagger}\widetilde{\Psi})^* \vec{a} = 0$$

has at least k linearly independent solutions $\vec{a}$'s so that

$$Rank(B^{\dagger}\widetilde{\Psi}) = Rank((B^{\dagger}\widetilde{\Psi})^*) \leq ((4-m) - k).$$

Here note that $(B^{\dagger}\widetilde{\Psi})^*$ is a matrix of $2 \times (4-m)$, and that $\vec{a}$ is a vector of $(4-m)$ dimensions.

Applied to prove the claim, the Green's formula, which is (2.1), gives, for $i = 1, \ldots, k$ and $j = 1, 2$,

$$0 = (L\varphi_i, \psi_j) - (\varphi_i, N\psi_j).$$

This equals, by the boundary formula (2.2),

$$(B\tilde{\varphi}_i, B_c^{\dagger}\widetilde{\psi_j}) + (B_c\tilde{\varphi}_i, B^{\dagger}\widetilde{\psi_j}).$$

Since $B\tilde{\varphi}_i = \vec{0}_m$, it follows that, for $j = 1, 2$,

$$(B^{\dagger}\widetilde{\psi_j})^*(B_c\tilde{\varphi}_i) = (B_c\tilde{\varphi}_i, B^{\dagger}\widetilde{\psi_j}) = 0$$

and then

$$
\begin{aligned}
(B^\dagger \widetilde{\Psi})^* (B_c \widetilde{\varphi_i}) &= (B^\dagger \begin{pmatrix} \widetilde{\psi_1} & \widetilde{\psi_2} \end{pmatrix})^* (B_c \widetilde{\varphi_i}) \\
&= \begin{pmatrix} B^\dagger \widetilde{\psi_1} & B^\dagger \widetilde{\psi_2} \end{pmatrix}^* (B_c \widetilde{\varphi_i}) \\
&= \begin{pmatrix} (B^\dagger \widetilde{\psi_1})^* \\ (B^\dagger \widetilde{\psi_2})^* \end{pmatrix} (B_c \widetilde{\varphi_i}) = \begin{pmatrix} (B^\dagger \widetilde{\psi_1})^* (B_c \widetilde{\varphi_i}) \\ (B^\dagger \widetilde{\psi_2})^* (B_c \widetilde{\varphi_i}) \end{pmatrix} \\
&= \begin{pmatrix} 0 \\ 0 \end{pmatrix}.
\end{aligned}
$$

Here note that

$$
\begin{aligned}
(\begin{pmatrix} a_1 \\ a_2 \end{pmatrix}, \begin{pmatrix} b_1 \\ b_2 \end{pmatrix}) &= \sum_{i=1}^{2} a_i \overline{b_i} = \begin{pmatrix} \overline{b_1} & \overline{b_2} \end{pmatrix} \begin{pmatrix} a_1 \\ a_2 \end{pmatrix} \\
&= \begin{pmatrix} b_1 \\ b_2 \end{pmatrix}^* \begin{pmatrix} a_1 \\ a_2 \end{pmatrix}.
\end{aligned}
$$

Hence, the equation

$$
(B^\dagger \widetilde{\Psi})^* \vec{a} = 0
$$

has at least the k linearly independent solutions $B_c \widetilde{\varphi_1}, \ldots, B_c \widetilde{\varphi_k}$, each of which is of $(4 - m)$ dimensions. It follows that

$$
Rank(B^\dagger \widetilde{\Psi}) = Rank((B^\dagger \widetilde{\Psi})^*) \le ((4 - m) - k),
$$

and then the claim is proved. Here used was the fact in linear algebra that if A is a matrix of $m \times n$, then $Rank(A) = Rank(A^*)$ and

$$
Rank(A) + dim(Ker(A)) = n,
$$

where $dim(Ker(A))$ is the dimension of the kernel $Ker(A)$ of A, which is a vector space defined by $\{x : Ax = \vec{0}_m\}$.

Step 3. Claim that

$$
(B^\dagger \widetilde{\Psi})^* \vec{b} = 0
$$

has exactly the k linearly independent solutions $B_c \widetilde{\varphi_1}, \ldots, B_c \widetilde{\varphi_k}$, so that

$$
r \equiv Rank(B^\dagger \widetilde{\Psi}) = (4 - m) - k.
$$

It then follows that the problem

$$
\pi_m^\dagger : \quad L^\dagger z = 0, z \in D(L^\dagger)
$$

has, by Lemma 4.9, exactly $(2 - Rank(B^\dagger \widetilde{\Psi})) = (2 - ((4 - m) - k)) = (k + m - 2)$ linearly independent solutions.

Suppose that the claim is not true, which means $r < ((4 - m) - k)$, and we seek a contradiction. Since $r = Rank(B^\dagger \widetilde{\Psi})$, it follows from Lemma 4.9 that the adjoint problem

$$
\pi_m^\dagger : \quad L^\dagger z = 0, z \in D(L^\dagger)
$$

has exactly $(2 - r)$ linearly independent solutions $z_1, \ldots, z_{(2-r)}$. Let ϕ_1, ϕ_2 be two linearly independent solutions of the equation $My = 0$, and let Φ be the corresponding associated matrix, in connection with $My = 0$. It follows from Step 2 that

$$
B_c^\dagger \widetilde{z_1}, \cdots, B_c^\dagger \widetilde{z_{(2-r)}}
$$

are $(2 - r)$ linearly independent solutions of

$$(B\widetilde{\Phi})^*\vec{b} = 0,$$

each of which being m-dimensional, and that

$$Rank(B\widetilde{\Phi}) = Rank((B\widetilde{\Phi})^*) \leq (m-(2-r)) = (m-2+r) < m-2+((4-m)-k) = (2-k)$$

is true. But this is a contradiction, since $Rank(B\widetilde{\Phi}) = (2 - k)$ by Lemma 4.9.

Step 4. It follows from Step 3 that if $z_1, \ldots, z_{(k+m-2)}$ are the $(k + m - 2)$ linearly independent solutions of the adjoint problem

$$\pi_m^\dagger : \quad L^\dagger z = 0, z \in D(L^\dagger),$$

then

$$(B\widetilde{\Phi})^*\vec{a} = 0$$

has exactly the $(k + m - 2)$ linearly independent solutions $B_c^\dagger \widetilde{z_1}, \ldots, B_c^\dagger \widetilde{z_{(k+m-2)}}$, where Φ is any associated matrix, in connection with the equation $My = 0$. $\quad\square$

In order to prove Theorem 2.9, we recall the following fact from linear algebra:

PROPOSITION 4.12. *If A is a matrix of $m \times n$ and $\vec{b} \in \mathbb{C}^n$ is a given vector, then*

$$A\vec{x} = \vec{b}$$

has a solution $\vec{x}$ if and only if

$$(\vec{b}, \vec{y}) = 0$$

for every solution $\vec{y}$ of the adjoint equation

$$A^*\vec{y} = \vec{0}_m.$$

REMARK 4.13. *It is easy to see the necessity part: Suppose that $A\vec{x} = \vec{b}$ has a solution $\vec{x}$, and that $\vec{y}$ is a solution to the adjoint equation*

$$A^*\vec{y} = \vec{0}_m.$$

Then using Lemma 4.2, we have

$$(\vec{b}, \vec{y}) = (A\vec{x}, \vec{y}) = (\vec{x}, A^*\vec{y}) = (\vec{x}, \vec{0}_m) = 0,$$

which is the necessity part.

For the sufficiency part, the following calculations provide essential ideas for a proof: Let

$$A = \begin{pmatrix} i & -i+1 & 3i \\ 1+i & 1-i & 2 \\ 1+2i & 3-i & 1 \end{pmatrix} \quad and \quad \vec{b} = \begin{pmatrix} b_1 \\ b_2 \\ b_3 \end{pmatrix}.$$

We shall find the condition on $\vec{b}$, so that $A\vec{x} = \vec{b}$ has a solution $\vec{x}$.

Applying row operations which are adding $(i-1)\times$ (row 1) to row 2 and adding $(i-2)\times$ (row 1) to row 3, we have

$$\left(A \mid \vec{b} \right) \longrightarrow \begin{pmatrix} i & -i+1 & 3i & | & b_1 \\ 0 & 1+i & -1-3i & | & (i-1)b_1 + b_2 \\ 0 & 2(1+i) & 2(-1-3i) & | & (i-2)b_1 + b_3 \end{pmatrix}.$$

Continuing to adding $(-2)\times$ (row 2) to row 3, we have

$$\longrightarrow \begin{pmatrix} 1 & -1-i & 3 & | & b_1 \\ 0 & 1+i & -1-3i & | & (i-1)b_1 + b_2 \\ 0 & 0 & 0 & | & -ib_1 - 2b_2 + b_3 \end{pmatrix}.$$

So the condition for $A\vec{x} = \vec{b}$ to have a solution is that the element $(-ib_1 - 2b_2 + b_3)$ at the position 3×4 equals zero. But this element is the same as

$$\vec{b}^T \begin{pmatrix} -i \\ -2 \\ 1 \end{pmatrix} = \begin{pmatrix} b_1 & b_2 & b_3 \end{pmatrix} \begin{pmatrix} -i \\ -2 \\ 1 \end{pmatrix} = \begin{pmatrix} b_1 \\ b_2 \\ b_3 \end{pmatrix} \cdot \begin{pmatrix} i \\ -2 \\ 1 \end{pmatrix}.$$

We claim that the vector $\begin{pmatrix} i \\ -2 \\ 1 \end{pmatrix}$ is a generating solution to the adjoint equation

$A^* \vec{y} = \begin{pmatrix} 0 \\ 0 \\ 0 \end{pmatrix}$, *so that the sufficiency part for this particular A is proved. Here note*

that every solution of $A^* \vec{y} = \begin{pmatrix} 0 \\ 0 \\ 0 \end{pmatrix}$ is of the form: $k \begin{pmatrix} i \\ -2 \\ 1 \end{pmatrix}$, $k \in \mathbb{C}$, which explains

the meaning of a generating solution.

The claim is seen as follows: If the above A is expressed as

$$A = \begin{pmatrix} \vec{\alpha} & \vec{\beta} & \vec{\gamma} \end{pmatrix},$$

where $\vec{\alpha}, \vec{\beta}$, and $\vec{\gamma}$ are the column 1, column 2, and column 3 of A, respectively, then the above calculations show that

$$0 = \vec{\alpha}^T \begin{pmatrix} -i \\ -2 \\ 1 \end{pmatrix} = \vec{\beta}^T \begin{pmatrix} -i \\ -2 \\ 1 \end{pmatrix} = \vec{\gamma}^T \begin{pmatrix} -i \\ -2 \\ 1 \end{pmatrix}.$$

That is,

$$\begin{pmatrix} 0 \\ 0 \\ 0 \end{pmatrix} = \begin{pmatrix} \vec{\alpha}^T \\ \vec{\beta}^T \\ \vec{\gamma}^T \end{pmatrix} \begin{pmatrix} -i \\ -2 \\ 1 \end{pmatrix} = A^T \begin{pmatrix} -i \\ -2 \\ 1 \end{pmatrix} = \bar{A}^T \begin{pmatrix} i \\ -2 \\ 1 \end{pmatrix} = A^* \begin{pmatrix} i \\ -2 \\ 1 \end{pmatrix}.$$

Thus $\begin{pmatrix} i \\ -2 \\ 1 \end{pmatrix}$ is a generating solution to the adjoint equation $A^* \vec{y} = \begin{pmatrix} 0 \\ 0 \\ 0 \end{pmatrix}$.

A proof of Proposition 4.12, of more generality, is given in [**11**, Pages 80, 87–88].

Proof of Theorem 2.9

PROOF. We consider two cases.

Case 1: The necessity part. Suppose that ϕ is a solution of (2.3), for which $M\phi = f$, and that z is a solution of the adjoint problem

$$\pi_m^\dagger : \quad L^\dagger z = 0, z \in D(L^\dagger),$$

for which $Nz = L^\dagger z = 0$. We need to prove the condition (2.4). By the boundary formula (2.2), we have

$$(f, z) - 0 = (M\phi, z) - (\phi, Nz) = (B\phi, B_c^\dagger z) + (B_c\phi, B^\dagger z) = (\vec{\gamma}, B_c^\dagger z) + 0,$$

which is (2.4).

Case 2: The sufficiency part. Suppose (2.4) holds. We need to show that (2.3) has a solution.

Let $\{\phi_1, \phi_2\}$ be a fundamental set of solutions for $My = 0$, which exist by Chapter 1. Let g be a particular solution of $My = f$, which exist by Chapter 1. Then every solution y of the equation $My = f$ is of the

$$y = \sum_{i=1}^{2} c_i \phi_i + g,$$

by Chapter 1. Here c_1 and c_2 are complex constants. It follows that (2.3) has a solution y if and only if c_1 and c_2 exist, such that

$$\vec{\gamma} = B(\tilde{y}) = B(\widetilde{\sum_{i=1}^{2} c_i \phi_i + g}) = \sum_{i=1}^{2} c_i B\tilde{\phi}_i + B\tilde{g} = (B\tilde{\Phi})\vec{c} + B\tilde{g}$$

or

$$(B\tilde{\Phi})\vec{c} = \vec{\gamma} - B\tilde{g}. \tag{4.5}$$

Here

$$\vec{c} = \begin{pmatrix} c_1 \\ c_2 \end{pmatrix}$$

and Φ is the associated matrix, corresponding to $\{\phi_1, \phi_2\}$.

It follows from Proposition 4.12 that (4.5) has a solution $\vec{c}$ if and only if

$$(\vec{\gamma} - B\tilde{g}, \vec{a}) = 0 \tag{4.6}$$

for every solution $\vec{a}$ of

$$(B\tilde{\Phi})^* \vec{a} = 0. \tag{4.7}$$

By Proposition 4.11, (4.7) has exactly the $(k + m - 2)$ linearly independent solutions $B_c^\dagger \tilde{z}_1, \cdots, B_c^\dagger z_{(k+m-2)}$, if $z_1, \cdots, z_{(k+m-2)}$ are the exact $(k + m - 2)$ linearly independent solutions of the adjoint problem

$$\pi_m^\dagger : \quad L^\dagger z = 0, z \in D(L^\dagger).$$

It follows that (4.6) is the same as

$$0 = (\vec{\gamma} - B\tilde{g}, B_c^\dagger \tilde{z}_i) = (\vec{\gamma}, B_c^\dagger \tilde{z}_i) - (B\tilde{g}, B_c^\dagger \tilde{z}_i) \tag{4.8}$$

for $i = 1, \cdots, (k + m - 2)$. But the boundary formula (2.2) yields

$$(f, z_i) - 0 = (Mg, z_i) - (g, Nz_i) = (B\tilde{g}, B_c^\dagger \tilde{z}_i) + (B_c \tilde{g}, B^\dagger \tilde{z}_i) = (B\tilde{g}, B_c^\dagger \tilde{z}_i) + 0.$$

This, together with (4.8), gives

$$(\vec{\gamma}, B_c^\dagger \tilde{z}_i) = (f, z_i),$$

which is (2.4).   □

We now prove the special case of Theorem 2.9, where $m = 2$ and the problem π_m only has the trivial zero solution:

Proof of Corollary 2.13

PROOF. By Proposition 4.11, the adjoint problem $\pi_m^\dagger$ has $(k + m - n) = (0 + m - 2) = 0$ solutions, that is, the adjoint problem only has the zero solution. Hence, the equation (2.3) has a solution by Theorem 2.9, since (2.4) holds:

$$(f, 0) = 0 = (\vec{\gamma}, B_c^\dagger \tilde{0}).$$

For uniqueness of solution, let y_1 and y_2 be two solutions to the equation (2.3). Substraction gives
$$L(y_1 - y_2) = 0, (y_1 - y_2) \in D(L).$$
By assumption that the problem
$$\pi_2 : \quad Ly = 0, y \in D(L)$$
only has the trivial zero solution, we have $y_1 - y_2 \equiv 0$, which is $y_1 \equiv y_2$, and so uniqueness of solution follows. $\qquad\square$

The result about when L is symmetric is proved below:

Proof of Proposition 1.5

PROOF. It is easy to see that
$$(My, z) = (y, Nz) + p(y'\overline{z} - y\overline{z'})|_a^b = (y, Nz) - (p(b)\tilde{W}(b) - p(a)\tilde{W}(a))$$
is true, derived by using integration by parts formula for continuously differentiable functions, where
$$My = (py')' + ry,$$
$$Nz = (pz')' + rz,$$
and
$$\tilde{W}(x) = \begin{vmatrix} y(x) & \overline{z}(x) \\ y'(x) & \overline{z'}(x) \end{vmatrix},$$
a variant of the Wronskian of $\{y(x), z(x)\}$.

Since $y, z \in D(L)$, there result
$$\alpha \begin{pmatrix} y(a) \\ y'(a) \end{pmatrix} + \beta \begin{pmatrix} y(b) \\ y'(b) \end{pmatrix} = 0$$
and
$$\alpha \begin{pmatrix} z(a) \\ z'(a) \end{pmatrix} + \beta \begin{pmatrix} z(b) \\ z'(b) \end{pmatrix} = 0,$$
and so
$$\alpha \begin{pmatrix} y(a) & \overline{z}(a) \\ y'(a) & \overline{z'}(a) \end{pmatrix} = -\beta \begin{pmatrix} y(b) & \overline{z}(b) \\ y'(b) & \overline{z'}(b) \end{pmatrix}.$$
Here note that $\overline{\alpha} = \alpha$ and $\overline{\beta} = \beta$. It follows that
$$|\alpha|\tilde{W}(a) = |-\beta|\tilde{W}(b) = |\beta|\tilde{W}(b), \tag{4.9}$$
where $|\alpha|$ and $|\beta|$ are the determinants of the matrices α and β, respectively. Here note that $|-\beta| = |\beta|$ since β is a matrix of 2×2.

Claim that $p(a)\tilde{W}(a) = p(b)\tilde{W}(b)$ if and only if $p(a)|\beta| = p(b)|\alpha|$, so that L is symmetric if and only if $p(a)|\beta| = p(b)|\alpha|$, and the proof is complete.

To this end, suppose that $p(a)\tilde{W}(a) = p(b)\tilde{W}(b)$ holds, where the functions y and z in $D(L)$ can be chosen, so that the corresponding $\tilde{W}(a) \neq 0$. Then by (4.9),
$$p(a)|\beta|\tilde{W}(a) = p(b)|\beta|\tilde{W}(b) = p(b)|\alpha|\tilde{W}(a),$$
and so $p(a)|\beta| = p(b)|\alpha|$, by cancelling $\tilde{W}(a)$ ($\neq 0$). Conversely, suppose that $p(a)|\beta| = p(b)|\alpha|$ holds. Then both $|\alpha|$ and $|\beta|$ are not zero, and, by (4.9),
$$p(a)\tilde{W}(a)|\beta| = p(b)\tilde{W}(a)|\alpha| = p(b)|\beta|\tilde{W}(b),$$

and so $p(a)\tilde{W}(a) = p(b)\tilde{W}(b)$, by cancelling $|\beta| \ (\neq 0)$.

However, if instead,

$$\alpha = \begin{pmatrix} c_1 & c_2 \\ 0 & 0 \end{pmatrix} \quad \text{and} \quad \beta = \begin{pmatrix} 0 & 0 \\ d_1 & d_2 \end{pmatrix},$$

where the complex numbers c_1, c_2, d_1, and d_2 are such that

$$c_1^2 + c_2^2 \neq 0 \quad \text{and} \quad d_1^2 + d_2^2 \neq 0,$$

then it is easy to see

$$\tilde{W}(a) = 0 = \tilde{W}(b),$$

and so L is still symmetric.   □

It follows from the proof of Proposition 1.5 that Proposition 1.5 also holds for a relaxing L with absolutely continuous coefficients:

PROPOSITION 4.14. *Proposition 1.5 continues to hold when the ordinary differential operator $L : D(L) \subset L^2(a, b) \longrightarrow L^2(a, b)$, is defined by*

$$Ly = (py')' + ry$$

for $y \in D(L)$, where the domain $D(L)$ of L is the set of functions $u \in L^2(a, b)$, such that $u(x), u'(x), p(x)$, and $p(x)u'(x)$ are absolutely continuous, and $(pu')'$ lies in $L^2(a, b)$.

PROOF. Here we refer to Chapter 5, Subsection 5.2 for the concepts of Lebesgue integration, almost everywhere, and to Chapter 6, Subsection 5.5, for the concept of absolute continuity and for the well-definedness of L.

Since the integration by parts formula (see [**8**, Page 103]):

$$\int_a^b (F(x)G'(x)\, dx = F(x)G(x)|_{x=a}^b - \int_a^b G(x)F'(x)\, dx$$

for absolutely continuous functions $F(x)$ and $G(x)$ on $[a, b]$, is true and since the Wronskian $W(x)$ is well-defined for the current relaxing L, the proof of Proposition 1.5 goes through, and Proposition 4.14 is proved.   □

The proof of Theorem 2.9 is based on Coddington and Levinson [**5**].

5. An Abstract Result

In this section, we shall present an abstract result, Theorem 5.25, that is close to Corollary 2.12. This theorem states that a nonhomogeneous equation $Ay = f$ in an abstract Hilbert space H is almost solvable if and only if f is orthogonal to the adjoint solutions z of the associated adjoint equation $A^\dagger z = 0$. Here $A^\dagger$ is the corresponding unique adjoint operator, associated with the operator A. We then show that a concrete example, Corollary 5.26, which is close to Corollary 2.12, follows as a special case of the abstract result here.

We prepare some background material first.

5.1. Linear Operators.

DEFINITION 5.1. *A non-empty set X with two operations $+$ and $\cdot$, is called a vector space over the field $\mathbb{C}$, the complex numbers, if the following are satisfied:*

- *$x + y \in X$ for $x, y \in X$.*
- *$\alpha \cdot x = \alpha x \in X$ for $x \in X$ and $\alpha \in \mathbb{C}$.*
- *$x + y = y + x$ for $x, y \in X$.*
- *$x + (y + z) = (x + y) + z$ for $x, y, z \in X$.*
- *There is a unique $0 \in X$, such that $0 + x = x$.*
- *For each $x \in X$, there is a unique $-x \in X$, such that $x + (-x) = 0$.*
- *$\alpha(x + y) = \alpha x + \alpha y$ for $x, y \in X$ and $\alpha \in \mathbb{C}$.*
- *$(\alpha + \beta)x = \alpha x + \beta x$ for $x \in X$ and $\alpha, \beta \in \mathbb{C}$.*
- *$\alpha(\beta x) = (\alpha \beta)x$ for $x \in X$ and $\alpha, \beta \in \mathbb{C}$.*
- *$1 \cdot x = x$ for $x \in X$.*
- *$0 \cdot x = 0 \in X$ for $x \in X$*

DEFINITION 5.2. *A norm $\| \cdot \|$ on a vector space X is a real-valued function, satisfying*

- *$\|x\| \geq 0$ for $x \in X$.*
- *$\|x\| \neq 0$ if $x \in X$ and $x \neq 0$.*
- *$\|\alpha x\| = |\alpha| \|x\|$ for $\alpha \in \mathbb{C}$ and $x \in X$.*
- *$\|x + y\| \leq \|x\| + \|y\|$ for $x, y \in X$.*

DEFINITION 5.3. *A vector space over the field $\mathbb{C}$, on which a norm is defined, is called a normed vector space.*

DEFINITION 5.4. *A normed vector space X with the norm $\|\cdot\|$, is called complete if every Cauchy sequence in X converges in X, that is, if every sequence x_n in X, with the property of $\lim_{n,m \to \infty} \|x_n - x_m\| = 0$, implies $\lim_{n \to \infty} \|x_n - x_0\| = 0$ for some element x_0 in X.*

DEFINITION 5.5. *A complete normed vector space is called a Banach space. For example, the field $\mathbb{C}$ of complex numbers, equipped with the norm of the usual distance function $|\cdot|$, is a Banach space. Here for $z = a + ib \in \mathbb{C}$, where $a, b \in \mathbb{R}$, and $i = \sqrt{-1}$, the complex unit, it is defined that $|z| = \sqrt{a^2 + b^2}$.*

Let $(X, \| \cdot \|_X)$ and $(Y, \| \cdot \|_Y)$ be two Banach spaces, with the norms $\| \cdot \|_X$ and $\| \cdot \|_Y$, respectively, over the field $\mathbb{C}$, the complex numbers. Assume that X, Y contain nonzero elements.

DEFINITION 5.6. *A function $T : D(T) \subset X \longrightarrow Y$, from its domain $D(T)$ to Y, is called a linear operator if*

$$T(x + y) = T(x) + T(y) = Tx + Ty$$

for $x, y \in X$, and if

$$T(\alpha x) = \alpha T x$$

for $x \in X$ and $\alpha \in \mathbb{C}$.
When the domain $D(T) = X$, we say that T is a linear operator from X to Y.

DEFINITION 5.7. *Assume that T is a linear operator from X to Y. T is continuous at $x_0 \in X$ if, for a given $\epsilon > 0$, there is a $\delta > 0$, such that*

$$\|Tx - Tx_0\|_Y < \epsilon$$

for all $x \in X$ with $\|x - x_0\|_X < \delta$.

LEMMA 5.8. *A linear operator T from X to Y is continuous at some $x_0 \in X$ if and only if it is uniformly continuous on X if and ony if there is an $M > 0$, such that*

$$\|Tx\|_Y \leq M\|x\|_X$$

for each $x \in X$.

In either case, we say that T is a continuous linear operator from X to Y. When $Y = \mathbb{C}$, we call T is a continuous linear functional on X.

DEFINITION 5.9. *If T is a continuous linear operator from X to Y, define the norm $\|T\| \equiv \|T\|_{X \to Y}$ of T by*

$$\|T\| \equiv \sup_{\|x\|_X \leq 1} \|Tx\|_Y,$$

LEMMA 5.10. *If T is a continuous linear operator from X to Y, then*

$$\|Tx\|_x \leq \|T\|\|x\|_X$$

and

$$\|T\| = \sup_{\|x\|_X = 1} \|Tx\|_Y \quad or \quad \sup_{x \neq 0} \frac{\|Tx\|_Y}{\|x\|_X}.$$

DEFINITION 5.11. *A vector space Z over the field $\mathbb{C}$, the complex numbers, is an inner product space if, to each pair $\{x, y\}, x, y \in X$, there is associated a complex number (x, y), called the inner product of x with y, such that, for $z \in X$ and $\alpha \in \mathbb{C}$, the following are true:*

$$(x + z, y) = (x, y) + (z, y),$$
$$(y, x) = \overline{(x, y)},$$
$$(\alpha x, y) = \alpha(x, y),$$

and

$$(x, x) \geq 0 \quad and \quad (x, x) \neq 0 \quad if \quad x \neq 0.$$

From this definition, it follows that

$$(x, \alpha y) = \overline{\alpha}(x, y)$$

and

$$(x, y + z) = (x, y) + (x, z).$$

LEMMA 5.12 (Cauchy-Schwarz inequality). *If $(Z, (\ , \ ))$ is an inner product space with the inner product $(\ , \ )$, then*

$$|(x, y)| \leq \sqrt{(x, x)}\sqrt{(y, y)}$$

for $x, y \in Z$.

LEMMA 5.13. *If $(Z, (\ , \ ))$ is an inner product space, then*

$$\|x\| \equiv \sqrt{(x, x)}$$

for $x \in Z$, defines a norm, called inner product norm, on Z and $(Z, \| \cdot \|)$ with the inner product norm $\| \cdot \|$ becomes a normed vector space.

DEFINITION 5.14. *An inner product space that is complete with respect to the inner product norm, is called a Hilbert space. Thus, a Hilbert space is a special Banach space. For Example, $L^2(a, b)$ with the inner product $(f, g) = \int_a^b f(x)\overline{g(x)}\, dx$, where $f, g \in L^2(a, b)$, is a Hilbert space.*

The material in this subsection is taken from Taylor and Lay [**21**].

5.2. Adjoint Operators. Suppose that

$$A : D(A) \subset H \longrightarrow H$$

is a linear operator in a Hilber space H. A linear operator

$$B : D(B) \subset H \longrightarrow H$$

is called adjoint to A if there is a $w \in H$, such that

$$(Ay, z) = (y, w)$$

for all $y \in D(A)$ and $z \in D(B)$. Here (y, w) is the inner product of y with w, in H.

If we take $D(B) = \{0\}$ and $B0 = 0$, then such a B is adjoint to A. In general, there are many B's that are adjoint to A. However, only one unique maximal B exists, which we denote as $A^\dagger$ instead, if A has a domain $D(A)$ that is dense in H. $A^\dagger$ is called the adjoint operator of A.

Here by $A^\dagger$ being maximal, it is meant that any operator B that is adjoint A, is a restriction of $A^\dagger$, that is, $D(A^\dagger) \supset D(B)$ and $A^\dagger z = Bz$ for $z \in D(B)$. In this case, we write $B \subset A^\dagger$.

Here that $D(A)$ is dense in H means that for each $f \in H$, there is a sequence $f_n \in D(A)$ that converges to f, i.e., $\lim_{n \to \infty} \|f_n - f\| = \lim_{n \to \infty} \sqrt{(f_n - f, f_n - f)} = 0$.

The adjoint operator $A^\dagger$ of A is defined as follows.

DEFINITION 5.15. *Suppose that*

$$A : D(A) \subset H \longrightarrow H$$

is a linear operator, whose domain $D(A)$ is dense in the Hilbert space H. An element $z \in H$ belongs to the domain $D(A^\dagger)$ of an operator

$$A^\dagger : D(A^\dagger) \subset H \longrightarrow H$$

if there is an $f \in H$, such that

$$(Ay, z) = (y, f)$$

for all $y \in D(A)$. The linear operator $A^\dagger$ defined by $A^\dagger z = f$, is called the adjoint operator of A.

REMARK 5.16. *We observe the following:*

- *The above f is unique, so that $A^\dagger$ is well-defined. To see this, let $f, g \in H$ be such that*

$$(Ay, z) = (y, f) = (y, g)$$

  for all $y \in D(A)$. This implies that

$$(y, f - g) = 0$$

  for $y \in D(A)$. We want to show that $f = g$. Let $w \in H$, arbitrarily. Since $D(A)$ is dense in H, there is a sequence $w_n \in D(A)$, such that

$$\lim_{n \to \infty} \|w_n - w\| = 0.$$

  Using $(w_n, f - g) = 0$ and the Cauchy-Schwarz inequality in Lemma 5.12, we have

$$|(w, f - g)| \leq |(w - w_n, f - g)| + |(w_n, f - g)|$$
$$\leq \|w - w_n\| \|f - g\| + 0$$
$$\longrightarrow 0,$$

and so $(w, f - g) = 0$. Since $w \in H$ is arbitrary, we can choose $w = f - h$ to obtain $\|f - g\|^2 = 0$. Thus $f = g$.

- $A^\dagger$ is linear, since, for $\alpha \in \mathbb{C}$, and for $z_1, z_2 \in D(A^\dagger)$ with $f_1 = A^\dagger z_1$ and $f_2 = A^\dagger z_2$, we have

$$
\begin{aligned}
(Ay, \alpha z_1) &= \overline{\alpha}(Ay, z_1) \\
&= \overline{\alpha}(y, f_1) \\
&= \overline{\alpha}(y, A^\dagger z_1) \\
&= (y, \alpha A^\dagger z_1,
\end{aligned}
$$

and so $\alpha z_1 \in D(A^\dagger)$ and $A^\dagger(\alpha z_1) = \alpha A^\dagger z_1$. Also we have

$$
\begin{aligned}
(Ay, z_1 + z_2) &= (Ay, z_1) + (Ay, z_2) \\
&= (y, f_1) + (y, f_2) \\
&= (y, A^\dagger z_1) + (y, A^\dagger z_2) \\
&= (y, A^\dagger z_1 + A^\dagger z_2),
\end{aligned}
$$

and so $(z_1 + z_2) \in D(A^\dagger)$ and $A^\dagger(z_1 + z_2) = A^\dagger z_1 + A^\dagger z_2$.

- $A^\dagger$ is maximal. To see this, let B be a linear operator that is adjoint to A, and let $z \in D(B)$. Then $(Ay, z) = (y, Bz)$ for $y \in D(A)$. Since $f \equiv Bz \in H$, it follows from the definition of $A^\dagger$ that $z \in D(A^\dagger)$ and $A^\dagger z = Bz$.

DEFINITION 5.17. *The null space $Null(A^\dagger)$ of $A^\dagger$ is defined by*

$$
Null(A^\dagger) = \{z \in D(A^\dagger) : A^\dagger z = 0\}.
$$

DEFINITION 5.18. *The range $Ran(A)$ of A is defined by*

$$
Ran(A) = \{w \in H : w = Ay \quad \text{for some} \quad y \in D(A)\}.
$$

DEFINITION 5.19. *The annihilator or the orthogonal complement $Ran(A)^\perp$ of $Ran(A)$ is defined by*

$$
Ran(A)^\perp = \{w \in H : (Ay, w) = 0 \quad \text{for all} \quad y \in D(A)\}.
$$

Here is a relation between A and $A^\dagger$:

PROPOSITION 5.20. *The null space $Null(A^\dagger)$ of $A^\dagger$ equals the orthogonal complement $Ran(A)^\perp$ of the range $Ran(A)$ of A.*

PROOF. We divide the proof into two steps.

Step 1. Let $z \in Null(A^\dagger)$. For $y \in D(A)$, we have

$$
0 = (y, A^\dagger z) = (Ay, z),
$$

and so $z \in Ran(A)^\perp$. Thus

$$
Null(A^\dagger) \subset Ran(A)^\perp.
$$

Step 2. Let $z \in Ran(A)^\perp$. Then for $y \in D(A)$, we have $0 = (Ay, z)$, and so there is an $f = 0 \in H$, such that $(y, f) = (y, 0) = 0 = (Ay, z)$. Hence by the definition of $A^\dagger$, we have $z \in D(A^\dagger)$ and $A^\dagger z = f = 0$. It follows that

$$
0 = (Ay, z) = (y, A^\dagger z)
$$

for all $y \in D(A)$. We claim that this continues to hold for all $y \in H$, so that

$$\|A^\dagger z\|^2 = 0,$$

by choosing $y = A^\dagger z$. From this, $z \in Null(A^\dagger)$ follows, and so

$$Ran(A)^\perp \subset Null(A^\dagger).$$

To prove the claim, let y_n be a sequence in $D(A)$, that converges to y, for which

$$\lim_{n \to \infty} \|y_n - y\| = 0.$$

This is possible, since $D(A)$ is dense in H. Then using the Cauchy-Schwarz inequality in Lemma 5.12, we have

$$|(y, A^\dagger z)| = |(y - y_n, A^\dagger z) + (y_n, A^\dagger z)|$$
$$\leq \|y - y_n\|\|A^\dagger z\| + 0$$
$$\longrightarrow 0,$$

and so $(y, A^\dagger z) = 0$, proving the claim. $\qquad\qquad\square$

We also need the following Definition 5.21 and Theorem 5.22.

DEFINITION 5.21. *The annihilator or the orthogonal complement $(Ran(A)^\perp)^\perp$ of $Ran(A)^\perp$ is defined by*

$$(Ran(A)^\perp)^\perp = \{f \in H : (f, w) = 0 \quad \text{for all} \quad w \in Ran(A)^\perp\}.$$

THEOREM 5.22 ([**16**], Page 136). *The following*

$$(Ran(A)^\perp)^\perp = \overline{Ran(A)}$$

is true, where $\overline{Ran(A)}$ is the closure of $Ran(A)$.

PROOF. We divide the proof into four steps.

Step 1. Claim that $(Ran(A)^\perp)^\perp$ is closed. To see this, let $y_n \in (Ran(A)^\perp)^\perp$ converge to y. Then for each $w \in Ran(A)^\perp$, we have

$$(y_n, w) = 0.$$

But by the Cauchy-Schwarz inequality in Lemma 5.12, we have

$$|(y_n - y, w)| \leq \|y_n - y\|\|w\|$$
$$\longrightarrow 0,$$

and so $(y_n, w) = 0$ converges to $(y, w) = 0$ as $n \longrightarrow \infty$. Thus $y \in (Ran(A)^\perp)^\perp$, and the claim is proved.

Step 2. Claim that $Ran(A) \subset (Ran(A)^\perp)^\perp$, and so $\overline{Ran(A)} \subset \overline{(Ran(A)^\perp)^\perp} = (Ran(A)^\perp)^\perp$ by Step 1. To see this, let $y \in Ran(A)$. Then for each $w \in Ran(A)^\perp$, we have

$$(y, w) = 0,$$

and so $y \in (Ran(A)^\perp)^\perp$. Thus the claim is proved.

Step 3. Claim that $(\overline{Ran(A)})^\perp = Ran(A)^\perp$. To see this, let $y \in (\overline{Ran(A)})^\perp$. Then

$$(y, w) = 0$$

for each $w \in \overline{Ran(A)} \supset Ran(A)$, and so $y \in Ran(A)^\perp$.

On the other hand, let $y \in Ran(A)^\perp$ and $w \in \overline{Ran(A)}$. Then there is a sequence $w_n \in Ran(A)$ that converges to w, and

$$(f, w_n) = 0$$

is true. This converges to $(f, w) = 0$, and so $f \in (\overline{Ran(A)})^\perp$. This is because

$$|(f, w - w_n)| \le \|f\| \|w - w_n\|$$
$$\longrightarrow 0$$

by the Cauchy-Schwarz inequality in Lemma 5.12.

Step 4. We utilize the following Riesz theorem and a corollary of Hahn-Banach theorem:

THEOREM 5.23 (Riesz Theorem). ([16, Pages 252–253]) *To each continuous linear functional x^* on H, there corresponds a unique $f \in H$, such that $X^*(x) = (x, f)$ and $\|x^*\|_{H \to H} = \|f\|$.*

THEOREM 5.24 (A Corollary of Hahn-Banach Theorem). ([16, Page 135]) *Let M be a proper closed subspace of H, for which $M \ne H$. Let x_0 be an element in H, that lies at a distance $d > 0$ from M. Then there is a continuous linear functional x^* on H, such that*

$$x^*(x_0) = 1, x^*(x) = 0 \quad for \quad x \in M, \quad and \quad \|x^*\|_{H \to H} = \frac{1}{d}.$$

We claim that $(Ran(A)^\perp)^\perp \subset \overline{Ran(A)}$, which, together with $\overline{Ran(A)} \subset (Ran(A)^\perp)^\perp$ in Step 2, completes the proof of the theorem. The claim is clearly true if $\overline{Ran(A)} = H$ and so, to prove the claim, suppose that $\overline{Ran(A)} \ne H$ and let $y \in H$ not belong to $\overline{Ran(A)}$. We need to show that y does not belong to $(Ran(A)^\perp)^\perp$.

Since $d \equiv dist(\overline{Ran(A)}, y) > 0$, the distance from y to $\overline{Ran(A)}$ being strictly greater than zero, and since $\overline{Ran(A)} \ne H$, $\overline{Ran(A)}$ being a proper subset of H, it follows from the above corollary to the Hahn-Banach theorem that there is a continuous linear functional x^* on H, such that

$$x^*(y) = 1$$

and $x^*(w) = 0$ for each $w \in \overline{Ran(A)}$. By the above Riesz theorem, there is a unique $f \in H$, such that

$$(f, y) = x^*(y) = 1$$

and $(f, w) = x^*(w) = 0$ for each $w \in \overline{Ran(A)}$. Thus $f \in (\overline{Ran(A)})^\perp$, and y does not belong to $((\overline{Ran(A)})^\perp)^\perp$ since $(f, y) = 1 \ne 0$. But

$$((\overline{Ran(A)})^\perp)^\perp = (Ran(A)^\perp)^\perp$$

by Step 3, and so the claim is proved. □

The material in this subsection is taken from Kato [16].

5.3. Solvability Condition. We now present here the abstract result that is close to Corollary 2.12.

THEOREM 5.25. *The following*

$$(Null(A^\dagger)^\perp = \overline{Ran(A)}$$

is true.

PROOF. The proof is completed by using Proposition 5.20 and Theorem 5.22.

□

From Theorem 5.25, a concrete example, close to Corollary 2.12, follows:

COROLLARY 5.26. *By choosing $A = L$, the differentiable operator defined in Section 1, to which corresponds the adjoint operator $A^\dagger = L^\dagger$ defined in Section 1 also, we have the following (i) and (ii) , close to Corollary 2.12:*

(i) *If $Ly = f$ for $f \in C[a, b]$ has a solution, then f is orthogonal to each adjoint solution z of $L^\dagger z = 0$, that is, $(f, z) = 0$ is true.*

(ii) *If $f \in C[a, b]$ is orthogonal to each adjoint solution z of $L^\dagger z = 0$, then f belongs to the closure $\overline{Ran(L)}$ of the range $Ran(L)$ of L; that is, there is a sequence $y_n \in D(L)$, such that Ly_n converges to f:*

$$\lim_{n \to \infty} \|Ly_n - f\| = 0.$$

PROOF. We divide the proof into three steps.

Step 1. We claim that $D(A) = D(L)$ is dense in $L^2(a, b)$, so that the adjoint operator $A^\dagger = L^\dagger$ is well-defined. To see this, we observe that Adams [**1**, Page 31], the set

$$C_0^\infty[a, b] \equiv \{u \in C^2[a, b] : u \quad \text{has compact support in but not equal to} \quad [a, b]\}$$

is dense in $(L^2(a, b), \|\cdot\|_2)$. That means that for each $f \in (L^2(a, b), \|\cdot\|_2)$, there is a sequence $f_n \in C_0^\infty[a, b]$ of functions, such that

$$\lim_{n \to \infty} \|f_n - f\|_2 = 0.$$

Here $C^\infty[a, b]$ is the set of functions on $[a, b]$ that are infinitely many times continuously differentiable, and that u has compact support in but not equal to $[a, b]$ means that the closure of the set of $x \in [a, b]$ on which $u(x)$ is not zero is a proper compact subset of $[a, b]$. Note that $C_0^\infty[a, b]$ is the same as the set of functions in $C^\infty[a, b]$ that vanish outside a proper compact subset of $[a, b]$.

Since $D(L)$ contains $C_0^2[a, b] \supset C_0^\infty[a, b]$, we have $D(L)$ is dense in $(L^2(a, b), \|\cdot\|_w)$.

Step 2. Suppose that $Ly = f$ for $f \in C[a, b]$ has a solution. Then $f \in Ran(A) = Ran(L) \subset \overline{Ran(L)}$, and so $f \in (Null(L^\dagger)^\perp$ clearly, by Theorem 5.25. Thus $(f, z) = 0$ is true for each solution z of the adjoint equation $L^\dagger z = 0$.

Step 3. Let $f \in C[a, b]$ be such that $(f, z) = 0$ for all solutions z to the adjoint equation $L^\dagger z = 0$. Thus $f \in (Null(L^\dagger)^\perp$. By Theorem 5.25, we have $f \in \overline{Ran(L)}$, and so there is a sequence $y_n \in D(L)$, such that $\lim_{n \to \infty} Ly_n = f$, or precisely,

$$\lim_{n \to \infty} \|Ly_n - f\| = 0.$$

□

REMARK 5.27. It is true that $(y_n, L^\dagger z)$ for $z \in D(L^\dagger)$, converges to (f, z). This follows from:

$$|(y_n, L^\dagger z) - (f, z)| = |(Ly_n - f, z)$$
$$\leq \|Ly_n - f\| \|z\| \longrightarrow 0,$$

where the Cauchy-Schwarz inequality in Lemma 5.12 was used.

6. Problems

Do as in Example 3.1, for the following boundary value problems:

1.
$$2y'' + 5y' - 3y = h(x), \quad 0 < x < 1,$$
$$y'(0) = a, \tag{6.1}$$
$$y'(1) = b,$$

where $h(x)$ is a continuous function on $[0,1]$, and a and b are given two constants.

2.
$$2y'' + 5y' - 3y = h(x), \quad 0 < x < 1,$$
$$y'(0) = a \in \mathbb{C}, \tag{6.2}$$

where $h(x)$ is a continuous function on $[0,1]$.

3.
$$2y'' + 5y' - 3y = h(x), \quad 0 < x < 1,$$
$$y'(0) = a,$$
$$y'(1) = b, \tag{6.3}$$
$$y(0) = c,$$

where a, b, and $c \in \mathbb{C}$, and $h(x)$ is a continuous function on $[0,1]$.

4.
$$2y'' + 5y' - 3y = h(x), \quad 0 < x < 1,$$
$$y'(0) = a,$$
$$y'(1) = b, \tag{6.4}$$
$$y(0) = c,$$
$$y(1) = d,$$

where a, b, c, and $d \in \mathbb{C}$, and $h(x)$ is a continuous function on $[0,1]$.

5.
$$x^2 y'' - 4xy' + 4y = h(x), \quad 1 < x < 2,$$
$$y(1) - y'(1) = a, \tag{6.5}$$
$$y(2) + y'(2) = b,$$

where $h(x)$ is a continuous function on $[1,2]$, and a and b are given two constants.

6.
$$x^2 y'' - 4xy' + 4y = h(x), \quad 1 < x < 2,$$
$$y(1) - y'(1) = a, \tag{6.6}$$

where $h(x)$ is a continuous function on $[1,2]$, and a is a given constant.

7.
$$x^2 y'' - 4xy' + 4y = h(x), \quad 1 < x < 2,$$
$$y(1) - y'(1) = a,$$
$$y(2) + y'(2) = b, \tag{6.7}$$
$$y(1) = c,$$

where $h(x)$ is a continuous function on $[1,2]$, and a, b, and c are given three constants.

8.

$$x^2 y'' - 4xy' + 4y = h(x), \quad 1 < x < 2,$$
$$y(1) - y'(1) = a,$$
$$y(2) + y'(2) = b, \tag{6.8}$$
$$y(1) = c,$$
$$y'(2) = d,$$

where $h(x)$ is a continuous function on $[1, 2]$, and $a, b, c,$ and d are given four constants.

9. Show that the corresponding operator L for the boundary value problem:

$$[(2 - (\sin(x))y']' + x(x - 3)y = h(x), \quad 0 < x < \pi,$$
$$2y(0) + \frac{5}{7}y'(0) + y(\pi) + 2y'(\pi) = 0$$
$$7y(0) + 3y'(0) + 6y(\pi) + 13y'(\pi) = 0,$$

is symmetric. Here $h(x)$ is a continuous function on $[0, \pi]$.

10. Prove Corollary 2.13 by using Lemma 4.9, equation (4.5), and the fact from linear algebra that the equation $A\vec{x} = \vec{b}$ has a unique solution $\vec{x}$ for $\vec{b} \in \mathbb{C}^2$ being given, if A is a square matrix of 2×2 with rank 2.

7. Solutions

1. Here

$$B = \begin{pmatrix} 0 & 1 & 0 & 0 \\ 0 & 0 & 0 & 1 \end{pmatrix},$$

$m = 2 = n$, and $\vec{\gamma} = \begin{pmatrix} a \\ b \end{pmatrix}$.

Let

$$B_c = \begin{pmatrix} 1 & 0 & 0 & 0 \\ 0 & 0 & 1 & 0 \end{pmatrix}, \hat{B}_c = \begin{pmatrix} 1 & 2 & 0 & 0 \\ 0 & 0 & 1 & 1 \end{pmatrix}.$$

Let $F = \begin{pmatrix} B \\ B_c \end{pmatrix}$. Applying row operations to compute the inverse F^{-1} of F, we have

$$F^{-1} = \begin{pmatrix} 0 & 0 & 1 & 0 \\ 1 & 0 & 0 & 0 \\ 0 & 0 & 0 & 1 \\ 0 & 1 & 0 & 0 \end{pmatrix}.$$

Similarly, the inverse $\hat{F}^{-1}$ of $\hat{F}$, where $\hat{F} = \begin{pmatrix} B \\ \hat{B}_c \end{pmatrix}$, is

$$\hat{F}^{-1} = \begin{pmatrix} -2 & 0 & 1 & 0 \\ 1 & 0 & 0 & 0 \\ 0 & -1 & 0 & 1 \\ 0 & 1 & 0 & 0 \end{pmatrix}.$$

Using Lemma 2.3, we have

$$S = \begin{pmatrix} -5 & -2 & 0 & 0 \\ 2 & 0 & 0 & 0 \\ 0 & 0 & 5 & 2 \\ 0 & 0 & -2 & 0 \end{pmatrix};$$

$$SF^{-1} = \begin{pmatrix} -2 & 0 & -5 & 0 \\ 0 & 0 & 2 & 0 \\ 0 & 2 & 0 & 5 \\ 0 & 0 & 0 & -2 \end{pmatrix};$$

$$\begin{pmatrix} B_c^\dagger \\ B^\dagger \end{pmatrix} = G = (SF^{-1})^* = \begin{pmatrix} -2 & 0 & 0 & 0 \\ 0 & 0 & 2 & 0 \\ -5 & 2 & 0 & 0 \\ 0 & 0 & 5 & -2 \end{pmatrix}, \quad \text{for which}$$

$$B_c^\dagger = \begin{pmatrix} -2 & 0 & 0 & 0 \\ 0 & 0 & 2 & 0 \end{pmatrix} \quad \text{and}$$

$$B^\dagger = \begin{pmatrix} -5 & 2 & 0 & 0 \\ 0 & 0 & 5 & -2 \end{pmatrix};$$

$$S\hat{F}^{-1} = \begin{pmatrix} 8 & 0 & -5 & 0 \\ -4 & 0 & 2 & 0 \\ 0 & -3 & 0 & 5 \\ 0 & 2 & 0 & -2 \end{pmatrix};$$

$$\begin{pmatrix} \hat{B}_c^\dagger \\ \hat{B}^\dagger \end{pmatrix} = \hat{G} = (S\hat{F}^{-1})^* = \begin{pmatrix} 8 & -4 & 0 & 0 \\ 0 & 0 & -3 & 2 \\ -5 & 2 & 0 & 0 \\ 0 & 0 & 5 & -2 \end{pmatrix}, \quad \text{for which}$$

$$\hat{B}_c^\dagger = \begin{pmatrix} 8 & -4 & 0 & 0 \\ 0 & 0 & -3 & 2 \end{pmatrix} \quad \text{and}$$

$$\hat{B}^\dagger = \begin{pmatrix} -5 & 2 & 0 & 0 \\ 0 & 0 & 5 & -2 \end{pmatrix}.$$

Using the Definition 2.6 of L and $L^\dagger$, we have

$$Ly = 2y'' + 5y' - 3y,$$
$$D(L) = \{y \in C^2[0,1] : y'(0) = 0, y'(1) = 0\};$$
$$L^\dagger z = 2z'' - 5z' - 3z,$$
$$D(L^\dagger) = \{z \in C^2[0,1] : B^\dagger \vec{z} = \vec{0}_2, \quad \text{or equivalently,}$$
$$\hat{B}^\dagger \vec{z} = \vec{0}_2, \quad \text{or equivalently,}$$
$$5z(0) - 2z'(0) = 0, 5z(1) - 2z'(1) = 0\}.$$

Thus the problem π_2 becomes

$$\pi_2 : \quad 2y'' + 5y' - 3y = 0,$$
$$y'(0) = 0, y'(1) = 0,$$

and the adjoint problem $\pi_2^\dagger$ becomes

$$\pi_2^\dagger: \quad 2z'' - 5z' - 3z = 0,$$
$$5z(0) - 2z'(0) = 0, 5z(1) - 2z'(1) = 0.$$

It follows that L is not symmetric, since

$$L^\dagger u = Lu$$

does not hold for $u \in D(L)$.

Here note that we can also obtain the $L^\dagger$, by using Definition 1.2. The expression

$$[p\bar{z}y' + q\bar{z}y - p\bar{z}'y - p'\bar{z}y]|_a^b = 0$$

for all $y \in D(L)$, in Definition 1.2, gives

$$[5\bar{z}(1) - 2\bar{z}'(1)]y(1) - [5\bar{z}(0) - 2\bar{z}'(0)]y(0) = 0.$$

Here $y \in C^2[0, \pi]$ and $y'(0) = 0 = y'(1)$. Thus choosing $y(1) = 1$ and $y(0) = 0$ leads to $5\bar{z}(1) - 2\bar{z}'(1) = 0$. But choosing $y(1) = 0$ and $y(0) = 1$ leads to $5\bar{z}(0) - 2\bar{z}'(0) = 0$. It follows that $D(L^\dagger) = \{z \in C^2[0, \pi] : 5z(0) - 2z'(0) = 0, 5z(1) - 2z'(1) = 0\}$.

Try $y = e^{rx}$ in the problem π_2, and we get the auxiliary equation $2r^2 + 5r - 3 = 0$. The two roots for this equation are: $r = \frac{1}{2}, -3$, and so the general solution for y is:

$$y = c_1 e^{\frac{1}{2}x} + c_2 e^{-3x}.$$

The boundary conditions $y'(0) = 0 = y'(1)$ give:

$$0 = c_1 = c_2,$$

and so $y \equiv 0$. Thus the problem π_2 has exactly 0 linearly independent solution. Similar calculations show that $z \equiv 0$, and so the adjoint problem $\pi_2^\dagger$ has also exactly 0 linearly independent solution. These two numbers 0 and 0 satisfy the relation:

$$0 = 0 + m - n = 0 + 2 - 2,$$

as desired.

Compute

$$B_c^\dagger \bar{z} = \begin{pmatrix} -2 & 0 & 0 & 0 \\ 0 & 0 & 2 & 0 \end{pmatrix} \begin{pmatrix} 0 \\ 0 \\ 0 \\ 0 \end{pmatrix}$$

$$= \begin{pmatrix} 0 \\ 0 \end{pmatrix},$$

$$\hat{B}_c^\dagger \bar{z} = \begin{pmatrix} 8 & -4 & 0 & 0 \\ 0 & 0 & -3 & 2 \end{pmatrix} \begin{pmatrix} 0 \\ 0 \\ 0 \\ 0 \end{pmatrix}$$

$$= \begin{pmatrix} 0 \\ 0 \end{pmatrix},$$

which are desired results.

By Theorem 2.9, (6.1) has a solution if and only if h satisfies

$$(h, z) = \int_0^1 h(x) 0 \, dx = 0$$

$$= (\vec{\gamma}, B_c^{\dagger}\tilde{z})$$

$$= (\begin{pmatrix} a \\ b \end{pmatrix}, \begin{pmatrix} 0 \\ 0 \end{pmatrix})$$

$$= 0.$$

But this is true for all h, and thus (6.1) has a solution for all given h.

2. Here

$$B = \begin{pmatrix} 0 & 1 & 0 & 0 \end{pmatrix},$$

$m = 1, n = 2$, and $\vec{\gamma} = a$.

Let

$$B_c = \begin{pmatrix} 1 & 0 & 0 & 0 \\ 0 & 0 & 1 & 0 \\ 0 & 0 & 0 & 1 \end{pmatrix}, \hat{B}_c = \begin{pmatrix} 1 & 1 & 0 & 0 \\ 0 & 0 & 1 & 2 \\ 0 & 2 & 1 & 1 \end{pmatrix}.$$

Let $F = \begin{pmatrix} B \\ B_c \end{pmatrix}$. Applying row operations to compute the inverse F^{-1} of F, we have

$$F^{-1} = \begin{pmatrix} 0 & 1 & 0 & 0 \\ 1 & 0 & 0 & 0 \\ 0 & 0 & 1 & 0 \\ 0 & 0 & 0 & 1 \end{pmatrix}.$$

Similarly, the inverse $\hat{F}^{-1}$ of $\hat{F}$, where $\hat{F} = \begin{pmatrix} B \\ \hat{B}_c \end{pmatrix}$, is

$$\hat{F}^{-1} = \begin{pmatrix} -1 & 1 & 0 & 0 \\ 1 & 0 & 0 & 0 \\ -4 & 0 & -1 & 2 \\ 2 & 0 & 1 & -1 \end{pmatrix}.$$

Using Lemma 2.3, we have

$$S = \begin{pmatrix} -5 & -2 & 0 & 0 \\ 2 & 0 & 0 & 0 \\ 0 & 0 & 5 & 2 \\ 0 & 0 & -2 & 0 \end{pmatrix};$$

$$SF^{-1} = \begin{pmatrix} -2 & -5 & 0 & 0 \\ 0 & 2 & 0 & 0 \\ 0 & 0 & 5 & 2 \\ 0 & 0 & -2 & 0 \end{pmatrix};$$

$$\begin{pmatrix} B_c^{\dagger} \\ B^{\dagger} \end{pmatrix} = G = (SF^{-1})^* = \begin{pmatrix} -2 & 0 & 0 & 0 \\ -5 & 2 & 0 & 0 \\ 0 & 0 & 5 & -2 \\ 0 & 0 & 2 & 0 \end{pmatrix}, \quad \text{for which}$$

$$B_c^{\dagger} = \begin{pmatrix} -2 & 0 & 0 & 0 \\ -5 & 2 & 0 & 0 \end{pmatrix} \quad \text{and}$$

$$B^{\dagger} = \begin{pmatrix} 0 & 0 & 5 & -2 \\ 0 & 0 & 2 & 0 \end{pmatrix};$$

$$S\hat{F}^{-1} = \begin{pmatrix} 3 & -5 & 0 & 0 \\ -2 & 2 & 0 & 0 \\ -16 & 0 & -3 & 8 \\ 8 & 0 & 2 & -4 \end{pmatrix};$$

$$\begin{pmatrix} \hat{B}_c^{\dagger} \\ \hat{B}^{\dagger} \end{pmatrix} = \hat{G} = (S\hat{F}^{-1})^* = \begin{pmatrix} 3 & -2 & -16 & 8 \\ -5 & 2 & 0 & 0 \\ 0 & 0 & -3 & 2 \\ 0 & 0 & 8 & -4 \end{pmatrix}, \quad \text{for which}$$

$$\hat{B}_c^{\dagger} = \begin{pmatrix} 3 & -2 & -16 & 8 \\ -5 & 2 & 0 & 0 \end{pmatrix} \quad \text{and}$$

$$\hat{B}^{\dagger} = \begin{pmatrix} 0 & 0 & -3 & 2 \\ 0 & 0 & 8 & -4 \end{pmatrix}.$$

Using the Definition 2.6 of L and $L^{\dagger}$, we have

$$Ly = 2y'' + 5y' - 3y,$$
$$D(L) = \{y \in C^2[0,1] : y'(0) = 0\};$$
$$L^{\dagger}z = 2z'' - 5z' - 3z,$$
$$D(L^{\dagger}) = \{z \in C^2[0,1] : \hat{B}^{\dagger}\vec{z} = \vec{0}_3, \quad \text{or equivalently,}$$
$$\hat{B}^{\dagger}\vec{z} = \vec{0}_3, \quad \text{or equivalently,}$$
$$-5z(0) + 2z'(0) = 0, 5z(1) - 2z'(1) = 0, z(1) = 0\}.$$

Thus the problem π_1 becomes

$$\pi_1: \quad 2y'' + 5y' - 3y = 0,$$
$$y'(0) = 0,$$

and the adjoint problem $\pi_1^{\dagger}$ becomes

$$\pi_1^{\dagger}: \quad 2z'' - 5z' - 3z = 0,$$
$$5z(0) - 2z'(0) = 0, 5z(1) - 2z'(1) = 0,$$
$$z(1) = 0.$$

It follows that L is not symmetric, since

$$L^{\dagger}u = Lu$$

does not hold for $u \in D(L)$.

Here note that we can also obtain the $L^{\dagger}$, by using Definition 1.2. The expression

$$[p\bar{z}y' + q\bar{z}y - p\bar{z}'y - p'\bar{z}y]\big|_a^b = 0$$

for all $y \in D(L)$, in Definition 1.2, gives

$$2\bar{z}(1)y'(1) + [5\bar{z}(1) - 2\bar{z}'(1)]y(1) - [5\bar{z}(0) - 2\bar{z}'(0)]y(0) = 0.$$

Here $y \in C^2[0,\pi]$ and $y'(0) = 0$. Thus choosing $y(1) = 1$ and $y(0) = 0 = y'(1)$ leads to $5\bar{z}(1) - 2\bar{z}'(1) = 0$. But choosing $y(1) = 0 = y'(1)$ and $y(0) = 1$ leads to $5\bar{z}(0) - 2\bar{z}(0) = 0$. Furthermore, choosing $y'(1) = 1$ and $y(1) = 0 = y(0)$ leads to $\bar{z}(1) = 0$. It follows that

$$D(L^{\dagger}) = \{z \in C^2[0,\pi] : 5z(0) - 2z'(0) = 0, 5z(1) - 2z'(1) = 0, z(1) = 0\}.$$

Try $y = e^{rx}$ in the problem π_2, and we get the auxiliary equation $2r^2 + 5r - 3 = 0$. The two roots for this equation are: $r = \frac{1}{2}, -3$, and so the general solution for y is:

$$y = c_1 e^{\frac{1}{2}x} + c_2 e^{-3x}.$$

The boundary conditions $y'(0) = 0$ give:

$$0 = \frac{1}{2}c_1 - 3c_2,$$

and so $c_1 = 6c_2$. Hence $y = c_2(6e^{\frac{-1}{2}x} + e^{-3x})$. Thus the problem π_2 has exactly 1 linearly independent solution. Similar calculations show that $z \equiv 0$, and so the adjoint problem $\pi_2^\dagger$ has also exactly 0 linearly independent solution. These two numbers 1 and 0 satisfy the relation:

$$0 = 1 + m - n = 1 + 1 - 2,$$

as desired.

Compute

$$B_c^\dagger \tilde{z} = \begin{pmatrix} -2 & 0 & 0 & 0 \end{pmatrix} \begin{pmatrix} 0 \\ 0 \\ 0 \\ 0 \end{pmatrix}$$

$$= 0,$$

$$\hat{B}_c^\dagger \tilde{z} = \begin{pmatrix} 3 & -2 & -16 & 8 \end{pmatrix} \begin{pmatrix} 0 \\ 0 \\ 0 \\ 0 \end{pmatrix}$$

$$= 0,$$

which are desired results.

By Theorem 2.9, (6.2) has a solution if and only if h satisfies

$$(h, z) = \int_0^1 h(x) 0 \, dx = 0$$
$$= (\vec{\gamma}, B_c^\dagger \tilde{z})$$
$$= (a, 0)$$
$$= 0.$$

But this is true for all h, and thus (6.2) has a solution for all given h.

3. Here

$$B = \begin{pmatrix} 0 & 1 & 0 & 0 \\ 0 & 0 & 0 & 1 \\ 1 & 0 & 0 & 0 \end{pmatrix},$$

$m = 3, n = 2$, and $\vec{\gamma} = \begin{pmatrix} a \\ b \\ c \end{pmatrix}$.

Let

$$B_c = \begin{pmatrix} 0 & 0 & 1 & 0 \end{pmatrix}, \hat{B}_c = \begin{pmatrix} 1 & 2 & 2 & 1 \end{pmatrix}.$$

Let $F = \begin{pmatrix} B \\ B_c \end{pmatrix}$. Applying row operations to compute the inverse F^{-1} of F, we have

$$F^{-1} = \begin{pmatrix} 0 & 0 & 1 & 0 \\ 1 & 0 & 0 & 0 \\ 0 & 0 & 0 & 1 \\ 0 & 1 & 0 & 0 \end{pmatrix}.$$

Similarly, the inverse $\hat{F}^{-1}$ of $\hat{F}$, where $\hat{F} = \begin{pmatrix} B \\ \hat{B}_c \end{pmatrix}$, is

$$\hat{F}^{-1} = \begin{pmatrix} 0 & 0 & 1 & 0 \\ 1 & 0 & 0 & 0 \\ -1 & -0.5 & -0.5 & 0.5 \\ 0 & 1 & 0 & 0 \end{pmatrix}.$$

Using Lemma 2.3, we have

$$S = \begin{pmatrix} -5 & -2 & 0 & 0 \\ 2 & 0 & 0 & 0 \\ 0 & 0 & 5 & 2 \\ 0 & 0 & -2 & 0 \end{pmatrix};$$

$$SF^{-1} = \begin{pmatrix} -2 & 0 & -5 & 0 \\ 0 & 0 & 2 & 0 \\ 0 & 2 & 0 & 5 \\ 0 & 0 & 0 & -2 \end{pmatrix};$$

$$\begin{pmatrix} B_c^\dagger \\ B^\dagger \end{pmatrix} = G = (SF^{-1})^* = \begin{pmatrix} -2 & 0 & 0 & 0 \\ 0 & 0 & 2 & 0 \\ -5 & 2 & 0 & 0 \\ 0 & 0 & 5 & -2 \end{pmatrix}, \quad \text{for which}$$

$$B_c^\dagger = \begin{pmatrix} -2 & 0 & 0 & 0 \\ 0 & 0 & 2 & 0 \end{pmatrix} \quad \text{and}$$

$$B^\dagger = \begin{pmatrix} -5 & 2 & 0 & 0 \\ 0 & 0 & 5 & -2 \end{pmatrix};$$

$$S\hat{F}^{-1} = \begin{pmatrix} -2 & 0 & -5 & 0 \\ 0 & 0 & 2 & 0 \\ -5 & -0.5 & -2.5 & 2.5 \\ 2 & 1 & 1 & -1 \end{pmatrix};$$

$$\begin{pmatrix} \hat{B}_c^\dagger \\ \hat{B}^\dagger \end{pmatrix} = \hat{G} = (S\hat{F}^{-1})^* = \begin{pmatrix} -2 & 0 & -5 & 2 \\ 0 & 0 & -0.5 & 1 \\ -5 & 2 & -2.5 & 1 \\ 0 & 0 & 2.5 & -1 \end{pmatrix}, \quad \text{for which}$$

$$\hat{B}_c^\dagger = \begin{pmatrix} -2 & 0 & -5 & 2 \\ 0 & 0 & -0.5 & 1 \\ -5 & 2 & -2.5 & 1 \end{pmatrix} \quad \text{and}$$

$$\hat{B}^\dagger = \begin{pmatrix} 0 & 0 & 2.5 & -1 \end{pmatrix}.$$

Using the Definition 2.6 of L and $L^\dagger$, we have

$$Ly = 2y'' + 5y' - 3y,$$

$$D(L) = \{y \in C^2[0,1] : y'(0) = 0, y'(1) = 0, y(0) = 0\};$$
$$L^\dagger z = 2z'' - 5z' - 3z,$$
$$D(L^\dagger) = \{z \in C^2[0,1] : B^\dagger \vec{z} = \vec{0}_2, \quad \text{or equivalently,}$$
$$\hat{B}^\dagger \vec{z} = \vec{0}_2, \quad \text{or equivalently,}$$
$$5z(1) - 2z'(1) = 0\}.$$

Thus the problem π_3 becomes

$$\pi_3 : \quad 2y'' + 5y' - 3y = 0,$$
$$y'(0) = 0, y'(1) = 0,$$
$$y(0) = 0,$$

and the adjoint problem $\pi_3^\dagger$ becomes

$$\pi_3^\dagger : \quad 2z'' - 5z' - 3z = 0,$$
$$5z(1) - 2z'(1) = 0.$$

It follows that L is not symmetric, since

$$L^\dagger u = Lu$$

does not hold for $u \in D(L)$.

Here note that we can also obtain the $L^\dagger$, by using Definition 1.2. The expression

$$[p\bar{z}y' + q\bar{z}y - p\bar{z}'y - p'\bar{z}y]|_a^b = 0$$

for all $y \in D(L)$, in Definition 1.2, gives

$$[5\bar{z}(1) - 2\bar{z}'(1)]y(1) = 0.$$

Here $y \in C^2[0,\pi]$ and $y'(0) = 0 = y'(1) = y(0)$. Thus choosing $y(1) = 1$ leads to $5\bar{z}(1) - 2\bar{z}'(1) = 0$. It follows that

$$D(L^\dagger) = \{z \in C^2[0,\pi] : 5z(1) - 2z'(1) = 0\}.$$

Try $y = e^{rx}$ in the problem π_2, and we get the auxiliary equation $2r^2 + 5r - 3 = 0$. The two roots for this equation are: $r = \frac{1}{2}, -3$, and so the general solution for y is:

$$y = c_1 e^{\frac{1}{2}x} + c_2 e^{-3x}.$$

The boundary conditions $y'(0) = 0 = y'(1) = y(0)$ give:

$$0 = c_1 = c_2,$$

and so $y \equiv 0$. Thus the problem π_2 has exactly 0 linearly independent solution. Similar calculations show that

$$z = c_1(e^{-\frac{1}{2}x} + 6e^{-\frac{7}{6}}e^{3x}),$$

and so the adjoint problem $\pi_2^\dagger$ has exactly 1 linearly independent solution. These two numbers 0 and 1 satisfy the relation:

$$1 = 0 + m - n = 0 + 3 - 2,$$

as desired.

Compute

$$B_c^\dagger \bar{z} = \begin{pmatrix} -2 & 0 & 0 & 0 \\ 0 & 0 & 2 & 0 \\ -5 & 2 & 0 & 0 \end{pmatrix} c_1 \begin{pmatrix} 1 + 6e^{-\frac{7}{6}} \\ -\frac{1}{2} + 18e^{-\frac{7}{6}} \\ 7e^{-\frac{1}{2}} \\ (17 + \frac{1}{2})e^{-\frac{1}{2}} \end{pmatrix}$$

$$= c_1 \begin{pmatrix} -2 - 12e^{-\frac{7}{6}} \\ 14e^{-\frac{1}{2}} \\ -6 + 6e^{-\frac{7}{6}} \end{pmatrix},$$

$$\hat{B}_c^\dagger \bar{z} = \begin{pmatrix} -2 & 0 & -5 & 2 \\ 0 & 0 & -0.5 & 1 \\ -5 & 2 & -2.5 & 1 \end{pmatrix} c_1 \begin{pmatrix} 1 + 6e^{-\frac{7}{6}} \\ -\frac{1}{2} + 18e^{-\frac{7}{6}} \\ 7e^{-\frac{1}{2}} \\ (17 + \frac{1}{2})e^{-\frac{1}{2}} \end{pmatrix}$$

$$= c_1 \begin{pmatrix} -2 - 12e^{-\frac{7}{6}} \\ 14e^{-\frac{1}{2}} \\ -6 + 6e^{-\frac{7}{6}} \end{pmatrix},$$

which are desired results.

By Theorem 2.9, (6.3) has a solution if and only if h satisfies

$$(h, z) = \overline{c_1} \int_0^1 h(x)(e^{-\frac{1}{2}x} + 6e^{-\frac{7}{6}}e^{3x})\, dx$$

$$= (\vec{\gamma}, B_c^\dagger \bar{z})$$

$$= \overline{c_1}\left(\begin{pmatrix} a \\ b \\ c \end{pmatrix}, \begin{pmatrix} -2 - 12e^{-\frac{7}{6}} \\ 14e^{-\frac{1}{2}} \\ -6 + 6e^{-\frac{7}{6}} \end{pmatrix} \right)$$

$$= \overline{c_1}[a(-2 - 12e^{-\frac{7}{6}}) + b(14e^{-\frac{1}{2}}) + c(-6 + 6e^{-\frac{7}{6}})],$$

that is, h satisfies

$$\int_0^1 h(x)(e^{-\frac{1}{2}x} + 6e^{-\frac{7}{6}}e^{3x})\, dx = a(-1 - 12e^{-\frac{7}{6}}) + b(14e^{-\frac{1}{2}}) + c(-6 + 6e^{-\frac{7}{6}}).$$

An admissible choice of h and a, b, and c is:

$$h \equiv 0, a = b = c = 0.$$

In this case, the unique solution y of (6.3) is:

$$y \equiv 0.$$

Another admissible choice is:

$$h(x) = e^{-3x}, a = -\frac{3}{7}, b = 0, c = \frac{2}{21}.$$

In this case, the unique solution y of (6.3) is:

$$y = \frac{2}{21}e^{-3x} - \frac{1}{7}xe^{-3x},$$

by using the method of undetermined coefficients or the method of parameters variation.

The choice of $h(x) = sin(x)$ leads to no solution, no matter what $\vec{\gamma} = \begin{pmatrix} a \\ b \\ c \end{pmatrix}$ is chosen. This can be checked by using the method of undetermined coefficients or the method of parameters variation.

4. Here

$$B = \begin{pmatrix} 0 & 1 & 0 & 0 \\ 0 & 0 & 0 & 1 \\ 1 & 0 & 0 & 0 \\ 0 & 0 & 1 & 0 \end{pmatrix},$$

$m = 4, n = 2$, and $\vec{\gamma} = \begin{pmatrix} a \\ b \\ c \\ d \end{pmatrix}$.

Let $B_c = 0_{0 \times 4}$, a zero matrix of 0×4, or equivalently, a zero vector $\vec{0}_4$ of four dimensions. This is the only complementary matrix to B.

Let $F = \begin{pmatrix} B \\ B_c \end{pmatrix} = B$. Applying row operations to compute the inverse F^{-1} of F, we have

$$F^{-1} = \begin{pmatrix} 0 & 0 & 1 & 0 \\ 1 & 0 & 0 & 0 \\ 0 & 0 & 0 & 1 \\ 0 & 1 & 0 & 0 \end{pmatrix}.$$

Using Lemma 2.3, we have

$$S = \begin{pmatrix} -5 & -2 & 0 & 0 \\ 2 & 0 & 0 & 0 \\ 0 & 0 & 5 & 2 \\ 0 & 0 & -2 & 0 \end{pmatrix};$$

$$SF^{-1} = \begin{pmatrix} -2 & 0 & -5 & 0 \\ 0 & 0 & 2 & 0 \\ 0 & 2 & 0 & 5 \\ 0 & 0 & 0 & -2 \end{pmatrix};$$

$$\begin{pmatrix} B_c^\dagger \\ B^\dagger \end{pmatrix} = G = (SF^{-1})^* = \begin{pmatrix} -2 & 0 & 0 & 0 \\ 0 & 0 & 2 & 0 \\ -5 & 2 & 0 & 0 \\ 0 & 0 & 5 & -2 \end{pmatrix}, \quad \text{for which}$$

$$B_c^\dagger = \begin{pmatrix} -2 & 0 & 0 & 0 \\ 0 & 0 & 2 & 0 \\ -5 & 2 & 0 & 0 \\ 0 & 0 & 5 & -2 \end{pmatrix} \quad \text{and}$$

$B^\dagger = 0_{0 \times 4}$,   a zero matrix of 0×4, or equivalently,

a zero vector $\vec{0}_4$ of four dimensions.

Using the Definition 2.6 of L and $L^\dagger$, we have

$$Ly = 2y'' + 5y' - 3y,$$

$$D(L) = \{y \in C^2[0,1] : y'(0) = 0, y'(1) = 0, y(0) = 0, y(1) = 0\};$$
$$L^\dagger z = 2z'' - 5z' - 3z,$$
$$D(L^\dagger) = \{z \in C^2[0,1] : B^\dagger \tilde{z} = 0, \quad \text{or equivalently,}$$
$$\text{no restriction on } \tilde{z}\}.$$

Thus the problem π_4 becomes

$$\pi_4 : \quad 2y'' + 5y' - 3y = 0,$$
$$y'(0) = 0, y'(1) = 0,$$
$$y(0) = 0, y(1) = 0,$$

and the adjoint problem $\pi_4^\dagger$ becomes

$$\pi_4^\dagger : \quad 2z'' - 5z' - 3z = 0.$$

It follows that L is not symmetric, since

$$L^\dagger u = Lu$$

does not hold for $u \in D(L)$.

Here note that we can also obtain the $L^\dagger$, by using Definition 1.2. The expression

$$[p\bar{z}y' + q\bar{z}y - p\bar{z}'y - p'\bar{z}y]|_a^b = 0$$

in Definition 1.2, is true for all $y \in D(L)$ and $z \in C^2[0,1]$, since

$$y'(0) = 0 = y'(1) = y(0) = y(1)$$

for $y \in D(L)$. It follows that

$$D(L^\dagger) = \in C^2[0,1].$$

Try $y = e^{rx}$ in the problem π_2, and we get the auxiliary equation $2r^2 + 5r - 3 = 0$. The two roots for this equation are: $r = \frac{1}{2}, -3$, and so the general solution for y is:

$$y = c_1 e^{\frac{1}{2}x} + c_2 e^{-3x}.$$

The boundary conditions $y'(0) = 0 = y'(1) = y(0) = y(1)$ give:

$$0 = c_1 = c_2,$$

and so $y \equiv 0$. Thus the problem π_2 has exactly 0 linearly independent solution. Similar calculations show that

$$z = c_1 e^{-\frac{1}{x}} + c_2 e^{3x},$$

and so the adjoint problem $\pi_2^\dagger$ has exactly 2 linearly independent solution. These two numbers 0 and 4 satisfy the relation:

$$2 = 0 + m - n = 0 + 4 - 2,$$

as desired.

Compute

$$B_c^\dagger \tilde{z} = \begin{pmatrix} -2 & 0 & 0 & 0 \\ 0 & 0 & 2 & 0 \\ -5 & 2 & 0 & 0 \\ 0 & 0 & 5 & -2 \end{pmatrix} \begin{pmatrix} c_1 + c_2 \\ -\frac{1}{2}c_1 + 3c_2 \\ c_1 e^{-\frac{1}{2}} + c_2 e^3 \\ -\frac{1}{2}c_1 e^{-\frac{1}{2}} + 3c_2 e^3 \end{pmatrix}$$

$$= \begin{pmatrix} -2c_1 - 2c_2 \\ 2c_1e^{-\frac{1}{2}} + 2c_2e^3 \\ -6c_1 + c_2 \\ 6c_1e^{-\frac{1}{2}} - c_2e^3 \end{pmatrix}.$$

By theorem 2.9, (6.4) has a solution if and only if h satisfies

$$(h, z) = \int_0^1 h(x)(\overline{c_1}e^{-\frac{1}{2}x} + \overline{c_2}e^{3x}\, dx$$

$$= \overline{c_1}(2 - 2e^{-\frac{1}{2}}) + \overline{c_2}(\frac{1}{3}e^3 - \frac{1}{3})$$

$$= (\vec{\gamma}, B_c^\dagger \tilde{z})$$

$$= (\begin{pmatrix} a \\ b \\ c \\ d \end{pmatrix}, \begin{pmatrix} -2c_1 - 2c_2 \\ 2c_1e^{-\frac{1}{2}} + 2c_2e^3 \\ -6c_1 + c_2 \\ 6c_1e^{-\frac{1}{2}} - c_2e^3 \end{pmatrix})$$

$$= \overline{a(-2c_1 - 2c_2)} + \overline{b(2c_1e^{-\frac{1}{2}} + 2c_2e^3)} + \overline{c(-6c_1 + c_2)} + \overline{d(6c_1e^{-\frac{1}{2}} - c_2e^3)}.$$

When $h(x) \equiv 1$, many choices can be made for $a, b, c,$ and d. For instance, we can choose

$$(a, b, c, d) = (\frac{1}{7}e^{-\frac{1}{2}} - \frac{1}{7}e^3, 0, \frac{1}{21}e^3 + \frac{2}{7}e^{-\frac{1}{2}} - \frac{1}{3}, 0) \quad \text{or}$$

$$= (b + e^{-\frac{1}{2}} - 1, \frac{(\frac{5}{3} - 2e^{-\frac{1}{2}} - \frac{1}{3}e^3)e^{-\frac{1}{2}}}{2(e^{-\frac{1}{2}} - e^3)}, 0, 0).$$

Under the first choice, the (unique solution) of (6.4) is:

$$y = \frac{2}{7}e^{-\frac{1}{2}}e^{\frac{1}{2}x} + \frac{1}{21}e^3e^{-3x} - \frac{1}{3},$$

by the method of undetermined coefficients or the method of parameters variation.

When $h(x) = \sin(x)$, there is no solution to (6.4), no matter what $\vec{\gamma} = (a, b, c, d)$ is chosen. This can be checked by the method of undetermined coefficients or the method of parameters variation.

5. Here

$$B = \begin{pmatrix} 1 & -1 & 0 & 0 \\ 0 & 0 & 1 & 1 \end{pmatrix},$$

$m = 2 = n, p(x) = x^2, q(x) = -4x, r(x) = 4,$ and

$$\vec{\gamma} = \begin{pmatrix} a \\ b \end{pmatrix}.$$

Let

$$B_c = \begin{pmatrix} 1 & 0 & 0 & 0 \\ 0 & 0 & 0 & 1 \end{pmatrix}, \hat{B}_c = \begin{pmatrix} 0 & 1 & 0 & 0 \\ 0 & 0 & 1 & 0 \end{pmatrix}.$$

Let $F = \begin{pmatrix} B \\ B_c \end{pmatrix}$. Applying row operations to compute the inverse F^{-1} of F, we have

$$F^{-1} = \begin{pmatrix} 0 & 0 & 1 & 0 \\ -1 & 0 & 1 & 0 \\ 0 & 1 & 0 & -1 \\ 0 & 0 & 0 & 1 \end{pmatrix}.$$

Similarly, the inverse $\hat{F}^{-1}$ of $\hat{F}$, where $\hat{F} = \begin{pmatrix} B \\ \hat{B}_c \end{pmatrix}$, is

$$\hat{F}^{-1} = \begin{pmatrix} 1 & 0 & 1 & 0 \\ 0 & 0 & 1 & 0 \\ 0 & 0 & 0 & 1 \\ 0 & 1 & 0 & -1 \end{pmatrix}.$$

Using Lemma 2.3, we have

$$S = \begin{pmatrix} 6 & -1 & 0 & 0 \\ 2 & 0 & 0 & 0 \\ 0 & 0 & -12 & 4 \\ 0 & 0 & -4 & 0 \end{pmatrix};$$

$$SF^{-1} = \begin{pmatrix} 1 & 0 & 5 & 0 \\ 0 & 0 & 1 & 0 \\ 0 & -12 & 0 & 16 \\ 0 & -4 & 0 & 4 \end{pmatrix};$$

$$\begin{pmatrix} B_c^\dagger \\ B^\dagger \end{pmatrix} = G = (SF^{-1})^* = \begin{pmatrix} 1 & 0 & 0 & 0 \\ 0 & 0 & -12 & -4 \\ 5 & 1 & 0 & 0 \\ 0 & 0 & 16 & 4 \end{pmatrix}, \quad \text{for which}$$

$$B_c^\dagger = \begin{pmatrix} 1 & 0 & 0 & 0 \\ 0 & 0 & -12 & -4 \end{pmatrix} \quad \text{and}$$

$$B^\dagger = \begin{pmatrix} 5 & 1 & 0 & 0 \\ 0 & 0 & 16 & 4 \end{pmatrix};$$

$$S\hat{F}^{-1} = \begin{pmatrix} 6 & 0 & 5 & 0 \\ 1 & 0 & 1 & 0 \\ 0 & 4 & 0 & -16 \\ 0 & 0 & 0 & -4 \end{pmatrix};$$

$$\begin{pmatrix} \hat{B}_c^\dagger \\ \hat{B}^\dagger \end{pmatrix} = \hat{G} = (S\hat{F}^{-1})^* = \begin{pmatrix} 6 & 1 & 0 & 0 \\ 0 & 0 & 4 & 0 \\ 5 & 1 & 0 & 0 \\ 0 & 0 & -16 & -4 \end{pmatrix}, \quad \text{for which}$$

$$\hat{B}_c^\dagger = \begin{pmatrix} 6 & 1 & 0 & 0 \\ 0 & 0 & 4 & 0 \end{pmatrix} \quad \text{and}$$

$$\hat{B}^\dagger = \begin{pmatrix} 5 & 1 & 0 & 0 \\ 0 & 0 & -16 & -4 \end{pmatrix}.$$

Using the Definition 2.6 of L and $L^\dagger$, we have

$$Ly = x^2 y'' - 4xy' + 4y,$$

$$D(L) = \{y \in C^2[1,2] : y(1) - y'(1) = 0, y(2) + y'(2) = 0\};$$
$$L^\dagger z = (x^2 z)'' - (-4xz)' + 4z,$$
$$= x^2 z'' + 8xz' + 10z$$
$$D(L^\dagger) = \{z \in C^2[1,2] : B^\dagger \vec{z} = \vec{0}_2, \quad \text{or equivalently,}$$
$$\hat{B}^\dagger \vec{z} = \vec{0}_2, \quad \text{or equivalently,}$$
$$5z(1) + z'(1) = 0, 4z(2) + z'(2) = 0\}.$$

Thus the problem π_2 becomes

$$\pi_2 : \quad x^2 y'' - 4xy' + 4y = 0,$$
$$y(1) - y'(1) = 0, y(2) + y'(2) = 0,$$

and the adjoint problem $\pi_2^\dagger$ becomes

$$\pi_2^\dagger : \quad x^2 z'' + 8xz' + 10z = 0,$$
$$5z(1) + z'(1) = 0, 4z(2) + z'(2) = 0.$$

It follows that L is not symmetric, since

$$L^\dagger u = Lu$$

does not hold for $u \in D(L)$.

Here note that we can also obtain the $L^\dagger$, by using Definition 1.2. The expression

$$[p\bar{z}y' + q\bar{z}y - p\bar{z}'y - p'\bar{z}y]|_a^b = 0$$

for all $y \in D(L)$, in Definition 1.2, gives

$$[-16\bar{z}(2) - 4\bar{z}'(2)]y(2) - [-5\bar{z}(1) - \bar{z}'(1)]y(1) = 0.$$

Here $y \in C^2[0, \pi]$ and

$$y(1) - y'(1) = 0, \quad y(2) + y'(2) = 0.$$

Thus choosing $y(2) = 0$ and $y(1) = 1$ leads to $5\bar{z}(1) + \bar{z}'(1) = 0$. But choosing $y(2) = 1$ and $y(1) = 0$ leads to $4\bar{z}(2) + \bar{z}(2) = 0$. It follows that

$$D(L^\dagger) = \{z \in C^2[1,2] : 5z(1) + z'(1) = 0, 4z(2) + z'(2) = 0\}.$$

Try $y = x^r$ in the problem π_2, and we get the auxiliary equation $r(r-1) - 4r + 4 = 0$. The two roots for this equation are: $r = 4, 1$, and so the general solution for y is:

$$y = c_1 x^4 + c_2 x.$$

The boundary conditions

$$y(1) - y'(1) = 0, \quad y(2) + y'(2) = 0,$$

give:

$$0 = c_1 = c_2,$$

and so $y \equiv 0$. Thus the problem π_2 has exactly 0 linearly independent solution. Similar calculations show that $z \equiv 0$, and so the adjoint problem $\pi_2^\dagger$ has also exactly 0 linearly independent solution. These two numbers 0 and 0 satisfy the relation:

$$0 = 0 + m - n = 0 + 2 - 2,$$

as desired.

Compute

$$B_c^\dagger \tilde{z} = \begin{pmatrix} 1 & 0 & 0 & 0 \\ 0 & 0 & -12 & -4 \end{pmatrix} \begin{pmatrix} 0 \\ 0 \\ 0 \\ 0 \end{pmatrix}$$

$$= \begin{pmatrix} 0 \\ 0 \end{pmatrix},$$

$$\hat{B}_c^\dagger \tilde{z} = \begin{pmatrix} 6 & 1 & 0 & 0 \\ 0 & 0 & 4 & 0 \end{pmatrix} \begin{pmatrix} 0 \\ 0 \\ 0 \\ 0 \end{pmatrix}$$

$$= \begin{pmatrix} 0 \\ 0 \end{pmatrix},$$

which are desired results.

By Theorem 2.9, (6.5) has a solution if and only if h satisfies

$$(h, z) = \int_1^2 h(x) 0 \, dx = 0$$
$$= (\vec{\gamma}, B_c^\dagger \tilde{z})$$
$$= (\begin{pmatrix} a \\ b \end{pmatrix}, \begin{pmatrix} 0 \\ 0 \end{pmatrix})$$
$$= 0.$$

But this is true for all h, and thus (6.5) has a (unique) solution for all given h and $\vec{\gamma}$.

6. Here

$$B = \begin{pmatrix} 1 & -1 & 0 & 0 \end{pmatrix},$$

$m = 1, n = 2, p(x) = x^2, q(x) = -4x, r(x) = 4$, and

$$\vec{\gamma} = a.$$

Let

$$B_c = \begin{pmatrix} 1 & 0 & 0 & 0 \\ 0 & 0 & 0 & 1 \\ 0 & 0 & 1 & 0 \end{pmatrix}, \hat{B}_c = \begin{pmatrix} 1 & 0 & 0 & 0 \\ 0 & 0 & 1 & 1 \\ 0 & 1 & 1 & 0 \end{pmatrix}.$$

Let $F = \begin{pmatrix} B \\ B_c \end{pmatrix}$. Applying row operations to compute the inverse F^{-1} of F, we have

$$F^{-1} = \begin{pmatrix} 0 & 1 & 0 & 0 \\ -1 & 1 & 0 & 0 \\ 0 & 0 & 0 & 1 \\ 0 & 0 & 1 & 0 \end{pmatrix}.$$

Similarly, the inverse $\hat{F}^{-1}$ of $\hat{F}$, where $\hat{F} = \begin{pmatrix} B \\ \hat{B}_c \end{pmatrix}$, is

$$\hat{F}^{-1} = \begin{pmatrix} 0 & 1 & 0 & 0 \\ -1 & 1 & 0 & 0 \\ 1 & -1 & 0 & 1 \\ -1 & 1 & 1 & -1 \end{pmatrix}.$$

Using Lemma 2.3, we have

$$S = \begin{pmatrix} 6 & -1 & 0 & 0 \\ 2 & 0 & 0 & 0 \\ 0 & 0 & -12 & 4 \\ 0 & 0 & -4 & 0 \end{pmatrix};$$

$$SF^{-1} = \begin{pmatrix} 1 & 5 & 0 & 0 \\ 0 & 1 & 0 & 0 \\ 0 & 0 & 4 & -12 \\ 0 & 0 & 0 & -4 \end{pmatrix};$$

$$\begin{pmatrix} B_c^\dagger \\ B^\dagger \end{pmatrix} = G = (SF^{-1})^* = \begin{pmatrix} 1 & 0 & 0 & 0 \\ 5 & 1 & 0 & 0 \\ 0 & 0 & 4 & 0 \\ 0 & 0 & -12 & -4 \end{pmatrix}, \quad \text{for which}$$

$$B_c^\dagger = \begin{pmatrix} 1 & 0 & 0 & 0 \end{pmatrix} \quad \text{and}$$

$$B^\dagger = \begin{pmatrix} 5 & 1 & 0 & 0 \\ 0 & 0 & 4 & 0 \\ 0 & 0 & -12 & -4 \end{pmatrix};$$

$$S\hat{F}^{-1} = \begin{pmatrix} 1 & 5 & 0 & 0 \\ 0 & 1 & 0 & 0 \\ -16 & 16 & 4 & -16 \\ -4 & 4 & 0 & -4 \end{pmatrix};$$

$$\begin{pmatrix} \hat{B}_c^\dagger \\ \hat{B}^\dagger \end{pmatrix} = \hat{G} = (S\hat{F}^{-1})^* = \begin{pmatrix} 1 & 0 & -16 & -4 \\ 5 & 1 & 16 & 4 \\ 0 & 0 & 4 & 0 \\ 0 & 0 & -16 & -4 \end{pmatrix}, \quad \text{for which}$$

$$\hat{B}_c^\dagger = \begin{pmatrix} 1 & 0 & -16 & -4 \\ 0 & 0 & 4 & 0 \end{pmatrix} \quad \text{and}$$

$$\hat{B}^\dagger = \begin{pmatrix} 5 & 1 & 16 & 4 \\ 0 & 0 & 4 & 0 \\ 0 & 0 & -16 & -4 \end{pmatrix}.$$

Using the Definition 2.6 of L and $L^\dagger$, we have

$$Ly = x^2 y'' - 4xy' + 4y,$$
$$D(L) = \{y \in C^2[0,1] : y(1) - y'(1) = 0\};$$
$$L^\dagger z = (x^2 z)'' - (-4xz)' + 4z$$
$$= x^2 z'' + 8xz' + 10z,$$
$$D(L^\dagger) = \{z \in C^2[0,1] : B^\dagger \vec{z} = \vec{0}_3, \quad \text{or equivalently,}$$

$$\hat{B}^\dagger \vec{z} = \vec{0}_3, \quad \text{or equivalently,}$$
$$5z(1) + z'(1) = 0,\, 3z(2) + z'(2) = 0,\, z(2) = 0\}.$$

Thus the problem π_1 becomes

$$\pi_1 : \quad x^2 y'' - 4xy' + 4y = 0,$$
$$y(1) - y'(1) = 0,$$

and the adjoint problem $\pi_1^\dagger$ becomes

$$\pi_1^\dagger : \quad x^2 z'' + 8xz' + 10z = 0,$$
$$5z(1) + z'(1) = 0,\, 3z(2) + z'(2) = 0,$$
$$z(2) = 0.$$

It follows that L is not symmetric, since

$$L^\dagger u = Lu$$

does not hold for $u \in D(L)$.

Here note that we can also obtain the $L^\dagger$, by using Definition 1.2. The expression

$$[p\bar{z}y' + q\bar{z}y - p\bar{z}'y - p'\bar{z}y]|_a^b = 0$$

for all $y \in D(L)$, in Definition 1.2, gives

$$4\bar{z}(2)y'(2) + [-12\bar{z}(2) - 4\bar{z}'(2)]y(2) - [-5\bar{z}(1) - \bar{z}'(1)]y(1) = 0.$$

Here $y \in C^2[0, \pi]$ and

$$y(1) - y'(1) = 0.$$

Thus choosing $y(2) = 0 = y'(2)$ and $y(1) = 1$ leads to $5\bar{z}(1) + \bar{z}'(1) = 0$. But choosing $y(2) = 1$ and $y(1) = 0 = y'(2)$ leads to $3\bar{z}(2) + \bar{z}(2) = 0$. Furthermore, choosing $y'(2) = 1$ and $y(2) = 0 = y(1)$ leads to $\bar{z}(2) = 0$. It follows that

$$D(L^\dagger) = \{z \in C^2[1, 2] : 5z(1) + z'(1) = 0,\, 3z(2) + z'(2) = 0,\, z(2) = 0\}.$$

Try $y = x^r$ in the problem π_2, and we get the auxiliary equation $r(r-1) - 4r + 4 = 0$. The two roots for this equation are: $r = 4, 1$, and so the general solution for y is:

$$y = c_1 x^4 + c_2 x.$$

The boundary condition

$$y(1) - y'(1) = 0$$

gives: $0 = c_1$, and so $y = c_2 x$. Thus the problem π_2 has exactly 1 linearly independent solution. Similar calculations show that $z \equiv 0$, and so the adjoint problem $\pi_2^\dagger$ has also exactly 0 linearly independent solution. These two numbers 1 and 0 satisfy the relation:

$$0 = 1 + m - n = 1 + 1 - 2,$$

as desired.

Compute

$$B_c^\dagger \bar{z} = \begin{pmatrix} 1 & 0 & 0 & 0 \end{pmatrix} \begin{pmatrix} 0 \\ 0 \\ 0 \\ 0 \\ 0 \end{pmatrix}$$

$$= 0,$$

$$\hat{B}_c^\dagger \bar{z} = \begin{pmatrix} 1 & 0 & -16 & -4 \end{pmatrix} \begin{pmatrix} 0 \\ 0 \\ 0 \\ 0 \\ 0 \end{pmatrix}$$

$$= 0,$$

which are desired results.

By Theorem 2.9, (6.6) has a solution if and only if h satisfies

$$(h, z) = \int_1^2 h(x) 0 \, dx = 0$$
$$= (\vec{\gamma}, B_c^\dagger \bar{z})$$
$$= (a, 0)$$
$$= 0.$$

But this is true for all h and a, and thus (6.6) has a solution for all given h and a.

7. Here

$$B = \begin{pmatrix} 1 & -1 & 0 & 0 \\ 0 & 0 & 1 & 1 \\ 1 & 0 & 0 & 0 \end{pmatrix},$$

$m = 3, n = 2, p(x) = x^2, q(x) = -4x, r(x) = 4$, and

$$\vec{\gamma} = \begin{pmatrix} a \\ b \\ c \end{pmatrix}.$$

Let

$$B_c = \begin{pmatrix} 0 & 0 & 0 & 1 \end{pmatrix}, \hat{B}_c = \begin{pmatrix} 1 & 2 & 2 & 1 \end{pmatrix}.$$

Let $F = \begin{pmatrix} B \\ B_c \end{pmatrix}$. Applying row operations to compute the inverse F^{-1} of F, we have

$$F^{-1} = \begin{pmatrix} 0 & 0 & 1 & 0 \\ -1 & 0 & 1 & 0 \\ 0 & 1 & 0 & -1 \\ 0 & 0 & 0 & 1 \end{pmatrix}.$$

Similarly, the inverse $\hat{F}^{-1}$ of $\hat{F}$, where $\hat{F} = \begin{pmatrix} B \\ \hat{B}_c \end{pmatrix}$, is

$$\hat{F}^{-1} = \begin{pmatrix} 0 & 0 & 1 & 0 \\ -1 & 0 & 1 & 0 \\ 2 & -1 & -3 & 1 \\ -2 & 2 & 3 & -1 \end{pmatrix}.$$

Using Lemma 2.3, we have

$$S = \begin{pmatrix} 6 & -1 & 0 & 0 \\ 2 & 0 & 0 & 0 \\ 0 & 0 & -12 & 4 \\ 0 & 0 & -4 & 0 \end{pmatrix};$$

$$SF^{-1} = \begin{pmatrix} 1 & 0 & 5 & 0 \\ 0 & 0 & 1 & 0 \\ 0 & -12 & 0 & 16 \\ 0 & -4 & 0 & 4 \end{pmatrix};$$

$$\begin{pmatrix} B_c^\dagger \\ B^\dagger \end{pmatrix} = G = (SF^{-1})^* = \begin{pmatrix} 1 & 0 & 0 & 0 \\ 0 & 0 & -12 & -4 \\ 5 & 1 & 0 & 0 \\ 0 & 0 & 16 & 4 \end{pmatrix}, \quad \text{for which}$$

$$B_c^\dagger = \begin{pmatrix} 1 & 0 & 0 & 0 \\ 0 & 0 & -12 & -4 \\ 5 & 1 & 0 & 0 \end{pmatrix} \quad \text{and}$$

$$B^\dagger = \begin{pmatrix} 0 & 0 & 16 & 4 \end{pmatrix};$$

$$S\hat{F}^{-1} = \begin{pmatrix} 1 & 0 & 5 & 0 \\ 0 & 0 & 1 & 0 \\ -32 & 20 & 48 & -16 \\ -8 & 4 & 12 & -4 \end{pmatrix};$$

$$\begin{pmatrix} \hat{B}_c^\dagger \\ \hat{B}^\dagger \end{pmatrix} = \hat{G} = (S\hat{F}^{-1})^* = \begin{pmatrix} 1 & 0 & -32 & -8 \\ 0 & 0 & 20 & 4 \\ 5 & 1 & 48 & 12 \\ 0 & 0 & -16 & -4 \end{pmatrix}, \quad \text{for which}$$

$$\hat{B}_c^\dagger = \begin{pmatrix} 1 & 0 & -32 & -8 \\ 0 & 0 & 20 & 4 \\ 5 & 1 & 48 & 12 \end{pmatrix} \quad \text{and}$$

$$\hat{B}^\dagger = \begin{pmatrix} 0 & 0 & -16 & -4 \end{pmatrix}.$$

Using the Definition 2.6 of L and $L^\dagger$, we have

$$Ly = x^2 y'' - 4xy' + 4y,$$
$$D(L) = \{y \in C^2[0,1] : y(1) - y'(1) = 0, y(2) + y'(2) = 0, y(1) = 0\};$$
$$L^\dagger z = (x^2 z)'' - (-4xz)' + 4z,$$
$$= x^2 z'' + 8xz' + 10z$$
$$D(L^\dagger) = \{z \in C^2[0,1] : B^\dagger \hat{z} = 0, \quad \text{or equivalently,}$$
$$\hat{B}^\dagger \hat{z} = 0, \quad \text{or equivalently,}$$
$$4z(2) + z'(2) = 0\}.$$

Thus the problem π_3 becomes

$$\pi_3 : \quad x^2 y'' - 4xy' + 4y = 0,$$
$$y(1) - y'(1) = 0, y(2) + y'(2) = 0,$$
$$y(1) = 0,$$

and the adjoint problem $\pi_3^\dagger$ becomes

$$\pi_3^\dagger : \quad x^2 z'' + 8xz' + 10z = 0,$$
$$4z(2) + z'(2) = 0.$$

It follows that L is not symmetric, since

$$L^\dagger u = Lu$$

does not hold for $u \in D(L)$.

Here note that we can also obtain the $L^\dagger$, by using Definition 1.2. The expression

$$[p\bar{z}y' + q\bar{z}y - p\bar{z}'y - p'\bar{z}y]|_a^b = 0$$

for all $y \in D(L)$, in Definition 1.2, gives

$$[-16\bar{z}(2) - 4\bar{z}'(2)]y(2) = 0.$$

Here $y \in C^2[0, \pi]$ and

$$y(1) - y'(1) = 0, \quad y(2) + y'(2) = 0, \quad y(1) = 0.$$

Thus choosing $y(2) = 1$ leads to $4\bar{z}(1) + \bar{z}(1) = 0$. It follows that

$$D(L^\dagger) = \{z \in C^2[1, 2] : 4z(2) + z'(2) = 0\}.$$

Try $y = x^r$ in the problem π_2, and we get the auxiliary equation $r(r-1) - 4r + 4 = 0$. The two roots for this equation are: $r = 4, 1$, and so the general solution for y is:

$$y = c_1 x^4 + c_2 x.$$

The boundary conditions

$$y(1) - y'(1) = 0, \quad y(2) + y'(2) = 0, \quad y(1) = 0,$$

give:

$$0 = c_1 = c_2,$$

and so $y \equiv 0$. Thus the problem π_2 has exactly 0 linearly independent solution. Similar calculations show that

$$z = c_1(x^{-2} - 16x^{-5}),$$

and so the adjoint problem $\pi_2^\dagger$ has exactly 1 linearly independent solution. These two numbers 0 and 1 satisfy the relation:

$$1 = 0 + m - n = 0 + 3 - 2,$$

as desired.

Compute

$$B_c^\dagger \bar{z} = \begin{pmatrix} 1 & 0 & 0 & 0 \\ 0 & 0 & -12 & -4 \\ 5 & 1 & 0 & 0 \end{pmatrix} c_1 \begin{pmatrix} -15 \\ 78 \\ -\frac{1}{4} \\ 1 \end{pmatrix}$$

$$= c_1 \begin{pmatrix} -15 \\ -1 \\ 3 \end{pmatrix},$$

$$\hat{B}_c^\dagger \bar{z} = \begin{pmatrix} 1 & 0 & -32 & -8 \\ 0 & 0 & 20 & 4 \\ 5 & 1 & 48 & 12 \end{pmatrix} c_1 \begin{pmatrix} -15 \\ 78 \\ -\frac{1}{4} \\ 1 \end{pmatrix}$$

$$= c_1 \begin{pmatrix} -15 \\ -1 \\ 3 \end{pmatrix},$$

which are desired results.

By Theorem 2.9, (6.7) has a solution if and only if h satisfies

$$(h, z) = \overline{c_1} \int_1^2 h(x)(x^{-2} - 16x^{-5}) \, dx$$

$$= (\vec{\gamma}, B_c^\dagger \tilde{z})$$

$$= \overline{c_1}\left(\begin{pmatrix} a \\ b \\ c \end{pmatrix}, \begin{pmatrix} -15 \\ -1 \\ 3 \end{pmatrix} \right)$$

$$= \overline{c_1}[a(-15) + b(-1) + c(3)],$$

That is, h satisfies

$$\int_1^2 h(x)(x^{-2} - 16x^{-5}) \, dx = a(-15) + b(-1) + c(3).$$

When $h(x) \equiv 1$, many choices can be made for a, b, and c. For instance, we can choose $(a, b, c) = (0, \frac{1}{4}, -1)$ or $(-\frac{1}{20}, 1, -1)$. Under the first choice, the (unique) solution of (6.7) is:

$$y = \frac{1}{12}x^4 - \frac{4}{3}x + \frac{1}{4},$$

by using the method of undetermined coefficients or the method of parameters variation.

When $h(x) = \sin(x)$, there is no solution for (6.7), no matter what $\vec{\gamma} = (a, b, c)$ is chosen. This can also be checked by using the method of undetermined coefficients or the method of parameters variation.

8. Here

$$B = \begin{pmatrix} 1 & -1 & 0 & 0 \\ 0 & 0 & 1 & 1 \\ 1 & 0 & 0 & 0 \\ 0 & 0 & 0 & 1 \end{pmatrix},$$

$m = 4, n = 2, p(x) = x^2, q(x) = -4x, r(x) = 4$, and

$$\vec{\gamma} = \begin{pmatrix} a \\ b \\ c \\ d \end{pmatrix}.$$

Let $B_c = 0_{0 \times 4}$, a zero matrix of 0×4, or equivalently, a zero vector $\vec{0}_4$ of four dimensions. This is the only complementary matrix to B.

Let $F = \begin{pmatrix} B \\ B_c \end{pmatrix} = B$. Applying row operations to compute the inverse F^{-1} of F, we have

$$F^{-1} = \begin{pmatrix} 0 & 0 & 1 & 0 \\ -1 & 0 & 1 & 0 \\ 0 & 1 & 0 & -1 \\ 0 & 0 & 0 & 1 \end{pmatrix}.$$

Using Lemma 2.3, we have

$$S = \begin{pmatrix} 6 & -1 & 0 & 0 \\ 2 & 0 & 0 & 0 \\ 0 & 0 & -12 & 4 \\ 0 & 0 & -4 & 0 \end{pmatrix};$$

$$SF^{-1} = \begin{pmatrix} 1 & 0 & 5 & 0 \\ 0 & 0 & 1 & 0 \\ 0 & -12 & 0 & 16 \\ 0 & -4 & 0 & 4 \end{pmatrix};$$

$$\begin{pmatrix} B_c^\dagger \\ B^\dagger \end{pmatrix} = G = (SF^{-1})^* = \begin{pmatrix} 1 & 0 & 0 & 0 \\ 0 & 0 & -12 & -4 \\ 5 & 1 & 0 & 0 \\ 0 & 0 & 16 & 4 \end{pmatrix}, \quad \text{for which}$$

$$B_c^\dagger = \begin{pmatrix} 1 & 0 & 0 & 0 \\ 0 & 0 & -12 & -4 \\ 5 & 1 & 0 & 0 \\ 0 & 0 & 16 & 4 \end{pmatrix} \quad \text{and}$$

$B^\dagger = 0_{0 \times 4}$, a zero matrix of 0×4, or equivalently,

a zero vector $\vec{0}_4$ of four dimensions.

Using the Definition 2.6 of L and $L^\dagger$, we have

$$Ly = x^2 y'' - 4xy' + 4y,$$
$$D(L) = \{y \in C^2[0,1] : y(1) - y'(1) = 0, y(2) + y'(2) = 0, y(1) = 0, y'(2) = 0\};$$
$$L^\dagger z = (x^2 z)'' - (-4xz)' + 4z$$
$$= x^2 z'' + 8xz' + 10z,$$
$$D(L^\dagger) = \{z \in C^2[0,1] : B^\dagger \tilde{z} = 0, \quad \text{or equivalently,}$$
no restriction on $\tilde{z}$\}.

Thus the problem π_4 becomes

$$\pi_4 : \quad x^2 y'' - 4xy' + 4y = 0,$$
$$y(1) - y'(1) = 0, y(2) + y'(2) = 0,$$
$$y(1) = 0, y'(2) = 0,$$

and the adjoint problem $\pi_4^\dagger$ becomes

$$\pi_4^\dagger : \quad x^2 z'' + 8xz' + 10z = 0.$$

It follows that L is not symmetric, since

$$L^\dagger u = Lu$$

does not hold for $u \in D(L)$.

Here note that we can also obtain the $L^\dagger$, by using Definition 1.2. The expression

$$[p\bar{z}y' + q\bar{z}y - p\bar{z}'y - p'\bar{z}y]|_a^b = 0$$

in Definition 1.2, is true for all $y \in D(L)$ and $z \in C^2[1, 2]$, since

$$y(1) = 0 = y'(1) = y'(2) = y(2)$$

for $y \in D(L)$. Here $y \in C^2[1, 2]$ and

$$y(1) - y'(1) = 0, \quad y(2) + y'(2) = 0, \quad y(1) = 0, \quad y'(2) = 0.$$

It follows that

$$D(L^\dagger) = C^2[1, 2].$$

Try $y = x^r$ in the problem π_2, and we get the auxiliary equation $r(r-1) - 4r + 4 = 0$. The two roots for this equation are: $r = 4, 1$, and so the general solution for y is:

$$y = c_1 x^4 + c_2 x.$$

The boundary conditions

$$y(1) - y'(1) = 0, \quad y(2) + y'(2) = 0, \quad y(1) = 0 = y'(2),$$

give:

$$0 = c_1 = c_2,$$

and so $y \equiv 0$. Thus the problem π_2 has exactly 0 linearly independent solution. Similar calculations show that

$$z = c_1 x^{-2} + c_2 x^{-5},$$

and so the adjoint problem $\pi_2^\dagger$ has exactly 2 linearly independent solution. These two numbers 0 and 2 satisfy the relation:

$$2 = 0 + m - n = 0 + 4 - 2,$$

as desired.

Compute

$$B_c^\dagger \tilde{z} = \begin{pmatrix} 1 & 0 & 0 & 0 \\ 0 & 0 & -12 & -4 \\ 5 & 1 & 0 & 0 \\ 0 & 0 & 16 & 4 \end{pmatrix} \begin{pmatrix} c_1 + c_2 \\ -2c_1 - 5c_2 \\ \frac{1}{4}c_1 + \frac{1}{32}c_2 \\ -\frac{1}{4}c_1 - \frac{5}{64}c_2 \end{pmatrix}$$

$$= \begin{pmatrix} c_1 + c_2 \\ -2c_1 - \frac{1}{16}c_2 \\ 3c_1 \\ 3c_1 + \frac{3}{16}c_2 \end{pmatrix}.$$

By Theorem 2.9, (6.8) has a solution if and only if h satisfies

$$(h, z) = \int_1^2 h(x)(\overline{c_1} x^{-2} + \overline{c_2} x^{-5})\, dx$$

$$= (\vec{\gamma}, B_c^\dagger \tilde{z})$$

$$= (\begin{pmatrix} a \\ b \\ c \\ d \end{pmatrix}, \begin{pmatrix} c_1 + c_2 \\ -2c_1 - \frac{1}{16}c_2 \\ 3c_1 \\ 3c_1 + \frac{3}{16}c_2 \end{pmatrix})$$

$$= a\overline{(c_1 + c_2)} + b\overline{(-2c_1 - \frac{1}{16}c_2)} + c\overline{(3c_1)} + d\overline{(3c_1 + \frac{3}{16}c_2)}.$$

When $h(x) \equiv 1$, many choices can be made for a, b, c and d. For instance, we can choose $(a, b, c, d) = (\frac{15}{64}, 0, \frac{17}{192}, 0)$ or $(0, -\frac{15}{4}, -\frac{7}{3}, 0)$. Under the first choice, the (unique) solution of (6.8) is:

$$y = \frac{1}{192}x^4 - \frac{1}{6}x + \frac{1}{4},$$

by using the method of undetermined coefficients or the method of parameters variation.

When $h(x) = \sin(x)$, there is no solution to (6.8), no matter what $\vec{\gamma} = (a, b, c, d)$ is chosen. This can also be checked by using the above either method mentioned.

9. Since

$$p(0) = 2 - \sin(0) = 2, p(\pi) = 2 - \sin(\pi) = 2,$$

$$\alpha = \begin{pmatrix} 2 & \frac{5}{7} \\ 7 & 3 \end{pmatrix}, \beta = \begin{pmatrix} 1 & 2 \\ 6 & 13 \end{pmatrix},$$

we have

$$|\alpha| = 1, |\beta| = 1,$$
$$p(0)|\alpha| = 2 = p(\pi)|\beta|,$$

and so L is symmetric by Proposition 1.5.

10. By equation (4.5), we need to show that

$$(B\tilde{\Phi})\vec{c} = \vec{\gamma} - B\tilde{g}$$

has a unique solution $\vec{c}$. But this is true, since $(\vec{\gamma} - B\tilde{g}) \in \mathbb{C}^2$ by the assumption of $m = 2$ and since $B\tilde{\Phi}$ has rank $(2 - k) = 2 - 0 = 2$ by Lemma 4.9. Here $k = 0$ follows from the assumption that the problem π_2 has only the trivial zero solution.

CHAPTER 5

Green Functions

1. Introduction

The functions that appear in this chapter are all assumed complex-valued. Consider the Sturm-Liouville boundary value problems:

$$(p(x)y')' + q(x)y = f(x), a < x < b$$
$$y(a) + \alpha_1 y'(a) = 0 \tag{1.1}$$
$$y(b) + \alpha_2 y'(b) = 0.$$

Here it is assumed that $p(x), p'(x), q(x)$, and $f(x)$ are continuous on $[a, b]$, that $|p(x)| \geq \delta > 0$ for some $\delta > 0$, and that $a < b$ are reals, and α_1 and α_2, complex numbers.

To rewrite (1.1) as a more compact form, let $L : D(L) \subset L^2(a, b) \longrightarrow L^2(a, b)$ be a linear differential operator defined by

$$Lu \equiv (p(x)u')' + q(x)u$$

for

$$u \in D(L) \equiv \{v \in C^2[a, b] : v(a) + \alpha_1 v'(a) = 0, v(b) + \alpha_2 v'(b) = 0\}.$$

Here $C^2[a, b]$ is the vector space of twice continuously differentiable functions on $[a, b]$, and $L^2(a, b)$ is the vector space of squarely Lebesgue integrable complex-valued functions on (a, b)(see Subsection 5.2). Using the L, the above boundary value problem can be written as the compact expression

$$Ly = f(x)$$
$$y \in D(L), \tag{1.2}$$

and the form

$$Lu = (p(x)u')' + q(x)$$

that L takes is called divergence form.

We assume that the homogeneous equation:

$$Ly = 0, y \in D(L)$$

only has the trivial solution $y \equiv 0$. Then it follows from Theorem 2.9 in Chapter 4 that (1.1) has a unique solution. It is the purpose of this chapter to find the so-called Green function $G(x, \xi)$ and express the unique solution of (1.1) in terms of this $G(x, \xi)$. In fact, more general (1.1) will also be studied.

The Green function $G(x, \xi)$ is a function that is defined on $[a, b] \times [a, b]$ and satisfies

(C1) $G(x, \xi)$ is jointly continuous in x, ξ, that is, $G(x, \xi)$ is continuous on the square $\{(x, \xi) : a \leq x \leq b, a \leq \xi \leq b\}$.

Using the Green function $G(x, \xi)$, we can represent the unique solution $u(x)$ of the equation (1.1) by

$$u(x) = \int_a^b G(x, \xi) f(\xi)\, d\xi.$$

To derive this Green function, the method presented here is to utilize the so-called Dirac delta function $\delta(x-\xi)$. This is a symbolic function, not a real function, that is created for convenience. This function is symbolically characterized by the two properties:

(D1) $\delta(x - \xi) = 0$ for $x \neq \xi$.

(D2) For a piecewise continuous function f on (c, d), where (c, d) is any subset of $(-\infty, \infty)$, it is true that

$$\int_c^d \delta(x - \xi) f(x)\, dx = \begin{cases} 0 & \text{if } \xi \notin (c, d); \\ f(\xi) & \text{if } \xi \in (c, d). \end{cases}$$

At $x = \xi$, $\delta(x - \xi)$ is thought of as ∞.

In order for

$$u(x) \equiv \int_{\xi=a}^b G(x, \xi) f(\xi)\, d\xi$$

to be a solution to the equation (1.1), we should have that u satisfies the boundary conditions and that $Lu = f$. The condition $Lu = f$ implies (see Subsection 5.3 for details) that

$$\int_{x=a}^b f(x)\phi(x)\, dx = \int_{x=a}^b Lu(x)\phi(x)\, dx$$

$$= \int_{x=a}^b [\int_{\xi=a}^b L_x G(x, \xi) f(\xi)\, d\xi]\phi(x)\, dx$$

is true for any twice continuously differentiable function ϕ on $[a, b]$, with ϕ, ϕ' vanishing at the end points $\{a, b\}$ of $[a, b]$. Here we use L_x to represent the L that acts on the variable x. If the above expression is compared to the second property (D2) of the delta function $\delta(x - \xi)$, we are led to require that

$$L_x G(x, \xi) = \delta(x - \xi).$$

This implies that

(C2) $L_x G(x, \xi) = 0$ for $a \leq x < \xi \leq b$ and $a \leq \xi < x \leq b$.

On the other hand, integrating, with respect to x, both sides of

$$L_x G(x, \xi) = \delta(x - \xi),$$

gives (see Subsection 5.3 for details) that

(C3) $\lim_{x \to \xi^+} G_x(x, \xi) - \lim_{x \to \xi^-} G_x(x, \xi) = \frac{1}{p(\xi)}$, where $G_x(x, \xi) = \frac{\partial}{\partial x} G(x, \xi)$.

To complete the finding of the $G(x, \xi)$, the last condition required by $G(x, \xi)$ is that the boundary conditions are met:

(C4)

$$G(a, \xi) + \alpha_1 G_x(a, \xi) = 0.$$

$$G(b, \xi) + \alpha_2 G_x(b, \xi) = 0.$$

Thus, the Green function $G(x, \xi)$ will be found by using the conditions $(C1), (C2),$ $(C3),$ and $(C4)$.

The idea for characterizing the symbolic delta function $\delta(x - \xi)$ is explained detailedly in Section 5.

The rest of this chapter is organized as follows. Section 2 states the main results, and Section 3 studies examples for illustration. Section 4 deals with the proof of the main results, and Section 5 provides the needed theoretical material for deriving the symbolic delta function $\delta(x - \xi)$. Finally, Section 6 contains a problems set, and their solutions are presented in Section 7.

2. Main Results

Define the linear differential operator

$$L : D(L) \subset L^2(a, b) \longrightarrow L^2(a, b),$$

by

$$\begin{cases} Lu = (p(x)u')' + q(x)u, \\ D(L) \equiv \{v \in C^2[a, b] : v(a) + \alpha_1 v'(a) = 0, v(b) + \alpha_2 v'(b) = 0\}, \end{cases}$$

where assumed are that the functions $p(x), p'(x),$ and $q(x)$ are continuous on $[a, b]$, that $p(x) \geq \delta > 0$ for some constant δ, and that the constants α_1 and α_2 are complex numbers.

THEOREM 2.1. *By assuming that the equation*

$$Ly = 0, \quad u \in D(L),$$

only has the trivial zero solution, it follows that there exists a so-called Green function $G(x, \xi) : [a, b] \times [a, b] \longrightarrow \mathbb{R}$, *which is determined uniquely by the conditions:*

(C1) *(Smoothness Condition)* $G(x, \xi)$ *is jointly continuous in* x, ξ, *that is,* $G(x, \xi)$ *is continuous on the square* $\{(x, \xi) : a \leq x \leq b, a \leq \xi \leq b\}$.

(C2) *(Solution of Homogeneous Equation)* $G(x, \xi)$ *satisfies:*

$$L_x G(x, \xi) = 0$$

for $a \leq x < \xi \leq b$ *and* $a \leq \xi < x \leq b$, *where* L_x *is the* L *that acts on* x.

(C3) *(Jump Condition)*

$$\lim_{x \to \xi^+} G_x(x, \xi) - \lim_{x \to \xi^-} G_x(x, \xi) = \frac{1}{p(\xi)}.$$

(C4) *(Boundary Conditions)*

$$G(a, \xi) + \alpha_1 G_x(a, \xi) = 0.$$

$$G(b, \xi) + \alpha_2 G_x(b, \xi) = 0.$$

Moreover, the Green function $G(x, \xi)$ *possesses the properties:*

(P1) *(Piecewise Smoothness)* $G_x(x, \xi)(\equiv \frac{\partial}{\partial x}G(x, \xi))$ *and* $G_{xx}(\equiv \frac{\partial^2}{\partial x^2}G(x, \xi))$ *are continuous on each of the triangles* $\{(x, \xi) : a \leq x \leq \xi \leq b\}$ *and* $\{(x, \xi) : a \leq \xi \leq x \leq b\}$, *except on the diagonal segment* $\{(x, \xi) : x = \xi, a \leq x, \xi \leq b\}$.

(P2) *(Dual Jump Condition)*

$$\lim_{\xi \to x^-} G_x(x, \xi) - \lim_{\xi \to x^+} G_x(x, \xi) = \frac{1}{p(x)}.$$

(P3) *(Symmetry Property)* $G(x, \xi) = G(\xi, x)$, *that is,* $G(x, \xi)$ *is symmetric in* x *and* ξ.

(P4) *(Uniqueness)* $G(x, \xi)$ *is unique.*

(P5) *(Solution of Nonhomogeneous Equation)* $u(x) \equiv \int_a^b G(\xi, x) f(\xi)\, d\xi$ *is the unique solution to the equation* (1.1) *for a given continuous function* f.

Boundary conditions that are more general than that in (1.1) are these:

THEOREM 2.2. *Theorem 2.1 remains true under the more general boundary condition of the form, Coddington and Levinson* [**5**, *Page* 297] *and Taylor and Lay* [**21**, *Page* 347]:

$$\begin{aligned}
\alpha_{11} y(a) + \alpha_{12} y'(a) + \beta_{11} y(b) + \beta_{12} y'(b) &= 0 \\
\alpha_{21} y(a) + \alpha_{22} y'(a) + \beta_{21} y(b) + \beta_{22} y'(b) &= 0,
\end{aligned} \tag{2.1}$$

where assumed is that α_{ij} *and* $\beta_{ij}, i, j = 1, 2$, *are complex numbers, such that*

$$|\alpha| \begin{vmatrix} u(a) & v(a) \\ u'(a) & v'(a) \end{vmatrix} = |\beta| \begin{vmatrix} u(b) & v(b) \\ u'(b) & v'(b) \end{vmatrix} \quad or \quad |\alpha| W(a) = |\beta| W(b) \tag{2.2}$$

for all $u, v \in C^2[a, b]$. *Here* $W(x)$ *is the Wronskian of* u *and* v, *and the determinants* $|\alpha|$ *and* $|\beta|$ *are given by*

$$|\alpha| = \begin{vmatrix} \beta_{11} & \beta_{12} \\ \beta_{21} & \beta_{22} \end{vmatrix} \quad and \quad |\beta| = \begin{vmatrix} \alpha_{11} & \alpha_{12} \\ \alpha_{21} & \alpha_{22} \end{vmatrix},$$

respectively. Furthermore, the boundary condition (2.1) *can be rewritten as the compact form:*

$$By \equiv (\alpha \quad \beta) \begin{pmatrix} y(a) \\ y'(a) \\ y(b) \\ y'(b) \end{pmatrix} = \alpha \begin{pmatrix} y(a) \\ y'(a) \end{pmatrix} + \beta \begin{pmatrix} y(b) \\ y'(b) \end{pmatrix} = \begin{pmatrix} 0 \\ 0 \\ 0 \\ 0 \end{pmatrix}. \tag{2.3}$$

The Step 3 in the proof of Theorem 2.2, given in Section 4, will show that

PROPOSITION 2.3. *The Green's function* $G(x, \xi)$ *in Theorem 2.2, once existing, is symmetric:* $G(x, \xi) = G(\xi, x)$, *if and only if the condition* (2.2) *is satisfied.*

REMARK 2.4. *This result seems new.*

Thus Proposition 2.3 can be used to check if the Green's function is symmetric, provided that the differential operator L in

$$Ly = (py')' + qy = f$$

is of divergence form. When

$$Ly = p(x) y'' + q(x) y' + r(x) y = f(x)$$

and L is not of divergence form, one can transform it to be so, by the multiplication $\exp\{\int \frac{q}{p}\, dx\} \frac{1}{p}$ [**5**, page 208]:

$$\begin{aligned}
\exp\{\int \frac{q}{p}\, dx\} \frac{1}{p} f &= \exp\{\int \frac{q}{p}\, dx\} \frac{1}{p} [py'' + qy' + ry] \\
&= [\exp\{\int \frac{q}{p}\, dx\} y']' + \exp\{\int \frac{q}{p}\, dx\} \frac{1}{p} ry \tag{2.4} \\
&= \tilde{L} y.
\end{aligned}$$

Here $\tilde{L}$ takes the divergence form, and Proposition 2.3 can then be used.

Here note that the condition (2.2) is not needed for the existence of the Green's function $G(x, \xi)$. It is instead used to ensure the symmetry property of $G(x, \xi)$ once $G(x, \xi)$ exists.

On the other hand, the corresponding linear differential operator L in Theorem 2.2 is also symmetric by Chapter 4, if either α_{ij} and $\beta_{ij}, i, j = 1, 2$, are reals, such that

$$p(a) \begin{vmatrix} \beta_{11} & \beta_{12} \\ \beta_{21} & \beta_{22} \end{vmatrix} = p(a)|\beta|$$

$$= p(b) \begin{vmatrix} \alpha_{11} & \alpha_{12} \\ \alpha_{21} & \alpha_{22} \end{vmatrix} = p(b)|\alpha|$$

and that both $|\alpha|$ and $|\beta|$ are not zero, or else α and β taking the forms

$$\alpha = \begin{pmatrix} c_1 & c_2 \\ 0 & 0 \end{pmatrix} \quad \text{and} \quad \beta = \begin{pmatrix} 0 & 0 \\ d_1 & d_2 \end{pmatrix},$$

respectively, are such that the complex numbers c_1, c_2, d_1, and d_2 satisfy

$$c_1^2 + c_2^2 \neq 0 \quad \text{and} \quad d_1^2 + d_2^2 \neq 0.$$

Here notice that the symmetry of L is different from that of $G(x, \xi)$.

Although the symmetry property of L is not necessarily needed here for the existence of the Green function $G(x, \xi)$, it indeed implies that, for some $c \in \mathbb{R}$, the equation

$$(L - c)y = 0, y \in D(L)$$

only has the trivial solution (see Chater 6, Section 4, Lemma 4.2). This is significant, in that the operator $(L - c)$ for this c will satisfy the requirements in Theorem 2.1 or 2.2, and give rise to the unique existence of the corresponding Green's function $G(x, \xi)$.

The above general boundary conditions in (2.1) or (2.3) includes the following as special cases:

- zero Dirichlet Boundary condition: $y(a) = 0 = y(b)$.
- zero Neumann boundary condition: $y'(a) = 0 = y'(b)$.
- zero Robin boundary condition:

$$y(a) + \alpha_1 y'(a) = 0, y(b) - \alpha_2 y'(b) = 0, \alpha_1, \alpha_2 > 0.$$

- Periodic boundary condition: $y(a) = y(b), y'(a) = y'(b)$.

PROPOSITION 2.5. *The above four special cases, except for the periodic one, all satisfy the condition* (2.2), *and so their corresponding Green's functions* $G(x, \xi)$*'s are symmetric. The periodic one will give rise to a symmetric* $G(x, \xi)$ *if and only if* $p(a) = p(b)$.

PROOF. See the Problems set in Section 6.   □

In fact, Theorem 2.2 is not only true for linear second order differential operator L; it is also true for higher order L, and L is allowed to take non-divergence form [5, Page 192] (see below). However, the symmetry property of the Green's function $G(x, \xi)$ need not hold here.

Assume that the linear n-th order $(n \geq 2)$ differential operator

$$L : D(L) \subset L^2(a, b) \longrightarrow L^2(a, b)$$

is defined by

$$Ly = p_0(x)y^{(n)} + p_1(x)y^{(n-1)} + \cdots + p_n(x)y$$

for $y \in D(L)$, the domain of L, which consists of n times continuously differentiable functions on $[a, b]$ that satisfies the boundary condition:

$$\begin{pmatrix} 0 \\ 0 \\ \vdots \\ 0 \end{pmatrix} = By \equiv (\alpha \quad \beta) \begin{pmatrix} y(a) \\ y'(a) \\ \vdots \\ y^{n-1}(a) \\ y(b) \\ y'(b) \\ \vdots \\ y^{n-1}(b) \end{pmatrix} \tag{2.5}$$

$$= \begin{pmatrix} \alpha_{11} & \alpha_{12} & \cdot & \cdot & \alpha_{1n} \\ \alpha_{21} & \alpha_{22} & \cdot & \cdot & \alpha_{2n} \\ \cdot & & & & \cdot \\ \alpha_{n1} & & & & \alpha_{nn} \end{pmatrix} \begin{pmatrix} y(a) \\ y'(a) \\ \cdot \\ \cdot \\ y^{(n-1)}(a) \end{pmatrix} + \begin{pmatrix} \beta_{11} & \beta_{12} & \cdot & \cdot & \beta_{1n} \\ \beta_{21} & \beta_{22} & \cdot & \cdot & \beta_{2n} \\ \cdot & & & & \\ \beta_{n1} & & & & \beta_{nn} \end{pmatrix} \begin{pmatrix} y(b) \\ y'(b) \\ \cdot \\ \cdot \\ y^{(n-1)}(b) \end{pmatrix}$$

Here α and β are given two $n \times n$ matrices of complex numbers, and the functions $p_i(x), i = 0, 1, \ldots, n$, are assumed continuous on $[a, b]$ with $p_0(x) \geq \delta > 0$ for some constant δ.

THEOREM 2.6. *By assuming that the equation*

$$Ly = 0, y \in D(L)$$

only has the trivial zero solution, it follows that there exists a unique Green function $G(x, \xi)$ [5, Page 192] which is determined by the conditions $(C1)' - (C4)'$ below, similar to the conditions $(C1)$-$(C4)$ in Theorem 2.1:

(C1)′ *(Smoothness Condition)* $\frac{\partial^k}{\partial x^k}G(x, \xi), k = 0, 1, \ldots, n - 2$, *are jointly continuous in x, ξ, that is, $\frac{\partial^k}{\partial x^k}G(x, \xi)$, $k = 0, 1, \ldots, n-2$), are continuous on the square $\{(x, \xi) : a \leq x \leq b, a \leq \xi \leq b\}$.*

(C2)′ *(Solution of Homogeneous Equation)*

$$L_x G(x, \xi) = 0$$

for $a \leq x < \xi \leq b$ and $a \leq \xi < x \leq b$;

(C3)′ *(Jump Condition)*

$$\lim_{x \to \xi^+} \frac{\partial^{n-1}}{\partial x^{n-1}}G(x, \xi) - \lim_{x \to \xi^-} \frac{\partial^{n-1}}{\partial x^{n-1}}G(x, \xi) = \frac{1}{p_0(\xi)};$$

(C4)′ *(Boundary Conditions) $G(x, \xi)$, with respect to the variable x, satisfies the boundary condition in (2.5).*

Moreover, the Green function $G(x, \xi)$ possesses the properties in Theorem 2.1, except for the symmetry one:

(P1)′ *(Piecewise Smoothness) $\frac{\partial^k}{\partial x^k}G(x, \xi), k = n - 1, n$, are continuous on each of the triangles $\{(x, \xi) : a \leq x \leq \xi \leq b\}$ and $\{(x, \xi) : a \leq \xi \leq x \leq b\}$, except on the diagonal segment $\{(x, \xi) : x = \xi, a \leq x, \xi \leq b\}$.*

(P2)' *(Dual Jump Condition)*

$$\lim_{\xi \to x^-} \frac{\partial^{n-1}}{\partial x^{n-1}} G(x,\xi) - \lim_{\xi \to x^+} \frac{\partial^{n-1}}{\partial x^{n-1}} G(x,\xi) = \frac{1}{p_0(x)}.$$

(P4)' *(Uniqueness)* $G(x,\xi)$ *is unique.*

(P5)' *(Solution of Nonhomogeneous Equation)* $u \equiv \int_a^b G(x,\xi) f(\xi)\, d\xi$ *is the unique solution to the equation*

$$Ly = f(x), y \in D(L)$$

for a given continuous function f.

Here note that the symmetry property for the $G(x,\xi)$ in Theorem 2.6 need not hold since L is of non-divergence form.

Theorem 2.1, 2.2, or 2.6, all assumes that the corresponding homogeneous equation

$$Ly = 0, \quad y \in D(L)$$

only has the zero solution, and in order for this to be true, an equivalent condition is:

PROPOSITION 2.7. *The equation*

$$Ly = 0, \quad y \in D(L)$$

in Theorem 2.6 only has the zero solution if and only if

$$\left| Bu_1 \quad Bu_2 \quad \cdots \quad Bu_n \right| = \left| (\alpha \quad \beta) \begin{pmatrix} u_1(a) & u_2(a) & \cdots & u_n(a) \\ u_1'(a) & u_2'(a) & \cdots & u_n'(a) \\ \vdots & & & \\ u_1^{(n-1)}(a) & u_2^{(n-1)}(a) & \cdots & u_n^{(n-1)}(a) \\ u_1(b) & u_2(b) & \vdots & u_n(b) \\ u_1'(b) & u_2'(b) & \cdots & u_n'(b) \\ \vdots & & & \\ u_1^{(n-1)}(b) & u_2^{(n-1)}(b) & \cdots & u_n^{(n-1)}(b) \end{pmatrix} \right|$$

$$\neq 0$$

for a fundamental set $\{u_1, u_2, \ldots, u_n\}$ *of solutions for* $Ly = 0$.

PROOF. Let $\{u_1, \ldots, u_n\}$ be a fundamental set of solutions for $Ly = 0$. Choose constants $\{c_i\}_{i=1}^n$, such that

$$\phi = c_1 u_1 + c_2 u_2 + \cdots + c_n u_n$$

is a solution of the homogeneous equation

$$Ly = 0, \quad y \in D(L).$$

In order this to be true, ϕ must satisfy, by (2.5),

$$\begin{pmatrix} 0 \\ 0 \\ \vdots \\ 0 \end{pmatrix} = B\phi = \sum_{i=1}^n c_i Bu_i$$

$$= \begin{pmatrix} Bu_1 & Bu_2 & \cdots & Bu_n \end{pmatrix} \begin{pmatrix} c_1 \\ c_2 \\ \vdots \\ c_n \end{pmatrix}.$$

This algebraic equation only has the zero solution, $0 = c_1 = \cdots = c_n$, that is,

$$Ly = 0, \quad y \in D(L)$$

only has the zero solution, if and only if, by using linear algebra,

$$\begin{vmatrix} Bu_1 & Bu_2 & \cdots & Bu_n \end{vmatrix} \neq 0.$$

$\square$

REMARK 2.8. *The domain of the differential operator L in Theorem 2.2 or 2.6 can be defined in terms of functions, which are absolutely continuous, instead of being twice continuously differentiable. In that case, a unique Green function exists similarly. See the end of Section 5 for a discussion of this.*

3. Examples

Before we prove the main results in Section 2, we look at some examples.

EXAMPLE 3.1. Consider the equation with zero Dirichlet boundary condition:

$$y'' + 4y = f(x), \quad 0 < x < 1,$$
$$y(0) = 0 = y(1),$$

where f is a continuous function on $[0, 1]$. Find the corresponding Green's function $G(x, \xi)$, and express the solution y in terms of $G(x, \xi)$.

Solution. By Theorem 2.1, the Green's function $G(x, \xi$, if existing, takes the form

$$G(x, \xi) = \begin{cases} d_1 \cos(2x) + d2 \sin(2x), & \text{if } x < \xi; \\ c_1 \cos(2x) + c_2 \sin(2x), & \text{if } x > \xi, \end{cases}$$

where $\{\cos(2x), \sin(2x)\}$ is a fundamental set of solutions for the homogeneous equation

$$y'' + 4y = 0,$$

and c_1, c_2, d_1, and d_2 satisfy

$$d_1 \cos(2\xi) + d_2 \sin(2\xi) = c_1 \cos(2\xi) + c_2 \sin(2\xi),$$
$$[-2c_1 \sin(2\xi) + 2c_2 \cos(2\xi)] - [-2d_1 \sin(2\xi) + 2d_2 \cos(2\xi)] = 1,$$
$$d_1 = 0,$$
$$c_1 \cos(2) + c_2 \sin(2) = 0.$$

After being solved, c_1, c_2, d_1, and d_2 are given by

$$c_1 = -\frac{1}{2} \sin(2\xi),$$
$$c_2 = \frac{1}{2}[\cot(2)] \sin(2\xi),$$
$$d_1 = 0,$$
$$d_2 = -\frac{1}{2} \cos(2\xi) + \frac{1}{2}[\cot(2)] \sin(2\xi).$$

It follows that

$$G(x,\xi) = \begin{cases} [-\frac{1}{2}\cos(2\xi) + \frac{1}{2}(\cot(2))\sin(2\xi)]\sin(2x), & \text{if } x < \xi; \\ -\frac{1}{2}[\sin(2\xi)]\cos(2x) + [\frac{1}{2}(\cot(2))(\sin(2\xi))]\sin(2x), & \text{if } x > \xi, \end{cases}$$

and that the solution y is expressed as

$$y(x) = \int_{\xi=0}^{1} G(x,\xi) f(\xi)\, d\xi.$$

It is readily checked that $G(x,\xi)$ is symmetric:

$$G(x,\xi) = G(\xi,x).$$

On the other hand, since

$$(\alpha \quad \beta) = \begin{pmatrix} 1 & 0 & 0 & 0 \\ 0 & 0 & 1 & 0 \end{pmatrix}, \{u_1, u_2\} = \{\cos(2x), \sin(2x)\},$$

we have

$$\begin{vmatrix} Bu_1 & Bu_2 \end{vmatrix} = \left| \begin{pmatrix} 1 & 0 & 0 & 0 \\ 0 & 0 & 1 & 0 \end{pmatrix} \begin{pmatrix} 1 & 0 \\ 0 & 2 \\ \cos(2) & \sin(2) \\ -2\sin(2) & 2\cos(2) \end{pmatrix} \right|$$

$$= \begin{vmatrix} 1 & 0 \\ \cos(2) & \sin(2) \end{vmatrix} = \sin(2) \neq 0,$$

and so the Green's function $G(x,\xi)$ exists by Proposition 2.7. $G(x,\xi)$ is symmetric by Proposition 2.5 since $|\alpha| = 0 = |\beta|$.

EXAMPLE 3.2. Consider the equation with zero Neumann boundary condition:

$$y'' + 4y = f(x), \quad 0 < x < 1,$$
$$y'(0) = 0 = y'(1).$$

where f is a continuous function on $[0,1]$. Find the Green's function $G(x,\xi)$, and express the solution y in terms of $G(x,\xi)$.

Solution. By Theorem 2.1, the Green's function $G(x,\xi$, if existing, takes the form

$$G(x,\xi) = \begin{cases} d_1\cos(2x) + d2\sin(2x), & \text{if } x < \xi; \\ c_1\cos(2x) + c_2\sin(2x), & \text{if } x > \xi, \end{cases}$$

where $\{\cos(2x), \sin(2x)\}$ is a fundamental set of solutions for the homogeneous equation

$$y'' + 4y = 0,$$

and c_1, c_2, d_1, and d_2 satisfy

$$d_1\cos(2\xi) + d_2\sin(2\xi) = c_1\cos(2\xi) + c_2\sin(2\xi),$$
$$[-2c_1\sin(2\xi) + 2c_2\cos(2\xi)] - [-2d_1\sin(2\xi) + 2d_2\cos(2\xi)] = 1,$$
$$2d_2 = 0,$$
$$-2c_1\sin(2) + 2c_2\cos(2) = 0.$$

After being solved, c_1, c_2, d_1, and d_2 are given by

$$c_1 = \frac{1}{2}[\cot(2)]\cos(2\xi),$$

$$c_2 = \frac{1}{2}\cos(2\xi),$$

$$d_1 = \frac{1}{2}\sin(2\xi) + \frac{1}{2}[\cot(2)]\cos(2\xi),$$

$$d_2 = 0.$$

It follows that

$$G(x,\xi) = \begin{cases} \frac{1}{2}[\sin(2\xi)]\cos(2x) + \frac{1}{2}(\cot(2))(\cos(2\xi))\cos(2x), & \text{if } x < \xi; \\ \frac{1}{2}[\cot(2)][\cos(2\xi)]\cos(2x) + \frac{1}{2}(\cos(2\xi))\sin(2x), & \text{if } x > \xi, \end{cases}$$

and that the solution y is expressed as

$$y(x) = \int_{\xi=0}^{1} G(x,\xi)f(\xi)\,d\xi.$$

It is readily checked that $G(x,\xi)$ is symmetric:

$$G(x,\xi) = G(\xi,x).$$

On the other hand, since

$$(\alpha \quad \beta) = \begin{pmatrix} 0 & 1 & 0 & 0 \\ 0 & 0 & 0 & 1 \end{pmatrix}, \{u_1, u_2\} = \{\cos(2x), \sin(2x)\},$$

we have

$$|Bu_1 \quad Bu_2| = \left| \begin{pmatrix} 0 & 1 & 0 & 0 \\ 0 & 0 & 0 & 1 \end{pmatrix} \begin{pmatrix} 1 & 0 \\ 0 & 2 \\ \cos(2) & \sin(2) \\ -2\sin(2) & 2\cos(2) \end{pmatrix} \right|$$

$$= \left| \begin{matrix} 0 & 2 \\ -2\sin(2) & 2\cos(2) \end{matrix} \right| = 4\sin(2) \neq 0,$$

and so the Green's function $G(x,\xi)$ exists by Proposition 2.7. $G(x,\xi)$ is symmetric by Proposition 2.5 since $|\alpha| = 0 = |\beta|$.

EXAMPLE 3.3. Consider the equation with zero Robin boundary condition:

$$y'' + 4y = f(x), \quad 0 < x < 1,$$

$$y(0) - y'(0) = 0,$$

$$y(1) + y'(1) = 0,$$

where f is a continuous function on $[0,1]$. Find the corresponding Green's function $G(x,\xi)$, and express the solution y in terms of $G(x,\xi)$.

Solution. By Theorem 2.1, the Green's function $G(x,\xi$, if existing, takes the form

$$G(x,\xi) = \begin{cases} d_1\cos(2x) + d2\sin(2x), & \text{if } x < \xi; \\ c_1\cos(2x) + c_2\sin(2x), & \text{if } x > \xi, \end{cases}$$

where $\{\cos(2x), \sin(2x)\}$ is a fundamental set of solutions for the homogeneous equation

$$y'' + 4y = 0,$$

and c_1, c_2, d_1, and d_2 satisfy

$$d_1\cos(2\xi) + d_2\sin(2\xi) = c_1\cos(2\xi) + c_2\sin(2\xi),$$

$$[-2c_1\sin(2\xi) + 2c_2\cos(2\xi)] - [-2d_1\sin(2\xi) + 2d_2\cos(2\xi)] = 1,$$

$$d_1 - 2d_2 = 0,$$

$$c_1[\cos(2) - 2\sin(2)] + c_2[\sin(2) + 2\cos(2)] = 0.$$

After being solved, c_1, c_2, d_1, and d_2 are given by

$$c_1 = \frac{\sin(2) + 2\cos(2)}{-4\cos(2) + 3\sin(2)}[\frac{1}{2}\sin(2\xi) + \cos(2\xi)],$$

$$c_2 = \frac{\cos(2) - 2\sin(2)}{4\cos(2) - 3\sin(2)}[\frac{1}{2}\sin(2\xi) + \cos(2\xi)],$$

$$d_1 = \frac{4\sin(2) - 2\cos(2)}{-4\cos(2) + 3\sin(2)}[\frac{1}{2}\sin(2\xi)] + \frac{\sin(2) + 2\cos(2)}{-4\cos(2) + 3\sin(2)}[\cos(2\xi)],$$

$$d_2 = \frac{2\sin(2) - \cos(2)}{-4\cos(2) + 3\sin(2)}[\frac{1}{2}\sin(2\xi)] + \frac{sin(2) + 2\cos(2)}{-4\cos(2) + 3\sin(2)}[\frac{1}{2}\cos(2\xi)].$$

It follows that

$$G(x,\xi)$$

$$= \begin{cases} \{\frac{4\sin(2)-2\cos(2)}{-4\cos(2)+3\sin(2)}[\frac{1}{2}\sin(2\xi)] + \frac{\sin(2)+2\cos(2)}{-4\cos(2)+3\sin(2)}[\cos(2\xi)]\}\cos(2x)+ \\ \{\frac{2\sin(2)-\cos(2)}{-4\cos(2)+3\sin(2)}[\frac{1}{2}\sin(2\xi)] + \frac{sin(2)+2\cos(2)}{-4\cos(2)+3\sin(2)}[\frac{1}{2}\cos(2\xi)]\}\sin(2x), & \text{if } x < \xi; \\ \{\frac{\sin(2)+2\cos(2)}{-4\cos(2)+3\sin(2)}[\frac{1}{2}\sin(2\xi) + \cos(2\xi)]\}\cos(2x)+ \\ \{\frac{\cos(2)-2\sin(2)}{4\cos(2)-3\sin(2)}[\frac{1}{2}\sin(2\xi) + \cos(2\xi)]\}\sin(2x), & \text{if } x > \xi, \end{cases}$$

and that the solution y is expressed as

$$y(x) = \int_{\xi=0}^{1} G(x,\xi)f(\xi)\,d\xi.$$

It is readily checked that $G(x,\xi)$ is symmetric:

$$G(x,\xi) = G(\xi,x).$$

On the other hand, since

$$(\alpha \quad \beta) = \begin{pmatrix} 1 & -1 & 0 & 0 \\ 0 & 0 & 1 & 1 \end{pmatrix}, \{u_1, u_2\} = \{\cos(2x), \sin(2x)\},$$

we have

$$|Bu_1 \quad Bu_2| = |\begin{pmatrix} 1 & -1 & 0 & 0 \\ 0 & 0 & 1 & 1 \end{pmatrix} \begin{pmatrix} 1 & 0 \\ 0 & 2 \\ \cos(2) & \sin(2) \\ -2\sin(2) & 2\cos(2) \end{pmatrix}|$$

$$= \begin{vmatrix} 1 & -2 \\ (\cos(2) - 2\sin(2)) & (\sin(2) + 2\cos(2)) \end{vmatrix} = 4\cos(2) - \sin(2) \neq 0,$$

and so the Green's function $G(x,\xi)$ exists by Proposition 2.7. $G(x,\xi)$ is symmetric by Proposition 2.5 since $|\alpha| = 0 = |\beta|$.

EXAMPLE 3.4. Consider the equation with periodic boundary condition:

$$y'' + 4y = f(x), \quad 0 < x < 1,$$

$$y(0) = y(1),$$

$$y'(0) = y'(1),$$

where f is a continuous function on $[0,1]$. Find the corresponding Green's function $G(x,\xi)$, and express the solution y in terms of $G(x,\xi)$.

Solution. By Theorem 2.1, the Green's function $G(x, \xi)$, if existing, takes the form

$$G(x, \xi) = \begin{cases} d_1 \cos(2x) + d2 \sin(2x), & \text{if } x < \xi; \\ c_1 \cos(2x) + c_2 \sin(2x), & \text{if } x > \xi, \end{cases}$$

where $\{\cos(2x), \sin(2x)\}$ is a fundamental set of solutions for the homogeneous equation

$$y'' + 4y = 0,$$

and c_1, c_2, d_1, and d_2 satisfy

$$d_1 \cos(2\xi) + d_2 \sin(2\xi) = c_1 \cos(2\xi) + c_2 \sin(2\xi),$$
$$[-2c_1 \sin(2\xi) + 2c_2 \cos(2\xi)] - [-2d_1 \sin(2\xi) + 2d_2 \cos(2\xi)] = 1,$$
$$d_1 = c_1 \cos(2) + c_2 \sin(2),$$
$$2d_2 = -2c_1 \sin(2) + 2c_2 \cos(2).$$

After being solved, c_1, c_2, d_1, and d_2 are given by

$$c_1 = -\frac{1}{4} \sin(2\xi) + \frac{1}{4} \frac{\sin(2)}{1 - \cos(2)} \cos(2\xi),$$

$$c_2 = \frac{1}{4} \cos(2\xi) + \frac{1}{4} \frac{\sin(2)}{1 - \cos(2\xi)} \sin(2\xi),$$

$$d_1 = \frac{1}{4} \sin(2\xi) + \frac{1}{4} \frac{\sin(2)}{1 - \cos(2)} \cos(2\xi),$$

$$d_2 = -\frac{1}{4} \cos(2\xi) + \frac{1}{4} \frac{\sin(2)}{1 - \cos(2)} \sin(2\xi).$$

It follows that

$$G(x, \xi) = \begin{cases} [\frac{1}{4} \sin(2\xi) + \frac{1}{4} \frac{\sin(2)}{1 - \cos(2)} \cos(2\xi)] \cos(2x) + \\ [-\frac{1}{4} \cos(2\xi) + \frac{1}{4} \frac{\sin(2)}{1 - \cos(2)} \sin(2\xi)] \sin(2x) & \text{if } x < \xi; \\ [-\frac{1}{4} \sin(2\xi) + \frac{1}{4} \frac{\sin(2)}{1 - \cos(2)} \cos(2\xi)] \cos(2x) + \\ [\frac{1}{4} \cos(2\xi) + \frac{1}{4} \frac{\sin(2)}{1 - \cos(2\xi)} \sin(2\xi)] \sin(2x) & \text{if } x > \xi, \end{cases}$$

and that the solution y is expressed as

$$y(x) = \int_{\xi=0}^{1} G(x, \xi) f(\xi) \, d\xi.$$

It is readily checked that $G(x, \xi)$ is symmetric:

$$G(x, \xi) = G(\xi, x).$$

On the other hand, since

$$(\alpha \quad \beta) = \begin{pmatrix} 1 & 0 & -1 & 0 \\ 0 & 1 & 0 & -1 \end{pmatrix}, \{u_1, u_2\} = \{\cos(2x), \sin(2x)\},$$

we have

$$|Bu_1 \quad Bu_2| = \left| \begin{pmatrix} 1 & 0 & -1 & 0 \\ 0 & 1 & 0 & -1 \end{pmatrix} \begin{pmatrix} 1 & 0 \\ 0 & 2 \\ \cos(2) & \sin(2) \\ -2\sin(2) & 2\cos(2) \end{pmatrix} \right|$$

$$= \begin{vmatrix} (1 - \cos(2)) & -\sin(2) \\ 2\sin(2) & (2 - 2\cos(2)) \end{vmatrix} = 4(1 - \cos(2)) \neq 0,$$

and so the Green's function $G(x, \xi)$ exists by Proposition 2.7. $G(x, \xi)$ is symmetric by Proposition 2.5 since $|\alpha| = 0 = |\beta|$.

EXAMPLE 3.5. Consider the equation with the boundary conditions:

$$y'' + 4y = f(x), \quad 0 < x < 1,$$
$$y(0) = 2y(1),$$
$$y'(0) = 2y'(1),$$

where f is a continuous function on $[0, 1]$. Find the corresponding Green's function $G(x, \xi)$, and express the solution y in terms of $G(x, \xi)$.

Solution. By Theorem 2.1, the Green's function $G(x, \xi$, if existing, takes the form

$$G(x, \xi) = \begin{cases} d_1 \cos(2x) + d2 \sin(2x), & \text{if } x < \xi; \\ c_1 \cos(2x) + c_2 \sin(2x), & \text{if } x > \xi, \end{cases}$$

where $\{\cos(2x), \sin(2x)\}$ is a fundamental set of solutions for the homogeneous equation

$$y'' + 4y = 0,$$

and c_1, c_2, d_1, and d_2 satisfy

$$d_1 \cos(2\xi) + d_2 \sin(2\xi) = c_1 \cos(2\xi) + c_2 \sin(2\xi),$$
$$[-2c_1 \sin(2\xi) + 2c_2 \cos(2\xi)] - [-2d_1 \sin(2\xi) + 2d_2 \cos(2\xi)] = 1,$$
$$d_1 = 2c_1 \cos(2) + 2c_2 \sin(2),$$
$$2d_2 = -4c_1 \sin(2) + 4c_2 \cos(2) = 0.$$

After being solved, c_1, c_2, d_1, and d_2 are given by

$$c_1 = \frac{\cos(2) - \frac{1}{2}}{5 - 4\cos(2)} \sin(2\xi) + \frac{\sin(2)}{5 - 4\cos(2)} \cos(2\xi),$$

$$c_2 = \frac{-\cos(2) + \frac{1}{2}}{5 - 4\cos(2)} \cos(2\xi) + \frac{\sin(2)}{5 - 4\cos(2)} \sin(2\xi),$$

$$d_1 = \frac{-\cos(2) + 2}{5 - 4\cos(2)} \sin(2\xi) + \frac{\sin(2)}{5 - 4\cos(2)} \cos(2\xi),$$

$$d_2 = \frac{\cos(2) - 2}{5 - 4\cos(2)} \cos(2\xi) + \frac{\sin(2)}{5 - 4\cos(2)} \sin(2\xi).$$

It follows that

$$G(x, \xi) = \begin{cases} [\frac{-\cos(2)+2}{5-4\cos(2)} \sin(2\xi) + \frac{\sin(2)}{5-4\cos(2)} \cos(2\xi)] \cos(2x) + \\ [\frac{\cos(2)-2}{5-4\cos(2)} \cos(2\xi) + \frac{\sin(2)}{5-4\cos(2)} \sin(2\xi)] \sin(2x), & \text{if } x < \xi; \\ [\frac{\cos(2)-\frac{1}{2}}{5-4\cos(2)} \sin(2\xi) + \frac{\sin(2)}{5-4\cos(2)} \cos(2\xi)] \cos(2x) + \\ [\frac{-\cos(2)+\frac{1}{2}}{5-4\cos(2)} \cos(2\xi) + \frac{\sin(2)}{5-4\cos(2)} \sin(2\xi)] \sin(2x), & \text{if } x > \xi, \end{cases}$$

and that the solution y is expressed as

$$y(x) = \int_{\xi=0}^{1} G(x, \xi) f(\xi) \, d\xi.$$

However, it is readily checked that $G(x, \xi)$ is not symmetric:

$$G(x, \xi) \neq G(\xi, x).$$

On the other hand, since

$$(\alpha \quad \beta) = \begin{pmatrix} 1 & 0 & -2 & 0 \\ 0 & 1 & 0 & -2 \end{pmatrix}, \{u_1, u_2\} = \{\cos(2x), \sin(2x)\},$$

we have

$$|Bu_1 \quad Bu_2| = \left| \begin{pmatrix} 1 & 0 & -2 & 0 \\ 0 & 1 & 0 & -2 \end{pmatrix} \begin{pmatrix} 1 & 0 \\ 0 & 2 \\ \cos(2) & \sin(2) \\ -2\sin(2) & 2\cos(2) \end{pmatrix} \right|$$

$$= \begin{vmatrix} (1 - 2\cos(2)) & -2\sin(2) \\ 4\cos(2) & (2 - 4\cos(2)) \end{vmatrix} = 2(5 - 4\cos(2)) \neq 0,$$

and so the Green's function $G(x, \xi)$ exists by Proposition 2.7. However, $G(x, \xi)$ is not symmetric by Proposition 2.5 since

$$|\alpha| = 1, \quad |\beta| = 4,$$

$$W(0) = \begin{vmatrix} 1 & 0 \\ 0 & 2 \end{vmatrix}, \quad W(1) = \begin{vmatrix} \cos(2) & \sin(2) \\ -2\sin(2) & 2\cos(2) \end{vmatrix} = 2,$$

$$|\alpha|W(0) = 2 \neq 4 = |\beta|W(1).$$

EXAMPLE 3.6. Consider the equation with the boundary conditions of simply supported(hinged) end [**3**, page 677, 8th edition]:

$$y^{(4)} + 4y'' = f(x), \quad 0 < x < 1,$$
$$y(0) = 0 = y''(0),$$
$$y(1) = 0 = y''(0),$$

where f is a continuous function on $[0, 1]$. Find the corresponding Green's function $G(x, \xi)$, and express the solution y in terms of $G(x, \xi)$.

Solution. By Theorem 2.1, the Green's function $G(x, \xi)$, if existing, takes the form

$$G(x, \xi) = \begin{cases} d_1 + d_2 x + d_3 \cos(2x) + d4 \sin(2x), & \text{if } x < \xi; \\ c_1 + c_2 x + c_3 \cos(2x) + c_4 \sin(2x), & \text{if } x > \xi, \end{cases}$$

where $\{1, x, \cos(2x), \sin(2x)\}$ is a fundamental set of solutions for the homogeneous equation

$$y^{(4)} + 4y'' = 0,$$

and c_i and $d_i, i = 1, 2, 3, 4$, satisfy

$$c_1 + c_2 \xi + c_3 \cos(2\xi) + c_4 \sin(2\xi) = d_1 + d_2 \xi + d_3 \cos(2\xi) + d_4 \sin(2\xi),$$
$$c_2 - 2c_3 \sin(2\xi) + 2d_4 \cos(2\xi) = d_2 - 2d_3 \sin(2\xi) + 2d_4 \cos(2\xi),$$
$$-4c_3 \cos(2\xi) - 4c_4 \sin(2\xi) = -4d_3 \cos(2\xi) - 4d_4 \sin(2\xi),$$
$$[8c_3 \sin(2\xi) - 8c_4 \cos(2\xi)] - [8d_3 \sin(2\xi) - 8d_4 \cos(2\xi)] = 1,$$
$$d_1 + d_3 = 0,$$
$$-4d_3 = 0,$$
$$c_1 + c_2 + c_3 \cos(2) + c_4 \sin(2) = 0,$$

$$- 4c_3 \cos(2) - 4c_4 \sin(2) = 0.$$

After being solved, c_i and $d_i, i = 1, 2, 3, 4$, are given by

$$c_1 = -\frac{1}{4}\xi,$$

$$c_2 = \frac{1}{4}\xi,$$

$$c_3 = \frac{1}{8}\sin(2\xi),$$

$$c_4 = -\frac{1}{8}[\cot(2)]\sin(2\xi),$$

$$d_1 = 0,$$

$$d_2 = \frac{1}{4}\xi - \frac{1}{4},$$

$$d_3 = 0,$$

$$d_4 = \frac{1}{8}\cos(2\xi) - \frac{1}{8}[\cot(2)]\sin(2\xi).$$

It follows that

$$G(x, \xi) = \begin{cases} (\frac{1}{4}\xi - \frac{1}{4})x + [\frac{1}{8}\cos(2\xi) - \frac{1}{8}(\cot(2))\sin(2\xi)]\sin(2x), & \text{if } x < \xi; \\ -\frac{1}{4}\xi + \frac{1}{4}\xi x + \frac{1}{8}[\sin(2\xi)]\cos(2x) + \\ [\frac{1}{8}\cos(2\xi) - \frac{1}{8}(\cot(2))(\sin(2\xi))]\sin(2x), & \text{if } x > \xi, \end{cases}$$

and that the solution y is expressed as

$$y(x) = \int_{\xi=0}^{1} G(x, \xi) f(\xi)\, d\xi.$$

On the other hand, since

$$(\alpha \quad \beta) = \begin{pmatrix} 1 & 0 & 0 & 0 & 0 & 0 & 0 & 0 \\ 0 & 0 & 1 & 0 & 0 & 0 & 0 & 0 \\ 0 & 0 & 0 & 0 & 1 & 0 & 0 & 0 \\ 0 & 0 & 0 & 0 & 0 & 0 & 1 & 0 \end{pmatrix},$$

$$\{u_1, u_2, u_3, u_4\} = \{1, x, \cos(2x), \sin(2x)\},$$

we have

$$|Bu_1 \quad Bu_2 \quad Bu_3 \quad Bu_4|$$

$$= \left| \begin{pmatrix} 1 & 0 & 0 & 0 & 0 & 0 & 0 & 0 \\ 0 & 0 & 1 & 0 & 0 & 0 & 0 & 0 \\ 0 & 0 & 0 & 0 & 1 & 0 & 0 & 0 \\ 0 & 0 & 0 & 0 & 0 & 0 & 1 & 0 \end{pmatrix} \begin{pmatrix} 1 & 0 & 1 & 0 \\ 0 & 1 & 0 & 2 \\ 0 & 0 & -4 & 0 \\ 0 & 0 & 0 & -8 \\ 1 & 1 & \cos(2) & \sin(2) \\ 0 & 1 & -2\sin(2) & 2\cos(2) \\ 0 & 0 & -4\cos(2) & -4\sin(2) \\ 0 & 0 & 8\sin(2) & -8\cos(2) \end{pmatrix} \right|$$

$$= \begin{vmatrix} 1 & 0 & 1 & 0 \\ 0 & 0 & -4 & 0 \\ 1 & 1 & \cos(2) & \sin(2) \\ 0 & 0 & -4\cos(2) & -4\sin(2) \end{vmatrix} = -16\sin(2) \neq 0,$$

and so the Green's function $G(x, \xi)$ exists by Proposition 2.7.

EXAMPLE 3.7. Consider the equation with zero boundary condition of periodic type:

$$x^2 y'' + xy' - y = f(x), \quad 1 < x < 2,$$
$$y(1) + 2y(2) = 0,$$
$$y'(1) + y'(2) = 0,$$

where f is a continuous function on $[1, 2]$. Find the corresponding Green's function $G(x, \xi)$, and express the solution y in terms of $G(x, \xi)$.

Solution. Method 1. By Theorem 2.1, the Green's function $G(x, \xi)$, if existing, takes the form

$$G(x, \xi) = \begin{cases} d_1 x + d_2 x^{-1}, & \text{if } x < \xi; \\ c_1 x + c_2 x^{-1}, & \text{if } x > \xi, \end{cases}$$

where $\{x, x^{-1}\}$ is a fundamental set of solutions for the homogeneous equation

$$x^2 y'' + xy' - y = 0,$$

and c_1, c_2, d_1, and d_2 satisfy

$$c_1 \xi + c_2 \xi^{-1} = d_1 \xi + d_2 \xi^{-1},$$
$$(c_1 - c_2 \xi^{-2}) - (d_1 - d_2 \xi^{-2}) = \frac{1}{\xi^2},$$
$$d_1 + d_2 + +4c_1 + c_2 = 0,$$
$$d_1 - d_2 + c_1 - \frac{1}{4} c_2 = 0.$$

After being solved, c_1, c_2, d_1, and d_2 are given by

$$c_1 = \frac{13}{82} \xi^{-2} + \frac{3}{82},$$
$$c_2 = \frac{6}{41} \xi^{-2} - \frac{14}{41},$$
$$d_1 = -\frac{14}{41} \xi^{-2} + \frac{3}{82},$$
$$d_2 = -\frac{6}{41} \xi^{-2} + \frac{13}{82}.$$

It follows that

$$G(x, \xi) = \begin{cases} (-\frac{14}{41} \xi^{-2} + \frac{3}{82}) x + (-\frac{6}{41} \xi^{-2} + \frac{13}{82}) x^{-1}, & \text{if } x < \xi; \\ (\frac{13}{82} \xi^{-2} + \frac{3}{82}) x + (-\frac{6}{41} \xi^{-2} - \frac{14}{41}) x^{-1}, & \text{if } x > \xi, \end{cases}$$

and that the solution y is expressed as

$$y(x) = \int_{\xi=0}^{1} G(x, \xi) f(\xi) \, d\xi.$$

It is readily checked that $G(x, \xi)$ is not symmetric:

$$G(x, \xi) \neq G(\xi, x).$$

On the other hand, since

$$(\alpha \quad \beta) = \begin{pmatrix} 1 & 0 & 0 & 0 \\ 0 & 0 & 1 & 0 \end{pmatrix}, \{u_1, u_2\} = \{x, x^{-1}\},$$

we have

$$|Bu_1 \quad Bu_2| = \left| \begin{pmatrix} 1 & 0 & 0 & 0 \\ 0 & 0 & 1 & 0 \end{pmatrix} \begin{pmatrix} 1 & 1 \\ 1 & -1 \\ 2 & \frac{1}{2} \\ 1 & -\frac{1}{4} \end{pmatrix} \right|$$

$$= \begin{vmatrix} 1 & 1 \\ 2 & \frac{1}{2} \end{vmatrix} = -\frac{3}{2} \neq 0,$$

and so the Green's function $G(x, \xi)$ exists by Proposition 2.7.

Method 2. By (2.4), the equation is transformed to

$$(xy')' - \frac{1}{x}y = \frac{f}{x},$$

and the corresponding L is of divergence form. By Theorem 2.1, the Green's function $G(x, \xi)$, if existing, takes the form

$$G(x, \xi) = \begin{cases} d_1 x + d2x^{-1}, & \text{if } x < \xi; \\ c_1 x + c_2 x^{-1}, & \text{if } x > \xi, \end{cases}$$

where $\{x, x^{-1}\}$ is a fundamental set of solutions for the homogeneous equation

$$(xy')' - \frac{1}{x}y = x^2 y'' + xy' - y$$

$$= 0,$$

and c_1, c_2, d_1, and d_2 satisfy

$$c_1\xi + c_2\xi^{-1} = d_1\xi + d_2\xi^{-1},$$

$$(c_1 - c_2\xi^{-2}) - (d_1 - d_2\xi^{-2}) = \frac{1}{\xi},$$

$$d_1 + d_2 + +4c_1 + c_2 = 0,$$

$$d_1 - d_2 + c_1 - \frac{1}{4}c_2 = 0.$$

After being solved, c_1, c_2, d_1, and d_2 are given by

$$c_1 = \frac{13}{82}\xi^{-1} + \frac{3}{82}\xi,$$

$$c_2 = \frac{6}{41}\xi^{-1} - \frac{14}{41}\xi,$$

$$d_1 = -\frac{14}{41}\xi^{-1} + \frac{3}{82}\xi,$$

$$d_2 = -\frac{6}{41}\xi^{-1} + \frac{13}{82}\xi.$$

It follows that

$$G(x, \xi) = \begin{cases} (-\frac{14}{41}\xi^{-1} + \frac{3}{82}\xi)x + (-\frac{6}{41}\xi^{-1} + \frac{13}{82}\xi)x^{-1}, & \text{if } x < \xi; \\ (\frac{13}{82}\xi^{-1} + \frac{3}{82}\xi)x + (-\frac{6}{41}\xi^{-1} - \frac{14}{41}\xi)x^{-1}, & \text{if } x > \xi, \end{cases}$$

and that the solution y is expressed as

$$y(x) = \int_{\xi=0}^{1} G(x, \xi)\frac{f(\xi)}{\xi} \, d\xi.$$

It is readily checked that $G(x, \xi)$ is symmetric:

$$G(x, \xi) = G(\xi, x).$$

On the other hand, since

$$(\alpha \quad \beta) = \begin{pmatrix} 1 & 0 & 0 & 0 \\ 0 & 0 & 1 & 0 \end{pmatrix}, \{u_1, u_2\} = \{x, x^{-1}\},$$

we have

$$|Bu_1 \quad Bu_2| = \left| \begin{pmatrix} 1 & 0 & 0 & 0 \\ 0 & 0 & 1 & 0 \end{pmatrix} \begin{pmatrix} 1 & 1 \\ 1 & -1 \\ 2 & \frac{1}{2} \\ 1 & -\frac{1}{4} \end{pmatrix} \right|$$

$$= \left| \begin{matrix} 1 & 1 \\ 2 & \frac{1}{2} \end{matrix} \right| = -\frac{3}{2} \neq 0,$$

and so the Green's function $G(x, \xi)$ exists by Proposition 2.7. $G(x, \xi)$ is symmetric by Proposition 2.5, since

$$|\alpha| = 1, \quad |\beta| = 2,$$

$$W(1) = \left| \begin{matrix} 1 & 1 \\ 1 & -1 \end{matrix} \right| = -2, \quad W(2) = \left| \begin{matrix} 2 & \frac{1}{2} \\ 1 & -\frac{1}{4} \end{matrix} \right| = -1,$$

$$|\alpha|W(1) = -2 = |\beta|W(2).$$

Here note that, while the Green's function by Method 1 is not symmetric, the Green's function by Method 2 is so.

4. Proof of the Main Results

Define the linear differential operator

$$L : D(L) \subset L^2(a, b) \longrightarrow L^2(a, b)$$

by

$$\begin{cases} Lu = (p(x)u')' + q(x)u, \\ D(L) \equiv \{v \in C^2[a, b] : v(a) + \alpha_1 v'(a) = 0, v(b) + \alpha_2 v'(b) = 0\}, \end{cases}$$

where assumed are that the functions $p(x), p'(x),$ and $q(x)$ are continuous on $[a, b]$, that $p(x) \geq \delta > 0$ for some constant δ, and that the constants α_1 and α_2 are complex numbers.

We now prove:

Theorem 2.1. *By assuming that the equation*

$$Ly = 0, \quad u \in D(L),$$

only has the trivial zero solution, it follows that there exists a so-called Green function $G(x, \xi) : [a, b] \times [a, b] \longrightarrow \mathbb{R}$, which is determined uniquely by the conditions:

(C1) *(Smoothness Condition) $G(x, \xi)$ is jointly continuous in x, ξ, that is, $G(x, \xi)$ is continuous on the square $\{(x, \xi) : a \leq x \leq b, a \leq \xi \leq b\}$.*

(C2) *(Solution of Homogeneous Equation) $G(x, \xi)$ satisfies:*

$$L_x G(x, \xi) = 0$$

for $a \leq x < \xi \leq b$ and $a \leq \xi < x \leq b$, where L_x is the L that acts on x.

(C3) *(Jump Condition)*

$$\lim_{x \to \xi^+} G_x(x, \xi) - \lim_{x \to \xi^-} G_x(x, \xi) = \frac{1}{p(\xi)}.$$

(C4) *(Boundary Conditions)*

$$G(a, \xi) + \alpha_1 G_x(a, \xi) = 0.$$

$$G(b, \xi) + \alpha_2 G_x(b, \xi) = 0.$$

Moreover, the Green function $G(x, \xi)$ possesses the properties:

(P1) *(Piecewise Smoothness)* $G_x(x, \xi)(\equiv \frac{\partial}{\partial x} G(x, \xi))$ *and* $G_{xx}(\equiv \frac{\partial^2}{\partial x^2} G(x, \xi))$ *are continuous on each of the triangles* $\{(x, \xi) : a \le x \le \xi \le b\}$ *and* $\{(x, \xi) : a \le \xi \le x \le b\}$, *except on the diagonal segment* $\{(x, \xi) : x = \xi, a \le x, \xi \le b\}$.

(P2) *(Dual Jump Condition)*

$$\lim_{\xi \to x^-} G_x(x, \xi) - \lim_{\xi \to x^+} G_x(x, \xi) = \frac{1}{p(x)}.$$

(P3) *(Symmetry Property)* $G(x, \xi) = G(\xi, x)$, *that is, $G(x, \xi)$ is symmetric in x and ξ.*

(P4) *(Uniqueness) $G(x, \xi)$ is unique.*

(P5) *(Solution of Nonhomogeneous Equation)* $u(x) \equiv \int_a^b G(\xi, x) f(\xi) \, d\xi$ *is the unique solution to the equation (1.1) for a given continuous function f.*

PROOF. The proof is divided into six steps.

Step 1. Since by the condition (C2),

$$L_x G(x, \xi) = 0$$

for $a \le x < \xi \le b$ and for $a \le \xi < x \le b$, it follows from the existence theorem, Corollary 2.2, in Chapter 8, for the linear ordinary differential equations with the initial conditions that there are two linearly independent solutions u_1, u_2 to the equation $Ly(x) = 0$, so that

$$G(x, \xi) = \begin{cases} c_1 u_1(x) + c_2 u_2(x) & \text{for } a \le x < \xi \le b; \\ d_1 u_1(x) + d_2 u_2(x) & \text{for } a \le \xi < x \le b \end{cases}$$

for some constants c_1, c_2, d_1, d_2. Those constants will be determined by the boundary conditions (C4), the continuity condition (C1) of $G(x, \xi)$ at $x = \xi$, and the jump condition (C3) for $G_x(x, \xi)$ at $x = \xi$.

Step 2. Choose c_1, c_2, d_1, d_2 such that

$$\lim_{x \to \xi^+} G(x, \xi) = \lim_{x \to \xi^-} G(x, \xi),$$

that

$$\lim_{x \to \xi^+} G_x(x, \xi) - \lim_{x \to \xi^-} G_x(x, \xi) = \frac{1}{p(\xi)},$$

and that $G(x, \xi)$ satisfies the boundary condition at $x = a$ and $x = b$.

Equivalently,

$$c_1 u_1(\xi) + c_2 u_2(\xi) = d_1 u_1(\xi) + d_2 u_2(\xi)$$

$$-[c_1 u_1'(\xi) + c_2 u_2'(\xi)] + [d_1 u_1'(\xi) + d_2 u_2'(\xi)] = \frac{1}{p(\xi)}$$

$$[c_1 u_1(a) + c_2 u_2(a)] + \alpha_1 [c_1 u_1'(a) + c_2 u_2'(a)] = 0$$
$$[d_1 u_1(b) d_2 u_2(b)] + \alpha_2 [d_1 u_1'(b) + d_2 u_2'(b)] = 0$$

or

$$(c_1 - d_1) u_1(\xi) + (c_2 - d_2) u_2(\xi) = 0$$
$$-(c_1 - d_1) u_1'(\xi) - (c_2 - d_2) u_2'(\xi) = \frac{1}{p(\xi)}$$

$$c_1 M + c_2 N = 0 \tag{4.1}$$
$$d_1 J + d_2 K = 0, \tag{4.2}$$

where

$$M = u_1(a) + \alpha_1 u_1'(a), \quad N = u_2(a) + \alpha_1 u_2'(a),$$
$$J = u_1(b) + \alpha_2 u_1'(b), \quad K = u_2(b) + \alpha_2 u_2'(b).$$

Since u_1 and u_2 are linearly independent, for which the Wronskian

$$W(u_1, u_2)(\xi) = \begin{vmatrix} u_1(\xi) & u_2(\xi) \\ u_1'(\xi) & u_2'(\xi) \end{vmatrix} \neq 0,$$

it follows that

$$(\lambda, \mu) \equiv (c_1 - d_1, c_2 - d_2), \tag{4.3}$$

by the Cramer's rule from the linear algebra, exists and is given by

$$
\lambda = \frac{\begin{vmatrix} 0 & u_2(\xi) \\ \frac{1}{p(\xi)} & -u_2'(\xi) \end{vmatrix}}{-W(u_1, u_2)(\xi)} = \frac{u_2(\xi)}{p(\xi) W(u_1, u_2)(\xi)}
$$

$$
\mu = \frac{\begin{vmatrix} u_1(\xi) & 0 \\ -u_1'(\xi) & \frac{1}{p(\xi)} \end{vmatrix}}{-W(u_1, u_2)(\xi)} = \frac{-u_1(\xi)}{p(\xi) W(u_1, u_2)(\xi)}. \tag{4.4}
$$

Substitution of $c_1 = \lambda + d_1$ and $c_2 = \mu + d_2$ into (4.1), in conjuction with (4.2), gives

$$d_1 M + d_2 N = -\lambda M - \mu N$$
$$d_1 J + d_2 K = 0. \tag{4.5}$$

It can be claimed that

$$\begin{vmatrix} M & N \\ J & K \end{vmatrix} \neq 0,$$

so that $d_1, d_2, c_1,$ and c_2 are solvable, and that Green's function $G(x, \xi)$ is constructed.

For if the claim is not true, then we have, by using linear algebra [9],

$$s \begin{pmatrix} N \\ K \end{pmatrix} + t \begin{pmatrix} M \\ J \end{pmatrix} = 0$$

for some constants s and t, not all zero. This, combined with the fact that u_1 satisfies

$$L u_1 = 0$$
$$u_1(a) + \alpha_1 u_1'(a) = M$$

$$u_1(b) + \alpha_2 u_1'(b) = J$$

and the fact that u_2 satisfies

$$Lu_2 = 0$$
$$u_2(a) + \alpha_2 u_2'(a) = N$$
$$u_2(b) + \alpha_2 u_2'(b) = K,$$

implies that $(tu_1 + su_2)$ is a solution to the equation

$$Ly = 0, y \in D(L).$$

Since zero is the only solution for that equation by assumption, it follows that

$$tu_1 + su_2 = 0,$$

and so u_1 and u_2 are linearly dependent. This is a contradiction to the linear independence of u_1 and u_2, and so the claim is proved.

Step 2' In fact, one can show

$$p(\xi)W(u_1, u_2)(\xi) = C, \quad \text{a constant for all } \xi, \tag{4.6}$$

and then it follows from (4.3), (4.4), and (4.5) that the numbers $c_1, c_2, d_1,$ and d_2 are given by

$$c_1 = \frac{u_1(\xi)NK - u_2(\xi)NJ}{CD}, c_2 = \frac{-u_1(\xi)MK + u_2(\xi)MJ}{CD},$$
$$d_1 = \frac{u_1(\xi)NK - u_2(\xi)MK}{CD}, d_2 = \frac{-u_1(\xi)NJ + u_2(\xi)MJ}{CD}, \tag{4.7}$$

where $D = \begin{vmatrix} M & N \\ J & K \end{vmatrix}$. To derive (4.6), one utilizes the two equations

$$(pu_1')' + qu_1 = 0,$$
$$(pu_2')' + qu_2 = 0,$$

that are satisfied by u_1 and u_2, respectively, by assumption. Multiply the top equation by u_2, and substract it from bottom equation that is multiplied by u_1. Then, by integration by parts formula, it follows that

$$p(x)W(u_1, u_2)(x)|_{x=x_1}^{x_2} = p(x)[u_1 u_2' - u_2 u_1']|_{x=x_1}^{x_2} = 0$$

for all $x_1, x_2 \in [a, b]$, and so $p(x)W(u_1, u_2)(x)$ is a constant.

Step 3. (Smoothness and Uniqueness of $G(x, \xi)$) Step 2 shows that the constructed $G(x, \xi)$ is jointly continuous in x, ξ, and that $G_x(x, \xi)$ is continuous in x, ξ, except for $x = \xi$, where a finite jump $\frac{1}{p(\xi)}$ exists. From this, it follows that

$$\max_{x, \xi \in [a,b]} |G_x(x, \xi)|$$

is uniformly finite.

Since

$$L_x G(x, \xi) = \frac{d}{dx}(p(x)G_x(x, \xi)) + q(x)G(x, \xi) = 0$$

for $x \neq \xi$, it results that $G_{xx} \equiv \frac{\partial^2}{\partial x^2}G(x, \xi)$ can be expressed in terms of $G_x(x, \xi)$ and $G(x, \xi)$:

$$G_{xx}(x, \xi) = \frac{-p'(x)G_x(x, \xi) - q(x)G(x, \xi)}{p(x)}$$

for $x \neq \xi$. This shows that $G_{xx}(x,\xi)$ is continuous in x,ξ, except for $x = \xi$, where $G_{xx}(x,\xi)$ has a finite jump, and so

$$\max_{x,\xi \in [a,b]} |G_{xx}(x,\xi)|$$

is uniformly finite.

To show uniqueness of the Green function, let $G(x,\xi)$ and $H(x,\xi)$ be two Green functions and let $F(x,\xi) = G(x,\xi) - H(x,\xi)$. Since the jump $\frac{1}{p(\xi)}$ of $G_x(x,\xi)$ at $x = \xi$ cancels that of $H_x(x,\xi)$, $F_x(x,\xi)$ has no jump at $x = \xi$, and so $F_x(x,\xi)$ is continuous everywhere. It follows that $F_{xx}(x,\xi)$ is also continuous everywhere, since $F_{xx}(x,\xi)$ can be expressed in terms of continuous $F_x(x,\xi)$ and $F(x,\xi)$. This, combined with the fact that

$$L_x F(x,\xi) = 0 \quad \text{for } a \leq x < \xi \leq b \text{ and } a \leq \xi < x \leq b,$$

yields

$$\lim_{x \to \xi} L_x F(x,\xi) = \lim_{x \to \xi^-} L_x F(x,\xi) = 0 = \lim_{x \to \xi^+} L_x F(x,\xi),$$

and so

$$L_x F(x,\xi) = 0 \quad \text{for } a \leq x \leq \xi \leq b \text{ and } a \leq \xi \leq x \leq b.$$

Because $F(x,\xi)$, as a function of x, also lies in $D(L)$ as $G(x,\xi)$ and $H(x,\xi)$ do, it follows that $\{F(x,\xi), 0\}$ are two solutions to the equation:

$$Ly(x) = 0, y \in D(L).$$

But this equation only has the zero solution by assumption, and so $F(x,\xi) \equiv 0$ or $G(x,\xi) \equiv H(x,\xi)$.

Step 4. (Symmetry of $G(x,\xi)$ in x,ξ) Substituting the c_1, c_2, d_1, and d_2 in (4.7), into the Green's function $G(x,\xi)$, one readily check that $G(x,\xi) = G(\xi,x)$, the desired symmetry property. The details are left to the reader (see the Problems set in Section 6).

Step 5. The jump condition

$$\lim_{x \to \xi^+} G_x(x,\xi) - \lim_{x \to \xi^-} G_x(x,\xi) = \frac{1}{p(\xi)}$$

produces, by the continuity of $p(x)$ in x,

$$\lim_{x \to \xi^-} G_x(x,\xi) = \lim_{x \to \xi^+} [G_x(x,\xi) - \frac{1}{p(\xi)}]$$

This proves (P2), since the right side of the above equation equals

$$\lim_{x = \xi + \epsilon, \epsilon \to 0^+} [G_x(x,\xi) - \frac{1}{p(\xi)}] = \lim_{\xi \to x^-} [G_x(x,\xi) - \frac{1}{p(\xi)}] = \lim_{\xi \to x^-} G_x(x,\xi) - \frac{1}{p(x)},$$

and the left side equals

$$\lim_{\xi \to x^+} G_x(x,\xi).$$

Step 6. (Claim that $u(x) \equiv \int_{\xi=a}^{b} G(x,\xi) f(\xi) \, d\xi$ is a solution to the equation (1.1).)

Let

$$G\{x > \xi\} \equiv G(x,\xi)|_{\{x > \xi\}}$$

be the restriction of $G(x, \xi)$ to the set

$$\{(x, \xi) : b \geq x > \xi \geq a\},$$

and similarly, let

$$G\{x < \xi\} \equiv G(x, \xi)|_{\{x < \xi\}}$$

be the restriction of $G(x, \xi)$ to the set

$$\{(x, \xi) : a \leq x < \xi \leq b\}.$$

Since

$$u(x) = \int_a^b G(x, \xi) f(\xi)\, d\xi = \int_{\xi=a}^x G(x, \xi) f(\xi)\, d\xi + \int_{\xi=x}^b G(x, \xi) f(\xi)\, d\xi$$

$$= \int_a^x G\{x > \xi\} f(\xi)\, d\xi + \int_x^b G\{x < \xi\} f(\xi)\, d\xi$$

by Apostol [**2**, Page 143], it follows from the fundamental theorem of calculus that

$$u'(x) = \int_{\xi=a}^x G_x\{x > \xi\} f(\xi)\, d\xi + \int_{\xi=x}^b G_x\{x < \xi\} f(\xi)\, d\xi$$

$$+ G\{x > \xi\}|_{\xi=x} f(x) - G\{x < \xi\}|_{\xi=x} f(x)$$

$$= \int_{\xi=a}^x G_x\{x > \xi\} f(\xi)\, d\xi + \int_{\xi=x}^b G_x\{x < \xi\} f(\xi)\, d\xi$$

$$+ [\lim_{\xi \to x^-} G(x, \xi) - \lim_{\xi \to x^+} G(x, \xi)] f(x),$$

the third term of which is zero by the continuity condition (C1) of $G(x, \xi)$ at $x = \xi$. Similarly, there results

$$u''(x) = \int_{\xi=a}^x G_{xx}\{x > \xi\} f(\xi)\, d\xi + \int_{\xi=x}^b G_{xx}\{x < \xi\} f(\xi)\, d\xi$$

$$+ G_x\{x > \xi\}|_{\xi=x} f(x) - G_x\{x < \xi\}|_{\xi=x} f(x)$$

$$= \int_{\xi=a}^x G_{xx}\{x > \xi\} f(\xi)\, d\xi + \int_{\xi=x}^b G_{xx}\{x < \xi\} f(\xi)\, d\xi$$

$$+ [\lim_{\xi \to x^-} G_x(x, \xi) - \lim_{\xi \to x^+} G_x(x, \xi)] f(x),$$

in which the third term equals $\frac{1}{p(x)} f(x)$ by the jump condition (C3).

It follows that

$$Lu(x) = \int_{\xi=a}^x L_x G(x, \xi) f(\xi)\, d\xi + \int_{\xi=x}^b L_x G(x, \xi) f(\xi)\, d\xi + f(x),$$

in which the first two terms equal zero by the condition (C2). Hence

$$Lu(x) = f(x)$$

results.

On the other hand, both

$$u(a) + \alpha_1 u'(a) = \int_{\xi=a}^b G(a, \xi) f(\xi)\, d\xi + \alpha_1 \int_{\xi=a}^b G_x(a, \xi)\, d\xi$$

$$= \int_{\xi=a}^b [G(a, \xi) + \alpha_1 G_x(a, \xi)] f(\xi)\, d\xi = 0$$

and, similarly,

$$u(b) + \alpha_2 u'(b) = 0$$

are true, and so $u(x)$ is a solution.   □

Boundary conditions that are more general than that in (1.1) are these:

Theorem 2.2. *Theorem 2.1 remains true under the more general boundary condition of the form, Coddington and Levinson* [**5**, *Page 297*] *and Taylor and Lay* [**21**, *Page 347*]:

$$
\begin{aligned}
\alpha_{11}y(a) + \alpha_{12}y'(a) + \beta_{11}y(b) + \beta_{12}y'(b) &= 0 \\
\alpha_{21}y(a) + \alpha_{22}y'(a) + \beta_{21}y(b) + \beta_{22}y'(b) &= 0,
\end{aligned}
\tag{4.8}
$$

where assumed is that α_{ij} and $\beta_{ij}, i, j = 1, 2$, are complex numbers, such that

$$
|\alpha| \begin{vmatrix} u(a) & v(a) \\ u'(a) & v'(a) \end{vmatrix} = |\beta| \begin{vmatrix} u(b) & v(b) \\ u'(b) & v'(b) \end{vmatrix} \quad or \quad |\alpha| W(a) = |\beta| W(b)
\tag{4.9}
$$

for all $u, v \in C^2[a, b]$. Here $W(x)$ is the Wronskian of u and v, and the determinants $|\alpha|$ and $|\beta|$ are given by

$$
|\alpha| = \begin{vmatrix} \beta_{11} & \beta_{12} \\ \beta_{21} & \beta_{22} \end{vmatrix} \quad and \quad |\beta| = \begin{vmatrix} \alpha_{11} & \alpha_{12} \\ \alpha_{21} & \alpha_{22} \end{vmatrix},
$$

respectively. Furthermore, the boundary condition (4.8) can be rewritten as the compact form:

$$
By \equiv \begin{pmatrix} \alpha & \beta \end{pmatrix} \begin{pmatrix} y(a) \\ y'(a) \\ y(b) \\ y'(b) \end{pmatrix} = \alpha \begin{pmatrix} y(a) \\ y'(a) \end{pmatrix} + \beta \begin{pmatrix} y(b) \\ y'(b) \end{pmatrix} = \begin{pmatrix} 0 \\ 0 \\ 0 \\ 0 \end{pmatrix}.
\tag{4.10}
$$

PROOF. All that need to be proved are the construction of the Green's function $G(x, \xi)$ and the symmetry property $G(x, \xi) = G(\xi, x)$, since the rest are the same as those in the proof of Theorem 2.1.

Step 1. (The construction of $G(x, \xi)$) As in the proof of Theorem 2.1, $G(x, \xi)$ takes the form

$$
G(x, \xi) = \begin{cases} d_1 u_1(x) + d_2 u_2(x), & \text{if } a \leq x < \xi \leq b; \\ c_1 u_1(x) + c_2 u_2(x), & \text{if } a \leq \xi < x \leq b. \end{cases}
$$

Here the constants c_1, c_2, d_1 and $, d_2$ are chosen, such that $G(x, \xi)$ is continuous at $x = \xi$, that $G_x(x, \xi)$ has the jump $\frac{1}{p(\xi)}$ at $x = \xi$, and that $G(x, \xi)$ satisfies the boundary conditions at $x = a$ and $x = b$. That is, $G(x, \xi)$ satisfies

$$
\lim_{x \to \xi^+} G(x, \xi) = \lim_{x \to \xi^-} G(x, \xi),
$$

$$
\lim_{x \to \xi^+} G_x(x, \xi) - \lim_{x \to \xi^-} G_x(x, \xi) = \frac{1}{p(\xi)},
$$

$$
\alpha \begin{pmatrix} G(a, \xi) \\ G_x(a, \xi) \end{pmatrix} + \beta \begin{pmatrix} G(b, \xi) \\ G_x(b, \xi) \end{pmatrix} = 0,
$$

where α is the matrix $(\alpha_{ij})_{2 \times 2}$, and β is the matrix $(\beta_{ij})_{2 \times 2}$. When the above being written out, c_1, c_2, d_1, and d_2 are conditioned by

$$c_1 u_1(\xi) + c_2 u_2(\xi) = d_1 u_1(\xi) + d_2 u_2(\xi),$$

$$[c_1 u_1'(\xi) + c_2 u_2'(\xi)] - [d_1 u_1'(\xi) + d_2 u_2'(\xi)] = \frac{1}{p(\xi)},$$

$$\begin{pmatrix} \alpha_{11} & \alpha_{12} \\ \alpha_{21} & \alpha_{22} \end{pmatrix} \begin{pmatrix} d_1 u_1(a) + d_2 u_2(a) \\ d_1 u_1'(a) + d_2 u_2'(a) \end{pmatrix} + \begin{pmatrix} \beta_{11} & \beta_{12} \\ \beta_{21} & \beta_{22} \end{pmatrix} \begin{pmatrix} c_1 u_1(b) + c_2 u_2(b) \\ c_1 u_1'(b) + c_2 u_2'(b) \end{pmatrix} = 0$$

or

$$(c_1 - d_1) u_1(\xi) + (c_2 - d_2) u_2(\xi) = 0,$$

$$(c_1 - d_1) u_1'(\xi) + (c_2 - d_2) u_2'(\xi) = \frac{1}{p(\xi)},$$

$$d_1 M_{11} + d_2 N_{11} + c_1 M_{12} + c_2 N_{12} = 0, \tag{4.11}$$

$$d_1 J_{11} + d_2 K_{11} + c_1 J_{12} + c_2 K_{12} = 0, \tag{4.12}$$

where

$$M_{11} = \alpha_{11} u_1(a) + \alpha_{12} u_1'(a), \quad N_{11} = \alpha_{11} u_2(a) + \alpha_{12} u_2'(a),$$

$$M_{12} = \beta_{11} u_1(b) + \beta_{12} u_1'(b), \quad N_{12} = \beta_{11} u_2(b) + \beta_{22} u_2'(b),$$

and

$$J_{11} = \alpha_{21} u_1(a) + \alpha_{22} u_1'(a), \quad K_{11} = \alpha_{21} u_2(a) + \alpha_{12} u_2(a),$$

$$J_{12} = \beta_{21} u_1(b) + \beta_{22} u_1'(b), \quad K_{12} = \beta_{21} u_2(b) + \beta_{22} u_2'(b).$$

Since u_1 and u_2 are linearly independent, for which the Wronskian

$$W(u_1, u_2)(\xi) = \begin{vmatrix} u_1(\xi) & u_2(\xi) \\ u_1'(\xi) & u_2'(\xi) \end{vmatrix} \neq 0,$$

it follows that

$$(\lambda, \mu) \equiv (c_1 - d_1, c_2 - d_2), \tag{4.13}$$

by the Cramer's rule from the linear algebra, exists and is given by

$$\lambda = \frac{\begin{vmatrix} 0 & u_2(\xi) \\ \frac{1}{p(\xi)} & u_2'(\xi) \end{vmatrix}}{W(u_1, u_2)(\xi)} = \frac{-u_2(\xi)}{p(\xi) W(u_1, u_2)(\xi)}, \tag{4.14}$$

$$\mu = \frac{\begin{vmatrix} u_1(\xi) & 0 \\ u_1'(\xi) & \frac{1}{p(\xi)} \end{vmatrix}}{W(u_1, u_2)(\xi)} = \frac{u_1(\xi)}{p(\xi) W(u_1, u_2)(\xi)}. \tag{4.15}$$

Substitution of $c_1 = \lambda + d_1$ and $c_2 = \mu + d_2$ into (4.11), in conjunction with (4.12), gives

$$d_1 M + d_2 N = -\lambda M_{12} - \mu N_{12},$$

$$d_1 J + d_2 K = -\lambda J_{12} - \mu K_{12}, \tag{4.16}$$

where

$$M = M_{11} + M_{12}, \quad N = N_{11} + N_{12},$$

$$J = J_{11} + J_{12}, \quad K = K_{11} + K_{12}.$$

It can be claimed that

$$\begin{vmatrix} M & N \\ J & K \end{vmatrix} \neq 0,$$

so that d_1, d_2, c_1, and c_2 are solvable and that $G(x, \xi)$ is constructed.

For if the claim is not true, then we have, by using linear algebra [9],

$$s \begin{pmatrix} N \\ K \end{pmatrix} + t \begin{pmatrix} M \\ J \end{pmatrix} = 0$$

for some constants s and t, not all zero. This, combined with the fact that u_1 satisfies

$$Lu_1 = 0,$$

$$\alpha \begin{pmatrix} u_1(a) \\ u_1'(a) \end{pmatrix} + \beta \begin{pmatrix} u_1(b) \\ u_1'(b) \end{pmatrix} = \begin{pmatrix} M \\ J \end{pmatrix}$$

and the fact that u_2 satisfies

$$Lu_2 = 0,$$

$$\alpha \begin{pmatrix} u_2(a) \\ u_2'(a) \end{pmatrix} + \beta \begin{pmatrix} u_2(b) \\ u_2'(b) \end{pmatrix} = \begin{pmatrix} N \\ K \end{pmatrix},$$

implies that $(tu_1 + su_2)$ is a solution to the equation

$$Ly = 0, y \in D(L).$$

Since zero is the only solution for that equation by assumption, it follows that

$$tu_1 + su_2 = 0,$$

and so u_1, u_2 are linearly dependent. This is a contradiction to the linear independence of u_1 and u_2, and so the claim is proved.

Step 2. As in **Step** $2'$ in the proof of Theorem 2.1, we have

$$p(\xi)W(u_1, u_2)(\xi) = C, \quad \text{a constant for all } \xi,$$

and so it follows from (4.13), (4.14), and (4.16) that the numbers c_1, c_2, d_1, and d_2 are given by

$$c_1 = -\frac{u_1(\xi)}{CD} \begin{vmatrix} N_{12} & N \\ K_{12} & K \end{vmatrix} - \frac{u_2(\xi)}{CD} \begin{vmatrix} M_{11} & N \\ J_{11} & K \end{vmatrix},$$

$$c_2 = \frac{u_1(\xi)}{CD} \begin{vmatrix} M & N_{11} \\ J & K_{11} \end{vmatrix} + \frac{u_2(\xi)}{CD} \begin{vmatrix} M & M_{12} \\ J & J_{12} \end{vmatrix},$$

$$d_1 = -\frac{u_1(\xi)}{CD} \begin{vmatrix} N_{12} & N \\ K_{12} & K \end{vmatrix} + \frac{u_2(\xi)}{CD} \begin{vmatrix} M_{12} & N \\ J_{12} & K \end{vmatrix}, \tag{4.17}$$

$$d_2 = -\frac{u_1(\xi)}{CD} \begin{vmatrix} M & N_{12} \\ J & K_{12} \end{vmatrix} + \frac{u_2(\xi)}{CD} \begin{vmatrix} M & M_{12} \\ J & J_{12} \end{vmatrix},$$

where

$$\begin{pmatrix} M \\ J \end{pmatrix} = \begin{pmatrix} M_{11} + M_{12} \\ J_{11} + J_{12} \end{pmatrix}, \quad \begin{pmatrix} N \\ K \end{pmatrix} = \begin{pmatrix} N_{11} + N_{12} \\ K_{11} + K_{12} \end{pmatrix},$$

$$D = \begin{vmatrix} M & N \\ J & K \end{vmatrix}.$$

Here used was the fact from the linear algebra that

$$\begin{vmatrix} c_{11} + \nu_{11} & c_{12} \\ c_{21} + \nu_{21} & c_{22} \end{vmatrix} = \begin{vmatrix} c_{11} & c_{12} \\ c_{21} & c_{22} \end{vmatrix} + \begin{vmatrix} \nu_{11} & c_{12} \\ \nu_{21} & c_{22} \end{vmatrix}. \tag{4.18}$$

Step 3. (Symmetry property $G(x,\xi) = G(\xi,x)$ of $G(x,\xi)$) Since

$$G(x,\xi) = \begin{cases} d_1(\xi)u_1(x) + d_2(\xi)u_2, & \text{if } x < \xi; \\ c_1(\xi)u_1(x) + c_2(\xi)u_2(x), & \text{if } x > \xi, \end{cases}$$

where $c_1(\xi)), c_2(\xi), d_1(\xi)$, and $d_2(\xi)$ are given in (4.17), it follows that

$$G(\xi,x) = \begin{cases} d_1(x)u_1(\xi) + d_2(x)u_2(\xi), & \text{if } x > \xi; \\ c_1(x)u_1(\xi) + c_2(x)u_2(\xi), & \text{if } x < \xi. \end{cases}$$

Calculations made on $G(\xi,x)$ show that, for $x > \xi$,

$$d_1(x)u_1(\xi) + d_2(x)u_2(\xi)$$

$$= [-\frac{u_1(x)}{CD}\begin{vmatrix} N_{12} & N \\ K_{12} & K \end{vmatrix} + \frac{u_2(x)}{CD}\begin{vmatrix} M_{12} & N \\ J_{12} & K \end{vmatrix}]u_1(\xi)+$$

$$[-\frac{u_1(x)}{CD}\begin{vmatrix} M & N_{12} \\ J & K_{12} \end{vmatrix} + \frac{u_2(x)}{CD}\begin{vmatrix} M & M_{12} \\ J & J_{12} \end{vmatrix}]u_2(\xi)$$

$$= u_1(x)\{-\frac{u_1(\xi)}{CD}\begin{vmatrix} N_{12} & N \\ K_{12} & K \end{vmatrix} - \frac{u_2(\xi)}{CD}[\begin{vmatrix} M_{11} & N_{12} \\ J_{11} & K_{12} \end{vmatrix} + \begin{vmatrix} M_{12} & N_{12} \\ J_{12} & K_{12} \end{vmatrix}]\}+$$

$$u_2(x)\{\frac{u_1(\xi)}{CD}[\begin{vmatrix} M_{12} & N_{11} \\ J_{12} & K_{11} \end{vmatrix} + \begin{vmatrix} M_{12} & N_{12} \\ J_{12} & K_{12} \end{vmatrix}] + \frac{u_2(\xi)}{CD}\begin{vmatrix} M & M_{12} \\ J & J_{12} \end{vmatrix}\},$$

Using the fact in (4.18), this equlas

$$c_1(\xi)u_1(x) + c_2(\xi)u_2(x)$$

$$= u_1(x)\{-\frac{u_1(\xi)}{CD}\begin{vmatrix} N_{12} & N \\ K_{12} & K \end{vmatrix} - \frac{u_2(\xi)}{CD}[\begin{vmatrix} M_{11} & N_{11} \\ J_{11} & K_{11} \end{vmatrix} + \begin{vmatrix} M_{11} & N_{12} \\ J_{11} & K_{12} \end{vmatrix}]\}+$$

$$u_2(x)\{\frac{u_1(\xi)}{CD}[\begin{vmatrix} M_{11} & N_{11} \\ J_{11} & K_{11} \end{vmatrix} + \begin{vmatrix} M_{12} & N_{11} \\ J_{12} & K_{11} \end{vmatrix}] + \frac{u_2(\xi)}{CD}\begin{vmatrix} M & M_{12} \\ J & J_{12} \end{vmatrix}\}$$

if and only if

$$\begin{vmatrix} M_{11} & N_{11} \\ J_{11} & K_{11} \end{vmatrix} = \begin{vmatrix} M_{12} & N_{12} \\ J_{12} & K_{12} \end{vmatrix}.$$

But the latter equation is the same as the condition (2.2), since

$$\begin{vmatrix} M_{11} & N_{11} \\ J_{11} & K_{11} \end{vmatrix} = \left| \alpha \begin{pmatrix} u_1(a) \\ u_1'(a) \end{pmatrix} \quad \alpha \begin{pmatrix} u_2(a) \\ u_2'(a) \end{pmatrix} \right| = |\alpha| \begin{vmatrix} u_1(a) & u_2(a) \\ u_1'(a) & u_2'(a) \end{vmatrix}$$

and

$$\begin{vmatrix} M_{12} & N_{12} \\ J_{12} & K_{12} \end{vmatrix} = \left| \beta \begin{pmatrix} u_1(b) \\ u_1'(b) \end{pmatrix} \quad \beta \begin{pmatrix} u_2(b) \\ u_2'(b) \end{pmatrix} \right| = |\beta| \begin{vmatrix} u_1(b) & u_2(b) \\ u_1'(b) & u_2'(b) \end{vmatrix}.$$

Similarly, calculations made on $G(\xi,x)$ show that, for $x < \xi$,

$$c_1(x)u_1(\xi) + c_2(x)u_2(\xi)$$

$$= [-\frac{u_1(x)}{CD}\begin{vmatrix} N_{12} & N \\ K_{12} & K \end{vmatrix} - \frac{u_2(x)}{CD}\begin{vmatrix} M_{11} & N \\ J_{11} & K \end{vmatrix}]u_1(\xi)+$$

$$[\frac{u_1(x)}{CD}\begin{vmatrix} M & N_{11} \\ J & K_{11} \end{vmatrix} + \frac{u_2(x)}{CD}\begin{vmatrix} M & M_{12} \\ J & J_{12} \end{vmatrix}]u_2(\xi)$$

$$= u_1(x)\{-\frac{u_1(\xi)}{CD} + \frac{u_2(\xi)}{CD}[\begin{vmatrix} M_{11} & N_{11} \\ J_{11} & K_{11} \end{vmatrix} + \begin{vmatrix} M_{12} & N_{11} \\ J_{12} & K_{11} \end{vmatrix}]\}+$$

$$u_2(x)\{-\frac{u_1(\xi)}{CD}[\begin{vmatrix} M_{11} & N_{11} \\ J_{11} & K_{11} \end{vmatrix} + \begin{vmatrix} M_{11} & N_{12} \\ J_{11} & K_{12} \end{vmatrix}] + \frac{u_2(\xi)}{CD}\begin{vmatrix} M & M_{12} \\ J & J_{12} \end{vmatrix}\},$$

and this equals

$$d_1(\xi)u_1(x) + d_2(\xi)u_2(x)$$

$$= u_1(x)\{-\frac{u_1(\xi)}{CD}\begin{vmatrix} N_{12} & N \\ K_{12} & K \end{vmatrix} + \frac{u_2(\xi)}{CD}[\begin{vmatrix} M_{12} & N_{11} \\ J_{12} & K_{11} \end{vmatrix} + \begin{vmatrix} M_{12} & N_{12} \\ J_{12} & K_{12} \end{vmatrix}]\} +$$

$$u_2(x)\{-\frac{u_1(\xi)}{CD}[\begin{vmatrix} M_{11} & N_{12} \\ J_{11} & K_{12} \end{vmatrix} + \begin{vmatrix} M_{12} & N_{12} \\ J_{12} & K_{12} \end{vmatrix}] + \frac{u_2(\xi)}{CD}\begin{vmatrix} M & M_{12} \\ J & J_{12} \end{vmatrix}\}$$

if and only if

$$\begin{vmatrix} M_{11} & N_{11} \\ J_{11} & K_{11} \end{vmatrix} = \begin{vmatrix} M_{12} & N_{12} \\ J_{12} & K_{12} \end{vmatrix}$$

or equivalently, the condition (2.2) holds.

Thus $G(\xi, x) = G(x, \xi)$ if and only if the condition (2.2) is satisfied. □

In fact, Theorem 2.2 is not only true for linear second order differential operator L; it is also true for higher order L, and L is allowed to take non-divergence form [5, Page 192] (see below).

Assume that the linear nth order ($n \geq 2$) differential operator

$$L : D(L) \subset L^2(a, b) \longrightarrow L^2(a, b)$$

is defined by

$$Ly = p_0(x)y^{(n)} + p_1(x)y^{(n-1)} + \cdots + p_n(x)y$$

for $y \in D(L)$, the domain of L, which consists of n times continuously differentiable functions on $[a, b]$ that satisfies the boundary condition:

$$\begin{pmatrix} 0 \\ 0 \\ \vdots \\ 0 \end{pmatrix} = By \equiv (\alpha \quad \beta) \begin{pmatrix} y(a) \\ y'(a) \\ \vdots \\ y^{n-1}(a) \\ y(b) \\ y'(b) \\ \vdots \\ y^{n-1}(b) \end{pmatrix} \tag{4.19}$$

$$= \begin{pmatrix} \alpha_{11} & \alpha_{12} & \cdot & \cdot & \alpha_{1n} \\ \alpha_{21} & \alpha_{22} & & & \alpha_{2n} \\ \vdots & & & & \vdots \\ \alpha_{n1} & & & & \alpha_{nn} \end{pmatrix} \begin{pmatrix} y(a) \\ y'(a) \\ \vdots \\ y^{(n-1)}(a) \end{pmatrix} + \begin{pmatrix} \beta_{11} & \beta_{12} & \cdot & \cdot & \beta_{1n} \\ \beta_{21} & \beta_{22} & & & \beta_{2n} \\ \vdots & & & & \vdots \\ \beta_{n1} & & & & \beta_{nn} \end{pmatrix} \begin{pmatrix} y(b) \\ y'(b) \\ \vdots \\ y^{(n-1)}(b) \end{pmatrix}.$$

Here α and β are given two $n \times n$ matrices of complex numbers, and the functions $p_i(x), i = 0, 1, \ldots, n$ are assumed continuous on $[a, b]$ with $p_0(x) \geq \delta > 0$ for some constant δ.

Theorem 2.6. *By assuming that the equation*

$$Ly = 0, y \in D(L)$$

only has the trivial zero solution, it follows that there exists a unique Green function $G(x, \xi)$ [5, Page 192] *which is determined by the conditions* $(C1)' - (C4)'$ *below, similar to the conditions* $(C1)$-$(C4)$ *in Theorem 2.1:*

(C1)' *(Smoothness Condition)* $\frac{\partial^k}{\partial x^k}G(x,\xi), k = 0, 1, \ldots, n-2$, *are jointly continuous in* x, ξ, *that is,* $\frac{\partial^k}{\partial x^k}G(x,\xi), k = 0, 1, \ldots, n-2$), *are continuous on the square* $\{(x,\xi) : a \le x \le b, a \le \xi \le b\}$.

(C2)' *(Solution of Homogeneous Equation)*

$$L_x G(x,\xi) = 0$$

for $a \le x < \xi \le b$ *and* $a \le \xi < x \le b$;

(C3)' *(Jump Condition)*

$$\lim_{x \to \xi+} \frac{\partial^{n-1}}{\partial x^{n-1}}G(x,\xi) - \lim_{x \to \xi-} \frac{\partial^{n-1}}{\partial x^{n-1}}G(x,\xi) = \frac{1}{p_0(\xi)};$$

(C4)' *(Boundary Conditions)* $G(x,\xi)$, *with respect to the variable* x, *satisfies the boundary condition in* (4.19).

Moreover, the Green function $G(x,\xi)$ *possesses the properties in Theorem 2.1, except for the symmetry one:*

(P1)' *(Piecewise Smoothness)* $\frac{\partial^k}{\partial x^k}G(x,\xi), k = n-1, n$, *are continuous on each of the triangles* $\{(x,\xi) : a \le x \le \xi \le b\}$ *and* $\{(x,\xi) : a \le \xi \le x \le b\}$, *except on the diagonal segment* $\{(x,\xi) : x = \xi, a \le x, \xi \le b\}$.

(P2)' *(Dual Jump Condition)*

$$\lim_{\xi \to x-} \frac{\partial^{n-1}}{\partial x^{n-1}}G(x,\xi) - \lim_{\xi \to x+} \frac{\partial^{n-1}}{\partial x^{n-1}}G(x,\xi) = \frac{1}{p_0(x)}.$$

(P4)' *(Uniqueness)* $G(x,\xi)$ *is unique.*

(P5)' *(Solution of Nonhomogeneous Equation)* $u \equiv \int_a^b G(x,\xi)f(\xi)\,d\xi$ *is the unique solution to the equation*

$$Ly = f(x), y \in D(L)$$

for a given continuous function f.

PROOF. In view of the proof of Theorem 2.1, it is readily seen that the construction of the Green's function $G(x,\xi)$ is all that needs to be proved, since the rest is the same as those in the proof of Theorem 2.1.

$G(x,\xi)$ will be constructed according to the conditions (C1)' to (C4)'. By (C2)', $G(x,\xi)$ takes the form

$$G(x,\xi) = \begin{cases} \sum_{i=1}^n d_i u_i(x), & \text{if } x < \xi; \\ \sum_{i=1}^n c_i u_i(x), & \text{if } x > \xi, \end{cases}$$

where $\{u_i(x)\}_{i=1}^n$, existing by Corollary 2.2, in Chapter 8 for the existence of solutions of linear ordinary differential equations with initial conditions, are n linearly independent solutions of

$$Ly = 0.$$

The constants d_i, c_j will be determined by $(C1)', (C3)'$, and $(C4)'$, and they satisfy

$$\sum_{i=1}^{n} c_i u_i(\xi) = \sum_{i=1}^{n} d_i u_i(\xi),$$

$$\sum_{i=1}^{n} c_i u_i'(\xi) = \sum_{i=1}^{n} d_i u_i'(\xi),$$

$$\vdots \tag{4.20}$$

$$\sum_{i=1}^{n} c_i u_i^{(n-2)}(\xi) = \sum_{i=1}^{n} d_i u_i^{(n-2)}(\xi),$$

$$\sum_{i=1}^{n} c_i u_i^{(n-1)}(\xi) - \sum_{i=1}^{n} d_i u_i^{(n-1)}(\xi) = \frac{1}{p_0(\xi)}$$

and

$$\begin{pmatrix} 0 \\ 0 \\ \vdots \\ 0 \end{pmatrix} = BG(x,\xi) = \begin{pmatrix} \alpha & \beta \end{pmatrix} \begin{pmatrix} \sum_{i=1}^{n} d_i u_i(a) \\ \sum_{i=1}^{n} d_i u_i'(a) \\ \vdots \\ \sum_{i=1}^{n} d_i u_i^{(n-1)}(a) \\ \sum_{i=1}^{n} c_i u_i(b) \\ \sum_{i=1}^{n} c_i u_i'(b) \\ \vdots \\ \sum_{i=1}^{n} c_i u_i^{(n-1)}(b) \end{pmatrix} \tag{4.21}$$

$$= \alpha \begin{pmatrix} \sum_{i=1}^{n} d_i u_i(a) \\ \sum_{i=1}^{n} d_i u_i'(a) \\ \vdots \\ \sum_{i=1}^{n} d_i u_i^{(n-1)}(a) \end{pmatrix} + \beta \begin{pmatrix} \sum_{i=1}^{n} c_i u_i(b) \\ \sum_{i=1}^{n} c_i u_i'(b) \\ \vdots \\ \sum_{i=1}^{n} c_i u_i^{(n-1)}(b) \end{pmatrix}.$$

It follows from (4.20) that

$$\begin{pmatrix} u_1(\xi) & \cdots & u_n(\xi) \\ u_1'(\xi) & \cdots & u_n'(\xi) \\ & \vdots & \\ u_1^{(n-1)}(\xi) & \cdots & u_n^{(n-1)}(\xi) \end{pmatrix} \begin{pmatrix} c_1 - d_1 \\ \vdots \\ c_n - d_n \end{pmatrix} = \begin{pmatrix} 0 \\ \vdots \\ 0 \\ \frac{1}{p_0(\xi)} \end{pmatrix},$$

and so

$$\lambda_i \equiv c_i - d_i, \quad i = 1, 2, \ldots, n \tag{4.22}$$

exist uniquely by linear algebra, since the Wronskian

$$W(u_1, \ldots, u_n)(\xi) = \begin{vmatrix} u_1(\xi) & \cdots & u_n(\xi) \\ u_1'(\xi) & \cdots & u_n'(\xi) \\ & \vdots & \\ u_1^{(n-1)}(\xi) & \cdots & u_n^{(n-1)}(\xi) \end{vmatrix} \neq 0.$$

Simplification shows that

$$\sum_{i=1}^{n} d_i \alpha \begin{pmatrix} u_i(a) \\ u_i'(a) \\ \vdots \\ u_i^{(n-1)}(a) \end{pmatrix} + \sum_{i=1}^{n} c_i \beta \begin{pmatrix} u_i(b) \\ u_i'(b) \\ \vdots \\ u_i^{(n-1)}(b) \end{pmatrix} = \begin{pmatrix} 0 \\ 0 \\ \vdots \\ 0 \end{pmatrix}, \tag{4.23}$$

and so, with $c_i = d_i + \lambda_i$ in (4.22) substituted into (4.23), it follows that

$$-\sum_{i=1}^{n} \lambda_i \beta \begin{pmatrix} u_i(b) \\ u_i'(b) \\ \vdots \\ u_i^{(n-1)}(b) \end{pmatrix} = \sum_{i=1}^{n} d_i [\alpha \begin{pmatrix} u_i(a) \\ u_i'(a) \\ \vdots \\ u_i^{(n-1)}(a) \end{pmatrix} + \beta \begin{pmatrix} u_i(b) \\ u_i'(b) \\ \vdots \\ u_i^{(n-1)}(b) \end{pmatrix}]$$

$$= \sum_{i=1}^{n} d_i B u_i$$

$$= \begin{pmatrix} B u_1 & B u_2 & \cdots & B u_n \end{pmatrix} \begin{pmatrix} d_1 \\ d2 \\ \vdots \\ d_n \end{pmatrix}.$$

Since $\begin{vmatrix} B u_1 & B u_2 & \cdots & B u_n \end{vmatrix} \neq 0$ is true, it results from the linear algebra that $d_i, i = 1, \ldots, n$, and then c_i, are solvable. Thus $G(x, \xi)$ is constructed. Indeed, if

$$\begin{vmatrix} B u_1 & B u_2 & \cdots & B u_n \end{vmatrix} = 0,$$

then $\{B u_i\}_{i=1}^{n}$ are linearly dependent, and so

$$\sum_{i=1}^{n} \nu_i B u_i = 0$$

with the constants $\nu_i, i = 1, \ldots, n$, not all zero. On the other hand,

$$L u_i = 0, i = 1, \ldots, n$$

by our choice of u_i, and so this, combined with the above equation, gives

$$L(\sum_{i=1}^{n} \nu_i u_i) = 0,$$

$$0 = \sum_{i=1}^{n} \nu_i B u_i = B(\sum_{i=1}^{n} \nu_i u_i).$$

But this equation has only the zero solution by assumption, and so

$$\sum_{i=1}^{n} \nu_i u_i = 0$$

with the constants ν_i's not all zero. This shows that $\{u_i\}_{i=1}^{n}$ are linearly dependent, a contradiction to the linear independence of $\{u_i\}_{i=1}^{n}$. $\square$

5. More Theoretical Material

In this section, the meaning of the Dirac function $\delta(x - \xi)$ will be explained in detail and the Remark 2.8 will be further discussed in the end. For this purpose, the Riemann integrals and Lebesgue integrals will be recalled.

REMARK 5.1. *The explanation of the Dirac delta function here seems a one that is presented for the first time.*

5.1. Riemann Integrals. As is recalled, the Riemann integral

$$\int_a^b f(x)\, dx$$

of f over $[a, b]$ for a bounded function f is defined [**2**, Page 174] by

$$\int_a^b f(x)\, dx \equiv \lim_{\|P\| \to 0} \sum_{i=1}^n f(c_i)\triangle x_i,$$

where P is a partition on $[a, b]$:

$$P : a = x_0 < x_1 < x_2 < \cdots < x_n = b,$$

$n \in \mathbb{N}, \triangle x_i \equiv x_i - x_{i-1}, \|P\| \equiv \max_{1 \le i \le n} \triangle x_i$, and $c_i \in [x_{i-1}, x_i]$ arbitrarily.

We have

$$\int_a^b f(x)\, dx \quad \text{exists if} \quad f : [a, b] \longrightarrow \mathbb{R} \quad \text{is piecewise continuous.}$$

More generally, a necessary and sufficient condition for a bounded function f on $[a, b]$ to be Riemann integrable is that the set of discontinuities of f in $[a, b]$ has measure zero [**2**, Page 171]. Thus if f is a bounded function on $[a, b]$, such that the set

$$\{x \in [a, b] : f(x) \quad \text{is not continuous at} \quad x\}$$

is countable, then f is Riemann integrable.

The meaning of measure zero will be made clear in the next subsection on Lebesgue integrals.

The meaning of

$$\lim_{\|P\| \to 0} \sum_{i=1}^n f(c_i)\triangle x_i = \int_a^b f(x)\, dx$$

is: for every $\epsilon > 0$, there is a $\delta > 0$ such that

$$\left| \lim_{\|P\| \to 0} \sum_{i=1}^n f(c_i)\triangle x_i - \int_a^b f(x)\, dx \right| < \epsilon$$

for all partitions P with $\|P\| < \epsilon$ and for all $c_i \in [x_{i-1}, x_i]$.

Another equivalent definition of Riemann integral is [**2**, Pages 141, 152–153, 174]: let

$$M_i \equiv \sup_{x \in [x_{i-1}, x_i]} f(x),$$

$$m_i \equiv \inf_{x \in [x_{i-1}, x_i]} f(x),$$

$$U(f, P) \equiv \sum_{i=1}^n M_i \triangle x_i, \quad L(f, P) \equiv \sum_{i=1}^n m_i \triangle x_i,$$

$$\int_a^{\overline{b}} f(x)\,dx \equiv \inf_P U(f,P),$$

and

$$\int_{\underline{a}}^b f(x)\,dx \equiv \sum_P L(f,P).$$

We say $\int_a^b f(x)\,dx$ exists if

$$\int_a^{\overline{b}} f(x)\,dx = \int_{\underline{a}}^b f(x)\,dx \equiv I$$

for some number I, and in this case, define

$$\int_a^b f(x)\,dx \equiv I.$$

5.2. Lebesgue Integrals. A generalization of Riemann integrals is the so-called Lebesgue integrals. It allows for more functions (in addition to continuous functions) and more sets (in addition to the set $[a,b]$) that can be used in its definition.

One definition of the Lebesgue integral to is to start with defining the measurable sets E and the Lebesgue measure m. Following that, define the measurable functions f. Finally, define the Lebesgue integral of f with respect to the Lebesgue measure m over the measurable set E:

$$\int_E f(x)\,d(m(x)).$$

The Measurable Sets And The Lebesgue Measure

Let $-\infty < a < b < +\infty$. Let I be an interval of the form:

$$(a,b), (a,b], [a,b), \quad \text{or } [a,b].$$

DEFINITION 5.2. *Define the length $l(I)$ of I to be*

$$l(I) \equiv b - a.$$

However, if instead, I is of the form:

$$(-\infty, a), (-\infty, a], (a, +\infty), ([a, +\infty), \quad \text{or } (-\infty, +\infty),$$

define

$$l(I) \equiv +\infty.$$

DEFINITION 5.3. *Let $P(\mathbb{R})$ be the set*

$$P(\mathbb{R}) \equiv \{A : A \quad \text{is a subset of} \quad \mathbb{R}\}.$$

DEFINITION 5.4. *For $A \in P(\mathbb{R})$, define the outer measure $m^*(A)$ of A to be*

$$m^*(A) \equiv \inf_{\cup_n I_n \supset A} \sum_n l(I_n),$$

where $I_n, n \in \mathbb{N}$, are open interval of $\mathbb{R}$. Here note that

$$\cup_n I_n \supset A$$

is possible since $(-\infty, +\infty)$ *is an open interval.*

Thus the outer measure

$$m^* : P(\mathbb{R}) \longrightarrow [0, +\infty]$$

is a set function, that is, a function that associates an extended real number to each set in in the collection $P(\mathbb{R})$ of subsets of $\mathbb{R}$.

It follows that $m^*(\emptyset) = 0$, that if $A \subset B$ then

$$m^*(A) \leq m^*(B),$$

and that for a single element set $\{a\}$, $a \in \mathbb{R}$,

$$m^*(\{a\}) = 0.$$

It follows that

$$m^*(I) = l(I) \quad \text{for an interval} \quad I$$

and that

$$m^*(A) = 0$$

for a countable set A.

DEFINITION 5.5. *A set* $E \in P(\mathbb{R})$ *is measurable if* E *satisfies*

$$m^*(A) = m^*(A \cap E) + m^*(A \cap E^c)$$

for each set $A \in P(\mathbb{R})$. *Here* E^c *is the complement set of* E.

It follows that

$$\text{if} \quad m^*(E) = 0, \quad \text{then} \quad E \quad \text{is measurable.}$$

It follows that the collection $\mathfrak{M}$ of measurable sets is a σ-algebra, that is, the complement of a measurable set is measurable and the union and intersection of a countable collection of measurable sets is measurable.

DEFINITION 5.6. *The collection* $\mathfrak{B}$ *of Borel sets in* $\mathbb{R}$ *is the smallest* σ*-algebra that contains all of the open sets in* $\mathbb{R}$.

It follows that

$$\mathfrak{B} \subset \mathfrak{M}.$$

Here note that each open set is the union of a countable number of open intervals.

DEFINITION 5.7. *Define the Lebesgue measure*

$$m : \mathfrak{M} \longrightarrow [0, +\infty]$$

by

$$m(E) \equiv m^*(E)$$

for $E \in \mathfrak{M}$.

Thus the Lebesgue measure m is a set function.

It follows that

$$m(\cup_{i=1}^{\infty} E_i) \leq \sum_{i=1}^{\infty} m(E_i)$$

for $E_i \in \mathfrak{M}$ and that

$$m(\cup_{i=1}^{\infty} E_i) = \sum_{i=1}^{\infty} m(E_i)$$

for mutually disjoint $E_i \in \mathfrak{M}$, that is,

$$E_i \cap E_j = \emptyset$$

for $i \neq j$.

Lebesgue Integrals

DEFINITION 5.8. *An extended real-valued (or a complex-valued) function*

$$f : D(f) \in \mathfrak{M} \longrightarrow [-\infty, +\infty] \quad or \ \mathbb{C}$$

is measurable if the set

$$\{x \in D(f) : f(x) > \alpha\}$$

is measurable for each $\alpha \in \mathbb{R}$ (or if the set

$$\{x \in D(f) : f(x) \in O\}$$

is measurable for each open set O in $\mathbb{C}$). Here $D(f)$ is the domain of f. Note the $>$ in the set

$$\{x \in D(f) : f(x) > \alpha\}$$

can be, equivalently, replaced by

$$\geq, <, \quad or \quad \leq.$$

DEFINITION 5.9. χ_A *is a measurable characteristic function with the characteristic set $A \subset \mathbb{R}$ if A is measurable and*

$$\chi_A(x) = \begin{cases} 1 & if \ x \in A \\ 0 & if \ x \notin A. \end{cases}$$

DEFINITION 5.10. *A real-valed function ϕ is a simple function if it is of the form*

$$\phi(x) = \sum_{i=1}^{k} \alpha_i \chi_{A_i}(x)$$

for some $k \in \mathbb{N}$, where $\alpha_i \neq 0, \alpha_i \in \mathbb{R}$ and χ_{A_i} are measurable characteristic functions. ϕ is a simple function of canonical form if $A_i = \{x : \phi(x) = \alpha_i\}$; in that case, the A_i are disjoint and the α_i are distinct. ϕ is a step function if A_i are intervals. Here note that a simple function can have many representations.

DEFINITION 5.11. **[8, pages 47–48]** *Define the Lebesgue integral*

$$\int_E \phi(x) \, d(m(x))$$

of a simple function $\phi(x) = \sum_{i=1}^{k} \alpha_i \chi_{A_i}(x)$ of canonical form, with respect to the Lebesgue measure m over the Lebesgue measurable set $E (\in \mathfrak{M})$, by

$$\int_E \phi(x) \, d(m(x)) \equiv \sum_{i=1}^{k} \alpha_i m(A_i \cap E).$$

Note that we also write

$$\int_E \phi(x) \, d(m(x)) = \int_E \phi(x) \, dx,$$

and that $\int_E \phi(x)\,d(m(x))$ may equal ∞.

DEFINITION 5.12. [8, page 48] *Define the Lebesgue integral $\int_E f(x)\,d(m(x)) = \int_E f(x)\,dx$ of a nonnegative extended real-valued measurable function f, by*

$$\int_E f(x)\,d(m(x)) = \int_e f(x)\,dx \equiv \sup_{0 \leq \phi \leq f} \int_E \phi(x)\,d(m(x)),$$

where ϕ are nonnegative simple functions and E is measurable. f is integrable over E if

$$\int_E f(x)\,d(m(x)) < \infty.$$

Remark. [8, page 48] Since $\int_E \phi(x)\,d(m(x)) \leq \int_E \psi(x)\,d(m(x))$ for two non-negative simple functions ϕ and ψ with $\phi \leq \psi$, it follows that the two definitions of $\int_E f(x)\,d(m(x))$ agree when f is a non-negative simple function. Note that an arbitrary real-valed function f can be written as

$$f = f^+ - f^-,$$

where

$$f^+(x) \equiv \max\{f(x), 0\} \geq 0,$$

the positive part of f, and

$$f^-(x) \equiv \max\{-f(x), 0\} \geq 0,$$

the negative part of f.

DEFINITION 5.13. *Define the Lebesgue integral $\int_E f(x)\,d(m(x)) = \int_E f(x)\,dx$ of an arbitrary measurable function f by*

$$\int_E f(x)\,d(m(x)) = \int_E f(x)\,dx \equiv \int_E f^+(x)\,d(m(x)) - \int_E f^-(x)\,d(m(x)).$$

f is integrable if both f^+ and f^- are integrable.

It follows that if a bounded real-valued function f on $[a, b]$ is Riemann integrable, then it is Lebesgue integrable and in this case,

$$\int_a^b f(x)\,dx = \int_{[a,b]} f(x)\,d(m(x)).$$

Thus Lebesgue integrals generalize Riemann integrals.

The above material about the Lebesgue integral is taken from [20].

DEFINITION 5.14. [8]*A complex-valued measurable function f is integrable if*

$$\int_E |f(x)|\,d(m(x)) < \infty.$$

Since

$$|f| \leq |Ref| + |Imf| \leq 2|f|,$$

f is integrable if and only if Ref and Imf are integrable. In that case, define

$$\int_E f(x)\,d(m(x)) \equiv \int_E Ref\,d(m(x)) + i\int_E Imf\,d(m(x)).$$

Here Ref is the real part of f and Imf is the imaginary part of f. Thus

$$f = Ref + iImf.$$

Recall that a complex-valued function

$$f : D(f) \subset \mathbb{R} \longrightarrow \mathbb{C}$$

is measurable if the set

$$\{x \in D(f) : f(x) \in O\}$$

is measurable for each open set O in $\mathbb{C}$.

Another equivalent definition of the Lebesgue integral is [12], [24, Pages 16–17]:

DEFINITION 5.15. *The Lebesgue integral* $\int_E \phi(x)\, d(m(x)) = \int_E \phi(x)\, dx$ *of a complex-valued simple function*

$$\phi(x) = \sum_{i=1}^{k} \alpha_i \chi_{A_i}(x), \alpha_i \in \mathbb{C}$$

of canonical form, is defined by

$$\int_E \phi(x)\, d(m(x)) = \int_E \phi(x)\, dx \equiv \sum_{i=1}^{k} \alpha_i m(A_i \cap E).$$

A real-valued (or complex-valued) measurable function f is integrable if there is a sequence of integrable simple functions ϕ_n, such that

$$\phi_n \longrightarrow f \quad \text{almost everywhere,}$$

that is,

$$m(\{x \in D(f) : \phi_n(x) \nrightarrow f(x)\}) = 0,$$

and that

$$\lim_{n,m \to \infty} \int_E |\phi_n(x) - \phi_m(x)|\, d(m(x)) = 0.$$

It follows that the limit

$$\lim_{n \to \infty} \int_E \phi_n(x)\, d(m(x))$$

exists and is independent of the choice of ϕ_n.

In the above case, define the Lebesgue integral

$$\int_E f(x)\, d(m(x)) = \int_E f(x)\, dx \equiv \lim_{n \to \infty} \int_E \phi_n(x)\, d(m(x)) = \lim_{n \to \infty} \int_E \phi_n(x)\, dx.$$

5.3. The Meaning of the Dirac Delta Function. Measures other than the Lebesgue measure exist [8], in which a special one, δ_ξ, called the Dirac measure centered at $\xi \in [a, b]$, is defined by

$$\delta_\xi(A) = \begin{cases} 0 & \text{if } \xi \notin A, \\ 1 & \text{if } \xi \in A \end{cases}$$

for a measurable set $A \subset [a, b]$. It has the properties: for $[c, d] \subset [a, b]$ and a piecewise continuous function f,

$$\int_{[c,d]} d(\delta_\xi(x)) = \begin{cases} 0 & \text{if } \xi \notin [c, d], \\ 1 & \text{if } \xi \in [c, d] \end{cases}$$

and

$$\int_{[c,d]} f(x)\, d(\delta_\xi(x)) = \begin{cases} 0 & \text{if } \xi \notin [c, d], \\ f(\xi) & \text{if } \xi \in [c, d]. \end{cases}$$

For notational convenience, we think of

$$\int_{[c,d]} d(\delta_\xi(x)) \quad \text{as} \quad \int_c^d \delta(x-\xi)\,dx \equiv \int_c^d \delta(x-\xi)\,d(m(x))$$

and

$$\int_{[c,d]} f(x)\,d(\delta_\xi(x)) \quad \text{as} \quad \int_c^d f(x)\delta(x-\xi)\,dx,$$

in which a symbolic non-negative function $\delta(x-\xi)$, not a real function, is created. This function is called the Dirac delta function. Mathematically speaking, it is not a function, but a symbol that is created for operational convenience.

It follows from

$$\int_c^d \delta(x-\xi)\,dx = \int_{[c,d]} d(\delta_\xi(x)) = 0 \quad \text{or} \quad 1$$

according as $\xi \notin [c,d]$ or $\xi \in [c,d]$, that $\delta(x-\xi)$ acts symbolically as a function which is zero on $[c,d]$, except for $x=\xi$, at which $\delta(x-\xi)$ is not defined. To see the symbolic behavior of $\delta(x-\xi)$ at $x=\xi$, choose $[c,d] = [\xi-\epsilon, \xi+\epsilon]$, $\epsilon > 0$ to have

$$\frac{1}{2\epsilon} = \frac{\int_{[\xi-\epsilon,\xi+\epsilon]} d(\delta_\xi(x))}{2\epsilon} = \frac{\int_{\xi-\epsilon}^{\xi+\epsilon} \delta(x-\xi)\,dx}{2\epsilon}$$

$$= \frac{\int_{-\epsilon}^{\epsilon} \delta(y)\,dy}{2\epsilon} = \frac{\int_0^\epsilon \delta(y)\,dy - \int_0^{-\epsilon} \delta(y)\,dy}{2\epsilon},$$

where $y \equiv x - \xi$. Letting $\epsilon \longrightarrow 0$, we have

$$+\infty = \frac{d}{d\epsilon}\Big(\int_0^\epsilon \delta(y)\,dy\Big)\big|_{\epsilon=0} = \delta(0)$$

by the fundamental theorem of calculus. Thus, symbolically

$$\delta(x-\xi) = +\infty$$

for $x = \xi$.

In order for

$$u(x) = \int_{\xi=a}^b G(x,\xi)f(\xi)\,d\xi$$

to be a solution of (1.1), we should have

$$Lu = L_x u = L_x\Big[\int_{\xi=a}^b G(x,\xi)f(\xi)\,d\xi\Big] = f(x).$$

This implies that

$$\int_{x=a}^b f(x)\phi(x)\,dx = \int_{x=a}^b [L_x \int_{\xi=a}^b G(x,\xi)f(\xi)\,d\xi]\phi(x)\,dx$$

$$= \int_{x=a}^b [\int_{\xi=a}^b L_x G(x,\xi)f(\xi)\,d\xi]\phi(x)\,dx$$

for $\phi \in C^2[a,b]$ with ϕ, ϕ' vanishing at end points: $\{a,b\}$. This follows because integration by parts and the Fubini theorem (see below) give:

$$\int_{x=a}^b [L_x \int_{\xi=a}^b G(x,\xi)f(\xi)\,d\xi]\phi(x)\,dx$$

$$= \int_{x=a}^{b} [q(x) \int_{\xi=a}^{b} G(x,\xi) f(\xi) \, d\xi] \phi(x) \, dx +$$

$$[p(x) \frac{d}{dx} \int_{\xi=a}^{b} G(x,\xi) f(\xi) \, d\xi] \phi(x)|_{x=a}^{b} -$$

$$\int_{x=a}^{b} [p(x) \frac{d}{dx} \int_{\xi=a}^{b} G(x,\xi) f(\xi) \, d\xi] \phi'(x) \, dx$$

$$= \int_{x=a}^{b} [q(x) \int_{\xi=a}^{b} G(x,\xi) f(\xi) \, d\xi] \phi(x) \, dx + 0 -$$

$$\int_{x=a}^{b} [p(x) \frac{d}{dx} \int_{\xi=a}^{b} G(x,\xi) f(\xi) \, d\xi] \phi'(x) \, dx,$$

whose third term equals

$$- [p(x) \int_{\xi=a}^{b} G(x,\xi) f(\xi) \, d\xi] \phi'(x)|_{x=a}^{b} +$$

$$\int_{x=a}^{b} [\int_{\xi=a}^{b} G(x,\xi) f(\xi) \, d\xi] \frac{d}{dx} (p(x) \phi'(x)) \, dx$$

$$= 0 + \int_{x=a}^{b} [\int_{\xi=a}^{b} G(x,\xi) f(\xi) \, d\xi] \frac{d}{dx} (p(x) \phi'(x)) \, dx$$

$$= \int_{\xi=a}^{b} [\int_{x=a}^{b} G(x,\xi) \frac{d}{dx} (p(x) \phi'(x)) \, dx] f(\xi) \, d\xi$$

$$= \int_{\xi=a}^{b} [\int_{x=a}^{b} \frac{d}{dx} (p(x) G_x(x,\xi)) \phi(x) \, dx] f(\xi) \, d\xi.$$

Here used was the fact that ϕ and ϕ' vanish at end points $\{a, b\}$. Thus

$$\int_{x=a}^{b} L_x [\int_{\xi=a}^{b} G(x,\xi) f(\xi) \, d\xi] \phi(x) \, dx = \int_{x=a}^{b} [\int_{\xi=a}^{b} L_x g(x,\xi) f(\xi) \, d\xi] \phi(x) \, dx$$

$$= \int_{x=a}^{b} f(x) \phi(x) \, dx,$$

and so we are led to require that

$$L_x G(x,\xi) = \delta(x - \xi).$$

This is the condition (C2).

Since

$$L_x G(x,\xi) = \delta(\xi - x),$$

integration by parts gives, for small $\epsilon > 0$,

$$1 = \int_{x=\xi-\epsilon}^{\xi+\epsilon} \delta(\xi - x) \, dx = \int_{x=\xi-\epsilon}^{\xi+\epsilon} L_x G(x,\xi) \, dx$$

$$= \int_{x=\xi-\epsilon}^{\xi+\epsilon} q(x) G(x,\xi) \, dx + p(x) G_x(x,\xi)|_{x=\xi-\epsilon}^{\xi+\epsilon}$$

$$= \int_{x=\xi-\epsilon}^{\xi+\epsilon} q(x) G(x,\xi) \, dx +$$

$$p(\xi + \epsilon) G_x(\xi + \epsilon, \xi) - p(\xi - \epsilon) G_x(\xi - \epsilon).$$

Letting $\epsilon \longrightarrow 0$, it follows that

$$\int_{x=\xi-\epsilon}^{\xi+\epsilon} q(x)G(x,\xi)\, dx \longrightarrow 0,$$

since $q(x)G(x,\xi)$ is continuous for $x \in [a,b]$. Because of this,

$$1 = p(x) \lim_{x\to\xi^+} G_x(x,\xi) - p(x) \lim_{x\to\xi^-} G_x(x,\xi),$$

since $p(x)$ is continuous. This is the jump condition (C3) for $G_x(x,\xi)$ at $x = \xi$.

Next, we deal with the previously cited Fubini Theorem, [**20**, Pages 303–308]. Recall the definition of the Lebesgue integrals in Subsection 5.2, in which m is the Lebesgue measure defined on the collection $\mathfrak{M}$ of the Lebesgue measurable subsets in $\mathbb{R}$. For $A, B \in \mathfrak{M}$, call $A \times B$ a measurable rectangle. Denote the collection of measurable rectangles by $\mathfrak{R}$. Define a set function

$$\lambda : \mathfrak{R} \longrightarrow [0, \infty]$$

by

$$\lambda(A \times B) \equiv m(A) \times m(B)$$

for $A \times B \in \mathfrak{R}$. λ can be extended to a complete measure on a σ-algebra $\mathfrak{S}$ containing $\mathfrak{R}$. This extended measure is called the product measure of m and m, and is denoted by $m \times m$. An element in $\mathfrak{S}$ is a subset in $\mathbb{R}^2 \equiv \mathbb{R} \times \mathbb{R}$, and is called a measurable set. $m \times m$ is called the two dimensional Lebesgue measure for $\mathbb{R}^2$.

An extended real-valued (or a complex-valued) function

$$f : D(f) \subset \mathbb{R}^2 \longrightarrow [-\infty, \infty](\text{or} \quad \mathbb{C})$$

is measurable if the domain $D(f)$ of f is measurable, that is, $D(f) \in \mathfrak{S}$, and if the set

$$\{(x,y) \in D(f) : f(x,y) > \alpha\}$$

is measurable for each $\alpha \in \mathbb{R}$ (or if the set

$$\{(x,y) \in D(f) : f(x,y) \in O\}$$

is measurable for each open set $O \subset \mathbb{C}$).

The Lebesgue integral of a measurable function $f(x,y)$ over a measurable set $S \in \mathfrak{S}$ is denoted by

$$\int_S f(x,y)\, d((m \times m)(x,y)).$$

THEOREM 5.16 (Fubini Theorem). *Suppose that $A, B \in \mathfrak{M}$, and that $f(x,y)$ is an integrable function on $A \times B$, that is $\int_{A\times B} |f(x,y)|\, d((m \times m)(x,y)) < \infty$. Then*

$$\int_{A\times B} f(x,y)\, d((m \times m)(x,y)) = \int_B [\int_A f(x,y)\, d(m(x))]\, d(m(y))$$

$$= \int_A [\int_B f(x,y)\, d(m(y))]\, d(m(x)).$$

Here note that we also write

$$\int_A f(x,y)\, d(m(x)) = \int_A f(x,y)\, dx$$

and

$$\int_B f(x,y)\, d(m(y)) = \int_B f(x,y)\, dy,$$

and note that $f(x, y)$ is integrable if $f(x, y)$ is continuous on $A \times B$, where A and B are bounded and closed in $\mathbb{R}$.

In what follows next, the Remark 2.8 that concerns the linear differential operator L with absolutely continuous coefficients, will be explored.

Let

$$L' : D(L') \subset (L^2(a,b), \| \cdot \|_w) \longrightarrow (L^2(a,b), \| \cdot \|_w)$$

be a linear differential operator, Taylor and Lay [21, Page 347], defined by

$$L'y = (p(x)y')' + q(x)y$$

for $y \in D(L')$

$$\equiv \{u \in L^2(a,b) : u, u', p, p(x)u' \in AC, (p(x)u')' \in L^2(a,b),$$

$$u(a) + \alpha_1 u'(a) = 0, u(b) + \alpha_2 u'(b) = 0\}.$$

Here the meanning of AC is defined below, $p(x), r(x) \geq \delta > 0$ is assumed for some $\delta > 0$ and for all $x \in [a,b]$, and $q(x), r(x)$ are measurable and bounded.

Since, for some $M > 0$, we have

$$0 < \delta \leq r(x) \leq M$$

for all $x \in [a,b]$, it follows as in Chapter 6 that

$$L^2(a,b) = L^2_w(a,b)$$

and that $\| \cdot \|_2$ and $\| \cdot \|_w$ are equivalent.

Here, Royden [20],

DEFINITION 5.17. *A complex-valued function f on $[a,b]$ is absolutely continuous (abbreviated as AC) if, for a given $\epsilon > 0$, there is a $\delta > 0$, such that*

$$\sum_{i=1}^{n} |f(b_i) - f(a_i)| < \epsilon$$

for all finite collection $\{(a_i, b_i)\}_{i=1}^{n}$ of nonoverlapping subintervals of $[a,b]$ with

$$\sum_{i=1}^{n} |b_i - a_i| < \delta.$$

LEMMA 5.18. *(Royden [20])A complex-valed function f on $[a,b]$ is absolutely continuous if and only if f' exists almost everywhere and*

$$f(x) = \int_a^x f'(t)\, dt + f(a).$$

Here the integral is taken with respect to the Lebesgue integral.

From the Definition 5.17 and Lemma 5.18, one can see that the domain $D(L')$ of L' is well-defined. It follows that proving that the unique Green function, corresponding to L', exists, is the same as proving Theorem 2.1. Indeed the only difference is that here we use Corollary 4.2 in Chapter 8 to obtain two linearly independent solutions $v_1(x)$ and $v_2(x)$, instead of using Corollary 2.1 in Chapter 8. Here likewise v_1 and v_2 are chosen to satisfy the initial conditions:

$$v_1(a) = 1, \quad v_1'(a) = 0,$$

$$v_2(a) = 0, \quad v_2'(a) = 1.$$

Such a choice makes $\{v_1, v_2\}$ linear independent, since their Wronskian at $x = a$ is not nonzero:

$$\begin{vmatrix} v_1(a) & v_2(a) \\ v_1'(a) & v_2'(a) \end{vmatrix} = \begin{vmatrix} 1 & 0 \\ 0 & 1 \end{vmatrix} \neq 0.$$

Here note that, instead of being twice continuously differentiable functions, as is the case for the proof of Theorem 2.1, the $v_i, i = 1, 2$, here are two continuously differentiable functions with $v_i', i = 1, 2$ being absolutely continuous, for which $v_i'', i = 1, 2$ exist almost everywhere. The details are left to the reader.

6. Problems

1. Consider the equation with zero Dirichlet boundary condition:

$$y'' = -x, \quad 0 < x < \pi,$$
$$y(0) = 0 = y(\pi).$$

Find the corresponding Green's function $G(x, \xi)$, and express the solution y in terms of $G(x, \xi)$.

2. Consider the equation with zero Dirichlet boundary condition:

$$y'' + 4y = f(x), \quad 0 < x < \pi,$$
$$y(0) = 0 = y(\pi),$$

where f is a continuous function on $[0, \pi]$. Find the corresponding Green's function $G(x, \xi)$, and express the solution y in terms of $G(x, \xi)$.

3. Consider the equation with zero Dirichlet boundary condition:

$$x^2 y'' + xy' - y = f(x), \quad 1 < x < 2,$$
$$y(1) = 0 = y(2),$$

where f is a continuous function on $[1, 2]$. Find the corresponding Green's function $G(x, \xi)$, and express the solution y in terms of $G(x, \xi)$.

4. Consider the equation with a boundary condition of left hinged end and a boundary condition of right clamped end [**3**, Pages 677–678, 8th edition]:

$$y^{(4)} + 4y'' = f(x), \quad 0 < x < 1,$$
$$y(0) = 0 = y''(0),$$
$$y(1) = 0 = y'(1),$$

where f is a continuous function on $[0, 1]$. Find the corresponding Green's function $G(x, \xi)$, and express the solution y in terms of $G(x, \xi)$.

5. Show that the Green's function $G(x, \xi) = G(\xi, x)$ in Theorem 2.1 is symmetric: $G(x, \xi) = G(\xi, x)$.

6. Prove Proposition 2.5.

Regarding Problems **7** to **13**, use Proposition 2.7 to show that the corresponding Green's functions exist, and then use Proposition 2.5 to show that the Green's functions are symmetric.

7.

$$y'' + 4y = f(x), \quad 0 < x < 1,$$
$$y(0) = 0,$$
$$y'(1) = 0.$$

8.

$$y'' + 4y = f(x), \quad 0 < x < 1,$$
$$y'(0) = 0,$$
$$y(1) = 0.$$

9.

$$y'' + 4y = f(x), \quad 0 < x < 1,$$
$$y(0) = 0,$$
$$y(1) + y'(1) = 0.$$

10.

$$y'' + 4y = f(x), \quad 0 < x < 1,$$
$$y(0) - y'(0) = 0,$$
$$y(1) = 0.$$

11.

$$y'' + 4y = f(x), \quad 0 < x < 1,$$
$$y'(0) = 0,$$
$$y(1) + y'(1) = 0.$$

12.

$$y'' + 4y = f(x), \quad 0 < x < 1,$$
$$y(0) - y'(0) = 0,$$
$$y'(1) = 0.$$

13.

$$(x^2 y')' - 6y = f(x), \quad 1 < x < 2,$$
$$\frac{1}{2}y(1) + y(2) = 0,$$
$$\frac{1}{2}y'(1) + y'(2) = 0.$$

14. Use Proposition 2.7 to show that the boundary value problem
$$y^{(4)} + 4y'' = f(x), \quad 0 < x < 1,$$
$$y(0) = 0 = y'(0),$$
$$y(1) = 0 = y'(1)$$
has a corresponding Green's function $G(x, \xi)$. Here f is a continuous function on $[0, 1]$.

15. Use Proposition 2.7 to show that the boundary value problem
$$y^{(4)} + 4y'' = f(x), \quad 0 < x < \pi,$$
$$y(0) = 0 = y''(0),$$
$$y(\pi) = 0 = y''(\pi)$$
does not have a correponding Green's function $G(x, \xi)$. Here f is a continuous function on $[0, \pi]$.

7. Solutions

1. By Theorem 2.1, the Green's function $G(x, \xi)$, if existing, takes the form

$$G(x, \xi) = \begin{cases} d_1 x + d_2, & \text{if } x < \xi; \\ c_1 x + c_2, & \text{if } x > \xi, \end{cases}$$

where $\{x, 1\}$ is a fundamental set of solutions for the homogeneous equation

$$y'' = 0,$$

and c_1, c_2, d_1, and d_2 satisfy

$$d_1 \xi + d_2 = c_1 \xi + c_2,$$
$$c_1 - d_1 = 1,$$
$$d_2 = 0,$$
$$c_1 \pi + c_2 = 0.$$

After being solved, c_1, c_2, d_1, and d_2 are given by

$$c_1 = \frac{\xi}{\pi},$$
$$c_2 = -\xi,$$
$$d_1 = \frac{\xi}{\pi} - 1,$$
$$d_2 = 0.$$

It follows that

$$G(x, \xi) = \begin{cases} (\frac{\xi}{\pi} - 1)x, & \text{if } x < \xi; \\ \frac{\xi}{\pi} x - \xi, & \text{if } x > \xi, \end{cases}$$

and that the solution y is expressed as

$$y(x) = \int_{\xi=0}^{1} G(x, \xi)(-\xi) \, d\xi$$
$$= \int_{0}^{x} (\frac{\xi}{\pi} x - \xi)(-xi) \, d\xi + \int_{x}^{\pi} (\frac{\xi}{\pi} - 1)x(-\xi) \, d\xi$$
$$= \frac{x}{6}(\pi^2 - x^2).$$

It is readily checked that $G(x, \xi)$ is symmetric:

$$G(x, \xi) = G(\xi, x).$$

On the other hand, since

$$(\alpha \quad \beta) = \begin{pmatrix} 1 & 0 & 0 & 0 \\ 0 & 0 & 1 & 0 \end{pmatrix}, \{u_1, u_2\} = \{x, 1\},$$

we have

$$|Bu_1 \quad Bu_2| = |\begin{pmatrix} 1 & 0 & 0 & 0 \\ 0 & 0 & 1 & 0 \end{pmatrix} \begin{pmatrix} 0 & 1 \\ 1 & 0 \\ \pi & 1 \\ 1 & 0 \end{pmatrix}|$$

$$= \begin{vmatrix} 0 & 1 \\ \pi & 1 \end{vmatrix} = -\pi \neq 0,$$

and so the Green's function $G(x, \xi)$ exists by Proposition 2.7. $G(x, \xi)$ is symmetric by Proposition 2.5 since $|\alpha| = 0 = |\beta|$.

2. Since

$$\begin{pmatrix} \alpha & \beta \end{pmatrix} = \begin{pmatrix} 1 & 0 & 0 & 0 \\ 0 & 0 & 1 & 0 \end{pmatrix}, \{u_1, u_2\} = \{\cos(2x), \sin(2x)\},$$

we have

$$\begin{vmatrix} Bu_1 & Bu_2 \end{vmatrix} = \begin{vmatrix} \begin{pmatrix} 1 & 0 & 0 & 0 \\ 0 & 0 & 1 & 0 \end{pmatrix} \begin{pmatrix} 1 & 0 \\ 0 & 1 \\ -1 & 0 \\ 0 & -1 \end{pmatrix} \end{vmatrix}$$

$$= \begin{vmatrix} 1 & 0 \\ -1 & 0 \end{vmatrix} = 0,$$

and so the Green's function $G(x, \xi)$ does not exist by Proposition 2.7.

3. Method 1. By Theorem 2.1, the Green's function $G(x, \xi)$, if existing, takes the form

$$G(x, \xi) = \begin{cases} d_1 x + d_2 x^{-1}, & \text{if } x < \xi; \\ c_1 x + c_2 x^{-1}, & \text{if } x > \xi, \end{cases}$$

where $\{x, x^{-1}\}$ is a fundamental set of solutions for the homogeneous equation

$$x^2 y'' + xy' - y = 0,$$

and c_1, c_2, d_1, and d_2 satisfy

$$c_1 \xi + c_2 \xi^{-1} = d_1 \xi + d_2 \xi^{-1},$$

$$(c_1 - c_2 \xi^{-2}) - (d_1 - d_2 \xi^{-2}) = \frac{1}{\xi^2},$$

$$d_1 + d_2 = 0,$$

$$2c_1 + \frac{1}{2} c_2 = 0.$$

After being solved, c_1, c_2, d_1, and d_2 are given by

$$c_1 = -\frac{1}{6} \xi^{-2} + \frac{1}{6},$$

$$c_2 = \frac{2}{3} \xi^{-2} - \frac{2}{3},$$

$$d_1 = -\frac{2}{3} \xi^{-2} + \frac{1}{6},$$

$$d_2 = \frac{2}{3} \xi^{-2} - \frac{1}{6}.$$

It follows that

$$G(x, \xi) = \begin{cases} (-\frac{2}{3}\xi^{-2} + \frac{1}{6})x + (\frac{2}{3}\xi^{-2} - \frac{1}{6})x^{-1}, & \text{if } x < \xi; \\ (-\frac{2}{3}\xi^{-2} + \frac{1}{6})x + (\frac{2}{3}\xi^{-2} - \frac{2}{3})x^{-1}, & \text{if } x > \xi, \end{cases}$$

and that the solution y is expressed as

$$y(x) = \int_{\xi=0}^{1} G(x, \xi) f(\xi)\, d\xi.$$

It is readily checked that $G(x, \xi)$ is not symmetric:

$$G(x, \xi) \neq G(\xi, x).$$

On the other hand, since

$$(\alpha \quad \beta) = \begin{pmatrix} 1 & 0 & 0 & 0 \\ 0 & 0 & 1 & 0 \end{pmatrix}, \{u_1, u_2\} = \{x, x^{-1}\},$$

we have

$$|Bu_1 \quad Bu_2| = \left| \begin{pmatrix} 1 & 0 & 0 & 0 \\ 0 & 0 & 1 & 0 \end{pmatrix} \begin{pmatrix} 1 & 1 \\ 1 & -1 \\ 2 & \frac{1}{2} \\ 1 & -\frac{1}{4} \end{pmatrix} \right|$$

$$= \begin{vmatrix} 1 & 1 \\ 2 & \frac{1}{2} \end{vmatrix} = -\frac{3}{2} \neq 0,$$

and so the Green's function $G(x, \xi)$ exists by Proposition 2.7.

Method 2. By (2.4), the equation is transformed to

$$(xy')' - \frac{1}{x}y = \frac{f}{x},$$

and the corresponding L is of divergence form. By Theorem 2.1, the Green's function $G(x, \xi)$, if existing, takes the form

$$G(x, \xi) = \begin{cases} d_1 x + d_2 x^{-1}, & \text{if } x < \xi; \\ c_1 x + c_2 x^{-1}, & \text{if } x > \xi, \end{cases}$$

where $\{x, x^{-1}\}$ is a fundamental set of solutions for the homogeneous equation

$$(xy')' - \frac{1}{x}y = x^2 y'' + xy' - y$$
$$= 0,$$

and $c_1, c_2, d_1,$ and d_2 satisfy

$$c_1 \xi + c_2 \xi^{-1} = d_1 \xi + d_2 \xi^{-1},$$

$$(c_1 - c_2 \xi^{-2}) - (d_1 - d_2 \xi^{-2}) = \frac{1}{\xi},$$

$$d_1 + d_2 = 0,$$

$$2c_1 + \frac{1}{2}c_2 = 0.$$

After being solved, $c_1, c_2, d_1,$ and d_2 are given by

$$c_1 = -\frac{1}{6}\xi^{-1} + \frac{1}{6}\xi,$$

$$c_2 = \frac{2}{3}\xi^{-1} - \frac{2}{3}\xi,$$

$$d_1 = -\frac{2}{3}\xi^{-1} + \frac{1}{6}\xi,$$

$$d_2 = \frac{2}{3}\xi^{-1} - \frac{1}{6}\xi.$$

It follows that

$$G(x, \xi) = \begin{cases} (-\frac{2}{3}\xi^{-1} + \frac{1}{6}\xi)x + (\frac{2}{3}\xi^{-1} - \frac{1}{6}\xi)x^{-1}, & \text{if } x < \xi; \\ (-\frac{2}{3}\xi^{-1} + \frac{1}{6}\xi)x + (\frac{2}{3}\xi^{-1} - \frac{2}{3}\xi)x^{-1}, & \text{if } x > \xi, \end{cases}$$

and that the solution y is expressed as

$$y(x) = \int_{\xi=0}^{1} G(x, \xi) \frac{f(\xi)}{\xi} \, d\xi.$$

It is readily checked that $G(x, \xi)$ is symmetric:

$$G(x, \xi) = G(\xi, x).$$

On the other hand, since

$$(\alpha \quad \beta) = \begin{pmatrix} 1 & 0 & 0 & 0 \\ 0 & 0 & 1 & 0 \end{pmatrix}, \{u_1, u_2\} = \{x, x^{-1}\},$$

we have

$$|Bu_1 \quad Bu_2| = \left| \begin{pmatrix} 1 & 0 & 0 & 0 \\ 0 & 0 & 1 & 0 \end{pmatrix} \begin{pmatrix} 1 & 1 \\ 1 & -1 \\ 2 & \frac{1}{2} \\ 1 & -\frac{1}{4} \end{pmatrix} \right|$$

$$= \begin{vmatrix} 1 & 1 \\ 2 & \frac{1}{2} \end{vmatrix} = -\frac{3}{2} \neq 0,$$

and so the Green's function $G(x, \xi)$ exists by Proposition 2.7. $G(x, \xi)$ is symmetric by Proposition 2.5 since $|\alpha| = 0 = |\beta|$.

4. By Theorem 2.1, the Green's function $G(x, \xi)$, if existing, takes the form

$$G(x, \xi) = \begin{cases} d_1 + d_2 x + d_3 \cos(2x) + d_4 \sin(2x), & \text{if } x < \xi; \\ c_1 + c_2 x + c_3 \cos(2x) + c_4 \sin(2x), & \text{if } x > \xi, \end{cases}$$

where $\{1, x, \cos(2x), \sin(2x)\}$ is a fundamental set of solutions for the homogeneous equation

$$y^{(4)} + 4y'' = 0,$$

and c_i and $d_i, i = 1, 2, 3, 4$, satisfy

$$c_1 + c_2\xi + c_3 \cos(2\xi) + c_4 \sin(2\xi) = d_1 + d_2\xi + d_3 \cos(2\xi) + d_4 \sin(2\xi),$$
$$c_2 - 2c_3 \sin(2\xi) + 2d_4 \cos(2\xi) = d_2 - 2d_3 \sin(2\xi) + 2d_4 \cos(2\xi),$$
$$- 4c_3 \cos(2\xi) - 4c_4 \sin(2\xi) = -4d_3 \cos(2\xi) - 4d_4 \sin(2\xi),$$
$$[8c_3 \sin(2\xi) - 8c_4 \cos(2\xi)] - [8d_3 \sin(2\xi) - 8d_4 \cos(2\xi)] = 1,$$
$$d_1 + d_3 = 0,$$
$$- 4d_3 = 0,$$
$$c_1 + c_2 + c_3 \cos(2) + c_4 \sin(2) = 0,$$
$$c_2 - 2c_3 \sin(2) + 2c_4 \cos(2) = 0.$$

After being solved, c_i and $d_i, i = 1, 2, 3, 4$, are given by

$$c_1 = -\frac{1}{4}\xi,$$

$$c_2 = \frac{1}{2} \frac{\cos(2)}{2 \cos(2) - \sin(2)} \xi - \frac{1}{4} \frac{1}{2 \cos(2) - \sin(2)} \sin(2\xi),$$

$$c_3 = \frac{1}{8}\sin(2\xi),$$

$$c_4 = -\frac{1}{4}\frac{1}{2\cos(2)-\sin(2)}\xi + \frac{1}{8}\frac{2\sin(2)+\cos(2)}{2\cos(2)-\sin(2)}\sin(2\xi),$$

$$d_1 = 0,$$

$$d_2 = \frac{1}{2}\frac{\cos(2)}{2\cos(2)-\sin(2)}\xi - \frac{1}{4}\frac{1}{2\cos(2)-\sin(2)}\sin(2\xi) - \frac{1}{4},$$

$$d_3 = 0,$$

$$d_4 = -\frac{1}{4}\frac{1}{2\cos(2)-\sin(2)}\xi + \frac{1}{8}\cos(2\xi) + \frac{1}{8}\frac{2\sin(2)+\cos(2)}{2\cos(2)-\sin(2)}\sin(2\xi).$$

It follows that

$$G(x,\xi) = \begin{cases} [\frac{1}{2}\frac{\cos(2)}{2\cos(2)-\sin(2)}\xi - \frac{1}{4}\frac{1}{2\cos(2)-\sin(2)}\sin(2\xi) - \frac{1}{4}]x+ \\ [-\frac{1}{4}\frac{1}{2\cos(2)-\sin(2)}\xi+ \\ \frac{1}{8}\cos(2\xi) + \frac{1}{8}\frac{2\sin(2)+\cos(2)}{2\cos(2)-\sin(2)}\sin(2\xi)]\sin(2x), & \text{if } x<\xi; \\ -\frac{1}{4}\xi + [\frac{1}{2}\frac{\cos(2)}{2\cos(2)-\sin(2)}\xi - \frac{1}{4}\frac{1}{2\cos(2)-\sin(2)}\sin(2\xi)]x+ \\ [\frac{1}{8}\sin(2\xi)]\cos(2x)+ \\ [-\frac{1}{4}\frac{1}{2\cos(2)-\sin(2)}\xi + \frac{1}{8}\frac{2\sin(2)+\cos(2)}{2\cos(2)-\sin(2)}\sin(2\xi)]\sin(2x), & \text{if } x>\xi, \end{cases}$$

and that the solution y is expressed as

$$y(x) = \int_{\xi=0}^{1} G(x,\xi)f(\xi)\,d\xi.$$

On the other hand, since

$$(\alpha \quad \beta) = \begin{pmatrix} 1 & 0 & 0 & 0 & 0 & 0 & 0 & 0 \\ 0 & 0 & 1 & 0 & 0 & 0 & 0 & 0 \\ 0 & 0 & 0 & 0 & 1 & 0 & 0 & 0 \\ 0 & 0 & 0 & 0 & 0 & 1 & 0 & 0 \end{pmatrix}, \{u_1, u_2, u_3, u_4\} = \{1, x, \cos(2x), \sin(2x)\},$$

we have

$$\left| Bu_1 \quad Bu_2 \quad Bu_3 \quad Bu_4 \right|$$

$$= \left| \begin{pmatrix} 1 & 0 & 0 & 0 & 0 & 0 & 0 & 0 \\ 0 & 0 & 1 & 0 & 0 & 0 & 0 & 0 \\ 0 & 0 & 0 & 0 & 1 & 0 & 0 & 0 \\ 0 & 0 & 0 & 0 & 0 & 1 & 0 & 0 \end{pmatrix} \begin{pmatrix} 1 & 0 & 1 & 0 \\ 0 & 1 & 0 & 2 \\ 0 & 0 & -4 & 0 \\ 0 & 0 & 0 & -8 \\ 1 & 1 & \cos(2) & \sin(2) \\ 0 & 1 & -2\sin(2) & 2\cos(2) \\ 0 & 0 & -4\cos(2) & -4\sin(2) \\ 0 & 0 & 8\sin(2) & -8\cos(2) \end{pmatrix} \right|$$

$$= \begin{vmatrix} 1 & 0 & 1 & 0 \\ 0 & 0 & -4 & 0 \\ 1 & 1 & \cos(2) & \sin(2) \\ 0 & 1 & -2\sin(2) & 2\cos(2) \end{vmatrix} = 4(2\cos(2)-\sin(2)) \neq 0,$$

and so the Green's function $G(x,\xi)$ exists by Proposition 2.7.

5.

$$G(x, \xi) = \begin{cases} c_1(\xi)u_1(x) + c_2(\xi)u_2(x), & \text{if } x < \xi; \\ d_1(\xi)u_1(x) + d_2(\xi)u_2(x), & \text{if } x > \xi, \end{cases}$$

where

$$c_1 = \frac{u_1(\xi)NK - u_2(\xi)NJ}{CD}, c_2 = \frac{-u_1(\xi)MK + u_2(\xi)MJ}{CD},$$

$$d_1 = \frac{u_1(\xi)NK - u_2(\xi)MK}{CD}, d_2 = \frac{-u_1(\xi)NJ + u_2(\xi)MJ}{CD},$$

$$D = \begin{vmatrix} M & N \\ J & K \end{vmatrix}.$$

On the other hand,

$$G(\xi, x) = \begin{cases} d_1(x)u_1(\xi) + d_2(x)u_2(\xi), & \text{if } x < \xi; \\ c_1(x)u_1(\xi) + c_2(x)u_2(\xi), & \text{if } x > \xi, \end{cases}$$

where calculations show that, for $x < \xi$,

$$d_1(x)u_1(\xi) + d_2(x)u_2(\xi)$$

$$= [-\frac{u_2(x)MK}{CD} + \frac{u_1(x)NK}{CD}]u_1(\xi) + [\frac{u_2(x)MJ}{CD} - \frac{u_1(x)NJ}{CD}]u_2(\xi)$$

$$= c_1(\xi)u_1(x) + c_2(\xi)u_2(x),$$

and that, for $x > \xi$,

$$c_1(x)u_1(\xi) + c_2(x)u_2(\xi)$$

$$= \frac{u_1(x)NK - u_2(x)NJ}{CD}u_1(\xi) + \frac{-u_1(x)MK + u_2(x)MJ}{CD}u_2(\xi)$$

$$= d_1(\xi)u_1(x) + d_2(\xi)u_2(x).$$

Thus $G(\xi, x) = G(x, \xi)$.

6. To the zero Dirichlet, Neumann, and Robin boundary conditions, the corresponding matrices $(\alpha \quad \beta)$ are

$$\begin{pmatrix} 1 & 0 & 0 & 0 \\ 0 & 0 & 1 & 0 \end{pmatrix}, \begin{pmatrix} 0 & 1 & 0 & 0 \\ 0 & 0 & 0 & 1 \end{pmatrix}, \text{ and } \begin{pmatrix} 0 & -1 & 0 & 0 \\ 0 & 0 & 1 & 1 \end{pmatrix},$$

respectively, and they all satisfy $|\alpha| = 0 = |\beta|$, for which (2.2) holds.
For the periodic condition,

$$(\alpha \quad \beta) = \begin{pmatrix} 1 & 0 & -1 & 0 \\ 0 & 1 & 0 & -1 \end{pmatrix},$$

and so $|\alpha| = 1 = |\beta|$. Since the condition (2.2) is the same as

$$|\alpha|W(u_1, u_2)(a) = |\beta|W(u_1, u_2)(b)$$

and $p(\xi)W(u_1, u_2)(\xi)$ is a constant by the Step 2 in the proof of Theorem 2.2, it follows that (2.2) holds if and only if $p(a) = p(b)$.

7. Since

$$(\alpha \quad \beta) = \begin{pmatrix} 1 & 0 & 0 & 0 \\ 0 & 0 & 0 & 1 \end{pmatrix}, \{u_1, u_2\} = \{\cos(2x), \sin(2x)\},$$

we have

$$|Bu_1 \quad Bu_2| = \left| \begin{pmatrix} 1 & 0 & 0 & 0 \\ 0 & 0 & 0 & 1 \end{pmatrix} \begin{pmatrix} 1 & 0 \\ 0 & 2 \\ \cos(2) & \sin(2) \\ -2\sin(2) & 2\cos(2) \end{pmatrix} \right|$$

$$= \left| \begin{matrix} 1 & 0 \\ -2\sin(2) & 2\cos(2) \end{matrix} \right| = 2\cos(2) \neq 0,$$

and so the Green's function $G(x, \xi)$ exists by Proposition 2.7. $G(x, \xi)$ is symmetric by Proposition 2.5 since $|\alpha| = 0 = |\beta|$.

8. Since

$$(\alpha \quad \beta) = \begin{pmatrix} 0 & 1 & 0 & 0 \\ 0 & 0 & 1 & 0 \end{pmatrix}, \{u_1, u_2\} = \{\cos(2x), \sin(2x)\},$$

we have

$$|Bu_1 \quad Bu_2| = \left| \begin{pmatrix} 0 & 1 & 0 & 0 \\ 0 & 0 & 1 & 0 \end{pmatrix} \begin{pmatrix} 1 & 0 \\ 0 & 2 \\ \cos(2) & \sin(2) \\ -2\sin(2) & 2\cos(2) \end{pmatrix} \right|$$

$$= \left| \begin{matrix} 0 & 2 \\ \cos(2) & \sin(2) \end{matrix} \right| = -2\cos(2) \neq 0,$$

and so the Green's function $G(x, \xi)$ exists by Proposition 2.7. $G(x, \xi)$ is symmetric by Proposition 2.5 since $|\alpha| = 0 = |\beta|$.

9. Since

$$(\alpha \quad \beta) = \begin{pmatrix} 1 & 0 & 0 & 0 \\ 0 & 0 & 1 & 1 \end{pmatrix}, \{u_1, u_2\} = \{\cos(2x), \sin(2x)\},$$

we have

$$|Bu_1 \quad Bu_2| = \left| \begin{pmatrix} 1 & 0 & 0 & 0 \\ 0 & 0 & 1 & 1 \end{pmatrix} \begin{pmatrix} 1 & 0 \\ 0 & 2 \\ \cos(2) & \sin(2) \\ -2\sin(2) & 2\cos(2) \end{pmatrix} \right|$$

$$= \left| \begin{matrix} 1 & 0 \\ (\cos(2) - 2\sin(2)) & (\sin(2) + 2\cos(2)) \end{matrix} \right| = \sin(2) + 2\cos(2) \neq 0,$$

and so the Green's function $G(x, \xi)$ exists by Proposition 2.7. $G(x, \xi)$ is symmetric by Proposition 2.5 since $|\alpha| = 0 = |\beta|$.

10. Since

$$(\alpha \quad \beta) = \begin{pmatrix} 1 & -1 & 0 & 0 \\ 0 & 0 & 1 & 0 \end{pmatrix}, \{u_1, u_2\} = \{\cos(2x), \sin(2x)\},$$

we have

$$|Bu_1 \quad Bu_2| = \left| \begin{pmatrix} 1 & -1 & 0 & 0 \\ 0 & 0 & 1 & 0 \end{pmatrix} \begin{pmatrix} 1 & 0 \\ 0 & 2 \\ \cos(2) & \sin(2) \\ -2\sin(2) & 2\cos(2) \end{pmatrix} \right|$$

$$= \left| \begin{matrix} 1 & -2 \\ \cos(2) & \sin(2) \end{matrix} \right| = \sin(2) + 2\cos(2) \neq 0,$$

and so the Green's function $G(x, \xi)$ exists by Proposition 2.7. $G(x, \xi)$ is symmetric by Proposition 2.5 since $|\alpha| = 0 = |\beta|$.

11. Since

$$(\alpha \quad \beta) = \begin{pmatrix} 0 & 1 & 0 & 0 \\ 0 & 0 & 1 & 1 \end{pmatrix}, \{u_1, u_2\} = \{\cos(2x), \sin(2x)\},$$

we have

$$|Bu_1 \quad Bu_2| = \left| \begin{pmatrix} 0 & 1 & 0 & 0 \\ 0 & 0 & 1 & 1 \end{pmatrix} \begin{pmatrix} 1 & 0 \\ 0 & 2 \\ \cos(2) & \sin(2) \\ -2\sin(2) & 2\cos(2) \end{pmatrix} \right|$$

$$= \left| \begin{matrix} 0 & 2 \\ (\cos(2) - 2\sin(2)) & (\sin(2) + 2\cos(2)) \end{matrix} \right| = 4\sin(2) - 2\cos(2) \neq 0,$$

and so the Green's function $G(x, \xi)$ exists by Proposition 2.7. $G(x, \xi)$ is symmetric by Proposition 2.5 since $|\alpha| = 0 = |\beta|$.

12. Since

$$(\alpha \quad \beta) = \begin{pmatrix} 1 & -1 & 0 & 0 \\ 0 & 0 & 0 & 1 \end{pmatrix}, \{u_1, u_2\} = \{\cos(2x), \sin(2x)\},$$

we have

$$|Bu_1 \quad Bu_2| = \left| \begin{pmatrix} 1 & -1 & 0 & 0 \\ 0 & 0 & 0 & 1 \end{pmatrix} \begin{pmatrix} 1 & 0 \\ 0 & 2 \\ \cos(2) & \sin(2) \\ -2\sin(2) & 2\cos(2) \end{pmatrix} \right|$$

$$= \left| \begin{matrix} 1 & -2 \\ -2\sin(2) & 2\cos(2) \end{matrix} \right| = 2\cos(2) - 4\sin(2) \neq 0,$$

and so the Green's function $G(x, \xi)$ exists by Proposition 2.7. $G(x, \xi)$ is symmetric by Proposition 2.5 since $|\alpha| = 0 = |\beta|$.

13. Since

$$(\alpha \quad \beta) = \begin{pmatrix} \frac{1}{2} & 0 & 1 & 0 \\ 0 & \frac{1}{2} & 0 & 1 \end{pmatrix}, \{u_1, u_2\} = \{x^2, x^{-3}\},$$

we have

$$|Bu_1 \quad Bu_2| = \left| \begin{pmatrix} \frac{1}{2} & 0 & 1 & 0 \\ 0 & \frac{1}{2} & 0 & 1 \end{pmatrix} \begin{pmatrix} 1 & 1 \\ 2 & -3 \\ 4 & \frac{1}{8} \\ 4 & -\frac{3}{16} \end{pmatrix} \right|$$

$$= \left| \begin{matrix} \frac{9}{2} & \frac{5}{8} \\ 5 & -\frac{27}{16} \end{matrix} \right| \neq 0,$$

and so the Green's function $G(x, \xi)$ exists by Proposition 2.7. $G(x, \xi)$ is symmetric by Proposition 2.5 since

$$|\alpha| = \frac{1}{4}, \quad |\beta| = 1,$$

$$W(1) = -5, \quad W(2) = -\frac{5}{4},$$

$$|\alpha|W(1) = -\frac{5}{4} = |\beta|W(2).$$

14. Since

$$(\alpha \quad \beta) = \begin{pmatrix} 1 & 0 & 0 & 0 & 0 & 0 & 0 & 0 \\ 0 & 1 & 0 & 0 & 0 & 0 & 0 & 0 \\ 0 & 0 & 0 & 0 & 1 & 0 & 0 & 0 \\ 0 & 0 & 0 & 0 & 0 & 1 & 0 & 0 \end{pmatrix}, \{u_1, u_2, u_3, u_4\} = \{1, x, \cos(2x), \sin(2x)\},$$

we have

$$|Bu_1 \quad Bu_2 \quad Bu_3 \quad Bu_4|$$

$$= \left| \begin{pmatrix} 1 & 0 & 0 & 0 & 0 & 0 & 0 & 0 \\ 0 & 1 & 0 & 0 & 0 & 0 & 0 & 0 \\ 0 & 0 & 0 & 0 & 1 & 0 & 0 & 0 \\ 0 & 0 & 0 & 0 & 0 & 1 & 0 & 0 \end{pmatrix} \begin{pmatrix} 1 & 0 & 1 & 0 \\ 0 & 1 & 0 & 2 \\ 0 & 0 & -4 & 0 \\ 0 & 0 & 0 & -8 \\ 1 & 1 & \cos(2) & \sin(2) \\ 0 & 1 & -2\sin(2) & 2\cos(2) \\ 0 & 0 & -4\cos(2) & -4\sin(2) \\ 0 & 0 & 8\sin(2) & -8\cos(2) \end{pmatrix} \right|$$

$$= \begin{vmatrix} 1 & 0 & 1 & 0 \\ 0 & 1 & 0 & 2 \\ 1 & 1 & \cos(2) & \sin(2) \\ 0 & 1 & -2\sin(2) & 2\cos(2) \end{vmatrix} = 4(1 - \cos(2) - \sin(2)) \neq 0,$$

and so the Green's function $G(x, \xi)$ exists by Proposition 2.7.

15. Since

$$(\alpha \quad \beta) = \begin{pmatrix} 1 & 0 & 0 & 0 & 0 & 0 & 0 & 0 \\ 0 & 0 & 1 & 0 & 0 & 0 & 0 & 0 \\ 0 & 0 & 0 & 0 & 1 & 0 & 0 & 0 \\ 0 & 0 & 0 & 0 & 0 & 0 & 1 & 0 \end{pmatrix}, \{u_1, u_2, u_3, u_4\} = \{1, x, \cos(2x), \sin(2x)\},$$

we have

$$|Bu_1 \quad Bu_2 \quad Bu_3 \quad Bu_4|$$

$$= \left| \begin{pmatrix} 1 & 0 & 0 & 0 & 0 & 0 & 0 & 0 \\ 0 & 0 & 1 & 0 & 0 & 0 & 0 & 0 \\ 0 & 0 & 0 & 0 & 1 & 0 & 0 & 0 \\ 0 & 0 & 0 & 0 & 0 & 0 & 1 & 0 \end{pmatrix} \begin{pmatrix} 1 & 0 & 1 & 0 \\ 0 & 1 & 0 & 2 \\ 0 & 0 & -4 & 0 \\ 0 & 0 & 0 & -8 \\ 1 & \pi & 1 & 0 \\ 0 & 1 & 0 & 2 \\ 0 & 0 & -4 & 0 \\ 0 & 0 & 0 & -8 \end{pmatrix} \right|$$

$$= \begin{vmatrix} 1 & 0 & 1 & 0 \\ 0 & 0 & -4 & 0 \\ 1 & \pi & 1 & 0 \\ 0 & 0 & -4 & 0 \end{vmatrix} = 0,$$

and so the Green's function $G(x, \xi)$ does not exist by Proposition 2.7.

CHAPTER 6

Eigenfunction Expansions

1. Introduction

Consider the eigenvalue problem:

$$
\begin{aligned}
(p(x)y')' + q(x)y &= \lambda r(x)y, \quad a < x < b \\
y(a) + \alpha_1 y'(a) &= 0, \\
y(b) + \alpha_2 y'(b) &= 0,
\end{aligned}
\tag{1.1}
$$

where $p(x), p'(x), q(x), r(x)$ are complex-valued continuous functions on $[a, b]$, with $\{|p(x)|, r(x)\} \geq \delta > 0$ for some $\delta > 0$ and for all x; $a < b$, $\alpha_1, \alpha_2 \in \mathbb{R}$, and λ is a complex number to be determined.

This problem will be studied through a certain linear differential operator L, which we define now. Let

$$
L : D(L) \subset (L^2(a, b), \| \cdot \|_2) \longrightarrow (C[a, b], \| \cdot \|_2) \subset (L^2(a, b), \| \cdot \|_2)
$$

be the linear differential operator defined by

$$
Lu \equiv (p(x)u')' + q(x)u
$$

for

$$
u \in D(L) \equiv \{v \in C^2[a, b] : v(a) + \alpha_1 v'(a) = 0, v(b) + \alpha_2 v'(b) = 0\}.
$$

Here $L^2(a, b)$ is the vector space of complex-valued functions on $[a, b]$, equipped with the norm $\| \cdot \|_2$, in which functions are absolutely square Lebesgue integrable (abbreviated as square integrable). That is,

DEFINITION 1.1. *A complex-valued function f is in $L^2(a, b)$ if*

$$
\int_a^b |f(x)|^2 \, dx < \infty,
$$

where the integral is taken with respect to the Lebesgue measure (see Chapter 5, Subsection 5.2).

$L^2(a, b)$ is an inner product space (see Definition 5.9) with respect to this inner product:

DEFINITION 1.2. *The inner product (f, g) of f with g is defined by*

$$
(f, g) \equiv \int_a^b f(x)\overline{g(x)} \, dx
$$

for $f, g \in L^2(a, b)$.

The norm $\| \cdot \|_2$ with which $L^2(a, b)$ is equipped is:

DEFINITION 1.3. *The norm* $\|\cdot\|_2$ *is defined by*

$$\|f\|_2 \equiv (f, f)^{\frac{1}{2}} = \left(\int_a |f(x)|^2 \, dx\right)^{\frac{1}{2}}$$

for $f \in L^2(a, b)$.

It follows from Chapter 4 that L is symmetric, that is,

$$(Lu, v) = (u, Lv)$$

for $u, v \in D(L)$.

In addition to $\|\cdot\|_2$, we shall also need a weighted norm $\|\cdot\|_w$. This is introduced by defining first the vector space $L^2_w(a, b)$, which relates to $L^2(a, b)$ in the following way:

DEFINITION 1.4. *Given the positive continuous function* $r(x)$, *a complex-valued function* f *on* (a, b) *is in* $L^2_w(a, b)$ *if*

$$\int_a^b r(x)|f(x)|^2 \, dx < \infty.$$

DEFINITION 1.5. *For* $f, g \in L^2_w(a, b)$, *the weighted inner product* $(f, g)_w$ *of* f *with* g, *is defined by*

$$(f, g)_w \equiv \int_a^b r(x)f(x)\overline{g(x)} \, dx.$$

Here $r(x)$ *is called the weight function for* $L^2_w(a, b)$.

Now we define

DEFINITION 1.6. *The weighted norm* $\|f\|_w$ *of* $f \in L^2_w(a, b)$ *is given by*

$$\|f\|_w \equiv \left(\int_a^b r(x)|f(x)|^2 \, dx\right)^{\frac{1}{2}}.$$

Since

$$0 < \delta \leq |r(x)| \leq \max_{x \in [a,b]} |r(x)| \equiv M,$$

we have

COROLLARY 1.7.

$$(L^2(a, b), \|\cdot\|_2) = (L^2_w(a, b), \|\cdot\|_w).$$

PROOF. See Problems set in Section 6.   □

It can be seen that

LEMMA 1.8. *The Cauchy-Schwarz inequality (see Section 5) is true:*

$$|(f, g)_w| = \left|\int_a^b r(x)f(x)\overline{g(x)} \, dx\right| \leq \int_a^b [\sqrt{r(x)}|f(x)|][\sqrt{r(x)}|g(x)|] \, dx$$

$$= (\sqrt{r}|f|, \sqrt{r}|g|) \leq \|\sqrt{r}|f|\|_2 \|\sqrt{r}|g|\|_2 = \|f\|_w \|g\|_w.$$

Here observe that, for $f, g \in L^2(a,b) = L_w^2(a,b)$, we have

$$(f,g)_w = (rf,g) = (f,rg)$$

and

$$(f,g) = (f,g)_w$$

when $r(x) \equiv 1$.

The norms $\| \cdot \|_2$ and $\| \cdot \|_w$ are related:

LEMMA 1.9. $\| \cdot \|_2$ *is equivalent to* $\| \cdot \|_w$, *in the sense that, for* $f \in L^2(a,b)$, *we have*

$$\epsilon_0 \|f\|_2 \leq \|f\|_w \leq \epsilon_1 \|f\|_2$$

for some constants $\epsilon_0, \epsilon_1 > 0$.

PROOF. This is readily checked by:

$$\|f\|_w^2 = (f,f)_w = (rf,f) \leq \max_{x \in [a,b]} r(x) \|f\|_2^2$$

and

$$\|f\|_w^2 = (f,f)_w = (rf,f) \geq \delta \|f\|_2^2.$$

$\square$

A result of this equivalence is:

COROLLARY 1.10. *The convergence in* $(L^2(a,b), \| \cdot \|_2)$ *is the same as that in* $(L^2(a,b), \| \cdot \|_w)$. *That is,*

$$\lim_{n \to \infty} \|f_n - f\|_2 = 0$$

for $f_n, f \in L^2(a,b)$, *is the same as*

$$\lim_{n \to \infty} \|f_n - f\|_w = 0.$$

PROOF. See Problems set in Section 6.

$\square$

By using L, the above eigenvalue problem (1.1) can be written as

$$Ly = \lambda r(x)y, \quad y \in D(L). \tag{1.2}$$

DEFINITION 1.11. λ *is called an eigenvalue for the operator* L *if there is a* $y \neq 0$ *which satisfies (1.2). In this case,* y *is called an eigenfunction or an eigenvector, associated with the eigenvalue* λ, *and* $\{\lambda, y\}$ *an eigenpair for* L. $\{\lambda, y\}$ *is a real eigenpair if* λ *is real and* y *real-valued. The eigenspace of* λ *is the set of eigenfunctions, associated with* λ. λ *is simple if its eigenspace is of one dimension.*

REMARK 1.12. *The eigenspace associated with the eigenvalue* λ, *is easily seen to be a vector space. For, if* y_1, y_2 *are two eigenfunctions associated with* λ *and* $\mu, \nu \in \mathbb{C}$ *are complex numbers, then*

$$L(\mu y_1) = \mu L y_1 = \mu(\lambda y_1) = \lambda(\mu y_1),$$

and so $\mu_1 y_1$ *lies in the eigenspace; also*

$$L(\mu y_1 + \nu y_2) = \mu L y_1 + \mu L y_2 = \mu \lambda y_1 + \nu \lambda y_2 = \lambda(\mu y_1 + \nu y_2),$$

and so $\mu y_1 + \nu y_2$ *lies in the eigenspace.*

We assume that $\lambda = 0$ is not an eigenvalue for the symmetric operator L, that is, we assume that the homogeneous equation: $Ly = 0, y \in D(L)$ only has the trivial solution $y \equiv 0$. This is no restriction, Coddington and Levinson [5, Page 193], since for a symmetric L, it is always possible (see Section 4, Lemma 4.2) to choose a real number $c_0 \in \mathbb{R}$, such that 0 is not an eigenvalue for the symmetric operator $L' \equiv (L - c_0)$ with $D(L') \equiv D(L)$. That is, the equation

$$L'y = 0, y \in D(L') = D(L)$$

only has the trivial solution $y \equiv 0$. In that case, we just work on L', instead of L, and the relation of eigenpairs between L' and L is: $\{c, y\}$ is an eigenpair for L' if and only if $\{c + c_0, y\}$ is an eigenpair for L.

By such an assumption that $\lambda = 0$ is not an eigenvalue for L, it follows that there is a unique Green function $G(x, \xi)$, associated with the operator L (see Chapter 5) and the nonhomogeneous equation

$$Ly = f, \quad y \in D(L) \tag{1.3}$$

has the solution

$$y = \mathbb{G}_0 f \equiv \int_{\xi = a}^{b} G(x, \xi) f(\xi) \, d\xi$$

for a continuous function f on $[a, b]$. Here

$$\mathbb{G}_0 : D(\mathbb{G}) \equiv C[a, b] \longrightarrow D(L) \subset C[a, b]$$

is introduced as a linear integral operator. Note that $\mathbb{G}_0 f$ still makes sense for a square integrable function f on (a, b) (see Section 5).

$\mathbb{G}_0$ can be seen to act as the inverse of L, for which

$$L\mathbb{G}_0 f = f \quad \text{for} \quad f \in C[a, b],$$
$$\mathbb{G}_0 L u = u \quad \text{for} \quad u \in D(L).$$

By using $\mathbb{G}_0$, the above eigenvalue problem (1.1) or (1.2) can be rewritten as

$$\mathbb{G}y = \frac{1}{\lambda} y,$$

where

$$\mathbb{G}y \equiv \mathbb{G}_0(ry) = \int_{\xi = a}^{b} G(x, \xi) r(\xi) y(\xi) \, d\xi.$$

Thus $\mathbb{G}$ has the same eigenfunction (if any) y as L does, but has the eigenvalue (if any) $\frac{1}{\lambda}$, the reciprocal of the eigenvalue λ for L.

It is the purpose of this chapter to show by using $\mathbb{G}$ that the eigenvalues and eigenfunctions of L exist and have special properties.

The rest of this chapter is organized as follows. Section 2 states the main results, and Section 3 studies examples for illustration. Section 4 deals with the proof of the main results, and Section 5 presents an abstract result that generalizes the main results in Section 2. Finally, Section 6 contains a problems set, and their solutions are placed in Section 7.

2. Main Results

THEOREM 2.1 (Coddington and Levinson, [**5**], Pages 193-201). *The following are true.*

(i) *The eigenvalues of L exist, are real numbers and simple, and form an increasing sequence of numbers:*

$$\lambda_0 < \lambda_1 < \lambda_2 < \cdots < \lambda_n < \cdots$$

with $\lim_{n \to \infty} \lambda_n = +\infty$. This says that the eigenvalues of L consists of $\{\lambda_i\}_{i=0}^{\infty}$.

(ii) *Eigenfunctions $\{\chi_i\}_{i=0}^{\infty}$ of L corresponding to real eigenvalues $\{\lambda_i\}_{i=0}^{\infty}$, exist, and are orthonormal in the sense that*

$$(\chi_i, \chi_j)_w = (\chi_i, r(x)\chi_j) = (r(x)\chi_i, \chi_j) = \delta_{ij}, i, j \in \mathbb{N},$$

where

$$\delta_{ij} = \begin{cases} 1, & \text{if } i = j, \\ 0, & \text{if } i \neq j; \end{cases}$$

that is,

$$(r(x)\chi_i, \chi_j) = (\chi_i, r(x)\chi_j) = 0$$

for $i \neq j$ and

$$(r(x)\chi_i, \chi_i) = (\chi_i, r(x)\chi_i) = 1.$$

(iii) *Orthonormal eigenfunctions $\{\chi_i\}_{i=0}^{\infty}$ are complete or span $D(L)$, with respect to the norm $\|\cdot\|_{\infty}$, in the sense that, for $f \in D(L)$, we have*

$$f = \sum_{i=0}^{\infty} (f, \chi_i)_w \chi_i = \sum_{i=1}^{\infty} (f, r\chi_i)\chi_i,$$

the convergence being uniform on $[a, b]$. In other words,

$$\left\| \sum_{i=0}^{m} (f, \chi_i)_w \chi_i - f \right\|_{\infty} \equiv \max_{x \in [a,b]} \left| \sum_{i=0}^{m} (f, \chi_i)_w \chi_i(x) - f(x) \right| \longrightarrow 0$$

as $m \longrightarrow \infty$.

(iv) *Eigenfunctions $\{\chi_i\}_{i=0}^{\infty}$ span $L^2(a, b)$ with respect to the norm $\|\cdot\|_2$, in the sense that, for $f \in L^2(a, b)$, we have*

$$f = \sum_{i=1}^{\infty} (f, \chi_i)_w \chi_i = \sum_{i=0}^{\infty} (f, r(x)\chi_i)\chi_i,$$

the convergence being taken with respect to the norm $\|\cdot\|_2$. In other words,

$$\left\| \sum_{i=0}^{m} (f, r(x)\chi_i)\chi_i - f \right\|_2 \equiv \left[\int_{x=a}^{b} \left| \sum_{i=0}^{m} (f, r(x)\chi_i)\chi_i(x) - f(x) \right|^2 dx \right]^{\frac{1}{2}} \longrightarrow 0$$

as $m \longrightarrow \infty$.

REMARK 2.2. *The eigenfunctions can be chosen to be real-valued if $p(x)$ and $q(x)$ are real-valued. We can see this by taking $\phi_i = \frac{\chi_i + \overline{\chi_i}}{2}$, and it follows readily that ϕ_i are real-valued eigenfunctions. See Lemma 4.4 in Subsection 4.2.*

REMARK 2.3. *Under boundary conditions of the form*

$$\begin{cases} \alpha_1 u(a) + u'(a) = 0 \\ \alpha_2 u(b) + u'(b) = 0, \quad or \end{cases}$$

$$\begin{cases} u(a) + \alpha_1 u'(a) = 0 \\ \alpha_2 u(b) + u'(b) = 0, \quad or \end{cases}$$

$$\begin{cases} \alpha_1 u(a) + u'(a) = 0 \\ u(b) + \alpha_2 u'(b) = 0, \quad or \end{cases}$$

the more general form (Coddington and Levinson [5, Page 297], and Taylor and Lay [21, Page 347])

$$\begin{cases} \alpha_{11} u(a) + \alpha_{12} u'(a) + \beta_{11} u(b) + \beta_{12} u'(b) = 0 \\ \alpha_{21} u(a) + \alpha_{22} u'(a) + \beta_{21} u(b) + \beta_{22} u'(b) = 0 \end{cases}$$

with

$$p(a) \begin{vmatrix} \beta_{11} & \beta_{12} \\ \beta_{21} & \beta_{22} \end{vmatrix} = p(b) \begin{vmatrix} \alpha_{11} & \alpha_{12} \\ \alpha_{21} & \alpha_{22} \end{vmatrix},$$

where $\alpha_{ij}, \beta_{ij} \in \mathbb{R}, i, j = 1, 2$, it is true that L is symmetric also (see Chapter 4) and that the corresponding Green function $G(x, \xi)$ possesses the same properties (see Chapter 5), Thus the proof below for Theorem 2.1 will still apply to those cases although, in the last case with the more general boundary conditions assumed, the eigenvalues of L are not necessarily simple. This last case includes the periodic boundary conditions:

$$\begin{cases} u(a) = u(b), \\ u'(a) = u'(b) \end{cases}$$

with $p(a) = p(b)$, for which see the examples in Section 2.

3. Examples

Before we prove Theorem 2.1, we look at some examples. Here notice that the eigenvalues in the examples are computed as the negative of those that are defined in Definition 1.11.

EXAMPLE 3.1. Solve the eigenvalue problem with the Dirichlet boundary condition:

$$y'' + \lambda y = 0, \quad 0 < x < b,$$
$$y(0) = 0 = y(b),$$

where $y = y(x), b > 0$, and λ is a constant to be determined by the boundary conditions.

Solution. Trying $y = e^{rx}$, where r is a constant to be found, we obtain

$$y' = re^{rx}, y'' = r^2 e^{rx}, 0 = y'' + \lambda y = (r^2 + \lambda)e^{rx}.$$

Since $e^{rx} \neq 0$, we have

$$r^2 + \lambda = 0.$$

Here note that $\lambda \in \mathbb{R}$, since the corresponding operator L is symmetric.

Case 1: $\lambda > 0$. Since $r = i\sqrt{\lambda}$ or $-i\sqrt{\lambda}$, we have the real-valued solution:

$$y = A\cos(\sqrt{\lambda}x) + B\sin(\sqrt{\lambda}x),$$

where A, B are two constants. The boundary conditions imply:

$$0 = y(0) = A \quad \text{and} \quad 0 = y(b) = B\sin(\sqrt{\lambda}b).$$

If $B = 0$, we only have the trivial zero solution: $y \equiv 0$, and so $B \neq 0$ and $\sin(\sqrt{\lambda}b) = 0$. It follows that $\lambda = \lambda_n = (\frac{n\pi}{b})^2, n \in \mathbb{Z}$, and that the eigenpairs are

$$\{\lambda_n, \phi_n\}_{n \in \mathbb{Z}} = \{(\frac{n\pi}{b})^2, \sin(\frac{n\pi}{b}x)\}_{n \in \mathbb{Z}}.$$

But it suffices to consider $n \in \mathbb{N}$, since $\lambda_{-n} = \lambda_n$ and $\sin(-\frac{n\pi}{b}x) = -\sin(\frac{n\pi}{b}x)$.

Case 2: $\lambda = 0$. We only obtain the trivial zero solution, since $y'' = 0, y = Cx + D$, and $C = D = 0$ by the boundary conditions.

Case 3: $\lambda < 0$. As in Case 1, we obtain

$$y = Ee^{\sqrt{-\lambda}x} + Fe^{-\sqrt{-\lambda}x},$$
$$0 = y(0) = E + F,$$
$$0 = y(b) = Ee^{\sqrt{-\lambda}b} + Fe^{-\sqrt{-\lambda}b}.$$

It follows from simple calculations that $E = 0 = F$, and that the trivial zero is the only solution for y.

REMARK 3.2. *Another way to see $\lambda \geq 0$ is:*

$$-\lambda \int_0^b y^2\, dx = \int_0^b y'' y\, dx$$
$$= y'y|_{x=0}^b - \int_0^b (y')^2\, dx$$
$$= -\int_0^b (y')^2\, dx$$

for $y \neq 0$, and so $\lambda \geq 0$.

The orthogonality: $(\phi_n, \phi_m) = 0$ for $n \neq m$, is true, since

$$(\phi_n, \phi_m) = \int_0^b \sin(\frac{n\pi}{b}x)\sin(\frac{m\pi}{b}x)\, dx$$
$$= \frac{1}{2}[\sin(\frac{(n-m)\pi}{b}x) - \sin(\frac{(n+m)\pi}{b}x)]|_{x=0}^b$$
$$= 0.$$

In order to have eigenfunctions of norm one, we normalize ϕ_n by choosing suitable c_n:

$$1 = (c_n\phi_n, c_n\phi_n)$$
$$= (c_n)^2 \int_0^b (\sin(\frac{n\pi}{b}x))^2\, dx$$
$$= c_n^2 \frac{b}{2},$$

and so $c_n = \sqrt{\frac{2}{b}}$. Here we used the formulas:

$$\sin^2(\frac{\theta}{2}) = \frac{1 - \cos(\theta)}{2},$$
$$2\sin(\theta_1)\sin(\theta_2) = \cos(\theta_1 - \theta_2) - \cos(\theta_1 + \theta_2).$$

Thus we obtain a complete orthonormal system as Theorem 2.1 ensures, which is:

$$\{\chi_n\}_{n\in\mathbb{N}} = \{\sqrt{\frac{2}{b}}\sin(\frac{n\pi}{b}x)\}_{n\in\mathbb{N}}.$$

EXAMPLE 3.3. Solve the eigenvalue problem with the Neumannn boundary conditions:

$$y'' + \lambda y = 0, \quad 0 < x < b,$$
$$y'(0) = 0 = y'(b),$$

where $y = y(x), b > 0$, and λ is constant to be determined by the boundary conditions.

Solution. As in Example 3.1, we have $\lambda \geq 0$, and

$$y = A\cos(\sqrt{\lambda}x) + B\sin(\sqrt{\lambda}x) \quad \text{for } \lambda > 0,$$
$$y = Cx + D \quad \text{for } \lambda = 0.$$

Case 1: $\lambda > 0$. The boundary condition gives:

$$0 = y'(0) = \sqrt{\lambda}B,$$
$$0 = y'(b) = -A\sqrt{\lambda}\sin(\sqrt{\lambda}b).$$

Since $A = 0$ gives only the trivial zero solution, we have $A \neq 0$ and $\sin(\sqrt{\lambda}b) = 0$. It follows that $\lambda = \lambda_n = (\frac{n\pi}{b})^2, n \in \mathbb{Z}$, and that the eigenpairs are

$$\{\lambda_n, \phi_n\}_{n\in\mathbb{Z}} = \{(\frac{n\pi}{b})^2, \cos(\frac{n\pi}{b}x)\}_{n\in\mathbb{Z}}.$$

But it suffices to consider only $n \in \mathbb{N} \cup \{0\}$, since $\lambda_{-n} = \lambda_n$ and $\cos(-\frac{n\pi}{b}x) = \cos(\frac{n\pi}{b}x)$.

Case 2: $\lambda = 0$. The boundary conditions give

$$0 = y'(0) = C,$$

and so $y = D$, a constant. But this is contained in Case 1.

As in Example 3.1, the normalization constants are $c_0 = \sqrt{\frac{1}{b}}$ and $c_n = \sqrt{\frac{2}{b}}, n \in \mathbb{N}$. Hence we obtain a complete orthonormal system as Theorem 2.1 ensures, which is

$$\{\chi_n\}_{n\in\mathbb{N}\cup\{0\}} = \{\sqrt{\frac{1}{b}}, \sqrt{\frac{2}{b}}\cos(\frac{n\pi}{b}x)\}_{n\in\mathbb{N}\cup\{0\}}.$$

EXAMPLE 3.4. Solve the eigenvalue problem with the periodic boundary conditions:

$$y'' + \lambda y = 0, \quad 0 < x < 2\pi,$$
$$y(0) = y(2\pi),$$
$$y'(0) = y'(2\pi),$$

where $y = y(x)$ and λ is a constant to be determined by the boundary conditions.

Solution. As in Example 3.1, we have $\lambda \geq 0$, and

$$y = A\cos(\sqrt{\lambda}x) + B\sin(\sqrt{\lambda}x) \quad \text{for } \lambda > 0,$$
$$y = Cx + D \quad \text{for } \lambda = 0.$$

Case 1: $\lambda > 0$. The boundary conditions give

$$A = A\cos(\sqrt{\lambda}2\pi) + B\sin(\sqrt{\lambda}2\pi),$$
$$\sqrt{\lambda}B = -\sqrt{\lambda}A\sin(\sqrt{\lambda}2\pi) + \sqrt{\lambda}B\cos(\sqrt{\lambda}2\pi).$$

To obtain A and B, we use: the top equation times B minus the bottom equation times $\frac{A}{\sqrt{\lambda}}$, and we have

$$0 = AB - AB = (A^2 + B^2)\sin(\sqrt{\lambda}2\pi).$$

It follows that $A^2 + B^2 \neq 0$ and

$$\sin(\sqrt{\lambda}2\pi) = 0.$$

For, if $A^2 + B^2 = 0$, then $A = 0 = B$ and we only have the trivial zero solution for y.

On the other hand, the top equation times A plus the bottom equation times $\frac{B}{\sqrt{\lambda}}$ gives

$$A^2 + B^2 = (A^2 + B^2)\cos(\sqrt{\lambda}2\pi).$$

Again, $A^2 + B^2 \neq 0$ and

$$\cos(\sqrt{\lambda}2\pi) = 1, \tag{3.1}$$

since $A^2 + B^2 = 0$ gives only the trivial zero solution. Thus (3.1), together with

$$\sin(\sqrt{\lambda}2\pi) = 0$$

gives

$$\lambda = \lambda_n = n^2, \quad n \in \mathbb{Z}.$$

It follows that the eigenpairs are

$$\{\lambda_n, \phi_n\}_{n \in \mathbb{Z}} = \{n^2, \{\cos(nx), \sin(nx)\}\}_{n \in \mathbb{Z}}.$$

But it suffices to consider only $n \in \mathbb{N} \cup \{0\}$, since

$$\lambda_{-n} = \lambda_n,$$
$$\cos(-nx) = \cos(nx), \sin(-nx) = -\sin(nx).$$

Here note that $\lambda_n, n \in \mathbb{N}$ is not simple, since, associated with it, there are two linearly independent eigenfunctions $\cos(nx), \sin(nx)$.

Case 2: $\lambda = 0$. The boundary condition $y(0) = y(2\pi)$ gives:

$$D = C(2\pi) + D,$$

and so $C = 0$ and the solution y is a constant. Hence this case is contained in Case 1.

The orthogonality: $(\phi_n, \phi_m) = 0$ for $n \neq m$, is true, since

$$\int_0^{2\pi} \sin(mx)\cos(nx)\,dx = \frac{1}{2}[-\frac{\cos((m+n)x)}{m+n} - \frac{\cos((m-n)x)}{m-n}]|_{x=0}^{2\pi}$$
$$= 0 \quad \text{for } n, m \in \mathbb{N}, n \neq m,$$

$$\int_0^{2\pi} \sin(nx)\cos(nx)\,dx = \frac{1}{2}\int_0^{2\pi} \sin(2nx)\,dx$$

$$= 0 \quad \text{for } n \in \mathbb{N},$$

$$(1, \cos(nx)) = \int_0^{2\pi} \cos(nx)\, dx = 0 \quad \text{for } n \in \mathbb{N}, \text{ and}$$

$$(1, \sin(nx)) = \int_0^{2\pi} \sin(nx)\, dx = 0 \quad \text{for } n \in \mathbb{N}.$$

Normalize ϕ_n by the calculations:

$$1 = (c_n \cos(nx), c_n \cos(nx))$$

$$= (c_n)^2 \int_0^{2\pi} \frac{1 + \cos(2nx)}{2}\, dx$$

$$= (c_n)^2 \pi,$$

and so choose $c_n = \sqrt{\frac{1}{\pi}}$; similarly, make the calculations:

$$1 = (d_n \sin(nx), d_n \sin(nx))$$

$$= (d_n)^2 \int_0^{2\pi} \sin^2(nx)\, dx$$

$$= (d_n)^2 \pi,$$

and so choose $d_n = \sqrt{\frac{1}{\pi}}$; finally, the calculations

$$1 = (c_0, c_0) = (c_0)^2 \int_0^{2\pi} 1\, dx = (c_0)^2 (2\pi)$$

choose $c_0 = \sqrt{\frac{1}{2\pi}}$.

Thus we obtain a complete orthonormal system as Theorem 2.1 ensures, which is

$$\{\sqrt{\frac{1}{2\pi}}, \sqrt{\frac{1}{\pi}} \cos(nx), \sqrt{\frac{1}{\pi}} \sin(nx)\}_{n \in \mathbb{N}}.$$

EXAMPLE 3.5. Solve the eigenvalue problem with the Robin boundary conditions:

$$y'' + \lambda y = 0, \quad 0 < x < \pi,$$

$$y(0) - y'(0) = 0,$$

$$y(\pi) + y'(\pi) = 0,$$

where $y = y(x)$ and λ is a constant to be determined by the boundary conditions.

Solution. As in Example 3.1, we have that $\lambda \geq 0$, and

$$y = A \cos(\sqrt{\lambda} x) + B \sin(\sqrt{\lambda} x) \quad \text{for } \lambda > 0,$$

$$y = Cx + D \quad \text{for } \lambda = 0.$$

Case 1: $\lambda > 0$. The boundary conditions give

$$A = \sqrt{\lambda} B,$$

$$A \cos(\sqrt{\lambda} \pi) + B \sin(\sqrt{\lambda} \pi) = \sqrt{\lambda} A \sin(\sqrt{\lambda} \pi) - \sqrt{\lambda} B \cos(\sqrt{\lambda} \pi),$$

and so

$$B(\sqrt{\lambda} \cos(\sqrt{\lambda} \pi) + \sin(\sqrt{\lambda} \pi)) = B(\lambda \sin(\sqrt{\lambda} \pi) - \sqrt{\lambda} \cos(\sqrt{\lambda} \pi)).$$

It follows that $B \neq 0$ and

$$(\lambda - 1)\sin(\sqrt{\lambda}\pi) = 2\sqrt{\lambda}\cos(\sqrt{\lambda}\pi)$$

or

$$\cot(\sqrt{\lambda}\pi) = \frac{\lambda - 1}{2\sqrt{\lambda}} = \frac{1}{2}\sqrt{\lambda} - \frac{1}{\sqrt{\lambda}}.$$

For, if $B = 0$, then $A = 0$ and the trivial zero is the only solution for y.

Letting $\mu = \sqrt{\lambda} > 0$, we see that eigenvalues λ can be obtained from the intersections of the two graphs:

$$y_1(\mu) = \cot(\mu\pi),$$

$$y_2(\mu) = \frac{1}{2}(\mu - \frac{1}{\mu}).$$

It follows from calculus that

$$y_2(1) = 0,$$

$$y_2' = \frac{1}{2}(1 + \frac{1}{\mu^2}) > 0,$$

$$y_2'' = -\frac{1}{\mu^3} < 0,$$

and so the graph of $y_2(\mu)$ passes through the point $(1, 0)$, increases, and concaves down. Similarly, we have

$$y_1(\frac{2n-1}{2}\pi) = 0, n \in \mathbb{N},$$

$$y_1'(\mu) = -\pi \csc^2(\mu\pi) < 0 \quad \text{for } 0 < \mu < 1,$$

and

$$y_1'' = 2\pi^2 \csc^2(\mu\pi)\cot(\mu\pi) \begin{cases} > 0, & \text{if } 0 < \mu < \frac{1}{2}; \\ < 0, & \text{if } \frac{1}{2} < \mu < 1. \end{cases}$$

Thus the graph of $y_1(\mu)$ for $0 < \mu < 1$, passes through the point $(\frac{1}{2}, 0)$, decreases, and concaves up for $0 < \mu < \frac{1}{2}$, and concaves down for $\frac{1}{2} < \mu < 1$. This graph of the graph of y_1 for $\mu > 1$ by periodic extension with the period 1.

Thus the graphs of $y_1(\mu)$ and $y_2(\mu)$ appear qualitatively as those in Figure 1:

Since each intersection point corresponds to an eigenvalue $\lambda = \lambda_n = (\mu_n)^2, n \in \mathbb{N}$, we have from the graphs that

$$\mu_1 = \sqrt{\lambda_1} \in (\frac{1}{2}, 1),$$

$$\mu_n = \sqrt{\lambda_n} \in (n - 1, n - \frac{1}{2}) \quad \text{for } n \geq 2, \quad \text{and}$$

$$\mu_n = \sqrt{\lambda_n} \approx n \quad \text{for large } n,$$

where the last property follows from

$$y_1(n) = \cot(n\pi) \approx \infty,$$

$$y_2(n) = \frac{1}{2}(n - \frac{1}{n}) \approx \frac{1}{2}n \approx \infty$$

for large n. Here $\approx$ means: approximately equals.

Thus the eigenpairs are

$$\{\lambda_n, \phi_n\}_{n \in \mathbb{N}} = \{\lambda_n, \sqrt{\lambda_n}\cos(\sqrt{\lambda_n}x) + \sin(\sqrt{\lambda_n}x)\}_{n \in \mathbb{N}},$$

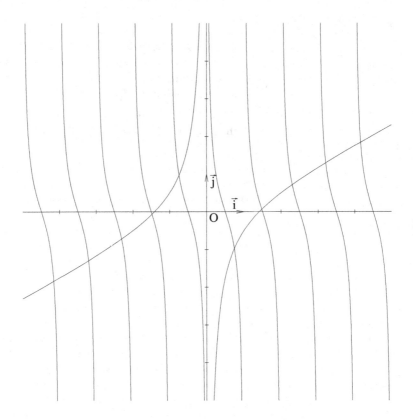

FIGURE 1. Intersection of two graphs

where λ_n satisfies

$$(\lambda_n - 1)\sin(\sqrt{\lambda_n}\pi) = 2\sqrt{\lambda_n}\cos(\sqrt{\lambda_n}\pi). \tag{3.2}$$

Case 2: $\lambda = 0$. The boundary condition gives:

$$D = y(0) = y'(0) = C,$$
$$C\pi + D = y(\pi) = -y'(\pi) = -C,$$

and so $C = 0 = D$ and the trivial zero is the only solution for y.

The orthogonality: $(\phi_n, \phi_m) = 0$ for $n \neq m$, is true, since

$$(\phi_n, \phi_m) = \int_0^\pi [\sqrt{\lambda_n}\sqrt{\lambda_m}\cos(\sqrt{\lambda_n}x)\cos(\sqrt{\lambda_m}x)$$
$$+ \sqrt{\lambda_m}\cos(\sqrt{\lambda_m}x)\sin(\sqrt{\lambda_n}x) + \sqrt{\lambda_n}\cos(\sqrt{\lambda_n}x)\sin(\sqrt{\lambda_m}x)$$
$$+ \sin(\sqrt{\lambda_n}x)\sin(\sqrt{\lambda_m}x)]\,dx$$
$$\equiv I_1 + I_2 + I_3 + I_4,$$

$$I_1 = \int_0^\pi \sqrt{\lambda_m}\sqrt{\lambda_n}\frac{1}{2}[\cos((\sqrt{\lambda_m} + \sqrt{\lambda_n})x) + \cos((\sqrt{\lambda_m} - \sqrt{\lambda_n})x)]\,dx$$
$$= \sqrt{\lambda_m}\sqrt{\lambda_n}\frac{1}{2}[\frac{\sin((\sqrt{\lambda_m} + \sqrt{\lambda_n})x)}{\sqrt{\lambda_m} + \sqrt{\lambda_n}} + \frac{\sin((\sqrt{\lambda_m} - \sqrt{\lambda_n})x)}{\sqrt{\lambda_m} - \sqrt{\lambda_n}}]|_{x=0}^\pi$$

$$= \sqrt{\lambda_m}\sqrt{\lambda_n}\frac{1}{2}[\frac{\sin((\sqrt{\lambda_m}+\sqrt{\lambda_n})\pi)}{\sqrt{\lambda_m}+\sqrt{\lambda_n}} + \frac{\sin((\sqrt{\lambda_m}-\sqrt{\lambda_n})\pi)}{\sqrt{\lambda_m}-\sqrt{\lambda_n}}$$

$$\equiv \frac{\sqrt{\lambda_m}\sqrt{\lambda_n}}{2}[\frac{J_1}{\sqrt{\lambda_m}+\sqrt{\lambda_n}} + \frac{J_2}{\sqrt{\lambda_m}-\sqrt{\lambda_n}}],$$

$$J_1 = \sin(\sqrt{\lambda_m}\pi)\cos(\sqrt{\lambda_n}\pi) + \cos(\sqrt{\lambda_m}\pi)\sin(\sqrt{\lambda_n}\pi)$$

$$= \frac{2\sqrt{\lambda_m}}{\lambda_m-1}\cos(\sqrt{\lambda_m}\pi)\cos(\sqrt{\lambda_n}\pi) + \cos(\sqrt{\lambda_m}\pi)\frac{2\sqrt{\lambda_n}}{\lambda_n-1}\cos(\sqrt{\lambda_n}\pi)$$

$$= \frac{2(\sqrt{\lambda_m}+\sqrt{\lambda_n})(\sqrt{\lambda_m}\sqrt{\lambda_n}-1)}{(\lambda_m-1)(\lambda_n-1)}\cos(\sqrt{\lambda_m}\pi)\cos(\sqrt{\lambda_n}\pi) \quad \text{by (3.2)},$$

$$J_2 = \sin(\sqrt{\lambda_m}\pi)\cos(\sqrt{\lambda_n}\pi) - \cos(\sqrt{\lambda_m}\pi)\sin(\sqrt{\lambda_n}\pi)$$

$$= \frac{2\sqrt{\lambda_m}}{\lambda_m-1}\cos(\sqrt{\lambda_m}\pi)\cos(\sqrt{\lambda_n}\pi) - \cos(\sqrt{\lambda_m}\pi)\frac{2\sqrt{\lambda_m}}{\lambda_m-1}\cos(\sqrt{\lambda_m}\pi)$$

$$= \frac{(-2)(\sqrt{\lambda_m}-\sqrt{\lambda_n})(\sqrt{\lambda_m}\sqrt{\lambda_n}+1)}{(\lambda_m-1)(\lambda_n-1)}\cos(\sqrt{\lambda_m}\pi)\cos(\sqrt{\lambda_n}\pi) \quad \text{by (3.2)},$$

$$I_1 = \frac{\sqrt{\lambda_m}\sqrt{\lambda_n}}{2}[\frac{J_1}{\sqrt{\lambda_m}+\sqrt{\lambda_n}} + \frac{J_2}{\sqrt{\lambda_m}-\sqrt{\lambda_n}}]$$

$$= \frac{-2\sqrt{\lambda_m}\sqrt{\lambda_n}}{(\lambda_m-1)(\lambda_n-1)}\cos(\sqrt{\lambda_m}\pi)\cos(\sqrt{\lambda_n}\pi),$$

$$I_2 + I_3 = \int_0^\pi \{\frac{1}{2}[\sin((\sqrt{\lambda_n}+\sqrt{\lambda_m})x) + \sin((\sqrt{\lambda_n}-\sqrt{\lambda_m})x)]\sqrt{\lambda_m}$$

$$+ \frac{1}{2}[\sin((\sqrt{\lambda_m}+\sqrt{\lambda_n})x) + \sin((\sqrt{\lambda_m}-\sqrt{\lambda_n})x)]\sqrt{\lambda_n}\} \, dx$$

$$= \int_0^\pi [\frac{1}{2}(\sqrt{\lambda_m}+\sqrt{\lambda_n})\sin((\sqrt{\lambda_m}+\sqrt{\lambda_n})x)$$

$$- \frac{1}{2}(\sqrt{\lambda_m}-\sqrt{\lambda_n})\sin((\sqrt{\lambda_m}-\sqrt{\lambda_n})x)] \, dx$$

$$= \frac{1}{2}[\cos((\sqrt{\lambda_m}-\sqrt{\lambda_n})x) - \cos((\sqrt{\lambda_m}+\sqrt{\lambda_n})x)]|_{x=0}^\pi$$

$$= \frac{1}{2}2\sin(\sqrt{\lambda_m}x)\sin(\sqrt{\lambda_n}x)|_{x=0}^\pi$$

$$= \sin(\sqrt{\lambda_m}\pi)\sin(\sqrt{\lambda_n}\pi)$$

$$= \frac{2\sqrt{\lambda_m}}{\lambda_m-1}\cos(\sqrt{\lambda_m}\pi)\frac{2\sqrt{\lambda_n}}{\lambda_n-1}\cos(\sqrt{\lambda_n}\pi), \quad \text{by (3.2)},$$

$$I_4 = \int_0^\pi \frac{1}{2}[\cos((\sqrt{\lambda_m}-\sqrt{\lambda_n})x) - \cos((\sqrt{\lambda_m}+\sqrt{\lambda_n})x)] \, dx$$

$$= \frac{1}{2}[\frac{J_2}{\sqrt{\lambda_m}-\sqrt{\lambda_n}} + \frac{J_1}{\sqrt{\lambda_m}+\sqrt{\lambda_n}}]$$

$$= \frac{-2\sqrt{\lambda_m}\sqrt{\lambda_n}}{(\lambda_m-1)(\lambda_n-1)}\cos(\sqrt{\lambda_m}\pi)\cos(\sqrt{\lambda_n}\pi), \quad \text{similarly to } I_1,$$

$$(\phi_n, \phi_m) = \sum_{i=1}^4 I_i$$

$$= [(-4) + 4]\frac{\sqrt{\lambda_m}\sqrt{\lambda_n}}{(\lambda_m - 1)(\lambda_n - 1)}\cos(\sqrt{\lambda_m}\pi)\cos(\sqrt{\lambda_n}\pi)$$

$$= 0.$$

Here we used (3.2):

$$\sin(\sqrt{\lambda_j}\pi) = \frac{2\sqrt{\lambda_j}}{\lambda_j - 1}\cos(\sqrt{\lambda_j}\pi), \quad j = m, n,$$

and the trigonometry formulas of converting product to sum and converting sum to product:

$$2\sin(\theta_1)\cos(\theta_2) = \sin(\theta_1 + \theta_2) + \sin(\theta_1 - \theta_2),$$
$$2\cos(\theta_1)\cos(\theta_2) = \cos(\theta_1 - \theta_2) + \cos(\theta_1 + \theta_2),$$
$$2\sin(\theta_1)\sin(\theta_2) = cos(\theta_1 - \theta_2) - \cos(\theta_1 + \theta_2).$$

Now we normalize ϕ_n by the calculations:

$$1 = (c_n\phi_n, c_n\phi_n)$$

$$= (c_n)^2\int_0^\pi [\lambda_n\cos^2(\sqrt{\lambda_n}x) + \sin^2(\sqrt{\lambda_n}x)$$

$$+ 2\sqrt{\lambda_n}\cos(\sqrt{\lambda_n}x)\sin(\sqrt{\lambda_n}x)]\,dx$$

$$\equiv (c_n)^2[K_1 + K_2 + K_3],$$

$$K_1 = \int_0^\pi \lambda_n\frac{1 + \cos(2\sqrt{\lambda_n}x)}{2}\,dx$$

$$= \lambda_n[\frac{1}{2}(\pi - 0) + \frac{1}{2}\frac{1}{2\sqrt{\lambda_n}}\sin(2\sqrt{\lambda_n}\pi)],$$

$$K_2 = \int_0^\pi \frac{1 - \cos(2\sqrt{\lambda_n}x)}{2}\,dx$$

$$= \frac{1}{2}(\pi - 0) - \frac{1}{2}\frac{1}{2\sqrt{\lambda_n}}\sin(2\sqrt{\lambda_n}\pi),$$

$$K_3 = \int_0^\pi \sqrt{\lambda_n}\sin(2\sqrt{\lambda_n}x)\,dx$$

$$= \frac{1}{2}[1 - \cos(2\sqrt{\lambda_n}\pi)]$$

$$= 1 - \cos^2(\sqrt{\lambda_n}\pi),$$

$$K_1 + K_2 = (\lambda_n + 1)\frac{\pi}{2} + \frac{\lambda_n - 1}{4\sqrt{\lambda_n}}\sin(2\sqrt{\lambda_n}\pi),$$

$$= (\lambda_n + 1)\frac{\pi}{2} + \frac{\lambda_n - 1}{2\sqrt{\lambda_n}}\sin(\sqrt{\lambda_n}\pi)\cos(\sqrt{\lambda_n}\pi)$$

$$= (\lambda_n + 1)\frac{\pi}{2} + \cos^2(\sqrt{\lambda_n}\pi),$$

$$1 = (c_n)^2\sum_{i=1}^3 K_i$$

$$= (c_n)^2[\frac{(\lambda_n + 1)\pi}{2} + 1 + (1 - 1)\cos^2(\sqrt{\lambda_n}\pi)]$$

$$= (c_n)^2\frac{(\lambda_n + 1)\pi + 2}{2};$$

so we choose

$$c_n = \sqrt{\frac{2}{(\lambda_n + 1)\pi + 2}}.$$

Here we used the trigonometry formulas:

$$\cos^2\left(\frac{\theta}{2}\right) = \frac{1 + \cos(\theta)}{2},$$

$$\sin^2\left(\frac{\theta}{2}\right) = \frac{1 - \cos(\theta)}{2},$$

$$\sin(2\theta) = 2\sin(\theta)\cos(\theta).$$

Thus we obtain a complete orthonormal system as Theorem 2.1 ensures, which is

$$\left\{\lambda_n, \sqrt{\frac{2}{(\lambda_n + 1)\pi + 2}}[\sqrt{\lambda_n}\cos(\sqrt{\lambda_n}x) + \sin(\sqrt{\lambda_n}x)]\right\}_{n \in \mathbb{N}},$$

where λ_n satisfies

$$\cot(\sqrt{\lambda_n}\pi) = \frac{1}{2}\left(\sqrt{\lambda_n} - \frac{1}{\sqrt{\lambda_n}}\right).$$

EXAMPLE 3.6. Solve the eigenvalue problem with the Dirichlet boundary conditions:

$$(xy')' + \lambda\frac{1}{x}y = 0, \quad 1 < x < e^2,$$

$$y(1) = 0 = y(e^2),$$

where $y = y(x)$ and λ is a constant to be determined by the boundary conditions.

Solution. Note that, as in Example 3.1, integration by parts applied to the equation gives:

$$-\lambda \int_1^{e^2} \frac{1}{x}y^2 \, dx = \int_0^{e^2} (xy')'y \, dx$$

$$= xyy'|_{x=1}^{e^2} - \int_1^{e^2} x(y')^2 \, dx$$

$$= 0 - \int_1^{e^2} x(y')^2 \, dx$$

$$\leq 0,$$

and so $\lambda \geq 0$.

Case 1: $\lambda = 0$. In this case, we have $(xy')' = 0$, and so

$$y = C\ln(x) + D.$$

Here the constants C, D are determined by the boundary conditions:

$$0 = y(1) = D,$$

$$0 = y(e^2) = C.$$

Thus the trivial zero is the only solution for y.

Case 2: $\lambda > 0$. The equation is the same as

$$x^2 y'' + xy' + \lambda y = 0,$$

which is a Cauchy-Euler equation. Trying $y = x^r$, where r is a constant to be determined, we have

$$0 = x^2 y'' + xy' + \lambda y$$
$$= [r(r-1) + r + \lambda]x^r.$$

Since $x^r \neq 0$ for $1 < x < e^2$, we have

$$r^2 + \lambda = 0,$$

and so $r = i\sqrt{\lambda}$ or $-i\sqrt{\lambda}$. Thus the real-valued solutions are:

$$y = A\cos(\sqrt{\lambda}\ln(x)) + B\sin(\sqrt{\lambda}\ln(x)),$$

where the constants A and B are determined by the boundary conditions:

$$0 = y(1) = A,$$
$$0 = y(e^2) = B\sin(\sqrt{\lambda}2).$$

Since $B = 0$ gives only the trivial zero solution for y, we have $B \neq 0$ and

$$\sin(\sqrt{\lambda}2) = 0, \sqrt{\lambda}2 = n\pi, n \in \mathbb{Z}.$$

It follows that the eigenpairs are

$$\{\lambda_n, \phi_n\}_{n\in\mathbb{Z}} = \{(\frac{n\pi}{2})^2, \sin(\frac{n\pi}{2}\ln(x))\}_{n\in\mathbb{Z}}$$

But it suffices to consider $n \in \mathbb{N}$ only, since $\lambda_{-n} = \lambda_n$ and

$$\sin(\frac{-n\pi}{2}\ln(x)) = -\sin(\frac{n\pi}{2}\ln(x)).$$

With the weight function $r(x) \equiv \frac{1}{x}$, the orthogonality: $(\phi_n, \phi_m)_w = 0$ for $n \neq m$, is true, since

$$(\phi_n, \phi_m)_w = \int_1^{e^2} \frac{1}{x}\sin(\frac{n\pi}{2}\ln(x))\sin(\frac{m\pi}{2}\ln(x))\,dx$$

$$= \frac{1}{2}\int_1^{e^2} [\cos(\frac{n-m}{2}\pi\ln(x)) - \cos(\frac{n+m}{2}\pi\ln(x))]\,d(\ln(x))$$

$$= \frac{1}{2}[\frac{\sin(\frac{n-m}{2}\pi\ln(x))}{\frac{n-m}{2}} - \frac{\sin(\frac{n+m}{2}\pi\ln(x))}{\frac{n+m}{2}}]|_{x=1}^{e^2}$$

$$= \frac{1}{2}[\frac{\sin((n-m)\pi)}{\frac{n-m}{2}} - \frac{\sin((n+m)\pi)}{\frac{n+m}{2}}]$$

$$= 0.$$

We now normalize ϕ_n by the calculations:

$$1 = (c_n\phi_n, c_n\phi_n)_w$$

$$= (c_n)^2 \int_1^{e^2} \frac{1}{x}\sin^2(\frac{n\pi}{2}\ln(x))\,dx$$

$$= (c_n)^2 \int_1^{e^2} \frac{1 - \cos(n\pi\ln(x))}{2}\,d(\ln(x))$$

$$= (c_n)^2 \frac{1}{2}[\ln(x) - \frac{\sin(n\pi\ln(x))}{n\pi}]|_1^{e^2}$$

$$= (c_n)^2 \frac{1}{2}[(2-0) - \frac{1}{n\pi}(0-0)]$$

$$= (c_n)^2.$$

Hence we choose $c_n = 1$.

Here the trigonometry formulas were used:

$$2\sin(\theta_1)\sin(\theta_2) = \cos(\theta_1 - \theta_2) - \cos(\theta_1 + \theta_2),$$

$$\sin^2(\frac{\theta}{2}) = \frac{1 - \cos(\theta)}{2}.$$

Thus we obtain a complete orthonormal system as Theorem 2.1 ensures, which is

$$\{\sin(\frac{n\pi}{2}\ln(x))\}_{n\in\mathbb{N}}.$$

EXAMPLE 3.7. Let

$$f(x) = x(x - \pi).$$

Expand f by using the orthonormal system in Example 3.1 with $b = \pi$.

Solution. Since $f \in C^2[0, \pi]$ and $f(0) = 0 = f(\pi)$, we have $f \in D(L)$. Thus Theorem 2.1 implies that

$$f(x) = \sum_{n=1}^{\infty}(f, \chi_n)\chi_n,$$

where

$$\chi_n = \sqrt{\frac{2}{\pi}}\sin(nx),$$

and that the convergence is uniform in $x \in [0, \pi]$.

The coefficients

$$(f, \chi_n) = \int_0^{\pi} x(x - \pi)\sqrt{\frac{2}{\pi}}\sin(nx)\,dx$$

can be calculated as follows:

$$\int_0^{\pi} x\sin(nx)\,dx = \int_0^{\pi} \frac{-x}{n}\,d(\cos(nx))$$

$$= \frac{-1}{n}[x\cos(nx)|_{x=0}^{\pi} - \int_0^{\pi}\cos(nx)\,dx]$$

$$= \frac{-1}{n}[\pi\cos(n\pi)]$$

$$= \frac{-1}{n}\pi(-1)^n,$$

$$\int_0^{\pi} x\cos(nx)\,dx = \int_0^{\pi} \frac{x}{n}\,d(\sin(nx))$$

$$= \frac{1}{n}[x\sin(nx)_{x=0}^{\pi} - \int_0^{\pi}\sin(nx)\,dx]$$

$$= \frac{(-1)^n - 1}{n^2},$$

$$\int_0^{\pi} x^2\sin(nx)\,dx = \int_0^{\pi} \frac{-x^2}{n}\,d(\cos(nx))$$

$$= \frac{-1}{n}[x^2 \cos(nx)|_{x=0}^{\pi} - \int_0^{\pi} 2x \cos(nx)\,dx]$$

$$= \frac{-1}{n}[\pi^2 \cos(n\pi) - 2\frac{(-1)^n - 1}{n^2}]$$

$$= \frac{-1}{n}(-1)^n \pi^2 + \frac{2[(-1)^n - 1]}{n^3}.$$

Thus

$$(f, \chi_n) = \sqrt{\frac{2}{\pi}} \frac{2[(-1)^n - 1]}{n^3}.$$

Since $f \in D(L)$, we have that $\sum_{n=1}^{\infty}(f, \chi_n)\chi_n$ converges to f uniformly by Theorem 2.1.

EXAMPLE 3.8. Solve the heat equation:

$$\frac{\partial}{\partial t}u(x, t) = \beta \frac{\partial^2}{\partial t^2}u(x, t), \quad 0 < x < \pi, t > 0,$$

$$u(0, t) = u(\pi, t) = 0,$$

$$u(x, 0) = f(x) = x(x - \pi),$$

where β is a positive constant.

Solution. By using the technique $u(x, t) = X(x)T(t)$ of separation of variables, we have

$$X\frac{d}{dt}T = \beta T \frac{d^2}{dx^2}X,$$

and so

$$\frac{1}{X}X'' = \frac{1}{\beta T}T' \equiv -\lambda,$$

where λ is a constant to be determined. Thus X satisfies the eigenvalue problem in Example 3.1:

$$X'' + \lambda X = 0,$$

$$X(0) = X(\pi) = 0,$$

and T satisfies the first order differential equation:

$$T' + \beta \lambda T = 0.$$

By the solution of Example 3.1,

$$\lambda = \lambda_n = n^2$$

and X is given by

$$X(x) = \sum_{n=1}^{\infty} a_n \sqrt{\frac{2}{\pi}} \sin(nx),$$

where the superposition principle was used and a_n's are constants to be determined. This yields easily

$$T(t) = e^{-\beta n^2 t},$$

and so

$$u(x, t) = \sum_{n=1}^{\infty} a_n e^{-\beta n^2 t} \sqrt{\frac{2}{\pi}} \sin(nx). \tag{3.3}$$

Because of the initial condition $u(x, 0) = f(x)$, we have

$$f(x) = x(x - \pi) = \sum_{n=1}^{\infty} a_n \sqrt{\frac{2}{\pi}} \sin(nx),$$

where, by the solution of Example 3.7, the convergence is in the supremum norm $\| \cdot \|_\infty$ and

$$a_n = \sqrt{\frac{2}{\pi}} \frac{2[(-1)^n - 1]}{n^3}.$$

Finally, we need to show that the series in (3.3) can be differentiated term by term, twice with respect to the variable x and once with respect to the variable t, so that this determined $u(x, t)$ is a solution of the heat equation. But this follows from the theorem about uniform convergence and differentiation, in advanced calculus [2, Page 230]:

THEOREM 3.9. *Assume that each function $f_n(x)$ defined on (a, b) has the derivative $f'_n(x)$ and that the series $\sum_{n=1}^{\infty} f_n(x)$ converges uniformly to $f(x)$ on (a, b). If the series $\sum_{n=1}^{\infty} f'_n(x)$ further converges uniformly to some function $g(x)$ on (a, b), then*

$$f'(x) = [\sum_{n=1}^{\infty} f_n(x)]' = \sum_{n=1}^{\infty} f'_n(x) \quad \text{for each } x \in (a, b).$$

To see this, we compute

$$\sum_{n=1}^{\infty} a_n \sqrt{\frac{2}{\pi}} e^{-\beta n^2 t} \sin(nx),$$

$$\sum_{n=1}^{\infty} a_n \sqrt{\frac{2}{\pi}} \frac{d}{dt}(e^{-\beta n^2 t}) \sin(nx) = \sum_{n=1}^{\infty} a_n \sqrt{\frac{2}{\pi}} (-\beta n^2) e^{-\beta n^2 t} \sin(nx),$$

$$\sum_{n=1}^{\infty} a_n \sqrt{\frac{2}{\pi}} e^{-\beta n^2 t} \frac{d}{dx} \sin(nx) = \sum_{n=1}^{\infty} a_n \sqrt{\frac{2}{\pi}} (n) e^{-\beta n^2 t} \cos(nx),$$

$$\sum_{n=1}^{\infty} a_n \sqrt{\frac{2}{\pi}} e^{-\beta n^2 t} \frac{d^2}{dx^2} \sin(nx) = \sum_{n=1}^{\infty} a_n \sqrt{\frac{2}{\pi}} (-n^2) e^{-\beta n^2 t} \sin(nx),$$

(3.4)

whose absolute values on

$$D \equiv \{(x, t) : 0 \le x \le \pi, t \ge t_0 > 0\}$$

are bounded, respectively, by

$$\sum_{n=1}^{\infty} M_0 \frac{1}{n^3}, \quad \sum_{n=1}^{\infty} M_0 \frac{e^{-\beta n^2 t}}{n},$$

$$\sum_{n=1}^{\infty} M_0 \frac{1}{n^2}, \quad \text{and} \quad \sum_{n=1}^{\infty} M_0 \frac{e^{-\beta n^2 t}}{n},$$

where M_0 is a positive constant. Since

$$\frac{e^{-\beta n^2 t}}{n} \le e^{-\beta n^2 t_0},$$

the ratio test shows that these series of upper bounds converge. Therefore, each series in (3.4) converges uniformly on D by the Weirstrass M-test [2, Page 223] (see Chapter 8), and so the cited Theorem above applies.

Next we consider an example where f does not lie in $D(L)$.

EXAMPLE 3.10. Let

$$f(x) = \begin{cases} -\frac{\pi}{2}x, & \text{if } 0 \leq x < \frac{\pi}{2}; \\ \frac{\pi}{2}x - \frac{\pi^2}{2}, & \text{if } \frac{\pi}{2} \leq x \leq \pi. \end{cases}$$

Expand f by using the orthonormal system in Example 3.1 with $b = \pi$.

Solution. Although f does not lie in $D(L)$, f does lie in $L^2(0, \pi)$, and so, by Theorem 2.1, we have

$$f(x) = \sum_{n=1}^{\infty} (f, \chi_n)\chi_n,$$

where

$$\chi_n = \sqrt{\frac{2}{\pi}} \sin(nx),$$

and that the convergence is in the $\| \cdot \|_2$ norm.

We compute the coefficients (f, χ_n):

$$(f, \chi_n) = \int_0^{\pi} f(x)\chi_n \, dx$$

$$= \int_0^{\frac{\pi}{2}} (-1)\frac{\pi}{2}x\sqrt{\frac{2}{\pi}} \sin(nx) \, dx$$

$$+ \int_{\frac{\pi}{2}}^{\pi} (\frac{\pi}{2}x - \frac{\pi^2}{2})\sqrt{\frac{2}{\pi}} \sin(nx) \, dx$$

$$= \frac{\sqrt{2\pi}}{n}[\frac{\pi}{2} \cos(\frac{n\pi}{2}) - \frac{1}{n} \sin(\frac{n\pi}{2})] - \frac{\pi\sqrt{\pi}}{\sqrt{2}n}(-1)^n$$

$$+ \frac{\pi\sqrt{\pi}}{\sqrt{2\pi}n}[(-1)^n - \cos(\frac{n\pi}{2})].$$

EXAMPLE 3.11. Solve the heat equation in Example 3.8, where $f(x)$ is the $f(x)$ in Example 3.10.

Solution. By the solution of Example 3.8, we have the solution $u(x, t)$ is given by

$$u(x, t) = \sum_{n=1}^{\infty} a_n\sqrt{\frac{2}{\pi}}e^{-\beta n^2 t} \sin(nx),$$

where a_n satisfies

$$f(x) = u(x, 0) = \sum_{n=1}^{\infty} \sqrt{\frac{2}{\pi}}e^{-\beta n^2 t},$$

whose convergence is, by the solution of Example 3.10, in the L_2 norm $\| \cdot \|_2$. Calculations yield

$$a_n = \frac{\sqrt{2\pi}}{n}[\frac{\pi}{2} \cos(\frac{n\pi}{2}) - \frac{1}{n} \sin(\frac{n\pi}{2})] - \frac{\pi\sqrt{\pi}}{\sqrt{2}n}(-1)^n$$

$$+ \frac{\pi\sqrt{\pi}}{\sqrt{2\pi}n}[(-1)^n - \cos(\frac{n\pi}{2})].$$

We need to show that the above series can be differentiated term by term, twice with respect to the variable x and once with respect to the variable t, so that this determined $u(x,t)$ is a solution of the heat equation. But this follows from the cited Theorem 3.9 in the solution of Example 3.8. To see this, we compute

$$\sum_{n=1}^{\infty} a_n \sqrt{\frac{2}{\pi}} e^{-\beta n^2 t} \sin(nx),$$

$$\sum_{n=1}^{\infty} a_n \sqrt{\frac{2}{\pi}} \frac{d}{dt}(e^{-\beta n^2 t}) \sin(nx) = \sum_{n=1}^{\infty} a_n \sqrt{\frac{2}{\pi}} (-\beta n^2) e^{-\beta n^2 t} \sin(nx),$$

$$\sum_{n=1}^{\infty} a_n \sqrt{\frac{2}{\pi}} e^{-\beta n^2 t} \frac{d}{dx} \sin(nx) = \sum_{n=1}^{\infty} a_n \sqrt{\frac{2}{\pi}} (n) e^{-\beta n^2 t} \cos(nx),$$

$$\sum_{n=1}^{\infty} a_n \sqrt{\frac{2}{\pi}} e^{-\beta n^2 t} \frac{d^2}{dx^2} \sin(nx) = \sum_{n=1}^{\infty} a_n \sqrt{\frac{2}{\pi}} (-n^2) e^{-\beta n^2 t} \sin(nx),$$

(3.5)

whose absolute values on

$$D \equiv \{(x,t) : 0 \le x \le \pi, t \ge t_0 > 0\}$$

are bounded, respectively, by

$$\sum_{n=1}^{\infty} M_0 \frac{e^{-\beta n^2 t}}{n}, \quad \sum_{n=1}^{\infty} M_0 n e^{-\beta n^2 t},$$

$$\sum_{n=1}^{\infty} M_0 e^{-\beta n^2 t}, \quad \text{and} \quad \sum_{n=1}^{\infty} M_0 n e^{-\beta n^2 t},$$

where M_0 is a positive constant. Since

$$e^{-\beta n^2 t}, \quad \frac{e^{-\beta n^2 t}}{n} \le e^{-\beta n^2 t_0} \quad \text{and}$$

$$n e^{-\beta n^2 t} \le e^{-\beta n^2 t_0 / 2} \quad \text{for large } n,$$

the ratio test shows that these series of upper bounds converge. Therefore, each series in (3.5) converges uniformly on D by the Weirstrass M-test [2, Page 223] (see Chapter 8), and so the cited Theorem 3.9 above applies.

4. Existence of Eigenvalues, and Completeness of Eigenfunctions

In order to prove Theorem 2.1, we need to make some preparations. Here we shall first study the properties of the eigenvalues of L stated in Theorem 2.1, and postpone the proof of the existence of eigenvalues until in Subsection 4.4.

4.1. Properties of Eigenvalues.

LEMMA 4.1. *The eigenvalues, if any, of the symmetric L are real numbers.*

PROOF. We assume that the eigenvalues of L exist.

Let λ be an eigenvalue of L and $u \ne 0$ be an eigenfunction, associated with λ. Since L is symmetric, we have

$$\overline{\lambda}(u, r(x)u) = (u, \lambda r(x)u) = (u, Lu)$$
$$= (Lu, u) = (\lambda r(x)u, u)$$
$$= \lambda(r(x)u, u),$$

and so

$$(\lambda - \overline{\lambda})(r(x)u, u) = 0.$$

Here

$$(r(x)u, u) = (u, r(x)u) = \int_a^b r(x)|u(x)|^2 \, dx.$$

Since $(r(x)u, u) \neq 0$, we have $\lambda = \bar{\lambda}$, and so λ is a real number. Here $\bar{\lambda}$ is the complex conjugate of λ. □

LEMMA 4.2. *(Coddington and Levinson [5, Pages 189-190]) The eigenvalues, if any, of the symmetric operator L is at most an enumerable set of real numbers, which can cluster only at ∞. That is, the set of the eigenvalues, if any, for the symmetric operator L is composed of a finite or infinite sequence $\lambda_m, m \in \mathbb{N}$, of real numbers, and if λ_m is an infinite sequence, then λ_m has no limit (or called cluster) points other than ∞.*

Therefore, it is always possible to choose a real number c, which is not an eigenvalue of the symmetric operator L.

PROOF. We assume that the eigenvalues of L exist.

(Coddington and Levinson [5], Page 190) The proof will be divided into seven steps.

Step 1. Let $\varphi_1 = \varphi_1(x, \lambda)$ be the unique solution for the equation with the initial conditions:

$$Ly(x) = \lambda r(x)y(x), \quad x \in (a, b),$$
$$y(a) = 1, \quad y'(a) = 0.$$

Let $\varphi_2 = \varphi_2(x, \lambda)$ be the unique solution to the equation with the initial conditions:

$$Ly(x) = \lambda r(x)y(x), \quad x \in (a, b),$$
$$y(a) = 0, \quad y'(a) = 1.$$

Here the unique existence of $\varphi_i, i = 1, 2$, follows from Chapter 8 on Existence and Uniqueness of solutions, Corollary 4.7.

Step 2. Also from Corollary 4.7, it follows that $\varphi_i(x, \lambda)$ and $\varphi'_i(x, \lambda) \equiv \frac{\partial}{\partial x}\varphi_i(x, \lambda), i = 1, 2$, are continuous functions of (x, λ) on $[a, b] \times \mathbb{C}$, and are entire functions of λ on $\mathbb{C}$ for each fixed $x \in [a, b]$. Since the Wronskian of φ_1 and φ_2 at $x = a$ equals

$$\begin{vmatrix} 1 & 0 \\ 0 & 1 \end{vmatrix},$$

which is not zero, φ_1 and φ_2 are linearly independent.

Step 3. Since the eigenvalues of the symmetric operator L are real numbers by Lemma 4.1, it follows that $\{\lambda_0, y_0\}$ is an eigenpair for L if and only if $\lambda_0 \in \mathbb{R}$ and

$$y_0 \equiv c_1\varphi_1(x, \lambda_0) + c_2\varphi_2(x, \lambda_0)$$

for some constants c_1 and c_2 which are not all zeros and are to be determined by the boundary conditions:

$$y(a) + \alpha_1 y'(a) = 0,$$
$$y(b) + \alpha_2 y'(b) = 0. \tag{4.1}$$

But (4.1) is true if and only if $\lambda_0 \in \mathbb{R}$ and

$$\sum_{i=1}^2 c_i\varphi_i(a, \lambda_0) + \alpha_1 \sum_{i=1}^2 c_i\varphi'_i(a, \lambda_0) = 0,$$

$$\sum_{i=1}^{2} c_i \varphi_i(b, \lambda_0) + \alpha_2 \sum_{i=1}^{2} c_i \varphi_i'(b, \lambda_0) = 0,$$

or equivalently,

$$\begin{pmatrix} \varphi_1(a, \lambda_0) + \alpha_1 \varphi_1'(a, \lambda_0) & \varphi_2(a, \lambda_0) + \alpha_1 \varphi_2'(a, \lambda_0) \\ \varphi_1(b, \lambda_0) + \alpha_2 \varphi_1'(b, \lambda_0) & \varphi_2(b, \lambda_0) + \alpha_2 \varphi_2'(b, \lambda_0) \end{pmatrix} \begin{pmatrix} c_1 \\ c_2 \end{pmatrix} = 0 \qquad (4.2)$$

has a nonzero solution $\begin{pmatrix} c_1 \\ c_2 \end{pmatrix}$.

Step 4. It follows from linear algebra that (4.2) has a nonzero solution $\begin{pmatrix} c_1 \\ c_2 \end{pmatrix}$ if and only if the determinant

$$\triangle = \triangle(\lambda_0) \equiv \begin{vmatrix} \varphi_1(a, \lambda_0) + \alpha_1 \varphi_1'(a, \lambda_0) & \varphi_2(a, \lambda_0) + \alpha_1 \varphi_2'(a, \lambda_0) \\ \varphi_1(b, \lambda_0) + \alpha_2 \varphi_1'(b, \lambda_0) & \varphi_2(b, \lambda_0) + \alpha_2 \varphi_2'(b, \lambda_0) \end{vmatrix}$$

$$= 0.$$

In other words, $\{\lambda_0, y_0\}$ is an eigenpair for L if and only if $\triangle(\lambda_0) = 0$, where $\lambda_0 \in \mathbb{R}$.

Step 5. Since $\varphi_i(x, \lambda)$ and $\varphi_i'(x, \lambda), i = 1, 2$ are entire function of λ on $\mathbb{C}$ by Step 2, we have that $\triangle(\lambda)$ is also an entire function of λ on $\mathbb{C}$. Since $\triangle(\lambda)$ equals 0 only when λ is a real eigenvalue by Step 4, $\triangle(\lambda)$ is an entire function of λ on $\mathbb{C}$, which is not identically 0.

Step 6. It is a fact (see e.g. Lang [**17**, Page 62] and Apostal [**2**, Page 452]) that a zero of a nonconstant analytic function is isolated. That is, if $w(z)$ is a nonconstant analytic function of z on an open disk S in $\mathbb{C}$ and $w(z_0) = 0$ for some $z_0 \in S$, then there is an open disk $\subset S$, centered at z_0, in which w has no further zeros; in other words, $w(z) \neq 0$, except for $z = z_0$. It follows from this fact and Step 5 that the eigenvalues of L are isolated real numbers. Therefore, the set of the eigenvalues for L is composed of a finite or infinite sequence $\lambda_m, m \in \mathbb{N}$ of real numbers.

Step 7. Claim that if λ_m is an infinite sequence of real numbers, then ∞ is the only possible limit (or called cluster) point. If the claim is false, then there is a subsequence λ_{k_m} of λ_m, which converges to some $\lambda_\infty \in \mathbb{R}$ as $m \longrightarrow \infty$ and $\lambda_\infty \neq \infty$. Since $\triangle(\lambda_{k_m}) = 0$ by Step 4, and since $\triangle(\lambda)$ is an entire (and so continuous) function of λ by Step 2, we have

$$0 = \triangle(\lambda_{k_m}) \longrightarrow 0 = \triangle(\lambda_\infty) \quad \text{as } m \longrightarrow \infty.$$

This shows that λ_∞ is also an eigenvalue of L by Step 4. Since $\lambda_{k_m} \longrightarrow \lambda_\infty$ as $m \longrightarrow \infty$ and λ_{k_m} are eigenvalues of L, it follows that every open disk of the eigenvalue λ_∞ contains an eigenvalue that is different from λ_∞. But this is a contradiction to the fact that every eigenvalue, especially the eigenvalue λ_∞, is isolated by Step 6. □

LEMMA 4.3. *The eigenvalues, if any, of L are simple.*

PROOF. We assume that the eigenvalues of L exist.

(Nagel and Saff [**19**]) Let λ be a real eigenvalue of L and let y_1, y_2 be two eigenfunctions, associated with λ. We will show that y_1 and y_2 are linearly dependent.

Since $y_1, y_2 \in D(L)$, we have

$$y_1(a) = -\alpha_1 y_1'(a),$$
$$y_2(a) = -\alpha_1 y_2'(a).$$

It follows that the Wronskian

$$
\begin{aligned}
W(y_1, y_1)(a) &= y_1(a)y_2'(a) - y_2(a)y_1'(a) \\
&= -\alpha_1 y_1'(a)y_2'(a) - (-1)\alpha_1 y_2'(a)y_1'(a) \\
&= 0,
\end{aligned}
$$

and so y_1 and y_2 are linearly dependent. $\qquad\qquad\qquad\square$

4.2. Properties of Eigenfunctions.

LEMMA 4.4. *The eigenfunctions, if any, of L can be chosen to be real-valued, if $p(x), q(x)$, and $r(x)$ are real-valued.*

PROOF. We assume that the eigenvalues of L exist.

Let $\lambda (\in \mathbb{R})$ be an eigenvalue for L and u be an eigenfunction associated with λ. Since the coefficient functions $p(x), q(x)$ of L and $r(x)$ are real-valued, we have

$$
L\bar{u} = \overline{Lu} = \overline{\lambda r(x)u} = \lambda r(x)\bar{u},
$$

and so the complex conjugate $\bar{u}$ of u, is another eigenfunction associated with λ. Since L is linear, we have that the real part

$$
\frac{u + \bar{u}}{2}
$$

of u is also an eigenfunction associated with λ, which is real-valued. $\qquad\square$

LEMMA 4.5. *The eigenfunctions, if any, of L associated with different eigenvalues are orthogonal, in the sense that, for ϕ, ψ, two eigenfunctions corresponding to two different real eigenvalues λ, μ, respectively, we have*

$$
(\psi, \phi)_w = (\psi, r(x)\phi) = 0.
$$

PROOF. We assume that the eigenvalues of L exist.

Let $\lambda, \mu \in \mathbb{R}$ be two different real eigenvalues for L, and ϕ and ψ eigenfunctions associated with λ and μ, respectively. Then we have

$$
\begin{aligned}
\lambda(\phi, \psi)_w = \lambda(r(x)\phi, \psi) &= (L\phi, \psi) = (\phi, L\psi) \\
&= (\phi, \mu r(x)\psi) = \mu(\phi, r(x)\psi) = \mu(r(x)\phi, \psi) \\
&= \mu(\phi, \psi)_w
\end{aligned}
$$

since λ, μ, and $r(x)$ are real, and so

$$
(\lambda - \mu)(\phi, \psi)_w = (\lambda - \mu)(r(x)\phi, \psi) = 0.
$$

Since $\lambda \neq \mu$ by assumption, it follows that

$$
(\phi, \psi)_w = (r(x)\phi, \psi) = 0,
$$

that is, ϕ and ψ are orthogonal. $\qquad\qquad\qquad\qquad\qquad\square$

4.3. Properties of Some Integral Operator.

LEMMA 4.6. *The set*

$$\{\mathbb{G}f_n : f_n \in C[a,b], \|f_n\|_w \leq 1, n \in \mathbb{N}\}$$

has a uniformly convergent subsequence in $C[a,b]$. *Here for* $f \in C[a,b]$, *we define*

$$\|f\|_w \equiv \left(\int_a^b r(x)|f(x)|^2 \, dx \right)^{\frac{1}{2}},$$

where the integral is taken with respect to the Lebesgue measure, and this equals the Riemann integral for the current $r, f \in C[a,b]$.

PROOF. We divide the proof into three steps.

Step 1. It is readily checked that

$$\|\mathbb{G}f_n\|_\infty \leq \gamma \int_a^b r(\xi) \cdot 1 \cdot |f_n(\xi)| \, d\xi$$

$$= \gamma(1, |f_n|)_w$$

$$\leq \gamma \|1\|_w \|f_n\|_w \gamma,$$

uniformly bounded for all n, where

$$\|f_n\|_w = 1,$$

$$\|1\|_w = \left(\int_a^b r(x) \, dx \right)^{\frac{1}{2}},$$

$$\gamma \equiv \max_{x, \xi \in [a,b]} |G(x, \xi)|,$$

$$\|f\|_\infty \equiv \max_{x \in [a,b]} |f(x)| \quad \text{for } f \in C[a,b].$$

Step 2. Let

$$G\{x \geq \xi\} = G(x, \xi)|_{\{x \geq \xi\}} \quad \text{be the restriction of } G(x, \xi) \text{ to the set}$$
$$\{(x, \xi) : a \leq \xi \leq x \leq b\},$$

$$G\{x \leq \xi\} = G(x, \xi)|_{\{x \leq \xi\}} \quad \text{be the restriction of } G(x, \xi) \text{ to the set}$$
$$\{(x, \xi) : a \leq x \leq \xi \leq b\}.$$

Since

$$(\mathbb{G}f_n)(x) = \int_a^x G(x, \xi) r(\xi) f_n(\xi) \, d\xi + \int_x^b G(x, \xi) r(\xi) f_n(\xi) \, d\xi$$

$$= \int_a^x G\{x \geq \xi\} r(\xi) f_n(\xi) \, d\xi + \int_x^b G\{x \leq \xi\} r(\xi) f_n(\xi) \, d\xi,$$

we have, as with Chapter 5 on the construction of the Green function $G(x, \xi)$, that

$$\frac{d}{dx}(\mathbb{G}f_n)(x) = \int_a^x G_x\{x \geq \xi\} r(\xi) f_n(\xi) \, d\xi + \int_x^b G_x\{x \leq \xi\} r(\xi) f_n(\xi) \, d\xi$$

$$+ G\{x \geq \xi\}|_{\xi=x} r(\xi) f_n(\xi)|_{\xi=x} + G\{x \leq \xi\}|_{\xi=x} r(\xi) f_n(\xi)|_{\xi=x}$$

$$= \int_a^x G_x\{x \geq \xi\} r(\xi) f_n(\xi) \, d\xi + \int_x^b G_x\{x \leq \xi\} r(\xi) f_n(\xi) \, d\xi$$

$$+ \lim_{\xi \to x^-} G(x, \xi) r(x) f_n(x) - \lim_{\xi \to x^+} G(x, \xi) r(x) f_n(x),$$

by the fundamental theorem of calculus. It follows that the last two terms on the right side of the above inequality equal zero, since

$$G_x\{x \geq \xi\} \equiv \frac{\partial}{\partial x}G\{x \geq \xi\},$$

$$G_x\{x \leq \xi\} \equiv \frac{\partial}{\partial x}G(x \leq \xi),$$

$G(x,\xi)$ is jointly continuous in x, ξ, and $r(\xi)f_n(\xi)$ is continuous.

Thus we have

$$\|\frac{d}{dx}\mathbb{G}f_n\|_\infty \leq 2\eta\|1\|_w\|f_n\|_w$$

as with Step 1, where $\|f_n\|_w = 1$ and

$$\eta \equiv \max\{\max_{x,\xi \in [a,b]} |G_x\{x \geq \xi\}|, \max_{x,\xi \in [a,b]} |G_x\{x \leq \xi\}|\}.$$

Here η is finite, since $G_x(x,\xi) \equiv \frac{\partial}{\partial x}G(x,\xi)$ is continuous on the triangles

$$\{(x,\xi) : a \leq x \leq \xi \leq b\} \quad \text{and}$$
$$\{(x,\xi) : a \leq \xi \leq x \leq b\},$$

and has a finite jump $\frac{1}{p(\xi)}$ at $x = \xi$.

As a result, we have that $\mathbb{G}f_n$ is equi-continuous in $C[a,b]$ (see below) by the mean value theorem, since

$$(\mathbb{G}f_n)(x_1) - (\mathbb{G}f_n)(x_2) = \frac{d}{dx}(\mathbb{G}f_n)(c)(x_1 - x_2)$$

for some c between $x_1, x_2 \in [a,b]$.

Step 3. Recall the Ascoli-Arzela Theorem:

THEOREM 4.7 (Ascoli-Arzela Theorem). *(Adams [1, Pages 10-11]) A sequence f_n of complex-valued functions in $C[a,b]$ is precompact, that is, f_n contains a uniformly convergent subsequence, if the two conditions are satisfied:*

- *There is a positive constant $M > 0$, such that*

$$\|f_n\|_\infty \equiv \max_{x \in [a,b]} |f_n(x)| \leq M \quad \text{(Uniformly boundedness)}$$

  for all n;
- *For every $\epsilon > 0$, there is a $\delta > 0$, such that*

$$|f_n(x) - f_n(y)| < \epsilon \quad \text{(Equi-continuity)}$$

  for n and all $x, y \in [a,b]$.

This theorem, combined with Steps 1 and 2, completes the proof.   □

LEMMA 4.8. *It is true that*

$$0 < \|\mathbb{G}\| \equiv \sup_{\|f\|_w=1, f \in C[a,b]} \|\mathbb{G}f\|_w < \infty \quad \text{and}$$

$$\|\mathbb{G}f\|_w \leq \|\mathbb{G}\|\|f\|_w.$$

PROOF. We divide the proof into three steps.

Step 1. Choose $u = \frac{1}{\sqrt{b-ar(x)}}$, so that $\|u\|_w = 1$. Since

$$L\mathbb{G}u = L\mathbb{G}_0(ru) = r(x)u(x)$$

$$= \frac{\sqrt{r(x)}}{\sqrt{b-a}}$$
$$> 0,$$

we have that $\mathbb{G}u$ is not identical to zero. For, otherwise $L\mathbb{G}u \equiv 0$, which is a contradiction. Thus

$$\|\mathbb{G}\| \equiv \sup_{\|f\|_w=1, f\in C[a,b]} \|\mathbb{G}f\|_w$$
$$\geq \|\mathbb{G}u\|_w (\neq 0)$$
$$\geq \delta > 0$$

for some $\delta > 0$.

Step 2. For $f \in C[a,b]$ with $\|f\|_w = 1$, we have

$$|\mathbb{G}f(x)| = \left| \int_{\xi=a}^{b} G(x,\xi)f(\xi)\,d\xi \right|$$
$$\leq \gamma \int_a^b r(\xi) \cdot 1 \cdot |f(\xi)|\,d\xi$$
$$= \gamma(|f|, 1)_w$$
$$\leq \gamma\|1\|_w\|f\|_w,$$

where

$$\gamma \equiv \max_{x,\xi\in[a,b]} |G(x,\xi)|.$$

So

$$\|\mathbb{G}\| \leq \gamma\|1\|_w < \infty.$$

Step 3. Since

$$\left\| \frac{u}{\|u\|_w} \right\|_w = 1$$

for nonzero $u \in C[a,b]$, we have

$$\|\mathbb{G}\| \equiv \sup_{\|f\|_w=1, f\in C[a,b]} \|\mathbb{G}f\|_w$$
$$\geq \left\| \mathbb{G}\frac{u}{\|u\|_w} \right\|_w = \frac{\|\mathbb{G}u\|_w}{\|u\|_w},$$

and so

$$\|\mathbb{G}u\|_w \leq \|\mathbb{G}\|\|u\|_w$$

for nonzero u; moreover, this is easily seen to hold also for zero u. □

LEMMA 4.9. *It is true that*

$$\|\mathbb{G}\| \equiv \sup_{\|f\|_w=1, f\in C[a,b]} \|\mathbb{G}f\|_w$$
$$= \sup_{\|f\|_w=1, f\in C[a,b]} |(\mathbb{G}f, f)_w|.$$

PROOF. The proof wil be divided into three steps.

Step 1. Let

$$\zeta = \sup_{\|h\|_w=1, h\in C[a,b]} |(\mathbb{G}h, h)_w|.$$

It follows that

$$|(\mathbb{G}f, f)_w| = \|f\|_w^2 |(\mathbb{G}\frac{f}{\|f\|_w}, \frac{f}{\|f\|_w})_w|$$

$$\leq \zeta \|f\|_w^2$$

for $f \in C[a, b]$ not identical to zero; clearly, this continues to hold for $f \equiv 0$.

Step 2. For $f \in C[a, b]$, we have

$$|(\mathbb{G}f, f)_w| \leq \|\mathbb{G}f\|_w \|f\|_w \leq \|\mathbb{G}\| \|f\|_w^2.$$

Taking the supremum over $f \in C[a, b]$ with $\|f\|_w = 1$ gives

$$\zeta \leq \|\mathbb{G}\|.$$

Step 3. To show $\zeta \geq \mathbb{G}$, which completes the proof by combining Step 2, let $f, g \in C[a, b]$. It follows from Step 1 that

$$\zeta \|f + g\|_w^2 \geq (\mathbb{G}(f + g), (f + g))_w$$

$$= (\mathbb{G}f, f)_w + (\mathbb{G}g, g)_w + 2Re(\mathbb{G}f, g)_w,$$

$$-\zeta \|f - g\|_w^2 \leq (\mathbb{G}(f - g), (f - g)_w)$$

$$= (\mathbb{G}f, f)_w + (\mathbb{G}g, g)_w - 2Re(\mathbb{G}f, g)_w.$$

By substracting, we have

$$Re(\mathbb{G}f, g)_w \leq \frac{1}{4}\zeta(\|f + g\|_w^2 + \|f - g\|_w^2$$

$$= \frac{1}{4}\zeta[(f + g, f + g)_w + (f - g, f - g)_w]$$

$$= \frac{1}{4}\zeta\{[(f, f)_w + (f, g)_w + (g, f)_w + (g, g)_w]$$

$$+ [(f, f)_w - (f, g)_w - (g, f)_w + (g, g)_w]\}$$

$$\leq \frac{1}{2}\zeta(\|f\|_w^2 + \|g\|_w^2).$$

Taking $f \in C[a, b]$ with $\|f\|_w = 1$ and $g = \frac{\mathbb{G}f}{\|\mathbb{G}f\|_w}$ for which $\|g\|_w = 1$, we have

$$\|\mathbb{G}f\|_w \leq \zeta.$$

Now taking supremum over $f \in C[a, b]$ with $\|f\|_w = 1$ gives

$$\|\mathbb{G}\| \leq \zeta.$$

Here note that $\|\mathbb{G}f\|_w \neq 0$ as with the Step 1 in proving Lemma 4.8. $\qquad\square$

4.4. Existence of Eigenvalues and Eigenfunctions. We now prove the existence of eigenvalues of the symmetric operator L:

LEMMA 4.10. *Either* $\|\mathbb{G}\| \neq 0$ *or* $-\|\mathbb{G}\| \neq 0$ *is an eigenvalue for* $\mathbb{G}$.

PROOF. Note that $(\mathbb{G}f, f)_w$ is real for $f \in C[a, b]$. This is because, for

$$u = \mathbb{G}f(= \mathbb{G}_0(rf)),$$

we have

$$Lu = rf,$$

$$(\mathbb{G}f, f)_w = (u, rf)$$

$$= (u, Lu) = (Lu, u) \quad \text{by the symmetry of } L$$

$$= (rf, \mathbb{G}f)$$
$$= (f, \mathbb{G}f)_w$$
$$= \overline{(f, \mathbb{G}f)_w}.$$

Thus

$$\|\mathbb{G}\| = \sup_{f \in C[a,b], \|f\|_w = 1} (\mathbb{G}f, f)_w \quad \text{or} \quad - \sup_{f \in C[a,b], \|f\|_w = 1} (\mathbb{G}f, f)_w$$

by Lemma 4.9.

We consider two cases.

Case 1. Suppose

$$\mu_0 \equiv \|\mathbb{G}\| = \sup_{f \in C[a,b], \|f\|_w = 1} (\mathbb{G}f, f)_w$$

is the case. It follows that there is a sequence of functions $f_n \in C[a,b]$ with $\|f_n\|_w = 1$, such that

$$(\mathbb{G}f_n, f_n)_w \longrightarrow \mu_0 \equiv \|\mathbb{G}\|. \tag{4.3}$$

By Lemma 4.6, $\mathbb{G}f_n$ has a uniformly convergent subsequence in $C[a,b]$, which we still call $\mathbb{G}f_n$, for convenience. Hence, $\mathbb{G}f_n$ converges to a $\phi_0 \in C[a,b]$, that is,

$$\|\mathbb{G}f_n - \phi_0\|_\infty \longrightarrow 0,$$

where $\|h\|_\infty \equiv \max_{x \in [a,b]} |f(x)|$ for $h \in C[a,b]$. This also implies

$$\|\mathbb{G}f_n - \phi_0\|_w^2 = \int_a^b |\mathbb{G}f_n(x) - \phi_0(x)|^2 \, dx$$

$$\leq \|\mathbb{G}f_n - \phi_0\|_\infty^2 \int_a^b 1 \, dx \tag{4.4}$$

$$\longrightarrow 0.$$

It will be proved that μ_0 is an eigenvalue of $\mathbb{G}$ and ϕ_0 is an eigenfunction associated with μ_0.

By triangularity inequality (Section 5), we have, for $f, g \in (L^2(a,b), \|\cdot\|_w)$, that

$$\|f\|_w = \|(f - g) + g\|_w \leq \|f - g\|_w + \|g\|_w,$$
$$\|g\|_w = \|(g - f) + f\|_w \leq \|g - f\|_w + \|f\|_w,$$

and so

$$-\|f - g\|_w \leq \|f\|_w - \|g\|_w \leq \|f - g\|_w,$$
$$\|\|f\|_w - \|g\|_w\| \leq \|f - g\|_w.$$

It follows then, in conjunction with (4.4), that

$$\|\|\mathbb{G}f_n\|_w - \|\phi_0\|_w\| \leq \|\mathbb{G}f_n - \phi_0\|_w$$

$$\longrightarrow 0. \tag{4.5}$$

Since

$$0 \leq \|\mathbb{G}f_n - \mu_0 f_n\|_w^2$$

$$= \|\mathbb{G}f_n\|_w^2 + \mu_0^2 \|f_n\|_w^2 - 2\mu_0 (\mathbb{G}f_n, f_n)_w$$

$$\longrightarrow \|\phi_0\|_w^2 - \mu_0^2$$

by (4.3) and (4.5), as $n \longrightarrow \infty$, we have

$$\|\phi_0\|_w^2 \geq \mu_0^2,$$

and so ϕ_0 is not identical to 0. On the other hand, since

$$\|\mathbb{G}f_n\|_w^2 \leq (\|\mathbb{G}\|\|f_n\|_w)^2 = \mu_0^2$$

by Lemma 4.8, it also follows from above that

$$\begin{aligned} 0 &\leq \|\mathbb{G}f_n - \mu_0 f_n\|_w^2 \\ &\leq 2\mu_0^2 - 2\mu_0(\mathbb{G}f_n, f_n)_w \longrightarrow 0, \end{aligned} \tag{4.6}$$

where (4.3) was used.

Therefore, by using triangularity inequality, Lemma 4.8, (4.4), and (4.6), we have

$$\begin{aligned} 0 &\leq \|\mathbb{G}\phi_0 - \mu_0\phi_0\|_w \\ &\leq \|\mathbb{G}\phi_0 - \mathbb{G}(\mathbb{G}f_n)\|_w + \|\mathbb{G}(\mathbb{G}f_n) - \mu_0(\mathbb{G}f_n)\|_w \\ &\quad + \|\mu_0\mathbb{G}f_n - \mu_0\phi_0\|_w \\ &\leq \|\mathbb{G}\|\|\phi_0 - \mathbb{G}f_n\|_w + \|\mathbb{G}(\mathbb{G}f_n) - \mu_0(\mathbb{G}f_n)\|_w \\ &\quad |\mu_0|\|\mathbb{G}f_n - \phi_0\|_w \\ &\longrightarrow 0 \end{aligned}$$

as $n \longrightarrow \infty$, and so

$$\|\mathbb{G}\phi_0 - \mu_0\phi_0\|_w = 0.$$

Since $\mathbb{G}\phi_0 - \mu_0\phi_0 \in C[a, b]$, we have

$$\mathbb{G}\phi_0 = \mu_0\phi_0,$$

and so $\mu_0 = \|\mathbb{G}\|$ is an eigenvalue of $\mathbb{G}$ and ϕ_0 is an eigenfunction associated with μ_0.

Case 2. The other case where

$$\|\mathbb{G}\| = \sup_{f \in C[a,b], \|f\|_w=1} (-1)(\mathbb{G}f, f)_w = - \inf_{f \in C[a,b], \|f\|_w=1} (\mathbb{G}f, f)_w$$

can be treated similarly. □

PROPOSITION 4.11. *There is a sequence of real numbers $\{\mu_n\}_{n=0}^{\infty}$ which are eigenvalues of $\mathbb{G}$ and satisfy*

$$|\mu_n| \longrightarrow 0,$$
$$0 \neq |\mu_0| \geq |\mu_1| \geq |\mu_2| \geq \cdots,$$

where $\mu_i \neq \mu_j$ for $i \neq j$. Moreover, a set of orthonormal eigenfunctions $\{\chi_n\}_{n=0}^{\infty}$ exist, associated with $\{\mu_n\}_{n=0}^{\infty}$.

PROOF. The proof will be divided into five steps.
Step 1. Let

$$\chi_0 = \frac{\phi_0}{\|\phi_0\|_w},$$

where ϕ_0 is from Lemma 4.9. Then

$$\|\chi_0\|_w = \left\|\frac{\phi_0}{\|\phi_0\|_w}\right\|_w = 1,$$

and χ_0 is an eigenfunction associated with the eigenvalue $\mu_0 \neq 0$ from Lemma 4.9.

Step 2. Let
$$G_1(x,\xi) = G(x,\xi)r(\xi) - \mu_0\chi_0(x)\chi_0(\xi)r(\xi),$$
and define
$$\begin{aligned}
\mathbb{G}_1 f &= \int_{\xi=a}^{b} G_1(x,\xi)f(\xi)\,d\xi \\
&= \mathbb{G}f - \mu_0(r\chi_0, f)\chi_0(x) \\
&= \mathbb{G}f - \mu_0(\chi_0, f)_w\chi_0(x)
\end{aligned}$$
for $f \in C[a,b]$. It is readily checked that $\mathbb{G}_1$ has the same properties as $\mathbb{G}$ had in Lemmas 4.6 and 4.9.

Suppose that $\mathbb{G}_1 \neq 0$, which will be justified in Step 4. It follows from Lemma 4.9 and Step 1 that $\mathbb{G}_1$ has a real eigenvalue $\mu_1 \neq 0$ and an eigenfunction χ_1, that is
$$\mathbb{G}_1\chi_1 = \mu_1\chi_1,$$
and that μ_1 and χ_1 satisfy
$$|\mu_1| = \sup_{f\in C[a,b], \|f\|_w=1} |(\mathbb{G}_1 f, f)_w|,$$
$$\|\chi_1\|_w = 1.$$

For $f \in C[a,b]$, $\mathbb{G}f$ is orthogonal to χ_0, and so, in particular,
$$(\mathbb{G}_1\chi_1, \chi_0)_w = 0. \tag{4.7}$$
This is because
$$\begin{aligned}
(\mathbb{G}_1 f, \chi_0)_w &= (\mathbb{G}_1 f, r\chi_0) \\
&= (\mathbb{G}f - \mu_0\chi_0(r\chi_0, f), r\chi_0) \\
&= (\mathbb{G}f, r\chi_0) - \mu_0(r\chi_0, f),
\end{aligned}$$
where $(\chi_0, r\chi_0) = (\chi_0, \chi_0)_w = 1$ was used, and because $(\mathbb{G}f, r\chi_0)$ equals, by the Fubini theorem (see Section 5),
$$\begin{aligned}
&\int_{x=a}^{b} [\int_{\xi=a}^{b} G(x,\xi)r(\xi)f(\xi)\,d\xi]r(x)\chi_0(x)\,dx \\
&= \int_{\xi=a}^{b} [\int_{x=a}^{b} G(x,\xi)r(x)\chi_0(x)\,dx]r(\xi)f(\xi)\,d\xi \\
&= \int_{\xi=a}^{b} (\mathbb{G}\chi_0)r(\xi)f(\xi)\,d\xi \\
&= \int_{\xi=a}^{b} \mu_0\chi_0(\xi)r(\xi)f(\xi)\,d\xi \\
&= (\mu_0 r\chi_0, f) \\
&= \mu_0(r\chi_0, f).
\end{aligned}$$

It follows that
$$\mu_1\chi_1 = \mathbb{G}_1\chi_1 = \mathbb{G}\chi_1 - \mu_0(r\chi_0, \chi_1)\chi_0 = \mathbb{G}\chi_1,$$
since
$$(r\chi_0, \chi_1) = (r\chi_0, \frac{1}{\mu_1}\mathbb{G}_1\chi_1) = \frac{1}{\mu_1}(r\chi_0, \mathbb{G}_1\chi_1)$$

$$= \frac{1}{\mu_1}(\chi_0, \mathbb{G}_1\chi_1)_w = 0 \quad \text{by (4.7)}.$$

Hence $\{\mu_1, \chi_1\}$ is also an eigenpair for $\mathbb{G}$, with $(\chi_1, \chi_0)_w = (r\chi_0, \chi_1) = 0$.

On the other hand, $|\mu_1| \le |\mu_0|$. This is because, for $f \in C[a, b]$, we have

$$\begin{aligned}
\|\mathbb{G}f\|_w^2 &= (\mathbb{G}f, \mathbb{G}f)_w \\
&= (\mathbb{G}_1 f + \mu_0(\chi_0, f)\chi_0, \mathbb{G}_1 f + \mu_0(\chi_0, f)\chi_0)_w \\
&= \|\mathbb{G}_1 f\|_w^2 + \mu_0^2 |(\chi_0, f)_w|^2 + 2\mu_0(\chi_0, f) Re(\mathbb{G}_1 f, \chi_0)_w \\
&\ge \|\mathbb{G}_1 f\|^2,
\end{aligned}$$

where $(\mathbb{G}_1 f, \chi_0)_w = 0$, and then

$$|\mu_0| \equiv \sup_{\|f\|_w = 1, f \in C[a,b]} \|\mathbb{G}f\|_w \ge \sup_{\|f\|_w = 1, f \in C[a,b]} \|\mathbb{G}_1 f\|_w = |\mu_1|.$$

It is also true that $\mu_1 \ne \mu_0$. For, otherwise χ_1 and χ_0 are two linearly independent eigenfunctions, associated with μ_0, a contradiction to μ_0 being simple by Lemma 4.3.

Step 3. Let

$$G_2(x, \xi) = G_1(x, \xi) - \mu_1 \chi_1(x) r(\xi)\chi_1(\xi)$$

and

$$\mathbb{G}_2 f \equiv \int_{\xi=a}^b G_2(x, \xi) f(\xi) \, d\xi$$

for $f \in C[a, b]$.

Suppose that $\mathbb{G}_2 \ne 0$, which will be justified in Step 4. It follows from Steps 1 and 2 that there is an eigenpair $\{\mu_2, \chi_2\}$ for $\mathbb{G}$, such that

$$\begin{aligned}
&\mu_2 \ne 0, \\
&|\mu_2| \le |\mu_1| \le |\mu_0|, \\
&\|\chi_2\|_w = 1, \\
&(\chi_2, \chi_i)_w = 0, i = 0, 1, \\
&\mu_i \ne \mu_j \quad \text{for } i \ne j, i, j = 0, 1, 2.
\end{aligned}$$

Step 4. Continuing in this way, we have a sequence of eigenpairs $\{\mu_i, \chi_i\}, i = 0, 1, 2, \ldots$, for $\mathbb{G}$, that satisfy

$$\begin{aligned}
&|\mu_0| \ge |\mu_1| \ge |\mu_2| \ge \cdots, \\
&\|\chi_i\|_w = 1, \\
&(\chi_i, \chi_j)_w = 0 \quad \text{for } i \ne j, \\
&\mu_i \ne \mu_j \quad \text{for } i \ne j.
\end{aligned}$$

This process for constructing the eigenpairs cannot be stopped. This is because, otherwise $\mathbb{G}_m = 0$ for some $m \in \mathbb{N}$, and then for $f \in C[a, b]$, we have

$$0 = \mathbb{G}_m f = \mathbb{G}f - \sum_{i=0}^{m-1} \mu_0(\chi_i, f)_w \chi_i,$$

$$0 = \frac{1}{r} L(\mathbb{G}f - \sum_{i=0}^{m-1} \mu_i(\chi_i, f)_w \chi_i)$$

$$= \frac{1}{r}(rf) - \sum_{i=0}^{m-1} \mu_i (\chi_i, f)_w \frac{1}{r} L\chi_i$$

$$= f - \sum_{i=0}^{m-1} \mu_i (\chi_i, f)_w \frac{1}{r} r\chi_i$$

$$= f - \sum_{i=0}^{m-1} (\chi_i, f)_w \frac{1}{\mu_i} G\chi_i;$$

this is a contradiction, since $f \in C[a,b]$ is a general continuous function on $[a,b]$ but

$$\sum_{i=0}^{m-1} (\chi_i, f) \frac{1}{\mu_i} G\chi_i$$

is a twice continuously differentiable function on $[a,b]$ and they cannot be equal in general.

Step 5. That $|\mu_n| \longrightarrow 0$ as $n \longrightarrow \infty$ will be proved in Proposition 4.13.   □

LEMMA 4.12. *If $f \in (L^2(a,b), \|\cdot\|_w)$ and $\{\chi_i\}_{i=0}^{\infty}$ is the orthonormal set from Proposition 4.11, then*

$$\sum_{i=0}^{m} |(f, \chi_i)_w|^2 \leq \|f\|_w^2$$

for $m = 0, 1, 2, \ldots$, and

$$\sum_{i=0}^{\infty} |(f, \chi_i)_w|^2 \leq \|f\|_w^2 \quad \text{(Bessel's inequality)}.$$

PROOF. For $m = 0, 1, 2, 3, \ldots$, we have

$$0 \leq \|f - \sum_{i=0}^{m} (f, \chi_i)_w \chi_i\|_w^2$$

$$= (f - \sum_{i=0}^{m} (f, \chi_i)_w \chi_i, f - \sum_{i=0}^{m} (f, \chi_i)\chi_i)_w$$

$$= \|f\|_w^2 - \sum_{i=0}^{m} |(f, \chi_i)_w|^2,$$

where note $(\chi_i, \chi_j)_w = 0$ for $i \neq j$ and $(\chi_i, \chi_i)_w = 1$. Hence

$$\sum_{i=0}^{m} |(f, \chi_i)_w|^2 \leq \|f\|_w^2,$$

from which the Bessel's inequality follows by letting $m \longrightarrow \infty$.   □

PROPOSITION 4.13. *Let*

$$\{\mu_i, \chi_i\}_{i=0}^{\infty}$$

be the eigenpairs from Proposition 4.11, and let $f \in C^2[a,b]$ satisfy the boundary conditions in $D(L)$. Then

$$\lim_{i \to \infty} |\mu_i| = 0$$

and, for $x \in [a, b]$, we have

$$f(x) = \sum_{i=0}^{\infty} (f, \chi_i)_w \chi_i(x),$$

where the series convergence is uniform in x.

Moreover,

$$\|f\|_w^2 = \sum_{i=0}^{\infty} |(f, \chi_i)_w|^2 \quad (Parseval\ equality).$$

PROOF. The proof will be divided into five steps.

Step 1. For each $x \in [a, b]$, let

$$G_{\{x\}}(\xi) \equiv G(x, \xi) r(\xi)$$

be a function of $\xi \in [a, b]$. Then

$$(G_{\{x\}}, \chi_i)_w = \int_{\xi=a}^{b} G(x, \xi) r(\xi) \chi_i(\xi) \, d\xi$$
$$= (\mathbb{G}\chi_i)(x) = \mu_i \chi_i(x).$$

The Bessel's inequality implies that, for each $x \in [a, b]$ and each $m = 0, 1, 2, 3, \ldots$, we have

$$\sum_{i=0}^{m} |\mu_i \chi_i(x)|^2 \le \|G_{\{x\}}\|_w^2$$

$$= \int_{\xi=a}^{b} |G(x, \xi) r(\xi)|^2 \, d\xi.$$

Multiplying both sides of the above by $r(x)$ and integrating both sides over x gives

$$\sum_{i=0}^{m} |\mu_i|^2 \le \gamma_0^2 (b-a)^2$$

for all m, where

$$\gamma_0 \equiv \max_{x, \xi \in [a, b]} |G(x, \xi) \sqrt{r(\xi)}|.$$

Letting $m \longrightarrow \infty$, we have

$$\sum_{i=0}^{\infty} |\mu_i|^2 \le \gamma^2 (b-a)^2.$$

Thus the series $\sum_{i=0}^{\infty} |\mu_i|^2$ converges, and so

$$\lim_{i \to \infty} |\mu_i|^2 = 0.$$

Step 2. Recall from Proposition 4.11 that, for $f \in C[a, b]$, we have

$$\mathbb{G}_m f \equiv \mathbb{G} f - \sum_{i=0}^{m-1} \mu_i (f, \chi_i)_w \chi_i$$

and

$$\|\mathbb{G}_m\| = |\mu_m|.$$

Since

$$\|\mathbb{G}_m f\|_w \le \|\mathbb{G}_m\| \|f\|_w \le |\mu_m| \|f\|_w \longrightarrow 0,$$

we have

$$\lim_{m \to \infty} \|\mathbb{G}f - \sum_{i=0}^{m-1} \mu_i(f, \chi_i)\chi_i\|_w = 0.$$

Step 3. We prove the assertion that

$$\sum_{i=0}^{\infty} \mu_i(f, \chi_i)_w \chi_i(x)$$

converges uniformly for $x \in [a, b]$. For $q > p, q, p \in \mathbb{N}$, we have

$$\sum_{i=p}^{q} \mu_i(f, \chi_i)_w \chi_i = \mathbb{G}(\sum_{i=p}^{q} (f, \chi_i)_w \chi_i).$$

Since

$$|(\mathbb{G}h)(x)| = |\int_{\xi=a}^{b} G(x, \xi) r(\xi) h(\xi) \, d\xi|$$
$$\leq \gamma |(1, |h|)_w| \leq \gamma \|1\|_w \|h\|_w,$$

for $h \in C[a, b]$, where

$$\gamma \equiv \max_{x, \xi \in [a,b]} |G(x, \xi)|,$$

it follows that

$$|\sum_{i=p}^{q} \mu_i(f, \chi_i)_w \chi_i(x)| \leq \gamma \|1\|_w \sum_{i=p}^{q} (f, \chi_i)_w \chi_i\|_w$$
$$= \gamma \|1\|_2 (\sum_{i=p}^{q} (f, \chi_i)_w \chi_i, \sum_{i=p}^{q} (f, \chi_i)_w \chi_i)_w^{\frac{1}{2}}$$
$$= \gamma \|1\|_2 [\sum_{i=p}^{q} |(f, \chi_i)_w|^2]^{\frac{1}{2}}.$$

The right side approaches zero as $q \geq p \longrightarrow \infty$. This is because

$$\sum_{i=0}^{\infty} |(f, \chi_i)_w|^2 \leq \|f\|_w^2$$

converges by the Bessel's inequality, where note that a convergent series

$$\sum_{i=0}^{\infty} a_i, \quad a_i \in \mathbb{C}$$

implies $\sum_{i=p}^{q} a_i \longrightarrow 0$ as $q \geq p \longrightarrow \infty$.

Thus

$$\sum_{i=0}^{m-1} \mu_i(f, \chi_i)_w \chi_i(x)$$

is a Cauchy sequence, uniformly for $x \in [a, b]$, and so converges uniformly to a continuous function on $[a, b]$ as $m \longrightarrow \infty$.

Step 4. Since $\mathbb{G}f \in C[a, b]$ also, it follows from Steps 2 and 3 that

$$\mathbb{G}f = \sum_{i=0}^{\infty} \mu_i(f, \chi_i)_w \chi_i(x)$$

for $f \in C[a,b]$, the convergence being uniform in $x \in [a,b]$.

Step 5. Thus if $u \in D(L)$, then $Lu, \frac{1}{r}Lu \in C[a,b]$, and so Step 4 gives

$$u = \mathbb{G}_0(Lu) = \mathbb{G}(\frac{1}{r}Lu)$$

$$= \sum_{i=0}^{\infty}(\frac{1}{r}Lu, \mu_i\chi_i)_w\chi_i = \sum_{i=0}^{\infty}(\frac{1}{r}Lu, \mathbb{G}\chi_i)_w\chi_i$$

$$= \sum_{i=0}^{\infty}(\mathbb{G}(\frac{1}{r}Lu), \chi_i)_w\chi_i = \sum_{i=0}^{\infty}(u, \chi_i)_w\chi_i.$$

Here we used

$$\mathbb{G}(\frac{1}{r}Lu) = \mathbb{G}_0(r\frac{1}{r}Lu) = u$$

for $u \in D(L)$, and used

$$(\mathbb{G}f, g)_w - (f, \mathbb{G}g)_w$$

for $f, g \in C[a,b]$, which follows from the symmetry of L:

$$(f, \mathbb{G}g)_w = (rf, \mathbb{G}g) = (L(\mathbb{G}_0 rf), \mathbb{G}_0(rg))$$

$$= (\mathbb{G}_0 rf, L(\mathbb{G}_0 rg)) = (\mathbb{G}f, rg)$$

$$= (\mathbb{G}f, g)_w.$$

$\square$

PROPOSITION 4.14. *Let $\{\mu_i, \chi_i\}_{i=0}^{\infty}$ be the eigenpairs from Proposition 4.11. If $f \in (L^2(a,b), \|\cdot\|_w)$, then*

$$f = \sum_{i=0}^{\infty}(f, \chi_i)_w\chi_i,$$

where the convergence is taken with respect to the norm $\|\cdot\|_w$:

$$\lim_{n\to\infty}\|\sum_{i=0}^{n}(f, \chi_i)\chi_i - f\|_w = 0.$$

PROOF. The proof will be divided into two steps.

Step 1. We will prove that $D(L)$ is dense in $(L^2(a,b), \|\cdot\|_w)$. Recall the fact, Adams [**1**, Page 31], that the set

$$C_0^2[a,b] \equiv \{u \in C^2[a,b] : u \text{ has compact support in but not equal to } [a,b]\}$$

is dense in $(L^2(a,b), \|\cdot\|_2)$. That means that, for each $f \in (L^2(a,b), \|\cdot\|_2)$, there is a sequence $f_n \in C_0^2[a,b]$ of functions, such that

$$\lim_{n\to\infty}\|f_n - f\|_2 = 0.$$

Since $\|\cdot\|_w$ is equivalent to $\|\cdot\|_2$, we have $C_0^2[a,b]$ is also dense in $(L^2(a,b), \|\cdot\|_w)$. In other words,

$$\lim_{n\to\infty}\|f_n - f\|_w = 0$$

holds also. Here that u has compact support in but not equal to $[a,b]$ means that the closure of the set of $x \in [a,b]$ on which $u(x)$ is not zero is a proper compact subset of $[a,b]$. It is true that $C_0^2[a,b]$ is the same as the set of functions in $C^2[a,b]$ that vanish outside a proper compact subset of $[a,b]$.

Since $D(L)$ contains $C_0^2[a,b]$, we have $D(L)$ is dense in $(L^2(a,b), \|\cdot\|_w)$.

Step 2. Let $\epsilon > 0$. For $f \in (L^2(a,b), \|\cdot\|_w)$, Step 1 says that there is a $u \in D(L)$, such that

$$\|f - u\|_w < \epsilon.$$

For this u, Proposition 4.13 implies that there is an $m_0 \in \mathbb{N}$, such that if $m \geq m_0$, then

$$\|u - \sum_{i=0}^m (u, \chi_i)_w \chi_i\|_w < \epsilon.$$

Using the triangularity inequality, we have

$$\|f - \sum_{i=0}^m (f, \chi_i)_w \chi_i\|_w \leq \|f - u\|_w + \|u - \sum_{i=0}^m (u, \chi_i)_w \chi_i\|_w$$

$$+ \|\sum_{i=0}^m (u - f, \chi_i)_w \chi_i\|_w,$$

the third term of which equals

$$(\sum_{i=0}^m (u - f, \chi_i)_w \chi_i, \sum_{i=0}^\infty (u - f, \chi_i)_w \chi_i)_w^{\frac{1}{2}}$$

$$= [\sum_{i=0}^m |(u - f, \chi_i)_w|^2]^{\frac{1}{2}} \leq \|u - f\|_w$$

by the Bessel's inequality. Thus for $m \geq m_0$, we have

$$\|f - \sum_{i=0}^m (f, \chi_i)_w \chi_i\|_w < 3\epsilon,$$

which completes the proof. □

4.5. Proof of the Main Results. We now prove Theorem 2.1:

PROOF. We divide the proof into two steps.

Step 1. [21] Let $\{\mu_i, \chi_i\}_{i=0}^\infty$ be the eigenpairs from Proposition 4.11. We will show that if $\mu \neq 0$ is not in the eigenvalue set

$$\{\mu_i\}_{i=0}^\infty,$$

then μ is not an eigenvalue of $\mathbb{G}$. If such a μ is an eigenvalue of $\mathbb{G}$, then

$$(y, \chi_i)_w = 0$$

for all i by Lemma 4.5, where $y \in D(L)$ is an eigenfunction, associated with μ. Thus, Proposition 4.13 implies that

$$y = \sum_{i=0}^\infty (y, \chi_i)_w \chi_i = 0,$$

a contradiction to $y \neq 0$ being an eigenfunction.

Step 2. Propositions 4.11, 4.13, 4.14, and Step 1 complete the proof. □

The proof of Theorem 2.1 is based on the book by Coddington and Levinson [5].

5. Abstract Expansion Results

We prepare some backgroud material first.

5.1. Linear Continuous Operators.

DEFINITION 5.1. *A non-empty set X with two operations $+$ and $\cdot$, is called a vector space over the field $\mathbb{C}$, the complex numbers, if the following are satisfied:*

- *$x + y \in X$ for $x, y \in X$.*
- *$\alpha \cdot x = \alpha x \in X$ for $x \in X$ and $\alpha \in \mathbb{C}$.*
- *$x + y = y + x$ for $x, y \in X$.*
- *$x + (y + z) = (x + y) + z$ for $x, y, z \in X$.*
- *There is a unique $0 \in X$, such that $0 + x = x$.*
- *For each $x \in X$, there is a unique $-x \in X$, such that $x + (-x) = 0$.*
- *$\alpha(x + y) = \alpha x + \alpha y$ for $x, y \in X$ and $\alpha \in \mathbb{C}$.*
- *$(\alpha + \beta)x = \alpha x + \beta x$ for $x \in X$ and $\alpha, \beta \in \mathbb{C}$.*
- *$\alpha(\beta x) = (\alpha\beta)x$ for $x \in X$ and $\alpha, \beta \in \mathbb{C}$.*
- *$1 \cdot x = x$ for $x \in X$.*
- *$0 \cdot x - 0 \in X$ for $x \in X$*

DEFINITION 5.2. *A norm $\| \cdot \|$ on a vector space X is a real-valued function that satisfies*

- *$\|x\| \geq 0$ for $x \in X$.*
- *$\|x\| \neq 0$ if $x \in X$ and $x \neq 0$.*
- *$\|\alpha x\| = |\alpha| \|x\|$ for $\alpha \in \mathbb{C}$ and $x \in X$.*
- *$\|x + y\| \leq \|x\| + \|y\|$ for $x, y \in X$.*

DEFINITION 5.3. *A vector space over the field $\mathbb{C}$, on which a norm is defined, is called a normed vector space.*

Let $(X, \| \cdot \|_X)$ and $(Y, \| \cdot \|_Y)$ be two normed vector spaces, with the norms $\| \cdot \|_X, \| \cdot \|_Y$, repestively, over the field $\mathbb{C}$, the complex numbers. Assume that X and Y contain nonzero elements.

DEFINITION 5.4. *A function $T : D(T) \subset X \longrightarrow Y$, from its domain $D(T)$ to Y, is called a linear operator if*

$$T(x + y) = T(x) + T(y) = Tx + Ty$$

for $x, y \in X$, and

$$T(\alpha x) = \alpha Tx$$

for $x \in X$ and $\alpha \in \mathbb{C}$.

When the domain $D(T) = X$, we say that T is a linear operator from X to Y.

DEFINITION 5.5. *Assume that T is a linear operator from X to Y. T is continuous at $x_0 \in X$ if, for a given $\epsilon > 0$, there is a $\delta > 0$, such that*

$$\|Tx - Tx_0\|_Y < \epsilon$$

for all $x \in X$ with $\|x - x_0\|_X < \delta$.

LEMMA 5.6. *A linear operator T from X to Y is continuous at some $x_0 \in X$ if and only if it is uniformly continuous on X if and ony if there is an $M > 0$, such that*

$$\|Tx\|_Y \leq M\|x\|_X$$

for each $x \in X$.

In either case, we say that T is a continuous linear operator from X to Y.

DEFINITION 5.7. *If T is a continuous linear operator from X to Y, define the norm $\|T\| \equiv \|T\|_{X \to Y}$ of T by*

$$\|T\| \equiv \sup_{\|x\|_X \leq 1} \|Tx\|_Y,$$

LEMMA 5.8. *If T is a continuous linear operator from X to Y, then*

$$\|Tx\|_x \leq \|T\| \|x\|_X$$

and

$$\|T\| = \sup_{\|x\|_X = 1} \|Tx\|_Y \quad or \quad \sup_{x \neq 0} \frac{\|Tx\|_Y}{\|x\|_X}.$$

DEFINITION 5.9. *A vector space Z over the field $\mathbb{C}$, the complex numbers, is an inner product space if, to each pair $\{x, y\}, x, y \in X$, there is associated a complex number (x, y), called the inner product of x with y, such that, for $z \in X$ and $\alpha \in \mathbb{C}$, the following are satisfied:*

$$(x + z, y) = (x, y) + (z, y),$$
$$(y, x) = \overline{(x, y)},$$
$$(\alpha x, y) = \alpha(x, y),$$

and

$$(x, x) \geq 0 \quad and \quad (x, x) \neq 0 \quad if \quad x \neq 0.$$

From this definition, it follows that

$$(x, \alpha y) = \overline{\alpha}(x, y)$$

and

$$(x, y + z) = (x, y) + (x, z).$$

LEMMA 5.10. *(Cauchy-Schwarz inequality)If $(Z, (\quad , \quad))$ is an inner product space with the inner product $(\quad , \quad)$, then*

$$|(x, y)| \leq \sqrt{(x, x)} \sqrt{(y, y)}$$

for $x, y \in Z$.

LEMMA 5.11. *If $(Z, (\quad , \quad))$ is an inner product space, then*

$$\|x\| \equiv \sqrt{(x, x)}$$

for $x \in Z$, defines a norm, called inner product norm, on Z and $(Z, \| \cdot \|)$ with the inner product norm $\| \cdot \|$ becomes a normed vector space.

DEFINITION 5.12. *Let $(Z, (\quad , \quad))$ be an inner product space. A linear operator*

$$A : D(A) \subset Z \longrightarrow Z$$

is called symmetric if

$$(Ax, y) = (x, Ay)$$

for $x, y \in D(A)$.

DEFINITION 5.13. *A linear operator A from X to Y is called compact if, for each bounded set $S \subset X$, $A(S)$ is relatively compact in Y; or equivalently, A is compact if, for each bounded sequence $x_n \in X$, Ax_n contains a convergent subsequence with the limit in Y.*

LEMMA 5.14. *If A is a linear compact operator from X to Y, then A is continuous.*

5.2. Expansion Results for Compact Symmetric Operators.

THEOREM 5.15. *Let* $(Z, (\quad, \quad))$ *be an inner product space with the inner product norm* $\| \cdot \|$. *If*

$$A : D(A) = Z \longrightarrow Z$$

is a nonzero linear compact symmetric operator, then the following are satisfied:

- *The eigenvalues of* A *exist, are real numbers, and either consist of* $\{0\} \cup \{\lambda_i\}_{i=1}^\infty$ *or consist of* $\{\lambda_i\}_{i=1}^\infty$, *where* $\{\lambda_i\}_{i=0}^\infty$ *is a non-increasing possibly terminating sequence of nonzero real numbers:*

$$0 \neq \lambda_1 \geq \lambda_2 \geq \cdots \geq \lambda_n \geq \cdots.$$

  Here that $\{\lambda_i\}_{i=1}^\infty$ *terminates means* $\{\lambda_i\}_{i=1}^\infty = \{\lambda_i\}_{i=1}^{i_0}$ *for some* $i_0 \in \mathbb{N}$. *If* $\{\lambda_i\}_{i=1}^\infty$ *does not terminate, then* $\lambda_i \neq 0$ *for all* $i = 1, 2, 3, \ldots$, *and* $\lim_{n\to\infty} |\lambda_n| = 0$.
- *The eigenvectors* $\{\chi_i\}_{i=1}^\infty$ *of* L *corresponding to the real eigenvalues* $\{\lambda_i\}_{i=1}^\infty$, *are orthonormal, in the sense that*

$$(\chi_i, \chi_j) = 0 \quad \text{for } i \neq j, \text{ and}$$
$$(\chi_i, \chi_i) = 1 \quad \text{for } i = 1, 2, \ldots.$$

- *The eigenvspace corresponding to each eigenvalue* λ_i *is of finite dimensions, and its dimension is the number of times that the eigenvalue* λ_i *is repeated in the sequence* $\{\lambda_i\}_{i=0}^\infty$.
- *For each* $x \in Z$, *we have*

$$Ax = \sum_{i=1}^{i_0} \lambda_i(x, \chi_i)\chi_i \quad or$$

$$= \sum_{i=1}^{\infty} \lambda_i(x, \chi_i)\chi_i,$$

according as $\{\lambda_i\}_{i=1}^\infty$ *terminates and equals, say* $\{\lambda_i\}_{i=1}^{i_0}$, *or not; that is,*

$$\left\| Ax - \sum_{i=1}^{i_0} \lambda_i(x, \chi_i)\chi_i \right\| = 0 \quad or$$

$$\lim_{m\to\infty} \left\| \sum_{i=1}^{m} \lambda_i(x, \chi_i)\chi_i - Ax \right\| = 0,$$

  accordingly.

Furthermore, if the inverse

$$T : D(T) \subset Z \longrightarrow Z$$

of A *exists, then*

$$z = \sum_{i=1}^{\infty} (z, \chi_i)\chi_i$$

for each $z \in D(T)$.

5.3. Properties of Compact Symmetric Operators. In order to prove Theorem 5.15, we need to prepare some Lemmas.

LEMMA 5.16. *The eigenvalues, if any, of A are real.*

PROOF. We assume that the eigenvalues of A exist.

Let λ be an eigenvalue of A and let $0 \neq x \in D(A) = Z$ be an eigenfunction, associated with λ. Then the symmetry of A implies

$$\lambda(x, x) = (\lambda x, x) = (Ax, x)$$
$$= (x, Ax) = (x, \lambda x)$$
$$= \overline{\lambda}(x, x).$$

Since $(x, x) = \|x\|^2 \neq 0$, we have $\lambda = \overline{\lambda}$, and so λ is real. $\qquad \square$

LEMMA 5.17. *The eigenvectors corresponding to different eigenvalues, if any, of A are orthogonal.*

PROOF. Assume that the eigenvalues of A exist.

Let λ_1, λ_2 be two different real eigenvalues of A, with a corresponding eigenvector x_1 and x_2, respectively. Then

$$(\lambda_1 - \lambda_2)(x_1, x_2) = (\lambda_1 x_1, x_2) - (x_1, \lambda_2 x_2)$$
$$= (Ax_1, x_2) - (x_1, Ax_2) = 0$$

by the symmetry of A. Since $\lambda_1 - \lambda_2 \neq 0$, we have

$$(x_1, x_2) = 0.$$

$\qquad \square$

LEMMA 5.18. *We have*

$$0 \neq \|A\| \equiv \sup_{\|x\|=1, x \in Z} \|Ax\| = \sup_{\|x\|=1, x \in Z} |(Ax, x)|.$$

PROOF. The proof will be divided into three steps.

Step 1. We will prove the assertion that $\|A\| \neq 0$. If

$$\sup_{x \neq 0} \frac{\|Ax\|}{\|x\|} = \|A\| = 0,$$

then

$$\|Ax\| = 0$$

for all $x \in Z$, and so $Ax = 0$ for all $x \in Z$. This leads to $A \equiv 0$, a contradiction to the assumption of $A \neq 0$.

Step 2. Let

$$\zeta \equiv \sup_{\|x\|=1, x \in Z} |(Ax, x)|.$$

For $x \in Z$, we have, by the Cauchy-Schwarz inequality in Lemma 5.10,

$$|(Ax, x)| \leq \|Ax\| \|x\|.$$

Taking the supremum over $x \in Z$ with $\|x\| = 1$ gives

$$\zeta \leq \|A\|.$$

Step 3. We shall show that $\zeta \geq \|A\|$. By Cauchy-Schwarz inequality, we have

$$\zeta \|x + y\|^2 \geq (A(x + y), (x + y))$$

$$= (Ax, x) + (Ay, y) + 2Re(Ax, y),$$
$$-\zeta\|x - y\|^2 \leq (A(x - y), (x - y))$$
$$= (Ax, x) + (Ay, y) - 2Re(Ax, y),$$

for $x, y \in Z$. By substracting, we have

$$Re(Ax, y) \leq \frac{1}{4}(\|x + y\|^2 + \|x - y\|^2)$$
$$= \frac{1}{4}[(x + y, x + y) + (x - y, x - y)]$$
$$\leq \frac{1}{2}\zeta(\|x\|^2 + \|y\|^2).$$

Taking $x \in Z$ with $\|x\| = 1$ and taking

$$y = \frac{Ax}{\|Ax\|},$$

we have

$$\|Ax\| \leq \zeta.$$

Taking supremum over $x \in Z$ with $\|x\| = 1$ gives

$$\|A\| \leq \zeta.$$

$\square$

LEMMA 5.19. *Either $\|A\| \neq 0$ or $-\|A\| \neq 0$ is an eigenvalue of A with a corresponding eigenvector x that satisfies*

$$\|x\| = 1,$$
$$|(Ax, x)| = \|A\|.$$

PROOF. Since

$$\overline{(Ax, x)} = (x, Ax) = (Ax, x)$$

by the symmetry of A, we have $(Ax, x) \in \mathbb{R}$, and so either

$$0 \neq \|A\| = \sup_{\|x\|=1, x \in Z} (Ax, x)$$

or

$$0 \neq \|A\| = \sup_{\|x\|=1, x \in Z} (-1)(Ax, x)$$

by Lemma 5.18.

We consider two cases.

Case 1. Suppose

$$0 \neq \lambda \equiv \|A\| = \sup_{\|x\|=1, x \in Z} (Ax, x)$$

is the case. From the definition of sup, there is a sequence x_n such that

$$\|x_n\| = 1,$$
$$(Ax_n, x_n) \longrightarrow \lambda = \|A\|$$

as $n \longrightarrow \infty$. Then

$$0 \leq \|Ax_n - \lambda x_n\|^2$$
$$= (Ax_n - \lambda x_n, Ax_n - \lambda x_n)$$
$$= (Ax_n, Ax_n) - 2\lambda(Ax_n, x_n) + (\lambda x_n, \lambda x_n)$$

$$= \|Ax_n\|^2 - 2\lambda(Ax_n, x_n) + \lambda^2 \|x_n\|^2$$
$$\leq \|A\|^2 - 2\lambda(Ax_n, x_n)) + \lambda^2$$
$$\longrightarrow 2\lambda^2 - 2\lambda^2 = 0$$

as $n \longrightarrow \infty$, we have

$$\|Ax_n - \lambda x_n\| \longrightarrow 0. \tag{5.1}$$

But the compactness of A implies that a subsequence Ay_n of Ax_n converges, where y_n is a subsequence of x_n. Hence (5.1) gives that y_n converges to a $x \in Z$, since $\lambda \neq 0$. It follows that

$$Ay_n \longrightarrow Ax,$$

since A is continuous by the compactness of A from Lemma 5.14. Thus $Ax = \lambda x$ and $\|x\| = 1$, which shows that $\lambda = \|A\|$ is an eigenvalue and x is a corresponding eigenvector.

Case 2. The case

$$\|A\| = \sup_{\|x\|=1, x \in Z} -(Ax, x) = - \inf_{\|x\|=1, x \in Z} (Ax, x)$$

can be treated similarly. $\qquad \square$

PROPOSITION 5.20. *There is a possibly terminating sequence $\{\lambda_i\}_{i=1}^{\infty}$ of real numbers which are eigenvalues of A and satisfy*

$$|\lambda_1| \geq |\lambda_2| \geq \cdots$$

with $\lambda_i \neq 0, i = 1, 2, \ldots$.

Furthermore, there is a set of orthonormal eigenvectors $\chi_i, i = 1, 2, \ldots$, corresponding to λ_i.

PROOF. The proof will be divided into four steps.

Step 1. Let $\lambda_1 \neq 0$ be the eigenvalue and χ_1 be the corresponding eigenvector from Lemma 5.16.

Step 2. Let $Z_1 = Z$ and

$$Z_2 \equiv \{x \in Z : (x, \chi_1) = 0\}.$$

Let $A_1 = A$, and let A_2 be the restriction $A_1|_{Z_2}$ of $A_1 = A$ to Z_2:

$$A_2 : D(A_2) = Z_2 \subset Z \longrightarrow Z,$$
$$A_2 x \equiv A_1 = Ax \quad \text{for } x \in Z_2.$$

It is easy to check that Z_2 is a closed subspace, and that Z_2 is invariant under A_2:

$$A_2 Z_2 \subset Z_2.$$

This is because

$$(A_2 x, \chi_1) = (Ax, \chi_1) = (x, A\chi_1)$$
$$= (x, \lambda_1 \chi_1) = \lambda_1 (x, \chi_1)$$
$$= 0$$

for $x \in Z_2$, and so $A_2 x \in Z_2$.

Moreover, we readily check that $A_2 : Z_2 \longrightarrow Z_2$ is compact and symmetric since $A : Z \longrightarrow Z$ is so. Thus if $A_2 \neq 0$, then Lemma 5.16 gives an eigenpair $\{\lambda_2, \chi_2\}$ for A_2. which satisfies $\chi_2 \in X_2$, Since

$$\chi_2 \in Z_2,$$

$$\lambda_2 \chi_2 = A_2 \chi_2 = A_1 \chi_2$$
$$= A\chi_2,$$

we have that $\{\lambda_2, \chi_2\}$ is also an eigenpair for A.

On the other hand, we have $|\lambda_2| \le |\lambda_1|$. This is because

$$|\lambda_2| = \|A_2\| = \sup_{\|x\|=1, x \in Z_2} \|A_2 x\|,$$

$$|\lambda_1| = \|A_1\| = \sup_{\|x\|=1, x \in Z_1} \|A_1 x\|,$$

and A_2 is the restriction of $A_1 = A$ to the subspace $Z_2 \subset Z_1 = Z$.

Step 3. As in Step 2, let

$$Z_3 = \{x \in Z : (x, \chi_1) = 0, (x, \chi_2) = 0\}$$

and A_3 be the restriction of $A_1 = A$ to Z_3. It follows as with Step 2 that Z_3 is a closed subspace invariant under A_3, and that there is an eigenpair $\{\lambda_3, \chi_3\}$ for A_3 and $A_1 = A$, with

$$|\lambda_3| \le |\lambda_2| \le |\lambda_1|,$$

provided that $A_3 \ne 0$.

Step 4. Continuing in this way, we obtain a sequence of eigenpairs $\{(\lambda_i, \chi_i)\}_{i=1}^{\infty}$ for A, a sequence of operators $\{A_i\}_{i=1}^{\infty}$, and a sequence of subspaces $\{Z_i\}_{i=1}^{\infty}$, such that

$$|\lambda_1| \ge |\lambda_2| \ge \cdots,$$
$$\lambda_i \ne 0,$$
$$(\chi_i, \chi_j) = 0 \quad \text{for } i \ne j,$$
$$(\chi_i, \chi_i) = 1,$$
$$Z = Z_1 \supset Z_2 \supset Z_3 \supset \cdots,$$

provided that $A_i \ne 0$ for all i. If

$$A_i \begin{cases} \ne 0 & \text{for } i = 1, 2, \cdots, i_0; \\ = 0 & \text{for } i = i_0 \end{cases}$$

for some $i_0 \ne 1$, then

$$0 = A_{i_0+1} = A_{i_0+2} = \cdots,$$

and the sequence λ_i terminates. $\qquad\square$

5.4. Proof of the Abstract Expansion Results. In this subsection, we will prove Theorem 5.15:

PROOF. We use the eigenpairs $\{\lambda_i, \chi_i\}$ from Proposition 5.20, and divide the proof into six steps.

Step 1. Since $|\lambda_i|$ is non-increasing, we have either

$$\lim_{i \to \infty} \lambda_i = 0$$

or

$$|\lambda_i| \ge \epsilon_0 > 0$$

for all i. If the former case occurs, then there is nothing to prove, and so we assume that the latter is the case and that the sequence λ_i does not terminate. Since A is compact,

$$\|\frac{\chi_i}{\lambda_i}\| = \frac{1}{|\lambda_i|} \leq \frac{1}{\epsilon_0},$$

bounded, and

$$A(\frac{\chi_i}{\lambda_i}) = \chi_i,$$

it follows that $\chi_i = A(\frac{\chi_i}{\lambda_i})$ contains a convergent subsequence; we still call this subsequence χ_i, for convenience. But this is a contradiction since

$$\|\chi_i - \chi_j\|^2 = (\chi_i - \chi_j, \chi_i - \chi_j)$$
$$= \|\chi_i\|^2 + \|\chi_j\|^2$$
$$= 1 + 1 = 2$$

for all i, j, showing that χ_i never converges. Here note that if χ_i converges, then

$$\|\chi_i - \chi_j\| \longrightarrow 0$$

as $i, j \longrightarrow \infty$.

Step 2. Suppose that the $A_i = A|_{Z_i}$, the restriction of A to the subspace Z_i of Z, in the Step 4 for the proof of Proposition 5.20, satisfes

$$A_i \begin{cases} \neq 0 & \text{for } i = 1, 2, \cdots, i_0; \\ = 0 & \text{for } i = i_0 + 1. \end{cases}$$

for some $i_0 > 1$, and let

$$z_{i_0} \equiv x - \sum_{i=1}^{i_0} (x, \chi_i)\chi_i$$

for $x \in Z$. Since

$$(z_{i_0}, \chi_i) = 0 \quad \text{for } i = 1, 2, \cdots, (n-1),$$

we have $z_{i_0} \in Z_{i_0+1}$, and so

$$0 = A_{i_0+1} z_{i_0} = A z_{i_0}$$
$$= Ax - \sum_{i=1}^{i_0} (x, \chi_i) A\chi_i$$
$$= Ax - \sum_{i=1}^{i_0} \lambda_i (x, \chi_i)\chi_i.$$

That is,

$$Ax = \sum_{i=1}^{n-1} \lambda_i (x, \chi_i)\chi_i$$

and λ_i terminates at λ_{i_0}.

Step 3. If the sequence λ_i does not terminate, define

$$z_n = x - \sum_{i=1}^{n-1} (x, \chi_i)\chi_i$$

for $x \in Z$. By using the orthonormality of χ_i, we have

$$
\|z_n\|^2 = (x - \sum_{i=1}^{n-1}(x, \chi_i)\chi_i, x - \sum_{i=1}^{n-1}(x, \chi_i)\chi_i)
$$

$$
= \|x\|^2 - \sum_{i=1}^{n-1}|(x, \chi_i)|^2 \tag{5.2}
$$

$$
\leq \|x\|^2.
$$

On the other hand, since $z_n \in Z_n$, we have

$$
\frac{\|Az_n\|}{\|z_n\|} = \frac{\|A_n z_n\|}{\|z_n\|} \leq \sup_{y \neq 0, y \in Z_n} \frac{\|A_n y\|}{\|y\|} = |\lambda_n|
$$

for $z_n \neq 0$. This, together with (5.2), shows

$$
\|Az_n\| \leq |\lambda_n| \|z_n\| \leq |\lambda_n| \|x\|
$$

for $z_n \neq 0$; this is easily seen to hold also for $z_n = 0$ since $Az_n = 0$ for $z_n = 0$.
Thus, using $|\lambda_n| \longrightarrow 0$ as $n \longrightarrow \infty$, it follows that

$$
Ax - \sum_{i=1}^{n-1}(x, \chi_i)A\chi_i = Ax - \sum_{i=1}^{n-1}(x, \chi_i)\lambda_i \chi_i
$$

$$
= Az_n \longrightarrow 0,
$$

and so

$$
Ax = \sum_{i=1}^{\infty}\lambda_i(x, \chi_i)\chi_i.
$$

Step 4. We prove the claim that either $\{0\} \cup \{\lambda_i\}_{i=1}^{\infty}$ or $\{\lambda_i\}_{i=1}^{\infty}$ constitute the eigenvalues of A. If $\lambda \neq 0$ and $\lambda \notin \{\lambda_i\}_{i=1}^{\infty}$ is an eigenvalue for A with a corresponding eigenvector $\chi \neq 0$, then

$$
(\chi, \chi_i) = 0
$$

for all i by Lemma 5.17. It follows from Step 3 that

$$
\lambda\chi = A\chi = \sum_{i=1}^{\infty}(\chi, \chi_i)\chi_i = 0,
$$

a contradiction to $\chi \neq 0$. This proves the claim. Here note $\lambda \neq 0$ by assumption.

Step 5. Each eigenvalue λ_i in the sequence $\{\lambda_i\}$, which is not zero, cannot be repeated infinitely many times. For, if so, say $\lambda_{i_0} \neq 0$ is repeated infinitely many times, then

$$
0 \neq \lambda_{i_0} = \lim_{i \to \infty} \lambda_i,
$$

a contradiction to $\lim_{i \to \infty} \lambda_i = 0$ from Step 1.

If λ_{i_0} is repeated p times, then the corresponding eigenspace contains a set of p orthonormal eigenvectors and is of p dimensions at least. We claim that it cannot be of dimensions greater than p. For, if so, then there is an eigenvector $\chi \neq 0$, such that

$$
A\chi = \lambda_{i_0}\chi,
$$

$$
\|\chi\| = 1,
$$

$$
(\chi, \chi_i) = 0 \quad \text{for all } i \text{ by Lemma 5.17.}
$$

But Step 3 implies that

$$\lambda_{i_0} \chi = A\chi = \sum_{i=1}^{\infty} (\chi, \chi_i)\chi_i$$
$$= 0,$$

a contradiction to $\chi \neq 0$. Here note that $\lambda_{i_0} \neq 0$.

Step 6. Suppose $T : D(T) \subset Z \longrightarrow Z$, is the inverse of A. Then T is symmetric. This is because, for $x, y \in D(T)$, we have $Tx, Ty \in D(A)$, and so

$$(Tx, y) = (Tx, ATy) = ((ATx, Ty)$$
$$= (x, Ty)$$

by the symmetry of A. Also, since

$$A\chi_i = \lambda_i \chi_i$$

is the same as

$$\chi_i = TA\chi_i = T(\lambda_i \chi_i)$$
$$= \lambda_i T\chi_i,$$

the eigenvalues of T consists of $\{\frac{1}{\lambda_i}\}$ with the corresponding eigenvectors $\{\chi_i\}$, and

$$\lim_{i \to \infty} |\frac{1}{\lambda_i}| = 0$$

if $\{\lambda_i\}$ does not terminate.

It follows from Step 3 that, for $z \in D(T)$,

$$z = ATz = \sum_{i=1}^{\infty} \lambda_i(Tz, \chi_i)\chi_i$$
$$= \sum_{i=1}^{\infty} (z, T\lambda_i \chi_i)\chi_i = \sum_{i=1}^{\infty} (z, \chi_i)\chi_i.$$

$\square$

The above material in this Section 5 is based on Taylor and Lay [21].

5.5. Examples from Ordinary Differential Operators. A corollary of Theorem 5.15 is:

COROLLARY 5.21. *Theorem 2.1 follows from Theorem 5.15.*

PROOF. Let $(Z, \|\cdot\|) = (C[a, b], \|\cdot\|_w)$, $A = \mathbb{G}$, and $T = L$, where $\mathbb{G}, L$, are the operators defined in Section 1, $C[a, b] \subset L^2(a, b)$, and $\|\cdot\|_w$ is the weighted inner product norm in $L^2(a, b)$. Here note that $C[a, b]$ is an inner product space with the weighted inner product $(,)_w$.

We now divide the proof into four steps.

Step 1. Since $A \equiv \mathbb{G}$ is symmetric, and compact by Lemma 4.6, we can apply the results in Theorem 5.15.

Step 2. We prove the assertion that the eigenvalues $\{\lambda_i\}_{i=1}^{\infty}$ from Theorem 5.15 do not terminate. By following the Step 4 in proving Proposition 4.11, we suppose otherwise $\{\lambda_i\}_{i=1}^{\infty} = \{\lambda_i\}_{i=1}^{m}$ for some m. Since, for $f \in D(A) = D(\mathbb{G}) = C[a, b]$, we have

$$\mathbb{G}f = \sum_{i=1}^{\infty} \lambda_i(f, \chi_i)_w \chi_i = \sum_{i=1}^{m} \lambda_i(f, \chi_i)\chi_i$$

by Theorem 5.15, it follows that

$$f = \frac{1}{r}L\mathbb{G}_0(rf) = \frac{1}{r}L\mathbb{G}f$$

$$= \sum_{i=1}^{m} \frac{1}{r}\lambda_i(f, \chi_i)_w \frac{1}{\lambda_i}r\chi_i$$

$$= \sum_{i=1}^{m} (f, \chi_i)_w \chi_i.$$

This is a contradiction, since f is a continuous function in general but the function

$$\sum_{i=1}^{m} (f, \chi_i)_w \chi_i \in C^2[a, b]$$

is twice continuously differentiable and they cannot be equal.

Step 3. The eigenvalues $\{\lambda_i\}_{i=1}^{\infty}$ are simple by Lemma 4.3, and so

$$\lambda_1 > \lambda_2 > \lambda_3 > \cdots \longrightarrow \infty.$$

Step 4. That

$$f = \sum_{i=1}^{\infty} (f, \chi_i)_w \chi_i$$

is true for $f \in L^2(a, b)$ follows from Proposition 4.14. $\qquad\square$

Another corollary of Theorem 5.15 is the following.
Let, Taylor and Lay [**21**, Pages 346-347],

$$L' : D(L') \subset (L^2(a, b), \|\cdot\|_w) \longrightarrow (L^2(a, b), \|\cdot\|_w)$$

be a linear differential operator, defined by

$$L'y = (p(x)y')' + q(x)y \quad \text{for } y \in D(L'),$$
$$D(L') \equiv \{u \in L^2(a, b) : u, u', p, p(x)u' \in AC, (p(x)u')' \in L^2(a, b),$$
$$u(a) + \alpha_1 u'(a) = 0, u(b) + \alpha_2 u'(b) = 0\},$$

where the meanning of AC is defined below; $p(x), r(x) \geq \delta > 0$ for some $\delta > 0$ and for all $x \in [a, b]$; and $q(x), r(x)$ are measurable and bounded.

Since, for some $M > 0$, we have

$$0 < \delta \leq r(x) \leq M$$

for all $x \in [a, b]$, it follows as before that

$$L^2(a, b) = L_w^2(a, b),$$

and that $\|\cdot\|_2$ and $\|\cdot\|_w$ are equivalent.
Here, Royden [**20**],

DEFINITION 5.22. *A complex-valued function f on $[a, b]$ is absolutely continuous (abbreviated as AC) if, for a given $\epsilon > 0$, there is a $\delta > 0$, such that*

$$\sum_{i=1}^{n} |f(b_i) - f(a_i)| < \epsilon$$

for all finite collection $\{(a_i, b_i)\}_{i=1}^{n}$ of nonoverlapping subintervals of $[a, b]$ with

$$\sum_{i=1}^{n} |b_i - a_i| < \delta.$$

LEMMA 5.23. *(Royden [20]) A complex-valed function f on $[a, b]$ is absolutely continuous if and only if f' exists almost everywhere and*

$$f(x) = \int_{a}^{x} f'(t)\, dt + f(a).$$

Here the integral is taken with respect to the Lebesgue integral.

From the Defintion 5.22 and Lemma 5.23, the domain $D(L')$ of L' is well-defined.

COROLLARY 5.24. *The results in Theorem 2.1 for the operator L also hold for the operator L'.*

In order to prove Corollary 5.24, we need to prepare an imbedding result, Adams [1, Pages 97–98] (compared to the Ascoli-Arzela Theorem 4.7), and some Lemmas.

THEOREM 5.25 (Sobolev Imbedding Theorem). *Let $f_n(x) \in L_w^2(a, b)$ be a sequence of complex-valued functions that satisfy, for all n and for some constant K,*

(i) $\|f_n\|_w \leq K$;
(ii) f_n' exists almost everywhere and $\|f_n'\|_w \leq k$;
(iii) f_n'' exists almost everywhere and $\|f_n''\|_w \leq K$.
Then $f_n \in C^1[a, b]$ and

$$\{\|f_n\|_\infty, \|f_n'\|_\infty\} \leq C$$

for some constant C and for all n.

As in Lemma 4.6, we have

LEMMA 5.26. *The set*

$$\{\mathbb{G}f_n : f_n \in L_w^2(a, b), \|f_n\|_w \leq 1, n \in \mathbb{N}\}$$

has a uniformly convergent subsequence with the limit in $C[a, b]$ and then in $L_w^2(a, b)$. Here for $f \in L_w^2(a, b)$, we define

$$\|f\|_w \equiv (\int_{a}^{b} r(x)|f(x)|^2\, dx)^{\frac{1}{2}},$$

where the integral is taken with respect to the Lebesgue measure, and this integral equals the Riemann integral if $f \in C[a, b]$.

REMARK 5.27. *Here note that the integral operator $\mathbb{G}$ is defined as before:*

$$\mathbb{G}f \equiv \int_a^b G(x,\xi)r(\xi)f(\xi)\,d\xi,$$

where the Green function is the Green function $G(x,\xi)$ for the operator L'. Recall, from Chapter 5 on the construction of Green function, that $G(x,\xi)$ is constructed by using two linearly independent solutions $u_1(x)$ and $u_2(x)$ of the equation $L'y = 0$, where, by being solutions, $u_i, i = 1, 2$, are meant to satisfy

$$L'u_i = 0$$

almost everywhere. It follows from Chapter 8, Corollary 4.2 that $u_i(x), i = 1, 2$ exist, u_i and u_i' are continuous, and u_i' is absolutely continuous. u_i'' exists almost everywhere by Lemma 5.23. It follows that

$$\{\|u_i\|_\infty, \|u_i'\|_\infty\} \le C$$

for some constant C, and that

$$\left\{ \max_{x,\xi \in [a,b]} |G(x,\xi)|, \ \max_{x,\xi \in [a,b]} |G_x(x,\xi)| \right\} \le M$$

for some constant M, where note the jump at $x = \xi$ of $G_x(x,\xi)$ is finite.

PROOF OF LEMMA 5.26. We divide the proof into four steps.
Step 1. We have

$$\|\mathbb{G}f_n\|_\infty \le \gamma \int_a^b r(\xi) \cdot 1 \cdot |f_n(\xi)|\,d\xi$$
$$= \gamma(1, |f_n|)_w \le \gamma\|1\|_w\|f_n\|_w$$
$$\le \gamma\|1\|_w,$$

uniformly bounded for all n, where

$$\|f_n\|_w \le 1 \quad \text{by assumption,}$$
$$\|1\|_w = \left(\int_a^b r(x)\,dx \right)^{\frac{1}{2}},$$
$$\gamma \equiv \max_{x,\xi \in [a,b]} |G(x,\xi)|.$$

Step 2. Let

$$G\{x \ge \xi\} = G(x,\xi)|_{\{x \ge \xi\}} \quad \text{be the restriction of } G(x,\xi) \text{ to the set}$$
$$\{(x,\xi) : a \le \xi \le x \le b\},$$
$$G\{x \le \xi\} = G(x,\xi)|_{\{x \le \xi\}} \quad \text{be the restriction of } G(x,\xi) \text{ to the set}$$
$$\{(x,\xi) : a \le x \le \xi \le b\}.$$

Since

$$(\mathbb{G}f_n)(x)$$
$$= \int_a^x G(x,\xi)r(\xi)f_n(\xi)\,d\xi + \int_x^b G(x,\xi)r(\xi)f_n(\xi)\,d\xi$$
$$= \int_a^x G\{x \ge \xi\}r(\xi)f_n(\xi)\,d\xi + \int_x^b G\{x \le \xi\}r(\xi)f_n(\xi)\,d\xi,$$

we have, as with Chapter 5, on the construction of the Green function $G(x, \xi)$, that

$$\frac{d}{dx}(\mathbb{G}f_n)(x)$$

$$= \int_a^x G_x\{x \geq \xi\}r(\xi)f_n(\xi)\,d\xi + \int_x^b G_x\{x \leq \xi\}r(\xi)f_n(\xi)\,d\xi$$
$$+ G\{x \geq \xi\}|_{\xi=x}r(x)f_n(x) - G\{x \leq \xi\}|_{\xi=x}r(x)f_n(x) \quad a.e.$$

$$= \int_a^x G_x\{x \geq \xi\}r(\xi)f_n(\xi)\,d\xi + \int_x^b G_x\{x \leq \xi\}r(\xi)f_n(\xi)\,d\xi$$
$$+ \lim_{\xi \to x^-} G(x, \xi)r(x)f_n(x) - \lim_{\xi \to x^+} G(x, \xi)r(x)f_n(x) \quad a.e.$$

by Lemma 5.23 on the fundamental theorem of calculus for Lebesgue integral. Here *a.e.* means almost everywhere. Since $G(x, \xi)$ is jointly continuous in x, ξ, the last two terms on the right side of the above inequality equal zero. It follows that

$$\frac{d}{dx}(\mathbb{G}f_n)(x)$$

$$= \int_a^x G_x(x \geq \xi)r(\xi)f_n(\xi)\,d\xi + \int_x^b G_x(x \leq \xi)r(\xi)f_n(\xi)\,d\xi \quad a.e.;$$

that

$$\{\| \int_a^x G_x\{x \geq \xi\}r(\xi)f(\xi)\,d\xi\|_\infty, \| \int_x^b G_x\{x \leq \xi\}r(\xi)f_n(\xi)\,d\xi\|_\infty\}$$
$$\leq \eta\|1\|_w\|f_n\|_w \leq \eta\|1\|_w$$

as with Step 1, where

$$\eta \equiv \max\{ \max_{x,\xi \in [a,b]} |G_x\{x \geq \xi\}|, \max_{x,\xi \in [a,b]} |G_x\{x \leq \xi\}|\}$$

and the jump $\frac{1}{p(\xi)}$ of $G_x(x, \xi)$ at $x = \xi$ is finite; and that

$$\|\frac{d}{dx}(\mathbb{G}f_n)(x)\|_w^2 \leq \int_a^b [2\eta\|1\|_w\|f_n\|_w]^2\,dx$$
$$= [2\eta\|1\|_w]^2(b-a), \quad \text{uniformly bounded.}$$

Step 3. As in Step 2, we have

$$\frac{d^2}{dx^2}(\mathbb{G}f_n)(x)$$

$$= \int_a^x G_{xx}\{x \geq \xi\}r(\xi)f_n(\xi)\,d\xi + \int_x^b G_{xx}\{x \leq \xi\}r(\xi)f_n(\xi)\,d\xi$$
$$+ \lim_{\xi \to x^-} G_x\{x, \xi\}r(x)f_n(x) - \lim_{\xi \to x^+} G_x\{x, \xi\}r(x)f_n(x) \quad a.e.$$

$$= \int_a^x G_{xx}\{x \geq \xi\}r(\xi)f_n(\xi)\,d\xi + \int_x^b G_{xx}\{x \leq \xi\}r(\xi)f_n(\xi)\,d\xi$$
$$+ \frac{r(x)f_n(x)}{p(x)} \quad a.e.,$$

where we used the jump $\frac{1}{p(x)}$ of $G_x(x, \xi)$ at $x = \xi$.

It follows as in Step 2 that $\|\frac{d^2}{dx^2}\mathbb{G}f_n\|_w$ is uniformly bounded, where note that

$$G_{xx}(x,\xi) \equiv \frac{\partial^2}{\partial x^2}G(x,\xi)$$
$$= \frac{-p'(x)G_x(x,\xi) - q(x)G(x,\xi)}{p(x)}$$

also has a finite jump at $x = \xi$, so that

$$\max_{x,\xi\in[a,b]} |G_{xx}\{x \geq \xi\}| \quad \text{and} \quad \max_{x,\xi\in[a,b]} |G_{xx}\{x \leq \xi\}|$$

are uniformly finite.

Step 4. The Sobolev imbedding Theorem 5.25 implies that

$$\|\mathbb{G}f_n\|_\infty \quad \text{and} \quad \|\frac{d}{dx}\mathbb{G}f_n\|_\infty$$

are uniformly bounded. Since, by the mean value theorem, we have

$$\mathbb{G}f_n(x_1) - \mathbb{G}f_n(x_2) = \frac{d}{dx}\mathbb{G}f_n(c)(x_1 - x_2)$$

for some c between x_1 and x_2, where $a \leq x_1, x_2 \leq b$, it follows that $\mathbb{G}f_n$ is equicontinuous. The proof is completed by applying the Ascoli-Arzela theorem 4.7. $\square$

We now prove Corollary 5.24

PROOF OF COROLLARY 5.24. We follow the proof of Corollary 5.21.

Let $(Z, \|\cdot\|) = (L^2(a,b), \|\cdot\|_w)$, $A = \mathbb{G}$, and $T = L'$, where $\mathbb{G}, L'$, are the operators defined in this Section 5 and $\|\cdot\|_w$ is the weighted inner product norm in $L^2(a,b)$.

Now we divide the proof into four stpes.

Step 1. Since $A \equiv \mathbb{G}$ is symmetric and compact by Lemma 5.26, we can apply the results in Theorem 5.15.

Step 2. We prove the assertion that the eigenvalues $\{\lambda_i\}_{i=1}^\infty$ from Theorem 5.15 do not terminate. By following the Step 4 in proving Proposition 4.11, we suppose otherwise $\{\lambda_i\}_{i=1}^\infty = \{\lambda_i\}_{i=1}^m$ for some m. Since, for $f \in D(A) = D(\mathbb{G}) = L^2(a,b)$, we have

$$\mathbb{G}f = \sum_{i=1}^\infty \lambda_i(f,\chi_i)_w\chi_i = \sum_{i=1}^m \lambda_i(f,\chi_i)\chi_i$$

by Theorem 5.15, it follows that

$$f = \frac{1}{r}L\mathbb{G}_0(rf) = \frac{1}{r}L\mathbb{G}f$$
$$= \sum_{i=1}^m \frac{1}{r}\lambda_i(f,\chi_i)_w\frac{1}{\lambda_i}r\chi_i$$
$$= \sum_{i=1}^m (f,\chi_i)_w\chi_i.$$

This is a contradiction, since f lies in $L^2(a,b)$ in general but the function

$$\sum_{i=1}^m (f,\chi_i)_w\chi_i \in D(L')$$

is absolutely continuous and they cannot be equal.

Step 3. The eigenvalues $\{\lambda_i\}_{i=1}^{\infty}$ are simple by Lemma 4.3, and so

$$\lambda_1 > \lambda_2 > \lambda_3 > \cdots \longrightarrow \infty.$$

Step 4. That

$$f = \sum_{i=1}^{\infty} (f, \chi_i)_w \chi_i$$

is true for $f \in L^2(a, b)$ follows from Proposition 4.14, since

$$D(L') \supset C_0^2[a, b].$$

$\square$

6. Problems

Solve the eigenvalue problems **1** to **9** by obtaining the normalized eigenpairs.

1.

$$2y'' + 4y' + \lambda y = 0, \quad 0 < x < 1,$$
$$y(0) = 0 = y(1).$$

2.

$$2y'' + 4y' + \lambda y = 0, \quad 0 < x < 1,$$
$$y'(0) = 0 = y'(1).$$

3.

$$2y'' + 4y' + \lambda y = 0, \quad 0 < x < 1,$$
$$y(0) = y(1),$$
$$y'(0) = y'(1).$$

4.

$$2y'' + 4y' + \lambda y = 0, \quad 0 < x < 1,$$
$$2y(0) + y'(0) = 0,$$
$$y(1) - 3y'(1) = 0.$$

5. Let

$$f(x) = \begin{cases} 2x, & \text{if } 0 \le x < \frac{1}{2}; \\ -2x + 2, & \text{if } \frac{1}{2} \le x \le 1. \end{cases}$$

Expand f by using the orthonormal system in Problem **3**.

6. Solve the heat equation

$$\frac{\partial}{\partial t} u(x, t) = \beta(2e^{-2x}) \frac{\partial}{\partial x} [e^{2x} \frac{\partial}{\partial x} u(x, t)], \quad 0 < x < 1, t > 0,$$
$$u(0, t) = u(1, t),$$
$$\frac{\partial}{\partial x} u(0, t) = \frac{\partial}{\partial x} u(1, 0),$$
$$u(x, 0) = f(x),$$

where β is a positive constant and $f(x)$ is the $f(x)$ in Problem **5**.

7.

$$(2xy')' + \frac{4}{x}y + \lambda \frac{1}{x}y = 0, \quad 1 < x < \sqrt{e},$$

$$y(1) = 0 = y(\sqrt{e}).$$

8.

$$(2xy')' + \frac{4}{x}y + \lambda\frac{1}{x}y = 0, \quad 1 < x < \sqrt{e},$$
$$y'(1) = 0 = y'(\sqrt{e}).$$

9.

$$(2xy')' + \frac{4}{x}y + \lambda\frac{1}{x}y = 0, \quad 1 < x < \sqrt{e},$$
$$y(1) = y(\sqrt{e}),$$
$$y'(1) = y'(\sqrt{e}).$$

10. Prove that the eigenvalues for the linear ordinary differential operator L in Section 1 are non-positive, if the coefficient function $q(x) \leq 0$, $\alpha_1 < 0$, and $\alpha_2 > 0$.
   11. Prove Corollary 1.7.
   12. Prove Corollary 1.10.

7. Solutions

1.

$$\{\lambda_n, \chi_n\}_{n\in\mathbb{N}} = \{2n^2\pi^2 - 4, \sqrt{2}\sin(n\pi x)\}_{n\in\mathbb{N}}.$$

2.

$$\{\lambda_n, \chi_n\}_{n=0}^{\infty} = \{-4, 1\} \cup \{2n^2\pi^2 - 4, \sqrt{2}\cos(n\pi x)\}_{n=1}^{\infty}.$$

3. The solution y is given by

$$y = c_1 \cos(\nu x) + c_2 \sin(\nu x),$$

where $\nu = \sqrt{\frac{4+\lambda}{2}}$. Using the boundary conditions, we have that c_1, c_2, and ν satisfy

$$c_1(1 - \cos(\nu)) = c_2 \sin(\nu),$$
$$\nu c_2(1 - \cos(\nu)) = -\nu c_1 \sin(\nu).$$

That implies

$$\sin(\nu) = 0,$$
$$(1 - \cos(\nu) = 0,$$

and so $\nu = \nu_n = 2n\pi$ or $\lambda = \lambda_n = 2(2n\pi)^2 - 4$. Here $n \in \mathbb{Z}$ but $n = 0, 1, \ldots$, is sufficient. Thus the normalized eigenpairs are

$$\{\lambda_n, \chi_n\}_{n=0}^{\infty} = \{-4, 1\} \cup \{2(2n\pi)^2 - 4, [\sqrt{2}\cos(2n\pi x), \sqrt{2}\sin(2n\pi x)]\}_{n=1}^{\infty}.$$

4. We discuss three cases.
   Case 1: $(4 + \lambda) > 0$. The solution y is given by

$$y = c_1 \cos(\nu x) + c_2 \sin(\nu x),$$

where $\nu = \sqrt{\frac{4+\lambda}{2}}$. Using the boundary conditions, we have that c_1, c_2, and ν satisfy

$$0 = 2c_1 + \nu c_2,$$
$$0 = (c_1 \cos(\nu) + c_2 \sin(\nu)) - 3(-\nu c_1 \sin(\nu) + \nu c_2 \cos(\nu)).$$

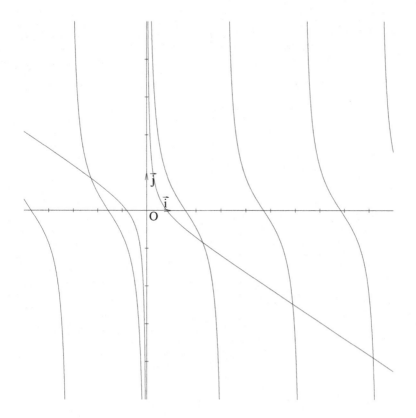

FIGURE 2. Intersection of two graphs

That implies

$$\cot(\nu) = \frac{2}{7\nu} - \frac{3\nu}{7},$$ (7.1)

and this equation has an infinite number of solutions $\nu = \nu_n, n = 1, 2, \ldots$, given by the intersection points of the two graphs (see Figure 2):

$$y_1 = \cot(\nu),$$

$$y_2 = \frac{2}{7\nu} - \frac{3\nu}{7}.$$

Those ν_n's give that $\lambda_n = 2\nu_n^2 - 4$ are the eigenvalues, and that $\phi_n = -\frac{\nu_n}{2}\cos(\nu_n x) + \sin(\nu_n x)$ are the associated eigenfunctions.

Case 2: $(4 + \lambda) < 0$. The solution y is given by

$$y = c_1 e^{\nu x} + c_2 e^{-\nu x},$$

where $\nu = \sqrt{\frac{-(4+\lambda)}{2}}$. Using the boundary conditions, we have that c_1, c_2, and ν satisfy

$$0 = 2(c_1 + c_2) + (c_1 \nu - c_2 \nu),$$

$$0 = (c_1 e^{\nu} + c_2 e^{-\nu}) - 3(c_1 \nu e^{\nu} - c_2 \nu e^{-\nu}).$$

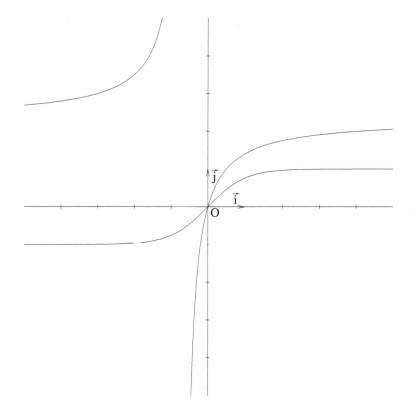

FIGURE 3. Intersection of two graphs

That implies

$$\tanh(\nu) = \frac{7\nu}{2 + 3\nu},$$

and this equation only has the solution $\nu = 0$ (see Figure 3), contradicting $\nu > 0$.

Case 3: $(4 + \lambda) = 0$. The solution y is given by

$$y = c_1 + c_2 x.$$

But the boundary conditions give $c_1 = 0 = c_2$, and so y is the trivial zero solution. Thus the only eigenvalues and eigenfuctions are those in Case 1.

We now verify the orthogonality: $(\phi_n, \phi_m) = 0$ if $n \neq m$, by the calculations:

$$(\phi_n, \phi_m) = \int_0^1 \frac{\nu_n \nu_m}{4} \cos(\nu_n x) \cos(\nu_m x)\, dx + (-1) \int_0^1 \frac{\nu_m}{2} \cos(\nu_m x) \sin(\nu_n x)\, dx$$

$$+ (-1) \int_0^1 \frac{\nu_n}{2} \cos(\nu_n x) \sin(\nu_m x)\, dx + \int_0^1 \sin(\nu_n x) \sin(\nu_m x)\, dx$$

$$\equiv I_1 + I_2 + I_3 + I_4,$$

$$I_1 = \frac{\nu_n \nu_m}{8} \int_0^1 [\cos((\nu_n + \nu_m)x) + \cos((\nu_n - \nu_m)x)]\, dx$$

$$= \frac{\nu_n \nu_m}{8} [\frac{\sin(\nu_n + \nu_m)}{\nu_n + \nu_m} + \frac{\sin(\nu_n - \nu_m)}{\nu_n - \nu_m}]$$

$$\equiv \frac{\nu_n \nu_m}{8} [\frac{J_1}{\nu_n + \nu_m} + \frac{J_2}{\nu_n - \nu_m}],$$

$$J_1 = \sin(\nu_n) \cos(\nu_m) + \cos(\nu_n) \sin(\nu_m)$$

$$= \sin(\nu_n)[\frac{2}{7\nu_m} - \frac{3\nu_m}{7}] \sin(\nu_m)$$

$$+ [\frac{2}{7\nu_n} - \frac{3\nu_n}{7}] \sin(\nu_n) \sin(\nu_m) \quad \text{by (7.1)}$$

$$= [\frac{2(\nu_n + \nu_m)}{7\nu_n \nu_m} - \frac{3(\nu_n + \nu_m)}{7}] \sin(\nu_n) \sin(\nu_m),$$

$$J_2 = \sin(\nu_n) \cos(\nu_m) - \cos(\nu_n) \sin(\nu_m)$$

$$= \sin(\nu_n)[\frac{2}{7\nu_m} - \frac{3\nu_m}{7}] \sin(\nu_m)$$

$$- [\frac{2}{7\nu_n} - \frac{3\nu_n}{7}] \sin(\nu_n) \sin(\nu_m) \quad \text{by (7.1)}$$

$$= [\frac{2(\nu_n - \nu_m)}{7\nu_n \nu_m} + \frac{3(\nu_n - \nu_m)}{7}] \sin(\nu_n) \sin(\nu_m),$$

$$I_1 = \frac{1}{14} \sin(\nu_n) \sin(\nu_m),$$

$$I_2 + I_3 = -\frac{\nu_m}{4} \int_0^1 [\sin((\nu_n + \nu_m)x) + \sin((\nu_n - \nu_m)x)] \, dx$$

$$- \frac{\nu_n}{4} \int_0^1 [\sin((\nu_n + \nu_m)x) - \sin((\nu_n - \nu_m)x)] \, dx$$

$$= -\frac{1}{4} \int_0^1 (\nu_n + \nu_m) \sin((\nu_n + \nu_m)x) \, dx + \frac{1}{4} \int_0^1 (\nu_n - \nu_m) \sin((\nu_n - \nu_m)x) \, dx$$

$$= -\frac{1}{2} \sin(\nu_n) \sin(\nu_m),$$

$$I_4 = \frac{1}{2} \int_0^1 [\cos((\nu_n - \nu_m)x) - \cos((\nu_n + \nu_m)x)] \, dx$$

$$= \frac{1}{2} [\frac{\sin(\nu_n - \nu_m)}{\nu_n - \nu_m} - \frac{\sin(\nu_n + \nu_m)}{\nu_n + \nu_m}]$$

$$= \frac{1}{2} [(\frac{2}{7\nu_n \nu_m} + \frac{3}{7}) \sin(\nu_n) \sin(\nu_m) - (\frac{2}{7\nu_n \nu_m} - \frac{3}{7}) \sin(\nu_n) \sin(\nu_m)] \quad \text{as for } I_1$$

$$= \frac{3}{7} \sin(\nu_n) \sin(\nu_m);$$

thus

$$(\phi_n, \phi_m) = \sum_{i=1}^{4} I_i$$

$$= (\frac{1}{14} - \frac{1}{2} + \frac{3}{7}) \sin(\nu_n) \sin(\nu_m)$$

$$= 0.$$

Next, we normalize the eigenfunctions $\phi_n = -\frac{\nu_n}{2} \cos(\nu_n x) + \sin(\nu_n x)$ by the calculations:

$$1 = (c_n \phi_n, c_n \phi_n)$$

$$= c_n^2 [\frac{\nu_n^2}{4} \int_0^1 (\cos(\nu_n x))^2 \, dx + \int_0^1 (\sin(\nu_n x))^2 \, dx$$

$$+ (\nu_n) \int_0^1 \cos(\nu_n x) \sin(\nu_n x) \, dx]$$

$$\equiv c_n^2 (I_1 + I_2 + I_3),$$

$$I_1 = \frac{1}{4} \nu_n^2 \int_0^1 \frac{1 + \cos(2\nu_n x)}{2} \, dx$$

$$= \frac{1}{4} \nu_n^2 [\frac{1}{2} + \frac{1}{4\nu_n} \sin(2\nu_n)],$$

$$I_2 = \int_0^1 \frac{1 - \cos(2\nu_n)}{2} \, dx$$

$$= \frac{1}{2} - \frac{1}{4\nu_n} \sin(2\nu_n),$$

$$I_3 = -\int_0^1 \sin(\nu_n x) \, d(\sin(\nu_n x))$$

$$= -\frac{1}{2} (\sin(\nu_n))^2;$$

thus

$$1 = c_n^2 (I_1 + I_2 + I_3)$$

$$= c_n^2 [(\frac{n_n^2}{8} + \frac{1}{2}) - (\frac{1}{4} + \frac{1}{7\nu_n^2} + \frac{3}{56} \nu_n^2)(\sin(\nu_n))^2]$$

$$\equiv c_n^2 d_n,$$

and so we choose $c_n = \frac{1}{\sqrt{d_n}}$. Here (7.1) was used.

Therefore, the normalized eigenpairs are

$$\{\lambda_n, \chi_n\}_{n=1}^\infty = \{2\nu_n^2 - 4, \frac{1}{\sqrt{d_n}} (-\frac{\nu_n}{2} \cos(\nu_n x) + \sin(\nu_n x))\}_{n=1}^\infty,$$

where ν_n satisfies (7.1).

5. By the solution to Problem **3**, the orthonormal system is

$$\{\chi_n\}_{n=0}^\infty = \{1, \sqrt{2} \cos(2n\pi x), \sqrt{2} \sin(2n\pi x)\}_{n=1}^\infty.$$

Although f does not lie in $D(L)$, f does lie in $L^2(0, 1)$, and so by Theorem 2.1, we have

$$f(x) = \sum_{n=0}^\infty (f, \chi_n) \chi_n$$

$$= (f, 1) + \sum_{n=1}^\infty (f, \sqrt{2} \cos(2n\pi x)) \sqrt{2} \cos(2n\pi x)$$

$$+ \sum_{n=1}^\infty (f, \sqrt{2} \sin(2n\pi x)) \sqrt{2} \sin(2n\pi x),$$

the convergence being in the norm $\| \cdot \|_2$. Here

$$(f, 1) = \frac{1}{2},$$

$$(f, \sqrt{2}\cos(2n\pi x)) = \frac{\sqrt{2}}{n^2\pi^2}[(-1)^n - 1],$$

$$(f, \sqrt{2}\sin(2n\pi x)) = 0.$$

6. By using the technique $u(x, t) = X(x)T(t)$ of separation of variables, we have

$$X\frac{d}{dt}T = \beta T\frac{d}{dx}(e^{2t}\frac{d}{dx}X),$$

and so

$$2e^{-2x}\frac{1}{X}(e^{2x}X')' = \frac{1}{\beta T}T' \equiv -\lambda,$$

where λ is a constant to be determined. Thus X satisfies the eigenvalue problem in Problem **3**:

$$2X'' + 4X' + \lambda X = 0,$$
$$X(0) = X(1),$$
$$X'(0) = X'(1)$$

and T satisfies the first order differential equation:

$$T' + \beta\lambda T = 0.$$

By the solution of Problem **3**, we have that

$$\lambda = \lambda_n = 2(2n\pi)^2 - 4, \quad n = 0, 1, \dots,$$

that the normalized eigenpairs are

$$\{\lambda_0, \phi_0\} \cup \{\lambda_n, [\chi_n, \varphi_n]\}_{n=1}^{\infty}$$
$$= \{-4, 1\} \cup \{2(2n\pi)^2 - 4, [\sqrt{2}\cos(2n\pi x), \sqrt{2}\sin(2n\pi x)]\}_{n=1}^{\infty},$$

and that, by superposition principle, X is given by

$$X(x) = c_0 + \sum_{n=1}^{\infty} a_n\chi_n(x) + \sum_{n=1}^{\infty} b_n\varphi_n(x),$$

where c_0, a_n's, and b_n's are constants to be determined. This yields easily

$$T(t) = e^{-\beta\lambda_n t},$$

and so

$$u(x, t) = c_0 e^{4\beta t} + \sum_{n=1}^{\infty} a_n e^{-\beta\lambda_n t}\chi_n(x) + \sum_{n=1}^{\infty} b_n e^{-\beta\lambda_n t}\varphi_n(x). \qquad (7.2)$$

Because of the initial condition $u(x, 0) = f(x)$, we have

$$f(x) = c_0 + \sum_{n=1}^{\infty} a_n\chi_n(x) + \sum_{n=1}^{\infty} b_n\varphi_n(x),$$

where, by the solution of Problem **5**, the convergence is in the L_2 norm $\|\cdot\|_2$ and

$$c_0 = (f, 1) = \frac{1}{2},$$

$$b_n = (f, \sqrt{2}\cos(2n\pi x)) = \frac{\sqrt{2}}{n^2\pi^2}[(-1)^n - 1],$$

$$a_n = (f, \sqrt{2}\sin(2n\pi x)) = 0.$$

Finally, we need to show that the series in (7.2) can be differentiated term by term, twice with respect to the variable x and once with respect to the variable t, so that this determined $u(x,t)$ is a solution of the heat equation. But this follows from the cited Theorem 3.9 in the solution of Example 3.8. To see this, we compute

$$\sum_{n=1}^{\infty} b_n \sqrt{2} e^{-\beta \lambda_n t} \sin(2n\pi x),$$

$$\sum_{n=1}^{\infty} b_n \sqrt{2} \frac{d}{dt} (e^{-\beta \lambda_n t}) \sin(2n\pi x) = \sum_{n=1}^{\infty} b_n \sqrt{2} (-\beta \lambda_n) e^{-\beta \lambda_n t} \sin(2n\pi x),$$

$$\sum_{n=1}^{\infty} b_n \sqrt{2} e^{-\beta \lambda_n t} \frac{d}{dx} \sin(2n\pi x) = \sum_{n=1}^{\infty} b_n \sqrt{2} (2n\pi) e^{-\beta \lambda_n t} \cos(2n\pi x),$$

$$\sum_{n=1}^{\infty} b_n \sqrt{2} e^{-\beta \lambda_n t} \frac{d^2}{dx^2} \sin(2n\pi x) = \sum_{n=1}^{\infty} b_n \sqrt{2} (-1)(2n\pi)^2 e^{-\beta \lambda_n t} \sin(2n\pi x),$$

$$(7.3)$$

whose absolute values on

$$D \equiv \{(x,t) : 0 \leq x \leq \pi, t \geq t_0 > 0\}$$

are bounded, respectively, by

$$\sum_{n=1}^{\infty} M_0 \frac{e^{-\beta \lambda_n t}}{n^2}, \quad \sum_{n=1}^{\infty} M_0 e^{-\beta \lambda_n t},$$

$$\sum_{n=1}^{\infty} M_0 \frac{e^{-\beta \lambda_n t}}{n}, \quad \text{and} \sum_{n=1}^{\infty} M_0 e^{-\beta \lambda_n t},$$

where M_0 is a positive constant. Since, for large n,

$$e^{-\beta \lambda_n t}, \frac{e^{-\beta \lambda_n t}}{n}, \frac{e^{-\beta \lambda_n t}}{n^2} \leq e^{-\beta (2n\pi)^2 t_0},$$

the ratio test shows that these series of upper bounds converge. Therefore, each series in (7.3) converges uniformly on D by the Weirstrass M-test [**2**, Page 223] (see Chapter 8), and so the cited Theorem 3.9 above applies.

7.

$$\{\lambda_n, \chi_n\}_{n\in\mathbb{N}} = \{2(2n\pi)^2 - 4, 2\sin(2n\pi \ln(x))\}_{n\in\mathbb{N}}.$$

8.

$$\{\lambda_n, \chi_n\}_{n=0}^{\infty} = \{-4, \sqrt{2}\} \cup \{2(2n\pi)^2 - 4, 2\cos(2n\pi \ln(x))\}_{n=1}^{\infty}.$$

9. The solution y is given by

$$y = c_1 \cos(\nu \ln(x)) + c_2 \sin(\nu \ln(x)),$$

where $\nu = \sqrt{\frac{4+\lambda}{2}}$. Using the boundary conditions, we have that $c_1, c_2,$ and ν satisfy

$$c_1 \left(1 - \cos(\frac{\nu}{2})\right) = c_2 \sin(\frac{\nu}{2}),$$

$$\nu c_2 \left(1 - \frac{1}{\sqrt{2}} \cos(\frac{\nu}{2})\right) = -\nu c_1 \frac{1}{\sqrt{e}} \sin(\frac{\nu}{2}).$$

That implies

$$\sin\left(\frac{\nu}{2}\right) = 0,$$

$$\left(1 - \cos\left(\frac{\nu}{2}\right)\right) = 0,$$

and so $\nu = \nu_n = 4n\pi$ or $\lambda = \lambda_n = 2(4n\pi)^2 - 4$. Here $n \in \mathbb{Z}$ but $n = 0, 1, \ldots$, is sufficient. Thus the normalized eigenpairs are

$$\{\lambda_n, \chi_n\}_{n=0}^{\infty} = \{-4, \sqrt{2}\} \cup \{2(4n\pi)^2 - 4, [2\cos(4n\pi \ln(x)), 2\sin(4n\pi \ln(x))]\}_{n=1}^{\infty}.$$

10. Suppose

$$Ly = (p(x)y')' + q(x)y = \lambda r(x)y.$$

Then using the integration by parts formula, we have

$$\lambda \int_a^b r(x)y^2\, dx = \int_a^b yLy\, dx$$

$$= [-\int_a^b p(x)(y')^2\, dx + p(x)y'y|_{x=a}^b]$$

$$+ \int_a^b q(x)y^2\, dx$$

$$\leq 0.$$

This is because, on the right side, the first term ≤ 0 clearly, the second term ≤ 0 by the boundary conditions in $D(L)$, and the third term ≤ 0 by $q(x) \leq 0$. Therefore, we have the eigenvalues $\lambda \leq 0$.

11. Suppose $f \in L^2(a, b)$. Then $\int_a^b |f(x)|^2\, dx < \infty$, finite, and so

$$\int_a^b |r(x)f(x)|^2\, dx \leq M^2 \int_a^b |f(x)|^2\, dx < \infty.$$

Thus $f \in L_w^2(a, b)$. Here $M \equiv \max_{x \in [a,b]} |r(x)|$. Conversely, suppose $f \in L_w^2(a, b)$. Then $\int_a^b |r(x)f(x)|^2\, dx < \infty$, and so

$$\int_a^b |f(x)|^2\, dx = \frac{1}{\delta^2}\delta^2 \int_a^b |f(x)|^2\, dx$$

$$\leq \frac{1}{\delta^2} \int_a^b |r(x)f(x)|^2\, dx < \infty.$$

Thus $f \in L^2(a, b)$. Here $\delta \leq |r(x)|$ for all $x \in [a, b]$, by assumption.

12. Suppose $\|f_n - f\|_2 \longrightarrow 0$ as $n \longrightarrow \infty$. By Lemma 1.9, we have

$$\|f_n - f\|_w \leq \epsilon_1 \|f_n - f\|_2 \longrightarrow 0$$

as $n \longrightarrow \infty$. Conversely, suppose $\|f_n - f\|_w \longrightarrow 0$ as $n \longrightarrow \infty$. By Lemma 1.9 again, we have

$$\|f_n - f\|_2 \leq \frac{1}{\epsilon_0} \|f_n - f\|_w \longrightarrow 0$$

as $n \longrightarrow \infty$.

CHAPTER 7

Long Time Behavior of Systems of Differential Equations

1. Introduction

Consider the homogeneous linear system of $n(\in \mathbb{N})$ first order ordinary differential equations

$$\frac{d}{dt}u(t) = Au, \quad t > 0,$$
$$u(0) = u_0 \in \mathbb{C}^n, \tag{1.1}$$

where A is an $n \times n$ matrix. From Chapter 2, (1.1) has the unique solution $\varphi(t) = e^{tA}u_0$, where e^{tA} is the exponential function of A, defined by

$$e^{tA} \equiv \lim_{m \to \infty} \sum_{k=0}^{m} \frac{(tA)^k}{k!}.$$

Chapter 2 also shows that there exist n unique matrices

$$M_{j,k}: \quad 1 \le j \le s \le n, 0 \le k \le m_j - 1,$$

such that

$$e^{tA} = \sum_{j=1}^{s} \sum_{k=0}^{m_j-1} t^k e^{t\lambda_j} M_{j,k}, \tag{1.2}$$

where $\lambda_j, j = 1, 2, \ldots, s$, are eigenvalues of A with multiplicity m_j. Here λ_j's are the roots of the characteristic polynomial equation

$$det(A - \lambda I) = 0$$

of A, and this equation is of the form

$$0 = det(A - \lambda I) = a_0(\lambda - \lambda_1)^{m_1}(\lambda - \lambda_2)^{m_2} \cdots (\lambda - \lambda_s)^{m_s}, \tag{1.3}$$

in which a_0 is some complex number and the natural numbers m_j satisfy

$$m_1 + m_2 + \cdots + m_s = n.$$

The form (1.3) taken by $det(A - \lambda I)$ is always true (see Chapter 2). Furthermore, after relabelling the n functions

$$t^k e^{t\lambda_i}, i = 1, 2, \ldots, s; k = 0, 1, \ldots, (m_i - 1),$$

as the n functions

$$z_i(t), i = 1, 2, \ldots, n,$$

and relabelling the n matrices

$$M_{i,k}, i = 1, 2, \ldots, s; k = 0, 1, \ldots, (m_i - 1),$$

as the n matrices

$$N_j, j = 1, 2, \ldots, n,$$

we can compute the n matrices $N_j, j = 1, \ldots, n$, as the n unique solutions of the algebraic equations:

$$
\begin{pmatrix}
z_1(0) & z_2(0) & \cdots & z_n(0) \\
z_1'(0) & z_2'(0) & \cdots & z_n'(0) \\
\vdots & & & \\
z_1^{(n-1)}(0) & z_2^{(n-1)}(0) & \cdots & z_n^{(n-1)}(0)
\end{pmatrix}
\begin{pmatrix}
N_1 \\ N_2 \\ \vdots \\ N_n
\end{pmatrix}
=
\begin{pmatrix}
I \\ A \\ \vdots \\ A^{n-1}
\end{pmatrix}.
\tag{1.4}
$$

If the initial condition for (1.1) is instead

$$
u(t_0) = u_0,
$$

where $t_0 \geq 0$, then the substitution $v(t) \equiv u(t - t_0)$ transforms (1.1) into

$$
\frac{d}{dt}v(t) = Av(t), \quad t > 0,
$$
$$
v(t_0) = u_0,
$$

again a form of (1.1). Thus, considering (1.1) is sufficient.

One of our two goals in this chapter is to show that the long time behavior of φ as $t \longrightarrow \infty$, is determined by the eigenvalues λ_j and the initial data u_0. Furthermore, the initial data u_0 can be conditioned by the matrices $M_{j,0}$. In particular, $\varphi(t)$ approaches zero as $t \longrightarrow \infty$, irrespective of u_0, provided that all the real parts of λ_j are negative. Similar results will also be obtained for the nonhomogeneous linear system

$$
\frac{d}{dt}u(t) = Au + g(t), \quad t > 0,
$$
$$
u(0) = u_0 \in \mathbb{C}^n,
\tag{1.5}
$$

to which the unique solution, by Chapter 2, is

$$
\varphi(t) = e^{tA}u_0 + \int_{\tau=0}^{t} e^{(t-\tau)A}g(\tau)\, d\tau.
\tag{1.6}
$$

In this case, the long time behavior of the solution $\varphi(t)$ also depends on $g(t)$. Here the vector-valued function $g(t)$ is continuous.

The other goal is to study the perturbed system of (1.1) by a small nonlinear term $g(t, u)$:

$$
\frac{d}{dt}u(t) = Au + g(t, u), \quad t > 0,
$$
$$
u(0) = u_0 \in \mathbb{C}^n,
\tag{1.7}
$$

where $g(t, u)$ is small in the sense that

$$
\lim_{|u| \to 0} \frac{|g(t, u)|}{|u|} = 0
\tag{1.8}
$$

uniformly for all $t \geq 0$. Here

DEFINITION 1.1. *The norm $|x|$ of a vector* $x = \begin{pmatrix} x_1 \\ x_2 \\ \vdots \\ x_n \end{pmatrix} \in \mathbb{C}^n$ *is defined by*

$$
|x| \equiv \sum_{i=1}^{n} |x_i|,
$$

where $|x_i|$ is the sum of the absolute values of the real part and the imaginary part of x_i. Similarly, the norm $|B|$ of a matrix $B = (b_{jk})_{n \times n}$ is defined by

$$|B| \equiv \sum_{j,k=1}^{n} |b_{jk}|.$$

Under the additional assumption that

(H1): $g(t, u)$ is continuous for small $|u|$ and $t \geq 0$,

Chapter 8 shows that (1.7) has a solution $\varphi(t)$ on a small t interval $[0, \beta]$, some $\beta > 0$. Here it will be shown that $\varphi(t)$ extends to exist even for all $t \geq 0$, and that not only $\varphi(t)$ but also other solutions that exist for all $t \geq 0$, approach zero as $t \longrightarrow \infty$, provided that all the eigenvalues of A have negative real parts. Additional aspects of long time behavior of solutions that exist on $[0, \infty)$ or $[T, \infty)$, some $T \geq 0$, are also examined. Here (H1) means

(H2):   There is a small $\delta > 0$, such that $g(t, u)$ is continuous on the region

$$\{(t, u) : t \geq 0 \quad \text{and} \quad 0 \leq |u| \leq \delta\}.$$

The rest of this chapter is organized as follows. Section 2 contains main results for the homogeneous linear system (1.1) and the nonhomogeneous linear system (1.5), and Section 3 contains those for the nonlinear system (1.7). Sections 4 and 5 illustrates, by examples, the main results for the linear system and the nonlinear system, respectively, and Sections 6 and 7 proves the respective main results. Finally, Section 8 presents a problems set, and solutions of the problems set are placed in Section 9.

2. Main Results for Linear Systems

Here are results in connection with the homogeneous linear system (1.1). To state those results, some definitions are required.

DEFINITION 2.1. *If M is an $n \times n$ matrix, $S(M) = S(\{u_j, j = 1, 2, \ldots, n\})$ is the subspace spanned by the column vectors $u_j, j = 1, \ldots, n$, in M. If M_1 and M_2 are two $n \times n$ matrices, $S(M_1 \cup M_2) = S(\{u_j, j = 1, \ldots, n; v_k, k = 1, \ldots, n\})$ is the subspace spanned by the column vectors $u_j, j = 1, \ldots, n$, and $v_k, k = 1, \ldots, n$, in the matrices M_1 amd M_2, respectively.*

DEFINITION 2.2. *A complex number λ is an eigenvalue of A if*

$$(A - \lambda I)x = 0$$

for some nonzero vector x in $\mathbb{C}^n$. In this case, x is an eigenvector of A, corresponding to λ, and the subspace $E(\lambda)$ consisting of all such x's is the eigenspace corresponding to λ.

Here note that $E(\lambda)$ equals the space spanned by the eigenvectors of λ.

DEFINITION 2.3. *The subspace*

$$K_\lambda \equiv \{x \in \mathbb{C}^n : (A - \lambda I)^\nu x = 0 \quad \text{but } (A - \lambda I)^{\nu-1} x \neq 0$$

$$\text{for some natural number } \nu \text{ and some nonzero vector } x\}$$

is the generalized eigenspace, corresponding to the generalized eigenvector λ. λ is an eigenvalue if $\nu = 1$.

In the case of (1.3), we have from linear algebra [**9**] that, for each $j = 1, 2, \ldots, s$,

$$K_{\lambda_j} = \{x \in \mathbb{C}^n : (A - \lambda_j I)^{m_j} x = 0\}$$
$$= \{x \in \mathbb{C}^n : (A - \lambda_j I)^{k-1} x \neq 0 \quad \text{but} \quad (A - \lambda_j I)^k x = 0, k = 1, 2, \ldots, m_j\}. \tag{2.1}$$

DEFINITION 2.4. *The difference $U \setminus V$ of set U from set V is defined by*

$$U \setminus V \equiv \{x : x \quad \text{is in } U \text{ but not in } V\}.$$

DEFINITION 2.5. *Let M be a set of vectors in $\mathbb{C}^n$. M is invariant under e^{tA} if*

$$e^{tA} M \subset M.$$

THEOREM 2.6. *The unique solution $\varphi(t) = e^{tA} u_0$ of (1.1) approaches zero as $t \longrightarrow \infty$, if u_0 is in $S(M_{j,0})$ for some $j \in \{1, 2, \ldots, s\}$, where the corresponding generalized eigenvalue λ_j has negative real part; conversely, the norm $|\varphi(t)|$ of the unique solution $\varphi(t)$ approaches $+\infty$ if u_0 is additionally nonzero and the corresponding generalized eigenvalue λ_j has positive real part; moreover, if the generalized eigenvalue λ_j has zero real part, then the unique solution $\varphi(t)$ either lies on a sphere for all $t \geq 0$ or its norm $|\varphi(t)|$ approaches $+\infty$, according as $0 \neq u_0 \in E(\lambda_j)$ or $0 \neq u_0 \in S(M_{j,0}) \setminus E(\lambda_j)$.*

Finally, the $\varphi(t) = e^{tA} u_0$ approaches zero for all u_0, as $t \longrightarrow \infty$, if all the generalized eigenvalues λ_j of A have negative real parts; conversely, if all λ_j's have positive real parts, then $|u(t)|$ approaches $+\infty$ for all u_0.

REMARK 2.7. *Since $S(\cup_{j=1}^s M_{j,0})$ equals the whole space $\mathbb{C}^n$ (see Lemma 6.1 in Section 6), u_0 equals some linear combination of the column vectors in the matrices $M_{j,0}, j = 1, 2, \ldots, s$. Thus if $0 \neq u_0$ has a component being some combination of the column vectors in some $M_{j,0}$, for which the corresponding eigenvalue λ_j has positive real part, then the norm $|\varphi(t)|$ of the unique solution $\varphi(t)$ approaches $+\infty$, by Theorem 2.6. This is particularly true, if u_0 is any nonzero column vector in the matrices*

$$M_{j,k} : \quad k = 1, 2, \ldots, m_j - 1,$$

as the columns of those matrices are in

$$K_{\lambda_j} (\supset E(\lambda_j)) = S(M_{j,0})$$

by Propositions 6.2 and 6.4 in Section 6. Similarly, for the case where λ_j has zero real part, the norm $|\varphi(t)|$ also approaches $+\infty$ by Theorem 2.6, if u_0 is a nonzero, non-eigenvalue column vector in the matrices

$$M_{j,k} : \quad k = 0, 1, \ldots, m_j - 2.$$

THEOREM 2.8. *Each subspace,*

$$S(M_{j,0}), \quad E(\lambda_j), \quad j = 1, 2, \ldots, s,$$

is invariant under e^{tA}.

COROLLARY 2.9. *The subspace M_0, spanned by the column vectors in the matrices*

$$M_{j,0}, j = 1, 2, \ldots, s,$$

for which the corresponding eigenvalues λ_j have negative real parts, is invariant under e^{tA}. Similarly, if the corresponding eigenvalues λ_j have positive real parts, then the spanned subspace M_∞ by the column vectors in the matrices

$$M_{j,0}, j = 1, 2, \ldots, s,$$

is also invariant under e^{tA}. Likewise, the subspace M_b, spanned by the eigenvectors in the eigenspaces

$$E(\lambda_j), j = 1, 2, \ldots, s,$$

in which the eigenvalues λ_j have zero real parts, is also invariant under e^{tA}.

PROOF. If $x \in M_0$, then, for each $j \in \{1, 2, \ldots, s\}$ with the corresponding eigenvalue λ_j having negative real part, there exist n complex numbers $\alpha_{j,k}, k = 1, 2, \ldots, s$, such that

$$x = \sum_{j=1}^{s} \left(\sum_{k=1}^{n} \alpha_{j,k} x_{j,k} \right),$$

where $x_{j,k}$'s are the n column vectors of the matrix $M_{j,0}$. Since

$$e^{tA} \left(\sum_{k=1}^{n} \alpha_{j,k} x_{j,k} \right) \in S(M_{j,0})$$

by Theorem 2.8, we have $x \in M_0$, and so M_0 is invariant under e^{tA}.

The cases for M_∞ and M_b, respectively, are similarly done. □

COROLLARY 2.10. *$e^{tA} u_0$ approaches 0 or $|e^{tA} u_0|$ approaches $+\infty$ according as $u_0 \in M_0$ or $0 \neq u_0 \in M_\infty \cup \triangle_\infty$. Here*

$$\triangle_\infty \equiv S(\cup_{j \in J} M_{j,0}) \setminus M_b,$$

where J is the set of the indices $j = 1, 2, \ldots, s$, for which the corresponding eigenvalues have zero real parts.

Moreover, if $u_0 \in M_b$, then $e^{tA} u_0$ is uniformly bounded for all $t \geq 0$ but $|e^{tA} u_0|$ does not approach zero as $t \longrightarrow \infty$.

PROOF. This follows easily from Theorem 2.6. □

DEFINITION 2.11. *M_0 and M_b are called the stable subspace with zero asymptote and the stable subspace with nonzero asymptote, respectively for (1.1). Conversely, M_∞ and $\triangle_\infty$ are called the unstable subspace and the unstable set, respectively for (1.1).*

Next are results in connection with the nonhomogeneous linear system (1.5).

REMARK 2.12. *These results and those above for the homogeneous linear system (1.1) are expressed in terms of the quantities $M_{j,0}$'s, in that this seems new.*

We first consider the case where the function $g(t)$ in (1.5) is constant in t.

THEOREM 2.13. *Let the function $g(t)$ in (1.5) be a constant g_0. Then the unique solution $\varphi(t)$ of (1.5) approaches a constant, as $t \longrightarrow \infty$:*

$$\begin{cases} \sum_{k=0}^{\mu-1} (-1)^{k+1} \lambda_j^{-(k+1)} (A - \lambda_j I)^k g_0, & \text{for some } 1 \leq \mu \leq m_j \text{ and for any } A; \\ -A^{-1} g_0, & \text{if, particularly, } A^{-1} \text{ exists,} \end{cases}$$

provided that $u_0 \in S(M_{i,0})$ and $g_0 \in S(M_{j,0})$ for some $i, j \in \{1, 2, \ldots, s\}$, where the corresponding generalized eigenvalues λ_i and λ_j have negative real parts. Here the μ is determined by the condition:

$$(A - \lambda_j I)^\mu g_0 = 0 \quad but \quad (A - \lambda_j I)^{\mu-1} g_0 \neq 0,$$

which is valid by Proposition 6.4.

Conversely, if $0 \neq u_0$ is in some $S(M_{i,0})$, where the generalized eigenvalue λ_i has positive real part, or else if $0 \neq g_0$ is in some $S(M_{j,0})$, where the generalized eigenvalue λ_j has positive real part, then the norm $|\varphi(t)|$ of the unique solution $\varphi(t)$ approaches $+\infty$, except for the case where $i = j$, u_0 and g_0 are in $E(\lambda_j)$, and $u_0 = -\frac{1}{\lambda_j} g_0$; in that case, it is true that

$$\varphi(t) = -\frac{1}{\lambda_j} g_0$$

for all t.

Moreover, if both the generalized eigenvalues λ_i and λ_j have zero real parts, then the unique solution $\varphi(t)$ either lies inside a sphere for all $t \geq 0$ or its norm $|\varphi(t)|$ approaches $+\infty$, according as the condition $(C1)$ is true or the condition $(C2)$ is true. Here

$(C1)$:   *u_0 is in $E(\lambda_i)$ and either $g_0 = 0$ or else $0 \neq g_0$ is in $E(\lambda_j)$ with $\lambda_j \neq 0$;*

$(C2)$:   *$0 \neq u_0$ is in $S(M_{i,0}) \setminus E(\lambda_i)$, or else $0 \neq g_0$ is in $E(\lambda_j)$ with $\lambda_j = 0$, or else $0 \neq g_0$ is in $S(M_{j,0}) \setminus E(\lambda_j)$.*

Finally, the solution $\varphi(t)$ approaches a constant for all u_0 and g_0, as $t \longrightarrow \infty$, if all the generalized eigenvalues λ_j of A have negative real parts.

We now consider the case where the function $g(t)$ in (1.5) is not necessarily a constant.

THEOREM 2.14. *The unique solution $\varphi(t)$ of (1.5) approaches a constant L_0, as $t \longrightarrow \infty$, provided that $u_0 \in S(M_{i,0})$ and $g(t) \in S(M_{j,0})$ for some $i, j \in \{1, 2, \ldots, s\}$, where the corresponding generalized eigenvalues λ_i and λ_j have negative real parts and $g(t)$ satisfies*

$(H3)$:   *$|g(t)| \leq K_0 e^{t\nu}$*

for some constants $K_0 \geq 0$ and $\nu \leq 0$. The constant L_0 is zero if $\nu < 0$.

Conversely, assume that $0 \neq u_0$ is in some $S(M_{i,0})$, where the generalized eigenvalue λ_i has positive real part; or else assume that $0 \neq g(t)$ is in some $S(M_{j,0})$, where the generalized eigenvalue λ_j has positive real part, and the condition $(H4)$ is true:

$(H4)$:   *$g(t)$ equals $c(t)g_0$, whose g_0 is in $S(M_{j,0})$ and whose $c(t)$, for some $1 \leq \mu \leq m_j$ and some $c_0 > 0$, satisfies:*

$$\lim_{t \to +\infty} \left| \sum_{k=0}^{\mu-1} \int_{\tau=0}^{t} e^{(t-\tau)\lambda_j} (t-\tau)^k c(\tau) \, d\tau \frac{(A - \lambda_j I)^k}{k!} g_0 \right| = +\infty;$$

Here the constant μ is determined by the condition:

$$(A - \lambda_j I)^\mu g_0 = 0 \quad but \quad (A - \lambda_j I)^{\mu-1} g_0 \neq 0,$$

which is valid by Proposition 6.4. Then the norm $|\varphi(t)|$ of the unique solution $\varphi(t)$ approaches $+\infty$, except for the case where $i = j$ and u_0 and g_0 are in $E(\lambda_j)$. In *that case,*

$$\varphi(t) = e^{t\lambda_j} u_0 + \int_{\tau=0}^{t} e^{(t-\tau)\lambda_j} c(\tau) g_0 \, d\tau.$$

In particular,

$$\varphi(t) = e^{t\lambda_j} u_0 +$$

$$e^{t\lambda_j} \frac{1}{(\alpha+1)\lambda_j} c_0 g_0 [1 - e^{-(\alpha+1)t\lambda_j}]$$

if $c(t) = e^{-\alpha\lambda_j t} c_0$ *where* c_0 *and* α *are constants and* $\alpha \neq -1$, *and such a* $\varphi(t)$

$$\begin{cases} \text{equals } e^{-\alpha t\lambda_j} u_0, & \text{if } u_0 = -\frac{1}{(\alpha+1)\lambda_j} c_0 g_0; \\ \text{has its norm } |\varphi(t)| \longrightarrow +\infty, & \text{otherwise;} \end{cases}$$

however, if α *equals* -1, $\varphi(t)$ *has its* $|\varphi(t)|$ *convergent to* $+\infty$.

 Moreover, if both the generalized eigenvalues λ_i *and* λ_j *have zero real parts, then the unique solution* $\varphi(t)$ *either lies inside a sphere for all* $t \geq 0$ *or its norm* $|\varphi(t)|$ *approaches* $+\infty$, *according as the condition* (C3) *is true or the condition* (C4) *is true. Here*

 (C3) : u_0 *is in* $E(\lambda_i)$ *and either* $g(t) \equiv 0$ *or else* $g(t) = c(t)g_0$, *whose* g_0 *is in* $E(\lambda_j)$ *with* $\lambda_j \neq 0$ *and whose* $c(t)$ *satisfies* $|\int_{\tau=0}^{t} e^{(t-\tau)\lambda_j} c(\tau) \, d\tau| < \infty$;

 (C4) : $0 \neq u_0$ *is in* $S(M_{i,0}) \setminus E(\lambda_i)$, *or else* $g(t) = c(t)g_0$, *whose* $g_0 \neq 0$ *is in* $E(\lambda_j)$ *with* $\lambda_j = 0$ *and whose* $c(t)$ *satisfies*

$$\lim_{t\to\infty} \left| \int_{\tau=0}^{t} c(\tau) \, d\tau \right| = +\infty;$$

or else $g(t)$ *satisfies* (H4) *but* $0 \neq g_0$ *is in* $S(M_{j,0}) \setminus E(\lambda_j)$.

 Finally, the solution $\varphi(t)$ *approaches a constant* L_1 *for all* u_0 *and for all* $g(t)$ *satisfying* (H3), *as* $t \longrightarrow \infty$, *if all the generalized eigenvalues* λ_j *of* A *have negative real parts. The constant* L_1 *is zero if* $\nu < 0$.

 PROPOSITION 2.15. *The function* $c(t)$ *in* $H(4)$, *where the generalized eigenvalue* λ_j *has positive real part, can be some linear combination or some product, of polynomial functions, sine functions, cosine function, and exponential functions. Furthermore,* $c(t)$ *can be a function, such that, for some constants* $c_0 > 0$ *and* $t_0 > 0$,

$$c(t) \geq c_0 > 0 \quad \text{for all } t \geq t_0,$$

provided that the generalized eigenvalue λ_j *is real and positive.*

 PROOF. When λ_j is real and positive and

$$c(t) \geq c_0 > 0 \quad \text{for all } t \geq t_0,$$

we have, for $t > t_0$,

$$\left| \int_{\tau=t_0}^{t} e^{(t-\tau)\lambda_j} (t-\tau)^k c(\tau) \, d\tau \right| = \left| \int_{\tau=0}^{t-t_0} e^{\tau\lambda_j} \tau^k c(t-\tau) \, d\tau \right|$$

$$\geq \int_{\tau=0}^{t-t_0} e^{\tau\lambda_j} \tau^k c_0 \, d\tau.$$

Since, for t large, the dominant term in

$$\int_{\tau=0}^{t-t_0} e^{\tau\lambda_j} \tau^k \, d\tau, \quad k = 1, 2, \dots, \mu - 1,$$

by (6.3) in Section 6, is the term with $k = \mu - 1$ and this dominant term, by (6.3), has its absolute value approaching $+\infty$ as $t \longrightarrow +\infty$, we have $c(t)$ satisfies $(H4)$.

On the other hand, since, for $p = \{0\} \cup \mathbb{N}$ and $\lambda_j \neq \nu \in \mathbb{C}$,

$$\int_{\tau=0}^{t} e^{(t-\tau)\lambda_j} (t-\tau)^k \tau^p e^{\tau\nu} \, d\tau$$

$$= \int_{\tau=0}^{t} e^{\tau\lambda_j} \tau^k (t-\tau)^p e^{(t-\tau)\nu} \, d\tau$$

$$= \sum_{m=0}^{p} \binom{p}{m} t^{p-m} e^{t\nu} (-1)^m \int_{\tau=0}^{t} e^{\tau(\lambda_j - \nu)} \tau^{k+m} \, d\tau,$$

we have that, by (6.3) in Section 6 again, the case is proved where λ_j has positive real part and $c(t)$ is some linear combination or some product, of polynomial functions, sine and cosine functions, and exponential functions. Here we used the fact that, for $\mu \in \mathbb{R}$, $\cos(\mu\tau)$ and $\sin(\mu\tau)$ equals the real part and the imaginary part, respectively, of $e^{i\mu\tau}$, the fact that

$$(a+b)^p = \sum_{m=0}^{p} \binom{p}{m} a^{p-m} b^m,$$

and the fact that the dominant term is the term with $k = \mu - 1$. However, when $\nu = \lambda_j$, the case is trivial.   □

3. Main Results for Nonlinear Systems

Here are results in connection with the nonlinear system (1.7). These results are based on [5], and their proofs modify those in [5] by using the quantities $\{M_{j,k}\}, j = 1, \ldots, s; k = 0, 1, \ldots, m_j - 1$.

THEOREM 3.1. *Let (1.8) and $(H1)$ hold, and let all the generalized eigenvalues of A have negative real parts. Then the initial value problem (1.7) with $|u_0|$ sufficiently small, has a solution $\varphi(t)$ on $[0, \infty)$, and this $\varphi(t)$, together with other solutions that also exist on $[0, \infty)$, approach zero as $t \longrightarrow +\infty$.*

REMARK 3.2. *(1.7) contains systems of the form*

$$\frac{d}{dt} v(t) = h(v), \quad t > 0,$$

where $h(x)$ satisfies $h(v_0) = 0$ and is differentiable at $x = v_0$. This can be easily seen by the substitution $u \equiv v - v_0$ and by the definition of differentiability [2, Page 346], in which the resulting new system is of the form

$$\frac{d}{dt} u = Dh(v_0)u + \tilde{h}(u).$$

Here the Jacobian matrix $Dh(x) = \begin{pmatrix} (h_1)_{x_1} & (h_1)_{x_2} & \cdots & (h_1)_{x_n} \\ & \vdots & & \\ (h_n)_{x_1} & (h_n)_{x_2} & \cdots & (h_n)_{x_n} \end{pmatrix}$ is the total derivative of $h(x)$ at x, $\tilde{h}(u)$ satisfies (1.8), and $(h_1)_{x_1}$ is the partial derivative of $h_1(x)$ with respective to the variable x_1. A sufficient condition for $h(x) = \begin{pmatrix} h_1 \\ \vdots \\ h_n \end{pmatrix}$

to be differentiable at $x = v_0$ is that all the partial derivatives $\frac{\partial}{\partial x_k} h_i(x)$ exist in the neighborhood of v_0 and are continuous at v_0, [2, Page 357].

Theorem 3.1 is still true if (1.8) is replaced by a weaker condition:

THEOREM 3.3. *Theorem 3.1 still holds if (1.8) is replaced by (A1) and (A2). Here*

$(A1):$ *There is a $k_0 > 0$, such that*

$$|g(t, u)| \leq k_0|u|$$

for all sufficiently small $|u|$ and for all $t \geq 0$.

$(A2):$ *Given $\epsilon > 0$, there are $\delta > 0$ and $T \geq 0$, such that*

$$|g(t, u)| \leq \epsilon|u|$$

if $|u| \leq \delta$ and $t \geq T$.

REMARK 3.4. *(A1) and (A2) are met if*

$$g(t, u) = B(t)u + f(t, u),$$

where $f(t, u)$ satisfies (1.8) and the matrix

$$B(t) \longrightarrow 0 \quad as \ t \longrightarrow \infty.$$

However,

THEOREM 3.5. *Let (A2) hold, and let A have at least one generalized eigenvalue with positive real part. Then not every solution of the initial value problem*

$$\frac{d}{dt}u = Au + g(t, u), \quad t > T, \tag{3.1}$$
$$u(T) = u_0,$$

with $|u(T)|$ sufficiently small, that exists on $[T, \infty)$, stays in the vicinity of the origin for all $t \geq T$, that is, there is an $\eta > 0$, such that those of solutions to (3.1) with $u(T)$ sufficiently small, that exist on $[T, \infty)$, possess one solution $\varphi(t)$ satisfying

$$|\varphi(t_1)| > \eta \quad for \ some \ t_1 \geq T.$$

Here the T in (A2) depends on the given ϵ.

REMARK 3.6. *It can be seen from the proof of Theorem 3.5 that $u(t) = \sum_{i=1}^{n} \varphi_i(t)d_i$ with $R(T) = 2\rho(T)$ or more generally, $R(T) > \rho(T)$, cannot stay in the vicinity of the origin, if it is a solution of the equation (3.1) with the initial value $\|\varphi(T)\|$ sufficiently small. Here*

$$\varphi(t) = \begin{pmatrix} \varphi_1(t) \\ \vdots \\ \varphi_n(t) \end{pmatrix},$$
$$R(T) = [\varphi_1^2(T) + \varphi_2^2(T) + \cdots + \varphi_m^2(T)]^{\frac{1}{2}},$$
$$\rho(T) = [\varphi_{m+1}^2(T) + \varphi_{m+2}^2(T) + \cdots + \varphi_n^2(T)]^{\frac{1}{2}},$$
$$\|\varphi(T)\| = \sqrt{R^2(T) + \rho^2(T)},$$

A has, with multiplicities counted, exactly m generalized eigenvalues whose real parts are positive, and $\{d_i\}_{i=1}^n$ is a special Jordan basis of $\mathbb{C}^n$, to which related are

$$\min_{j=1,\ldots,s_0} \{Re(\lambda_j)\} > \sigma \quad and \quad \max_{i=2,\ldots,n} \{\gamma_i\} < \frac{\sigma}{20}.$$

The details of these quantities are referred to the proof of Theorem 3.5.

In spite of Theorem 3.5, solutions of (1.7) for $|u_0|$ sufficiently small and u_0 restricted within a particular set, still approach zero as $t \longrightarrow +\infty$, provided that further assumptions are imposed. In order to state this result, we need to make some preparations.

First of all, we single out the equation in (1.7):

$$\frac{d}{dt}u = Au + g(t,u), \quad t > 0. \tag{3.2}$$

Secondly, we recall: Let $\{e_1, e_2, \ldots, e_n\}$ be the standard basis for $\mathbb{C}^n$, that is, let e_j be the column vector whose components are all zero, except for the j-th component which equals 1. Let $\{d_j\}_{j=1}^n$ be another basis for $\mathbb{C}^n$, and for $x \in \mathbb{C}^n$, let

$$x = \sum_{j=1}^n \alpha_j e_j \quad and$$

$$x = \sum_{i=1}^n \beta_i d_i,$$

relative to basis $\{e_j\}_{j=1}^n$ and $\{d_i\}_{i=1}^n$, respectively. Then it follows from linear algebra [9] that

$$d_i = \sum_{j=1}^n p_{ji} e_j \tag{3.3}$$

for some unique constants p_{ji}, and so

$$x = \sum_{j=1}^n \alpha_j e_j = \sum_{j=1}^n (\sum_{i=1}^n p_{ji}\beta_i) e_j.$$

Thus $\alpha_j = \sum_{i=1}^n p_{ji}\beta_i$, that is,

$$\begin{pmatrix} \alpha_1 \\ \vdots \\ \alpha_n \end{pmatrix} = P \begin{pmatrix} \beta_1 \\ \vdots \\ \beta_n \end{pmatrix} \tag{3.4}$$

for the invertible matrix $P = \begin{pmatrix} p_{11} & \cdots & p_{1n} \\ \vdots & & \\ p_{n1} & \cdots & p_{nn} \end{pmatrix}$.

Thirdly, we define:

DEFINITION 3.7. *A set S of points in $\mathbb{C}^n$, in a restricted sense, is an $m(\in \mathbb{N})$-dimensional manifold (or surface), relative to a basis $\{d_j\}_{j=1}^n$ in $\mathbb{C}^n$, if there are continuous functions $\Psi_j, j = m+1, m+2, \ldots, n$, such that a point $x \in S$ taking the form $x = \sum_{j=1}^n \beta_j d_j$, satisfies*

$$\beta_j = \Psi_j(\beta_1, \beta_2, \ldots, \beta_m), \quad j = m+1, \ldots, n.$$

If the partial derivatives of Ψ_j exist, then , for each $j = m+1, \ldots, n$,

$$\frac{\partial}{\partial \beta_j} x$$

is the tangent vector to the surface S at the point x and along the direction d_j. Those tangent vectors

$$\frac{\partial}{\partial d_j} x, \quad j = m+1, \ldots, n$$

span the tangent plane to the S at x.

Thus, if the above x takes the form

$$x = \sum_{j=1}^{n} x_j e_j, \quad \text{relative to the standard basis } \{e_j\}_{j=1}^{n},$$

then the above m-dimensional manifold (or surface) S can also be defined by:

$$x = \begin{pmatrix} x_1 \\ \vdots \\ x_m \\ \vdots x_n \end{pmatrix} = P \begin{pmatrix} \beta_1 \\ \vdots \\ \beta_m \\ \Psi_{m+1}(\beta_1, \ldots, \beta_m) \\ \vdots \\ \Psi_n(\beta_1, \ldots, \beta_m) \end{pmatrix},$$

in terms of the curvilinear coordinates $\beta_1, \ldots, \beta_m$.

Finally, a further assumption needed is:

For each $\epsilon > 0$, let there exist constants $\delta > 0$ and $T \geq 0$, such that the function

$$g(t, x) = \begin{pmatrix} g_1(t, x) \\ \vdots \\ g_n(t, x) \end{pmatrix} = \begin{pmatrix} g_1 \\ \vdots \\ g_n \end{pmatrix}$$

satisfies:

$$|g(t, \tilde{x}) - g(t, x)| \leq \epsilon |\tilde{x} - x| \tag{3.5}$$

for all $t \geq T$ and for $|\tilde{x}|, |x| \leq \delta$.

REMARK 3.8. *A sufficient condition for (3.5) to hold is that the Jacobian matrix*

$$g_x(t, x) = \begin{pmatrix} (g_1)_{x_1} & (g_1)_{x_2} & \cdots & (g_1)_{x_n} \\ \vdots & & & \\ (g_n)_{x_1} & (g_n)_{x_2} & \cdots & (g_n)_{x_n} \end{pmatrix}$$

exists and satisfies:

$$\lim_{|x| \to 0, t \to +\infty} |g_x(t, x)| = 0.$$

Now,

THEOREM 3.9. *Let $(H1)$ and (3.5) hold, and let $g(t, 0) \equiv 0$ for $t \geq 0$. Further, let*

$$m = m_1 + m_2 + \cdots + m_{s_0}$$

eigenvalues λ_j's with multiplicity m_j's, where $j = 1, 2, \ldots, s_0 < s$, of A, have negative real parts, and let the remaining $(n - m)$ eigenvalues have positive real parts. Then, for any large enough t_0, there exists an m-dimensional manifold (or surface) S containing the origin, such that a unique solution $\varphi(t) = \varphi(t, t_0; \tilde{s})$ of (3.2) with the initial value $\varphi(t_0) = \tilde{s}$ on S and sufficiently near the origin, exists and satisfies:

$$\varphi(t) \to 0 \quad as\ t \to +\infty.$$

This solution $\varphi(t) = \varphi(t, t_0; \tilde{s})$ also lies on S for $t \geq t_0$, and the manifold (or surface) S is, at the origin, tangent to the subspace M_0 spanned by the column vectors in the matrices:

$$M_{j,0}, \quad j = 1, 2, \ldots, s_0.$$

Moreover, $\varphi(t, t_0; \tilde{s})$ is also the unique solution of (3.2) with the initial vaue $\varphi(\tau, t_0; \tilde{s})$ at $t = \tau \geq t_0$, and, by defining $V(t, \tau)\varphi(\tau, t_0; \tilde{s})$ to be this solution, it follows that $V(t, \tau), t \geq \tau \geq t_0$, is a family of exponentially decaying evolution operators:

$$V(t, \tau) : D \equiv \{\varphi(\rho, t_0; \tilde{s}) : \rho \geq t_0, \tilde{s} \in S\} \to D,$$
$$V(\tau, \tau)\varphi(\tau, t_0; \tilde{s}) = \varphi(\tau, t_0; \tilde{s}),$$
$$V(t, \tau)V(\tau, r)\varphi(r, t_0; \tilde{s}) = V(t, r)\varphi(r, t_0; \tilde{s}) \quad for\ t \geq \tau \geq r \geq t_0,$$
$$|V(t, \tau)\varphi(\tau, t_0; \tilde{s})| \leq Ke^{-k_0(t-t_0)} \quad for\ some\ positive\ constants\ K, k_0.$$

Finally, no solution $\varphi(t)$ of (3.2) with $\varphi(t_0)$ sufficiently near the origin but not on S, can stay in the vicinity of the origin for all $t \geq t_0$. That is, there is an $\eta > 0$, such that, if $\varphi(t)$ is a solution of (3.2) with $\varphi(t_0)$ sufficiently near the origin but not on S, then

$$|\varphi(t_1)| > \eta \quad for\ some\ t_1 \geq t_0.$$

It will follow from the proof of Theorem 3.9 that

THEOREM 3.10. *Let the assumptions in Theorem 3.9 hold. That is: let $H(1)$ and (3.5) hold, and let $g(t, 0) \equiv 0$ for $t \geq 0$; further, let*

$$m = m_1 + m_2 + \cdots + m_{s_0}$$

eigenvalues λ_j's with multiplicity m_j's, where $j = 1, 2, \ldots, s_0 < s$, of A, have negative real parts, and let the remaining $(n - m)$ eigenvalues have positive real parts.

Then, for any large enough t_0, there exists, for each $j = 1, 2, \ldots, s_0$, an m_j-dimensional manifold (or surface) S_j containing the origin, such that a unique solution $\varphi(t) = \varphi(t, t_0; \tilde{s})$ of (3.2) with the initial value $\varphi(t_0) = \tilde{s}$ on S_j and sufficiently near the origin, exists and satisfies:

$$\varphi(t) \to 0 \quad as\ t \to +\infty.$$

The manifold (or surface) S_j equals

$$S \cap S(M_{j,0} \cup (\cup_{k=s_0+1}^{s} M_{k,0})),$$

where S is the manifold (or surface) S in Theorem 3.9 and $S(M_{j,0} \cup (\cup_{k=s_0+1}^{s} M_{j,0}))$ is the subspace spanned by the column vectors in the matrices:

$$M_{k,0}, \quad k = j, s_0 + 1, s_0 + 2, \ldots, s.$$

Moreover, S_j is, at the origin, tangent to the subspace $S(M_{j,0})$.

THEOREM 3.11. *Following Theorem 3.10, if the function $g(t, u)$ in (3.2) has no components in the subspace $S(\cup_{k=s_0+1}^s M_{k,0})$ spanned by the column vectors in the matrices*

$$M_{k,0}, \quad k = s_0 + 1, s_0 + 2, \ldots, s,$$

then

$$S_j = S(M_{j,0}) \quad \text{for each } j = 1, \ldots, s_0;$$
$$S = M_0, \quad \text{the } M_0 \text{ in Theorem 3.9.}$$

If $g(t, u)$ has a nozero component in $S(\cup_{j=s_0+1,\ldots,s} M_{j,0})$ but this component is zero for $u \neq 0$ in some $S(M_{j_0,0})$ where $j_0 \in \{1, \ldots, s_0\}$, then nevertheless,

$$S_{j_0} = S(M_{j_0,0}).$$

If $S_{j_0} \neq S(M_{j_0,0})$ for some $j_0 \in \{1, 2, \ldots, s_0\}$, then there is a $\tilde{s}$ in $S(M_{j_0,0})$ that is sufficiently near the origin, such that no solution $\varphi(t)$ of (3.2) with $\varphi(t_0) = \tilde{s}$ can stay in the vicinity of the origin for all $t \geq t_0$.

REMARK 3.12. *However, if $g(t,u)$ has a nonzero component in $S(\cup_{j=s_0+1,\ldots,s} M_{j,0})$ and this component is still nonzero so long as u has a nonzero component in some $S(M_{j_0,0})$ where $j \in \{1, \ldots, s_0\}$, then it can be conjectured that*

$$S_{j_0} \neq S(M_{j_0,0}) \quad \text{and so } S \neq M_0.$$

REMARK 3.13. *The above Theorems 3.10 and 3.11 seem new.*

4. Examples for Linear Systems

Before we prove the main results, we look at some examples. These examples study the long time behavior, as $t \longrightarrow +\infty$, of the solutions of the given initial value problems. They also find the corresponding stable subspaces M_0's with zero asymptotes, the stable subspaces M_b's with nonzero asymptotes, the unstable subspaces M_∞'s, and the unstable subsets $\triangle_\infty$'s.

Here are examples related to the homogeneous system (1.1).

EXAMPLE 4.1.

$$\frac{d}{dt} u = Au, \quad t > 0,$$
$$u(0) = u_0, u_1, u_2, u_3, u_4, u_5, \text{and } u_6, \text{respectively,}$$

where

$$u_0 = \alpha_0 \begin{pmatrix} 12 \\ 4 \\ 3 \end{pmatrix}, \quad u_1 = \alpha_1 \begin{pmatrix} 6 \\ 4 \\ 6 \end{pmatrix}, \quad u_2 = \alpha_2 \begin{pmatrix} 18 \\ -4 \\ 2 \end{pmatrix},$$
$$u_3 = u_0 + u_1, \quad u_5 = u_0 + u_2, \quad u_6 = u_1 + u_2,$$
$$u_7 = u_0 + u_1 + u_2,$$

α_0, α_1, and α_2 are constants, and

$$A = \begin{pmatrix} -3 & 9 & 0 \\ 0 & -3 & 4 \\ -1 & 0 & 4 \end{pmatrix}.$$

Solution. The eigenvalues of A, by (1.3), are $0, 3$, and -5. Using (1.2) and (1.4), or equivalently, using (1.2), the formula $\frac{d}{dt}e^{tA} = Ae^{tA}$, and letting $t = 0$, we have

$$e^{tA} = M_{1,0} + e^{3t}M_{2,0} + e^{-5t}M_{3,0},$$
$$Ae^{tA} = 3e^{3t}M_{2,0} - 5e^{5t}M_{3,0},$$
$$A^2e^{tA} = 9e^{3t}M_{2,0} + 25e^{-5t}M_{3,0},$$

and

$$I = M_{1,0} + M_{2,0} + M_{3,0},$$
$$A = 3M_{2,0} - 5M_{3,0},$$
$$A^2 = 9M_{2,0} + 25M_{3,0},$$

where $M_{j,0}, j = 1, 2, 3$, are solved as:

$$M_{1,0} = \frac{1}{15}\begin{pmatrix} 12 & 36 & -36 \\ 4 & 12 & -12 \\ 3 & 9 & -9 \end{pmatrix},$$

$$M_{2,0} = \frac{1}{24}\begin{pmatrix} -6 & -9 & 36 \\ -4 & -6 & 24 \\ -6 & -9 & 36 \end{pmatrix},$$

$$M_{3,0} = \frac{1}{40}\begin{pmatrix} 18 & -81 & 36 \\ -4 & 18 & -8 \\ 2 & -9 & 4 \end{pmatrix}.$$

Thus, it follows from Theorem 2.6 that, as $t \longrightarrow +\infty$,

$$\begin{cases} \text{the solution } u(t) \text{ lies on a sphere,} & \text{if } u(0) = u_0; \\ |u(t)| \text{ approaches } +\infty, & \text{if } u(0) = u_1; \\ u(t) \text{ approaches zero,} & \text{if } u(0) = u_2; \\ |u(t)| \text{ approaches } +\infty, & \text{if } u(0) = u_3; \\ u(t) \text{ lies on a sphere,} & \text{if } u(0) = u_4; \\ |u(t)| \text{ approaches } +\infty, & \text{if } u(0) = u_5; \\ |u(t)| \text{ approaches } +\infty, & \text{if } u(0) = u_6. \end{cases}$$

This is because $u_0 \in S(M_{1,0})$, $u_1 \in S(M_{2,0})$, and $u_2 \in S(M_{3,0})$.

Moreover, the subspaces M_0, M_b, and M_∞, and the subset $\triangle_\infty$ are:

$$M_0 = S(M_{3,0}) = \{\alpha\begin{pmatrix} 18 \\ -4 \\ 2 \end{pmatrix} : \alpha \in \mathbb{C}\},$$

$$M_b = S(M_{1,0}) = E(0) = \{\alpha\begin{pmatrix} 12 \\ 4 \\ 3 \end{pmatrix} : \alpha \in \mathbb{C}\},$$

$$M_\infty = S(M_{2,0}) = \{\alpha\begin{pmatrix} -6 \\ -4 \\ -6 \end{pmatrix} : \alpha \in \mathbb{C}\},$$

$$\triangle_\infty = S(M_{1,0}) \setminus M_b = \emptyset,$$

respectively.

EXAMPLE 4.2.

$$\frac{d}{dt}u = Au, \quad t > 0,$$
$$u(0) = u_0,$$

where

$$A = \begin{pmatrix} 2 & 1 & 1 \\ 1 & 2 & 1 \\ -2 & -2 & -1 \end{pmatrix}, \text{ and } u_0 \text{ is any vector in } \mathbb{C}^3.$$

Solution. The eigenvalues of A, by (1.3), are $1, 1$, and 1, which are positive, and so, by Theorem 2.6, the norm $|u(t)|$ of the solution $u(t)$ approaches $+\infty$ for all u_0, as $t \longrightarrow +\infty$.

Moreover, as in solving Example 4.1, we have

$$e^{tA} = e^t M_{1,0} + t e^t M_{1,1} + t^2 e^t M_{1,2},$$

where

$$M_{1,0} = I, \quad \text{the identity, } M_{1,1} = \begin{pmatrix} 1 & 1 & 1 \\ 1 & 1 & 1 \\ -2 & -2 & -2 \end{pmatrix},$$
$$M_{1,2} = 0,$$

and so

$$M_0 = \emptyset, M_b = \emptyset,$$
$$M_\infty = S(M_{1,0}) = S(I) = \mathbb{C}^3, \triangle_\infty = \emptyset.$$

EXAMPLE 4.3.

$$\frac{d}{dt}u = Au, \quad t > 0,$$
$$u(0) = u_0, u_1, \quad \text{and } u_2, \text{ respectively,}$$

where $A = \begin{pmatrix} 0 & 1 \\ 0 & 0 \end{pmatrix}$, and

$$u_0 = \begin{pmatrix} 1 \\ 0 \end{pmatrix}, \quad u_1 = \begin{pmatrix} 0 \\ 1 \end{pmatrix}, \quad \text{and } u_2 = u_0 + u_1.$$

Solution. As in Example 4.1, we have that the eigenvalues of A are 0 and 0, and that

$$M_{1,0} = I = \begin{pmatrix} 1 & 0 \\ 0 & 1 \end{pmatrix} \quad \text{and } M_{1,1} = A.$$

Therefore, it follows from Theorem 2.6 that, as $t \longrightarrow +\infty$,

$$\begin{cases} u(t) \text{ lies on a circle,} & \text{if } u(0) = u_0; \\ |u(t)| \text{ approaches } +\infty, & \text{if } u(0) = u_1 \text{ or } u_2. \end{cases}$$

This is because $u_0 \in E(0)$, the eigenspace of the eigenvalue 0, and $u_1, u_2 \in S(M_{1,0}) \backslash E(0)$.

Moreover, we have

$$M_0 = \emptyset, M_b = E(0) = \{\alpha \begin{pmatrix} 1 \\ 0 \end{pmatrix} : \alpha \in \mathbb{C}\},$$

$$M_\infty = \emptyset, \Delta_\infty = S(M_{1,0}) \setminus M_b = \mathbb{C}^2 \setminus E(0).$$

EXAMPLE 4.4.

$$\frac{d}{dt}u = Au, \quad t > 0,$$
$$u(0) = u_0 \in \mathbb{C}^2,$$

where $A = \begin{pmatrix} 0 & 1 \\ -2 & -3 \end{pmatrix}$.

Solution. The eigenvalues of A, by (1.3), are -1 and -2, which are negative, and so, it follows from Theorem 2.6 that $u(t)$ approaches zero for all u_0, as $t \longrightarrow +\infty$.

Moreover, as in solving Example 4.1, we have

$$e^{tA} = e^{-t}M_{1,0} + e^{-2t}M_{2,0},$$

where

$$M_{1,0} = \begin{pmatrix} 2 & 1 \\ -2 & -1 \end{pmatrix}, \quad M_{2,0} = \begin{pmatrix} -1 & -1 \\ 2 & 2 \end{pmatrix},$$

and so

$$M_0 = S(M_{1,0} \cup M_{2,0}) = \left\{ \alpha \begin{pmatrix} 1 \\ -1 \end{pmatrix} + \beta \begin{pmatrix} -1 \\ 2 \end{pmatrix} : \alpha, \beta \in \mathbb{C} \right\},$$

$$M_b = \emptyset, M_\infty = \emptyset,$$

$$\Delta_\infty = \emptyset.$$

Next are examples related to the nonhomogeneous linear system (1.5), where $g(t) = g_0$ is a constant.

EXAMPLE 4.5.

$$\frac{d}{dt}u = Au + g_0, \quad t > 0,$$
$$u(0) = u_0,$$

where u_0 and g_0 are constant vectors, and A is the A in Example 4.1:

$$A = \begin{pmatrix} -3 & 9 & 0 \\ 0 & -3 & 4 \\ -1 & 0 & 4 \end{pmatrix}.$$

Solution. As in solving Example 4.1, we have that the eigenvalues of A are $0, 3$, and -5, and that

$$e^{tA} = M_{1,0} + e^{3t}M_{2,0} + e^{-5t}M_{3,0},$$

where

$$M_{1,0} = \frac{1}{15} \begin{pmatrix} 12 & 36 & -36 \\ 4 & 12 & -12 \\ 3 & 9 & -9 \end{pmatrix},$$

$$M_{2,0} = \frac{1}{24} \begin{pmatrix} -6 & -9 & 36 \\ -4 & -6 & 24 \\ -6 & -9 & 36 \end{pmatrix},$$

$$M_{3,0} = \frac{1}{40} \begin{pmatrix} 18 & -81 & 36 \\ -4 & 18 & -8 \\ 2 & -9 & 4 \end{pmatrix}.$$

Thus, it follows from Theorem 2.13 that, as $t \longrightarrow +\infty$,

$$\begin{cases} \text{the solution } u(t) \text{ lies on a sphere,} & \text{if } u(0) = u_0 \text{ and } g_0 = 0; \\ |u(t)| \text{ approaches } +\infty, & \text{if } u(0) = u_1 \text{ or } g_0 = u_1; \\ u(t) \longrightarrow \frac{1}{5}g_0 = -A^{-1}g_0, & \text{if } u(0) = g_0 = u_2; \\ |u(t)| \text{ approaches } +\infty, & \text{if } u(0) = u_2 \text{ and } g_0 = u_0; \\ u(t) = -\frac{1}{3}g_0 \text{ for all } t, & \text{if } g_0 = u_1 \text{ and } u_0 = -\frac{1}{3}g_0. \end{cases}$$

Here

$$u_0 = \begin{pmatrix} 12 \\ 4 \\ 3 \end{pmatrix} \in S(M_{1,0}), \quad u_1 = \begin{pmatrix} -6 \\ -4 \\ -6 \end{pmatrix} \in S(M_{2,0}),$$

$$u_2 = \begin{pmatrix} 18 \\ -4 \\ 2 \end{pmatrix} \in S(M_{3,0}), \quad \text{an eigenvalue of } A \text{ by Propositions 6.2 and 6.3, and so}$$

$(A + 5)u_2 = 0 \quad \text{but } u_2 \neq 0.$

EXAMPLE 4.6.

$$\frac{d}{dt}u = Au + g_0, \quad t > 0,$$
$$u(0) = u_0,$$

where u_0 and g_0 are constant vectors, and A is the A in Example 4.2:

$$A = \begin{pmatrix} 2 & 1 & 1 \\ 1 & 2 & 1 \\ -2 & -2 & -1 \end{pmatrix}.$$

Solution. As in solving Example 4.2, we have that the eigenvalues of A are $1, 1$, and 1, and that

$$e^{tA} = e^t M_{1,0} + te^t M_{1,1} + t^2 e^t M_{1,2},$$

where

$$M_{1,0} = I, \quad \text{the identity, } M_{1,1} = \begin{pmatrix} 1 & 1 & 1 \\ 1 & 1 & 1 \\ -2 & -2 & -2 \end{pmatrix},$$

$$M_{1,2} = 0.$$

To find the eigenspace $E(1)$ of the eigenvalue 1, we form the augmented matrix

$$F \equiv \begin{pmatrix} M_{1,0} \\ M_{1,1} \end{pmatrix}.$$

By performing column operations on F, in which $(-1)\times$ column 1 is added to columns 2 and 3, we have

$$G = \begin{pmatrix} 1 & -1 & -1 \\ 0 & 1 & 0 \\ 0 & 0 & 1 \\ * & * & * \\ 1 & 0 & 0 \\ 1 & 0 & 0 \\ -2 & 0 & 0 \end{pmatrix}.$$

Thus,

$$E(1) = \{\alpha \begin{pmatrix} -1 \\ 1 \\ 0 \end{pmatrix} + \beta \begin{pmatrix} -1 \\ 0 \\ 1 \end{pmatrix} : \alpha, \beta \in \mathbb{C}\},$$

by Propositions 6.2 and 6.3.

Therefore, in view of

$$S(M_{1,0}) = S(I) = \mathbb{C}^3,$$

it follows from Theorem 2.13 that, as $t \longrightarrow +\infty$, the norm $|u(t)|$ of the solution $u(t)$ approaches $+\infty$ for all u_0 and g_0, except for the case u_0 and g_0 are in $E(1)$ and $u_0 = -g_0$. In that case,

$$u(t) = -g_0$$

for all t.

EXAMPLE 4.7.

$$\frac{d}{dt}u = Au + g_0, \quad t > 0,$$

$$u(0) = u_0,$$

where u_0 and g_0 are two constant vectors, and A is the A in Example 4.3:

$$A = \begin{pmatrix} 0 & 1 \\ 0 & 0 \end{pmatrix}.$$

Solution. As in solving Example 4.3, we have that the eigenvalues of A are 0 and 0, and that

$$M_{1,0} = I = \begin{pmatrix} 1 & 0 \\ 0 & 1 \end{pmatrix} \quad \text{and} \quad M_{1,1} = A.$$

The eigenspace $E(0)$, by Propositions 6.2 and 6.3, is

$$\{\alpha \begin{pmatrix} 1 \\ 0 \end{pmatrix} : \alpha \in \mathbb{C}\}.$$

Therefore, it follows from Theorem 2.13 that, as $t \longrightarrow +\infty$,

$$\begin{cases} u(t) \text{ lies inside a circle,} & \text{if } u(0) = u_0 \text{ and } g_0 = 0; \\ |u(t)| \text{ approaches } +\infty, & \text{if } u(0) = u_1 \text{ or } u_2; \\ & \text{or if } g_0 = u_0, \\ & \text{or if } g_0 = u_1, \\ & \text{or if } g_0 = u_2, \end{cases}$$

where $u_0 = \begin{pmatrix} 1 \\ 0 \end{pmatrix} \in E(0)$, and $u_1 = \begin{pmatrix} 0 \\ 1 \end{pmatrix}$ and $u_2 = u_0 + u_1$ are in $S(M_{1,0}) \setminus E(0)$.
In fact, $|u(t)| \longrightarrow +\infty$ for any $g_0 \neq 0$ or any $u(0)$ which is not in $E(0)$.

EXAMPLE 4.8.

$$\frac{d}{dt}u = Au + g_0, \quad t > 0,$$
$$u(0) = u_0,$$

where u_0 and g_0 are constant vectors, and A is the A in Example 4.4:

$$A = \begin{pmatrix} 0 & 1 \\ -2 & -3 \end{pmatrix}.$$

Solution. As in Solving Example 4.4, the eigenvalues of A, by (1.3), are -1 and -2, which are negative, and so, it follows from Theorem 2.13 that, as $t \longrightarrow +\infty$, $u(t)$ approaches a constant for all u_0 and g_0.

Next are examples related to the nonhomogeneous linear system (1.5), where $g(t)$ is not necessarily a constant.

EXAMPLE 4.9.

$$\frac{d}{dt}u = Au + c(t)g_0, \quad t > 0,$$
$$u(0) = u_0,$$

where u_0 and g_0 are constant vectors, $c(t)$ is a continuous function, and A is the A in Example 4.1:

$$A = \begin{pmatrix} -3 & 9 & 0 \\ 0 & -3 & 4 \\ -1 & 0 & 4 \end{pmatrix}.$$

Solution. As in solving Example 4.1, we have that the eigenvalues of A are $0, 3$, and -5, and that

$$e^{tA} = M_{1,0} + e^{3t}M_{2,0} + e^{-5t}M_{3,0},$$

where

$$M_{1,0} = \frac{1}{15}\begin{pmatrix} 12 & 36 & -36 \\ 4 & 12 & -12 \\ 3 & 9 & -9 \end{pmatrix},$$

$$M_{2,0} = \frac{1}{24}\begin{pmatrix} -6 & -9 & 36 \\ -4 & -6 & 24 \\ -6 & -9 & 36 \end{pmatrix},$$

$$M_{3,0} = \frac{1}{40}\begin{pmatrix} 18 & -81 & 36 \\ -4 & 18 & -8 \\ 2 & -9 & 4 \end{pmatrix}.$$

Thus, it follows from Theorem 2.14 and Proposition 2.15 that, as $t \longrightarrow +\infty$,

$$
\begin{cases}
\text{the solution } u(t) \text{ lies inside a sphere,} & \text{if } u(0) = u_0 \text{ and } g_0 = 0; \\
|u(t)| \text{ approaches } +\infty, & \text{if } u(0) = u_1 \text{ or } c(t)g_0 = [\alpha_{-1}\frac{t+2}{t+1} + \alpha_0 + \\
& \quad \alpha_1 e^{(-1+i)t} + \alpha_2 \cos(2t) + \alpha_3 \sin(3t)]u_1; \\
u(t) \longrightarrow \text{ a constant,} & \text{if } u(0) = u_2 \text{ and } c(t)g_0 = \frac{1}{t+1}u_2; \\
|u(t)| \text{ approaches } +\infty, & \text{if } u(0) = u_2 \text{ and } c(t)g_0 = [\alpha_0 + \alpha_1\frac{1}{t+1}]u_0; \\
u(t) = e^{-3t}u_0 \text{ for all } t, & \text{if } c(t)g_0 = e^{-3t}u_1 \text{ and } u_0 = -\frac{1}{12}g_0.
\end{cases}
$$

Here $\alpha_{-1}, \alpha_0, \ldots,$ and α_3 are constants, and

$$
u_0 = \begin{pmatrix} 12 \\ 4 \\ 3 \end{pmatrix} \in S(M_{1,0}), \quad u_1 = \begin{pmatrix} -6 \\ -4 \\ -6 \end{pmatrix} \in S(M_{2,0}),
$$

$$
u_2 = \begin{pmatrix} 18 \\ -4 \\ 2 \end{pmatrix} \in S(M_{3,0}), \quad \text{an eigenvalue of } A \text{ by Propositions 6.2 and 6.3, and so}
$$

$$
(A+5)u_2 = 0 \quad \text{but } u_2 \neq 0,
$$

and

$$
\left|\frac{1}{t+1}\right| \leq 1, \quad \int_0^{+\infty} \frac{1}{t+1}\, dt = +\infty,
$$

$$
\frac{t+2}{t+1} \geq 1.
$$

EXAMPLE 4.10.

$$
\frac{d}{dt}u = Au + c(t)g_0, \quad t > 0,
$$

$$
u(0) = u_0,
$$

where u_0 and g_0 are constant vectors, $c(t)$ is a continuous function, and A is the A in Example 4.2:

$$
A = \begin{pmatrix} 2 & 1 & 1 \\ 1 & 2 & 1 \\ -2 & -2 & -1 \end{pmatrix}.
$$

Solution. As in solving Example 4.6, we have that the eigenvalues of A are $1, 1,$ and $1,$ that

$$
e^{tA} = e^t M_{1,0} + te^t M_{1,1} + t^2 e^t M_{1,2},
$$

where

$$
M_{1,0} = I, \quad \text{the identity, } M_{1,1} = \begin{pmatrix} 1 & 1 & 1 \\ 1 & 1 & 1 \\ -2 & -2 & -2 \end{pmatrix},
$$

$$
M_{1,2} = 0,
$$

and that the eigenspace $E(1)$ of 1 is:

$$
E(1) = \{\alpha \begin{pmatrix} -1 \\ 1 \\ 0 \end{pmatrix} + \beta \begin{pmatrix} -1 \\ 0 \\ 1 \end{pmatrix} : \alpha, \beta \in \mathbb{C}\}.
$$

Therefore, in view of
$$S(M_{1,0}) = S(I) = \mathbb{C}^3,$$
it follows from Theorem 2.14 and Proposition 2.15 that, as $t \longrightarrow +\infty$,

$$\begin{cases} |u(t)| \text{ approaches } +\infty, & \text{for all } u_0, g_0, \text{ and } c(t) = \alpha_{-1}\frac{t+2}{t+1} + \alpha_0 + \\ & \quad \alpha_1 e^{(-1+i)t} + \alpha_2 \cos(2t) + \alpha_3 \sin(3t); \\ u(t) = e^{-3t}u_0 \text{ for all } t, & \text{if } c(t)g_0 = e^{-3t}u_1 \text{ and } u_0 = -\frac{1}{4}g_0. \end{cases}$$

Here $\alpha_{-1}, \ldots,$ and α_3 are constants, and

$$u_1 = \begin{pmatrix} -1 \\ 1 \\ 0 \end{pmatrix} + \begin{pmatrix} -1 \\ 0 \\ 1 \end{pmatrix} \in E(1).$$

EXAMPLE 4.11.

$$\frac{d}{dt}u = Au + c(t)g_0, \quad t > 0,$$
$$u(0) = u_0,$$

where u_0 and g_0 are two constant vectors, $c(t)$ is a continuous function, and A is the A in Example 4.3:

$$A = \begin{pmatrix} 0 & 1 \\ 0 & 0 \end{pmatrix}.$$

Solution. As in solving Example 4.7, we have that the eigenvalues of A are 0 and 0, that

$$M_{1,0} = I = \begin{pmatrix} 1 & 0 \\ 0 & 1 \end{pmatrix} \quad \text{and } M_{1,1} = A,$$

and the eigenspace $E(0)$ of 0 is:

$$E(0) = \{\alpha \begin{pmatrix} 1 \\ 0 \end{pmatrix} : \alpha \in \mathbb{C}\}.$$

Therefore, it follows from Theorem 2.14 and Proposition 2.15 that, as $t \longrightarrow +\infty$,

$$\begin{cases} u(t) \text{ lies inside a circle}, & \text{if } u(0) = u_0 \text{ and } g_0 = 0; \\ |u(t)| \text{ approaches } +\infty, & \text{if } u(0) = u_1 \text{ or } u_2; \\ & \text{or if } c(t)g_0 = [\alpha_0 + \alpha_1 \frac{1}{t+1}]u_0, \\ & \text{or if } g_0 = e^{(-1+i)t}u_1, \\ & \text{or if } g_0 = u_2 \cos(2t), \end{cases}$$

where α_0 and α_1 are constants, $u_0 = \begin{pmatrix} 1 \\ 0 \end{pmatrix} \in E(0)$, and

$$u_1 = \begin{pmatrix} 0 \\ 1 \end{pmatrix} \text{ and } u_2 = u_0 + u_1 \text{ are in } S(M_{1,0}) \setminus E(0).$$

EXAMPLE 4.12.

$$\frac{d}{dt}u = Au + c(t)g_0, \quad t > 0,$$
$$u(0) = u_0,$$

where u_0 and g_0 are constant vectors, $c(t)$ is a continuous function, and A is the A in Example 4.4:

$$A = \begin{pmatrix} 0 & 1 \\ -2 & -3 \end{pmatrix}.$$

Solution. As in Solving Example 4.8, the eigenvalues of A are -1 and -2, which are negative, and so, it follows from Theorem 2.14 that, as $t \longrightarrow +\infty$, $u(t)$ approaches a constant for all u_0 and $c(t)g_0$ whose $c(t)$ satisfies $(H3)$:

$$|c(t)| \leq K_0 e^{t\nu},$$

where $K_0 \geq 0$ and $\nu \leq 0$. This constant is zero if $\nu < 0$.

5. Examples for Nonlinear Systems

Here are examples related to the nonlinear system 1.7.

EXAMPLE 5.1.

$$\frac{d}{dt}u = Au + f(u), \quad t > 0,$$
$$u(0) = u_0,$$

where

$$u = \begin{pmatrix} u_1 \\ u_2 \\ u_3 \end{pmatrix}, \quad A = \begin{pmatrix} -2 & -1 & -1 \\ -1 & -2 & -1 \\ 2 & 2 & 1 \end{pmatrix},$$

$$f(u) = \begin{pmatrix} 3u_1 u_2 u_3 \\ u_1^2 + u_2^2 + u_3^2 \\ u_1 u_2 + u_2 u_3 + u_3 u_1 \end{pmatrix} \equiv g(u) \quad \text{and}$$

$$\begin{pmatrix} 3|u_1 u_2 u_3| \\ |u_1^2 + u_2^2 + u_3^2| \\ |u_1 u_2 + u_2 u_3 + u_3 u_1| \end{pmatrix} \equiv h(u), \quad \text{respectively,}$$

and u_0 is any vector in $\mathbb{C}^3$.

Solution. Whether $f(u) = g(u)$ or $f(u) = h(u)$, we have that $f(u)$ satisfies (1.8), because

$$|u_1 u_2 u_3| \leq (|u_1| + |u_2| + |u_3|)^3,$$
$$|u_1^2 + u_2^2 + u_3^2| \leq (|u_1| + |u_2| + |u_3|)^2,$$
$$|u_1 u_2 + u_2 u_3 + u_3 u_1| \leq (|u_1| + |u_2| + |u_3|)^2.$$

In the case where $f(u) = g(u)$, this can also be seen by using Remark 3.2 because $Au + f(u)$ is differentiable at 0 and its Jacobian matrix at 0 is A.

The eigenvalues of A, by (1.3), are $-1, -1$, and -1, which are negative. Thus, it follows from Theorem 3.1 that this initial value problem with $|u_0|$ sufficiently small has a solution $\varphi(t)$, and that this $\varphi(t)$, together with other solutions which also exist on $[0, \infty)$, approaches zero as $t \longrightarrow +\infty$.

EXAMPLE 5.2.

$$\frac{d}{dt}u = Au + B(t)u + f(u), \quad t > 0,$$
$$u(0) = u_0,$$

where

$$u = \begin{pmatrix} u_1 \\ u_2 \\ u_3 \end{pmatrix}, \quad A = \begin{pmatrix} -2 & -1 & -1 \\ -1 & -2 & -1 \\ 2 & 2 & 1 \end{pmatrix},$$

$$B(t) = \begin{pmatrix} e^{-t} & e^{-t}t & e^{-t}t^2 \\ \frac{1}{1+t^2} & \frac{\sin(t)}{1+t^2} & \frac{\cos(2t)}{1+t^4} \\ e^{-2t} & e^{-3t}t^5 & e^{-t}t^8 \end{pmatrix},$$

$f(u)$ is as in Example 5.1, and u_0 is any vector in $\mathbb{C}^3$.

Solution. Since $B(t) \to 0$ as $t \to +\infty$, it follows from the solution of Example 5.1, Remark 3.4, and Theorem 3.3 that this initial value problem with $|u_0|$ sufficiently small has a solution $\varphi(t)$, and that this $\varphi(t)$, together with other soltions which also exist on $[0, \infty)$, approaches zero as $t \longrightarrow +\infty$.

EXAMPLE 5.3.

$$\frac{d}{dt}u = Au + B(t)u + f(u), \quad t > t_0 \geq 0,$$
$$u(t_0) = u_0,$$

where

$$u = \begin{pmatrix} u_1 \\ u_2 \\ u_3 \end{pmatrix}, \quad A = \begin{pmatrix} -3 & 9 & 0 \\ 0 & -3 & 4 \\ -1 & 0 & 4 \end{pmatrix},$$

and $B(t)$ and $f(u)$ are as in Example 5.2.

Solution. By the solution of Example 5.2, the condition $(A2)$ is met. Since the eigenvalues of A are, by (1.3), $0, 3$, and -5, where 3 is positive, it follows from Theorem 3.5 that not every solution $u(t)$ of this initial value problem with $|u(T)| = |u_0|$ sufficiently small, which exists on $[T, \infty)$, stays in the vicinity of the origin for all $t \geq T$. Moreover, such a solution $u(t)$ which cannot stay in the vicinity of the origin has its initial value $u(T)$ conditioned as follows.

Using (1.2) and (1.4) or equivalently, using the formula $\frac{d}{dt}e^{tA} = Ae^{tA}$ in (1.2) and letting $t = 0$, we have

$$e^{tA} = M_{1,0} + e^{3t}M_{2,0} + e^{-5t}M_{3,0},$$
$$Ae^{tA} = 3e^{3t}M_{2,0} - 5e^{5t}M_{3,0},$$
$$A^2 e^{tA} = 9e^{3t}M_{2,0} + 25e^{-5t}M_{3,0},$$

and

$$I = M_{1,0} + M_{2,0} + M_{3,0},$$
$$A = 3M_{2,0} - 5M_{3,0},$$
$$A^2 = 9M_{2,0} + 25M_{3,0},$$

where $M_{j,0}, j = 1, 2, 3$, are solved as:

$$M_{1,0} = \frac{1}{15} \begin{pmatrix} 12 & 36 & -36 \\ 4 & 12 & -12 \\ 3 & 9 & -9 \end{pmatrix},$$

$$M_{2,0} = \frac{1}{24} \begin{pmatrix} -6 & -9 & 36 \\ -4 & -6 & 24 \\ -6 & -9 & 36 \end{pmatrix},$$

$$M_{3,0} = \frac{1}{40} \begin{pmatrix} 18 & -81 & 36 \\ -4 & 18 & -8 \\ 2 & -9 & 4 \end{pmatrix}.$$

It follows from Theorem 5.14 in Chapter 3 that a Jordan basis $\{d_1, d_2, d_3\}$ is

$$\{ \frac{1}{15} \begin{pmatrix} 12 \\ 4 \\ 3 \end{pmatrix}, \frac{1}{24} \begin{pmatrix} -6 \\ -4 \\ -6 \end{pmatrix}, \frac{1}{40} \begin{pmatrix} 18 \\ -4 \\ 2 \end{pmatrix} \},$$

that is, the first column of the augmented matrix $\begin{pmatrix} M_{1,0} \\ M_{2,0} \\ M_{3,0} \end{pmatrix}$.

Suppose that $u(t)$ is a solution, and that its initial value $u(T)$ is sufficiently near the origin and takes the form

$$u(T) = \beta_1 d_1 + \beta_2 d_2 + \beta_3 d_3,$$

relative to the above basis. Then it follows from Remark 3.6 that $u(t)$ cannot stay in the vicinity of the origin if $R(T) = 2\rho(T)$ or more generally, $R(T) > \rho(T)$, where

$$R(T) = \beta_2 \quad \text{and} \quad \rho(T) = \sqrt{\beta_1^2 + \beta_3^2}.$$

EXAMPLE 5.4.

$$\frac{d}{dt} u(t) = Bu + f(u), \quad t > 0,$$

$$u(0) = u_0,$$

where

$$B = \begin{pmatrix} -3 & -1 & -1 \\ -1 & -3 & -1 \\ 2 & 2 & 0 \end{pmatrix} \quad \text{and} \quad f(u) = \begin{pmatrix} \sin(u_1 - 2) + \cos(u_1 - 2) \\ \sin(u_2 - 4) + \cos(u_2 - 4) \\ \sin(u_3 - 1) + \cos(u_3 - 1) \end{pmatrix}.$$

Solution. (1.8) is satisfied by Remark 3.2 because $Bu + f(u)$ is differentiable at $v_0 = \begin{pmatrix} 2 \\ 4 \\ 1 \end{pmatrix}$ and its Jacobian matrix at v_0 is

$$A = \begin{pmatrix} -2 & -1 & -1 \\ -1 & -2 & -1 \\ 2 & 2 & 1 \end{pmatrix}.$$

Since A has, by the solution of Example 5.1, negative eigenvalues $-1, -1$, and -1, it follows from Remark 3.2 and Theorem 3.1 that this initial value problem with $|u_0|$ sufficiently small has a solution $\varphi(t)$, and that this $\varphi(t)$, together with other solutions which also exist on $[0, \infty)$, approaches v_0 as $t \to +\infty$.

EXAMPLE 5.5.

$$\frac{d}{dt} u = Au + B(t)u + f(u), \quad t > t_0 \geq 0,$$

$$u(t_0) = u_0,$$

where

$$u = \begin{pmatrix} u_1 \\ u_2 \\ u_3 \end{pmatrix}, \quad A = \begin{pmatrix} 0 & 1 & 0 \\ 0 & 1 & -1 \\ 1 & -1 & 0 \end{pmatrix},$$

and $B(t)$ and $f(u)$ are as in Example 5.2.

Solution. By the solution of Example 5.2, the condition $(A2)$ is met. Since the eigenvalues of A are, by (1.3), $1, 1$, and -1, where 1 is positive, it follows from Theorem 3.5 that not every solution $u(t)$ of this initial value problem with $|u(T)| = |u_0|$ sufficiently small, which exists on $[T, \infty)$, stays in the vicinity of the origin for all $t \geq T$. Moreover, such a solution $u(t)$ which cannot stay in the vicinity of the origin has its initial value $u(T)$ conditioned as follows.

Using (1.2) and (1.4) or equivalently, using the formula $\frac{d}{dt}e^{tA} = Ae^{tA}$ in (1.2) and letting $t = 0$, we have

$$I = M_{1,0} + M_{2,0},$$
$$A = M_{1,0} + M_{1,1} - M_{2,0},$$
$$A^2 = M_{1,0} + 2M_{1,1} + M_{2,0},$$

where $M_{1,0}, M_{1,1}$, and $M_{2,0}$ are solved as:

$$M_{1,0} = \frac{1}{2} \begin{pmatrix} \frac{3}{2} & \frac{1}{2} & \frac{1}{2} \\ \frac{1}{2} & \frac{3}{2} & -\frac{1}{2} \\ 1 & -1 & 1 \end{pmatrix},$$

$$M_{1,1} = \frac{1}{2} \begin{pmatrix} -1 & 1 & -1 \\ -1 & 1 & -1 \\ 0 & 0 & 0 \end{pmatrix}, \quad \text{and}$$

$$M_{2,0} = \frac{1}{2} \begin{pmatrix} \frac{1}{2} & -\frac{1}{2} & -\frac{1}{2} \\ -\frac{1}{2} & \frac{1}{2} & \frac{1}{2} \\ -1 & 1 & 1 \end{pmatrix}.$$

It follows from Theorem 5.14 in Chapter 3 that a Jordan basis $\{d_1, d_2, d_3\}$ is

$$\left\{ \begin{pmatrix} \frac{3}{4} \\ \frac{1}{4} \\ \frac{1}{4} \\ \frac{1}{2} \end{pmatrix}, \begin{pmatrix} -\frac{1}{2} \\ -\frac{1}{2} \\ 0 \end{pmatrix}, \begin{pmatrix} \frac{1}{4} \\ -\frac{1}{4} \\ -\frac{1}{2} \end{pmatrix} \right\},$$

that is, the first column of the augmented matrix $\begin{pmatrix} M_{1,0} \\ M_{2,0} \\ M_{3,0} \end{pmatrix}$. By referring to Remark 3.6, a special Jordan basis is

$$\{\gamma d_1, d_2, d_3\}, \quad \text{where } \gamma \equiv \frac{0.1}{20} < \frac{\sigma}{20} \equiv \frac{0.5}{20} \text{ and } 1 > 0.5 \equiv \sigma.$$

Suppose that $u(t)$ is a solution, and that its initial value $u(T)$ is sufficiently near the origin and takes the form

$$u(T) = \beta_1 \left(\frac{0.1}{20}\right) d_1 + \beta_2 d_2 + \beta_3 d_3,$$

relative to the above special Jordan basis. Then it follows from Remark 3.6 that $u(t)$ cannot stay in the vicinity of the origin if $R(T) = 2\rho(T)$ or more generally,

$R(T) > \rho(T)$, where

$$R(T) = \sqrt{\beta_1^2 + \beta_2^2} \quad \text{and} \quad \rho(T) = \beta_3.$$

EXAMPLE 5.6.

$$\frac{d}{dt}u = Au + B(t)u + f(u), \quad t > 0,$$

where

$$u = \begin{pmatrix} u_1 \\ u_2 \\ u_3 \end{pmatrix}, \quad A = \begin{pmatrix} 0 & 1 & 0 \\ 0 & 1 & -1 \\ 1 & -1 & 0 \end{pmatrix}, \quad \text{the } A \text{ in Example 5.5,}$$

$$f(u) = \begin{pmatrix} 3u_1 u_2 u_3 \\ u_1^2 + u_2^2 + u_3^2 \\ u_1 u_2 + u_2 u_3 + u_3 u_1 \end{pmatrix},$$

and $B(t)$ is as in Example 5.2.

Solution. The condition (3.5) is met by Remark 3.8. This is because the Jacobian matrix $g_x(t, u)$ of $g(t, u) \equiv B(t)u + f(u)$ is

$$B(t) + \begin{pmatrix} 3u_2 u_3 & 3u_1 u_3 & 3u_1 u_2 \\ u_1 & u_2 & u_3 \\ u_2 + u_3 & u_3 + u_1 & u_1 + u_2 \end{pmatrix}$$

and clearly, $|g_x(t, u)| \to 0$ when both $t \to +\infty$ and $|u| \to 0$. Since $g(t, 0) = 0$ and the eigenvalues of A are, by the solution of Example 5.5, $1, 1$, and -1, we have that the conclusion of Theorem 3.9 hold, where the one-dimensional manifold (or surface) $S = S_2$ is, at the origin, tangent to, by the solution of Example 5.5,

$$M_0 = S(M_{2,0}) = S(\{\begin{pmatrix} \frac{1}{2} \\ -\frac{1}{2} \\ -1 \end{pmatrix}\}), \quad \text{where}$$

$$M_{2,0} = \frac{1}{2} \begin{pmatrix} \frac{1}{2} & -\frac{1}{2} & -\frac{1}{2} \\ -\frac{1}{2} & \frac{1}{2} & \frac{1}{2} \\ -1 & 1 & 1 \end{pmatrix}.$$

EXAMPLE 5.7.

$$\frac{d}{dt}u = Au + f(u), \quad t > 0,$$

where

$$u = \begin{pmatrix} u_1 \\ u_2 \\ u_3 \end{pmatrix}, \quad A = \begin{pmatrix} 0 & 1 & 0 \\ 0 & 1 & -1 \\ 1 & -1 & 0 \end{pmatrix}, \quad \text{the } A \text{ in Example 5.5,}$$

$$f(u) = u_1 u_2 u_3 \begin{pmatrix} 1/4 \\ -1/4 \\ -1/2 \end{pmatrix} \equiv u_1 u_2 u_3 d_3 \quad \text{and}$$

$$u_1 u_2 u_3 \begin{pmatrix} 3/2 & 1/2 & 1/2 \\ 1/2 & 3/2 & -1/2 \\ 1 & -1 & 1 \end{pmatrix} \begin{pmatrix} u_1 \\ u_2 \\ u_3 \end{pmatrix} \equiv u_1 u_2 u_3 v = u_1 u_2 u_3 \begin{pmatrix} v_1 \\ v_2 \\ v_3 \end{pmatrix}, \quad \text{respectively.}$$

Solution. The condition (3.5) is met by Remark 3.8. This is because the Jacobian matrix $f_x(u)$ of $f(u)$ is

$$\begin{pmatrix} \frac{1}{4}u_2u_3 & \frac{1}{4}u_1u_3 & \frac{1}{4}u_1u_2 \\ \frac{-1}{4}u_2u_3 & \frac{-1}{4}u_1u_3 & \frac{-1}{4}u_1u_2 \\ \frac{-1}{2}u_2u_3 & \frac{-1}{2}u_1u_3 & \frac{-1}{2}u_1u_2 \end{pmatrix} \quad \text{and}$$

$$\begin{pmatrix} u_2u_3v_1 + u_1u_2u_3\frac{3}{2} & u_1u_3v_1 + u_1u_2u_3\frac{1}{2} & u_1u_2v_1 + u_1u_2u_3\frac{1}{2} \\ u_2u_3v_2 + u_1u_2u_3\frac{1}{2} & u_1u_3v_2 + u_1u_2u_3\frac{3}{2} & u_1u_2v_2 + u_1u_2u_3\frac{-1}{2} \\ u_2u_3v_3 + u_1u_2u_3 & u_1u_3v_3 + u_1u_2u_3(-1) & u_1u_2v_3 + u_1u_2u_3 \end{pmatrix},$$

respectively, and clearly, $|f_x(u)| \to 0$ when $|u| \to 0$. Since $f(0) = 0$ and the eigenvalues of A are, by the solution of Example 5.5, $1, 1$, and -1, we have that the conclusion of Theorem 3.9 hold, where the one-dimensional manifold (or surface) $S = S_2$ is, at the origin, tangent to, by the solution of Example 5.5,

$$M_0 = S(M_{2,0}) = S(\{ \begin{pmatrix} \frac{1}{2} \\ -\frac{1}{2} \\ -1 \end{pmatrix} \}), \quad \text{where}$$

$$M_{2,0} = \frac{1}{2} \begin{pmatrix} \frac{1}{2} & -\frac{1}{2} & -\frac{1}{2} \\ -\frac{1}{2} & \frac{1}{2} & \frac{1}{2} \\ -1 & 1 & 1 \end{pmatrix}.$$

Furthermore, it follows from Theorem 3.11 that $S = M_0$. This is because $f(u) = u_1u_2u_3d_3$ has no components in $S(M_{1,0})$ where $M_{1,0} = \begin{pmatrix} 3/2 & 1/2 & 1/2 \\ 1/2 & 3/2 & -1/2 \\ 1 & -1 & 1 \end{pmatrix}$ from the solution of Example 5.5, and because $f(u) = u_1u_2u_3v = 0$ when

$$u = cd_3 \in S(M_{2,0}), \quad c \text{ a constant.}$$

EXAMPLE 5.8.

$$\frac{d}{dt}u = Au + g(u), \quad t > 0,$$

$$u(0) = u_0,$$

where

$$u = \begin{pmatrix} u_1 \\ u_2 \\ u_3 \end{pmatrix}, \quad A = \begin{pmatrix} -1 & 0 & 0 \\ 0 & -2 & 0 \\ 0 & 0 & 3 \end{pmatrix},$$

$$g(u) = \begin{pmatrix} 0 \\ u_1^3 \\ -u_1^2 \end{pmatrix}, \quad \text{and } u_0 = \begin{pmatrix} \alpha_1 \\ \alpha_2 \\ \alpha_3 \end{pmatrix}, \text{ a constant vector in } \mathbb{C}^3.$$

Solution. As in the solution of Example 5.6, the condition (3.5) is satisfied, $g(0) = 0$, the eigenvalues of A are $-1, -2$, and 3, and

$$e^{tA} = e^{-t}M_{1,0} + e^{-2t}M_{2,0} + e^{3t}M_{3,0}$$

$$= e^{-t} \begin{pmatrix} 1 & 0 & 0 \\ 0 & 0 & 0 \\ 0 & 0 & 0 \end{pmatrix} + e^{-2t} \begin{pmatrix} 0 & 0 & 0 \\ 0 & 1 & 0 \\ 0 & 0 & 0 \end{pmatrix} + e^{3t} \begin{pmatrix} 0 & 0 & 0 \\ 0 & 0 & 0 \\ 0 & 0 & 1 \end{pmatrix}.$$

Because of its simplicity where

$$u_1' = -u_1,$$

$$u'_2 = -2u_2 + u_1^3,$$
$$u'_3 = 3u_3 - u_1^2,$$

this equation has the solution which is easily obtained as

$$u(t) = \begin{pmatrix} \alpha_1 e^{-t} \\ (\alpha_2 + \alpha_1^3)e^{-2t} - \alpha_1^3 e^{-3t} \\ (\alpha_3 - \frac{\alpha_1^2}{5})e^{3t} + \alpha_1^2 e^{-2t}/5 \end{pmatrix} \quad \text{with } u(0) = \begin{pmatrix} \alpha_1 \\ \alpha_2 \\ \alpha_3 \end{pmatrix}.$$

In order for $u(t) \to 0$ as $t \to +\infty$, we require

$$\alpha_3 - \frac{\alpha_1^2}{5} = 0,$$

and so the two-dimensional manifold (or surface) S predicted by Theorem 3.9 is

$$S = \{ \begin{pmatrix} \alpha_1 \\ \alpha_2 \\ \alpha_3 \end{pmatrix} \in \mathbb{C}^3 : \alpha_3 - \frac{\alpha_1^2}{5} = 0 \}.$$

Clearly, S is, at the origin, tangent to the plane

$$M_0 = S(M_{1,0} \cup M_{2,0}) = \{ \alpha_1 \begin{pmatrix} 1 \\ 0 \\ 0 \end{pmatrix} + \alpha_2 \begin{pmatrix} 0 \\ 1 \\ 0 \end{pmatrix} : \alpha_1, \alpha_2 \in \mathbb{C} \}.$$

Furthermore, the one-dimensional manifolds (or surfaces) predicted by Theorem 3.10 are

$$S_1 = S \cap S(M_{3,0} \cup M_{1,0}) = \{ \begin{pmatrix} \alpha_1 \\ 0 \\ \frac{\alpha_1^2}{5} \end{pmatrix} : \alpha_1 \in \mathbb{C} \},$$

$$S_2 = S \cap S(M_{3,0} \cup M_{2,0}) = \{ \alpha_2 \begin{pmatrix} 0 \\ 1 \\ 0 \end{pmatrix} : \alpha_2 \in \mathbb{C} \},$$

and S_1 and S_2 are, at the origin, tangent to

$$S(M_{1,0}) = \{ \alpha_1 \begin{pmatrix} 1 \\ 0 \\ 0 \end{pmatrix} : \alpha_1 \in \mathbb{C} \} \quad \text{and} \quad S(M_{2,0}) = \{ \alpha_2 \begin{pmatrix} 0 \\ 1 \\ 0 \end{pmatrix} : \alpha_2 \in \mathbb{C} \},$$

respectively. It is obvious that

$$S_1 \neq S(M_{1,0}) \quad \text{and} \quad S_2 = S(M_{2,0}),$$

as conjectured and predicted by Remark 3.12 and Theorem 3.11, respectively, since

$$g(u) = \begin{cases} \begin{pmatrix} 0 \\ \alpha_1^3 \\ -\alpha_1^2 \end{pmatrix} \neq 0, & \text{if } \alpha_1 \neq 0 \text{ and } u = \begin{pmatrix} \alpha_1 \\ \alpha_2 \\ \alpha_3 \end{pmatrix}; \\ \\ 0, & \text{if } \alpha_2 \neq 0 \text{ and } u = \begin{pmatrix} 0 \\ \alpha_2 \\ 0 \end{pmatrix}. \end{cases}$$

The solution $\varphi(t)$ of this initial value problem with $\varphi(0) = u_0 = \begin{pmatrix} 1 \\ 0 \\ 0 \end{pmatrix}$ is

$$\begin{pmatrix} e^{-t} \\ e^{-2t} - e^{-3t} \\ -\frac{1}{5}(e^{3t} - e^{-2t}) \end{pmatrix},$$

and so it cannot stay in the vicinity of the origin for all $t \geq 0$. This is predicted by Theorem 3.11 since $S_1 \neq S(M_{1,0})$ and $u_0 \in S(M_{1,0}) \setminus S_1$.

EXAMPLE 5.9.

$$\frac{d}{dt}u = Au + g(u), \quad t > 0,$$
$$u(0) = u_0,$$

where

$$u = \begin{pmatrix} u_1 \\ u_2 \\ u_3 \end{pmatrix}, \quad A = \begin{pmatrix} -1 & 0 & 0 \\ 0 & -2 & 0 \\ 0 & 0 & 3 \end{pmatrix},$$

$$g(u) = \begin{pmatrix} u_3^2 \\ u_3^2 \\ 0 \end{pmatrix}, \quad \text{and } u_0 = \begin{pmatrix} \alpha_1 \\ \alpha_2 \\ \alpha_3 \end{pmatrix}, \text{ a constant vector in } \mathbb{C}^3.$$

Solution. As in the solution of Example 5.6, the condition (3.5) is satisfied, $g(0) = 0$, the eigenvalues of A are $-1, -2$, and 3, and

$$e^{tA} = e^{-t}M_{1,0} + e^{-2t}M_{2,0} + e^{3t}M_{3,0}$$

$$= e^{-t}\begin{pmatrix} 1 & 0 & 0 \\ 0 & 0 & 0 \\ 0 & 0 & 0 \end{pmatrix} + e^{-2t}\begin{pmatrix} 0 & 0 & 0 \\ 0 & 1 & 0 \\ 0 & 0 & 0 \end{pmatrix} + e^{3t}\begin{pmatrix} 0 & 0 & 0 \\ 0 & 0 & 0 \\ 0 & 0 & 1 \end{pmatrix}.$$

Because of its simplicity where

$$u_1' = -u_1 + u_3^2,$$
$$u_2' = -2u_2 + u_3^2,$$
$$u_3' = 3u_3,$$

this equation has the solution which is easily obtained as

$$u(t) = \begin{pmatrix} (\alpha_1 - \frac{\alpha_3^2}{7})e^{-t} + \frac{\alpha_3^2}{7}e^{6t} \\ (\alpha_2 - \frac{\alpha_3^2}{8})e^{-2t} + \frac{\alpha_3^2}{8}e^{6t} \\ \alpha_3 e^{3t} \end{pmatrix} \quad \text{with } u(0) = \begin{pmatrix} \alpha_1 \\ \alpha_2 \\ \alpha_3 \end{pmatrix}.$$

In order for $u(t) \to 0$ as $t \to +\infty$, we require

$$\alpha_3 = 0,$$

and so the two-dimensional manifold (or surface) S predicted by Theorem 3.9 is

$$S = \left\{ \begin{pmatrix} \alpha_1 \\ \alpha_2 \\ 0 \end{pmatrix} : \alpha_1, \alpha_2 \in \mathbb{C}^3 \right\}.$$

Clearly, S is, at the origin, tangent to the plane

$$M_0 = S(M_{1,0} \cup M_{2,0}) = \{\alpha_1 \begin{pmatrix} 1 \\ 0 \\ 0 \end{pmatrix} + \alpha_2 \begin{pmatrix} 0 \\ 1 \\ 0 \end{pmatrix} : \alpha_1, \alpha_2 \in \mathbb{C}\}.$$

Furthermore, the one-dimensional manifolds (or surfaces) predicted by Theorem 3.10 are

$$S_1 = S \cap S(M_{3,0} \cup M_{1,0}) = \{\begin{pmatrix} \alpha_1 \\ 0 \\ 0 \end{pmatrix} : \alpha_1 \in \mathbb{C}\},$$

$$S_2 = S \cap S(M_{3,0} \cup M_{2,0}) = \{\alpha_2 \begin{pmatrix} 0 \\ 1 \\ 0 \end{pmatrix} : \alpha_2 \in \mathbb{C}\},$$

and S_1 and S_2 are, at the origin, tangent to

$$S(M_{1,0}) = \{\alpha_1 \begin{pmatrix} 1 \\ 0 \\ 0 \end{pmatrix} : \alpha_1 \in \mathbb{C}\} \quad \text{and} \quad S(M_{2,0}) = \{\alpha_2 \begin{pmatrix} 0 \\ 1 \\ 0 \end{pmatrix} : \alpha_2 \in \mathbb{C}\},$$

respectively. It is obvious that

$$S = M_0, \quad S_1 = S(M_{1,0}) \quad \text{and} \quad S_2 = S(M_{2,0}),$$

as predicted by Theorem 3.11 since $g(u)$ has no components in $S(M_{3,0})$.

6. Proof of the Main Results for Linear Systems

To prove the results in Section 2, we first need to study some important properties from our article [13] of the matrices $M_{i,k}$ in (1.2).

LEMMA 6.1. *The spanned subspace $S(\cup_{j=1}^{s} M_{j,0})$ by the column vectors in the matrices*

$$M_{j,0} : \quad j = 1, 2, \ldots, s,$$

equals the vector space $\mathbb{C}^n$.

PROOF. Evaluating (1.2) at $t = 0$, we have

$$I = \sum_{j=1}^{s} M_{j,0},$$

and so the column vectors in the matrices

$$M_{j,0} : \quad j = 1, 2, \ldots, s,$$

span the vector space $\mathbb{C}^n$. This is because the column vectors in the identity matrix I does so. $\square$

PROPOSITION 6.2. *The following are true for $i = 1, \ldots, s$:*

$$M_{i,k} = \frac{1}{k!}(A - \lambda_i I)^k M_{i,0}, \quad k = 0, 1, \ldots, (m_i - 1),$$
$$(A - \lambda_i I)^{m_i} M_{i,0} = 0, \quad \text{and}$$
$$A M_{i,m_i - 1} = \lambda_i M_{i,m_i - 1}.$$

Thus we have that each nonzero column vector in the matrix M_{i,m_i-1}, is an eigenvector, that each column vector in the matrix $M_{i,k}$ equals its corresponding column *vector in the matrix $M_{i,0}$, pre-multiplied by the factor $\frac{1}{k!}(A - \lambda_i I)^k$, and that each column vector in any of the matrices*

$$M_{j,k} : \quad k = 0, 1, \ldots, m_j - 1,$$

is in K_{λ_j}.

PROOF. Applying the identity

$$\frac{d}{dt} e^{tA} = A e^{tA}$$

(in Chapter 2) to (1.2), it follows that

$$\sum_{k=0}^{m_i-1} \sum_{i=1}^{s} t^k e^{t\lambda_i} A M_{i,k}$$

$$= \sum_{i=1}^{s} e^{t\lambda_i} [A M_{i,0} + t A M_{i,1} + \cdots + t^{m_i-1} A M_{i,m_i-1}]$$

$$= \sum_{k=0}^{m_i-1} \sum_{i=1}^{s} [t^k \lambda_i e^{t\lambda_i} + k t^{k-1} e^{t\lambda_i}] M_{i,k}$$

$$= \sum_{i=1}^{s} e^{t\lambda_i} [(\lambda_i M_{i,0} + M_{i,1}) + t(\lambda_i M_{i,1} + 2 M_{i,2}) + \cdots$$

$$+ t^{m_1-2}(\lambda_i M_{i,m_i-2} + (m_i - 1) M_{i,m_i-1}) + t^{m_i-1} M_{i,m_i-1}],$$

and so

$$A M_{i,0} = \lambda_i M_{i,0} + M_{i,1},$$
$$A M_{i,1} = \lambda_i M_{i,1} + 2 M_{i,2},$$

$$\vdots \tag{6.1}$$

$$A M_{i,m_i-2} = \lambda_i M_{i,m_i-2} + (m_i - 1) M_{i,m_i-1},$$
$$A M_{i,m_i-1} = \lambda_i M_{i,m_i-1}.$$

Here used was the fact in Chapter 8 that the set

$$\{t^k e^{t\lambda_i} : i = 1, \ldots, s; k = 0, \ldots, m_i - 1\}$$

is linearly independent.

From (6.1), it results that

$$M_{i,k} = \frac{1}{k!}(A - \lambda_i I)^k M_{i,0}, \quad i = 1, \ldots, s; k = 1, \ldots, (m_i - 1), \quad \text{and}$$
$$(A - \lambda_i I)^{m_i} M_{i,0} = 0, \quad i = 1, \ldots, s.$$

$\square$

PROPOSITION 6.3. *If v is a nonzero column vector in M_{i,m_i-1}, then v is an eigenvector of A. However, if v is instead a zero vector, then its corresponding nonzero column vector x in M_{i,m_i-2} is an eigenvector; if x is zero again, then its corresponding nonzero column vector y in M_{i,m_i-3} is an eigenvector. This process* *can be continued, and each $M_{i,k}$ can be the new $M_{i,k}$, obtained by performing column operations on the augmented matrix*

$$F_i \equiv \begin{pmatrix} M_{i,0} \\ M_{i,1} \\ M_{i,2} \\ \vdots \\ M_{i,m_i-1} \end{pmatrix}.$$

Eigenvectors that are found in the end of this process span the eigenspace $E(\lambda_i)$.

PROOF. This is readily seen, by using Proposition 6.2 and Lemma 6.1.   □

PROPOSITION 6.4. *For each $i = 1,\ldots,s$, the spanned subspace $S(M_{i,0})$ by the column vectors in the matrix $M_{i,0}$ equals the generalized eigenspace K_{λ_i}, corresponding to the generalized eigenvalue λ_i.*

PROOF. Let $0 \neq p \in K_{\lambda_i}$. Then there is a natural number $1 \leq r \leq m_i$, such that $(A - \lambda_i I)^r p = 0$. but $(A - \lambda_i I)^{r-1} p \neq 0$. Since

$$e^{tA} = \sum_{j=1}^{s} \sum_{k=0}^{m_j-1} t^k e^{t\lambda_j} M_{j,k}$$

from (1.2), we have

$$e^{tA} p = e^{t\lambda_i} p + \sum_{k=1}^{r-1} e^{t\lambda_i} \frac{t^k}{k!} (A - \lambda_i)^k p$$

$$= \sum_{j=1}^{s} \sum_{k=0}^{m_j-1} t^k e^{t\lambda_j} M_{j,k} p$$

$$= e^{t\lambda_i} M_{i,0} p + \sum_{j=1,j\neq i}^{s} \sum_{k=0}^{m_j-1} t^k e^{t\lambda_j} M_{j,k} p + \sum_{k=1}^{m_i-1} t^k e^{t\lambda_i} M_{i,k} p.$$

It follows that

$$p = M_{i0} p,$$

as the set

$$\{t^k e^{t\lambda_j} : j = 1,\ldots,s; k = 0,\ldots,(m_j - 1)\}$$

is linearly independent, a fact in Chapter 8. Thus $K_{\lambda_i} \subset S(M_{i,0})$.

That $S(M_{i,0}) \subset K_{\lambda_i}$ follows from Proposition 6.2, and so $S(M_{i,0}) = K_{\lambda_i}$.   □

PROPOSITION 6.5. *The $M_{i,k}$ satisfies:*

$$M_{i,l} M_{j,m} = 0 \quad \text{if } i \neq j;$$

$$M_{i,m} M_{i,l} = M_{i,l} M_{i,m} = \frac{(A - \lambda_i)^l}{l!} M_{i,m} = \frac{(A - \lambda_i)^{m+l}}{l!m!} M_{i,0}$$

$$\text{if } 0 \leq l, m \leq m_i - 1 \text{ and } l + m \leq m_i - 1;$$

$$M_{i,m} M_{i,l} = 0 \quad \text{if } l + m > m_i - 1; \quad \text{and}$$

$$M_{i,k} M_{i,0} = M_{i,k}, \quad i = 1,\ldots,s.$$

PROOF. Making use of (1.2) and Proposition 6.2, we have

$$e^{tA}M_{j,m} = e^{t\lambda_j} \sum_{l=0}^{m_j-1} \frac{[t(A-\lambda_j)]^l}{l!} M_{j,m}$$

$$= \sum_{l=0}^{m_j-1} t^l e^{t\lambda_j} M_{j,l} M_{j,m} + \sum_{i\neq j, i=1}^{s} \sum_{l=0}^{m_i-1} t^l e^{t\lambda_i} M_{i,l} M_{j,m}.$$

Since

$$\{t^k e^{t\lambda_i} : i = 1,\ldots,s; k = 0,\ldots,(m_i-1)\}$$

is linearly independent, a fact in Chapter 8, we have

$$M_{i,l} M_{j,m} = 0 \quad \text{for} \quad i \neq j,$$

$$M_{j,l} M_{j,m} = \frac{(A-\lambda_j)^l}{l!} M_{j,m}.$$

The rest follows from Proposition 6.2. $\square$

We now prove Theorem 2.6:

PROOF. If $u_0 \in S(M_{j,0})$ for some j, then $u_0 \in K_{\lambda_j}$ by Proposition 6.4, and so, for some natural number $1 \leq \nu \leq m_j$,

$$(A-\lambda_j I)^\nu u_0 = 0, \quad \text{but} \quad (A-\lambda_j I)^{\nu-1} u_0 \neq 0,$$

where λ_j is the corresponding generalized eigenvalue. It follows that

$$\varphi(t) \equiv e^{tA} u_0$$

$$= e^{t\lambda_j} e^{t(A-\lambda_j I)} u_0$$

$$= e^{t\lambda_j} [I + \frac{(A-\lambda_j I)t}{1!} + \cdots + \frac{((A-\lambda_j I)t)^{\nu-1}}{(\nu-1)!}] u_0$$

$$\equiv \phi_\nu(t),$$

and so, as $t \longrightarrow \infty$,

$$\begin{cases} \varphi(t) \longrightarrow 0, & \text{if } \lambda_j \text{ has negative real part;} \\ |\varphi(t)| \longrightarrow +\infty, & \text{if } \lambda_j \text{ has positive real part.} \end{cases}$$

Furthermore, since $(A-\lambda_j I)u_0 = 0$ for $u_0 \in E(\lambda_j)$, we have

$$\varphi(t) \equiv e^{tA} u_0 = \begin{cases} e^{t\lambda_j} u_0, & \text{if } u_0 \in E(\lambda_j); \\ \text{The above } \phi_\nu(t) \text{ with } \nu \geq 2, & \text{if } u_0 \in S(M_{j,0}) \setminus E(\lambda_j). \end{cases}$$

Thus if λ_j has zero real part, then $\varphi(t)$ lies on a sphere for all $t \geq 0$ or approaches $+\infty$ as $t \longrightarrow +\infty$, according as

$$u_0 \in E(\lambda_j) \quad \text{or} \quad S(M_{j,0}) \setminus E(\lambda_j).$$

Finally, (1.2) gives

$$\varphi(t) = e^{tA} u_0 = \sum_{j=1}^{s} \sum_{k=0}^{m_j-1} t^k e^{t\lambda_j} M_{j,k} u_0,$$

and so, for all u_0, we have $\lim_{t\to+\infty} \varphi(t) = 0$ or $\lim_{t\to+\infty} |u(t)| = +\infty$ according as all the eigenvalues λ_j have negative real parts or all λ_j's have positive real parts. $\square$

We continue to prove Theorem 2.8:

PROOF. If $u_0 \in S(M_{j,0})$ for some j, then $u_0 \in K_{\lambda_j}$ by Proposition 6.4, and so, for some natural number $1 \le \nu \le m_j$, we have

$$(A - \lambda_j I)^\nu u_0 = 0, \quad \text{but } (A - \lambda_j I)^{\nu-1} u_0 \ne 0 \tag{6.2}$$

where λ_j is the corresponding generalized eigenvalue. It follows that

$$
\begin{aligned}
\varphi(t) &\equiv e^{tA} u_0 \\
&= e^{t\lambda_j} e^{t(A - \lambda_j I)} u_0 \\
&= e^{t\lambda_j} [I + \frac{(A - \lambda_j I)t}{1!} + \cdots + \frac{((A - \lambda_j I)t)^{\nu-1}}{(\nu-1)!}] u_0 \\
&\equiv \phi_\nu(t),
\end{aligned}
$$

and so

$$(A - \lambda_j I)^\nu \varphi(t) = 0 \quad \text{but } (A - \lambda_j I)^{\nu-1} \varphi(t) \ne 0$$

by (6.2). This shows $\varphi(t) = e^{tA} u_0 \in S(M_{j,0})$, and so $S(M_{j,0})$ is invariant under e^{tA}.

If $u_0 \in E(\lambda_j)$, it is readily seen that the above arguments with $\nu = 1$ apply. □

We next prove Theorem 2.13:

PROOF. We divide the proof into five steps.

Step 1. By repeated use of the formula of integration by parts, we have, for $m \in \{0\} \cup \mathbb{N}$ and $\lambda \ne 0$,

$$
\begin{aligned}
\int e^{t\lambda} t^m \, dt &= \lambda^{-1}(t^m e^{t\lambda} - m \int t^{m-1} e^{t\lambda} \, dt) \\
&= \lambda^{-1}[t^m e^{t\lambda} - \lambda^{-1} m(t^{m-1} e^{t\lambda} - (m-1) \int t^{m-2} e^{t\lambda} \, dt)] \\
&\vdots \\
&= \sum_{l=0}^{m} (-1)^l m(m-1)(m-2)\cdots(m-l+1)\lambda^{-(l+1)} t^{m-l} e^{t\lambda},
\end{aligned}
$$

where $(-1)^l m(m-1)\cdots(m-l+1) \equiv 1$ is defined. Hence

$$
\int_{t=0}^{\tau} e^{t\lambda} t^m \, dt = -(-1)^m m! \lambda^{-(m+1)} + \\
\sum_{l=0}^{m} (-1)^l m(m-1)\cdots(m-l+1)\lambda^{-(l+1)} \tau^{m-l} e^{\tau\lambda}, \tag{6.3}
$$

which approaches

$$(-1)^m m! \lambda^{-(m+1)},$$

for negative real part of λ, as $\tau \longrightarrow +\infty$.

Step 2. By (1.6), the unique solution to the nonhomogeneous linear system (1.5) is

$$\varphi(t) = e^{tA} u_0 + \int_{\tau=0}^{t} e^{(t-\tau)A} g_0 \, d\tau.$$

Thus, if $u_0 \in S(M_{i,0})$ for some i, then, as in proving Theorem 2.6, we have

$$e^{tA} u_0 = e^{t\lambda_i}[I + (A - \lambda_i I)t + \cdots + \frac{((A - \lambda_j I)t)^{\nu-1}}{(\nu - 1)!}]u_0 \qquad (6.4)$$

for some $1 \le \nu \le m_i$. Similarly, if $g_0 \in S(M_{j,0})$ for some j, we have

$$e^{(t-\tau)A} g_0 = e^{(t-\tau)\lambda_j}[I + (t - \tau)(A - \lambda_j I) + \cdots + \frac{((t - \tau)(A - \lambda_j I))^{\mu-1}}{(\mu - 1)!}]g_0 \qquad (6.5)$$

for some $1 \le \mu \le m_j$. Hence, the statement in the first paragraph of Theorem 2.13 is proved. This is due to the fact that

$$e^{tA} u_0 \longrightarrow 0$$

by (6.4) for negative real part of λ_i as $t \longrightarrow +\infty$, and to the fact that

$$\int_{\tau=0}^{t} e^{(t-\tau)A} g_0 \, d\tau = -A^{-1}e^{(t-\tau)A}g_0|_{\tau=0}^{t} = -A^{-1}(I - e^{tA})g_0$$

approaches $-A^{-1}g_0$ for negative real part of λ_j as $t \longrightarrow +\infty$ if A^{-1} exists, or else to the fact that, by (6.3) and (6.5),

$$\int_{\tau=0}^{t} e^{(t-\tau)A} g_0 \, d\tau \longrightarrow \sum_{k=0}^{\mu-1}(-1)^{k+1}\lambda_j^{-(k+1)}(A - \lambda_j I)^k g_0$$

for negative real part of λ_j as $t \longrightarrow +\infty$.

Step 3. This above argument in Step 1 and Step 2 with positive real parts of λ_i and λ_j, also proves the statement in the second paragraph of Theorem 2.13, except for the case where $u_0, g_0 \in E(\lambda_j)$ and $u_0 = -\frac{1}{\lambda_j}g_0$; but this case is also true since, then,

$$\varphi(t) = e^{t\lambda_j} u_0 + \int_{\tau=0}^{t} e^{(t-\tau)\lambda_j} g_0 \, d\tau$$

$$= -\frac{1}{\lambda_j}g_0.$$

Step 4. The above argument in Step 1 and Step 2, with zero real parts of λ_i and λ_j, also proves the statement in the third paragraph of Theorem 2.13, except for the case where $0 \ne g_0 \in E(\lambda_j)$ with $\lambda_j = 0$; but this case is also true since, then,

$$\int_{\tau=0}^{t} e^{(t-\tau)A} g_0 \, d\tau = \int_{\tau=0}^{t} e^{(t-\tau)\lambda_j} g_0 \, d\tau$$

$$= \int_{\tau=0}^{t} g_0 \, d\tau = tg_0,$$

whose norm approaches $+\infty$ as $t \longrightarrow +\infty$.

Step 5. Finally, using

$$S(\cup_{j=1}^{s} M_{j,0}) = \mathbb{C}^n$$

from Lemma 6.1, it is readily seen that the above argument in Step 1 and Step 2 with negative real parts of all generalized eigenvalues of A, also proves the statement in the last paragraph of Theorem 2.13. $\qquad \square$

Lastly, we deal with the proof of Theorem 2.14. However, we need the following lemma first (cf. [**2**, Page 276]):

LEMMA 6.6. *Let* $f(t,\tau) = h_1(t,\tau) + ih_2(t,\tau)$, *where* $t, \tau \geq 0$, $i \equiv \sqrt{-1}$ *with* $i^2 = -1$, *and, for each* $t \geq 0$, $h_1(t,\tau)$ *and* $h_2(t,\tau)$ *are two Riemann-integrable functions on* $[0, b]$ *for every* $b \geq 0$. *Assume that there is a constant* $M > 0$ *such that*

$$\int_{\tau=0}^{t} |f(t,\tau)| \, d\tau \leq M \quad \text{for every } t \geq 0.$$

Then both the limits $\lim_{t \to +\infty} \int_{\tau=0}^{t} f(t,\tau) \, d\tau$ *and* $\lim_{t \to +\infty} \int_{\tau=0}^{t} |f(t,\tau)| \, d\tau$ *exist. Here* $|f(t,\tau)| \equiv |h_1(t,\tau)| + |h_2(t,\tau)|$ *is defined.*

PROOF. We use the proof in [**2**, Page 276]. Let $F(t) = \int_{\tau=0}^{t} |f(t,\tau)| \, d\tau$. Then $\lim_{t \to +\infty} F(t)$ exists, since $F(t)$ is an increasing function of t and bounded above by M. Similarly, the limits

$$H_j \equiv \lim_{t \to +\infty} \int_{\tau=0}^{t} |h_j(t,\tau)| \, d\tau, j = 1, 2$$

exist, since $|h_j(t,\tau)| \leq |f(t,\tau)|$.

It is easy to see

$$0 \leq |h_1(t,\tau)| - h_1(t,\tau) \leq 2|h_1(t,\tau)|,$$

and so

$$\lim_{t \to +\infty} \int_{\tau=0}^{t} \{|h_1(t,\tau)| - h_1(t,\tau)\} \, d\tau$$

and then

$$\lim_{t \to +\infty} \int_{\tau=0}^{t} h_1(t,\tau) \, d\tau = \lim_{t \to +\infty} \int_{\tau=0}^{t} |h_1(t,\tau)| \, d\tau -$$

$$\lim_{t \to +\infty} \int_{\tau=0}^{t} \{|h_1(t,\tau)| - h_1(t,\tau)\} \, d\tau$$

exists. This is because $\int_{\tau=0}^{t} \{|h_1(t,\tau)| - h_1(t,\tau)\} \, d\tau$ is then an increasing function of t and bounded above by $\lim_{t \to +\infty} \int_{\tau=0}^{t} 2|h_1(t,\tau)| \, d\tau$. The case for $h_2(t,\tau)$ is likewise treated.

Therefore the limit $\lim_{t \to +\infty} \int_{\tau=0}^{t} f(t,\tau) \, d\tau$ exists. $\square$

We finally prove Theorem 2.14:

PROOF. We divide the proof into five steps.
Step 1. By (6.3), we have

$$\int_{\tau=0}^{t} e^{(t-\tau)\lambda}(t-\tau)^m e^{\tau\nu} \, d\tau = -(-1)^m m! \lambda^{-(m+1)} e^{t\nu} +$$

$$\sum_{l=0}^{m} (-1)^l m(m-1) \cdots (m-l+1)\lambda^{-(l+1)} t^{m-l} e^{t\lambda},$$

$$(6.6)$$

which, for negative real part of λ_j, approaches

$$(-1)^m m! \lambda^{-(m+1)} \quad \text{or } 0$$

as $t \longrightarrow +\infty$, according as $\nu = 0$ or $\nu < 0$.

Step 2. By (1.6), the unique solution to the nonhomogeneous linear system (1.5) is

$$\varphi(t) = e^{tA}u_0 + \int_{\tau=0}^{t} e^{(t-\tau)A}g(t)\,d\tau.$$

Thus, if $u_0 \in S(M_{i,0})$ for some i, then, as in proving Theorem 2.6, we have (6.4) holds:

$$e^{tA}u_0 = e^{t\lambda_i}[I + (A - \lambda_i I)t + \cdots + \frac{((A - \lambda_j I)t)^{\nu-1}}{(\nu-1)!}]u_0$$

for some $1 \le \nu \le m_i$. Similarly, if $g(t) \in S(M_{j,0})$ for some j, we have the formula, similar to (6.5):

$$e^{(t-\tau)A}g(t) = e^{(t-\tau)\lambda_j}[I + (t-\tau)(A - \lambda_j I) + \cdots + \frac{((t-\tau)(A - \lambda_j I))^{\mu-1}}{(\mu-1)!}]g(t) \quad (6.7)$$

for some $1 \le \mu \le m_j$. Hence the statement in the first paragraph of Theorem 2.14 is proved. This is due to the fact that

$$e^{tA}u_0 \longrightarrow 0$$

by (6.4) for negative real part of λ_i as $t \longrightarrow +\infty$, to the fact that, by (6.7) and (6.6), we have

$$|\int_{\tau=0}^{t} e^{(t-\tau)A}g(\tau)\,d\tau| \le \int_{\tau=0}^{t} |e^{(t-\tau)A}g(\tau)|\,d\tau$$

$$\le \sum_{k=0}^{\mu-1} \int_{\tau=0}^{t} e^{(t-\tau)Re(\lambda_j)}(t-\tau)^k \frac{|A - \lambda_j I|^k}{k!}|g(t)|\,d\tau$$

$$\le \sum_{k=0}^{\mu-1} K_0 \frac{|A - \lambda_j I|^k}{k!} \int_{\tau=0}^{t} e^{(t-\tau)Re(\lambda_j)}(t-\tau)^k e^{\nu\tau}\,d\tau$$

$$\longrightarrow \begin{cases} \sum_{k=0}^{\mu-1}(-1)^{k+1}\lambda_j^{-(k+1)}|(A - \lambda_j I)|^k K_0, & \text{if } \nu = 0; \\ 0, & \text{if } \nu < 0, \end{cases}$$

for negative real part of λ_j as $t \longrightarrow +\infty$, and to the fact that Lemma 6.6 can be applied. Here $Re(\lambda_j)$ is the real part of λ_j, and the following inequality was used, [5, Page 62]:

$$|Bx + Cy| \le |B||x| + |C||y|$$

is true for matrices $B = (b_{jk})_{n \times n}$ and $C = (c_{jk})_{n \times n}$ and vectors

$$x = \begin{pmatrix} x_1 \\ \vdots \\ x_n \end{pmatrix}, y \in \mathbb{C}^n,$$

in which

$$|B| \equiv \sum_{j,k=1}^{n} |b_{ij}| \quad \text{and} \quad |x| \equiv \sum_{k=1}^{n} |x_k|$$

are defined.

Step 3. To prove the statement in the second paragraph of Theorem 2.13, we compute, for $0 \le k \le \mu - 1$,

$$|\int_{\tau=0}^{t} e^{(t-\tau)A}g(\tau)\,d\tau| = |\sum_{k=0}^{\mu-1} \int_{\tau=0}^{t} e^{(t-\tau)\lambda_j}\frac{(t-\tau)^k}{k!}(A - \lambda_j I)^k c(t)g_0\,d\tau|.$$

Thus, the argument in Step 1 and Step 2 with positive real parts of λ_i and λ_j, proves the statement in the second paragraph of Theorem 2.14, except for the case where $i = j$ and $u_0, g_0 \in E(\lambda_j)$; but this case is readily seen to be true by simple calculations.

Step 4. The above argument in Step 1, Step 2, and Step 3 with zero real parts of $\lambda_i \lambda_j$, also proves the statement in the third paragraph of Theorem 2.14, except for the case where $g(t)$ satisfies $(H4)$ with $0 \neq g_0 \in E(\lambda_j)$ and $\lambda_j = 0$; but this case is also true since, then,

$$\left| \int_{\tau=0}^{t} e^{(t-\tau)A} g(t) \, d\tau \right| = \left| \int_{\tau=0}^{t} e^{(t-\tau)\lambda_j} c(t) \, d\tau \|g_0| \right|,$$

which approaches $+\infty$ as $t \longrightarrow +\infty$.

Step 5. Finally, using

$$S(\cup_{j=1}^{s} M_{j,0}) = \mathbb{C}^n$$

from Lemma 6.1, it is readily seen that the above argument in Step 1 and Step 2, with negative real parts of all generalized eigenvalues of A, also proves the statement in the last paragraph of Theorem 2.14. □

7. Proof of the Main Results for Nonlinear Systems

Here we prove the results in Section 3.
Proof of Theorem 3.1.

PROOF. Since all the generalized eigenvalues of A have negative real parts, it follows from (1.2) that there are positive constants $K > 1$ and $\sigma > 0$, such that

$$|e^{tA}| \leq Ke^{-t\sigma} \quad \text{for all } t \geq 0. \tag{7.1}$$

Using the assumptions of (1.8) and $(H1)$, we have that, given $\epsilon > 0$, there is a $\delta > 0$ such that, for $|u| \leq \delta$ and $t \geq 0$, we have

$$|g(t, u)| \leq \frac{\epsilon}{K}|u|, \tag{7.2}$$

$$g(t, u) \quad \text{is continuous.}$$

Let the initial value u_0 in (1.5) be small enough that

$$|u_0| \leq \frac{\delta}{K} < \delta.$$

Thus, Corollary 2.8 for local existence in Chapter 8 gives a solution $\varphi(t)$ of (1.5) on a small t-interval $[0, \eta]$ for $\eta > 0$. It is easily seen that the $\varphi(t)$ satisfies the integral equation

$$\varphi(t) = e^{tA} u_0 + \int_0^t e^{(t-\tau)A} g(\tau, \varphi(\tau)) \, d\tau,$$

similar to (1.6), and this gives, by (7.1) and (7.2),

$$|\varphi(t)| \leq Ku_0 e^{-t\sigma} + e^{-t\sigma} \int_0^t e^{\tau\sigma} |\varphi(\tau)| \, d\tau$$

or

$$e^{t\sigma} |\varphi(t)| \leq Ku_0 + \epsilon \int_0^t e^{\tau\sigma} |\varphi(\tau)| \, d\tau.$$

Then it follows from the Gronwall's inequality in Chapter 8 that

$$e^{t\sigma}|\varphi(t)| \le Ku_0 e^{t\epsilon}$$

or

$$|\varphi(t)| \le Ku_0 e^{-t(\sigma-\epsilon)} \quad \text{for } t \in [0, \tau]. \tag{7.3}$$

By choosing $\epsilon < \sigma$, it follows from (7.3) that $|\varphi(t)| \le Ku_0 \le \delta$, and so the $\varphi(t)$ can be extended from $[0, \eta]$ to $[0, \infty)$ by repeatedly using Theorem 2.9 for continuation of solutions in Chapter 8. This extended $\varphi(t)$ still satisfies (7.3) even for all $t \ge 0$. Since

$$|\varphi(t)| \longrightarrow 0 \quad \text{as } t \longrightarrow \infty,$$

the proof is complete. □

Proof of Theorem 3.3.

PROOF. We divide the proof into three steps.

Step 1. The two norms $\|x\|$ and $|x|$ of $x = \begin{pmatrix} x_1 \\ \vdots \\ x_n \end{pmatrix} \in \mathbb{C}^n$ are related by

$$\frac{1}{\sqrt{n}}|x| \le \|x\| \le |x|, \tag{7.4}$$

where $\|x\|^2 \equiv \sum_{j=1}^n |x_j|^2$ and $|x| \equiv \sum_{j=1}^n |x_j|$. To see this, calculate

$$[|x_1| + \cdots + |x_n|]^2 \ge |x_1|^2 + \cdots + |x_n|^2 = \|x\|^2,$$

which proves the inequality: $\|x\| \le |x|$. To prove the other inequality, we use mathematical induction. The case $n = 1$ is trivially true: $\|x\| = |x_1|$. The case $n = 2$ is also true:

$$2\|x\|^2 - |x|^2 = 2(|x_1|^2 + |x_2|^2) - (|x_1| + |x_2|)^2$$
$$= (|x_1| - x_2|)^2 \ge 0.$$

Now assume the case $n = k \in \mathbb{N}$ is true: $k\|x\|^2 \ge |x|^2$ or

$$k(|x_1|^2 + \cdots + |x_k|^2) \ge (|x_1| + \cdots + |x_k|)^2.$$

Then, for $n = k + 1$, we have

$$(k + 1)[(|x_1|^2 + \cdots + |x_k|^2) + |x_{k+1}|^2]$$
$$= k(|x_1|^2 + \cdots + |x_k|^2) + k|x_{k+1}|^2 + (|x_1|^2 + \cdots + |x_k|^2 + |x_{k+1}|^2)$$
$$\ge (|x_1| + \cdots + |x_k|)^2 + k|x_{k+1}|^2 + (|x_1|^2 + \cdots + |x_k|^2 + |x_{k+1}|^2),$$

from which, by substracting

$$[(|x_1| + \cdots + |x_k|) + |x_{k+1}|]^2 = (|x_1| + \cdots + |x_k|)^2 + |x_{k+1}|^2 + 2|x_{k+1}|(|x_1| + \cdots + |x_k|),$$

we derive

$$(|x_{k+1}| - |x_1|)^2 + (|x_{k+1}| - |x_2|)^2 + \cdots + (|x_{k+1}| - |x_k|)^2 \ge 0.$$

Thus the proof of induction is complete.

Step 2. As in the proof of Theorem 3.1, (3.2) has a solution $\varphi(t)$ on a small interval $[0, \eta], \eta > 0$. Thus, on $[0, \eta]$, we have

$$2\|\varphi\| \|\varphi\|' = \frac{d}{dt}\|\varphi\|^2 = \frac{d}{dt}(\varphi, \varphi)$$

$$
\begin{aligned}
&= (\varphi', \varphi) + (\varphi, \varphi') \\
&= [(A\varphi, \varphi) + (\varphi, A\varphi)] + [(g(t, \varphi), \varphi) + (\varphi, g(t, \varphi))] \\
&= 2Re(A\varphi, \varphi) + 2Re(g(t, \varphi), \varphi) \\
&\leq 2|(A\varphi, \varphi)| + 2|(g(t, \varphi), \varphi)| \\
&\leq 2\|A\varphi\|\|\varphi\| + 2\|g(t, \varphi)\|\|\varphi\|,
\end{aligned}
$$

where (φ', φ) is the inner product of φ' with φ, $Re(A\varphi, \varphi)$ is the real part of $(A\varphi, \varphi)$, and the last inequality follows from the Cauchy-Schwarz inequality in Lemma 5.12 of Chapter 4. This gives, by assumimg $\|\varphi(t)\| \neq 0$ for each $t \in [0, \eta]$ for a moment, that

$$
\begin{aligned}
\|\varphi\|' &\leq \|A\varphi\| + \|g(t, \varphi)\| \\
&\leq \|A\|\|\varphi\| + |g(t, \varphi)| \\
&\leq |A|\|\varphi\| + k_0|\varphi|ne \\
&\leq (|A| + k_0\sqrt{n})\|\varphi\|,
\end{aligned}
$$

where Step 1 was used and the matrix $A = (a_{ij})_{n \times n}$ satisfies

$$
\begin{aligned}
\|Ax\|^2 &= \sum_{i=1}^{n} |\sum_{j=1}^{n} a_{ij}x_j|^2 \\
&\leq \sum_{i=1}^{n} (\sum_{j=1}^{n} |a_{ij}|^2 \sum_{j=1}^{n} |x_j|^2) \quad \text{by the Cauchy-Schwarz inequality} \\
&\equiv \|A\|^2\|x\|^2, \\
\|A\|^2 &= \sum_{i,j=1}^{n} |a_{ij}|^2 \leq (\sum_{i,j=1}^{n} |a_{ij}|)^2 \equiv |A|.
\end{aligned}
$$

By multiplying the above differential inequality by $e^{-(|A|+k_0\sqrt{n})t}$, we have

$$
\frac{d}{dt}[e^{-(|A|+k_0\sqrt{n})t}\|\varphi\|] \leq 0,
$$

and so

$$
\|\varphi\| \leq \|\varphi(0)\|e^{(|A|+k_0\sqrt{n})t}
$$

or, by Step 1,

$$
|\varphi| \leq \sqrt{n}|\varphi(0)|e^{(|A|+k_0\sqrt{n})t}.
$$

This shows that, by choosing $|\varphi(0)|$ sufficiently small, $|\varphi(t)|$, for $t \in [0, \eta]$, is also small enough. Hence, $\varphi(t)$ can be extended beyond $t = T$ and $|\varphi(T)|$ is accordingly small enough, by repeatedly using Theorem 2.9 for continuation of solutions in Chapter 8. Thus the proof of Theorem 3.1 can be applied to the interval $[T, \infty)$, in view of $(A2)$, and this proves Theorem 3.3.

Step 3. Now suppose $\|\varphi(t_0)\| = 0$ for some $t_0 \in [0, \eta]$. Then $\varphi(t)$ satisfies the integral equation

$$
\begin{aligned}
\varphi(t) &= e^{tA}\varphi(t_0) + \int_{\tau=t_0}^{t} e^{(t-\tau)A}g(\tau, \varphi)\, d\tau \\
&= \int_{\tau=t_0}^{t} e^{(t-\tau)A}g(\tau, \varphi)\, d\tau,
\end{aligned}
$$

and so, by the use of $(A1)$ and (7.1),

$$e^{t\sigma}|\varphi| \leq k_0 \int_{n e \tau = t_0}^{t} e^{\tau\sigma}|\varphi|\, d\tau.$$

This yields, by the Gronwall inequality in Chapter 8,

$$\varphi(t) \equiv 0 \quad \text{on } [t_0, \eta],$$

which, together with the arguments in Step 2, ensures that $\varphi(t)$ is extendable beyond $t = T$ and the entended $|\varphi(T)|$ is zero also. Thus, the conclusion in Step 2 still holds. $\qquad\square$

Proof of Theorem 3.5.

PROOF. We divide the proof into six steps.

Step 1. It is true that

$$A = \sum_{j=1}^{s} \lambda_j M_{j,0} + M_{j,1},$$

where $M_{j,0}$ and $M_{j,1}$ are from (1.2) and $M_{j,1}$ might be zero. To see this, apply $\frac{d}{dt}e^{tA} = Ae^{tA}$ to (1.1). It follows that

$$A = e^{-tA}\frac{d}{dt}e^{tA}$$

$$= \sum_{l=1}^{s}\sum_{m=0}^{m_l-1}(-t)^m e^{-t\lambda_l} M_{l,m}[\sum_{j=1}^{s}\sum_{k=1}^{m_j-1} kt^{k-1}e^{t\lambda_j} M_{j,k} +$$

$$\sum_{j=1}^{s}\sum_{k=0}^{m_j-1}\lambda_j t^k e^{t\lambda_j} M_{j,k}]$$

$$= \sum_{j=1}^{s}(\lambda_j M_{j,0} + M_{j,1}),$$

where Proposition 6.5 was used.

Step 2. Let $\{c_i\}_{i=1}^{n}$ be a Jordan basis of $\mathbb{C}^n$ that comes from Theorem 5.14 in Chapter 3. Then, by this Theorem 5.14, $\{c_i\}_{i=1}^{n}$ consists of cycles

$$S_{p,j}^{(l)} = \{p, (A - \lambda_j)p, \ldots, (A - \lambda_j)^{(l-1)}p\}, \quad j = 1, 2, \ldots, s \leq n$$

of generalized vectors that have various lengths $1 \leq l \leq m_{j-1}, j = 1, 2, \ldots, s \leq n$, and each cycle is a nonzero column vector in the augmented matrices

$$F_j = \begin{pmatrix} M_{j,0} \\ 1!M_{j,1} \\ 2!M_{j,2} \\ \vdots \\ (m_j - 1)M_{j,m_j-1} \end{pmatrix}, \quad j = 1, 2, \ldots, s \leq n$$

on which the needed column operations have been performed, as is stated in this Theorem 5.14. Here $\{c_i\}_{i=1}^{n}$ is so indexed that its elements which lie in a cycle $S_{p,j}^{(l)}$ appear in the same order as the elements of $S_{p,j}^{(l)}$ do.

Let $\{d_i\}_{i=1}^n$ be a basis of $\mathbb{C}^n$ that is obtained from the Jordan basis $\{c_i\}_{i=1}^n$ by replacing each cycle

$$S_{p,j}^{(l)} = \{p, (A - \lambda_j)p, \ldots, (A - \lambda_j)^{(l-1)}p\}$$

in $\{c_i\}_{i=1}^n$ by the cycle

$$U_{p,j}^{(l)} = \{\gamma^{(l-1)}p, \gamma^{(l-2)}(A - \lambda_j)p, \ldots, \gamma(A - \lambda_j)^{(l-2)}p, (A - \lambda_j)^{(l-1)}p\},$$

where γ is an arbitrarily positive constant. Observe that $U_{p,j}^{(1)} = p = S_{p,j}^{(1)}$, for which γ is not needed

Step 3. Relative to the basis $\{d_i\}_{i=1}^n$ and the standard basis $\{e_i\}_{i=1}^n$, respectively, assume that a continuous vector-valued function u on $[0, \infty)$ takes the form

$$u = \sum_{i=1}^n \varphi_i(t)d_i \quad \text{and } u = \sum_{i=1}^n \phi_i(t)e_i, \text{ respectively.}$$

Similarly, assume that, for this u, $g(t, u)$ takes the form

$$g(t, u) = \sum_{i=1}^n h_i(t, u)d_i \quad \text{and } g(t, u) = \sum_{i=1}^n f_i(t, u)e_i, \text{ respectively.}$$

Here e_i is a column vector with each component equal to 0, except for the i-th component which is 1.

Let

$$\varphi(t) = \begin{pmatrix} \varphi_1(t) \\ \varphi_2(t) \\ \vdots \\ \varphi_n(t) \end{pmatrix}, \quad \phi(t) = \begin{pmatrix} \phi_1(t) \\ \phi_2(t) \\ \vdots \\ \phi_n(t) \end{pmatrix},$$

$$h(t, u) = \begin{pmatrix} h_1(t, u) \\ h_2(t, u) \\ \vdots \\ h_n(t, u) \end{pmatrix}, \quad f(t, u) = \begin{pmatrix} f_1(t, u) \\ f_2(t, u) \\ \vdots \\ f_n(t, u) \end{pmatrix}.$$

Thus $u(t) = \phi(t)$ and $g(t, u) = f(t, u)$.

By $(A2)$, we have, for $\epsilon > 0$, there are $\eta > 0$ and $T > 0$, so that

$$\|h(t, \phi)\| \le \epsilon\|\varphi\|, \tag{7.5}$$

if $|P|\sqrt{n}\|\varphi\| < \eta$ and $t \ge T$. This is a result of the calculations:

$$h(t, \phi) = Pf(t, \phi) \quad \text{for some invertible matrix } P \text{ by (3.4)},$$

$$|P|^{-1}\|h(t, \phi)\| \le |f(t, \phi)| = |g(t, \phi)| \le \epsilon \frac{1}{|P|^2\sqrt{n}}|\phi|,$$

$$|u| = |\phi| \le |P|\|\varphi\|$$

$$\le |P|\sqrt{n}\|\varphi\| \quad \text{by (7.4).}$$

Step 4. It suffices to show that there is a solution of (3.1) on $[T, \infty)$, which is, we assume, the above $u(t) = \phi(t) = \sum_{i=1}^n \varphi_i(t)d_i$, such that the quantity $\|\varphi(t)\|$ of this solution for sufficiently small $\|\varphi(T)\|$ cannot stay in the vicinity of zero for all $t \ge T$. This is because

$$|\varphi(t)| \ge \|\varphi(t)\| \ge \frac{1}{\sqrt{n}}|\varphi(t)| \quad \text{by (7.4),}$$

$\phi = P\varphi$ for the above invertible matrix P by (3.4),

$$|\phi| \geq \frac{1}{\|P^{-1}\|}|\varphi| \geq \frac{1}{\|P^{-1}\|}\frac{1}{\|P\|}|\phi|.$$

Step 5. Now we estimate $\|\varphi\|$. From (3.1), we have by making use of Step 1 and Proposition 6.5 that

$$\sum_{i=1}^{n} \varphi_i' d_i = A \sum_{i=1}^{n} \varphi_i(t) d_i + \sum_{i=1}^{n} h_i(t) d_i$$

$$= [\sum_{i=1}^{m_1} \lambda_1 \varphi_i(t) d_i + \sum_{i=m_1+1}^{m_1+m_2} \lambda_2 \varphi_i(t) d_i + \cdots$$

$$+ \sum_{i=m_{s-1}+1}^{m_{s-1}+m_s} \lambda_s \varphi_i(t) d_i] \tag{7.6}$$

$$+ \sum_{i=2}^{n} \gamma_i \varphi_{i-1}(t) d_i + \sum_{i=1}^{n} h_i(t) d_i,$$

where the constant $\gamma_i \geq 0$ is either 0 or else can be made arbitrarily small by Step 2.

Assume that, for some $1 \leq s_0 \leq s$, the generalized eigenvalues

$$\lambda_j, \quad j = 1, 2, \ldots, s_0,$$

each with multiplicity m_j, constitute $m \geq 1$ ones that have positive real parts, and assume that the remaining generalized values have nonnegative real parts. Further, assume that the first m d_i's in the basis $\{d_i\}_{i=1}^{n}$ correspond to those m generalized eigenvalues, and assume that the real parts of those m generalized eigenvalues exceed some $\sigma > 0$.

Let

$$R^2 = \sum_{i=1}^{m} |\varphi_i(t)|^2 \quad \text{and} \quad \rho^2 = \sum_{i=m+1}^{n} |\varphi_i(t)|^2.$$

Now suppose that the conclusion in Theorem 3.5 is false, and we seek a contradiction. Thus, for the chosen η and T, there is a $\delta > 0$ such that the above $u(t) = \phi(t) = \sum_{i=1}^{n} \varphi_i d_i$ satisfies

$$\|\varphi(t)\| < \eta \quad \text{for } t \geq T$$

if $\|\varphi(T)\| < \delta$. Choose this $u(t)$ with small $\|\varphi(T)\| < \delta$, so that $R(T) = 2\rho(T) > 0$ or more generally, $R(T) > \rho(T)$, and for a moment, assume that $\varphi(t) \neq 0$ for all $t \geq T$.

By choosing

$$\max_{i=2,\ldots,n} \{\gamma_i\} < \frac{\sigma}{20}$$

and by noting

$$\min_{j=1,\ldots,s_0} \{Re(\lambda_j)\} > \sigma,$$

it follows from (7.6) that

$$2RR' = \frac{d}{dt}R^2 = \sum_{i=1}^{n} [(\bar{\varphi}_i)'\varphi + \bar{\varphi}\varphi']$$

$$= \{\sum_{i=1}^{m_1}(\lambda_1 + \bar{\lambda}_1)|\varphi_i|^2 + \sum_{i=m_1+1}^{m_1+m_2}(\lambda_2 + \bar{\lambda}_2)|\varphi_i|^2 + \cdots$$

$$+ \sum_{i=m_{s_0-1}+1}^{m_{s_0-1}+m_{s_0}}(\lambda_{s_0} + \bar{\lambda}_{s_0})|\varphi_i|^2\} + [\sum_{i=2}^{m}\gamma_i\overline{\varphi_{i-1}}\varphi_i + \sum_{i=2}^{m}\gamma_i\varphi_{i-1}\bar{\varphi}_i] +$$

$$[\sum_{i=1}^{m}h_i\bar{\varphi}_i + \sum_{i=1}^{m}\bar{h}_i\varphi_i]$$

$$\geq 2\sigma R^2 - 2(\frac{\sigma}{20})R^2 - 2\|h\|R.$$

Here $Re(\lambda_j)$ is the real part of λ_j, and the Cauchy-Schwarz inequality was used. This continues to yield, by choosing $\epsilon < \frac{\sigma}{10}$ and by noting $-\|h\| \geq -\epsilon\|\varphi\| \geq -\epsilon(R + \rho)$ from (7.5),

$$RR' \geq (\frac{1}{2}(2\sigma - \frac{1}{10}\sigma - \frac{1}{5}\sigma)R^2 - \epsilon\rho R$$

$$\geq \frac{1}{2}\sigma R^2 - \frac{\sigma}{10}\rho R,$$

or, by dividing R,

$$R' \geq \frac{1}{2}\sigma R - \frac{\sigma}{10}\rho. \tag{7.7}$$

In the same way, we have, by noting $\max_{i=s_0+1,\ldots,n}\{Re(\lambda_i)\} \leq 0$,

$$2\rho\rho' = \frac{d}{dt}\rho^2 = \sum_{i=m+1}^{n}(\bar{\varphi}_i'\varphi_i + \bar{\varphi}_i\varphi')$$

$$\leq 2(\frac{\sigma}{20})\rho^2 + 2\epsilon(R + \rho)\rho$$

$$\leq \frac{1}{10}\sigma\rho^2 n + \frac{1}{5}\sigma(R + \rho)\rho,$$

or, by dividing ρ,

$$\rho' \leq \frac{\sigma}{10}R + \frac{3}{20}\sigma\rho. \tag{7.8}$$

Combining the above two differential inequalities (7.7) and (7.8), we have

$$(R - \rho)' \geq \frac{4}{10}R - \frac{1}{4}\sigma$$

$$\geq \frac{1}{4}(R - \rho),$$

and this gives, after being multiplied by $e^{-\frac{1}{4}\sigma t}$,

$$\frac{d}{dt}[(R(t) - \rho(t)e^{-\frac{1}{\sigma}t}] \geq 0$$

or

$$R(t) - \rho(t) \geq (R(T) - \rho(T))e^{\frac{\sigma(t-T)}{4}}.$$

Since $R(T) - \rho(T) > 0$, we have

$$\|\varphi(t)\| \geq R(t)$$

can be made arbitrarily large by choosing large t, and this is a contradiction to $\|\varphi(t)\| < \eta$ for all $t \geq T$.

Step 6. Now suppose that $\varphi(t_1) = 0$ for some $t_1 \geq T$, and we seek a contradiction. Then $t_1 > T$ follows, since $t_1 = T$ contradicts $\varphi(T) \neq 0$. Thus $\varphi(t_2) = 0$ for some $T \leq t_2 < t_1$, otherwise, on $[T, t_1)$,

$$R(t) - \rho(t) \geq (R(T) - \rho(T))e^{\frac{\sigma(t-T)}{4}} \quad \text{by Step 5,}$$

which, by letting $t \to t_1$ and by using the continuity of $\varphi(t)$ in t, gives $\varphi(t_1) \neq 0$, a contradiction. In the same way, we have

$$\varphi(t_3) = 0$$

for some $T \leq t_3 < t_2 < t_1$, and eventually,

$$\varphi(t_0) = 0$$

for some $t_0 = T$. This is a contradiction to $\varphi(T) \neq 0$. $\qquad\square$

REMARK 7.1. *It can be seen from the above proof that* $u(t) = \sum_{i=1}^{n} \varphi_i(t)d_i$ *with* $R(T) = 2\rho(T)$ *or more generally,* $R(T) > \rho(T)$, *cannot stay in the vicinity of the origin if it is a solution of the equation* (3.1) *with the initial value* $\|\varphi(T)\|$ *sufficiently small. Here* σ *and* γ_i *are chosen so that*

$$\min_{j=1,\ldots,s_0} \{Re(\lambda_j)\} > \sigma \quad \text{and} \quad \max_{i=2,\ldots,n} \{\gamma_i\} < \frac{\sigma}{20}.$$

Proof of Theorem 3.9.

PROOF. We divide the proof into ten steps.

Step 1. Since the generalized eigenspace K_{λ_j} corresponding to the generalized eigenvector λ_j has the dimension m_j, a fact from linear algebra [9], it follows from Lemma 6.1 and Proposition 6.4 that $\mathbb{C}^n$ has a basis

$$\{d_j\}_{j=1}^{n},$$

whose first m_1 d_j's come from the columns of $M_{1,0}$, next m_2 d_j's from those of $M_{2,0}$, $\ldots$, and last m_s d_j's from those of $M_{s,0}$. Here note that

$$n = m_1 + \cdots + m_{s_0} + \cdots + m_s,$$
$$m = m_1 + \cdots + m_{s_0}.$$

Step 2. Relative to the basis $\{d_j\}_{j=1}^{n}$, let $a = \sum_{j=1}^{m} a_j d_j$ be a vector in $\mathbb{C}^n$ and consider the integral equation

$$\theta(t,a) = U_1(t-t_0)a + \int_{q=t_0}^{t} U_1(t-q)g(q,\theta(q,a))\,dq$$
$$- \int_{q=t}^{\infty} U_2(t-q)g(q,\theta(q,a))\,dq. \tag{7.9}$$

Here $t_0 \geq T$ is a fixed number where T is the T in (3.5), and $U_1(t)$ and $U_2(t)$ are, respectively,

$$U_1(t) = \sum_{j=1}^{s_0} \sum_{k=0}^{m_j-1} t^k e^{t\lambda_j} M_{j,k}, \tag{7.10}$$

$$U_2(t) = \sum_{j=s_0+1}^{s} \sum_{k=0}^{m_j-1} t^k e^{t\lambda_j} M_{j,k} \tag{7.11}$$

and satisfy

$$e^{tA} = U_1(t) + U_2(t) \tag{7.12}$$

in view of (1.2).

Step 3. We shall solve (7.9) by the method of successive approximations in Chapter 8. Let $\alpha > 0$ be chosen so that $-\alpha$ is larger than the real parts of the generalized eigenvalues $\{\lambda_j\}_{j=1}^{s_0}$. Then it follows from the above $U_1(t)$ and U_2 that there are positive constants K and σ, such that

$$|U_1(t)| \le Ke^{-(\alpha+\sigma)t} \quad \text{for } t \ge 0 \tag{7.13}$$

$$|U_2(t)| \le Ke^{\sigma t} \quad \text{for } t \le 0. \tag{7.14}$$

Here note by assumption that the real parts of the generalized eigenvalues $\{\lambda_j\}_{j=1}^{s_0}$ of A are negative and the real parts of $\{\lambda_j\}_{j=s_0+1}^{n}$ are positive. Let the small ϵ in (3.5) be chosen so that $\frac{2\epsilon K}{\sigma} < \frac{1}{2}$, and let $|a|$ be small so that $2K|a| < \delta$ where δ is the δ related to (3.5). Using the successive approximations to (7.9):

$$\theta_{(0)}(t,a) = 0,$$

$$\theta_{(l+1)}(t,a) = U_1(t-t_0)a + \int_{q=t_0}^{t} U_1(t-q)g(q,\theta_{(l)}(q,a))\,dq$$

$$- \int_{q=t}^{\infty} U_2(t-q)g(q,\theta_{(l)}(q,a))\,dq, \quad l = 0,1,2,\dots,$$

it will follow that

$$|\theta_{(l+1)}(t,a) - \theta_{(l)}(t,a)| \le \frac{K|a|e^{-\alpha(t-t_0)}}{2^l}. \tag{7.15}$$

This is a result of the calculations that use

$$g(t,0) = 0, \quad \frac{2\epsilon K}{\sigma} < \frac{1}{2}, \quad 2K|a| < \delta,$$

(3.5), (7.13), and (7.14):

$$|\theta_{(1)} - \theta_{(0)}| = |\theta_{(1)} - 0|$$

$$= |U_1(t-t_0)a + \int_{q=t_0}^{t} U_1(t-q)g(q,0)\,dq - \int_{q=t}^{\infty} U_2(t-q)g(q,0)\,dq|$$

$$\le K|a|e^{-(\alpha+\sigma)(t-t_0)}$$

$$\le K|a|e^{-\alpha(t-t_0)} \le \frac{\delta}{2} \le \delta;$$

$$|\theta_{(2)} - \theta_{(1)}|$$

$$\le \int_{q=t_0}^{t} Ke^{-(\alpha+\sigma)(t-q)}\epsilon|\theta_{(1)}|\,dq + \int_{q=t}^{\infty} Ke^{\sigma(t-q)}\epsilon|\theta_{(1)}|\,dq$$

$$\le \int_{q=t_0}^{t} Ke^{-(\alpha+\sigma)(t-q)}\epsilon K|a|e^{-\alpha(q-t_0)}\,dq + \int_{q=t}^{\infty} Ke^{\sigma(t-q)}\epsilon K|a|e^{-\alpha(q-t_0)}\,dq$$

$$\le K(\frac{\epsilon K|a|}{\sigma})e^{-\alpha(t-t_0)} + K(\frac{\epsilon K|a|}{\sigma})e^{-\alpha(t-t_0)} \quad (\text{Here } \int_{t_0}^{t} e^{\sigma q}\,dq \le \frac{e^{\sigma t}}{\sigma})$$

$$= K(\frac{2\epsilon K}{\sigma})|a|e^{-\alpha(t-t_0)}$$

$$\leq \frac{K|a|e^{-\alpha(t-t_0)}}{2} \leq \frac{\delta}{4};$$

$$|\theta_{(2)}| \leq |\theta_{(1)}| + \frac{\delta}{4}$$

$$\leq \frac{\delta}{2} + \frac{\delta}{4} \leq \delta;$$

$$|\theta_{(3)} - \theta_{(2)}|$$

$$\leq \int_{q=t_0}^{t} Ke^{-(\alpha+\sigma)(t-q)}\epsilon|\theta_{(2)} - \theta_{(1)}|\,dq + \int_{q=t}^{\infty} Ke^{\sigma(t-q)}\epsilon|\theta_{(2)} - \theta_{(1)}|\,dq$$

$$\leq \frac{1}{2}K(\frac{2\epsilon K}{\sigma})|a|e^{-\alpha(t-t_0)}$$

$$\leq \frac{K|a|e^{-\alpha(t-t_0)}}{2^2} \leq \frac{\delta}{8};$$

$$|\theta_{(3)}| \leq |\theta_{(2)}| + \frac{\delta}{8}$$

$$\leq \frac{\delta}{2} + \frac{\delta}{4} + \frac{\delta}{8} \leq \delta;$$

$$\vdots,$$

$$|\theta_{(l+1)} - \theta_{(l)}|$$

$$\leq \frac{K|a|e^{-\alpha(t-t_0)}}{2^l} \leq \frac{\delta}{2^{l+1}}, \quad l = 0, 1, 2, \ldots; \quad \text{(by induction)}$$

$$|\theta_{(l+1)}| \leq \frac{\delta}{2} + \frac{\delta}{4} + \cdots + \frac{\delta}{2^{l+1}} \leq \delta.$$

Step 4. Since

$$\theta_{(l+1)} = \theta_{(0)} + \sum_{j=0}^{l}(\theta_{(j+1)} - \theta_{(j)}),$$

it follows from (7.15) that

$$|\theta_{(l+1)}| \leq \sum_{j=0}^{l} \frac{K|a|e^{-\alpha(t-t_0)}}{2^j}$$

$$\leq \sum_{j=0}^{\infty} \frac{K|a|e^{-\alpha(t-t_0)}}{2^j} \tag{7.16}$$

$$= 2K|a|e^{-\alpha(t-t_0)} \quad \text{for all } l.$$

Thus the Weirstrass M-test from Advanced Calculus, Apostol [2, Page 223] (See Chapter 8), implies that, as $l \longrightarrow \infty$, $\theta_{(l+1)}$ converges uniformly for $t \geq t_0 \geq T$ to a continuous function $\theta(t, a)$ for large $t \geq t_0$ and for small $|a| < \frac{\delta}{2K}$. This $\theta(t, a)$ satisfies

$$|\theta(t, a)| \leq 2K|a|e^{-\alpha(t-t_0)} \tag{7.17}$$

by (7.16) where l is let go to ∞, and is a solution to (7.9) by (7.14), and is a solution of (3.2) by the fundamental theorem of calculus for improper integrals (See the NOTE and the NOTE, respectively in [2, Page 162 and Page 277]), in which (7.14) is used again. That $\theta(t, a) \longrightarrow 0$ as $t \longrightarrow \infty$ follows from (7.17).

Step 5. From (7.9), we have that the initial value $\theta(t_0, a)$ of $\theta(t, a)$ satisfies

$$\theta(t_0, a) = a - \int_{q=t_0}^{\infty} U_2(t_0 - q)g(q, \theta(q, a)) \, dq, \tag{7.18}$$

where, relative to the basis $\{d_j\}_{j=1}^{n}$ in $\mathbb{C}^n$ (See Steps 1 and 2), $a = \sum_{j=1}^{m} a_j d_j$ and $\int_{q=t_0}^{\infty} U_2(t_0 - q)g(q, \theta(q, a)) \, dq$, by using Proposition 6.5, takes the form

$$-\int_{q=t_0}^{\infty} U_2(t_0 - q)g(q, \theta(q, a)) \, dq = \sum_{j=m+1}^{n} \Psi_j(a) d_j. \tag{7.19}$$

Thus, relative to that basis, the first m components $\theta_j, j = 1, 2, \ldots, m$ of $\theta(t_0, a)$ are

$$\theta_j = a_j, \quad j = 1, 2, \ldots, m, \tag{7.20}$$

and the later $(n - m)$ components $\{\theta_j\}_{j=m+1}^{n}$ are

$$\theta_j = \Psi_j(a) = \Psi_j(\theta_1, \theta_2, \ldots, \theta_m), \quad j = m + 1, m + 2, \ldots, n. \tag{7.21}$$

It is readily seen that (7.21) defines an m-dimensional manifold (or surface) S, where a is sufficiently near the origin:

$$2K|a| < \delta.$$

Step 6. Because of (3.5), the uniqueness result in Chapter 8 gives the uniqueness of solutions of (3.2) which start sufficiently near the origin. Therefore, if $\varphi(t)$ is any solution of (3.2) with $\varphi(t_0)$ on S and $|\varphi(t_0)|$ sufficiently small, then $\varphi(t) = \theta(t, a)$ for some a with $|a|$ sufficiently small. Here $\theta(t, a)$ is the solution of (3.2) in Step 5 that satisfies $\theta(t_0, a) = \varphi(t_0)$. Thus

$$\varphi(t) = \theta(t, a) \longrightarrow 0 \quad \text{as } t \longrightarrow \infty.$$

Step 7. Assume that, relative to the standard basis $\{e_j\}_{j=1}^{n}$ in $\mathbb{C}^n$, the vector

$$\int_{q=t_0}^{\infty} U_2(t_0 - q)g(q, \theta(q, a)) \, dq$$

takes the form

$$\int_{q=t_0}^{\infty} U_2(t_0 - q)g(q, \theta(q, a)) \, dq = \sum_{j=1}^{n} \mu_j(a) e_j.$$

Then in view of (7.19), it follows from linear algebra [9] that

$$\begin{pmatrix} 0 \\ \vdots \\ 0 \\ \Psi_{m+1}(a) \\ \vdots \\ \Psi_n(a) \end{pmatrix} = Q \begin{pmatrix} \mu_1(a) \\ \vdots \\ \mu_m(a) \\ \vdots \\ \mu_n(a) \end{pmatrix}$$

for some invertible matrix Q, and so, for $j = m + 1, \ldots, n$,

$$|\Psi_j(a)| \leq |Q| \sum_{j=1}^{n} |\mu_j(a)| = |Q| \left| \int_{q=t_0}^{\infty} U_2(t_0 - q)g(q, \theta(q, a)) \, dq \right|. \tag{7.22}$$

Using (7.14), $g(t,0) \equiv 0$, and (3.5), it follows from (7.22) that, for $j = m+1, \ldots, n$,

$$|\Psi_j(a) - \Psi_j(0)| = |\Psi_j(a)|$$

$$\leq 2K^2|Q||\epsilon||a| \int_{q=t_0}^{\infty} e^{\sigma(t_0-q)} \, dq \quad \text{(Here } e^{-\alpha(q-t_0)} \leq 1\text{)}$$

$$\leq \frac{2K^2|a||Q|}{\sigma} \epsilon.$$

This shows that, for $\epsilon > 0$ given, there is a $\delta > 0$ such that, if $2K|a| < \delta$, then

$$\frac{|\Psi_j(a) - \Psi_j(0)|}{|a|} \leq \frac{2K^2|Q|}{\sigma} \epsilon.$$

Therefore

$$\lim_{|a| \to 0} \frac{|\Psi_j(a) - \Psi_j(0)|}{|a|} = 0,$$

and so, by choosing $a = a_k d_k$, $k = 1, \ldots, m$, respectively, we have

$$\frac{\partial}{\partial a_k} \Psi_j(0) = 0, \quad j = m+1, \ldots, n.$$

Applying this to (7.18), it follows that

$$\frac{\partial}{\partial a_k} \theta(t_0, a)|_{a=0} = \begin{cases} d_k, & \text{if } k = 1, \ldots, m; \\ 0, & \text{otherwise.} \end{cases}$$

That is, the tangent vectors to the surface S at the origin:

$$\frac{\partial}{\partial a_k} \theta(t_0, 0), \quad k = 1, \ldots, m$$

are on the subspace spanned by $\{d_k\}_{k=1}^m$ which equals M_0; or equivalently, the surface S is tangent to M_0 at the origin.

Step 8. We will show that if $\varphi(t)$ is a solution of (3.2) with $\varphi(t_0)$ sufficiently near the origin but not on S, then

$$|\varphi(t_1)| > \delta \quad \text{for some } t_1 \geq t_0,$$

where δ is the δ in Step 3, that is, the δ related to (3.5). Otherwise, suppose $|\varphi(t)| \leq \delta$ for all $t \geq t_0$, and we seek a contradiction. It follows immediately from (3.2) that $\varphi(t)$ satisfies

$$\varphi(t) = e^{(t-t_0)A} \varphi(t_0) + \int_{q=t_0}^{t} e^{(t-q)A} g(q, \varphi(q)) \, dq,$$

similar to the formula (1.6). This can be rewritten, by using (7.12), as

$$\varphi(t) = U_1(t - t_0)\varphi(t_0) + U_2(t - t_0)\mu + \int_{q=t_0}^{t} U_1(t - q)g(q, \varphi(q)) \, dq$$

$$- \int_{q=t}^{\infty} U_2(t_0 - q)g(q, \varphi(q)) \, dq,$$
(7.23)

where

$$\mu = \int_{q=t_0}^{\infty} U_2(t_0 - q)g(q, \varphi(q)) \, dq + \varphi(t_0),$$

a constant vector. This μ exists finitely because $|g(q, \varphi(q))|$ is finite for finite $|\varphi(q)|$ by (3.5) and $g(q, 0) \equiv 0$, for which (7.14) can be applied.

Furthermore, the μ should not have components in the subspace spanned by $\{d_k\}_{k=m+1}^n$, so that the term $U_2(t-t_0)\mu$ on the right side of (7.23) should be zero. For, otherwise, $U_2(t-t_0)\mu$ is unbounded in view of (7.11) as $t \longrightarrow \infty$, and this contradicts the boundedness of $|\varphi(t)|$ on the left side of (7.23). Thus $\varphi(t)$ satisfies (7.9). This $\varphi(t)$ will be equal to $\theta(t,a)$ constructed above, so that $\varphi(t_0)$ is on S as $\theta(t_0, a)$ is, contradicting the fact that $\varphi(t_0)$ is not on S. To see this, observe that

$$|\theta(t,a) \leq 2K|a| \leq \delta \quad \text{for } t \geq t_0$$

by (7.17). Since $\varphi(t)$ and $\theta(t,a)$ are two solutions of (7.9), we have, by using (7.13), (7.14), and $|\theta(t,a)|, |\varphi(t)| \leq \delta$ for $t \geq t_0$,

$$|\varphi(t) - \theta(t,a)| \leq K\epsilon e^{-\alpha t} \int_{t_0}^t e^{\sigma q}|\varphi(q) - \theta(q,a)|\, dq$$
$$+ K\epsilon e^{\sigma t} \int_t^\infty e^{-\sigma q}|\varphi(q) - \theta(q,a)|\, dq.$$

By letting $M \equiv \sup_{t \geq t_0}|\varphi(t) - \theta(t,a)|$, which is possible because of $|\varphi(t)|, |\theta(t,a)| \leq \delta$, it follows that

$$M \leq 2K\epsilon \frac{M}{\sigma}.$$

Thus $M = 0$, which is $\varphi(t) = \theta(t,a)$, otherwise, by dividing by M, we have

$$\frac{2K\epsilon}{\sigma} \geq 1,$$

which contradicts $\frac{2K\epsilon}{\sigma} < \frac{1}{2}$.

Step 9. We will show that the above $\theta(\tau, a)$ also lies on S for each $\tau \geq t_0$. Consider the equation (3.2) equipped with the initial value $u(\tau) = \theta(\tau, a)$. Corollary 2.8 for local existence in Chapter 8 gives a unique solution $\varphi(t)$ on some small t-interval $[\tau, \tau + \eta]$ for $\eta > 0$, where uniqueness of $\varphi(t)$ follows from the uniqueness result in Chapter 8. As in Step 8, $\varphi(t)$ satisfies the integral equation

$$\varphi(t) = e^{(t-\tau)A}\theta(\tau, a) + \int_{q=\tau}^t e^{(t-q)A}g(q, \varphi(q))\, dq. \tag{7.24}$$

Since

$$e^{(t-\tau)A}\theta(\tau, a) = [U_1(t-\tau) + U_2(t-\tau)][U_1(\tau - t_0)a+$$
$$\int_{q=t_0}^\tau U_1(\tau - q)g(q, \theta(q,a))\, dq - \int_{q=\tau}^\infty U_2(\tau - q)g(q, \theta(q,a))\, dq]$$
$$= U_1(t - t_0)a + \int_{q=t_0}^\tau U_1(t-q)g(q, \theta(q,a))\, dq -$$
$$\int_{q=\tau}^\infty U_2(t-q)g(q, \theta(q,a))\, dq$$
$$= \theta(t,a) - \int_{q=\tau}^t U_1(t-q)g(q, \theta(q,a))\, dq -$$
$$\int_{q=\tau}^t U_2(t-q)g(q, \theta(q,a))\, dq$$
$$= \theta(t,a) - \int_{q=\tau}^t e^{(t-q)A}g(q, \theta(q,a))\, dq,$$

(7.24) becomes

$$\varphi(t) - \theta(t, a) = \int_{q=\tau}^{t} e^{(t-q)A} [g(q, \varphi) - g(q, \theta(q, a))] \, dq,$$

and then, by (3.5) and

$$|e^{tA}| \leq M e^{\beta t} \quad \text{for some } M, \beta > 0,$$

we have

$$|\varphi(t) - \theta(t, a)| \leq M \epsilon e^{\beta t} \int_{q=\tau}^{t} |\varphi(q) - \theta(q, a)| \, dq$$

or

$$e^{-\beta t} |\varphi(t) - \theta(t, a)| \leq M \epsilon \int_{q=\tau}^{t} e^{-\beta q} |\varphi(q) - \theta(q, a)| \, dq.$$

Here we used

$$U_1(t - \tau)U_1(\tau - q) = U_1(t - q),$$
$$U_2(t - \tau)U_2(\tau - q) = U_2(t - q), \quad \text{and}$$
$$U_1(t - \tau)U_2(\tau - q) = 0 = U_2(t - \tau)U_1(t - q),$$

as a result of Proposition 6.5 and

$$\begin{aligned}
U_1(t - q) + U_2(t - q) &= e^{(t-q)A} \\
&= e^{(t-\tau)A} e^{(\tau - q)A} \\
&= [U_1(t - \tau) + U_2(t - \tau)][U_1(\tau - q) + U_2(\tau - q)] \\
&= U_1(t - \tau)U_1(\tau - q) + U_2(t - \tau)U_2(\tau - q).
\end{aligned}$$

It follows from the Gronwall inequality in Chapter 8 that

$$e^{-\beta t} |\varphi(t) - \theta(t, a)| = 0 \quad \text{or} \quad \varphi(t) = \theta(t, a), \tag{7.25}$$

and then

$$|\varphi(t)| \leq |\theta(t, a)| \leq 2K |a| e^{-\alpha(t - t_0)} \leq 2K |a| < \delta.$$

Thus $|\varphi(\tau)| < \delta$, and $\varphi(t) = \varphi(t, \theta(\tau, a))$ can be extended from $[\tau, \tau + \eta]$ to $[\tau, \infty)$ by repeatedly using Theorem 2.9 for continuation of solutions in Chapter 8, and this extended $\varphi(t)$ by (7.25) equals $\theta(t, a)$ for $t \geq \tau$. This $\varphi(t) = \theta(t, a)$ becomes a unique solution of (3.2) for $t \geq \tau$ with the initial value sufficiently near the origin:

$$|\varphi(\tau)| = |\theta(\tau, a)| < \delta.$$

Since

$$|\varphi(t)| = |\theta(t, a)| \leq 2K |a| e^{-\alpha(t - t_0)}$$
$$\leq 2K |a| < \delta,$$

Step 8 shows that $\varphi(\tau) = \theta(\tau, a)$ must lie on S.

Step 10. Let $\tilde{s} = \theta(t_0, a) \in S$, where $2K |a| < \delta$, and write $\theta(t, a) = \theta(t, t_0; \tilde{s})$ to indicate the dependence on t_0 and $\tilde{s}$. Here $\theta(t, t_0; \tilde{s})$ is the unique solution to the equation (3.2) with the initial value $\tilde{s}$ at $t = t_0$. Step 9 shows that $\theta(t, t_0; \tilde{s})$ is the unique solution of the equation (3.2) with the initial value $\theta(\tau, t_0; \tilde{s})$ at $t = \tau \geq t_0$, and so, by defining $V(t, \tau)\theta(\tau, t_0; \tilde{s})$ to be this solution, it follows that $V(t, \tau), t \geq \tau \geq t_0$, is a family of exponentially decaying evolution operators:

$$V(t, \tau) : D \equiv \{\theta(\rho, t_0; \tilde{s}) : \rho \geq t_0, \tilde{s} \in S\} \longrightarrow D,$$

$$V(\tau, \tau)\theta(\tau, t_0; \tilde{s}) = \theta(\tau, t_0; \tilde{s}),$$
$$V(t, \tau)[V(\tau, r)\theta(r, t_0; \tilde{s})] = V(t, \tau)\theta(\tau, t_0; \tilde{s}) = \theta(t, t_0; \tilde{s})$$
$$= V(t, r)\theta(r, t_0; \tilde{s}) \quad \text{for } t \geq \tau \geq r \geq t_0.$$
$$|V(t, \tau)\theta(\tau, t_0; \tilde{s})| = |\theta(t, t_0; \tilde{s})| \leq 2K|a|e^{-\alpha(t-t_0)} \longrightarrow 0 \quad \text{as } t \longrightarrow +\infty.$$

$\square$

REMARK 7.2. *The proof in Steps 9 and 10 seems new.*

Proof of Theorem 3.10.

PROOF. In (7.9) and (7.21), by letting $a = \sum_{k=1}^m a_k d_k$, where, for each $j = 1, \ldots, s_0$,

$$a_k \begin{cases} \neq 0, & \text{if } d_k \text{ is one of those } m_j \text{ column vectors } d_j\text{'s from the matrix } M_{j,0}; \\ = 0, & \text{otherwise,} \end{cases}$$

the proof of Theorem 3.9 gives the needed results. $\square$

Proof of Theorem 3.11.

PROOF. The first paragraph of Theorem 3.11 follows from using (7.21) and the fact that the matrix $U_2(t)$ annihilates all the vectors that are not in the subspace $S(\cup_{k=s_0+1}^s M_{k,0})$. This is because, then, the θ_j's in (7.21) where $j = m + 1, \ldots, n$, are all zero.

To prove the second paragraph, we use (7.9) and (7.21) where choose a be the nonzero a in the above proof of Theorem 3.10 in which $j = j_0$. Then $a \in S(M_{j_0,0})$, and $U_1(t - t_0)a$ is also a nonzero vector in $S(M_{j_0,0})$ because $U_1(t)$ leaves $S(M_{j_0,0})$ invariant. Since a nonzero component in

$$S(\cup_{j=s_0+1,\ldots,s} M_{j,0})$$

of $g(t, u)$ for $u \neq 0$ in $S(M_{j_0,0})$ is zero by the assumption on g, it follows from the construction of the solution $\theta(t, a)$ in the proof of Theorem 3.9 that $\theta(t, a)$ has only components in $S(M_{j_0,0})$ and that $g(t, \theta(t, a))$ has no components in

$$S(\cup_{j=s_0+1,\ldots,s} M_{j,0}).$$

Therefore, the θ_j's in (7.21) where $j = m + 1, \ldots, n$, are all zero. This proves the second paragraph.

The third paragraph follows from applying Theorem 3.9. $\square$

8. Problems

Study the long time behavior, as $t \longrightarrow +\infty$, of the solutions of the given initial value problems. In the case of linear homogeneous systems, also find the corresponding stable subspaces M_0's with zero asymptotes, the stable subspaces M_b's with nonzero asymptotes, the unstable subspaces M_∞'s, and the unstabe subsets $\triangle_\infty$'s. However, in the case of nonlinear systems, study those problems by following Section 5.

1.

$$\frac{d}{dt} u = Au, \quad t > 0,$$
$$u(0) = u_0, u_1, u_2, u_3, u_4, u_5, \quad \text{and } u_6, \text{ respectively,}$$

where

$$u_0 = \alpha_0 \begin{pmatrix} 1 \\ 0 \\ 0 \end{pmatrix}, \quad u_1 = \alpha_1 \begin{pmatrix} 1 \\ 5 \\ 20 \end{pmatrix}, \quad u_2 = \alpha_2 \begin{pmatrix} 1 \\ -1 \\ 2 \end{pmatrix},$$

$$u_3 = u_0 + u_1, \quad u_4 = u_0 + u_2, \quad u_5 = u_2 + u_3,$$

$$u_6 = u_0 + u_1 + u_2,$$

$\alpha_0, \alpha_1,$ and α_2 are constants, and

$$A = \begin{pmatrix} 0 & 1 & 0 \\ 0 & 1 & 1 \\ 0 & 8 & 3 \end{pmatrix}.$$

2.

$$\frac{d}{dt}u = Au, \quad t > 0,$$

$$u(0) = u_0, u_1, \quad \text{and } u_2, \text{ respectively,}$$

where

$$u_0 = \begin{pmatrix} \frac{3}{4} \\ 1 \\ 0 \end{pmatrix}, u_1 = \begin{pmatrix} \frac{1}{2} \\ 1 \\ 1 \end{pmatrix}, \quad \text{and } u_2 = \begin{pmatrix} 1 \\ 0 \\ 0 \end{pmatrix},$$

and

$$A = \begin{pmatrix} -2 & 1 & 0 \\ 0 & -1 & 1 \\ 0 & -1 & 1 \end{pmatrix}.$$

3.

$$\frac{d}{dt}u = Au, \quad t > 0,$$

$$u(0) = u_0 \quad \text{and } u_1, \text{ respectively,}$$

where

$$u_0 = \alpha_1 \begin{pmatrix} 1 \\ 0 \\ 0 \\ 0 \\ 0 \end{pmatrix} + \alpha_2 \begin{pmatrix} 0 \\ 1 \\ 0 \\ 0 \\ 0 \end{pmatrix} + \alpha_3 \begin{pmatrix} 0 \\ 1 \\ 1 \\ 1 \\ 0 \end{pmatrix} + \alpha_4 \begin{pmatrix} 0 \\ -1 \\ -1 \\ 0 \\ 1 \end{pmatrix}, u_1 = \begin{pmatrix} 0 \\ 1 \\ 1 \\ 0 \\ 0 \end{pmatrix},$$

$\alpha_j, j = 1, \ldots, 4$, are constants, and

$$A = \begin{pmatrix} 1 & 0 & 0 & 1 & -1 \\ 0 & 1 & -2 & 3 & -3 \\ 0 & 0 & -1 & 2 & -2 \\ 1 & -1 & 1 & 0 & 1 \\ 1 & -1 & 1 & -1 & 2 \end{pmatrix}$$

from [**10**, Page 172].

4.

$$\frac{d}{dt}u = Au + g_0, \quad t > 0,$$

$$u(0) = u_0, u_1, \quad \text{and } u_2, \text{ respectively,}$$

where g_0 is a constant vector,

$$u_0 = \begin{pmatrix} 1 \\ 0 \\ 0 \end{pmatrix}, u_1 = \begin{pmatrix} 1 \\ 5 \\ 20 \end{pmatrix}, u_2 = \begin{pmatrix} 1 \\ -1 \\ 2 \end{pmatrix},$$

and A is the A in Problem **1**:

$$A = \begin{pmatrix} 0 & 1 & 0 \\ 0 & 1 & 1 \\ 0 & 8 & 3 \end{pmatrix}.$$

5.

$$\frac{d}{dt}u = Au + g_0, \quad t > 0,$$
$$u(0) = u_0, u_1, u_2, \quad \text{and } u_3, \text{ respectively,}$$

where g_0 is a constant vector,

$$u_0 = \begin{pmatrix} \frac{3}{4} \\ 1 \\ 0 \end{pmatrix}, u_1 = \begin{pmatrix} -\frac{1}{4} \\ 0 \\ 1 \end{pmatrix}, u_2 = \begin{pmatrix} \frac{1}{2} \\ 1 \\ 1 \end{pmatrix}, \quad \text{and } u_3 = \begin{pmatrix} 1 \\ 0 \\ 0 \end{pmatrix},$$

and A is the A in Problem **2**:

$$A = \begin{pmatrix} -2 & 1 & 0 \\ 0 & -1 & 1 \\ 0 & -1 & 1 \end{pmatrix}.$$

6.

$$\frac{d}{dt}u = Au + g_0, \quad t > 0,$$
$$u(0) = u_0 \quad \text{and } u_1, \text{ respectively,}$$

where g_0 is a constant vector,

$$u_0 = \alpha_1 \begin{pmatrix} 1 \\ 0 \\ 0 \\ 0 \\ 0 \end{pmatrix} + \alpha_2 \begin{pmatrix} 0 \\ 1 \\ 0 \\ 0 \\ 0 \end{pmatrix} + \alpha_3 \begin{pmatrix} 0 \\ 1 \\ 1 \\ 1 \\ 0 \end{pmatrix} + \alpha_4 \begin{pmatrix} 0 \\ -1 \\ -1 \\ 0 \\ 1 \end{pmatrix}, u_1 = \begin{pmatrix} 0 \\ 1 \\ 1 \\ 0 \\ 0 \end{pmatrix},$$

$\alpha_j, j = 1, \dots, 4$, are constants, and A is the A in Problem **3**:

$$A = \begin{pmatrix} 1 & 0 & 0 & 1 & -1 \\ 0 & 1 & -2 & 3 & -3 \\ 0 & 0 & -1 & 2 & -2 \\ 1 & -1 & 1 & 0 & 1 \\ 1 & -1 & 1 & -1 & 2 \end{pmatrix}$$

from [**10**, Page 172].

7.

$$\frac{d}{dt}u = Au + c(t)g_0, \quad t > 0,$$
$$u(0) = u_0, u_1, \quad \text{and } u_2, \text{ respectively,}$$

where g_0 is a constant vector, $c(t)$ is a continuous function,

$$u_0 = \begin{pmatrix} 1 \\ 0 \\ 0 \end{pmatrix}, u_1 = \begin{pmatrix} 1 \\ 5 \\ 20 \end{pmatrix}, u_2 = \begin{pmatrix} 1 \\ -1 \\ 2 \end{pmatrix},$$

and A is the A in Problem **1**:

$$A = \begin{pmatrix} 0 & 1 & 0 \\ 0 & 1 & 1 \\ 0 & 8 & 3 \end{pmatrix}.$$

8.

$$\frac{d}{dt}u = Au + c(t)g_0, \quad t > 0,$$

$$u(0) = u_0, u_1, u_2, \quad \text{and } u_3, \text{ respectively,}$$

where g_0 is a constant vector, $c(t)$ is a continuous function,

$$u_0 = \begin{pmatrix} \frac{3}{4} \\ 1 \\ 0 \end{pmatrix}, u_1 = \begin{pmatrix} -\frac{1}{4} \\ 0 \\ 1 \end{pmatrix}, u_2 = \begin{pmatrix} \frac{1}{2} \\ 1 \\ 1 \end{pmatrix}, \quad \text{and } u_3 = \begin{pmatrix} 1 \\ 0 \\ 0 \end{pmatrix},$$

and A is the A in Problem **2**:

$$A = \begin{pmatrix} -2 & 1 & 0 \\ 0 & -1 & 1 \\ 0 & -1 & 1 \end{pmatrix}.$$

9.

$$\frac{d}{dt}u = Au + c(t)g_0, \quad t > 0,$$

$$u(0) = u_0 \quad \text{and } u_1, \text{ respectively,}$$

where g_0 is a constant vector, $c(t)$ is a continuous function,

$$u_0 = \alpha_1 \begin{pmatrix} 1 \\ 0 \\ 0 \\ 0 \\ 0 \end{pmatrix} + \alpha_2 \begin{pmatrix} 0 \\ 1 \\ 0 \\ 0 \\ 0 \end{pmatrix} + \alpha_3 \begin{pmatrix} 0 \\ 1 \\ 1 \\ 1 \\ 0 \end{pmatrix} + \alpha_4 \begin{pmatrix} 0 \\ -1 \\ -1 \\ 0 \\ 1 \end{pmatrix}, u_1 = \begin{pmatrix} 0 \\ 1 \\ 1 \\ 0 \\ 0 \end{pmatrix},$$

$\alpha_j, j = 1, \ldots, 4$, are constants, and A is the A in Problem **3**:

$$A = \begin{pmatrix} 1 & 0 & 0 & 1 & -1 \\ 0 & 1 & -2 & 3 & -3 \\ 0 & 0 & -1 & 2 & -2 \\ 1 & -1 & 1 & 0 & 1 \\ 1 & -1 & 1 & -1 & 2 \end{pmatrix}$$

from [**10**, Page 172].

10.

$$\frac{d}{dt}u = Au + f(u), \quad t > 0,$$

$$u(0) = u_0,$$

where

$$u = \begin{pmatrix} u_1 \\ u_2 \\ u_3 \\ u_4 \end{pmatrix}, \quad A = \begin{pmatrix} -2 & 4 & -2 & -2 \\ 2 & 0 & -1 & -3 \\ 2 & 2 & -3 & -3 \\ 2 & 6 & -3 & -7 \end{pmatrix},$$

$$f(u) = \begin{pmatrix} 4u_1 u_2 u_3 \\ u_1^2 u_2 + u_2^2 u_3 + u_3^2 u_4 + u_4^2 u_1 \\ u_1 u_2 + u_2 u_3 + u_3 u_4 + u_4 u_1 \\ u_1^3 + u_2^3 + u_3^3 + u_4^3 \end{pmatrix} \equiv g(u) \quad \text{and}$$

$$\begin{pmatrix} 4|u_1 u_2 u_3| \\ |u_1^2 u_2 + u_2^2 u_3 + u_3^2 u_4 + u_4^2 u_1| \\ |u_1 u_2 + u_2 u_3 + u_3 u_4 + u_4 u_1| \\ |u_1^3 + u_2^3 + u_3^3 + u_4^3| \end{pmatrix} \equiv h(u), \quad \text{respectively,}$$

and u_0 is any vector in $\mathbb{C}^4$.

11.

$$\frac{d}{dt} u = Au + B(t)u + f(u), \quad t > 0,$$
$$u(0) = u_0,$$

where

$$u = \begin{pmatrix} u_1 \\ u_2 \\ u_3 \\ u_4 \end{pmatrix}, \quad B(t) = \begin{pmatrix} e^{-t} & e^{-t}t & e^{-t}t^2 & e^{-t}t^3 \\ \frac{1}{1+t^2} & \frac{\sin(t)}{1+t^2} & \frac{\cos(2t)}{1+t^4} & \frac{\sin(3t)}{1+t^4} \\ e^{-2t} & e^{-3t}t^5 & e^{-t}t^8 & e^{-4t}t^9 \\ \frac{1}{1+e^t} & \frac{t+1}{1+e^t} & \frac{t^2+2t}{1+e^t} & \frac{t^3+3t^2+1}{1+e^t} \end{pmatrix},$$

A and $f(u)$ are as in Problem **10**, and u_0 is any vector in $\mathbb{C}^4$.

12.

$$\frac{d}{dt} u = Au + B(t)u + f(u), \quad t > t_0 \geq 0,$$
$$u(t_0) = u_0,$$

where $u = \begin{pmatrix} u_1 \\ u_2 \\ u_3 \end{pmatrix}$, A is as in Problem **1**, $B(t)$ is the $B(t)$ in Problem **11** whose first column is deleted, and

$$f(u) = \begin{pmatrix} u_1^2 u_2 + u_2^2 u_3 + u_3^2 u_1 \\ u_1 u_2 + u_2 u_3 + u_3 u_1 \\ u_1^3 + u_2^3 + u_3^3 \end{pmatrix} \equiv g(u) \quad \text{and}$$

$$\begin{pmatrix} |u_1^2 u_2 + u_2^2 u_3 + u_3^2 u_1| \\ |u_1 u_2 + u_2 u_3 + u_3 u_1| \\ |u_1^3 + u_2^3 + u_3^3| \end{pmatrix} \equiv h(u), \quad \text{respectively.}$$

13.

$$\frac{d}{dt} u(t) = Bu + f(u), \quad t > 0,$$
$$u(0) = u_0,$$

where

$$B = \begin{pmatrix} -3 & 4 & -2 & -2 \\ 2 & -1 & -1 & -3 \\ 2 & 2 & -4 & -3 \\ 2 & 6 & -3 & -8 \end{pmatrix} \quad \text{and} \quad f(u) = \begin{pmatrix} \tan(u_1 - 3) \\ \tan(u_2 - 5) \\ \tan(u_3 - 4) \\ \tan(u_4 - 6) \end{pmatrix}.$$

14.

$$\frac{d}{dt} u = Au + B(t)u + f(u), \quad t > t_0 \geq 0,$$
$$u(t_0) = u_0,$$

where

$$u = \begin{pmatrix} u_1 \\ u_2 \\ u_3 \end{pmatrix}, \quad A = \begin{pmatrix} 0 & 1 & 0 \\ 0 & 1 & 1 \\ 0 & -1 & 3 \end{pmatrix},$$

and $B(t)$ and $f(u)$ are as in Problem **12**.

15.

$$\frac{d}{dt} u = Au + B(t)u + f(u), \quad t > 0,$$

where

$$u = \begin{pmatrix} u_1 \\ u_2 \\ u_3 \\ u_4 \end{pmatrix}, \quad A = \begin{pmatrix} 3 & -1 & -5 & 1 \\ 1 & 1 & -1 & 0 \\ 0 & 0 & -2 & -1 \\ 0 & 0 & 1 & 0 \end{pmatrix},$$

$$f(u) = \begin{pmatrix} 4u_1 u_2 u_3 \\ u_1^2 u_2 + u_2^2 u_3 + u_3^2 u_4 + u_4^2 u_1 \\ u_1 u_2 + u_2 u_3 + u_3 u_4 + u_4 u_1 \\ u_1^3 + u_2^3 + u_3^3 + u_4^3 \end{pmatrix},$$

and $B(t)$ is as in Problem **11**.

16.

$$\frac{d}{dt} u = Au + g(u), \quad t > 0,$$
$$u(0) = u_0,$$

where

$$u = \begin{pmatrix} u_1 \\ u_2 \\ u_3 \\ u_4 \end{pmatrix}, \quad A = \begin{pmatrix} -2 & 0 & 0 & 0 \\ 0 & -3 & 0 & 0 \\ 0 & 0 & 4 & 0 \\ 0 & 0 & 0 & 5 \end{pmatrix},$$

$$g(u) = \begin{pmatrix} 0 \\ u_1^3 \\ -u_1^4 \\ -u_3^2 \end{pmatrix}, \quad \text{and} \quad u_0 = \begin{pmatrix} \alpha_1 \\ \alpha_2 \\ \alpha_3 \\ \alpha_4 \end{pmatrix}, \text{ a constant vector in } \mathbb{C}^4.$$

17.

$$\frac{d}{dt} u = Au + f(u), \quad t > 0,$$

where

$$u = \begin{pmatrix} u_1 \\ u_2 \\ u_3 \\ u_4 \end{pmatrix}, \quad A = \begin{pmatrix} 3 & -1 & -5 & 1 \\ 1 & 1 & -1 & 0 \\ 0 & 0 & -2 & -1 \\ 0 & 0 & 1 & 0 \end{pmatrix}, \quad \text{the } A \text{ in Problem } \mathbf{14},$$

$$f(u) = u_1 u_2 u_3 u_4 \left[\begin{pmatrix} 25 \\ 4 \\ 27 \\ 0 \end{pmatrix} + \begin{pmatrix} -14 \\ 10 \\ 0 \\ 27 \end{pmatrix} \right] \equiv u_1 u_2 u_3 u_4 (d_3 + d_4) \quad \text{and}$$

$$u_1 u_2 u_3 u_4 \begin{pmatrix} 27 & 0 & -25 & 4 \\ 0 & 27 & -4 & -10 \\ 0 & 0 & 0 & 0 \\ 0 & 0 & 0 & 0 \end{pmatrix} \begin{pmatrix} u_1 \\ u_2 \\ u_3 \\ u_4 \end{pmatrix} \equiv u_1 u_2 u_3 u_4 v = u_1 u_2 u_3 u_4 \begin{pmatrix} v_1 \\ v_2 \\ v_3 \\ v_4 \end{pmatrix},$$

respectively.

18.

$$\frac{d}{dt} u = Au + g(u), \quad t > 0,$$
$$u(0) = u_0,$$

where

$$u = \begin{pmatrix} u_1 \\ u_2 \\ u_3 \\ u_4 \end{pmatrix}, \quad A = \begin{pmatrix} -2 & 0 & 0 & 0 \\ 0 & -3 & 0 & 0 \\ 0 & 0 & 4 & 0 \\ 0 & 0 & 0 & 5 \end{pmatrix}, \quad \text{the } A \text{ in Problem } \mathbf{16}$$

$$g(u) = \begin{pmatrix} u_4^3 \\ -u_4^3 \\ u_4^2 \\ 0 \end{pmatrix}, \quad \text{and } u_0 = \begin{pmatrix} \alpha_1 \\ \alpha_2 \\ \alpha_3 \\ \alpha_4 \end{pmatrix}, \quad \text{a constant vector in } \mathbb{C}^4.$$

9. Solutions

1. As in solving Example 4.1, we have that the eigenvalues of A are $0, 5$, and -1, and that

$$e^{tA} = M_{1,0} + e^{5t} M_{2,0} + e^{-t} M_{3,0},$$

where

$$M_{2,0} = \frac{1}{30}(A + A^2) = \frac{1}{30}\begin{pmatrix} 0 & 2 & 1 \\ 0 & 10 & 5 \\ 0 & 40 & 20 \end{pmatrix},$$

$$M_{1,0} = I - \frac{1}{5}(A^2 - 4A) = \begin{pmatrix} 1 & 0.6 & -0.2 \\ 0 & 0 & 0 \\ 0 & 0 & 0 \end{pmatrix},$$

$$M_{3,0} = \frac{1}{6}(A^2 - 5A) = \frac{1}{6}\begin{pmatrix} 0 & -4 & 1 \\ 0 & 4 & -1 \\ 0 & -8 & 2 \end{pmatrix}.$$

Thus, it follows from Theorem 2.6 that, as $t \longrightarrow +\infty$,

$$\begin{cases} \text{the solution } u(t) \text{ lies on a sphere,} & \text{if } u(0) = u_0; \\ |u(t)| \text{ approaches } +\infty, & \text{if } u(0) = u_1; \\ u(t) \text{ approaches zero,} & \text{if } u(0) = u_2; \\ |u(t)| \text{ approaches } +\infty, & \text{if } u(0) = u_3; \\ u(t) \text{ lies on a sphere,} & \text{if } u(0) = u_4; \\ |u(t)| \text{ approaches } +\infty, & \text{if } u(0) = u_5; \\ |u(t)| \text{ approaches } +\infty, & \text{if } u(0) = u_6. \end{cases}$$

This is because $u_0 \in S(M_{1,0}), u_1 \in S(M_{2,0})$, and $u_2 \in S(M_{3,0})$.

Moreover, the subspaces M_0, M_b, and M_∞, and the subset $\triangle_\infty$ are:

$$M_0 = S(M_{3,0}) = \{\alpha \begin{pmatrix} 1 \\ -1 \\ 2 \end{pmatrix} : \alpha \in \mathbb{C}\},$$

$$M_b = S(M_{1,0}) = E(0) = \{\alpha \begin{pmatrix} 1 \\ 0 \\ 0 \end{pmatrix} : \alpha \in \mathbb{C}\},$$

$$M_\infty = S(M_{2,0}) = \{\alpha \begin{pmatrix} 1 \\ 5 \\ 20 \end{pmatrix} : \alpha \in \mathbb{C}\},$$

$$\triangle_\infty = S(M_{1,0}) \setminus M_b = \emptyset,$$

respectively.

2. As in solving Example 4.1, we have that the eigenvalues of A are $0, 0$, and -2, and that

$$e^{tA} = M_{1,0} + tM_{1,1} + e^{-2t}M_{2,0},$$

where

$$M_{1,0} = I - \frac{A^2}{4} = \begin{pmatrix} 0 & \frac{3}{4} & -\frac{1}{4} \\ 0 & 1 & 0 \\ 0 & 0 & 1 \end{pmatrix},$$

$$M_{1,1} = A + \frac{A^2}{2} = \begin{pmatrix} 0 & -\frac{1}{2} & \frac{1}{2} \\ 0 & -1 & 1 \\ 0 & -1 & 1 \end{pmatrix},$$

$$M_{2,0} = \frac{A^2}{4} = \begin{pmatrix} 1 & -\frac{3}{4} & \frac{1}{4} \\ 0 & 0 & 0 \\ 0 & 0 & 0 \end{pmatrix}.$$

Thus, it follows from Theorem 2.6 that, as $t \longrightarrow +\infty$,

$$\begin{cases} |u(t)| \text{ approaches } +\infty, & \text{if } u(0) = u_0; \\ u(t) \text{ lies on a sphere,} & \text{if } u(0) = u_1; \\ u(t) \longrightarrow 0, & \text{if } u(0) = u_2. \end{cases}$$

This is because $u_0 \in S(M_{1,0}) \setminus E(0), u_1 \in E(0)$, and $u_2 \in S(M_{2,0})$.

Moreover, we have

$$M_0 = S(M_{2,0}) = \{\alpha \begin{pmatrix} 1 \\ 0 \\ 0 \end{pmatrix} : \alpha \in \mathbb{C}\},$$

$$M_b = E(0) = \{\alpha \begin{pmatrix} \frac{1}{2} \\ 1 \\ 1 \end{pmatrix} : \alpha \in \mathbb{C}\},$$

$$M_\infty = \emptyset,$$
$$\triangle_\infty = S(M_{1,0}) \setminus E(0)$$

$$= \{\alpha \begin{pmatrix} \frac{3}{4} \\ 1 \\ 0 \end{pmatrix} + \beta \begin{pmatrix} -\frac{1}{4} \\ 0 \\ 1 \end{pmatrix} : \alpha, \beta \in \mathbb{C}\} \setminus \{\alpha \begin{pmatrix} \frac{1}{2} \\ 1 \\ 1 \end{pmatrix} : \alpha \in \mathbb{C}\}.$$

3. As in solving Example 4.1, we have that the eigenvalues of A are $1, 1, 1, 1$, and -1, and that

$$e^{tA} = e^t M_{1,0} + t e^t M_{1,1} + t^2 e^t M_{1,2} + t^3 e^t M_{1,3} + e^{-t} M_{2,0},$$

where

$$M_{1,0} = \begin{pmatrix} 1 & 0 & 0 & 0 & 0 \\ 0 & 1 & -1 & 1 & -1 \\ 0 & 0 & 0 & 1 & -1 \\ 0 & 0 & 0 & 1 & 0 \\ 0 & 0 & 0 & 0 & 1 \end{pmatrix}, \quad M_{1,1} = \begin{pmatrix} 0 & 0 & 0 & 1 & -1 \\ 0 & 0 & 0 & 1 & -1 \\ 0 & 0 & 0 & 0 & 0 \\ 1 & -1 & 1 & -1 & 1 \\ 1 & -1 & 1 & -1 & 1 \end{pmatrix},$$

$$M_{1,2} = 0, \quad M_{1,3} = 0$$

$$M_{2,0} = \begin{pmatrix} 0 & 0 & 0 & 0 & 0 \\ 0 & 0 & 1 & -1 & 1 \\ 0 & 0 & 1 & -1 & 1 \\ 0 & 0 & 0 & 0 & 0 \\ 0 & 0 & 0 & 0 & 0 \end{pmatrix}.$$

Thus, it follows from Theorem 2.6 that, as $t \longrightarrow +\infty$,

$$\begin{cases} |u(t)| \text{ approaches } +\infty, & \text{if } u(0) = u_0; \\ u(t) \longrightarrow 0, & \text{if } u(0) = u_1. \end{cases}$$

This is because $u_0 \in S(M_{1,0})$ and $u_1 \in S(M_{2,0})$.

Moreover, we have

$$M_0 = S(M_{2,0}) = \{\alpha u_1 : \alpha \in \mathbb{C}\},$$
$$M_b = \emptyset,$$
$$M_\infty = S(M_{1,0}) = \{u_0 : \alpha_j \in \mathbb{C}, j = 1, \ldots, 4\},$$
$$\triangle_\infty = \emptyset.$$

4. As in solving Problem **1**, we have that the eigenvalues of A are $0, 5$, and -1, and that

$$e^{tA} = M_{1,0} + e^{5t} M_{2,0} + e^{-t} M_{3,0},$$

where

$$M_{2,0} = \frac{1}{30}(A + A^2) = \frac{1}{30}\begin{pmatrix} 0 & 2 & 1 \\ 0 & 10 & 5 \\ 0 & 40 & 20 \end{pmatrix},$$

$$M_{1,0} = I - \frac{1}{5}(A^2 - 4A) = \begin{pmatrix} 1 & 0.6 & -0.2 \\ 0 & 0 & 0 \\ 0 & 0 & 0 \end{pmatrix},$$

$$M_{3,0} = \frac{1}{6}(A^2 - 5A) = \frac{1}{6}\begin{pmatrix} 0 & -4 & 1 \\ 0 & 4 & -1 \\ 0 & -8 & 2 \end{pmatrix}.$$

Thus, it follows from Theorem 2.13 that, as $t \longrightarrow +\infty$,

$$\begin{cases} \text{the solution } u(t) \text{ lies on a sphere,} & \text{if } u(0) = u_0 \text{ and } g_0 = 0; \\ |u(t)| \text{ approaches } +\infty, & \text{if } u(0) = u_1 \text{ or } g_0 = u_1; \\ u(t) \longrightarrow g_0 = -A^{-1}g_0, & \text{if } u(0) = g_0 = u_2; \\ |u(t)| \text{ approaches } +\infty, & \text{if } u(0) = u_2 \text{ and } g_0 = u_0; \\ u(t) = -\frac{1}{5}g_0 \text{ for all } t, & \text{if } g_0 = u_1 \text{ and } u_0 = -\frac{1}{5}g_0. \end{cases}$$

Here

$$u_0 = \begin{pmatrix} 1 \\ 0 \\ 0 \end{pmatrix} \in S(M_{1,0}), \quad u_1 = \begin{pmatrix} 1 \\ 5 \\ 20 \end{pmatrix} \in S(M_{2,0}),$$

$$u_2 = \begin{pmatrix} 1 \\ -1 \\ 2 \end{pmatrix} \in S(M_{3,0}), \quad \text{an eigenvalue of } A \text{ by Propositions 6.2 and 6.3, and so}$$

$(A + 1)u_2 = 0$ but $u_2 \neq 0$.

5. As in solving Problem **2**, we have that the eigenvalues of A are $0, 0$, and -2, and that

$$e^{tA} = M_{1,0} + tM_{1,1} + e^{-2t}M_{2,0},$$

where

$$M_{1,0} = I - \frac{A^2}{4} = \begin{pmatrix} 0 & \frac{3}{4} & -\frac{1}{4} \\ 0 & 1 & 0 \\ 0 & 0 & 1 \end{pmatrix},$$

$$M_{1,1} = A + \frac{A^2}{2} = \begin{pmatrix} 0 & -\frac{1}{2} & \frac{1}{2} \\ 0 & -1 & 1 \\ 0 & -1 & 1 \end{pmatrix},$$

$$M_{2,0} = \frac{A^2}{4} = \begin{pmatrix} 1 & -\frac{3}{4} & \frac{1}{4} \\ 0 & 0 & 0 \\ 0 & 0 & 0 \end{pmatrix}.$$

The eigenspace $E(0)$ of 0, by Propositions 6.2 and 6.3, is:

$$E(0) = \{\alpha \begin{pmatrix} \frac{1}{2} \\ 1 \\ 1 \end{pmatrix} : \alpha \in \mathbb{C}\}.$$

Thus, it follows from Theorem 2.13 that, as $t \longrightarrow +\infty$,

$$
\begin{cases}
|u(t)| \text{ approaches } +\infty, & \text{if } u(0) = u_0 \text{ or } u_1; \\
& \text{or if } g_0 = u_0; \\
& \text{or if } g_0 = u_1; \\
& \text{or if } g_0 = u_2; \\
u(t) \text{ lies inside a sphere,} & \text{if } u(0) = u_2 \text{ and } g_0 = 0; \\
u(t) \longrightarrow 0, & \text{if } u(0) = u_3,
\end{cases}
$$

where

$$
u_3 = \begin{pmatrix} 1 \\ 0 \\ 0 \end{pmatrix} \in S(M_{2,0}), \quad u_2 = \begin{pmatrix} \frac{1}{2} \\ 1 \\ 1 \end{pmatrix} \in E(0),
$$

$$
u_0 = \begin{pmatrix} \frac{3}{4} \\ 4 \\ 1 \\ 0 \end{pmatrix} \text{ and } u_1 = \begin{pmatrix} -\frac{1}{4} \\ 4 \\ 0 \\ 1 \end{pmatrix} \text{ are in } S(M_{1,0}) \setminus E(0).
$$

6. As in solving Problem **3**, we have that the eigenvalues of A are $1, 1, 1, 1$, and -1, and that

$$
e^{tA} = e^t M_{1,0} + t e^t M_{1,1} + t^2 e^t M_{1,2} + t^3 e^t M_{1,3} + e^{-t} M_{2,0},
$$

where

$$
M_{1,0} = \begin{pmatrix} 1 & 0 & 0 & 0 & 0 \\ 0 & 1 & -1 & 1 & -1 \\ 0 & 0 & 0 & 1 & -1 \\ 0 & 0 & 0 & 1 & 0 \\ 0 & 0 & 0 & 0 & 1 \end{pmatrix}, \quad M_{1,1} = \begin{pmatrix} 0 & 0 & 0 & 1 & -1 \\ 0 & 0 & 0 & 1 & -1 \\ 0 & 0 & 0 & 0 & 0 \\ 1 & -1 & 1 & -1 & 1 \\ 1 & -1 & 1 & -1 & 1 \end{pmatrix}
$$

$$
M_{1,2} = 0, \quad M_{1,3} = 0
$$

$$
M_{2,0} = \begin{pmatrix} 0 & 0 & 0 & 0 & 0 \\ 0 & 0 & 1 & -1 & 1 \\ 0 & 0 & 1 & -1 & 1 \\ 0 & 0 & 0 & 0 & 0 \\ 0 & 0 & 0 & 0 & 0 \end{pmatrix}.
$$

To find the eigenspace $E(1)$ of 1, we form the augmented matrix

$$
F \equiv \begin{pmatrix} M_{1,0} \\ M_{1,1} \end{pmatrix} = \begin{pmatrix} 1 & 0 & 0 & 0 & 0 \\ 0 & 1 & -1 & 1 & -1 \\ 0 & 0 & 0 & 1 & -1 \\ 0 & 0 & 0 & 1 & 0 \\ 0 & 0 & 0 & 0 & 1 \\ * & * & * & * & * \\ 0 & 0 & 0 & 1 & -1 \\ 0 & 0 & 0 & 1 & -1 \\ 0 & 0 & 0 & 0 & 0 \\ 1 & -1 & 1 & -1 & 1 \\ 1 & -1 & 1 & -1 & 1 \end{pmatrix}
$$

Performing column operations on F by adding column 1 to column 2, adding $(-1)\times$ column 1 to column 3, and adding column 5 to column 4, we obtain

$$
\begin{pmatrix}
1 & 1 & -1 & 0 & 0 \\
0 & 1 & -1 & 0 & -1 \\
0 & 0 & 0 & 0 & -1 \\
0 & 0 & 0 & 1 & 0 \\
0 & 0 & 0 & 1 & 1 \\
* & * & * & * & * \\
0 & 0 & 0 & 0 & -1 \\
0 & 0 & 0 & 0 & -1 \\
0 & 0 & 0 & 0 & 0 \\
1 & 0 & 0 & 0 & 1 \\
1 & 0 & 0 & 0 & 1
\end{pmatrix},
$$

and so, by Propositions 6.2 and 6.3,

$$
E(1) = \{\beta_1 \begin{pmatrix} 1 \\ 1 \\ 0 \\ 0 \\ 0 \end{pmatrix} + \beta_2 \begin{pmatrix} 0 \\ 0 \\ 0 \\ 1 \\ 1 \end{pmatrix} : \beta_1, \beta_2 \in \mathbb{C}\}.
$$

Thus, it follows from Theorem 2.13 that, as $t \longrightarrow +\infty$,

$$
\begin{cases}
|u(t)| \text{ approaches } +\infty, & \text{if } u(0) = u_0 \text{ and } g_0 \text{ is not in } E(1); \\
& \text{or if } g_0 = u_0 \text{ and } u(0) \text{ is not in } E(1); \\
u(t) = -g_0, & \text{if } u(0), g_0 \in E(1) \text{ and } u_0 = -g_0; \\
u(t) \longrightarrow 0, & \text{if } u(0) = u_1,
\end{cases}
$$

where $u_0 \in S(M_{1,0})$ and $u_1 \in S(M_{2,0})$.

7. As in solving Problem **1**, we have that the eigenvalues of A are $0, 5$, and -1, and that

$$
e^{tA} = M_{1,0} + e^{5t} M_{2,0} + e^{-t} M_{3,0},
$$

where

$$
M_{2,0} = \frac{1}{30}(A + A^2) = \frac{1}{30} \begin{pmatrix} 0 & 2 & 1 \\ 0 & 10 & 5 \\ 0 & 40 & 20 \end{pmatrix},
$$

$$
M_{1,0} = I - \frac{1}{5}(A^2 - 4A) = \begin{pmatrix} 1 & 0.6 & -0.2 \\ 0 & 0 & 0 \\ 0 & 0 & 0 \end{pmatrix},
$$

$$
M_{3,0} = \frac{1}{6}(A^2 - 5A) = \frac{1}{6} \begin{pmatrix} 0 & -4 & 1 \\ 0 & 4 & -1 \\ 0 & -8 & 2 \end{pmatrix}.
$$

Thus, it follows from Theorem 2.14 and Proposition 2.15 that, as $t \longrightarrow +\infty$,

$$
\begin{cases}
\text{the solution } u(t) \text{ lies inside a sphere,} & \text{if } u(0) = u_0 \text{ and } g_0 = 0; \\
|u(t)| \text{ approaches } +\infty, & \text{if } u(0) = u_1 \text{ or } c(t)g_0 = [\alpha_{-1}\frac{t+2}{t+1} + \alpha_0 + \\
& \qquad \alpha_1 e^{(-1+i)t} + \alpha_2 \cos(2t) + \alpha_3 \sin(3t)]u_1; \\
u(t) \longrightarrow \text{ a constant,} & \text{if } u(0) = u_2 \text{ and } c(t)g_0 = \frac{1}{t+1}u_2; \\
|u(t)| \text{ approaches } +\infty, & \text{if } u(0) = u_2 \text{ and } c(t)g_0 = [\alpha_0 + \alpha_1\frac{1}{t+1}]u_0; \\
u(t) = e^{-3t}u_0 \text{ for all } t, & \text{if } c(t)g_0 = e^{-3t}u_1 \text{ and } u_0 = -\frac{1}{20}g_0.
\end{cases}
$$

Here $\alpha_{-1}, \alpha_0, \ldots$, and α_3 are constants, and

$$
u_0 = \begin{pmatrix} 1 \\ 0 \\ 0 \end{pmatrix} \in S(M_{1,0}), \quad u_1 = \begin{pmatrix} 1 \\ 5 \\ 20 \end{pmatrix} \in S(M_{2,0}),
$$

$$
u_2 = \begin{pmatrix} 1 \\ -1 \\ 2 \end{pmatrix} \in S(M_{3,0}), \quad \text{an eigenvalue of } A \text{ by Propositions 6.2 and 6.3, and so}
$$

$$
(A+1)u_2 = 0 \quad \text{but } u_2 \neq 0,
$$

and

$$
|\frac{1}{t+1}| \leq 1, \qquad \int_0^{+\infty} \frac{1}{t+1}\,dt = +\infty,
$$

$$
\frac{t+2}{t+1} \geq 1.
$$

8. As in solving Problem **2**, we have that the eigenvalues of A are $0, 0$, and -2, and that

$$
e^{tA} = M_{1,0} + tM_{1,1} + e^{-2t}M_{2,0},
$$

where

$$
M_{1,0} = I - \frac{A^2}{4} = \begin{pmatrix} 0 & \frac{3}{4} & -\frac{1}{4} \\ 0 & 1 & 0 \\ 0 & 0 & 1 \end{pmatrix},
$$

$$
M_{1,1} = A + \frac{A^2}{2} = \begin{pmatrix} 0 & -\frac{1}{2} & \frac{1}{2} \\ 0 & -1 & 1 \\ 0 & -1 & 1 \end{pmatrix},
$$

$$
M_{2,0} = \frac{A^2}{4} = \begin{pmatrix} 1 & -\frac{3}{4} & \frac{1}{4} \\ 0 & 0 & 0 \\ 0 & 0 & 0 \end{pmatrix}.
$$

The eigenspace $E(0)$ of 0, by Propositions 6.2 and 6.3, is:

$$
E(0) = \{\alpha \begin{pmatrix} \frac{1}{2} \\ 1 \\ 1 \end{pmatrix} : \alpha \in \mathbb{C}\}.
$$

Thus, it follows from Theorem 2.14 and Proposition 2.15 that, as $t \longrightarrow +\infty$,

$$
\begin{cases}
|u(t)| \text{ approaches } +\infty, & \text{if } u(0) = u_0 \text{ or } u_1; \\
& \text{or if } g_0 = e^{(-1+i)t}u_0; \\
& \text{or if } g_0 = u_1 \cos(2t); \\
& \text{or if } g_0 = (\alpha_0 + \alpha_1 \frac{1}{t+1})u_2; \\
u(t) \text{ lies inside a sphere}, & \text{if } u(0) = u_2 \text{ and } g_0 = 0; \\
u(t) \longrightarrow 0, & \text{if } u(0) = u_3,
\end{cases}
$$

where

$$
u_3 = \begin{pmatrix} 1 \\ 0 \\ 0 \end{pmatrix} \in S(M_{2,0}), \quad u_2 = \begin{pmatrix} \frac{1}{2} \\ 1 \\ 1 \end{pmatrix} \in E(0),
$$

$$
u_0 = \begin{pmatrix} \frac{3}{4} \\ 4 \\ 1 \\ 0 \end{pmatrix} \text{ and } u_1 = \begin{pmatrix} -\frac{1}{4} \\ \frac{4}{0} \\ 0 \\ 1 \end{pmatrix} \text{ are in } S(M_{1,0}) \setminus E(0).
$$

9. As in solving Problem **3**, we have that the eigenvalues of A are $1, 1, 1, 1$, and -1, that

$$
e^{tA} = e^t M_{1,0} + te^t M_{1,1} + t^2 e^t M_{1,2} + t^3 e^t M_{1,3} + e^{-t} M_{2,0},
$$

where

$$
M_{1,0} = \begin{pmatrix} 1 & 0 & 0 & 0 & 0 \\ 0 & 1 & -1 & 1 & -1 \\ 0 & 0 & 0 & 1 & -1 \\ 0 & 0 & 0 & 1 & 0 \\ 0 & 0 & 0 & 0 & 1 \end{pmatrix}, \quad M_{1,1} = \begin{pmatrix} 0 & 0 & 0 & 1 & -1 \\ 0 & 0 & 0 & 1 & -1 \\ 0 & 0 & 0 & 0 & 0 \\ 1 & -1 & 1 & -1 & 1 \\ 1 & -1 & 1 & -1 & 1 \end{pmatrix}
$$

$$
M_{1,2} = 0, \quad M_{1,3} = 0
$$

$$
M_{2,0} = \begin{pmatrix} 0 & 0 & 0 & 0 & 0 \\ 0 & 0 & 1 & -1 & 1 \\ 0 & 0 & 1 & -1 & 1 \\ 0 & 0 & 0 & 0 & 0 \\ 0 & 0 & 0 & 0 & 0 \end{pmatrix},
$$

and that the eigenspace $E(1)$ of 1 is:

$$
E(1) = \left\{ \beta_1 \begin{pmatrix} 1 \\ 1 \\ 0 \\ 0 \\ 0 \end{pmatrix} + \beta_2 \begin{pmatrix} 0 \\ 0 \\ 0 \\ 1 \\ 1 \end{pmatrix} : \beta_1, \beta_2 \in \mathbb{C} \right\}.
$$

Thus, it follows from Theorem 2.14 and Proposition 2.15 that, as $t \longrightarrow +\infty$,

$$
\begin{cases}
|u(t)| \text{ approaches } +\infty, & \text{if } u(0) = u_0 \text{ and } g_0 \text{ is not in } E(1); \\
& \text{or if } c(t)g_0 = [\beta_{-1}\frac{t+2}{t+1} + \beta_0 + \beta_1 e^{(-1+i)t}+ \\
& \beta_2 \cos(2t) + \beta_3 \sin(3t)]u_0 \text{ and} \\
& u(0) \text{ is not in } E(1); \\
u(t) = e^{-3t}u_0, & \text{if } u(0), g_0 \in E(1), c(t) = e^{-3t}, \text{ and } u_0 = -\frac{1}{4}g_0; \\
u(t) \longrightarrow 0, & \text{if } u(0) = u_1,
\end{cases}
$$

where $\beta_{-1},\ldots,$ and β_3 are constants, $u_0 \in S(M_{1,0})$, and $u_1 \in S(M_{2,0})$.

10. Whether $f(u) = g(u)$ or $f(u) = h(u)$, we have that $f(u)$ satisfies (1.8), because

$$|u_1 u_2 u_3| \le (|u_1| + |u_2| + |u_3| + |u_4|)^3,$$
$$|u_1^2 u_2 + u_2^2 u_3 + u_3^2 u_4 + u_4^2 u_1| \le (|u_1| + |u_2| + |u_3| + |u_4|)^3,$$
$$|u_1 u_2 + u_2 u_3 + u_3 u_4 + u_4 u_1| \le (|u_1| + u_2| + |u_3||u_4|)^2,$$
$$|u_1^3 + u_2^3 + u_3^3 + u_4^3| \le (|u_1| + |u_2| + |u_3| + |u_4|)^3.$$

In the case where $f(u) = g(u)$, this can also be seen by using Remark 3.2 because $Au + f(u)$ is differentiable at 0 and its Jacobian matrix at 0 is A.

The eigenvalues of A, by (1.3), are $-2, -2, -4$, and -4, which are negative. Thus, it follows from Theorem 3.1 that this initial value problem with $|u_0|$ sufficiently small has a solution $\varphi(t)$, and that this $\varphi(t)$, together with other solutions which also exist on $[0, \infty)$, approaches zero as $t \longrightarrow +\infty$.

11. Since $B(t) \to 0$ as $t \to +\infty$, it follows from the solution of Problem **10**, Remark 3.4, and Theorem 3.3 that this initial value problem with $|u_0|$ sufficiently small has a solution $\varphi(t)$, and that this $\varphi(t)$, together with other soltions which also exist on $[0, \infty)$, approaches zero as $t \longrightarrow +\infty$.

12. By the solution of Problem **11**, the condition $(A2)$ is met. Since the eigenvalues of A are, by (1.3), $0, 5$, and -1, where 5 is positive, it follows from Theorem 3.5 that not every solution $u(t)$ of this initial value problem with $|u(T)| = |u_0|$ sufficiently small, which exists on $[T, \infty)$, stays in the vicinity of the origin for all $t \ge T$. Moreover, such a solution $u(t)$ which cannot stay in the vicinity of the origin has its initial value $u(T)$ conditioned as follows.

From the solution of Problem **1**, we have

$$e^{tA} = M_{1,0} + e^{5t}M_{2,0} + e^{-t}M_{3,0},$$

where

$$
M_{2,0} = \frac{1}{30}(A + A^2) = \frac{1}{30}
\begin{pmatrix}
0 & 2 & 1 \\
0 & 10 & 5 \\
0 & 40 & 20
\end{pmatrix},
$$

$$
M_{1,0} = I - \frac{1}{5}(A^2 - 4A) =
\begin{pmatrix}
1 & 0.6 & -0.2 \\
0 & 0 & 0 \\
0 & 0 & 0
\end{pmatrix},
$$

$$
M_{3,0} = \frac{1}{6}(A^2 - 5A) = \frac{1}{6}
\begin{pmatrix}
0 & -4 & 1 \\
0 & 4 & -1 \\
0 & -8 & 2
\end{pmatrix}.
$$

It follows from Theorem 5.14 in Chapter 3 that a Jordan basis $\{d_1, d_2, d_3\}$ is

$$\{\begin{pmatrix} 1 \\ 0 \\ 0 \end{pmatrix}, \frac{1}{30}\begin{pmatrix} 1 \\ 5 \\ 20 \end{pmatrix}, \frac{1}{6}\begin{pmatrix} 1 \\ -1 \\ 2 \end{pmatrix}\},$$

that is, the first column of $M_{1,0}$, the third column of $M_{2,0}$, and the third column of $M_{3,0}$.

Suppose that $u(t)$ is a solution, and that its initial value $u(T)$ is sufficiently near the origin and takes the form

$$u(T) = \beta_1 d_1 + \beta_2 d_2 + \beta_3 d_3,$$

relative to the above basis. Then it follows from Remark 3.6 that $u(t)$ cannot stay in the vicinity of the origin if $R(T) = 2\rho(T)$ or more generally, $R(T) > \rho(T)$, where

$$R(T) = \beta_2 \quad \text{and} \quad \rho(T) = \sqrt{\beta_1^2 + \beta_3^2}.$$

13. (1.8) is satisfied by Remark 3.2 because $Bu + f(u)$ is differentiable at

$$v_0 = \begin{pmatrix} 3 \\ 5 \\ 4 \\ 6 \end{pmatrix}$$ and its Jacobian matrix at v_0 is

$$A = \begin{pmatrix} -2 & 4 & -2 & -2 \\ 2 & 0 & -1 & -3 \\ 2 & 2 & -3 & -3 \\ 2 & 6 & -3 & -7 \end{pmatrix}.$$

Since A has, by the solution of Problem **10**, negative eigenvalues $-2, -2, -4$, and -4, it follows from Remark 3.2 and Theorem 3.1 that this initial value problem with $|u_0|$ sufficiently small has a solution $\varphi(t)$, and that this $\varphi(t)$, together with other solutions which also exist on $[0, \infty)$, approaches v_0 as $t \to +\infty$.

14. By the solution of Problem **12**, the condition (A2) is met. Since the eigenvalues of A are, by (1.3), $2, 2$, and 0, where 2 is positive, it follows from Theorem 3.5 that not every solution $u(t)$ of this initial value problem with $|u(T)| = |u_0|$ sufficiently small, which exists on $[T, \infty)$, stays in the vicinity of the origin for all $t \geq T$. Moreover, such a solution $u(t)$ which cannot stay in the vicinity of the origin has its initial value $u(T)$ conditioned as follows.

Using (1.2) and (1.4) or equivalently, using the formula $\frac{d}{dt}e^{tA} = Ae^{tA}$ in (1.2) and letting $t = 0$, we have

$$I = M_{1,0} + M_{2,0},$$
$$A = 2M_{1,0} + M_{1,1},$$
$$A^2 = 4M_{1,0} + 4M_{1,1},$$

where $M_{j,0}, j = 1, 2, 3$, are solved as:

$$M_{1,0} = \begin{pmatrix} 0 & \frac{3}{4} & -\frac{1}{4} \\ 0 & 1 & 0 \\ 0 & 0 & 1 \end{pmatrix},$$

$$M_{1,1} = \begin{pmatrix} 0 & -\frac{1}{2} & \frac{1}{2} \\ 0 & -1 & 1 \\ 0 & -1 & 1 \end{pmatrix}$$

$$M_{2,0} = \begin{pmatrix} 1 & -\frac{3}{4} & \frac{1}{4} \\ 0 & 0 & 0 \\ 0 & 0 & 0 \end{pmatrix}.$$

It follows from Theorem 5.14 in Chapter 3 that a Jordan basis $\{d_1, d_2, d_3\}$ is

$$\left\{ \begin{pmatrix} -\frac{1}{4} \\ 0 \\ 1 \end{pmatrix}, \begin{pmatrix} \frac{1}{2} \\ 1 \\ 1 \end{pmatrix}, \begin{pmatrix} 1 \\ 0 \\ 0 \end{pmatrix} \right\},$$

that is, the third column of the augmented matrix $\begin{pmatrix} M_{1,0} \\ M_{1,1} \end{pmatrix}$ and the first column of $M_{2,0}$. By referring to Remark 3.6, a special Jordan basis is

$$\{\gamma d_1, d_2, d_3\}, \quad \text{where } \gamma \equiv \frac{1}{20} < \frac{\sigma}{20} \equiv \frac{1.5}{20} \text{ and } 2 > 1.5 \equiv \sigma.$$

Suppose that $u(t)$ is a solution, and that its initial value $u(T)$ is sufficiently near the origin and takes the form

$$u(T) = \beta_1 (\frac{1}{20}) d_1 + \beta_2 d_2 + \beta_3 d_3,$$

relative to the above special Jordan basis. Then it follows from Remark 3.6 that $u(t)$ cannot stay in the vicinity of the origin if $R(T) = 2\rho(T)$ or more generally, $R(T) > \rho(T)$, where

$$R(T) = \sqrt{\beta_1^2 + \beta_2^2} \quad \text{and } \rho(T) = \beta_3.$$

15. The condition (3.5) is met by Remark 3.8. This is because the Jacobian matrix $g_x(t, u)$ of $g(t, u) \equiv B(t)u + f(u)$ is

$$B(t) + \begin{pmatrix} 4u_2 u_3 & 4u_1 u_3 & 4u_1 u_2 & 0 \\ 2u_1 u_2 + u_4^2 & u_1^2 + 2u_2 u_3 & u_2^2 + 2u_3 u_4 & u_4^2 + 2u_1 u_2 \\ u_2 + u_4 & u_3 + u_1 & u_4 + u_2 & u_3 + u_1 \\ 3u_1^2 & 3u_2^2 & 3u_3^2 & 3u_4^2 \end{pmatrix}$$

and clearly, $|g_x(t, u)| \to 0$ when both $t \to +\infty$ and $|u| \to 0$. Since $f(0) = 0$ and the eigenvalues of A are, by (1.3), $2, 2, -1$, and -1, we have that the conclusion of Theorem 3.9 hold, where the two-dimensional manifold (or surface) $S = S_2$ is, at the origin, tangent to, by the solution of Example 5.23 in Chapter 3,

$$M_0 = S(M_{2,0}) = S(\{ \begin{pmatrix} 25 \\ 4 \\ 27 \\ 0 \end{pmatrix}, \begin{pmatrix} -14 \\ 10 \\ 0 \\ 27 \end{pmatrix} \}), \quad \text{where}$$

$$M_{2,0} = \frac{1}{27} \begin{pmatrix} 0 & 0 & 25 & -14 \\ 0 & 0 & 4 & 10 \\ 0 & 0 & 27 & 0 \\ 0 & 0 & 0 & 27 \end{pmatrix}.$$

16. As in the solution of Problem **15**, the condition (3.5) is satisfied, $g(0) = 0$, the eigenvalues of A are $-2, -3, 4$, and 5, and

$$e^{tA} = e^{-2t}M_{1,0} + e^{-3t}M_{2,0} + e^{4t}M_{3,0} + e^{5t}M_{4,0}$$

$$= e^{-2t}\begin{pmatrix} 1 & 0 & 0 & 0 \\ 0 & 0 & 0 & 0 \\ 0 & 0 & 0 & 0 \\ 0 & 0 & 0 & 0 \end{pmatrix} + e^{-3t}\begin{pmatrix} 0 & 0 & 0 & 0 \\ 0 & 1 & 0 & 0 \\ 0 & 0 & 0 & 0 \\ 0 & 0 & 0 & 0 \end{pmatrix} +$$

$$e^{4t}\begin{pmatrix} 0 & 0 & 0 & 0 \\ 0 & 0 & 0 & 0 \\ 0 & 0 & 1 & 0 \\ 0 & 0 & 0 & 0 \end{pmatrix} + e^{5t}\begin{pmatrix} 0 & 0 & 0 & 0 \\ 0 & 0 & 0 & 0 \\ 0 & 0 & 0 & 0 \\ 0 & 0 & 0 & 1 \end{pmatrix}.$$

Because of its simplicity where

$$u_1' = -2u_1,$$
$$u_2' = -3u_2 + u_1^3,$$
$$u_3' = 4u_3 - u_1^4,$$
$$u_4' = 5u_4 - u_3^2,$$

this equation has the solution which is easily obtained as

$$u(t) = \begin{pmatrix} c_1 e^{-2t} \\ c_2 e^{-3t} - \frac{c_1^3}{3}e^{-6t} \\ c_3 e^{4t} + \frac{c_1^4}{12}e^{-8t} \\ c_4 e^{5t} - \frac{c_3^2}{3}e^{8t} + \frac{c_3 c_1^4}{54}e^{-4t} + \frac{c_1^8}{3024}e^{-16t} \end{pmatrix},$$

where

$$u(0) = \begin{pmatrix} \alpha_1 \\ \alpha_2 \\ \alpha_3 \\ \alpha_4 \end{pmatrix} \quad \text{and}$$

$$\begin{pmatrix} c_1 \\ c_2 \\ c_3 \\ c_4 \end{pmatrix} = \begin{pmatrix} \alpha_1 \\ \alpha_2 + \frac{\alpha_1^3}{3} \\ \alpha_3 - \frac{\alpha_1^4}{12} \\ \alpha_4 + \frac{1}{3}(\alpha_3 - \frac{\alpha_1^4}{12})^2 - \frac{(\alpha_3 - \frac{\alpha_1^4}{12})\alpha_1^4}{54} - \frac{\alpha_1^8}{3024} \end{pmatrix}.$$

In order for $u(t) \to 0$ as $t \to +\infty$, we require

$$\alpha_3 - \frac{\alpha_1^4}{12} = 0 \quad \text{and} \quad \alpha_4 - \frac{\alpha_1^8}{3024} = 0,$$

and so the two-dimensional manifold (or surface) S predicted by Theorem 3.9 is

$$S = \{\begin{pmatrix} \alpha_1 \\ \alpha_2 \\ \alpha_3 \\ \alpha_4 \end{pmatrix} \in \mathbb{C}^4 : \alpha_3 - \frac{\alpha_1^4}{12} = 0 \quad \text{and} \quad \alpha_4 - \frac{\alpha_1^8}{3024} = 0\}.$$

Clearly, S is, at the origin, tangent to the plane

$$M_0 = S(M_{1,0} \cup M_{2,0}) = \{\alpha_1 \begin{pmatrix} 1 \\ 0 \\ 0 \\ 0 \end{pmatrix} + \alpha_2 \begin{pmatrix} 0 \\ 1 \\ 0 \\ 0 \end{pmatrix} : \alpha_1, \alpha_2 \in \mathbb{C}\}.$$

Furthermore, the one-dimensional manifolds (or surfaces) predicted by Theorem 3.10 are

$$S_1 = S \cap S(M_{3,0} \cup M_{4,0} \cup M_{1,0}) = \{\begin{pmatrix} \alpha_1 \\ 0 \\ \frac{\alpha_1^4}{12} \\ \frac{\alpha_1^8}{3024} \end{pmatrix} : \alpha_1 \in \mathbb{C}\},$$

$$S_2 = S \cap S(M_{3,0} \cup M_{4,0} \cup M_{2,0}) = \{\alpha_2 \begin{pmatrix} 0 \\ 1 \\ 0 \\ 0 \end{pmatrix} : \alpha_2 \in \mathbb{C}\},$$

and S_1 and S_2 are, at the origin, tangent to

$$S(M_{1,0}) = \{\alpha_1 \begin{pmatrix} 1 \\ 0 \\ 0 \\ 0 \end{pmatrix} : \alpha_1 \in \mathbb{C}\} \quad \text{and} \quad S(M_{2,0}) = \{\alpha_2 \begin{pmatrix} 0 \\ 1 \\ 0 \\ 0 \end{pmatrix} : \alpha_2 \in \mathbb{C}\},$$

respectively. It is obvious that

$$S_1 \neq S(M_{1,0}) \quad \text{and} \quad S_2 = S(M_{2,0}),$$

as conjectured and predicted by Remark 3.12 and Theorem 3.11, respectively, since

$$g(u) = \begin{cases} \begin{pmatrix} 0 \\ \alpha_1^3 \\ -\alpha_1^4 \\ -\alpha_3^2 \end{pmatrix} \neq 0, & \text{if } \alpha_1 \neq 0 \text{ and } u = \begin{pmatrix} \alpha_1 \\ \alpha_2 \\ \alpha_3 \\ \alpha_4 \end{pmatrix}; \\ \\ 0, & \text{if } \alpha_2 \neq 0 \text{ and } u = \begin{pmatrix} 0 \\ \alpha_2 \\ 0 \\ 0 \end{pmatrix}. \end{cases}$$

The solution $\varphi(t)$ of this initial value problem with $\varphi(0) = u_0 = \begin{pmatrix} 1 \\ 0 \\ 0 \\ 0 \end{pmatrix}$ cannot stay

in the vicinity of the origin for all $t \geq 0$, because in this case,

$$c_3 = -\frac{1}{12} \neq 0 \quad \text{and} \quad c_4 = \frac{1}{432} - \frac{1}{648} - \frac{1}{3024} \neq 0.$$

This is predicted by Theorem 3.11 since $S_1 \neq S(M_{1,0})$ and $u_0 \in S(M_{1,0}) \setminus S_1$.

17. The condition (3.5) is met by Remark 3.8. This is because the Jacobian matrix $f_x(u)$ of $f(u)$ is

$$\begin{pmatrix} 11u_2u_3u_4 & 11u_1u_3u_4 & 11u_1u_2u_4 & 11u_1u_2u_3 \\ 14u_2u_3u_4 & 14u_1u_3u_4 & 14u_1u_2u_4 & 14u_1u_2u_3 \\ 27u_2u_3u_4 & 27u_1u_3u_4 & 27u_1u_2u_4 & 27u_1u_2u_3 \\ 27u_2u_3u_4 & 27u_1u_3u_4 & 27u_1u_2u_4 & 27u_1u_2u_3 \end{pmatrix} \quad \text{and}$$

$$\begin{pmatrix} f_{11} & \cdots & f_{14} \\ & \vdots & \\ f_{41} & \cdots & f_{44} \end{pmatrix}, \quad \text{respectively,}$$

where

$$f_{11} = u_2u_3u_4v_1 + 27u_1u_2u_3u_4, \quad f_{12} = u_1u_3u_4v_1,$$
$$f_{13} = u_1u_2u_4v_1 - 25u_1u_2u_3u_4, \quad f_{14} = u_1u_2u_3v_1 + 4u_1u_2u_3u_4$$
$$f_{21} = u_2u_3u_4v_2, \quad f_{22} = u_1u_3u_4v_2 + 27u_1u_2u_3u_4,$$
$$f_{23} = u_1u_2u_4v_2 - 4u_1u_2u_3u_4, \quad f_{24} = u_1u_2u_3v_2 - 10u_1u_2u_3u_4$$
$$f_{31} = u_2u_3u_4v_3, \quad f_{32} = u_1u_3u_4v_3,$$
$$f_{33} = u_1u_2u_4v_3, \quad f_{34} = u_1u_2u_3v_3,$$
$$f_{41} = u_2u_3u_4v_4, \quad f_{42} = u_1u_3u_4v_4,$$
$$f_{43} = u_1u_2u_4v_4, \quad f_{44} = u_1u_2u_3v_4.$$

Clearly, $|f_x(u)| \to 0$ when $|u| \to 0$. Since $f(0) = 0$ and the eigenvalues of A are, by the solution of Example 5.23 in Chapter 3, $2, 2, -1$, and -1, we have that the conclusion of Theorem 3.9 hold, where the two-dimensional manifold (or surface) $S = S_2$ is, at the origin, tangent to, by the solution of Example 5.23 in Chapter 3,

$$M_0 = S(M_{2,0}) = \{c_3d_3 + c_4d_4 = c_3 \begin{pmatrix} 25 \\ 4 \\ 27 \\ 0 \end{pmatrix} + c_4 \begin{pmatrix} -14 \\ 10 \\ 0 \\ 27 \end{pmatrix} : c_3, c_4 \in \mathbb{C}\}, \quad \text{where}$$

$$M_{2,0} = \frac{1}{27} \begin{pmatrix} 0 & 0 & 25 & -14 \\ 0 & 0 & 4 & 10 \\ 0 & 0 & 27 & 0 \\ 0 & 0 & 0 & 27 \end{pmatrix}.$$

Furthermore, it follows from Theorem 3.11 that $S = M_0$. This is because $f(u) = u_1u_2u_3u_4(d_3 + d_4)$ has no components in $S(M_{1,0})$ where

$$M_{1,0} = \frac{1}{27} \begin{pmatrix} 27 & 0 & -25 & 4 \\ 0 & 27 & -4 & -10 \\ 0 & 0 & 0 & 0 \\ 0 & 0 & 0 & 0 \end{pmatrix}$$

from the solution of Example 5.23 in Chapter 3, and because $f(u) = u_1u_2u_3v = 0$ when

$$u = c_3d_3 + c_4d_4 \in S(M_{2,0}), \quad c_3, c_4, \text{ constants.}$$

18. As in the solution of Problem **15**, the condition (3.5) is satisfied, $g(0) = 0$, the eigenvalues of A are $-2, -3, 4$, and 5, and

$$e^{tA} = e^{-2t}M_{1,0} + e^{-3t}M_{2,0} + e^{4t}M_{3,0} + e^{5t}M_{4,0}$$

$$= e^{-2t} \begin{pmatrix} 1 & 0 & 0 & 0 \\ 0 & 0 & 0 & 0 \\ 0 & 0 & 0 & 0 \\ 0 & 0 & 0 & 0 \end{pmatrix} + e^{-3t} \begin{pmatrix} 0 & 0 & 0 & 0 \\ 0 & 1 & 0 & 0 \\ 0 & 0 & 0 & 0 \\ 0 & 0 & 0 & 0 \end{pmatrix} +$$

$$e^{4t} \begin{pmatrix} 0 & 0 & 0 & 0 \\ 0 & 0 & 0 & 0 \\ 0 & 0 & 1 & 0 \\ 0 & 0 & 0 & 0 \end{pmatrix} + e^{5t} \begin{pmatrix} 0 & 0 & 0 & 0 \\ 0 & 0 & 0 & 0 \\ 0 & 0 & 0 & 0 \\ 0 & 0 & 0 & 1 \end{pmatrix}.$$

Because of its simplicity where

$$u_1' = -2u_1 + u_4^3,$$
$$u_2' = -3u_2 - u_4^3,$$
$$u_3' = 4u_3 + u_4^2,$$
$$u_4' = 5u_4,$$

this equation has the solution which is easily obtained as

$$u(t) = \begin{pmatrix} c_1 e^{-2t} + \frac{c_4^3}{17} e^{15t} \\ c_2 e^{-3t} - \frac{c_4^3}{18} e^{15t} \\ c_3 e^{4t} + \frac{c_4^2}{6} e^{10t} \\ c_4 e^{5t} \end{pmatrix},$$

where

$$u(0) = \begin{pmatrix} \alpha_1 \\ \alpha_2 \\ \alpha_3 \\ \alpha_4 \end{pmatrix} \quad \text{and} \quad \begin{pmatrix} c_1 \\ c_2 \\ c_3 \\ c_4 \end{pmatrix} = \begin{pmatrix} \alpha_1 - \dfrac{\alpha_4^3}{17} \\ \alpha_2 + \dfrac{\alpha_4^3}{18} \\ \alpha_3 - \dfrac{\alpha_4^2}{6} \\ \alpha_4 \end{pmatrix}.$$

In order for $u(t) \to 0$ as $t \to +\infty$, we require

$$\alpha_3 = 0 = \alpha_4,$$

and so the two-dimensional manifold (or surface) S predicted by Theorem 3.9 is

$$S = \left\{ \begin{pmatrix} \alpha_1 \\ \alpha_2 \\ 0 \\ 0 \end{pmatrix} : \alpha_1, \alpha_2 \in \mathbb{C}^3 \right\}.$$

Clearly, S is, at the origin, tangent to the plane

$$M_0 = S(M_{1,0} \cup M_{2,0}) = \left\{ \alpha_1 \begin{pmatrix} 1 \\ 0 \\ 0 \\ 0 \end{pmatrix} + \alpha_2 \begin{pmatrix} 0 \\ 1 \\ 0 \\ 0 \end{pmatrix} : \alpha_1, \alpha_2 \in \mathbb{C} \right\}.$$

Furthermore, the one-dimensional manifolds (or surfaces) predicted by Theorem 3.10 are

$$S_1 = S \cap S(M_{3,0} \cup M_{4,0} \cup M_{1,0}) = \left\{ \begin{pmatrix} \alpha_1 \\ 0 \\ 0 \\ 0 \end{pmatrix} : \alpha_1 \in \mathbb{C} \right\},$$

$$S_2 = S \cap S(M_{3,0} \cup M_{4,0} \cup M_{2,0}) = \left\{ \alpha_2 \begin{pmatrix} 0 \\ 1 \\ 0 \\ 0 \end{pmatrix} : \alpha_2 \in \mathbb{C} \right\},$$

and S_1 and S_2 are, at the origin, tangent to

$$S(M_{1,0}) = \left\{ \alpha_1 \begin{pmatrix} 1 \\ 0 \\ 0 \\ 0 \end{pmatrix} : \alpha_1 \in \mathbb{C} \right\} \quad \text{and} \quad S(M_{2,0}) = \left\{ \alpha_2 \begin{pmatrix} 0 \\ 1 \\ 0 \\ 0 \end{pmatrix} : \alpha_2 \in \mathbb{C} \right\},$$

respectively. It is obvious that

$$S = M_0, \quad S_1 = S(M_{1,0}) \quad \text{and} \quad S_2 = S(M_{2,0}),$$

as predicted by Theorem 3.11 since $g(u) = 0$ for

$$u = \alpha_1 \begin{pmatrix} 1 \\ 0 \\ 0 \\ 0 \end{pmatrix} \quad \text{or} \quad \alpha_2 \begin{pmatrix} 0 \\ 1 \\ 0 \\ 0 \end{pmatrix}$$

where $\alpha_1 \ \alpha_2$ are not zero.

CHAPTER 8

Existence and Uniqueness Theorems

1. Introduction

Unless otherwise stated, all the functions in this chapter are assumed real-valued. Although the main results in Section 2 are stated for the case with real-valued functions, they are also valid for the case with complex-valued functions. This can be seen from the proofs in Section 3.

Now consider the linear second order ordinary differential equation

$$y''(t) + p(t)y'(t) + q(t)y(t) = r(t), \quad a \leq t \leq b,$$
$$y(\tau) = y_0, y'(\tau) = y_1, \quad \tau \in [a, b], \tag{1.1}$$

where $y_0, y_1, a, b \in \mathbb{R}$ are given four real numbers with $a < b$, $y = y(\tau)$ is a twice continuously differentiable function on $[a, b]$, and $p(\tau), q(\tau)$, and $r(\tau)$ are given three continuous functions on $[a, b]$.

By using the substitutions:

$$\begin{cases} u_1 = y, \\ u_2 = y', \\ u = \begin{pmatrix} u_1 \\ u_2 \end{pmatrix}, \end{cases}$$

we have

$$\frac{d}{dt} u = \begin{pmatrix} u_1' \\ u_2' \end{pmatrix} = \begin{pmatrix} y' \\ y'' \end{pmatrix}$$
$$= \begin{pmatrix} y' \\ -py' - qy + r \end{pmatrix}$$
$$= \begin{pmatrix} u_2 \\ -pu_2 - qu_1 + r \end{pmatrix}.$$

It follows that (1.1) can be rewritten as the system of two linear first order differential equations with the initial conditions:

$$\frac{d}{dt} u = B(t)u + h(t), \quad a \leq t \leq b,$$
$$u(\tau) = u_0, \quad \tau \in [a, b], \tag{1.2}$$

where

$$u = \begin{pmatrix} u_1 \\ u_2 \end{pmatrix} \in \mathbb{R}^2, \quad B(x) = \begin{pmatrix} 0 & 1 \\ -q & -p \end{pmatrix},$$
$$h(t) = \begin{pmatrix} 0 \\ r(t) \end{pmatrix}, \quad u_0 = \begin{pmatrix} y_0 \\ y_1 \end{pmatrix}.$$

It is readily checked that if y is a (unique) solution to (1.1), then $\begin{pmatrix} y \\ y' \end{pmatrix}$ is a (unique) solution to the first order linear system (1.2). Conversely, if

$$u = \begin{pmatrix} u_1 \\ u_2 \end{pmatrix}$$

is a (unique) solution to the first order linear system (1.2), then u_1 is a (unique) solution to the equation (1.1). Uniqueness of a solution also follows from the same reasoning. Thus solving (1.1) is the same as solving (1.2).

Here, by a solution y to the equation (1.1), it is meant that y is a twice continuously differentiable function on $[a, b]$ and y satisfies the equation (1.1). And, by a solution

$$u = \begin{pmatrix} u_1 \\ u_2 \end{pmatrix}$$

to the first order linear system (1.2), it is meant that $u_1, u_2 \in C[a, b]$ are continuously differentiable functions on $[a, b]$ and $u = \begin{pmatrix} u_1 \\ u_2 \end{pmatrix}$ satisfies (1.2).

The above idea also applies to the higher order linear ordinary differential equation with the initial conditions:

$$y^{(n)} + p_1(t)y^{(n-1)} + p_2(t)y^{(n-3)} + \cdots + p_n(t)y = p_0(t), \quad a \le t \le b,$$
$$y(\tau) = y_0, y'(\tau) = y_1, \ldots, y^{(n-1)}(\tau) = y_{n-1}, \quad \tau \in [a, b], \tag{1.3}$$

where $n \in \mathbb{N}$; $y_i, i = 0, 1, \ldots, n-1, a, b$, are given real numbers with $a < b$; $y = y(t)$ is a n times continuously differentiable function on $[a, b]$; and $p_i(t), i = 0, 1, \ldots, n$, are given continuous functions on $[a, b]$. By using the substitutions: $u_1 = y, u_2 = y', \ldots, u_n = y^{(n-1)}$, and the substitution:

$$u = \begin{pmatrix} u_1 \\ u_2 \\ \vdots \\ u_n \end{pmatrix},$$

we have

$$\frac{d}{dt} u = \begin{pmatrix} u_1' \\ u_2' \\ \vdots \\ u_n' \end{pmatrix} = \begin{pmatrix} y' \\ y'' \\ \vdots \\ y^{(n)} \end{pmatrix}$$

$$= \begin{pmatrix} u_2 \\ u_3 \\ \vdots \\ u_n \\ -p_1 y^{(n-1)} - p_2 y^{(n-2)} \cdots -p_{n-1} y' - p_n y + p_0 \end{pmatrix}$$

$$= \begin{pmatrix} u_2 \\ u_3 \\ \vdots \\ u_n \\ -p_n u_1 - p_{n-1} u_2 \cdots -p_1 u_n \end{pmatrix} + \begin{pmatrix} 0 \\ 0 \\ \vdots \\ 0 \\ p_0 \end{pmatrix},$$

and so equation (1.3) becomes the first order linear system of n differential equations with the initial conditions:

$$\frac{d}{dt}u = B(t)u + h(t), \quad a \leq t \leq b,$$

$$u(\tau) = u_0, \quad \tau \in [a, b],$$

(1.4)

where

$$B(t) = \begin{pmatrix} 0 & 1 & 0 & \cdot & \cdots & & \cdot & 0 & 0 \\ 0 & 0 & 1 & 0 & 0 & \cdots & & \cdot & 0 \\ \cdot & & \cdot & \cdot & \cdot & \cdots & & \cdot & \cdot \\ \vdots & \vdots & \vdots & \vdots & \cdots & \vdots & \vdots & 0 \\ 0 & 0 & \cdot & \cdot & \cdots & 0 & 0 & 1 \\ -p_n & -p_{n-1} & \cdot & \cdot & \cdots & -p_3 & -p_2 & -p_1 \end{pmatrix}, \quad h(t) = \begin{pmatrix} 0 \\ 0 \\ \vdots \\ 0 \\ p_0 \end{pmatrix}.$$

As is with the case of linear second order equation (1.1), solving the equation (1.3) is the same as solving the first order linear system (1.4).

One of our two purposes in this chapter to study the more general first order linear system of 2 or $n \in \mathbb{N}$ differential equations with the initial conditions:

$$\frac{d}{dt}u = A(t)u + f(t), \quad a \leq t \leq b,$$

$$u(\tau) = u_0 \in \mathbb{R}^2 \quad \text{or} \quad \mathbb{R}^n, \quad \tau \in [a, b],$$

(1.5)

from which solutions of the equation (1.1) or (1.3) are deduced. Here $A(t) = (a_{ij}(t))$ is a matrix of 2×2 or $n \times n$ with $a_{ij}(t), i, j = 1, 2$ or $i, j = 1, 2, \ldots, n$, continuous functions on $[a, b]$; $f(x)$ is a vector of 2 or n components with each component, a continuous function on $[a, b]$; and u_0 is a constant vector in $\mathbb{R}^2$ or $\mathbb{R}^n$.

The other purpose of this chapter to study the general first order nonlinear system of $n \in \mathbb{N}$ differential equations with the initial conditions:

$$\frac{d}{dt}u = g(t, u) \in \mathbb{R}^n, \quad \tau - a < t < \tau + a,$$

$$u(\tau) = u_0 \in \mathbb{R}^n,$$

(1.6)

where τ is some given real number in $\mathbb{R}$; u_0 is some given point in $\mathbb{R}^n$ with $n \in \mathbb{N}$; $g(t, x) \in \mathbb{R}^n$ is a continuous vector-valued function on the region

$$R \equiv \{(t, x) : |t - \tau| \leq a, |x - \xi| \leq b, t \in \mathbb{R}, x \in \mathbb{R}^n\}$$

and satisfies the Lipschitz condition in x:

$$|g(t, x_1) - g(t, x_2)| \leq L_0 |x_1 - x_2|$$

for all $(t, x_1), (t, x_2) \in R$. Here for $x = \begin{pmatrix} x_1 \\ x_2 \\ \vdots \\ x_n \end{pmatrix}$ and $y = \begin{pmatrix} y_1 \\ y_2 \\ \vdots \\ y_n \end{pmatrix} \in \mathbb{R}^n$, we define $|x - y|$ by

$$|x - y| \equiv \sum_{i=1}^{n} |x_i - y_i|.$$

This is the distance in $\mathbb{R}^n$, between $x \in \mathbb{R}^n$ and $y \in \mathbb{R}^n$, which generalizes that in $\mathbb{R}$.

The rest of this chapter is organized as follows. Section 2 states the main results, and section 3 deals with the proof of the main results. Finally, Section 4 presents some results with Riemann or Lebesgue integrable coefficients.

The material in this chapter is based on Coddington and Levinson [5].

2. The Main Results with Continuous Coefficients

Although the following results are stated for the case with real-valued functions, they are also valid for the case with complex-valued functions. This can be seen by the proofs in Section 3.

THEOREM 2.1. *There is a unique solution* $u = \begin{pmatrix} u_1 \\ u_2 \end{pmatrix}$ *with* $u_i, i = 1, 2$, *continuously differentiable functions on* $[a, b]$, *to the first order linear system of two differential equations with the initial conditions:*

$$\frac{d}{dt} u = A(t)u + f(t), \quad a \le t \le b,$$
$$u(\tau) = u_0 \in \mathbb{R}^2, \quad \tau \in [a, b]. \tag{2.1}$$

Here $0 < a < b$ *are given two real numbers, and* u_0 *is a given constant real vector in* $\mathbb{R}^2$; $A(t) = (a_{ij}(t))$ *is a given matrix of* 2×2 *with* $a_{ij}(t), i, j = 1, 2$, *continuous functions on* $[a, b]$, *and* $f(t) = \begin{pmatrix} f_1(t) \\ f_2(t) \end{pmatrix}$ *is a given vector with each component, a continuous function on* $[a, b]$.

COROLLARY 2.2. *There is a unique solution* $y \in C^2[a, b]$ *to the linear second order ordinary differential equation with the initial conditions:*

$$y''(t) + p(t)y'(t) + q(t)y(t) = r(t), \quad a \le t \le b,$$
$$y(\tau) = y_0, y'(\tau) = y_1, \quad \tau \in [a, b]. \tag{2.2}$$

Here y_0, y_1, *and* $0 < a < b \in \mathbb{R}$ *are given four real numbers;* $p(t), q(t)$, *and* $r(t)$ *are given three continuous functions on* $[a, b]$; *and* $C^2[a, b]$ *is the vector space of all twice continuously differentiable functions on* $[a, b]$.

THEOREM 2.3. *There is a unique vector solution* $u \in \mathbb{R}^n$ *with its components, continuously differentiable functions on* $[a, b]$, *to the first order linear system of* n *differential equations with the initial conditions:*

$$\frac{d}{dt} u = A(t)u + f(t), \quad a \le t \le b,$$
$$u(\tau) = u_0 \in \mathbb{R}^n, \quad \tau \in [a, b]. \tag{2.3}$$

Here $n \in \mathbb{N}$ *is a natural number, and* $0 < a < b$ *and* τ *are given three real numbers;* u_0 *is a given real constant vector in* $\mathbb{R}^n$, *and* $A(t) = (a_{ij}(t))$ *is a given matrix of* $n \times n$ *with* $a_{ij}(t), i, j = 1, 2, \ldots, n$, *continuous functions on* $[a, b]$; *finally,* $f(t)$ *is a given vector of* n *components, with each component* $f_i, i = 1, 2, \ldots, n$, *a continuous function on* $[a, b]$.

COROLLARY 2.4. *There is a unique solution $y \in C^n[a,b]$ to the n-th ($n \in \mathbb{N}$) order linear ordinary differential equation with the initial condition:*

$$y^{(n)} + p_1(t)y^{(n-1)} + p_2(t)y^{(n-3)} + \cdots + p_n(t)y = p_0(x), \quad a \le t \le b,$$

$$y(\tau) = y_0, y'(\tau) = y_1, \ldots, y^{(n-1)}(\tau) = y_{n-1}, \quad \tau \in [a,b].$$

(2.4)

Here $0 < a < b$ and $y_i, i = 0, 1, 2, \ldots, (n-1)$, are given real numbers; $p_i(t), i = 0, 1, \ldots, n$, are given continuous functions on $[a,b]$; and $C^n[a,b]$ is the vector space of all n times continuously differentiable functions on $[a,b]$.

Concerning nonlinear systems, we have:

THEOREM 2.5 (Picard-Lindelof local unique existence theorem). *(Coddington and Levinson [5, Page 12]) There is a unique solution $u \in \mathbb{R}^n, n \in \mathbb{N}$, with each component of u, a continuously differentiable function on $[\tau - \alpha, \tau + \alpha]$, to the first order nonlinear system (1.6) on a small t interval $[\tau - \alpha, \tau + \alpha]$, of n differential equations with the initial conditions:*

$$\frac{d}{dt}u = g(t,u), \quad \tau - \alpha \le t \le \tau + \alpha,$$

$$u(\tau) = u_0 = \xi \in \mathbb{R}^n.$$

(2.5)

Here $0 < a < b$ and τ are three given real numbers; $u_0 = \xi$ is some given point in $\mathbb{R}^n$; and $g(t,x) \in \mathbb{R}^n$ is a given continuous function on the region

$$R \equiv \{(t,x) : |t - \tau| \le a, x \in \mathbb{R}^n, |x - u_0| = |x - \xi| \le b\}$$

with

$$M_0 \equiv \max_{(t,x) \in R} |g(t,x)| < \infty$$

and satisfies the Lipschitz condition in x:

$$|g(t,x_1) - g(t,x_2)| \le L_0|x_1 - x_2|$$

for all $(t,x_1), (t,x_2) \in R$, with $0 < L_0 < \infty$ and

$$\alpha \equiv \min\{a, \frac{b}{M_0}\}.$$

If the Lipschitz condition for g in Theorem 2.5 is dropped, the uniqueness result might not hold and we have the following existence result without uniqueness:

THEOREM 2.6 (Cauchy-Peano local existence theorem). *(Coddington and Levinson [5, Page 6]) There is a solution $u \in \mathbb{R}^n, n \in \mathbb{N}$, with each component of u, a continuously differentiable function on $[\tau - \alpha, \tau + \alpha]$, to the first order nonlinear system (1.6) on a small t interval $[\tau - \alpha, \tau + \alpha]$, of $n \in \mathbb{N}$ differential equations with the initial conditions:*

$$\frac{d}{dt}u = g(t,u), \quad \tau - \alpha \le t \le \tau + \alpha,$$

$$u(\tau) = u_0 = \xi \in \mathbb{R}^n.$$

(2.6)

Here $0 < a < b$ and τ are three given real numbers in $\mathbb{R}$; $u_0 = \xi$ is some given point in $\mathbb{R}^n$; and $g(t,x) \in \mathbb{R}^n$ is a given continuous function on the region

$$R \equiv \{(t,x) : |t - \tau| \le a, x \in \mathbb{R}^n, |x - u_0| = |x - \xi| \le b\}$$

with

$$M_0 \equiv \max_{(t,x) \in R} |g(t,x)| < \infty,$$

$$\alpha \equiv \min\{a, \frac{b}{M_0}\}.$$

To illustrate Theorem 2.6, consider the example, Coddington and Levinson [**5**, Page 7],

EXAMPLE 2.7.

$$\frac{d}{dt}u(t) = \sqrt{u}, \quad t \in [0, 1],$$
$$u(0) = 0.$$

By separating the variables u and t and integrating both sides, we have

$$\int u^{-\frac{1}{2}} \, du = \int 1 \, dt,$$

and so $2u^{\frac{1}{2}} = t + \lambda$. Thus, for each $0 \leq \lambda \leq 1$, the function

$$u_\lambda(t) = \begin{cases} 0, & \text{if } 0 \leq t \leq \lambda; \\ (\frac{t-\lambda}{2})^2, & \text{if } \lambda < t \leq 1, \end{cases}$$

is a solution. Hence this problem lacks uniqueness of a solution. This is due to the fact that the function $g(t, u) = \sqrt{u}$ is not Lipschitz continuous with respect to the variable u on any $[0, \alpha]$ where $\alpha > 0$. g being not Lipschitz continuous can be readily seen, for, otherwise, there is an $L_0 > 0$, such that

$$|\sqrt{u} - 0| \leq L_0|u - 0| \quad \text{for small } u > 0.$$

This would imply

$$\frac{1}{\sqrt{u}} \leq L_0 \quad \text{for small } u > 0,$$

which is impossible.

COROLLARY 2.8 (Local Existence). *If, not on the closed rectangle R, but on a domain D in the (t, x) plane, the function $g(t, x)$ is continuous, then we still have:*

The equation 2.5 has a solution for any (τ, ξ) in D, which exists locally in some t inteveral $[\tau - \alpha, \tau + \alpha]$ for some $\alpha > 0$. Here a domain D in the (t, x) plane is an open connected region.

This corollary will be used in Chapter 7.

PROOF. Since D is open, we can find an open disk $B_{r_0}(\tau, \xi)$ in D, which has the radius $r_0 > 0$ and the center (τ, ξ). Let R be any closed rectangle in $B_{r_0}(\tau, \xi)$, that contains the point (τ, ξ). The result follows by applying Theorem 2.6 to the R. □

When a solution for (2.5) exists only on a small open t interval $(\tau - \alpha, \tau + \alpha)$, we might possibly enlarge this small interval:

THEOREM 2.9 (Continuation of Solutions). *Suppose additionally that the $g(t, x)$ in Corollary 2.8 is bounded on D and that (2.5) has a solution u on the open interval $(a, b) \equiv (\tau - \alpha, \tau + \alpha)$. Then the limits $L_l \equiv \lim_{t \to a+} u(t)$ and $L_r \equiv \lim_{t \to b-} u(t)$ exist, and u can be continued to the right $(a, b + \epsilon]$ (or to the left $[a - \epsilon, b)$) of (a, b) for some $\epsilon > 0$ if the point (b, L_r) (or the point (a, L_l)) is in D.*

This theorem will be used in Chapter 7.

Before proving Theorem 2.9, consider the example, Coddington and Levinson, [**5**, Page 14],

EXAMPLE 2.10.

$$\frac{d}{dt}u(t) = u^3, \quad [-2, 0),$$

$$u(-2) = \frac{1}{2},$$

By separating the variables u and t and integrating both sides, we have

$$\int u^{-3} \, du = \int 1 \, dt,$$

and so $\frac{u^{-2}}{-2} = t + \lambda$. Thus this problem has the unique solution

$$u(t) = \sqrt{-\frac{1}{2t}}$$

on $[-2, 0)$, but it cannot be extended to $[-2, 0]$. This is because $\lim_{t \to 0-} u(t) = +\infty$, which is not in a region D where $g(t, u) = u^3$ is bounded.

Proof of Theorem 2.9:

PROOF. Since the solution u satisfies, by integrating (2.5),

$$u(t) = \xi + \int_{s=\tau}^{t} g(s, u(s)) \, ds, \quad t \in (a, b), \tag{2.7}$$

we have, for $a < \lambda, \nu < b$,

$$|u(\nu) - u(\lambda)| \leq \int_{s=\lambda}^{\nu} |g(s, u(s))| \, ds \leq M_0 |\nu - \mu|, \tag{2.8}$$

where $M_0 \equiv \max_{(t,x) \in R} |g(t, x)|$. Thus $u(t)$ is Cauchy in t and so converges by Advanced Calculus, Apostol [**2**, Page 73], as $t \longrightarrow b^-$ or a^+. That is, L_l and L_r exist. Here, as $\lambda, \mu \longrightarrow b^-$ or a^+, we have, by (2.8),

$$|u(\nu) - u(\lambda)| \longrightarrow 0,$$

and so $u(t)$ is Cauchy in t.

Suppose that the pint (b, L_r) lies in D, then the solution u can be continued to the right $(a, b + \epsilon]$ for some $\epsilon > 0$. To see this, we divide the proof into two steps:

Step 1. By defining the function $\tilde{u}$ by

$$\tilde{u}(t) = \begin{cases} u(t), & \text{if } t \in (a, b); \\ L_r, & \text{if } t = b, \end{cases}$$

a continuous extension of u to the point $t = b$, we have $\tilde{u}$ is a solution on $(a, b]$. This follows from the reasoning: Since $\tilde{u}$ is continuous on $(a, b]$ and g is bounded and continuous on D, we have from Advanced Calculus, Apostol [**2**], that

$$\int_{s=\tau}^{t} g(s, \tilde{u}) \, ds, \quad t \in (a, b]$$

exists and is continuous. Thus (2.7) extends to

$$\tilde{u}(t) = \xi + \int_{s=\tau}^{t} g(s, \tilde{u}) \, ds, \quad t \in (a, b], \tag{2.9}$$

and so

$$\lim_{t \to b^-} \tilde{u}(t)' = g(b, \tilde{u}(b))$$

follows by the fundamental theorem of calculus, (the **NOTE** in [**2**, Page 162]). This shows that $\tilde{u}$ is a solution on $(a, b]$.

Step 2. Since the point (b, L_r) is in D, we have by Theorem 2.6 that (2.5) has a solution ψ passing through the point (b, L_r) that exists on some interval $[b, b+\epsilon])$, $\epsilon > 0$. Then the function $\hat{u}$ defined by

$$\hat{u}(t) = \begin{cases} \tilde{u}(t), & \text{if } t \in (a, b]; \\ \psi(t), & \text{if } t \in [b, b+\epsilon], \end{cases}$$

is a solution of (2.5) on $(a, b + \epsilon]$. This follows from the fundamental theorem of calculus if we can show

$$\hat{u}(t) = \xi + \int_{s=\tau}^{t} g(s, \hat{u}(s)) \, ds, \quad t \in (a, b+\epsilon]. \tag{2.10}$$

But (2.10) is true for $t \in (a, b]$, since $\hat{u} = \tilde{u}$ and $\tilde{u}$ satisfies (2.10). Further, (2.10) is also true for $t \in [b, b+\epsilon]$ because on $[b, b+\epsilon]$, we have $\hat{u} = \psi$ where ψ is a solution of (2.5) passing through the point (b, L_r), and so

$$\hat{u}(t) = L_r + \int_{s=b}^{t} g(s, \hat{u}(s)) \, ds;$$

but

$$L_r = \xi + \int_{s=\tau}^{b} g(s, \hat{u}(s)) \, ds$$

by (2.9).

It follows similarly that u can be continued to the left $[a - \epsilon, a)$ for some $\epsilon > 0$ if the point (a, L_l) lies in D.

See the book by Coddington and Levinson [**5**, Pages 13-15]. □

REMARK 2.11. *The $\hat{u}$ in the above proof is called a continuation of u to $(a, b+\epsilon]$.*

3. Proof of the Main Results

Although the proofs below are given for the case with real-valued functions, they are also valid for the case with complex-valued functions. This is done by simply defining

$$|z(t)| = |v(t)| + |h(t)|$$

for a complex-valued function $z(t) = v(t) + ih(t)$ on $[c, d], c < d$, where $v(t)$ and $h(t)$ are real-valued functions on $[c, d]$, and i is the complex unit satisfying

$$i = \sqrt{-1}, i^2 = -1, i^3 = -i, i^4 = 1.$$

3.1. The Linear Case with Continuous Coefficients.

PROOF. We use the method of successive approximations, Coddington and Levinson [**5**, Pages 12 and 98]. Integrating the equation (2.1) gives

$$u(t) = u_0 + \int_{\tau}^{t} A(s)u(s) \, ds + \int_{\tau}^{t} f(s) \, ds, \tag{3.1}$$

the integrated form of (2.1), to which we define the successive approximations

$$\phi_i = \begin{pmatrix} (\phi_i)_1 \\ (\phi_i)_2 \end{pmatrix}, \quad i = 0, 1, 2, \ldots,$$

by

$$\phi_0 \equiv u_0,$$

$$\phi_i \equiv u_0 + \int_\tau^t A(s)\phi_{i-1}(s)\, ds + \int_\tau^t f(s)\, ds, \quad i = 1, 2, \ldots,$$

where $\int_\tau^t f(s)\, ds$ is defined by $\begin{pmatrix} \int_\tau^t f_1(s)\, ds \\ \int_\tau^t f_2(s)\, ds \end{pmatrix}$, and $\int_\tau^t A(s)\phi_{i-1}(s)\, ds$ is similarly defined.

Since $\phi_0 = u_0$ is a constant continuous function on $[a, b]$, and since $a_{jk}(s), (A(s)u_0)_j$, the jth component of $A(s)u_0$, and $f_j(s), j, k = 1, 2$, are all continuous also, we have that

$$\phi_1(t) = u_0 + \int_\tau^t A(s)u_0\, ds + \int_\tau^t f(s)\, ds$$

is likewise continuous on $[a, b]$. It follows by mathematical induction that $\phi_i, i = 1, 2, \ldots$, is well-defined, and is a continuous function on $[a, b]$.

We now divide the proof into ten steps.

Step 1. By defining

$$|A(t)| \equiv \sum_{j,k=1}^{2} |a_{jk}(t)|,$$

we have, for $x = \begin{pmatrix} x_1 \\ x_2 \end{pmatrix} \in \mathbb{R}^2$,

$$|A(t)x| \le |A(t)||x|.$$

This is because

$$|x_j| \le |x|, j = 1, 2,$$

$$A(t)x = \begin{pmatrix} \sum_{j=1}^{2} a_{1j}(t)x_j \\ \sum_{j=1}^{2} a_{2j}(t)x_j \end{pmatrix},$$

$$|A(t)x| = \sum_{k=1}^{2} |\sum_{j=1}^{2} a_{kj}(t)x_j|$$

$$\le \sum_{k=1}^{2} \sum_{j=1}^{2} |a_{kj}(t)||x| = |A(t)||x|.$$

Step 2. Since $a_{jk}(t)$ and $f_j, j, k = 1, 2$ are all continuous on $[a, b]$, we have

$$\max\{|A(t)|, |f(t)|\} = \max\{\sum_{j,k=1}^{2} |a_{jk}(t)|, |f(t)|\} \le k(t)$$

for some positive continuous function $k(t)$ on $[a, b]$. For example, we can take

$$k(t) = |A(t)| + |f(t)|.$$

It follows from Step 1 that

$$|A(t)\phi_{i-1}| \le k(t)|\phi_{i-1}|.$$

Step 3. For $t \ge \tau$, claim

$$|\phi_i - \phi_{i-1}| \le (1 + |u_0|)\frac{K(t)^i}{i!},$$

where $K(t) = \int_\tau^t k(s)\,ds$. We are lead to this claim by the calculations:

$$|\phi_1 - \phi_0| = |\int_\tau^t A(s)\phi_0(s)\,ds + \int_\tau^t f(s)\,ds|$$

$$= \sum_{j=1}^{2}|\int_\tau^t (A(s)\phi_0(s))_j\,ds + \int_\tau^t f_j(s)\,ds|$$

$$\le \sum_{j=1}^{2}\int_\tau^t |(A(s)\phi_0(s))_j|\,ds + \sum_{j=1}^{2}\int_\tau^t |f_j(s)|\,ds$$

$$= \int_\tau^t |A(s)\phi_0(s)|\,ds + \int_\tau^t |f(s)|\,ds$$

$$\le \int_\tau^t k(s)|u_0|\,ds + \int_\tau^t k(s)\,ds$$

$$= (1 + |u_0|)\int_\tau^t k(s)\,ds = (1 + |u_0|)K(t),$$

$$|\phi_2 - \phi_1| = |\int_\tau^t A(s)[\phi_1(s) - \phi_0(s)]\,ds|$$

$$\le \int_\tau^t k(s)[(1 + |u_0|)K(s)\,ds$$

$$= (1 + |u_0|)\int_\tau^t \frac{dK(s)}{ds}K(s)\,ds$$

$$= (1 + |u_0|)\int_{s=\tau}^t K(s)\,dK(s)$$

$$= (1 + |u_0|)\frac{K(s)^2}{2!}|_{s=\tau}^t$$

$$= (1 + |u_0|)\frac{K(t)^2}{2!},$$

where $K(\tau) = 0$. Here we used the facts that, for a Riemann integrable function $h(t)$ on $[a, b]$,

$$|\int_a^b h(t)\,dt| \le \int_a^b |h(t)|\,dt$$

is true (Apostal [**2**, Page 155]), that the Fundamental Theorem of Calculus holds for continuous $k(t)$ (Apostal [**2**, Pages 161-162]):

$$\frac{d}{dt}K(t) = \frac{d}{dt}\int_\tau^t k(s)\,ds = k(t),$$

and that the formula of change of variables holds (Apostal [2], Page 163):

$$\int_\tau^t K(s)\frac{d}{ds}K(s)\,ds = \int_\tau^t K(s)\,dK(s).$$

The claim will be proved by mathematical induction on i. It is clearly true for $i = 1$, and so now, assume that it is true for $i = m \in \mathbb{N}$:

$$|\phi_m - \phi_{m-1}| \le (1 + |u_0|)\frac{K(t)^m}{m!}.$$

It will also be true for $i = m + 1$ by the calculations:

$$|\phi_{m+1} - \phi_m| \le \int_\tau^t k(s)|\phi_m - \phi_{m-1}|\,ds$$

$$\le (1 + |u_0|)\int_\tau^t \frac{dK(s)}{ds}\frac{K(s)^m}{m!}\,ds$$

$$= (1 + |u_0|)\int_{s=\tau}^t \frac{K(s)^m}{m!}\,dK(s)$$

$$= (1 + |u_0|)\frac{K(s)^{m+1}}{(m+1)!}\Big|_{s=\tau}^t$$

$$= (1 + |u_0|)\frac{K(t)^{m+1}}{(m+1)!},$$

where $K(\tau) = 0$.

Step 4. For $t \le \tau$, we have

$$|\phi_i - \phi_{i-1}| \le (1 + |u_0|)\frac{G(t)^i}{i!},$$

where $G(t) = \int_t^\tau k(s)\,ds$. This is proved similarly as in Step 3, where note:

$$|\phi_1 - \phi_0| = |\int_\tau^t A(s)\phi_0(s)\,ds + \int_\tau^t f(s)\,ds|$$

$$\le \int_t^\tau k(s)|u_0|\,ds + \int_t^\tau k(s)\,ds$$

$$= (1 + |u_0|)\int_t^\tau k(s)\,ds,$$

$$|\phi_i - \phi_{i-1}| \le (1 + |u_0|)\int_t^\tau |\phi_{i-1} - \phi_{i-2}|\,ds.$$

Step 5. Steps 3 and 4 give

$$|(\phi_i)_j - (\phi_{i-1})_j| \le \sum_{j=1}^2 |(\phi_i)_j - (\phi_i)_j| = |\phi_i - \phi_{i-1}|$$

$$\le (1 + |u_0|)\frac{K_0^i}{i!}$$

for each $j = 1, 2$, and for all $t \in [a, b]$, where $K_0 \equiv \int_a^b k(a)\,ds$, and

$$\max\{\max_{b \ge t \ge \tau}\int_\tau^t k(s)\,ds, \max_{a \le t \le \tau}\int_t^\tau k(s)\,ds\}$$

$$\leq \int_a^b k(s)\, ds < \infty,$$

and $(\phi_i)_1$ and $(\phi_i)_2$ are the first and the second component of the vector ϕ_i, respectively.

Step 6. It is a fact (Weirstrass M-test) from Advanced Calculus, Apostal [**2**, Page 223] that if a sequence $f_m(t)$ of functions on $[a, b]$ and a sequence M_m of nonnegative numbers, $m \in \mathbb{N}$, satisfy

$$0 \leq |f_m(t)| \leq M_m$$

for each m and for all $t \in [a, b]$, then we have $\sum_{m=1}^\infty f_m(t)$ converges uniformly for all $t \in [a, b]$, provided that $\sum_{m=1}^\infty M_m$ converges also.

Since

$$\sum_{i=1}^\infty (1 + |u_0|)\frac{K_0^i}{i!} = (1 + |u_0|)e^{K_0}$$

converges, it follows from the above Weirstrass M-test that each component function

$$(\phi_m - \phi_0)_j = \sum_{j=1}^m (\phi_i - \phi_{i-1})_j, j = 1, 2,$$

of the vector function $(\phi_m - \phi_0)$, converges uniformly for all $t \in [a, b]$. In this case, we say that the vector function

$$\phi_m(t) - \phi_0 = \sum_{i=1}^m (\phi_i - \phi_{i-1}),$$

converges uniformly for all $t \in [a, b]$. This implies that

$$\phi_m = (\phi_m - \phi_0) + \phi_0$$

converges uniformly to some vector function $\phi(t)$.

Step 7. It is a fact from Advanced Calculus, Apostal [**2**, Page 221] that if $f_m(t)$, a sequence of continuous functions on $[a, b]$, converges uniformly for all $t \in [a, b]$, to $f_0(t)$, then $f_0(t)$ is also a continuous function on $[a, b]$. This fact gives that each component $(\phi)_i$ of ϕ is a continuous function on $[a, b]$, and in this case, we say that ϕ is a continuous function on $[a, b]$.

Step 8. It is a fact (Uniform Convergence Theorem) from Adavanced Calculus, Apostal [**2**, Page 225] that if $f_m(t)$, a sequence of functions on $[a, b]$, is Riemann integrable on $[a, b]$ for each $m \in \mathbb{N}$, and converges uniformly to $f_0(t)$, then

$$\int_a^t f_n(s)\, ds$$

converges uniformly to

$$\int_a^t f(s)\, ds.$$

Since ϕ_{i-1} converges uniformly to ϕ by Step 6, as $i \to \infty$, and since

$$|(A(t)\phi_{i-1}(t) - A(t)\phi(t))_j| \leq \sum_{j=1}^2 |(A(t)\phi_{i-1}(t) - A(t)\phi(t))_j|$$

$$= |A(t)\phi_{i-1}(t) - A(t)\phi(t)| \leq k(t)|\phi_{i-1}(t) - \phi(t)|$$

$$\leq k_0|\phi_{i-1}(t) - \phi(t)|$$

for $j = 1, 2$, where $k_0 \equiv \max_{t \in [a,b]} k(t) < \infty$, we have that as $i \to \infty$, each component $(A(t)\phi_{i-1})_j, j = 1, 2$, of the vector function $A(t)\phi_{i-1}(t)$ converges uniformly to the corresponding component $(A(t)\phi(t))_j, j = 1, 2$, of the vector function $A(t)\phi(t)$. It follows from the above Uniform Convergence Theorem that, for $j = 1, 2$, we have

$$(\phi_i)_j = (\phi_0)_j + \int_\tau^t (A(s)\phi_{i-1}(s))_j \, ds + \int_\tau^t f_j(s) \, ds$$

converges uniformly to the function

$$(\phi)_j = (\phi_0)_j + \int_\tau^t (A(s)\phi(s))_j \, ds + \int_\tau^t f_j(s) \, ds,$$

where $(\phi)_j, (A(s)\phi(s))_j$, and f_j are the jth component of the vectors $\phi, A(s)\phi$, and f, respectively.

Step 9. It is a fact (the Fundamental Theorem of Calculus) from Calculus, Apostal [**2**, Page 162], that, for a continuous function $f_0(t)$ on $[a, b]$, we have $\int_a^t f_0(s) \, ds$ is differentiable, and

$$\frac{d}{dt} \int_a^t f_0(s) \, ds = f_0(t).$$

Since ϕ and $A(t)$ are continuous on $[a, b]$, we have that $A(t)\phi(t)$ is continuous on $[a, b]$. Since f is also continuous on $[a, b]$, it follows from the above Fundamental Theorem of calculus that, for $j = 1, 2,$, we have $(\phi)_j$ is differentiable, and

$$(\phi)_j(\tau) = (\phi_0)_j(\tau) = (u_0)_j,$$

$$\frac{d}{dt}(\phi)_j(t) = (A(t)\phi(t))_j + f_j(t),$$

where $(A(t)\phi(t))_j, j = 1, 2$, is the jth component of the vector $A(t)\phi(t)$. Equivalently, ϕ satisfies

$$\phi(\tau) = u_0,$$

$$\frac{d}{dt}\phi = \begin{pmatrix} \frac{d}{dt}(\phi)_1 \\ \frac{d}{dt}(\phi)_2 \end{pmatrix}$$

$$= A(t)\phi(t) + f(t).$$

Thus ϕ is a solution to (2.1).

Step 10. A solution to (2.1) is unique. This will be proved by applying the fact (Gronwall Inequality), Coddington and Levinson [**5**, Page 37] or Hartman [**15**, Page 24], that for two nonnegative, continuous functions φ and ψ on $[a, b]$, and for a nonnegative constant $C \geq 0$, we have the integral inequality

$$\varphi(t) \leq C + \int_a^t \varphi(s)\psi(s) \, ds$$

implies

$$\varphi(t) \leq C e^{\int_a^t \psi(s) \, ds}.$$

Suppose that φ and ψ are two solutions to the equation (2.1). Integrations give

$$\varphi(t) = u_0 + \int_\tau^t A(s)\varphi(s) \, ds + \int_\tau^t f(s) \, ds,$$

$$\psi(t) = u_0 + \int_\tau^t A(s)\psi(s) \, ds + \int_\tau^t f(s) \, ds,$$

and substraction gives

$$|\varphi(t) - \psi(t)| = |\int_\tau^t A(s)(\varphi(s) - \psi(s))\,ds|,$$

$$\sum_{j=1}^2 |\int_\tau^t (A(s)(\varphi(s) - \psi(s)))_j\,ds| \le \sum_{j=1}^2 \int_a^t |(A(s)(\varphi(s) - \psi(s)))_j|\,ds$$

$$= \int_a^t |A(s)(\varphi(s) - \psi(s))|\,ds$$

$$\le \int_a^t k(s)|\varphi(s) - \psi(s)|\,ds,$$

where Step 1 was used. It follows from the above Gronwall Inequality that for $j = 1, 2$, we have

$$0 \le |(\varphi(t) - \psi(t))_j| \le |\varphi(t) - \psi(t)| \le 0,$$

and so $\varphi \equiv \psi$. Here note that

$$\int_a^t k(s)\,ds \le \int_a^b k(s)\,ds < \infty.$$

□

Proof of Corollary 2.2.

PROOF. As stated in Section 1, solving the equation (1.1) is the same as solving the first order system (1.2), and so Corollary 2.2 is proved by applying Theorem 2.1. □

Proof Theorem 2.3.

PROOF. The proof is the same as that for Theorem 2.1, where note that $|A(t)|$ is defined as

$$|A(t)| \equiv \sum_{i,j=1}^n |a_{ij}(t)|,$$

and, for $x = \begin{pmatrix} x_1 \\ x_2 \\ \vdots \\ x_n \end{pmatrix} \in \mathbb{R}^n$, $|x|$ is defined as

$$|x| \equiv \sum_{i=1}^n |x_i|,$$

and the inequality

$$|A(t)x| \le |A(t)||x|$$

is also true. □

Proof of Corollary 2.4.

PROOF. As stated in Section 1, solving the equation (1.3) is the same as solving the first order system (1.4), and so Corollary 2.4 is proved by applying Theorem 2.3. □

3.2. The Nonlinear Case with Continuous Coefficients and with Uniqueness Result.

Proof of Theorem 2.5.

PROOF. We use the method of successive approximations, Coddington and Levinson [**5**, Pages 12 and 98]. Define the successive approximations

$$\phi_i = \begin{pmatrix} (\phi_i)_1 \\ (\phi_i)_2 \\ \vdots \\ (\phi_i)_n \end{pmatrix}, \quad i = 0, 1, 2, \ldots,$$

of (2.5) by

$$\phi_0 \equiv u_0,$$

$$\phi_i \equiv u_0 + \int_\tau^t g(s, \phi_{i-1}(s))\, ds, \quad i = 1, 2, \ldots,$$

where $|t - \tau| \leq \alpha$, and $(\phi_i)_j, j = 1, 2, \ldots, n$, is the jth component of the vector function ϕ_i.

Note that $\alpha > 0$ is restricted by $\alpha \leq a$, in order that the set $\{t : |t - \tau| \leq \alpha\}$ stays within the set $\{t : |t - \tau| \leq a\}$. The other restriction that $\alpha \leq \frac{b}{M_0}$ will be used in Step 1 below.

We now divide the proof into nine steps.

Step 1. Claim that $(t, \phi_i) \in R$ is true for all i and for all t with $|t - \tau| \leq \alpha$, so that, as $|g(t, \phi_i)| \leq M_0$ is true for all t with $|t - \tau| \leq \alpha$, the Lipschitz condition for $g(t, \phi_i)$ can be used.

The claim will be proved by mathematical induction on i. It is clearly true for $i = 0$:

$$|u_0 - \xi| = |\xi - \xi| = 0 \leq b,$$

and so assume it is true for $i = m \in \{0\} \cup \mathbb{N}$, that is, assume

$$|\phi_m - \xi| \leq b.$$

(for which $|g(t, \phi_m)| \leq M_0$) It will also be true for $i = m + 1$ by the calcalations:

$$|\phi_{m+1} - \xi| = \Big| \int_\tau^t g(s, \phi_m(s))\, ds \Big|$$

$$= \sum_{j=1}^n \Big| \int_\tau^t (g(s, \phi_m(s)))_j\, ds \Big|$$

$$\leq \begin{cases} \sum_{j=1}^n \int_\tau^t |(g(s, \phi_m(s)))_j|\, ds & \text{if } t - \tau \geq 0; \\ \sum_{j=1}^n \int_t^\tau |(g(s, \phi_m(s)))_j|\, ds & \text{if } t - \tau \leq 0: \end{cases}$$

$$= \begin{cases} \int_\tau^t |g(s, \phi_m(s))|\, ds, & \text{if } t - \tau \geq 0: \\ \int_t^\tau |g(s, \phi_m(s))|\, ds, & \text{if } t - \tau \leq 0; \end{cases}$$

$$\leq \begin{cases} \int_\tau^t M_0\, ds, & \text{if } t - \tau \geq 0; \\ \int_t^\tau M_0\, ds, & \text{if } t - \tau \leq 0; \end{cases}$$

$$= M_0|t - \tau| \leq M_0\alpha \leq M_0\frac{b}{M_0} = b.$$

Step 2. For $t - \tau \geq 0$, claim that

$$|\phi_i - \phi_{i-1}| \leq M_0\frac{L_0^i}{L_0}\frac{(t - \tau)^i}{i!}$$

is true, where $M_0 = \max_{(t,x)\in R}|g(t, x)|$ and L_0 is the Lipschitz constant for $g(t, x)$ on the region R. We are lead to this claim by the calculations:

$$|\phi_1 - \phi_0| = |\int_\tau^t g(s, \phi_0(s))\, ds| = \sum_{j=1}^n |\int_\tau^t (g(s, \phi_0(s)))_j\, ds|$$

$$\leq \sum_{j=1}^n \int_\tau^t |(g(s, \phi_0(s)))_j|\, ds = \int_\tau^t |g(s, \phi_0(s))|\, ds$$

$$\leq \int_\tau^t M_0\, ds = M_0(t - \tau),$$

$$|\phi_2 - \phi_1| = |\int_\tau^t [g(s, \phi_1(s)) - g(s, \phi_0(s))]\, ds|$$

$$= \sum_{j=1}^n |\int_\tau^t (g(s, \phi_1(s)) - g(s, \phi_0(s)))_j\, ds|$$

$$\leq \int_\tau^t |g(s, \phi_1(s)) - g(s, \phi_0(s))|\, ds$$

$$\leq \int_\tau^t L_0|\phi_1 - \phi_0|\, ds = \int_\tau^t L_0 M_0(s - \tau)\, ds$$

$$= M_0 L_0\frac{(t - \tau)^2}{2!}.$$

The claim will be proved by mathematical induction on i. It is clearly true for $i = 1$ by the above calculations, and so assume it is true for $i = m \in \mathbb{N}$, that is, assume

$$|\phi_m - \phi_{m-1}| \leq M_0\frac{L_0^m}{L_0}\frac{(t - \tau)^m}{m!}.$$

It will also be true for $i = m + 1$ by the calculations:

$$|\phi_{m+1} - \phi_m| \leq \int_\tau^t L_0|\phi_m - \phi_{m-1}|\, ds$$

$$\leq M_0\frac{L_0^{m+1}}{L_0}\int_\tau^t \frac{(s - \tau)^m}{m!}\, ds$$

$$= M_0\frac{L_0^{m+1}}{L_0}\frac{(t - \tau)^{m+1}}{(m + 1)!}.$$

Step 3. For $t - \tau \leq 0$, we have

$$|\phi_i - \phi_{i-1}| \leq M_0\frac{L_0^i}{L_0}\frac{(\tau - t)^i}{i!}.$$

This is proved similarly as in Step 2, where note:

$$|\phi_1 - \phi_0| = |\int_\tau^t g(s, \phi_0(s))\, ds|$$

$$\leq \int_t^\tau |g(s, \phi_0(s))|\, ds,$$

$$|\phi_i - \phi_{i-1}| \leq \int_t^\tau |g(s, \phi_{i-1}) - g(s, \phi_{i-2})|\, ds.$$

Step 4. It follows from Steps 2 and 3 that

$$|(\phi_i - \phi_{i-1})_j| \leq \sum_{j=1}^n |(\phi_i - \phi_{i-1})_j| = |\phi_i - \phi_{i-1}|$$

$$\leq M_0 \frac{L_0^i}{L_0} \frac{|t - \tau|^i}{i!} \leq M_0 \frac{L_0^i}{L_0} \frac{a^i}{i!}$$

for each $j = 1, 2, \ldots, n$, and for all $|t - \tau| \leq \alpha \leq a$, where note that $(\phi_i - \phi_{i-1})_j$ is the jth component of the vector $\phi_i - \phi_{i-1}$.

Step 5. Since

$$\sum_{i=1}^\infty M_0 \frac{L_0^i}{L_0} \frac{a^i}{i!} = \frac{M_0}{L_0} e^{L_0 a}$$

converges, it follows from the Weirstrass M-test (cited before) that each component

$$(\phi_m - \phi_0)_j = \sum_{i=1}^m (\phi_i - \phi_{i-1})_j, j = 1, 2, \ldots, n$$

of the vector function $(\phi_m - \phi_0)$ converges uniformly for all $t \in [a, b]$ as $m \longrightarrow \infty$, and in that case, we say that the vector function

$$\phi_m(t) - \phi_0 = \sum_{i=1}^m (\phi_i - \phi_{i-1}),$$

converges uniformly for all $t \in [a, b]$. This implies that

$$\phi_m = ((\phi_m - \phi_0) + \phi_0$$

converges uniformly to some vector function $\phi(t)$.

Step 6. As the Step 7 in the proof of Theorem 2.1, each component $(\phi)_i$ of ϕ is a continuous function on $[a, b]$, and then so is ϕ.

Step 7. Since ϕ_{i-1} converges uniformly to ϕ by Step 5 as $i \to \infty$, and since

$$|(g(t, \phi_{i-1}(t)) - g(t, \phi(t)))_j| \leq \sum_{j=1}^n |(g(t, \phi_{i-1}(t)) - g(t, \phi(t)))_j|$$

$$= |g(t, \phi_{i-1}(t)) - g(t, \phi(t))|$$

$$\leq L_0 |\phi_{i-1}(t) - \phi(t)|$$

for $j = 1, 2, \ldots, n$, where L_0 is the Lipschitz constant for g, we have that as $i \to \infty$, each component $(g(t, \phi_{i-1}))_j, j = 1, 2, \ldots, n$, of the vector function $g(t, \phi_{i-1}(t))$ converges uniformly to the corresponding component function $(g(t, \phi(t)))_j, j = 1, 2, \ldots, n$, of the vector function $g(t, \phi(t))$. It follows from the Uniform Convergence Theorem (cited before) that

$$(\phi_i)_j = (\phi_0)_j + \int_\tau^t (g(s, \phi_{i-1}(s)))_j\, ds$$

converges uniformly to

$$(\phi)_j = (\phi_0)_j + \int_\tau^t (g(s, \phi(s)))_j \, ds,$$

where $(\phi)_j$ and $(g(s, \phi(s)))_j$ are the jth component of the vectors ϕ and $g(s, \phi)$, respectively.

Step 8. Since ϕ is continuous on $\{t : |t - \tau| \le a\}$ and $g(t, x)$ continuous on R, we have that $g(t, \phi(t))$ is continuous on R. It follows from the Fundamental Theorem of calculus (cited before) that $(\phi)_j$ is differentiable, and

$$(\phi)_j(\tau) = (\phi_0)_j(\tau) = (u_0)_j,$$

$$\frac{d}{dt}(\phi)_j(t) = (g(t, \phi(t)))_j,$$

where $(g(t, \phi(t)))_j, j = 1, 2, \ldots, n$, is the jth component of the vector $g(t, \phi(t))$. Equivalently, ϕ satisfies

$$\phi(\tau) = u_0,$$

$$\frac{d}{dt}\phi = \begin{pmatrix} \frac{d}{dt}(\phi)_1 \\ \frac{d}{dt}(\phi)_2 \\ \vdots \\ \frac{d}{dt}(\phi)_n \end{pmatrix} = g(t, \phi(t)).$$

Thus ϕ is a solution to (2.5).

Step 9. A solution to (2.5) is unique. This will be proved by using Gronwall Inequality (cited before). Suppose that φ and ψ are two solutions to the equation (2.5). Then integrations give

$$\varphi(t) = u_0 + \int_\tau^t g(s, \varphi(s)) \, ds$$

$$\psi(t) = u_0 + \int_\tau^t g(s, \psi(s)) \, ds,$$

and substraction gives

$$|\varphi(t) - \psi(t)| = \left| \int_\tau^t (g(s, \varphi(s)) - g(s, \psi(s))) \, ds \right|,$$

$$\sum_{j=1}^n \left| \int_\tau^t (g(s, \varphi(s)) - g(s, \psi(s)))_j \, ds \right|$$

$$\le \sum_{j=1}^n \int_a^t |(g(s, \varphi(s)) - g(s, \psi(s)))_j| \, ds$$

$$= \int_a^t |g(s, \varphi(s)) - g(s, \psi(s))| \, ds \le \int_a^t L_0 |\varphi(s) - \psi(s)| \, ds.$$

Thus, the Gronwall Inequality gives

$$0 \le |(\varphi(t) - \psi(t))_j| \le |\varphi(t) - \psi(t)| \le 0, \quad j = 1, 2, \ldots, n,$$

and so $\varphi \equiv \psi$. $\qquad \qquad \qquad \qquad \qquad \qquad \qquad \qquad \qquad \square$

3.3. The Nonlinear Case with Continuous Coefficients but without Uniqueness Result.

Proof of Theorem 2.6.

PROOF. We divide the proof into ten steps.

Step 1. Let $m \in \mathbb{N}$, and divide the integrval $[\tau - \alpha, \tau + \alpha]$ into $2m$ equal sub-intervals, each with the length $\frac{\alpha}{m}$. This is done by the partition

$$P : \tau - \alpha = t_{-m} < \cdots < t_{-1} < t_0 \equiv \tau < t_1 < \cdots < t_m \equiv \tau + \alpha,$$

where

$$t_i - t_{(i-1)} = \frac{\alpha}{m}, \quad i = -(m-1), \ldots, -1, 0, 1, \ldots, m,$$

$$\alpha \equiv \min\{a, \frac{b}{M_0}\}.$$

Step 2. For each $m \in \mathbb{N}$, define the polygonal line vector-valued functions $\varphi_m(t)$ by the following Cases 1 to 4:

Case 1: $t \in [t_{-1}, t_1]$. Let

$$\varphi_m(t_0) = \varphi_m(\tau) \equiv \xi.$$

Since $|\varphi_m(t_0) - \xi| = |\xi - \xi| = 0 \leq b$, we have $g(t_0, \varphi_m(t_0))$ is well-defined, and then

$$|g(t_0, \varphi_m(t_0))| \leq M_0$$

by the definition of M_0. Thus we can define

$$\varphi_m(t) \equiv \varphi_m(t_0) + g(t_0, \varphi_m(t_0))(t - t_0) \quad \text{for} \quad t \in [t_{-1}, t_1]. \tag{3.2}$$

From this, we have $g(t, \varphi_m(t))$ is well-defined for $t \in [t_{-1}, t_1]$, and then

$$|g(t, \varphi_m(t))| \leq M_0 \quad \text{for } t \in [t_{-1}, t_1]$$

by the definition of M_0. This is because

$$|\varphi_m(t) - \xi| = |g(t_0, \varphi_m(t_0))(t - t_0)| \leq M_0 |t - t_0|$$

$$\leq M_0 \frac{\alpha}{m} \leq M_0 \alpha \leq b \quad \text{for} \quad t \in [t_{-1}, t_1], \tag{3.3}$$

where note $\frac{1}{m} \leq 1$ by $1 \leq m$.

Case 2: $t \in [t_1, t_2]$. Since $|\varphi_m(t_1) - \xi| \leq b$ by Case 1, we have $g(t_1, \varphi_m(t_1))$ is well-defined, and then

$$|g(t_1, \varphi_m(t_1))| \leq M_0$$

by the definition of M_0. Thus we can define

$$\varphi_m(t) \equiv \varphi_m(t_1) + g(t_1, \varphi_m(t_1))(t - t_1) \quad \text{for} \quad t \in [t_1, t_2]. \tag{3.4}$$

From this, we have $g(t, \varphi_m(t))$ is well-defined for $t \in [t_1, t_2]$, and then

$$|g(t, \varphi_m(t))| \leq M_0 \quad \text{for } t \in [t_1, t_2]$$

by the definition of M_0. This is because

$$|\varphi_m(t) - \xi| = |(\varphi_m(t_1) - \xi) + g(t_1, \varphi_m(t_1))(t - t_1)|$$

$$\leq |\varphi_m(t_1) - \xi| + |g(t_1, \varphi_m(t_1))||t - t_1|$$

$$\leq M_0 |t_1 - t_0| + M_0 |t_2 - t_1| \quad \text{by (3.2)} \tag{3.5}$$

$$\leq M_0 (2\frac{\alpha}{m}) \leq M_0 \alpha \leq M_0 \frac{b}{M_0} = b \quad \text{for } t \in [t_1, t_2],$$

where note $\frac{2}{m} \leq 1$ by $2 \leq m$.

Case 3: $t \in [t_{-2}, t_{-1}]$. Since $|\varphi_m(t_{-1}) - \xi| \leq b$ by Case 1, we have $g(t_{-1}, \varphi_m(t_{-1}))$ is well-defined, and then

$$|g(t_{-1}, \varphi_m(t_{-1}))| \leq M_0$$

by the definition of M_0. Thus we can define

$$\varphi_m(t) = \varphi_m(t_{-1}) + g(t_{-1}, \varphi_m(t_{-1}))(t - t_{-1}) \quad \text{for} \quad t \in [t_{-2}, t_{-1}]. \tag{3.6}$$

From this, we have $g(t, \varphi_m(t))$ is well-defined for $t \in [t_{-2}, t_{-1}]$, and then

$$|g(t, \varphi_m(t))| \leq M_0 \quad \text{for } t \in [t_{-2}, t_{-1}]$$

by the definition of M_0. This is because

$$
\begin{aligned}
|\varphi_m(t) - \xi| &= |(\varphi_m(t_{-1}) - \xi) + g(t_{-1}, \varphi_m(t_{-1}))(t - t_{-1})| \\
&\leq |\varphi_m(t_{-1}) - \xi| + |g(t_{-1}, \varphi_m(t_{-1}))||t - t_{-1}| \\
&\leq M_0|t_{-1} - t_0| + M_0|t_{-2} - t_{-1}| \quad \text{by (3.4)} \\
&\leq M_0(2\frac{\alpha}{m}) \leq M_0\alpha \leq M_0\frac{b}{M_0} = b \quad \text{for } t \in [t_{-2}, t_{-1}],
\end{aligned}
\tag{3.7}
$$

where note $\frac{2}{m} \leq 1$ by $2 \leq m$.

Case 4. Continuing in the above way, we are led to the result: for each $i = 1, 2, \cdots, m$, we can define

$$
\begin{aligned}
\varphi_m(t) &\equiv \varphi_m(t_{i-1}) + g(t_{i-1}, \varphi_m(t_{i-1}))(t - t_{i-1}) \\
&\text{for } t \in [t_{i-1}, t_i] \text{ and } t \in [t_{-i}, t_{-(i-1)}],
\end{aligned}
\tag{3.8}
$$

and then

$$|\varphi_m(t) - \xi| \leq M_0 \sum_{k=1}^{i} |t_k - t_{k-1}| \quad \text{or } M_0 \sum_{k=1}^{i} |t_{-(i-1)} - t_{-i}|$$

$$\leq M_0(i\frac{\alpha}{m}) \leq M_0\alpha \tag{3.9}$$

$$\leq M_0\frac{b}{M_0} = b,$$

where note $\frac{i}{m} \leq 1$ by $i \leq m$.

Here note that $\alpha > 0$ is restricted by $\alpha < a$, in order that the set $\{t : |t - \tau| \leq \alpha\}$ stays within the set $\{t : |t - \tau| \leq a\}$. The other restriction that $\alpha \leq \frac{b}{M_0}$ is used, in order that the set $\{\varphi_m(t) \in \mathbb{R}^n\}$ stays within the set $\{x \in \mathbb{R}^n : |x - \xi| \leq b\}$, that is,

$$|\varphi_m(t) - \xi| \leq b,$$

as Step 2 derived.

Step 3. It follows from Case 4 of Step 2 that $(t, \varphi_m(t)) \in R$ is true for each m and for all t with $|t - \tau| \leq \alpha$, and

$$|g(t, \varphi_m(t))| \leq M_0$$

is true for all $m \in \mathbb{N}$.

Step 4. It is a fact (Ascoli-Arzela Theorem) from Analysis, Adams [**1**], that for a sequence of real-valued (or complex-valued) continuous functions $f_k(t)$ on $[c, d], c < d$, it is true that $f_k(t)$ has a uniformly convergent subsequence with the limit being continuous also on $[c, d]$, provided that f_k satisfies the conditions of Uniform boundedness and Equi-continuity:

$$\|f_k\| \equiv \max_{t \in [c,d]} |f_k(t)| \leq K_0 \quad \text{(Uniform boundedness)}$$

for some $K_0 > 0$ and for all $k \in \mathbb{N}$.

$$|f_k(t) - f_k(s)| \le K_1|t - s| \quad \text{(Equi-continuity)}$$

for some constant $K_1 > 0$ and for all $k \in \mathbb{N}$ and all $t, s \in [c, d]$.

Step 5. Claim that $(\varphi_m)_j(t)$, the j-th component of the vector function $\varphi_m(t)$, $j = 1, 2, \cdots, n$, satisfies the conditions of Uniform boundedness and Equi-continuity, so that the Ascoli-Arzela Theorem in Step 4 applies. The claim is proved by the following Cases 1 to 3:

Cases 1. If $t, s \in [a, b]$ lies in the same sub-interval $[t_{i-1}, t_i]$ for some $i \in [-m, m] \cap \mathbb{Z}$, then

$$\varphi_m(t) = \varphi_m(t_{i-1}) + g(t_{i-1}, \varphi_m(t_{i-1}))(t - t_{i-1}),$$
$$\varphi_m(s) = \varphi_m(t_{i-1}) + g(t_{i-1}, \varphi_m(t_{i-1}))(s - t_{i-1}).$$

Thus substraction gives

$$|(\varphi_m)_j(t) - (\varphi_m)_j(s)| \le \sum_{j=1}^{n} |(\varphi_m)_j(t) - (\varphi_m)_j(s)| = |\varphi_m(t) - \varphi_m(s)|$$
$$= |\varphi_m(t_{i-1}, \varphi_m(t_{i-1}))(t - s)| \le M_0|t - s|,$$

the equi-continuity.

Case 2. Suppose that $t, s \in [a, b]$ lies in different sub-intervals, say, $s \in [t_{i-1}, t_i]$ and $t \in [t_{j-1}, t_j]$ for some $j > i$. Here $i, j \in [-m, m] \cap \mathbb{Z}$. Then

$$|t - s| = |t - t_{j-1}| + |t_{j-1} - t_{j-2}| + \cdots + |t_i - s|$$

and

$$|(\varphi_m)_k(t) - (\varphi_m)_k(s)| \le \sum_{k=1}^{n} |(\varphi_m)_k(t) - (\varphi_m)_k(s)| = |\varphi_m(t) - \varphi_m(s)|$$
$$\le |\varphi_m(t) - \varphi_m(t_{j-1})| + |\varphi_m(t_{j-1}) - \varphi_m(t_{j-2})| + \cdots + |\varphi_m(t_i) - \varphi_m(s)|$$
$$\le M_0|t - t_{j-1}| + M_0|t_{j-1} - t_{j-2}| + \cdots + M_0|t_i - s|$$
$$= M_0|t - s|,$$

the equi-continuously, where Case 1 was used.

Case 3. By letting $s = \tau$ in Case 2, for which $\varphi_m(s) = \varphi_m(\tau) = \xi$, it follows that

$$|(\varphi_m)_k(t)| \le \sum_{k=1}^{n} |(\varphi_m)_k(t)| = |\varphi_m(t)| \le |\xi| + |(\varphi_m)_k(t) - \xi|$$
$$\le |\xi| + M_0|t - \tau| \le |\xi| + M_0\alpha$$
$$\le |\xi| + M_0\frac{b}{M_0},$$

uniformly bounded for all $m \in \mathbb{N}$ and all $|t - \tau| \le \alpha$.

Step 6. Thus the Ascoli-Arzela Theorem above implies that $(\varphi_m)_1$, the first component of φ_m, has a convergent subsequence, say $(\varphi_{m_1}), m_1 \in \mathbb{N}$, with the uniform limit, say, $(\varphi)_1$, a continuous function also. The Ascoli-Arzela Theorem applied to $(\varphi_{m_1})_2$, the second component of φ_m, gives that $(\varphi_{m_1})_2$ has a convergent subsequence, say, $(\varphi_{m_2})_2, m_2 \in \mathbb{N}$, with the limit, say, $(\varphi)_2$, a continuous function also. Here note that $(\varphi_{m_2})_1$, a subsequence of $(\varphi_m)_1$ also, converges uniformly to $(\varphi)_1$ also.

Continuing in this way, in a finite process of n times, we have that $(\varphi_{m_n})_n, m_n \in \mathbb{N}$, a subsequence of $(\varphi_m)_n$, converges uniformly to, say, φ_n, a continuous functnion. Here note that n is fixed and is the dimension of the real vectore space $\mathbb{R}^n$. It follows that

$$\varphi_{m_n} = \begin{pmatrix} (\varphi_{m_n})_1 \\ (\varphi_{m_n})_2 \\ \vdots \\ (\varphi_{m_n})_n \end{pmatrix} \longrightarrow \varphi \equiv \begin{pmatrix} (\varphi)_1 \\ (\varphi)_2 \\ \vdots \\ (\varphi)_n \end{pmatrix}$$

uniformly as $m_n \longrightarrow \infty$.

Step 7. From the construction of the polygonal straight line functions $\varphi_m(t)$, we have

$$\varphi_{m_n}(t) = \varphi_{m_n}(t_{i-1}) + g(t_{i-1}, \varphi_{m_n}(t_{i-1}))(t - t_{i-1})$$

for $t \in [t_{i-1}, t_i]$ or $t \in [t_{-i}, t_{-(i-1)}]$, where $i = 1, 2, \ldots, m_n$, and m_n is a subsequence of $m \in \mathbb{N}$. Note that m_n could be, say, $(2m+1)^n$ or $(m^2+6)n$ or $\ldots$, etc.

It follows by differentiation that

$$\frac{d}{dt}\varphi_{m_n}(t) = g(t_{i-1}, \varphi_{m_n}(t_{i-1}))$$

for $t \in (t_{i-1}, t_i)$, where $\frac{d}{dt}\varphi_{m_n}(t)$ does not exist and has a jump at

$$t \in \{t_{-(m_n-1)}, \ldots, t_{m_n}\}.$$

This equation can be written as the convenient form

$$\frac{d}{dt}\varphi_{m_n}(t) = g(\mu_{m_n}(t), \chi_{m_n}(t))$$

for $t \in [\tau - \alpha, \tau + \alpha] \setminus \{t_{-m_n}, \ldots, t_{m_n}\}$ (the node points being taken out), where the step functions $\mu_{m_n}(t)$ and $\chi_{m_n}(t)$ are defined by:

$$\mu_{m_n}(\tau) = t_0 = \tau,$$
$$\mu_{m_n}(t) = t_i \quad \text{for} \quad t \in (t_{i-1}, t_i]$$

and

$$\chi_{m_n}(\tau) = \xi,$$
$$\chi_{m_n}(t) = \varphi_{m_n}(t_i) \quad \text{for} \quad t \in (t_{i-1}, t_i].$$

Integration gives that

$$\varphi_{m_n}(t) = \xi + \int_\tau^t g(\mu_{m_n}(t), \chi_{m_n}(s))\, ds \quad \text{for} \quad t \in [\tau - \alpha, \tau + \alpha]. \tag{3.10}$$

Step 8. The steps functions $\mu_{m_n}(t)$ and $\chi_{m_n}(t)$ satisfy:

$$|\mu_{m_n}(t) - t| \leq \max_{i \in [-m_n, m_n] \cap \mathbb{Z}} |t_i - t_{i-1}|$$
$$= \frac{\alpha}{m_n} \longrightarrow 0$$

as m_n (or m) $\longrightarrow \infty$;

$$|\chi_{m_n}(t) - \varphi_{m_n}(t)| \leq \max_{i \in [-m_n, m_n] \cap \mathbb{Z}} |g(t_{i-1}, \varphi_{m_n}(t_{i-1})||t_i - t_{i-1}|$$
$$\leq M_0 \frac{\alpha}{m_n} \longrightarrow 0$$

as m_n (or m) $\longrightarrow \infty$.

Since $\varphi_{m_n}(t) \longrightarrow \varphi(t)$ uniformly, we have $\chi_{m_n}(t) \longrightarrow \varphi(t)$ uniformly also.

Step 9. Since, by Step 8, φ_{m_n} and $\chi_{m_n}(t)$ both converge uniformly to $\varphi(t)$, and $\mu_{m_n}(t)$ converges uniformly to t as m_n (or m) $\longrightarrow \infty$, and since $g(t,x)$ is uniformly continuous on R (because $g(t,x)$ is continuous on R and R is bounded and closed in $\mathbb{R}^2$, *for which we say that R is compact in $\mathbb{R}^2$*), we have, Apostal [2], that each component $(g(\mu_{m_n}(t), \varphi_{m_n})_j, j = 1, 2, \cdots, n$, of the vector function $g(\mu_{m_n}(t), \varphi_{m_n}(t))$ converges uniformly to the corresponding component $(g(t, \varphi(t)))_j, j = 1, 2, \cdots, n$, of the vector function $g(t, \varphi(t))$. It follows from the Uniform Convergence Theorem (cited before) that for $j = 1, 2, \cdots, n$, the jth component

$$(\varphi_{m_n})_j = (\xi)_j + \int_\tau^t (g(\mu_{m_n}(s), \varphi_{m_n}(s)))_j \, ds$$

of the equation (3.10), converges uniformly to the equation

$$(\varphi)_j = (\xi)_j + \int_\tau^t (g(s, \varphi(s)))_j \, ds,$$

where $(\xi)_j, (\varphi)_j$, and $(g(s, \varphi(s)))_j$ are the jth component of the vectors ξ, φ, and $g(s, \varphi)$, respectively.

Step 10. Since φ is continuous on $\{t : |t - \tau| \leq \alpha\}$ and $g(t, x)$ continuous on R, we have that $g(t, \varphi(t))$ is continuous on R. It follows from the Fundamental Theorem of calculus (cited before) that for $j = 1, 2, \cdots, n$, $(\varphi)_j$ is differentiable, $(\varphi)_j(\tau) = (\xi)_j$, and

$$\frac{d}{dt}(\varphi)_j(t) = (g(t, \varphi(t)))_j,$$

where $(g(t, \varphi(t)))_j, j = 1, 2, \cdots, n$, is the jth component of the vector $g(t, \varphi(t))$. Equivalently, ϕ satisfies

$$\phi(\tau) = \xi,$$

$$\frac{d}{dt}\varphi = \begin{pmatrix} \frac{d}{dt}(\varphi)_1 \\ \frac{d}{dt}(\varphi)_2 \\ \vdots \\ \frac{d}{dt}(\varphi)_n \end{pmatrix} = g(t, \varphi(t).$$

Thus φ is a solution to (2.5). □

4. The Results with Riemann or Lebesgue Integrable Coefficients

Although Theorems 2.1 and 2.3 are only stated for continuous $a_{ij}(t)$ and $f_i(t)$ on $[a, b]$, they are also true for Riemann integrable (Apostal [2, Pages 174, 141, 152–153]) or Lebesgue integrable (Royden [20, Page 75]) $a_{ij}(t)$ and $f_i(t)$.

These two kinds of integrals are reviewed in Chapter 5 on Green functions. Roughly speaking, the definition of Lebesgue integrals, compared to that of Riemann integrals, allow for more functions (including continuous, piece-wise continuous functions) and more sets (including inteverals) that can be used. Thus a Riemann integrable function on $[a, b]$ is also Lebesgue integrable on (a, b), but not vice versa, and the class of Lebesgue integrable functions is larger than that of Riemann integrable functions. One concept, much used here, is that of *almost everywhere*: We say that a property, depending on $x \in [a, b]$, is true *almost everywhere* in x, if the set of all $x \in [a, b]$, for which that property is false, has the measure equal to zero. Here the measure of a set is sort of the length of that set; for example, the measure of the set $[a, b]$ or the set $(a, b]$ or the set $[a, b)$ or the set (a, b), equals $(b - a)$, and the measure of the point set $\{a\}$, $\{b\}$ or the points set $\{a, b\}$, all equals zero.

On the other hand, if, in Theorem 2.1 (or in Theorem 2.3), we replace $f(t)$ by 0, and $A(t)$ by $B(t) + \mu C(t)$, where μ is a complex parameter, and $B(t) = (b_{ij}(t)$ and $C(t) = (c_{ij}(t)$ are two matrices with their elements $a_{ij}(t)$ and $c_{ij}(t)$, continuous complex-valued functions on $[a, b]$, then similar results follow, in which the unique solution $\varphi = \varphi(t, \xi, \mu)$ will depend on the initial data $\varphi(\tau) = u_0 = \xi$ and the complex parameter μ.

Therefore, in this section we shall prove Theorems 2.1 and 2.3 for Riemann or Lebesgue integrable $a_{ij}(t)$ and $f_i(t)$, and further prove the result and its corollary when the $f(t)$ and $A(t)$ in Theorem 2.1 are replaced by 0 and $B(t) + \mu C(t)$, respectively, in which the dependence of the solution on the initial data and the complex-valued parameter will be studied. We include this section because of its use in other chapters.

4.1. Existence and Uniqueness Theorems. Theorems 2.1 and 2.3 assumed that $a_{ij}(t)$ and $f_i(t)$ are continuous on $[a, b]$, so that

$$\max\{|A(t)|, |f(t)|\} \leq k(t), \quad \text{where}$$

$$\int_a^b k(t)\, dt < \infty \tag{*}$$

for some positive, continuous function $k(t)$, and it is the condition (*) that is eventually needed for their proofs. Therefore, if we assume that $a_{ij}(t)$ and $f_i(t)$ are Riemannn integrable or Lebesgue integrable, and that they are bounded by some Riemann integrable or Lebesgue integrable function $k(t)$, then Theorems 2.1 and 2.3 might still be true. In fact, this is the case as the following results show.

Note that, although these results below are stated for real-valued functions, they are also valid for complex-valued functions. This can be seen from their proofs.

THEOREM 4.1. *(Coddington and Levinson* [**5**, *Pages 97-98]) There is a unique solution* $u = \begin{pmatrix} u_1 \\ u_2 \end{pmatrix} \in \mathbb{R}^2$, *to the first order linear system* (2.1) *of 2 differential equations with the initial conditions:*

$$\frac{d}{dt} u = A(t)u + f(t), \quad a < t < b,$$

$$u(\tau) = u_0 \in \mathbb{R}^2, \tau \in [a, b], \tag{4.1}$$

in the sense that $u_i, i = 1, 2$, *is absolutely continuous (so that* u_1 *and* u_2 *are differentiable almost everywhere),* $(A(t)u(t))_i, i = 1, 2$, *is Lebesgue integrable (see Chapter 5 for the concept of Absolute Continuity, Almost Everywhere, and Lebesgue Integrable), and*

$$u_i(t) = (u_0)_i + \int_\tau^t (A(s)u(s))_i\, ds + \int_\tau^t f_i(s)\, ds,$$

for which we write

$$u(t) = u_0 + \int_\tau^t A(s)u(s)\, ds + \int_\tau^t f(s)\, ds. \tag{3.1}$$

Or equivalently, $u(t)$ satisfies (2.1) for almost every t. Here $(u_0)_i, (A(t)u(t))_i$, and $f_i(s)$ are the ith component of $u_0, A(s)u(s)$, and $f(t)$, respectively, and

$$\int_\tau^t f(s)\, ds = \begin{pmatrix} \int_\tau^t f_1(s)\, ds \\ \int_\tau^t f_2(s)\, ds \end{pmatrix},$$

and then $\int_\tau^t A(s)u(s)\, ds$ is similarly defined. Here $0 < a < b$ are given two real numbers, and u_0 is a given constant real vector in $\mathbb{R}^2$; $A(t) = (a_{ij}(t))$ is a given matrix of 2×2, and $a_{ij}(t), i, j = 1, 2$, are Riemann integrable or Lebesgue integrable, real-valued functions on $[a, b]$; finally, $f(t) = \begin{pmatrix} f_1(t) \\ f_2(t) \end{pmatrix}$ is a given vector function with each component, a Riemann integrable or Lebesgue integrable, real-valued function on $[a, b]$.

Note that under the assumptions on $a_{jk}(t)$ and f_j, we have $A(t)$ and $f(t)$ satisfy

$$\max\{|A(t)|, |f(t)|\} \le k(t),$$

$$\int_a^b k(t)\, dt < \infty$$

for some $k(t) > 0$, Riemann integrable or Lebesgue integrable on $[a, b]$. For example, we can take

$$k(t) = \max\{|A(t)|, |f(t)|\} \equiv \max\{\sum_{j,k=1}^2 |a_{jk}(t)|, \sum_{j=1}^2 |f_j(t)|\}.$$

COROLLARY 4.2. *The linear second order ordinary differential equation with the initial conditions:*

$$\begin{aligned} y''(t) + p(t)y'(t) + q(t)y(t) &= r(t), \quad a < t < b, \\ y(\tau) = y_0, \quad y'(\tau) &= y_1, \quad \tau \in [a, b] \end{aligned} \tag{4.2}$$

has a unique solution $y(t)$, in the sense that $y(t)$ is continuously differentiable with y' absolutely continuous (so that y'' is differentiable almost everywhere), and $y(t)$ satisfies (1.1) for almost every t. Here y_0, y_1, and $0 < a < b$ are given four real numbers, and $p(t), q(t)$, and $r(t)$ are given Riemann integrable or Lebesgue integrable, real-valued functions on $[a, b]$.

Note that $p(t), q(t)$, and $r(t)$ satisfy

$$\max\{|p(t)|, |q(t)|, |r(t)|\} \le k(t),$$

$$\int_a^b k(t)\, dt < \infty$$

for some $k(t) > 0$, Riemann integrable or Lebesgue integrable on $[a, b]$.

THEOREM 4.3. *(Coddington and Levinson [**5**, Pages 97-98]) There is a unique solution $u \in \mathbb{R}^n$, to the first order system (2.3) of $n \in \mathbb{N}$ differential equations with the initial conditions:*

$$\begin{aligned} \frac{d}{dt}u &= A(t)u + f(t), \quad a < t < b, \\ u(\tau) &= u_0 \in \mathbb{R}^n, \quad \tau \in [a, b], \end{aligned} \tag{4.3}$$

in the sense that each component $(u)_i, i = 1, 2, \ldots, n,$ *of u is absolutely continuous (so that* $(u)_i$ *is differentiable almost everywhere), each component* $(A(t)u)_i, i = 1, 2, \ldots, n,$ *of A(t)u is Lebesgue integrable, and*

$$u(t) = u_0 + \int_\tau^t A(s)u, ds + \int_\tau^t f(s)\, ds. \qquad (4.4)$$

Or equivalently, u satisfies (2.3) for almost every t. Here $n \in \mathbb{N}$, *and* $0 < a < b$ *and* τ *are given three real numbers;* u_0 *is a given real constant vector in* $\mathbb{R}^n$, *and* $A(t) = (a_{ij}(t))$ *is a given matrix of* $n \times n$, *with* $a_{ij}(t)$, *a Riemann integrable or Lebesgue integrable, real-valued function on* $[a, b]$; *finally,* $f(t) = \begin{pmatrix} f_1(t) \\ \vdots \\ f_n(t) \end{pmatrix}$ *is a given vector function, with each component, a Riemann integrable or Lebesgue integrable, real-valued function on* $[a, b]$.

Note that $A(t)$ and $f(t)$ satisfy

$$\max\{|A(t)|, |f(t)|\} \le k(t),$$

$$\int_a^b k(t)\, dt < \infty$$

for some $k(t) > 0$, Riemann integrable or Lebesgue integrable on $[a, b]$.

COROLLARY 4.4. *The nth* $(n \in \mathbb{N})$ *order linear ordinary differential equation with the initial condition:*

$$y^{(n)} + p_1(t)y^{(n-1)} + p_2(t)y^{(n-3)} + \cdots + p_n(t)y = p_0(x), \quad a < t < b,$$
$$y(\tau) = y_0, y'(\tau)y_1, \ldots, y^{(n-1)}(\tau) = y_{n-1}, \quad \tau \in [a, b] \qquad (4.5)$$

has a unique solution $y(t)$, *in the sense that* $y^{(i)}(t), i = 0, 1, 2, \ldots, (n-2),$ *is continuously differentiable with* $y^{(n-1)}$ *absolutely continuous (so that* $y^{(n-1)}$ *is differentiable almost everywhere), and* $y(t)$ *satisfies (1.3) for almost every t. Here* $0 < a < b$ *and* $y_i, i = 0, 1, 2, \ldots, (n-1)$ *are given real numbers, and* $p_i(t), i = 0, 1, \ldots, n$ *are given Riemann integrable or Lebesgue integrable, real-valued functions on* $[a, b]$.

Note that $p_i(t), i = 0, 1, \ldots, n$, satisfy

$$\max_{i \in \{0,1,\ldots,n\}} \{|p_i|(t)\} \le k(t),$$

$$\int_a^b k(t)\, dt < \infty$$

for some $k(t) > 0$, Riemann integrable or Lebesgue integrable on $[a, b]$.

Proof of Theorems 4.1 and 4.3.

Although the following proofs are given for the case with real-valued functions, they are also valid for the case with complex-valued functions. This is done by simply defining

$$|z(t)| = |v(t)| + |h(t)|$$

for a complex-valued function $z(t) = v(t) + ih(t)$ on $[c, d], c < d$, where $v(t)$ and $h(t)$ are real-valued on $[c, d]$, and i is the complex unit satisfying

$$i = \sqrt{-1}, i^2 = -1, i^3 = -i, i^4 = 1.$$

Proof of Theorem 4.1.

PROOF. The proof is much like that for Theorem 2.1.

Since $a_{jk}(t), j, k = 1, 2$ and $f_j(t), j = 1, 2$ are Riemann integrable or Lebesgue integrable, we have that $A(t)$ and $f(t)$ satisfy

$$\max\{|A(t)|, |f(t)|\} \le k(t),$$

$$\int_a^b k(t)\, dt < \infty$$

for some Riemann integrable or Lebesgue integrable function $k(t) > 0$, respectively. For example, we can take

$$k(t) = |A(t)| + |f(t)| \equiv \sum_{j,k=1}^{2} |a_{jk}(t)| + \sum_{j=1}^{2} |f_j(t)|.$$

We now divide the proof into eleven steps.

Step 1. We use the method of successive approximations, Coddington and Levinson [5, Pages 12, 98]. Integrating the equation (2.1) gives

$$u(t) = u_0 + \int_\tau^t A(s)u(s)\, ds + \int_\tau^t f(s)\, ds, \qquad (3.1)$$

the integrated form of (2.1), to which we define the successive approximations

$$\phi_i = \begin{pmatrix} (\phi_i)_1 \\ (\phi_i)_2 \end{pmatrix}, \quad i = 0, 1, 2, \cdots,$$

by

$$\phi_0 \equiv u_0,$$

$$\phi_i \equiv u_0 + \int_\tau^t A(s)\phi_{i-1}(s)\, ds + \int_\tau^t f(s)\, ds, \quad i = 1, 2, \cdots,$$

where $\int_\tau^t f(s)\, ds \equiv \begin{pmatrix} \int_\tau^t f_1(s)\, ds \\ \int_\tau^t f_2(s)\, ds \end{pmatrix}$, and $\int_\tau^t A(s)\phi_{i-1}(s)\, ds$ is similarly defined. Here $\phi_0 = u_0$ is a well-defined, constant function on $[a, b]$.

It is a fact, Apostal [2, Pages 161-162] and Royden [20, Page 105], that for a Riemann integrable or Lebesgue integrable function $h(t)$ on $[a, b]$, we have $\int_a^t h(s)\, ds$ is a continuous function on $[a, b]$. It follows that $(\phi_1)_j, j = 1, 2$, and then ϕ_1, where

$$\phi_1(t) = u_0 + \int_\tau^t A(s)u_0\, ds + \int_\tau^t f(s)\, ds,$$

is continuous on $[a, b]$, since $a_{jk}(s), (A(s)u_0)_j$ (the jth component of $A(s)u_0$), and $f_j(s), j, k = 1, 2$, are Riemann integrable or Lebesgue integrable.

Inductively, ϕ_i is well-defined and continuous on $[a, b]$.

Step 2. By defining

$$|A(t)| \equiv \sum_{j,k=1}^{2} |a_{jk}(t)|,$$

we have, for $x = \begin{pmatrix} x_1 \\ x_2 \end{pmatrix} \in \mathbb{R}^2$,

$$|A(t)x| \leq |A(t)||x|.$$

This follows from:

$$|x_j| \leq |x|, \quad j = 1, 2,$$

$$A(t)x = \begin{pmatrix} \sum_{j=1}^{2} a_{1j}(t)x_j \\ \sum_{j=1}^{2} a_{2j}(t)x_j \end{pmatrix},$$

$$|A(t)x| = \sum_{k=1}^{2} |\sum_{j=1}^{2} a_{kj}(t)x_j| \leq \sum_{k=1}^{2} \sum_{j=1}^{2} |a_{kj}(t)||x|$$

$$= |A(t)||x|.$$

Step 3. For $t \geq \tau$, claim

$$|\phi_i - \phi_{i-1}| \leq (1 + |u_0|)\frac{K(t)^i}{i!},$$

where $K(t) = \int_\tau^t k(s)\, ds$. We are led to this claim by the calculations:

$$|\phi_1 - \phi_0| = |\int_\tau^t A(s)\phi_0(s)\, ds + \int_\tau^t f(s)\, ds|$$

$$= \sum_{j=1}^{2} |\int_\tau^t (A(s)\phi_0(s))_j\, ds + \int_\tau^t f_j(s)\, ds|$$

$$\leq \sum_{j=1}^{2} \int_\tau^t |A(s)\phi_0(s))_j|\, ds + \sum_{j=1}^{2} \int_\tau^t |f_j(s)|\, ds$$

$$= \int_\tau^t |A(s)\phi_0(s)|\, ds + \int_\tau^t |f(s)|\, ds$$

$$\leq \int_\tau^t k(s)|u_0|\, ds + \int_\tau^t k(s)\, ds$$

$$= (1 + |u_0|)\int_\tau^t k(s)\, ds = (1 + |u_0|)K(t);$$

$$|\phi_2 - \phi_1| = |\int_\tau^t A(s)[\phi_1(s) - \phi_0(s)]\, ds|$$

$$\leq \int_\tau^t |A(s)[\phi_1(s) - \phi_0(s)]|\, ds$$

$$\leq \int_\tau^t k(s)[(1 + |u_0|)K(s)\, ds$$

$$= (1 + |u_0|)\int_\tau^t \frac{dK(s)}{ds}K(s)\, ds$$

$$= (1 + |u_0|)\int_{s=\tau}^t K(s)\, dK(s)$$

$$= (1 + |u_0|)\frac{K(s)^2}{2!}\Big|_{s=\tau}^t$$

$$= (1 + |u_0|)\frac{K(t)^2}{2!},$$

where $K(\tau) = 0$. Here we used the facts that, Apostal [2, Page 155] and Royden [20, Page 93],

$$\left| \int_a^t h(t)\, dt \right| \le \int_a^t |h(t)|\, dt$$

for a Riemann integrable or Lebesgue integrable function $h(t)$ on $[a, b]$, that, Royden [20, Page 107],

$$\frac{d}{dt} K(t) = \frac{d}{dt} \int_\tau^t k(s)\, ds = k(t) \quad \text{for almost every } t,$$

and that the formula of change of variable is true, Royden [20, Page 279] and Folland [8, Page 86]:

$$\int_{s=\tau}^t K(s)\frac{d}{ds}K(s)\, ds = \int_{s=\tau}^t K(s)\, dK(s).$$

Here also note, from above calculations,

$$\left| \int_\tau^t A(s)\phi_{i-1}(s)\, ds + \int_\tau^t f(s)\, ds \right| \le \int_\tau^t |A(s)\phi_{i-1}(s)|\, ds + \int_\tau^t |f(s)|\, ds.$$

The claim follows from mathematical induction on i. Clearly, it is true for $i = 1$, and so assume it is true for $i = m \in \mathbb{N}$:

$$|\phi_m - \phi_{m-1}| \le (1 + |u_0|)\frac{K(t)^m}{m!}.$$

It will also be true for $i = m + 1$ by the calculations:

$$|\phi_{m+1} - \phi_m| \le \int_\tau^t k(s)|\phi_m - \phi_{m-1}|\, ds$$

$$\le (1 + |u_0|) \int_\tau^t \frac{dK(s)}{ds}\frac{K(s)^m}{m!}\, ds$$

$$= (1 + |u_0|) \int_{s=\tau}^t \frac{K(s)^m}{m!}\, dK(s)$$

$$= (1 + |u_0|)\frac{K(s)^{m+1}}{(m+1)!}\Big|_{s=\tau}^t$$

$$= (1 + |u_0|)\frac{K(t)^{m+1}}{(m+1)!},$$

where $K(\tau) = 0$.

Step 4. For $t \le \tau$, we have

$$|\phi_i - \phi_{i-1}| \le (1 + |u_0|)\frac{G(t)^i}{i!},$$

where $K(t) = \int_t^\tau k(s)\, ds$. This is proved similarly as in Step 3, where note:

$$|\phi_1 - \phi_0| = \left| \int_\tau^t A(s)\phi_0(s)\, ds + \int_\tau^t f(s)\, ds \right|$$

$$\le \int_t^\tau |A(s)\phi_0(s)|\, ds + \int_t^\tau |f(s)|\, ds$$

$$\leq \int_t^\tau k(s)|u_0|\,ds + \int_t^\tau k(s)\,ds$$

$$= (1+|u_0|)\int_t^\tau k(s)\,ds,$$

$$|\phi_i - \phi_{i-1}| \leq (1+|u_0|)\int_t^\tau |\phi_{i-1} - \phi_{i-2}|\,ds.$$

Step 5. It follows from Steps 3 and 4 that

$$|(\phi_i)_j - (\phi_{i-1})_j| \leq \sum_{j=1}^2 |(\phi_i)_j - (\phi_i)_j| = |\phi_i - \phi_{i-1}|$$

$$\leq (1+|u_0|)\frac{K_0^i}{i!}$$

for each $j = 1, 2$ and for all $t \in [a, b]$, where $K_0 \equiv \int_a^b k(a)\,ds$, and

$$\max\{\max_{b \geq t \geq \tau} \int_\tau^t k(s)\,ds,\ \max_{a \leq t \leq \tau} \int_t^\tau k(s)\,ds\}$$

$$\leq \int_a^b k(s)\,ds = K_0 < \infty,$$

and $(\phi_i)_1$ and $(\phi_i)_2$ are the first and the second component of the vector function ϕ_i, respectively.

Step 6. Since

$$\sum_{i=1}^\infty (1+|u_0|)\frac{K_0^i}{i!} = (1+|u_0|)e^{K_0}$$

converges, it follows from the Weirstrass M-test (cited before) that each component

$$(\phi_m - \phi_0)_j = \sum_{j=1}^m (\phi_i - \phi_{i-1})_j, \quad j = 1, 2$$

of the vector function $(\phi_m - \phi_0)$ converges uniformly for all $t \in [a, b]$. In this case, we say that the vector function

$$\phi_m(t) - \phi_0 = \sum_{i=1}^m (\phi_i - \phi_{i-1}),$$

converges uniformly for all $t \in [a, b]$. This implies that

$$\phi_m = ((\phi_m - \phi_0) + \phi_0$$

converges uniformly to some vector function $\phi(t)$.

Step 7. As cited before, it is a fact from Advanced Calculus, Apostal [**2**, Page 221], that if $f_m(t)$, a sequence of continuous functions on $[a, b]$, converges uniformly for all $t \in [a, b]$, to $f_0(t)$, then $f_0(t)$ is also continuous on $[a, b]$. This fact gives that each component $(\phi)_i$ of ϕ is a continuous function on $[a, b]$, and in this case, we say that ϕ is a continuous fuction on $[a, b]$.

Step 8. In the case that $a_{jk}(t)$ and $f_j, j, k = 1, 2$, are Riemann integrable on $[a, b]$, we have from the definition of Riemann integrals, Apostal [**2**, Page 141] that $a_{jk}(t), f_j(t)$, and $k(t)$ are all bounded by some constant $N_0 > 0$.

Since ϕ_{i-1} converges uniformly to ϕ by Step 6 as $i \to \infty$, and since

$$|(A(t)\phi_{i-1}(t) - A(t)\phi(t))_j| \leq \sum_{j=1}^{2} |(A(t)\phi_{i-1}(t) - A(t)\phi(t))_j|$$
$$= |A(t)\phi_{i-1}(t) - A(t)\phi(t)| \leq k(t)|\phi_{i-1}(t) - \phi(t)|$$
$$\leq N_0|\phi_{i-1}(t) - \phi(t)|$$

for $j = 1, 2$, we have that as $i \to \infty$, each component function $(A(t)\phi_{i-1})_j$ of the vector function $A(t)\phi_{i-1}(t)$ converges uniformly to the corresponding component function $(A(t)\phi(t))_j$ of the vector function $A(t)\phi(t)$. It follows from the Uniform Convergence Theorem (cited before) that, for $j = 1, 2$,

$$(\phi_i)_j = (\phi_0)_j + \int_{\tau}^{t} (A(s)\phi_{i-1}(s))_j \, ds + \int_{\tau}^{t} f_j(s) \, ds$$

converges uniformly to the function

$$(\phi)_j = (\phi_0)_j + \int_{\tau}^{t} (A(s)\phi(s))_j \, ds + \int_{\tau}^{t} f_j(s) \, ds.$$

Or equivalently,

$$\phi = \phi_0 + \int_{\tau}^{t} A(s)\phi(s) \, ds + \int_{\tau}^{t} f(s) \, ds.$$

Thus ϕ is a solution to (3.1).

Step 8'. In the case that $a_{jk}(t)$ and $f_j(t)$ are Lebesgue integrable, we use the fact that for a Lebesgue integrable function $h(t)$ on $[a, b]$,

$$|h(t)| < \infty$$

is true for almost every $t \in [a, b]$, and then we have

$$\max\{|A(t)|, |f(t)|\} \leq k(t) < \infty$$

for almost every $t \in [a, b]$. Since $\phi_{i-1}(t) \longrightarrow \phi(t)$ uniformly as $i \longrightarrow \infty$, we have that

$$|(A(s)\phi_{i-1}(s) - A(s)\phi(s))_j| \leq |A(s)\phi_{i-1}(s) - A(s)\phi(s)|$$
$$\leq |A(s)||\phi_{i-1}(s) - \phi(s)|$$
$$\leq k(s)|\phi_{i-1}(s) - \phi(s)| \longrightarrow 0$$

for almost every $t \in [a, b]$.

Next we will use the fact (the Lebesgue Convergence Theorem, Royden [20, Page 91]) that for a sequence $h_m(t)$ of Lebesgue integrable functions on $[a, b]$ which converges to $h_0(t)$ for almost every $t \in [a, b]$, we have

$$\int_{a}^{t} h_m(s) \, ds \longrightarrow \int_{a}^{t} h_0(s) \, ds$$

if $|h_m(t)| \leq v(t)$ for some Lebesgue integrable function $v(t)$ and for all m. Since

$$|\phi_{i-1}| = |\phi_0 + \sum_{k=1}^{i} (\phi_k - \phi_{k-1})|$$
$$\leq |\phi_0| + \sum_{k=1}^{i} |\phi_k - \phi_{k-1}|$$

$$\leq |\phi_0| + (1 + |u_0|)e^{K_0} \leq N_1 < \infty$$

for some constant $N_1 > 0$ by Step 6, and since

$$|(A(s)\phi_{i-1}(s))_j| \leq |A(s)\phi_{i-1}(s)| \leq k(s)N_1,$$

where $k(s)N_1$ is Lebesgue integrable, it follows from the above Lebesgue Convergence Theorem that

$$(\phi_i)_j = (\phi_0)_j + \int_\tau^t (A(s)\phi_{i-1}(s))_j\, ds + \int_\tau^t f_j(s)\, ds$$

converges to the function

$$(\phi)_j = (\phi_0)_j + \int_\tau^t (A(s)\phi(s))_j\, ds + \int_\tau^t f(s)\, ds.$$

Or equivalently,

$$\phi = \phi_0 + \int_\tau^t A(s)\phi(s)\, ds + \int_\tau^t f(s)\, ds.$$

Thus ϕ is a solution to (3.1).

Step 9. It is a fact (the Fundamental Theorem of Calculus for Lebesgue integrable functions) from Royden [20, Page 107] that for a Lebesgue integrable function $f_0(t)$ on $[a, b]$, we have $\int_a^t f_0(s)\, ds$ is differentiable for almost every $t \in [a, b]$ and

$$\frac{d}{dt} \int_a^t f_0(s)\, ds = f_0(t)$$

for almost every $t \in [a, b]$. Since $A(t)\phi(t)$ and $f(t)$ are Lebesgue integrable on $[a, b]$, it follows from the above Fundamental Theorem of calculus that, for $j = 1, 2$, we have $(\phi)_j$ is differentiable for almost every $t \in [a, b]$, and

$$(\phi)_j(\tau) = (\phi_0)_j(\tau) = (u_0)_j,$$

$$\frac{d}{dt}(\phi)_j(t) = (A(t)\phi(t))_j + f_j(t)$$

for almost every $t \in [a, b]$. Or equivalently, ϕ satisfies

$$\phi(\tau) = u_0,$$

$$\frac{d}{dt}\phi = \begin{pmatrix} \frac{d}{dt}(\phi)_1 \\ \frac{d}{dt}(\phi)_2 \end{pmatrix}$$

$$= A(t)\phi(t) + f(t)$$

for almost every $t \in [a, b]$.

Step 10. A solution to (3.1) is unique. This will be proved by applying the fact (Gronwall Inequality), Coddington and Levinson [5, Page 37] or Hartman [15, Page 24], that, for two nonnegative, continuous or piece-wise continuous or Rieman integrable or Lebesgue integrable functions φ and ψ on $[a, b]$, and for a nonnegative constant $C \geq 0$, we have the integral inequality

$$\varphi(t) \leq C + \int_a^t \varphi(s)\psi(s)\, ds$$

implies

$$\varphi(t) \leq Ce^{\int_a^t \psi(s)\, ds}.$$

Suppose that φ and ψ are two solutions to the equation (3.1). Integrations give

$$\varphi(t) = u_0 + \int_\tau^t A(s)\varphi(s)\,ds + \int_\tau^t f(s)\,ds,$$

$$\psi(t) = u_0 + \int_\tau^t A(s)\psi(s)\,ds + \int_\tau^t f(s)\,ds,$$

and substraction gives

$$|\varphi(t) - \psi(t)| = |\int_\tau^t A(s)(\varphi(s) - \psi(s))\,ds|,$$

$$\sum_{j=1}^2 |\int_\tau^t (A(s)(\varphi(s) - \psi(s))_j\,ds| \le \sum_{j=1}^2 \int_a^t |(A(s)(\varphi(s) - \psi(s))_j|\,ds$$

$$= \int_a^t |A(s)(\varphi(s) - \psi(s))|\,ds$$

$$\le \int_a^t k(s)|\varphi(s) - \psi(s)|\,ds,$$

where Step 2 was used. It follows from the above Gronwall Inequality that, for $j = 1, 2$, we have

$$0 \le |(\varphi(t) - \psi(t))_j| \le |\varphi(t) - \psi(t)| \le 0,$$

and so $\varphi \equiv \psi$. Here note that

$$\int_a^t k(s)\,ds \le \int_a^b k(s)\,ds < \infty.$$

□

Proof of Corollary 4.2.

PROOF. As stated in Section 1, solving the equation (1.1) is the same as solving the first order system (1.2), and so Corollary 4.2 is proved by applying Theorem 4.1. Here we used the fact, Royden [**20**, Page 110], that a complex-valued function $z(t)$ on $[a, b]$ is absolutely continuous if and only if z' exists almost everywhere and

$$z(t) = \int_a^t z'(s)\,ds + z(a),$$

where the integral is taken with respect to the Lebesgue integral. □

Proof of Theorem 4.3.

PROOF. The proof is the same as that for Theorem 4.1, where note that $|A(t)|$ is defined as

$$|A(t)| \equiv \sum_{i,j=1}^n |a_{ij}(t)|,$$

and, for $x = \begin{pmatrix} x_1 \\ x_2 \\ \vdots \\ x_n \end{pmatrix} \in \mathbb{R}^n$, $|x|$ is defined as

$$|x| \equiv \sum_{i=1}^{n} |x_i|,$$

and the inequality

$$|A(t)x| \leq |A(t)||x|$$

is also true. $\square$

Proof of Corollary 4.4.

PROOF. As in Section 1, solving the equation (1.3) is the same as solving the first order system (1.4), and so Corollary 4.4 is proved by applying Theorem 4.3. Here we used the fact, Royden [20, Page 110], that a complex-valued function $z(t)$ on $[a, b]$ is absolutely continuous if and only if z' exists almost everywhere and

$$z(t) = \int_a^t z'(s)\, ds + z(a),$$

where the integral is taken with respect to the Lebesgue integral. $\square$

4.2. Dependence on Initial Conditions and Parameters. As is explained at the beginning of this Section 4, we will not only study the unique existence problem for Riemann or Lebesgue integrable (a_{ij}) and f_i, but also study the dependence of the solution φ on the initial data u_0 and the complex-valued parameter μ. Here are the results:

THEOREM 4.5. *(Coddington and Levinson [5, Page 37]) There is a unique vector-valued solution $u = \begin{pmatrix} u_1 \\ u_2 \end{pmatrix} \in \mathbb{C}^2$, with u_1 and u_2 complex-valued, analytic functions on $[a, b]$, to the first order linear system of two differential equations with the initial conditions:*

$$\frac{d}{dt} u = C(t)u + \mu D(t)u, \quad a < t < b, \tag{4.6}$$
$$u(\tau) = u_0 \in \mathbb{C}^2, \quad \tau \in [a, b],$$

where $0 < a < b$ are given two real numbers, and u_0 is a given constant complex-valued vector in $\mathbb{C}^2$; $B(t) = (a_{ij}(t))$ and $C(t) = (c_{ij}(t))$ are given two matrices of 2×2, and $a_{ij}(t)$ and $c_{ij}(t)$ are continuous complex-valued functions on $[a, b]$.

Further, $u = u(t, u_0, \mu)$ is a continuous function of (t, u_0, μ) on $[a, b] \times \mathbb{C}^2 \times \mathbb{C}$, and, for each fixed $t \in [a, b]$, we have $u(t, u_0, \mu)$ is a vector-valued analytic function of (u_0, μ) on $\mathbb{C}^2 \times \mathbb{C}$. In particular, for each fixed $(t, u_0) \in [a, b] \times \mathbb{C}^2$, we have $u(t, u_0, \mu)$ is a vector-valued entire function of $\mu \in \mathbb{C}$.

Recall the following:

DEFINITION 4.6. *Let $w = w(z)$ be a complex-valued function defined on an open subset S in $\mathbb{C}$.*

*(i) (Lange [**17**, Pages 30, 68, 129]) w is called analytic at $z_0 \in S$ if $w(z)$ is differentiable (or called holomorphic) at z_0, or equivalently, if $w(z)$ equals some convergent power series*

$$w(z) = \sum_{n=0}^{\infty} a_n (z - z_0)^n$$

for z in every neighborhood of z_0, contained in S.

*(ii) (Lang [**17**, Page 129]) The w in (i) is called an entire function of z if $S = \mathbb{C}$.*

COROLLARY 4.7. *The linear second order ordinary differential equation with the initial conditions:*

$$y''(t) + p(t)y'(t) + q(t)y(t) = \mu r(t)y(t), \quad a < t < b,$$
$$y(\tau) = y_0, y'(\tau) = y_1, \quad \tau \in [a, b] \tag{4.7}$$

has a unique complex-valued solution, where y_0, y_1, are given complex numbers, and $0 < a < b \in \mathbb{R}$ are given real numbers; $p(t), q(t), r(t)$ are given continuous complex-valued functions on $[a, b]$, and $\mu \in \mathbb{C}$ is a complex parameter.

Further, $y = y(t, y_0, y_1, \mu)$ and $y' = y'(y, y_0, y_1, \mu) = \frac{d}{dt} y(t, y_0, y_1, \mu)$ are continuous functions of (t, y_0, y_1, μ) on $[a, b] \times \mathbb{C}^2 \times \mathbb{C}$, and, for each fixed $t \in [a, b]$, we have y and y' are analytic functions of (y_0, y_1, μ) on $\mathbb{C}^2 \times \mathbb{C}$. In particular, for each fixed $(t, y_0, y_1) \in [a, b] \times \mathbb{C}^2$, we have y and y' are entire functions of $\mu \in \mathbb{C}$.

The following proofs for Theorem 4.5 and Corollary 4.7 will be less detailed, compared to those for Theorem 2.1 and Corollary 2.2.

Proof of Theorem 4.5.

PROOF. We follow the proof of Theorem 2.1, and divide the proof into six steps.

Step 1. Define the successive approximations

$$\phi_i, \quad i = 0, 1, \ldots,$$

by

$$\phi_0 = \phi_0(t, u_0, \mu) \equiv u_0,$$
$$\phi_i = \phi_i(t, u_0, \mu)$$
$$\equiv \phi_0 + \int_{\tau}^{t} (C(s)\phi_{i-1}(s, u_0, \mu) + \mu D(s)\phi_{i-1}(s, u_0, \mu)) \, ds, \quad i = 1, 2, \ldots.$$

Step 2. It is readily seen that $\phi_i(t, u_0, \mu)$ are well-defined, continuous functions of $(t, u_0, \mu) \in [a, b] \times \mathbb{C}^2 \times \mathbb{C}$, and, for each fixed $t \in [a, b]$, they are analytic functions of $(u_0, \mu) \in \mathbb{C}^2 \times \mathbb{C}$.

Step 3. The proof of Theorem 2.1 shows that, for $t \geq \tau$, we have

$$|\phi_i(t, u_0, \mu) - \phi_{i-1}(t, u_0, \mu)| \leq (1 + |u_0|) \frac{K(t, u_0, \mu)^i}{i!},$$

where

$$K(t, u_0, \mu) = \int_{\tau}^{t} k(s, u_0, \mu) \, ds$$

with $k(s, u_0, \mu)) = |C(s)| + |\mu||D(s)|$.

Step 3. The proof of Theorem 2.1 shows that, for $t \leq \tau$, we have

$$|\phi_i(t, u_0, \mu) - \phi_{i-1}(t, u_0, \mu)| \leq (1 + |u_0|) \frac{K(t, u_0, \mu)^i}{i!},$$

where

$$K(t, u_0, \mu) = \int_t^\tau k(s, u_0, \mu) \, ds$$

with $k(s, u_0, \mu) = |C(s)| + |\mu||D(s)|$.

Step 4. Steps 2 and 3 show

$$|\phi_i(t, u_0, \mu) - \phi_{i-1}(t, u_0, \mu)| \leq (1 + |u_0|) \frac{K_0(t, u_0, \mu)^i}{i!},$$

where

$$K_0(t, u_0, \mu) = \int_a^b k(s, u_0, \mu) \, ds.$$

Step 5. The proof of Theorem 2.1 shows that, for (t, u_0, μ) satisfying $t \in [a, b]$ and $|u_0| + |\mu| \leq R_0$, where $R_0 > 0$ is an arbitrarily given positive number, we have K_0 is finite,

$$\sum_{i=1}^\infty |\phi_i(t, u_0, \mu) - \phi_{i-1}(t, u_0, \mu)|$$

$$\leq \sum_{i=1}^\infty (1 + |u_0|) \frac{K_0^i}{i!} = (1 + |u_0|) e^{K_0} < \infty,$$

and

$$\phi_m(t, u_0, \mu) = \phi_0 + \sum_{i=1}^m (\phi_i(t, u_0, \mu) - \phi_{i-1}(t, u_0, \mu))$$

converges uniformly to some continuous complex-valued function $\phi(t, u_0, \mu)$ as $m \longrightarrow \infty$. This $\phi(t, u_0, \mu)$ is the unique solution to the equation (4.6). Since $R_0 > 0$ is arbitrary, $\phi(t, u_0, \mu)$ is continuous on $[a, b] \times \mathbb{C}^2 \times \mathbb{C}$. Here we used the fact that the uniform limit function of a sequence of continuous complex-valued functions is continuous also, Lang [**17**, Page 51].

Step 6. We use fact that the uniform limit function of a sequence of analytic functions is analytic also, Lang [**17**, Page 156], and then we have that, for each fixed $t \in [a, b]$, $\phi(t, u_0, \mu)$ is an analytic function of (u_0, μ) on

$$\{(u_0, \mu) \in \mathbb{C}^2 : |u_0| + |\mu| \leq R_0\},$$

where $R_0 > 0$ is an arbitrarily given positive number. This is because each $\phi_i(t, u_0, \mu)$ is such a function. Since $R_0 > 0$ is arbitrary, we have $\phi(t, u_0, \mu)$, for each fixed $t \in [a, b]$, is analytic on $\mathbb{C}^2 \times \mathbb{C}$. In particular, $\phi(t, u_0, \mu)$, for each fixed $(t, u_0) \in [a, b] \times \mathbb{C}^2$, is an entire function of μ. $\square$

Proof of Corollary 4.7.

PROOF. By choosing

$$C(t) = \begin{pmatrix} 0 & 1 \\ -q(t) & -p(t) \end{pmatrix} \quad \text{and} \quad D(t) = \begin{pmatrix} 0 & 0 \\ r(t) & 0 \end{pmatrix},$$

it follows from Theorem 4.5 that the equation

$$\frac{d}{dt}u = C(t)u + \mu D(t)u, \quad a < t < b, \tag{4.6}$$

$$u(\tau) = u_0, \quad \tau \in [a, b]$$

has a unique solution $u = u(t, u_0, \mu)$. By letting

$$u_0 = \begin{pmatrix} y_0 \\ y_1 \end{pmatrix},$$

$$u = \begin{pmatrix} y(t, y_0, y_1, \mu) \\ \frac{d}{dt}y(t, y_0, y_1, \mu) \end{pmatrix},$$

we have y is the unique solution to the equation (4.7) that satisfies the required properties. $\qquad\square$

Bibliography

[1] R. A. Adams, *Sobolev Spaces*, Academic Press, New York, (1975).

[2] T. M. Apostol, *Mathematical Analysis*, second edition, Addision-Wesley Publishing Company, Inc., (1974).

[3] W. E. Boyce and R. C. DiPrima, *Elemetary Differential Equations and Boundary Value Problems*, Fifth Edition, John Wiley & Sons, Inc., New York, 1992.

[4] R. V. Churchill, *Operational Mathematics*, 3rd ed., New York, McGraw-Hill, (1971).

[5] E. A. Coddington and N. Levision, *Theory of Ordinary Differential Equations*, McGraw-Hill Book Company Inc., New York, (1955).

[6] E. A. Coddington. *An Introduction to Ordinary Differential Equations*, Prentice-Hall, Inc., Englewood Cliffs, New Jersey, (1961).

[7] C. G. Cullen, *Linear Algebra and Differential Equations*, Second edition, PWS-KENT Publishing Company, Boston, Massachussetts, 1991.

[8] G. B. Folland, *Real Analysis, Modern Techniques and Their Applications*, John-Wiley & Sons, New York, (1984).

[9] S. H. Friedberg, A. J. Insel, and L. E. Spence. *Linear Algebra*, Second Edition, Prentice-Hall Inc., New Jersey, 1989.

[10] F. R. Gantmacher, *The Theory of Matrices*, Volume one, Chelsea, New York, 1977.

[11] , W. H. Greub, *Linear Algebra*, Fourth Edition, Springer-Verlag, New York, 1981.

[12] P. R. Halmos, *Measure Theory*, Springer-Verlag, New York, (1974).

[13] J. -Y. Han and C. -Y. Lin, *A new proof of Jordan canonical forms of a square matrix*, Linear and Multilinear Algebra, Vol. 57, No. 4, (2009), 369-386.

[14] J. -Y. Han and C. -Y. Lin, *A direct approach to power series solutions of weakly singular linear systems*, submitted.

[15] P. Hartman, *Ordinary Differential Equations*, Second Edition, Birkhauser, Boston, (1982).

[16] T. Kato, *Perturbation Theory for Linear Operators*, Springer, New York, (1966).

[17] S. Lang, *Complex Analysis*, Fourth Edition, Springer-Verlag, New York, (1999).

[18] P. Lancaster and M. Tismenetsky, *The Theory of Matrices with Applications*, Second edition, Academic Press, New York, 1985.

[19] R. K. Nagel and E. B. Saff, *Fundamentals of Differential Equations and Boundary Value Problems*, Addison-Wesley Publishing Company, New York, (1996).

[20] H. L. Royden, *Real Analysis*, Macmillan Publishing Company, New York, (1988).

[21] A. E. Taylor and D. C. Lay, *Introduction to Functional Analysis*, Wiley, New York, (1980).

[22] D. V. Widder, *Advanced Calculus*, 2nd ed., Englewood Cliffs, New Jersey: Prentice-Hall, (1961).

[23] W. Walter, *Ordinary Differential Equations*, Springer-Verlag, New York, 1998.

[24] K. Yosida, *Functional Analysis*, Springer, New York, (1980).

Index